PHYSIOLOGICAL BASIS OF AGING AND GERIATRICS

SECOND EDITION

ARROWE PARK HOSPITAL
MCARDLE
LIBRARY

PHYSIOLOGICAL BASIS OF AGING AND GERIATRICS

SECOND EDITION

EDITED BY
PAOLA S. TIMIRAS

CRC Press
Boca Raton London New York Washington, D.C.

Library of Congress Cataloging-in-Publication Data

Physiological basis of aging and geriatrics / edited by Paola S. Timiras
p. cm.
Rev. ed. of: Physiological basis of geriatrics. c1988
Includes bibliographical references and index.
ISBN 0-8493-8979-8
1. Aging—Physiological aspects. 2. Geriatrics. I. Timiras, Paola S.
II. Physiological basis of geriatrics.
QP86.P557 1994
612.6′7—dc20

93-44066
CIP

Direct all inquiries to CRC Press LLC, 2000 N.W. Corporate Blvd., Boca Raton, Florida 33431.

International Standard Book Number 8493-8979-8
Library of Congress Card Number 93-44066
Printed in the United States of America 5 6 7 8 9 0
Printed on acid-free paper

PREFACE

Given the rapidly increasing advances in the areas of gerontology and geriatrics, one of the major purposes of this Second Edition is to update the material presented. Another goal is to attract a broad readership by providing a book that will be useful to a spectrum of individuals with varying biological backgrounds. Although this may seem too ambitious a task, it is necessary in view of the continuing progress in this expanding field and a need for a common understanding of the biology of aging by individuals from various disciplines. To fulfill these goals, the original text has been streamlined with more explicit and concise explanations. Focus is on the established facts of physiologic aging rather than on speculative considerations of mechanisms.

However, some of the goals of the First Edition (1988) — e.g., viewing aging from several perspectives — have been preserved. Using physiology as the unifying concept, the present work attempts to assimilate and distill information from multiple sources in a concise, comprehensive text. According to the traditional physiologic approach, aging is examined in the various organs and systems, focusing predominantly on humans but drawing, when appropriate, on animal research.

Generally, the meaning of aging is not perceived at its own value; rather, what is meaningful is "to reach old age". The concepts of *continuity* and *change* are fundamental to all biologic processes. As one ages, continuity of prior physiologic events may provide "a usable past" that can shape future functions. When one goes back in time looking for continuity, one is also aware that continuing change occurs simultaneously. In a previous text — *Developmental and Physiological Aging* (1972) — identity in old age was viewed as being a dynamic process, as it is at younger ages. Similarly, in this edition, reports of studies of cellular and molecular events throughout the lifespan may provide a better understanding of the continuum of changes with aging.

One of the characteristics of aging is the increasing incidence of disease. Clinical correlations are included both as practical reference for the geriatrician and as a comparison with normal aging for the gerontologist.

Aging has been viewed as a "journey taking place in a community setting" from which the individual cannot be separated. Therefore, aspects of demographic, comparative, and differential aging are included together with a discussion of the several theories of aging to provide a more complete profile of the aging of individuals and populations.

As more is learned of the aging process, the possibility of intervention is moving within reach. When aging is viewed through the eyes of medicine, the emphasis is on "the elderly at risk" and on the need for assessing, managing, and reducing "risk" factors. Equally important is the strengthening of physiologic competence. The bases for future interventions to improve the quality of life are discussed with updates of those currently in use.

ACKNOWLEDGMENTS

I have been greatly encouraged in my task of preparing this Second Edition by support from the collaborators of the First Edition. All have willingly accepted the task of reviewing and updating their respective chapters or sections within very stringent deadlines. To them, I offer my most heartfelt thanks. I also address a special thought to the late Dr. Doherty B. Hudson, whose friendship and competent collaboration I miss sorely.

Special thanks go to Dr. Mary L. Timiras for providing pertinent clinical correlations throughout and for preparing, with Dr. Judith L. Beizer, Chapter 23, which was previously written by Dr. Hudson; to Dr. Sally Oklund for her many drawings, which effectively visualize complex material; and to Dr. L. L. Rosenberg for his substantive and constructive editorial criticisms.

I am indebted to Ana Dueñas, who competently prepared the manuscript and dealt with the word processing and editing. I would also like to thank the students Karen Ho and Paul Lee, bibliographic assistant Ms. Anelia Popnikolova, and librarian Ms. Ingrid Radkey for their assistance with the bibliography.

THE EDITOR

Dr. Paola S. Timiras is a Professor in the Department of Molecular and Cell Biology at the University of California at Berkeley (UCB). A native of Rome, Italy, Dr. Timiras is a graduate of the University of Grenoble, France (B.A.), the University of Rome (B.A., M.D., Summa cum Laude), and the University of Montréal, Quebec, Canada (Ph.D.).

Before joining the University of California in 1956, she was a faculty member of the University of Montréal (1950–1953) and the University of Utah (1953–1955). She chaired the Department of Physiology–Anatomy at the UCB from 1978 to 1984.

Her interest in development and aging has led her to participate in several related societies. She directed an innovative medical program (Health and Medical Science, 1973–1975) and a training program in developmental physiology and aging (1965–1976) at UCB. She was one of the founders and president (1978–1981) of the International Society of Developmental Neuroscience, as well as vice president and president of the International Society of Psychoneuroendocrinology (1974–1982). Her research on the neuroendocrinology of development and aging has resulted in the publication of more than 350 papers and several books.

Dr. Timiras is also a consultant for many government agencies concerned with biological research, and she is on the editorial boards of several journals. Her contributions to teaching and research have been recognized by several awards in the United States and abroad.

CONTRIBUTORS

Robert W. Atherton
Department of Zoology and Physiology
University of Wyoming
Laramie, Wyoming

Judith L. Beizer
College of Pharmacy and
Allied Health Professions
St. John's University
Jamaica, New York

Trudy M. Forte
Life Sciences Division
Lawrence Berkeley National Laboratory
University of California–Berkeley
Berkeley, California

Anne M. Holehan
Institute of Human Ageing
University of Liverpool
Liverpool, England

Rolf J. Mehlhorn
Heavy Metal and Free Radical Toxicology
Lawrence Berkeley National Laboratory
University of California–Berkeley
Berkeley, California

Esmail Meisami
Department of Physiology and Biophysics
University of Illinois
Urbana, Illinois

Brian J. Merry
Institute of Human Ageing
University of Liverpool
Liverpool, England

Paul Segall
Bio Time, Inc.
Berkeley, California

Ramesh Sharma
Department of Biochemistry
North Eastern Hill University
Shillong, India

Hal Sternberg
Bio Time, Inc.
Berkeley, California

Paola S. Timiras
Department of Molecular and Cell Biology
University of California–Berkeley
Berkeley, California

Mary L. Timiras
Center for Geriatric Health Care
Beth Israel Medical Center
Newark, New Jersey

TABLE OF CONTENTS

I GENERAL PERSPECTIVES

Chapter 1 Introduction: Aging as a Stage in the Life Cycle 1
Paola S. Timiras

Chapter 2 Demographic, Comparative, and Differential Aging 7
Paola S. Timiras

Chapter 3 Aging and Disease 23
Paola S. Timiras

Chapter 4 Theories of Aging 37
Ramesh Sharma

II MOLECULAR AND CELLULAR AGING

Chapter 5 Degenerative Changes in Cells and Cell Death 47
Paola S. Timiras

Chapter 6 Oxidants and Antioxidants in Aging 61
Rolf J. Mehlhorn

Chapter 7 Aging of the Immune System 75
Hal Sternberg

III SYSTEMIC AND ORGANISMIC AGING

Chapter 8 Aging of the Nervous System: Structural and Biochemical Changes 89
Paola S. Timiras

Chapter 9 Aging of the Nervous System: Functional Changes 103
Paola S. Timiras

Chapter 10 Aging of the Sensory Systems 115
Esmail Meisami

Chapter 11 Aging of the Adrenals and Pituitary 133
Paola S. Timiras

Chapter 12 Aging of the Female Reproductive System: The Menopause 147
Brian J. Merry and Anne M. Holehan

Chapter 13 Aging of the Male Reproductive System 171
Brian J. Merry and Anne M. Holehan

Chapter 14 Aging of the Thyroid Gland and Basal Metabolism 179
Paola S. Timiras

Chapter 15 The Endocrine, Pancreas, and Carbohydrate Metabolism 191
Paola S. Timiras

Chapter 16 Cardiovascular Alterations with Age . . . 199
Paola S. Timiras

Chapter 17 Plasma Lipoproteins: Their Metabolism and Role in Atherosclerosis . . . 215
Trudy M. Forte

Chapter 18 Aging of Respiration, Erythrocytes, and the Hematopoietic System . . . 225
Paola S. Timiras and Richard W. Atherton

Chapter 19 The Kidney, the Lower Urinary Tract, the Prostate, and Body Fluids . . . 235
Mary L. Timiras

Chapter 20 Aging of the Gastrointestinal Tract and Liver . . . 247
Paola S. Timiras

Chapter 21 Aging of the Skeleton, Joints, and Muscles . . . 259
Paola S. Timiras

Chapter 22 Aging of the Skin and Connective Tissue . . . 273
Mary L. Timiras

IV PREVENTION AND REHABILITATION

Chapter 23 Pharmacology and Drug Management in the Elderly . . . 279
Judith L. Beizer and Mary L. Timiras

Chapter 24 Effects of Diet on Aging . . . 285
Brian J. Merry and Anne M. Holehan

Chapter 25 An Agenda for Healthful Aging . . . 311
Paola S. Timiras and Paul E. Segall

I
GENERAL
PERSPECTIVES

1 INTRODUCTION: AGING AS A STAGE IN THE LIFE CYCLE

Paola S. Timiras

As we move towards the 21st century, one of the major concerns of the biomedical community, and of society at large, is the increasing proportion of the elderly in the population, both in the developed and developing nations of the world. In the U.S., the percentage of individuals 65 years of age and older was 4% in 1900, 7% in 1940, and 11% in 1980. In 1990, it has risen to 13%, an increase that is even higher in France, Japan, and Sweden. It is expected that by the year 2030 about one fifth (18%) of the population of the U.S. will live 65 years and longer. Such impressive and persistent increase in the elderly (65+) population underlies the pressing need for a better understanding of the aging process and of the problems of the elderly.

Although the pathology of older people has been studied extensively for combating specific diseases associated with this "high-risk" group, the physiology of aging has not been of primary interest, partly because of the difficulty of isolating "normal" from "abnormal" aging processes. Old age in humans is conventionally accepted as the stage of the life cycle that starts around 65 years of age and terminates with death, but it is difficult to circumscribe its temporal boundaries in physiologic terms.

Aging has, so far, defied all attempts to establish objective landmarks that would precisely signal its earlier stages. It lacks specific markers characteristic of other life periods, such as menarche at puberty. Rather, its onset occurs at some "indeterminate" point following maturity, and its progression follows timetables that differ with each individual. Thus, "physiologic heterogeneity" is one of the consistent characteristics of the elderly population.

1 BASES AND ORGANIZATION OF SUBJECT MATERIAL

The breadth of the included material reflects the wealth of information on aging since publication of the earlier text by the same author.[1] Historically, physiology focuses on the structure and function of bodily biological systems. A prominent part of the book is therefore concerned with systemic and organismic aging. Herein, we attempt to present the aging of the various systems in the traditional manner, focusing on the human but drawing, where necessary, on animal research. To provide a more comprehensive understanding of these basic aspects, clinical correlations are included. This book is not considered to be a textbook of medicine but will, however, include therapies that employ physiological concepts.

Any consideration of aging should not ignore the psychological and social components, particularly important for the geriatrician who deals with the individual as a whole. These fields require extensive and competent presentation well beyond the scope of this book; therefore, such subject material has not been included.

Understanding systemic and organismic aging can only be achieved based on current knowledge of molecular and cellular aging. The various theories of aging are also necessary background for comprehending aging processes. All of these, together with demographics and comparative and differential aspects of aging, are considered in the first four chapters.

Our anticipated goal is to provide a book that will be useful to a broad spectrum of individuals with different degrees of biologic background. Each chapter is organized to begin with *structure,* then move to *function,* considered first in the adult (a brief synopsis) and then in the aged. Essential material is presented in normal text size; smaller type furnishes additional information, in-depth discussion, or less well-accepted or controversial matters.

The intent to attract a broad readership may seem too ambitious. The studies of aging and geriatrics are fast expanding and attracting people from many disciplines who require a common understanding of the biology of aging. Current texts are directed to very specific aspects of aging or are designed for a particular group of readers. The present work, using physiology as the unifying concept, assimilates and distills information from multiple sources to produce an understandable, comprehensive text.

2 PRE- AND POSTNATAL LIFE STAGES

In the broadest sense, the lifespan of the individual is divided into two main periods: prenatal and postnatal. For a long time, attention was focused on the prenatal period when the most striking physiologic events occur. Important changes continue to occur even in the adult state and systematic studies have now been extended to embrace old age. The necessity for caution in establishing "chronologic boundaries" for life stages is abundantly clear; not only are there numerous examples of individual variation in any species, but the limits of life periods also shift forward and backward.

Although many animal species are capable of an independent existence at relatively immature stages, others, including mammals, are not. The human newborn, fully formed anatomically, is still utterly dependent on adults for food and care. Throughout infancy, childhood, and adolescence, a gradual remolding of body shape continues, together with the acquisition of new functions and the perfecting of already established functions. At about 25 years, the last of these changes is completed and the body is stabilized in the adult condition. The mature adult period lasts for approximately another 40 years and encompasses the period of maximal physiologic competence. The progressive decline of physiologic potentialities with age leads to senescence and death.

Both prenatal and postnatal stages can be subdivided into several periods, each of which may be distinguished by morphologic, physiologic, biochemical, and psychologic features. The main divisions and the approximate time periods of the lifespan in humans are listed in Table 1-1.

0-8493-8979-8/94/$0.00+$.50

TABLE 1-1
Stages of the Lifespan

Stage	Duration
Prenatal life	
Ovum	Fertilization through week 1
Embryo	Weeks 2-8
Fetus	Months 3-10
Birth	
Postnatal life	
Neonatal period	Newborn; birth through week 2
Infancy	3 weeks until end of first year
Childhood	
Early	Years 2-6
Middle	Years 7-10
Later	Prepubertal; females 9-15; males 12-16
Adolescence	The 6 years following puberty
Adulthood	Between 20 and 65 years
Senescence	From 65 years on
Death	

The prenatal period includes three main stages: ovum, embryonic, and fetal stages. The postnatal period commences with birth and continues into stages of neonatal life, infancy, childhood, adolescence, adulthood, and old age. The practical necessity of research to fragment the biologic study of living organisms, however, has a tendency to obscure the dynamic relationships that are obtained at many levels of organization. This is true whether we speak of the morphology and function of a single system or of the effects of environmental stress from one age period to another.

The physiologic profile of a given individual must be assessed with regard to his particular life history in all its biosocial complexity. Succeeding stages of the lifespan cannot help but reflect on the events that have occurred in the preceding stages. The concept that attained growth tells of past growth and foretells growth yet to be achieved can even more appropriately be applied to the later stages af the lifespan: in old age, function most certainly depends on past physical history. Indeed, as the saying goes, growing old gracefully is the work of a lifetime.

2.1 Heredity and Environment

Developmental processes and their regulation by heredity and environment have engaged biologists for many decades. At all stages of life, the directing force of heredity and the molding influences of the internal and external environment interplay in determining physiologic competence and the length of the lifespan; to weigh the value of one against the other is to disturb the integrity of the whole.

Heredity operates through internal factors present in the fertilized egg itself. Chief among these are the genes, or hereditary determiners, located in the chromosomes, which contain the genetic contribution of each parent. Environment supplies the external factors that make development possible and allow inherited potentials to find expression. Environmental factors (1) include temperature, humidity, atmospheric gases, nutrition, drugs, infections, radiation, and so forth; (2) can condition the appearance and modify the type of genetic characters, influence their expression, and alter their composition to make possible the creation of new inheritable characters (mutation); and (3) are operative throughout the lifespan.

One of the basic characteristics of humans and animals is their capacity to adapt to an ever-changing environment. Adaptation is attained through a series of physiologic adjustments that serve to restore the normal state once it has been disrupted. These adjustments are grouped under the term *homeostasis,* and a large part of physiology is concerned with regulatory mechanisms (such as negative and positive feedback) that act to maintain constant the internal environment. By the time of maturity, numerous such adjustments already have been accomplished.

Homeostatic regulation, together with inherited traits, defines the competence with which the individual will continue to respond to environmental challenges; it is in this sense that aging has been viewed as a general decline in the capacity to adapt to continuing stress and that external factors have been related to the length of the lifespan. After consideration of all these influences, the best assurance for a long lifespan remains the thoughtful selection of long-lived parents.

3 STAGES OF MATURITY AND AGING

The mature years are considered a major life stage. They are characterized by great functional stability as connoted by the attainment of optimal, integrated function of all body systems. Function in adulthood is taken as a standard against which to measure any degree of physiologic or pathologic deviation. Most textbooks in human physiology take the mature 25-year-old, 77-kg, 170-cm-tall man as their model. In the present text, this period and this average man will serve mainly as a reference point for the discussion of aging and the aged.

Functional competence is multifaceted, and optimal performance may differ from age to age and from one parameter to another. In other words, it would be unsound, if not physiologically incorrect, to assume that a function is maximally efficient only during adulthood and that differences in the earlier or later years necessarily represent functional immaturity or deterioration, respectively. Rather, one must view physiologic competence as having several levels of integration, depending on the requirements of the organism at any specific age and the type and severity of the challenges to which the organism is exposed. So much negativity has already been attached to physiologic changes with aging that we have elected in this text to focus, whenever possible, on more positive aspects such as the persistence of adequate function of several organs and systems into old age.

Evidence for conceptualizing physiologic competence along a continuum throughout the lifespan emerges from the study of developmental physiology and aging. Although the age of 65 has been accepted as the demarcation line between maturity and old age, a person of 65 may be quite healthy and a long way from "retiring" from life — a situation about which social scientists frequently express concern.

If the adult years are difficult to subdivide into physiologic stages, aging or senescence is even more of a challenge. Inasmuch as the "winter years" are increasingly besieged by pathologic events, the study of aging is diversely approached, i.e., in terms of the extent to which physiologic competence is retained.

Despite the absence of sequential stages of decline from maturity to death, in aging as in development, the changes that do take place do not occur simultaneously at all levels of functional organization, nor do they involve all functions to the same degree. It appears more likely that each change, whether at the molecular, cellular, tissular, or organismic level, follows its own timetable and that the latter is differentially susceptible to specific intrinsic and extrinsic factors.

Gerontology is not restricted to the identification of pathologic factors. Old age cannot be regarded simplistically as a "disease", but involves a complexity of physiologic and pathologic phenomena, all of which are subject to numerous environmental influences. Thus, as a life stage, aging is presented here not only as it affects the whole organism, but also on the basis of alterations occurring at the level of cellular and subcellular elements, tissues, organs, and systems.

TABLE 1-2

Brief Glossary of Aging-Related Terms

Term	Definition
Aging	Latin "aetas", age or lifetime The condition of becoming old
Geriatrics	Greek "geron", old man, and "iatros", healer A medical specialty dealing with the problems and diseases of the elderly
Gerontology	Greek "geron", old man, and "logos", knowledge A branch of biology studying aging and the problems of old age
Senescence	Latin "senex", old man. The condition of being old, used interchangeably with aging
Lifespan	The duration of the life of an individual/organism in a particular environment and/or under specific circumstances
Average lifespan	The average of individual lifespans for members of a group (cohort) of the same birth date
Life expectancy	The average amount of time of life remaining for a definite population with the same birth date
Active life expectancy	As above with the addition that the remaining life be free of a specific level of disability
Longevity	Long duration of an individual's life; the condition of being "long-lived"
Maximum lifespan	The length of life of the longest-lived individual member of a species

4 DEFINITIONS

The interpretation of aging as a physiologic process upon which pathology and disease have been superimposed has been formalized under the separate disciplines of *gerontology* — the study of aging processes — and *geriatrics* — the treatment of the debilitating symptoms and diseases associated with old age (Table 1-2). The terms aging and senescence are often used interchangeably despite some minor interpretative differences; *aging* refers more appropriately to the process of growing old regardless of chronologic age, whereas *senescence* is restricted to the state of old age characteristic of the late years of the lifespan (Table 1-2). *Lifespan* refers to the duration of the life of an individual. It is genetically determined for each species, and, within the species, for each individual. The *average lifespan* represents the average of individual lifespans of a cohort (i.e., a group of individuals having a common demographic factor such as age) born at the same date (Table 1-2). Both terms represent concepts that may be viewed as theoretical since the lifespan implies optimal life duration in an environment free of risk factors (a very unrealistic event!), and average lifespan is characterized by considerable heterogeneity and variations.

One of the most important ways in which aging processes may be evaluated is in terms of life survivorship of individuals of the same age (or birth cohort). This is described as *life expectancy,* that is, total years of survival from the moment (age) under study. As the number and severity of disabilities increase with advancing age, *active life expectancy* refers to years of survival without disability (Table 1-2). Still another definition concerns *longevity,* meaning the "long" duration of individual life and maximum lifespan, that is, the *maximum lifespan or the (longest) life of the oldest individuals* of a species (Table 1-2). As we have already indicated, the average lifespan in humans has considerably lengthened in this century, but the age reached by the longest-lived individual has not significantly changed from previous reports.

Definitions of *aging* are numerous; indeed, there are as many definitions as there are theories of aging (Table 1-3). Those presented here are arbitrarily selected on the basis of their relevance to physiology.

Accordingly, a first definition of aging is *the sum of all changes that occur in an organism with the passage of time.* As already discussed in this chapter, this definition visualizes the lifespan as an orderly unfolding of precisely timed events from fertilization to death, with aging as a stage of the lifespan.

Despite the many differences that are the consequence of heredity, sex, past history, and life experience, it is undeniable that older persons share certain traits with respect to the onset, rate, and characteristics of the aging process. Indeed, we cannot ignore the fact that aging is a deteriorative process characterized by increased vulnerability and decreased viability. The first definition, therefore, must be amended to indicate that, with advancing age, functional impairment increases as does the probability of death. Thus, in humans, aging takes the form of morphologic and functional involution, always progressive and often silent, which affects most organs, tissues, and cells and results in a gradual decline in performance, leading to *functional impairment and death.*

Adaptation to environmental stress is one of the major functions of the organism, and, as the capacity for adaptation progressively fails with age, so aging may be further defined as *a decreasing ability to survive stress.* At the organismic and systemic levels, then, aging is recognized as the progressive deterioration of the physiologic processes necessary for the maintenance of a constant "milieu interieur" and death as the ultimate failure of the organism to maintain this constancy, that is, to sustain homeostasis. Other definitions of aging focus on *cellular and molecular changes,* and aging and death would result from alterations therein. Proponents of this view ascribe aging to progressive cellular changes (the expression of "gerontogenes") in the plasma and intracellular membranes (fluidity, permeability, transport), cytoplasm (accumulation of free radicals, cross-linkages, lipofuscin), and nucleus (DNA damage and DNA repair failure, mRNA error catastrophe, alterations in replication, and histocompatibility complex).

5 DEATH VERSUS IMMORTALITY

While aging is approached gradually without any specific physiologic markers of its onset, death is the terminal event of this stage in the lifespan. Various types of death terminate the aging process. In broad terms, these include trauma, accidents, and disease. Trauma and accidents (e.g., high-speed vehicle crashes, dangerous occupations, drug abuse, smoking) are the major causes of death in young adulthood. Disease processes that overwhelm the defense or repair systems of the body affect all ages but are particularly life endangering in the very young and in the old. Many diseases that lead to death in the perinatal period — period of high risk — have been conquered in developed countries, including the U.S. Today most deaths from disease occur in the elderly, in whom diminished function makes the accumulation of pathology less tolerable than in the young. Indeed, some diseases occur almost exclusively in the old, and this linkage of pathology with old age justifies the argument of some investigators that aging itself is a disease.

Senescence is not an accepted cause of death and is not so noted on the death certificate. Whether death is a natural event has never been validated scientifically. The attitude of tacit acceptance of a debilitating old age, prevalent even among biologists, is now being replaced by one that regards senescence as the "subversion of function", the inevitability of which is open to question. Nostalgically we consider immortality in the light of extending our life through our offspring; however, this sense of immortality is no longer adequate. It is frustrating that in a time when humans have gone into and returned from outer space and can manipulate DNA, they have not conquered death. Death, indeed, remains the last "sacred" enemy.

6 PHYSIOLOGIC, CLINICAL, AND INTERVENTIVE CONSIDERATIONS

Biologists working in the area of aging have defined two major goals to guide their research programs: one entails the effort to prolong human life, and the other the effort to significantly enhance viability, the so-called quality of life, throughout the lifespan. Organ transplantation, perhaps, encompasses both objectives, but it represents a "crude and clumsy" approach in either case. Pharmacologic and genetic research, on the other hand, has startling potential for the amelioration and prolongation of life and is currently attracting increasing support. Nevertheless, the fact that the answers remain obscure need not deter our search, for only by continually posing bold new questions can we hope to accomplish the spirited goals of the gerontologists, whose motto "to add years to life and life to years" emphasizes for all of us the importance of the quality of life. Whether specific age-related changes are the cause or consequence of the aging process remains to be established. The concept of a single causative factor pervades many areas of biologic study, and gerontology is no exception. It is unlikely that a single "triggering" event is responsible for the aging of the organism, but, rather, aging probably entails numerous and complex interactions at different levels. Viewing aging, therefore, as the consequence of many diverse factors, we shall consider, first, the aging of human and animal populations. This comparative and demographic section will be followed by consideration of the molecular and cellular aspects of aging. Inasmuch as our major concern is with the human, the aging of systems and organs will be discussed along with selected relevant clinical aspects. And finally, in the desire to actively improve the quality of life of the elderly, we shall emphasize means for "healthful" aging.

TABLE 1-3

Definitions of Aging

Aging as a stage of the lifespan	The sum of all changes occurring in an organism with the passage of time
Aging as a deteriorative process	The sum of all changes occurring with time and leading to functional impairment and death
Aging as cellular and molecular damage	Changes in membranes, cytoplasm, and/or nucleus

REFERENCE

1. Timiras, P.S., *Developmental Physiology and Aging,* Macmillan, New York, 1972.

ADDITIONAL READINGS

Bergener, M., Ermini, M., and Stahelin, H.B., Eds., *Thresholds in Aging,* Academic Press, New York, 1985.

Butler, R.N., Schneider, E.L., Spratt, R.L., and Warner, H.R., *Modern Biological Theories of Aging,* Vol. 31, Raven Press, New York, 1987.

Calow, P., *Life Cycles: An Evolutionary Approach to the Physiology of Reproduction, Development and Aging,* Chapman and Hall, London, 1978.

Carter, N. (Ed.), *Development, Growth and Aging,* Croom Helm, London, 1980.

Comfort, A., *The Biology of Senescence,* 3rd ed., Elsevier, New York, 1979.

Coni, N., Davison, W., and Webster, S., *Ageing: The Facts,* Oxford University Press, Oxford, 1984.

Danon, D., Shock, N.W., and Marois, M., *Aging: A Challenge of Science and Society,* Vol. 1, Biology, Oxford University Press, Oxford, 1981.

Davis, B.B. and Wood, W.A., Eds., *Homeostatic Function and Aging,* Vol. 30, Raven Press, New York, 1985.

Finch, C.E., *Longevity, Senescence and the Genome,* University of Chicago Press, Chicago, 1990.

Finch, C.E. and Johnson, T.E., *Molecular Biology of Aging,* Wiley-Liss, New York, 1990.

Fries, J.F. and Crapo, L.M., *Vitality and Aging,* W.H. Freeman, San Francisco, 1981.

Frolkis, V.V., *Aging and Life-Prolonging Processes,* Springer-Verlag, New York, 1982.

Johnson, H.A., Ed., *Relations Between Normal Aging and Disease,* Vol. 28, Aging, Raven Press, New York, 1985.

Kenney, R.A., *Physiology of Aging: A Synopsis,* Year Book Medical Publishers, Chicago, 1982.

Kimmel, D.C., *Adulthood and Aging. An Interdisciplinary Developmental View,* 2nd ed., John Wiley & Sons, New York, 1980.

Masoro, E.J., Adelman, R.C., and Roth, G.S., *CRC Handbook of Physiology of Aging,* CRC Press, Boca Raton, FL, 1981.

Perlmutter, M. and Hall, E., *Adult Development and Aging,* John Wiley & Sons, New York, 1985.

Rockstein, M. and Sussman, M., *Biology of Aging,* Wadsworth Publishing Co., Belmont, CA, 1979.

Rothstein, M., Adler, W.H., Cristofalo, V.J., Finch, C.E., Florini, J.R., and Martin, G.M., *Review of Biological Research in Aging,* Vol. 4, Wiley-Liss, New York, 1990.

Smith, D.W., Bierman, E.L., and Robinson, N.M., Eds., *The Biologic Ages of Man: From Conception Through Old Age,* 2nd ed., W.B. Saunders, Philadelphia, 1978.

Wallace, R.B. and Woodson, F.W., Eds., *The Epidemiologic Study of the Elderly,* Oxford University Press, New York, 1992.

Young, J.Z., *An Introduction to the Study of Man,* Clarendon Press, London, 1971.

I
GENERAL
PERSPECTIVES

2 DEMOGRAPHIC, COMPARATIVE, AND DIFFERENTIAL AGING

Paola S. Timiras

1 DEMOGRAPHY OF AGING

As indicated in the previous chapter, it is a major achievement of civilization that life expectancy of humans has increased throughout history. It is in this century that this increase, previously quite slow, has undergone a dramatic acceleration. In the U.S., average life expectancy has increased from about 50 years in 1900 to about 76 years in 1990, with women living about 5 to 7 more years than men (Figure 2-1). Longer life expectancy results in part from a reduction of mortality. This reduction may be ascribed to biomedical advances; but, perhaps, more than progress in medical technology and molecular biology, improvement of sanitation, life habits, and of socioeconomic conditions may account for this trend.

Not only are the elderly living longer, but they also represent the most rapidly growing segment of the population in developed countries. Considering world population, the percentage of individuals 65 years of age and older in developed countries was about 8% in 1960 and 11% in 1980; it is expected to reach 13% and 16% in the years 2000 and 2020, respectively. In the less-developed countries, the rise in the proportion of 65-year-old individuals and older has been slower but may almost double between 1960 (4%) and 2020 (7%) (Table 2-1).[1] In the U.S. in 1990, persons 65 years and older represented 13% of the population. By the year 2030, this proportion will increase to 18%. This so-called *"graying"* of the population is also reflected in the higher proportion of the old-old (75 years and older) who represent the fastest growing segment of the population,[2] increasing from 1% in 1910 to an expected 8% in 2030 (Figure 2-2). It is predicted that by the year 2000, almost half of the deaths in the U.S. will occur after age 80.[3] It is probably safe to state that most of these elderly individuals are in need of some type of assistance. The social, economic, and psychologic as well as the medical problems of the elderly segment of the population demand attention and should be dealt with by expeditious and judicious planning.

The demographic study of a population (that is, size and density, distribution, and vital statistics) shows that aging of the population may be due to (1) longer survival and (2) declining birth rates and consequent higher proportion of individuals in older age groups. Both ways seem to be operative at this time in the U.S. and developed countries; birth rates continue to remain low and more people survive to old age. *"Demographic transition"* is the shift from high fertility/high mortality to low fertility/low mortality and, consequently, from a low to a high proportion of older people within the general population. This unprecedented progression of the elderly population is expected to continue and to increase throughout the world over the next century.

1.1 Life Tables

One way of describing mortality is to construct life tables from census and vital statistics.[4] A life table represents the action of mortality on a conventional cohort of newborn individuals as they proceed through successive ages. It indicates, therefore, the number of survivors at each birthday out of an initial cohort of size 10n (commonly 10,000 or 100,000) at age 0, until the last survivor is eliminated by death. The life table consists of several columns. Each column gives (in abbreviated form): (1) age (x = 0 is the point of birth, and from then on age is calculated in years); (2) the mortality rate, i.e., the probability of dying between two birthdays; (3) the number of deaths between two birthdays; (4) the number of survivors at each birthday; and (5) expectations of life at each birthday. Thus, a life table reveals a series of probabilities of death, or the risk that a group of individuals, upon reaching their *x*th birthday, will die before the *x*th + 1 birthday. Other functions of the life table (such as the number of deaths in each interval or the life expectancy at each birthday) can be easily derived. Thus, the life table is a precise tool for measuring the level of mortality — and therefore the length of life — of any population.

Life tables, however, do not offer any indication as to the process of aging of a real population, since this depends on the past history (i.e., during the preceding 100 years) of fertility, mortality, and migration of the population. If we rule out migration and consider a population that gains or loses elements only because of birth and death, the age structure of a population (or, in other words, the percentage of the total population in the various age groups) is determined more by the changing level of the birth rate than by the trend of mortality.

A decline in the number of births, for instance, determines the reduction of the size of each newborn cohort and, if sustained for one or more decades, induces a reduction of the proportional importance of infants, children, and the young, and, conversely, an increased proportional importance

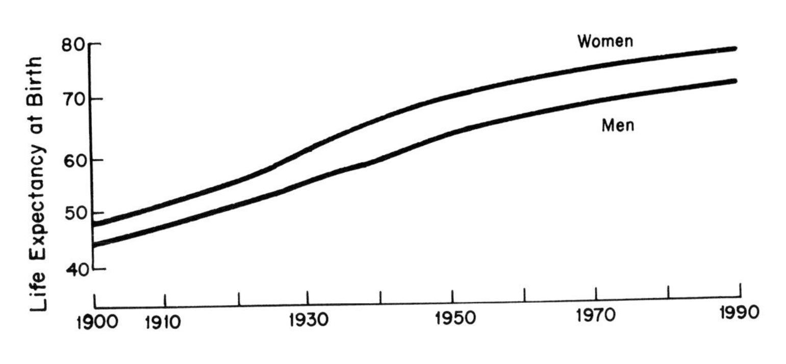

FIGURE 2-1 *Life expectancy at birth (U.S., 1900–1990).* Note progressive increase in life expectancy since 1900 and the greater life expectancy of women.

0-8493-8979-8/94/$0.00+$.50

TABLE 2-1

Total Population in the Older Ages: 1960–2020

	1960			1980		
	Total population[a]	**65 Years and over**	**80 Years and over**	**Total population**	**65 Years and over**	**80 Years and over**
World	3037.0	165.3	19.9	4432.1	295.5	35.3
More developed regions (MDR)	944.9	80.3	11.7	1131.3	127.8	20.9
Less developed regions (LDR)	2092.3	85.0	8.1	3300.8	131.7	14.4
	2000[b]			**2020**		
	Total population	**65 Years and over**	**80 Years and over**	**Total population**	**65 Years and over**	**80 Years and over**
World	6118.9	402.9	59.6	7813.0	649.2	101.6
More developed regions (MDR)	1272.2	166.0	30.2	1360.2	212.4	43.3
Less developed regions (LDR)	4846.7	236.9	29.4	6452.8	436.9	58.2

Note: MDR: Northern America, Europe, Japan, Australia/New Zealand, and former USSR. LDR: Africa, Latin America, China and other East Asia (excluding Japan), Oceania (excluding Australia and New Zealand).

[a] Population in millions.
[b] Projections are medium variant.[1]

Source: Based on United Nations, *1982 Demographic Indicators of Countries; Estimates and Projections as Assessed in 1980, Population Studies,* series A, number 82, United Nations, New York. (Adapted with permission from Reference 1.)

of the aged. Thus, aging of a population would occur when its age structure is altered in such a way as to favor an increased proportion of the older versus the younger individuals.

A decline in mortality, on the other hand — such as the continuous one witnessed by the developed countries during the last 100 years — proportionally enhances the chances of survival from birth to all ages, it does not greatly change the age structure of the population concerned if, in the meantime, fertility has remained constant. Therefore, increasing longevity may depend on the decline of the risks of mortality at the various ages, as well as the process of population aging produced by reduced birth rates.

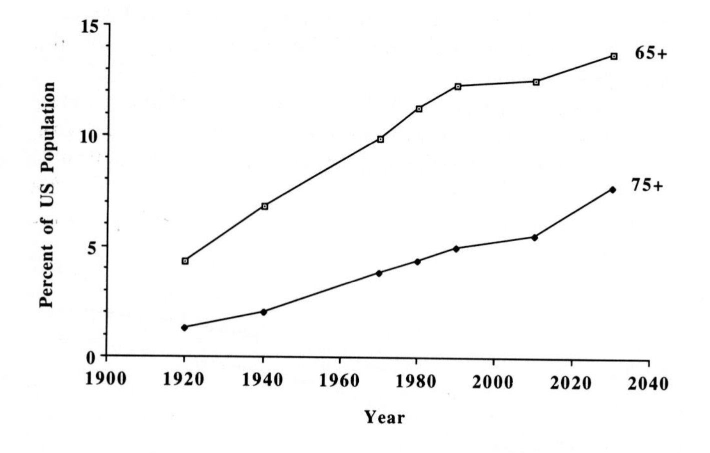

FIGURE 2-2 *Percentage of elderly in U.S. population (1900–2030).* Note approximately sixfold increase (real and projected) in percentage of population 75 years of age and older between 1910 and 2030.

1.2 Survival Curves

A number of survival curves, obtained from plots of life table survivorship, are portrayed graphically in Figure 2-3, where the abscissa measures chronologic age and the ordinate indicates that portion of the original population surviving (percentage of survivors). The straight portion of Curve 2 indicates a very low rate of death — approximately 100% survival — at early ages, but, with increasing age, the number of survivors decreases, as shown by the shoulder of the curve, and the 50% survival is the median age at death; the tail of the curve indicates the maximum lifespan or maximum longevity in the population.

Cumulative genetic or environmental hazards may lead to a more rapid decrease in the number of survivors. In this case the curve will shift to the left, as shown in Curve 1. Although the average lifespan is shortened, the maximum lifespan does not necessarily change. If conditions are improved, more people will survive longer, the curve becomes more "rectangular", and the average lifespan is shifted to the right, as shown in Curve 3.

This "rectangularization" or "squaring" of the survival curve (linear decelerating curve) is postulated by a number of gerontologists; it implies that (1) the optimal life expectancy is achieved, (2) all persons live healthy and active lives until they reach the maximum lifespan age of 110 years, and (3) then die suddenly (preferably in their sleep!) without having suffered any serious disease or disability in the last years of their life.[5-9]

Despite the progressive rise in average life expectancy, maximum lifespan has been little affected. To increase the maximum lifespan — the length of life of the oldest members of a population (Curve 4) — is one of the goals of gerontologists, but to date is the most difficult to achieve.

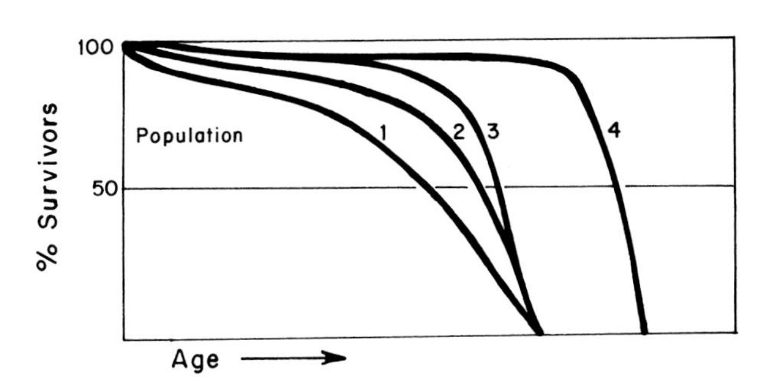

FIGURE 2-3 *Population survival curves (plotted from life table survivorship data).* Flat portion of curve indicates very low death rates; 50% survival is median age at death; tail of curve indicates maximum lifespan (longevity). Curve 2, representative survival curve; Curve 1, increased genetic or environmental hazards; Curve 3, improved health or living conditions; Curve 4, goal of gerontologists ("rectangularization" of curve and lengthening of maximum lifespan).

Survival data provide a concrete measure against which any theory of aging eventually must be measured. Removal of a particular cause or a number of factors inducing aging may result in a shift in the average longevity of a population, producing a more rectangular curve. In contrast, adding risk factors results in decreased survivorship at early ages, and the curve becomes less rectangular. The value of a given intervention to delay aging and improve the quality of life of the elderly must be weighed against its side effects. Practical consequences are involved in any suggested remedies or preventives adopted on a wide scale; if the removal of a damaging factor doubled the lifespan but rendered people impotent, not many people would choose longer life.

1.3 Sex Differences

Demographic studies in humans show that today, primarily in developed countries, females live longer than males (Figure 2-1).[10,11] This was not always the case: throughout most of history, women had a shorter lifespan than men. This shorter lifespan was attributed primarily to the danger of pregnancy and especially childbirth, but women were also more susceptible to a large number of diseases (e.g., infectious diseases, cancer, heart disease). However, this trend started to be reversed about 150 years ago and, currently, women continue to live longer than men.

Several causes have been implicated in the so-called "gender gap" in favor of women;[12] they are either related to the environment (e.g., geography, country, income) or the genome (e.g., XX versus XY phenotype). Sex-dependent longevity for females has been attributed alternately to:

1. Lesser life stress in females and, therefore, lower incidence of cardiovascular diseases
2. Less smoking (although the number of women smokers is on the increase in several countries)
3. Protecting action of estrogens (with respect to cardiovascular diseases, but primarily before menopause)
4. Longer gene recombination frequency
5. Lesser accumulation of mitochondrial DNA deletions/mutations with better protection against oxidative damage (crucial at the embryonal stage when one of the two female X has not yet been inactivated)

Studies in animals demonstrate a higher survival of females as compared with males in several species.[13] However, given the genetic sex differences and the possible role of environmental factors (biologic, behavioral, socioeconomic), none of the above explanations nor comparisons with other animal species have so far provided any satisfactory explanation for the gender gap in humans.

1.4 Differences in Ethnic Groups and in Populations Historically Understudied in the U.S.

The study of aging in populations whose life experience may be expected to differ from that of the "majority" is important not only (1) to characterize the functional status and health of these populations and (2) to plan for specific health services for them, but also (3) to provide useful insights into aging in general.[14] These populations may include Afro-Americans, Native Americans, and Hispanics or may include individuals with chronic illnesses (e.g., end-stage renal disease) and congenital diseases (e.g., Down's syndrome). Differences within these groups and with the white population are confounded by differences in cultural, socioeconomic, and environmental conditions, often disadvantageous, and superimposed on genetic factors (e.g., greater susceptibility of Afro-Americans to hypertension and non-insulin-dependent diabetes and of Native Americans to renal insufficiency).

Life expectancy at birth continues to be substantially greater for whites than for Afro-Americans and, in general, most minority groups. Some recent (1990) statistics show a life expectancy of 75.4 for whites and 69.4 for Afro-Americans (Table 2-2). This difference is attributable to the lower mortality of whites younger than 65 years. Currently, the difference in mortality between the two major U.S. racial groups (whites and blacks) is narrowing with the improvement of socioeconomic conditions (e.g., education, occupation, income) for the Afro-Americans.

After the age of 80, blacks and people of other races combined seem to have lower mortality than whites.[15-17] This "crossover effect" is not completely understood. One suggestion is that after 80 years of age, socioeconomic variables are weaker determinants of mortality than at younger ages. Another suggestion is the "selective survival" of the fittest minority group members: those individuals who have survived severe stress throughout life represent a selected subpopulation genetically endowed with exceptional life vigor.[17,18]

Another group that may be included in this category are the homeless. Although they are typically young, the number of homeless elderly is growing.[16,17] The homeless have higher than expected mortality rates and numerous health problems.[19-21] They represent a unique group to study the interaction of aging with a variety of illnesses and exposure to unfavorable environments. Similarly challenging is the study of aging of prison inmates among whom drug abuse is frequent.

Studies of the demographics and epidemiology of minority groups are just beginning. They should be pursued actively as they offer unique opportunities to address important issues such as the role of genetic factors in chronic illnesses, the interactions of drug abuse and aging, and the optimization of long-term, intensive delivery of health care services.[14]

TABLE 2-2

Life Expectancy (in Years) at Birth and at Age 65 According to Race and Gender

	Black		White	
Age	**Men**	**Women**	**Men**	**Women**
Birth	65.2	73.5	72.0	78.8
65 years	13.4	17.0	14.8	18.7

Data from U.S. Bureau of Commerce (1990).

1.5 Aging in Primitive and Ancient Populations

Comparison of the aging status of ancient and primitive populations with populations in historical times (including modern and contemporary periods) discloses a consistent but slow progression. Extremely ancient human populations take us back to the neolithic and paleolithic eras of the old world, millennia before the beginning of history. The information we have concerning survivorship in the prehistoric old world prior to 10,000 years ago is derived entirely from skeletal material and can be expressed only in terms of age at death.[22] According to the limited number (less than 100 specimens) of studies, it is evident that few persons survived past the age of 40. As shown in Figure 2-4, in the neolithic period, a 5-year-old had a life expectancy of 20 more years; at 20 years of age, perhaps 14 more years; and at 60 there were no survivors. Despite the sketchy data, the findings of several investigators show that in the Stone Age, before the rise of the Mesopotamian and Egyptian civilizations, aging, as we know it today, simply did not exist.

Not all investigators agree with this point of view. In their opinion, the evidence for the paleolithic and neolithic eras would be little more than a guess. Many, perhaps with reason, say that the incidence of infectious diseases has grown with the increased density of population made possible by the development of agriculture. Early humans, living under simpler rural conditions, may have had a longer lifespan than revealed by the few fossils available.

There is little evidence of progress in lowering mortality until perhaps the 18th century. Nevertheless, some reports suggest that, with continually improving conditions of life, life expectancy progressively increased. A Roman at 10 years of age had 30 more years of life to look forward to and, at 60, perhaps 5 more years of life to enjoy (Figure 2-4).

Taking only a few selected examples throughout modern times, we see that in the France of 1800 and up to 1900 in the U.S., life expectancy was relatively short until 10 years of age, a sign of high mortality due to childhood diseases. In the U.S., life expectancy started to increase in the second half of the 19th century, probably due to the improved hygiene, more efficient prevention and treatment of infectious diseases, and higher income (hence, better nutrition). This progress has rapidly accelerated since the beginning of the 20th century, and it is expected that it will continue into the 21st century.

In contrast to the historical trend in lengthening of the average lifespan, the maximum lifespan has remained essentially unchanged throughout the centuries. In ancient, as in contemporary times, the number of centenarians remains limited and only a few among the longest-lived individuals reach the maximum age (see below).

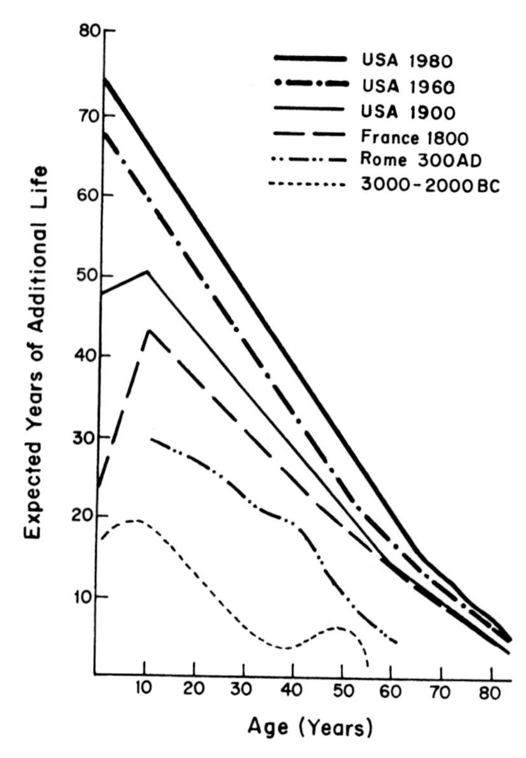

FIGURE 2-4 *Lifespan through the ages in various populations.* Life expectancy at birth has increased dramatically since neolithic times. The conquest of infectious diseases has raised the life expectancy during childhood years, but nothing has effectively prolonged lifespan well beyond age 65, although more people now reach this age.

1.6 Current Comparative Lifespan in Various Countries

Prolongation of life expectancy from prehistoric through historic times and its marked increase in this century are not equally shared today by all world populations. As indicated earlier, the standard of living, particularly in terms of sanitation and nutrition, is very important in the survivorship of human populations. Indeed, the socioeconomic and public health conditions of a population can be inferred from data regarding early childhood mortality and the average length of lifespan: the lower the life conditions, the higher the childhood mortality and the shorter the lifespan.

Essential differences among countries are manifested in early age rather than later in the lifespan. Even in populations with a shorter average lifespan, a few individuals are capable of living as long a life as those from populations with a longer average lifespan. Individuals with unusually long lifespans have been found in populations living in a rather hostile environment and under marginal economic conditions. It is to be noted that in some of the developing countries, life expectancy, although still remaining well below that of economically advanced countries, has shown extraordinary progress since the 1950s (Table 2-1).[1] In India, for example, life expectancy — which at the end of World War II ranged from 27 to 30 years — has now increased to 55 to 58 years. The difficulties of collecting reliable census data in such a large country with such varied populations are obviously enormous, and any demographic study should be carefully evaluated. Nevertheless, some of the available figures are striking. According to some estimates of the sample registration system from the 1981 census,[23,24] life expectancy at birth for 1970 to 1975 was 50.5 years for males and 49 years for females, and this has increased in 1981 to 58 years for males and 55 for females. Keeping in mind that India's total population is about 700 million, and that this population will soon witness a marked increase in the number of individuals 50 years old and older, one can appreciate the enormous impact that such an increase will have on that country's resources.

Although we can view with some satisfaction the considerable progress in sanitation and medical care that this increased life expectancy represents for India and other developing countries, we can only be overwhelmed by the economic and social implications that such a lengthening of the lifespan implies for such a large population. Clearly, it is none too soon for planning a vigorous and effective program for the support of this so far minor, but rapidly expanding, segment of the population. While the major concern remains survival of the young during their early years (together with reduction in the number of births) and improvement of the health of the general population, the realization of an increased number of individuals over 50 years who will compete for medical and welfare benefits must be addressed immediately.

1.7 Maximum Lifespan in Humans

Among mammals, humans represent the longest-living species.[25,26] Documented records have fixed the longest human lifetime at 114 years, although there are current reports (not yet verified) of lifetimes of 117 to 120 years. Accurate records of human longevity are difficult to collect; not only are they subject to unavoidable recording errors, especially in countries where demographic records are not kept consistently, but it is well know that, beyond a certain age, elderly individuals show a tendency to exaggerate their age.[27]

Claims of extreme longevity abound in particular regions.[28] For example, Russian investigators have frequently reported the celebrated longevity of some inhabitants of the Caucasus and of Siberia who claim to be 130 years old. Other unusually long-lived populations have been studied in the high valley of Ecuador and in the Karakoram Mountains, west of Tibet. The subjects interviewed were very old individuals, even centenarians, but their exact age and their proportion in the total population have been questioned.[29,30] In these populations, conditions are generally extremely arduous and caloric intake is low; genetic isolation and separation from urban viruses and pollution may be some of the key factors in the longevity of these populations. Again, a note of caution is needed. It has been clearly demonstrated that age in these populations has been grossly overstated and, therefore, it is dangerous to attempt to find any cause-effect relationship.[31]

2 COMPARATIVE PHYSIOLOGY OF AGING

2.1 Selective Longevity

Senescence may be viewed as a manifestation of the process of adaptation. In evolutionary terms, the problem of adaptation has been solved by the theory of natural selection, that is, (1) by the selection of inherited characteristics most favorable for survival and reproduction in a particular environment, and (2) by passing the genes specifying these characteristics to the succeeding generations. In this sense then, aging would represent the declining force of natural selection with age. The survival of the species demands the aging and death of its members. Survival beyond the period of reproductive activity represents a luxury that few species can afford. Contrasting the immortality of the germ line with the mortality of somatic cells (the soma) suggests that, at least in some species, limiting the individual lifespan may be a positively beneficial adaptation. The so-called "disposable soma theory" estimates that the investment humans have made in protecting somatic cells (beyond the energy required for maximizing fertility) would provide enough energy until about 40 years; from that age on, the ability to respond to stress would be progressively lost until death would ensue.[32,33]

In many species, however, and certainly in humans, longevity can theoretically be subject to positive selection based on criteria of fitness other than reproductive capacity. Although much has been written about such reasoning, the question still remains — why is the lifespan of most species finite? One approach to this question is to compare life duration among several phyla and, from this comparison, to extract principles applicable to the aging process in general (Table 2-3).

Although the interest of gerontologists in obtaining accurate information on the lifespan of animals has increased considerably in the last few decades, much of the literature and knowledge of the nature of aging in animal and plant species is still incomplete.[34,35] Senescence is assumed to occur in all vertebrates and leads to death, hence its qualification as being *universal, progressive, deleterious, and irreversible*.[36] It is characterized by an exponential increase in mortality and is terminated by death. However, there are certain life forms — particularly plants and invertebrates — in which aging may not occur at all.[37]

2.2 Physiologic Correlates of Longevity

Among common factors governing aging in animals is *brain weight and its relation to body weight*. Data from several orders of placental mammals show a highly significant relationship between lifespan and *body weight:* the bigger the animal, the longer the lifespan. For example, the elephant may reach or exceed 70 years in captivity, whereas the rat seldom lives more than 3 years. There are, however, many exceptions to this generalization: humans may reach 114 years of age, whereas other, larger mammals show a shorter potential longevity (horse: approximately 60 years; hippopotamus and rhinoceros: approximately 50 years; bear: 30 years; camel: 25 years). Among domestic carnivores, cats, although generally smaller in size than dogs, live longer.

The same data show an even more significant relationship between lifespan and *brain weight*. For example, insectivores with a smaller, simpler brain have a shorter lifespan than ungulates, and these have a simpler brain and shorter lifespan than humans. Humans with the longest lifespan have the heaviest brain in relation to body weight and also the most complicated structurally.[38] Among 12 major brain regions, the neocortex (the largest and most recently developed part of the cerebral cortex) shows the strongest correlation with the lifespan.[39]

TABLE 2-3

Physiologic Correlates with Longevity

Index Studied	Correlation
Body weight	Direct
Brain/body weight	Direct
Basal metabolic rate	Inverse
Stress	Inverse
Reproductive function/fecundity	Inverse
Length of growth period	Direct
Evolution	Uncertain

Important physiologic implications can be extrapolated from these data, such as the relation between brain size and longevity. The higher the brain weight is relative to the body weight (the higher the brain/body weight ratio) and, particularly, the greater the degree of cerebral cortical expansion (i.e., the process of encephalization), the more precise are the physiologic regulations and thus the greater chance for longer survival.[40] Such an interpretation seems well justified by the essential role the nervous system plays in regulating vital physiologic adjustments, especially responses to environmental demands. Maintenance of the physiologic optimum and reduction of the magnitude of the fluctuations occurring over time diminish the probability of irreversible changes per unit time and, thus, the rate of aging and the incidence of death.

Appealing as this interpretation may be to a physiologist, some reservations must be kept in mind when formulating direct correlations between body and brain size and lifespan. The correlations themselves are not always appropriate when the terms of comparison represent different entities, one essentially anatomical (stature, brain weight) and the other essentially evolutionary, functional, and biochemical (duration of life). Another limitation is the paucity of data on the maximum lifespan of most animal species. While the brain is a functionally most important organ, a similarly positive relationship exists between size of several organs (e.g., adrenal, liver, spleen) and lifespan.[41]

Another criterion of the relationship between a physiologic parameter and the length of the lifespan is *basal metabolic rate* (BMR): the higher the metabolic rate, the shorter the lifespan (Table 2-3).[42] BMR represents the amount of energy liberated per unit time by the catabolism of food and physiologic processes in the body. It is measured at rest, in a comfortable ambient temperature and at least 12 h from the last meal. BMR is expressed in kilocalories per unit of body surface area and can be compared among different individuals. The energy thus liberated appears as external work, heat, and energy storage species.

Comparison of shrews and bats shows that shrews, with the highest metabolic rate of all mammals, have a lifespan of 1 year compared with bats with a much lower metabolism and a lifespan of over 15 years.[32,43] The higher metabolic rate would accelerate the accumulation of nuclear errors (DNA damage) or of cellular damage (accumulation of damage by free radicals) and thereby shorten the lifespan.

The Pacific salmon and the Atlantic eel are prime examples of the inverse relationship between accumulation of damage, stress, and length of lifespan. Both fish age rapidly at the time of spawning and die shortly thereafter. However, if spawning and the associated stress are prevented, they will continue for several years.[44] Lower temperature would increase longevity by decreasing immune responsiveness and thereby preventing autoimmune disorders.[45]

Fecundity, as an expression of reproductive function and measured by the number of young born per year of mature life, appears also to be inversely related to longevity. Shrews, with a short lifespan, have a large litter size and produce two litters per year, whereas the longer-lived bats have only one young per year. *Duration of growth* also has been related to the lifespan. For example, comparison of the chimpanzee to humans shows that growth periods last approximately 10 years in the chimpanzee and 20 years in humans, and their respective lifespans are approximately 40 and 100 years.

Duration of the growth period can be prolonged in experimental animals, primarily rodents, and the onset of maturation can be delayed by restricting food intake in terms of total calories or of some specific dietary components. With these dietary manipulations, not only is the lifespan prolonged, but some specific functions, such as reproduction and thermoregulation, are maintained until advanced age; also, the onset of aging-related pathology is delayed and its severity is reduced.[46,47]

Several factors superimposed on the genetic makeup — including body and brain size, metabolic rate, fecundity, duration of growth, as well as stress and conditions of life — undoubtedly contribute to the difference in longevity among species. As discussed above, natural selection suggests that the individual member of a group must survive through the reproductive period to ensure continuation of the species; thereafter, survival of the postreproductive individual becomes indifferent or detrimental (e.g., food competition) to the group. In this sense, a gene that acted to ensure a maximum number of offspring in youth, but produced disease at later ages, might be positively selected. In modern times, not only is life expectancy increased, but humans live well beyond the reproductive years, women some 40 years beyond the childbearing period. Longevity of a species beyond the reproductive years must be incidental to some earlier events, or the infertile individual must confer some advantage to the fertile. In our society, this must be the case, for older members do contribute to the maintenance of the entire population structure and to the development and progress of our society.[48]

Selective Immortality

There are certain life forms — particularly plants and invertebrates — in which aging may not occur at all. Among the unicellular organisms, protozoa have received considerable attention, but no one has definitively demonstrated or negated their immortality. For example, certain strains of *Paramecium* maintained in culture may be capable of surviving indefinitely without conjugation and cross-fertilization, whereas other strains do not survive unless periodically cross-fertilized. Other unicellular organisms, such as some *ciliates,* undergo age-related changes that involve not only cell size and a change in cell shape but also alterations in cell skeleton and cell behavior.

Fruit flies may be made to live longer by selective breeding of long-lived progenitors.[49] The difference between long-lived and normal specimens would be the presence of a variant of a normal enzyme-specifying gene for the antioxidant superoxide dismutase. The enzyme would be particularly active in the long-lived flies and provide a stronger protection against the damage induced by accumulation of free radicals (Chapter 6). Bacteria, yeast, and clones of many kinds of plants may not exhibit senescence. Similarly, metazoa, such as sea anemones, appear to be ageless; in these jellyfish, the individual "jellies" have a fixed lifespan, but the larval "stub" produces a constant supply of new blanks as old ones are removed and the specimens remain vigorous indefinitely. Even in some vertebrates, such as large fish and tortoises, the process of aging is so slow as to be almost undetectable. From these and similar studies, it may be concluded that there is no inherent factor common to all forms of life that automatically produces senescence. *"Gerontogenes"*, that is, genes inducing senescence, have been described, but firm evidence of their presence is not yet available. Rather, *"longevity-determining genes"* or gene mutations responsible for increasing the lifespan have been identified in worms and yeast (Chapter 4).

The absence of signs of aging cannot be equated with immortality; the advantage of an ageless organism consists simply in the fact that its lifespan is not affected by the passage of time; that is, it still may die, but from a cause that can kill it at any age. In contrast, among the animals that undergo the characteristic age-dependent loss of vigor with advancing age, the mortality rate steadily increases at a rate typical for each species, thus producing a finite lifespan.

3 DIFFERENTIAL AGING IN HUMANS

3.1 "Successful" versus "Usual" Aging

Chronologic age (age in number of years) and *physiologic age* (age in terms of functional capacity) do not always coincide, and physical appearance and health status often belie chronologic age. Although no specialized knowledge is required to estimate an individual's age, in many cases a person may look younger or older than his/her chronologic age. From a physiologic standpoint, such disparities in the timetable of aging may occur among individuals or among selected populations; they result from complex interactions between genetic and environmental factors that operate on the individual as the child of her/his parents and also as a member of a societal group. Studies on aging have often ignored or attributed differences among individuals of the same age to specific genetic endowment, yet one of the characteristics of the aging human population is its substantive heterogeneity; that is, some individuals "age" at a much slower or faster rate than others. Accordingly, aging processes have been divided into *"usual aging"*, referring to the average physiologic changes, and *"successful aging"*, referring to those individuals who show minimal functional decrements with aging.[50]

One of the major difficulties in the study of aging is *the lack of biomarkers of aging*. Although the search for such biomarkers is an ongoing process, its ultimate outcome is uncertain. It is assumed that it is possible to define the phenotype of aging. However, according to some gerontologists,[51] impediment to progress in the study of aging is due to the very paucity of reliable and valid biomarkers. Rather, aging appears to involve numerous processes and to be highly polygenic in its determination. Possibly, a number of different mechanisms are responsible for the phenotypic alterations: different individuals would exhibit various patterns of aging, based on their specific inheritance and environmental experiences.[52]

The distinction between usual, i.e., average aging, and successful, i.e., better than average aging, supports the hypothesis that extrinsic factors play an important role in age-associated functional decline. As we consider the many cross-cultural differences that greatly influence aging, it is difficult to ascribe change to age alone. Attributing change to age *per se* may often be exaggerated, and factors such as diet, exercise, drugs, and psychosocial environment should not be underestimated or ignored as potential moderators of the aging process. Taking these elements into account, the prospects for avoidance, or eventual reversal, of functional loss with age are vastly improved and the risks of adverse consequences are reduced.[50]

3.2 Heterogeneity of Physiologic Aging

Changes with aging lack uniformity, not only among individuals of the same species but also within the same individual; onset, rate, and magnitude of changes vary depending on the cell, tissue, organ, system, or laboratory value considered.[53-55] Traditionally, in humans, the physiologic "norm" is represented by the sum of all functions in a 25-year-old man, with a weight of 70 kg and a height of 170 cm, who is free of any disease. Comparison with this "ideal man" inevitably discloses a range of functional decrements with advancing age. Early studies were conducted in selected samples of "representative" elderly.[53] As the prevalence of chronic disease increases with age (Chapter 3), a large part of the data of functional loss with age in these studies may have been due to the effects of disease rather than the natural concomitants of aging itself. In these earlier studies, comparison of several functions from young to old age focused on a gradation of decrements with old age.

More recent research has challenged the inevitability of functional impairment with aging. Significant laboratory changes may be erroneously attributed to aging, and normal aging changes may be misinterpreted as evidence of disease. An example of the heterogeneity of aging changes is illustrated by a number of laboratory values, many of which remain unchanged with aging, while a few decrease or increase. Laboratory values are always given within a normal range, and many of the age-related levels do not change in absolute terms but only alter value distribution within this normal range (Table 2-4).[54]

Successful aging is a demonstration that aging can occur with little loss of function. Thus, in certain functions, regulation remains quite efficient until advanced age; in others, it may decline at an early age. Examples of this type of differential aging include fasting blood glucose levels and acid-base balance, which remain stable as late as 70 to 90 years. In contrast, the basal metabolic rate declines continuously throughout the lifespan. Certain sensory modalities, such as vision and hearing, show functional decrements beginning in early adulthood. One classic example of a unique timetable involving an organ that develops and ages during a specific period of the lifespan is the ovary: it begins to function at adolescence (in humans, approximately, 10 to 12 years) and ceases to function at menopause (in humans, as long as 40 years before death; Chapter 12). Other classic examples include embryonal structures, such as the placenta, which develop and age within a relatively short period of time as compared to the total length of the lifespan. Whichever the organ or tissue considered, timetables of aging represent an approximation, for the onset of aging cannot be pinpointed precisely by a specific physiologic sign.

Because aging is a slow and continuous process, some of its effects can be observed only when they have progressed sufficiently to induce alterations that can be identified and validated by available testing methods. An illustrative example is atherosclerosis, the consequences of which become manifest

TABLE 2-4

Laboratory Values in Old Age

Unchanged	Decreased	Increased
Hepatic function tests	Serum albumin	Alkaline phosphatase
Serum bilirubin	HDL cholesterol (women)	Uric acid
Aspartate aminotransferase (AST)	Serum B_{12}	Total cholesterol
Alanine aminotransferase (ALT)	Serum Magnesium	HDL cholesterol (men)
Gamma-glutamyltransferase (GGTP)	PaO_2	Triglycerides
Coagulation tests	Creatinine clearance	TSH (?)
Biochemical tests	T_3 (?)	Glucose tolerance tests
Serum electrolytes	White blood count	Fasting blood sugar (within normal range)
Total protein		Postprandial blood sugar
Calcium		
Phosphorus		
Serum folate		
Arterial blood test		
pH		
$PaCO_2$		
Renal function tests		
Serum creatinine		
Thyroid function tests		
T_4		
Complete blood count		
Hematocrit		
Hemoglobin		
Red blood cells		
Platelets		

Adapted from Cavalieri, T. A., Chopra, A., and Bryman, P. N., *Geriatrics,* 47, 66, 1992. With permission.

in middle and old age even though the atherosclerotic lesion may start early in infancy (Chapter 16).

As mentioned above, several functions, as is the case of fasting blood sugar values, are minimally affected by aging. However, when blood sugar levels are tested after increased physiologic demand (e.g., a sugar load as in the sugar tolerance test), the efficiency with which the organism is capable of maintaining levels within normal limits and the rapidity with which these levels return to normal demonstrate marked differences between adult and old. Similarly, conduction velocity in nerves, cardiac index (cardiac output per minute per square meter of body), renal function (filtration rate, blood flow), and respiratory function (vital capacity and maximum breathing capacity) are less capable of withstanding stress in old rather than in young individuals. Thus, placing a system under stress brings to light age differences not otherwise detectable. It also clearly demonstrates the declining ability of the aging organisms to withstand or respond adequately to stress (Chapter 11).

Other functions of the body begin to age relatively early in adult life and fall to a minimum before the age of 65, officially heralding the stage of senescence. For example, in the eye, accommodation begins to decline in the teens and regresses to a minimum in the mid-fifties. Similarly, the auditory function begins to deteriorate at adolescence and deterioration continues steadily thereafter, culminating around 50 years of age. This auditory deterioration may also be hastened by the environmental noise to which, in our civilization, individuals are continuously exposed from young age. Some comparative studies in isolated populations living in a quiet environment (e.g., forest-dwelling African tribes) and maintaining a good auditory function into old age seem to support the view that continuous exposure to noise may be harmful to the auditory function. However, many other factors in the life-style may also be operative. A more complete discussion of changes with aging in various functions is presented in the corresponding chapters of this book.

Changes in Body and Organ Weight

Anatomical changes vary greatly with age of onset, structure examined, and the individual's sex and history. Some prominent changes that occur in body morphology with aging include stature (or standing height), sitting height, breadth of shoulder, and depth of chest — all of which show a progressive reduction with aging in contrast to cephalic diameters, which remain practically unchanged. The *considerable and progressive diminution in stature* that appears to occur in all humans may be ascribed, at least in part, to alterations (osteoporosis) in bone structure to be discussed in Chapters 12 and 21.

Other age-related changes in body morphology are represented by an *increase in fat deposition* particularly evident in the subcutaneous layer, where it is easily quantifiable by measuring the thickness of the skinfold at a number of points in the body. Fat deposition increases in parallel with body weight, reaches a peak at approximately 50 years, and progressively declines thereafter. Although body weight usually increases at the initial stages of senescence, it decreases in later stages.

Weight loss begins in humans between the fifth and seventh decades, a period of increased mortality, and involves primarily the skeletal muscles, liver, kidney, and adrenal. In association with the decrease in body weight, the *weight of organs declines* with aging. Although a comprehensive tabulation of changes in organ weight with age exists only for humans, features such as skeletal muscle atrophy are typical of mammalian aging. In humans, as in other mammals, this skeletal atrophy may be due not only to degenerative changes in the muscle itself but also to the decreased use (disuse) of the muscles due to impairment of other functions (e.g., sensory, peripheral, and central nervous motor) as well as the changing life style of the older individual (Chapters 3 and 21).

Brain weight declines significantly, but less than that of the above organs except in pathologic conditions in which brain atrophy may be quite severe, as in most cases of senile dementia of the Alzheimer type (Chapter 8). In contrast, *heart weight* does not change and, in fact, may even increase, especially when measured relative to whole-body weight. This is an indication of the attempt by cardiac muscle to compensate for its declining functional activity and the increased peripheral resistance (as a consequence of atherosclerosis). Compensation is achieved through hypertrophy of contractile cells (myocytes) and/or of interstitial cells (connective, conductive, and endothelial cells). The weight of some *bones,* such as the ribs, decreases with aging, whereas that of others, such as the sternum, increases with age.

In view of the high incidence of pathology with aging, sudden or progressive changes in body weight, abnormal in severity and timing, may also be indicative of onset and course of disease. In this manner, body weight represents a relatively simple measure of the physiologic and, eventually, pathologic assessment of the health status of the elderly (as it is also an indicator of the growth progress during development).

3.3 Persistence of Physiologic Competence with Aging

So much negativity has been attached to physiologic decline with aging that it may be appropriate here to underline the positive aspects as well. A positive tack should consider the many compensatory responses that go into effect to successfully offset such decline and thereby permit efficient adaptation and survival well into old age.

Successful aging implies persistence of adequate function of several organs and tissues into old age. Many mechanisms emerge in later years that compensate for specific types of functional loss (as discussed throughout this book). The potential for rehabilitation at advanced ages is much greater than previously supposed.[55,56] At all levels of disability, there is a significant probability of regaining function.

Undoubtedly, a number of functions decline in individuals 65 years and older as compared with those of 25 years. As people live longer, the classic profile of the 25-year-old man, taken as the standard of optimal physiologic competence against which all other physiologic patterns must be compared, appears unrealistic and outdated. Heterogeneity among old persons increases with age. The more than 2 million Americans who are now (1993) 85 years of age and older show numerous differences in their physiologic and pathologic profiles. This extreme heterogeneity of functional status even in the oldest-old supports the view that aging must be evaluated on an individual basis.[52] In addition to genetic make-up, health, social status, economic and environmental conditions are responsible for the demographic and epidemiologic history of the elderly.

Competence of the elderly must take into account the increasing demands imposed by the pathology associated with old age (Chapter 3). Even in those cases where the disease process is not currently curable (e.g., Alzheimer's disease), appropriate management may slow the functional loss associated with the disease by maximizing environmental input and social support.[57]

The association of aging with pathology affects the orientation of research in this area. Two main approaches are currently extant:

1. Some physiologists follow the traditional view that research in aging should be aimed at specific functional and clinical entities (e.g., cardiovascular aging and atherosclerosis; brain aging and Alzheimer's disease) rather than at aging *per se.*
2. Other physiologists argue that treatment and even cure of many diseases would prolong life only a few years; they strongly support research on the aging process itself.

Both research goals should be pursued simultaneously.

As 25 years no longer represent half of the lifespan, the physiologic profile of the old individual must be evaluated against a standard, optimal for each specific age. Such a standard should serve as a measure against which to assess the physiologic competence of individuals of comparable age, at different stages of aging. Standards of competence should parallel progressive age. If physiologic competence represents the capacity for survival, then the old individual may well be considered competent.

3.4 Assessment of Physiologic Age in Humans

Assessment of physiologic competence in humans, at any age, is a multifactorial process. It requires quantitative measurements of numerous parameters selected as indices of physical, neurologic, and behavioral competence at progressive ages.[58-61] To establish an accurate profile that will reflect the different age-related timetables for body systems and combine them to represent the health status of the individual, a number of criteria must be satisfied; these criteria should account for many variables, among which some of the prerequisites are as follows:

- The variables must be indicative of a function important to the competence or general health of the individual and capable of influencing the rate of aging.
- They must correlate with chronologic age.
- They must change sufficiently and with discernible regularity over time to reveal significant differences over a 3- to 5-year interval between tests.
- They must be practically measurable in an individual or cohort of individuals without hazard, discomfort, or expense to the participant or excessive labor or expense for the investigator.

The validity of any assessment lies in the choice of the right test or battery of tests best qualified to provide an overall picture of current health and to eventually serve as a basis for prediction of future health and length of the lifespan. Such a choice is complicated by the need to take into account also the financial feasibility and the facilities available for the testing. The relatively large number of tests for assessing physiologic competence and health status in the elderly reflects the current failure to reach a consensus on the best checklist.

A global measure of physiologic status may be derived from many different combinations of tests; selection will depend on the purpose of the assessment and who will use these data. *With respect to purpose of assessment,* measurements may be expected to:

- Describe the physiologic status of an individual at progressive chronologic ages
- Screen a selected population for assessment of overall physiologic competence or competence of specific functions using either cross-sectional or longitudinal sampling methods (see below)
- Monitor the efficacy of specific treatments, drugs, exercise, diet

- Predict persistence or loss of physiologic competence, to determine incidence of disease, and to evaluate life expectancy

With respect to the user, the choice of tests will depend on whether he/she is a health provider, specialist, researcher, or case manager. Even with a precise identification of the nature of the assessment needed and of the user who needs it, it still remains difficult to choose the most significant and feasible tests.

Some tests, although relatively innocuous in young and healthy individuals, may be troublesome for the elderly.[62-65] Yet, it may be contrary to the interest of the elderly to assert that *they represent a vulnerable group* needing special protection. Rather, benefits may accrue for the elderly from their participation in medical and psychosocial survey research. Not only may they lead to the discovery of an unsuspected illness and its eventual treatment, but they may also provide altruistic commitment and mental satisfaction.[66]

3.5 Geriatric Assessment

Geriatric assessment involves a multidimensional diagnostic process (including physiologic assessment) designed to qualify an elderly individual not only in terms of functional capabilities and disabilities but also of medical and psychosocial characteristics (Table 2-5). It is usually conducted by a multidisciplinary team with the intent of formulating a comprehensive plan for therapy and long-term follow-up. Major purposes are (1) to improve diagnosis (medical and psychosocial), (2) to plan appropriate rehabilitation and other therapy, (3) to determine optimal living location, arrange for high-quality follow-care and case management, and (4) to establish baseline information useful for future comparison. Most of these assessment programs include several tests, which have been grouped into three categories:

- Tests that examine general physical health as represented by physiologic competence and the absence of disease
- Tests that measure the ability to perform basic self-care activities, the so-called *activities of daily living* (ADL) (Table 2-6)
- Tests that measure, in addition to basic activities, the ability to perform more complex *instrumental activities daily living* (IADL) (Table 2-6)

The incidence of disease and pathology are discussed in Chapter 3. ADLs and IADLs are widely utilized as representative of measures useful in home-dwelling populations and as representative of capability for independent living or, *vice versa,* as indicators of disability (Figure 2-5). *Disability is the inability to perform a specific function because of health or age and results from impaired functional performance.* In most testing, the degree of wellness, i.e., the absence of disability/disease, is recorded and, reciprocally, the presence and severity of disability/disease are recorded as well. The severity of the disability may be measured in terms of whether a person (1) does not perform the activity at all, (2) can only perform it with the help of another person or if a person is available (but does not actually give aid), or (3) can perform it with the help of special equipment. Disability is coded according to five degrees of severity: (1) no disability; (2) at least one IADL disability but no ADL; (3) one or two ADL disabilities; (4) three to four ADL disabilities; and (5) five to six ADL disabilities.[67]

With advancing age, disability intensity increases (Figure 2-5), with the highest disability at 85 years and older (85+). It is to be noted that the greater intensity of disability of women than men becomes manifest at the later ages of 75 to 85+ years. Thus, females with a longer average lifespan

TABLE 2-5

"Simple" Functional Assessment of Ambulatory Elderly

History

Physical Examination

Including neurologic and musculoskeletal evaluation of arm and leg, evaluation of vision, hearing, and speech

Urinary Incontinence (eventually fecal incontinence)
  presence and degree of severity

Nutrition
  dental evaluation
  body weight
  laboratory tests depending on nutritional status and diet

Mental Status
  Folstein Mini-mental Status Score
  If score < 24, search for causes of cognitive impairment

Depression

If Geriatric Depression scale is positive:
  Check for adverse medications
  Initiate appropriate treatment

ADL and IADL (see Table 2-6)

Home Environment and Social Support
  Evaluation of home safety and family and community resources

Adapted from Lachs, M. S., Feinstein, A. R., Cooney, L. M., Jr., et al., *Ann. Int. Med.,* 112, 699, 1990. With permission.

TABLE 2-6

Categories of Physical Health Index Measuring Physical Competence

Physical health	Activities of daily living	Instrumental activities of daily living
Bed days	Feeding	Cooking
Restricted-activity days	Bathing	Cleaning
Hospitalization	Toileting	Using telephone
Physician visits	Dressing	Writing
Pain and discomfort	Ambulation	Reading
Symptoms	Transfer from bed	Shopping
Signs on physical exam	Transfer from toilet	Laundry
Physiologic indicators (e.g., lab tests, X-rays, pulmonary and cardiac functions)	Bowel and bladder control	Managing medications
Permanent impairments (e.g., vision, hearing, speech, paralysis, amputations, dental)	Grooming	Using public transportation
Diseases/diagnosis	Communication	Walking outdoors
Self-rating of health	Visual acuity	Climbing stairs
Physician's ratings of health	Upper extremities (e.g., grasping and picking up objects)	Outside work (e.g., gardening, snow shoveling)
	Range of motion of limbs	Ability to perform in paid employment
		Managing money
		Traveling out of town

Note: Many of the items presented are components of several measures of physical and functional health as discussed in a number of geriatric screening and assessment programs.[58-61]

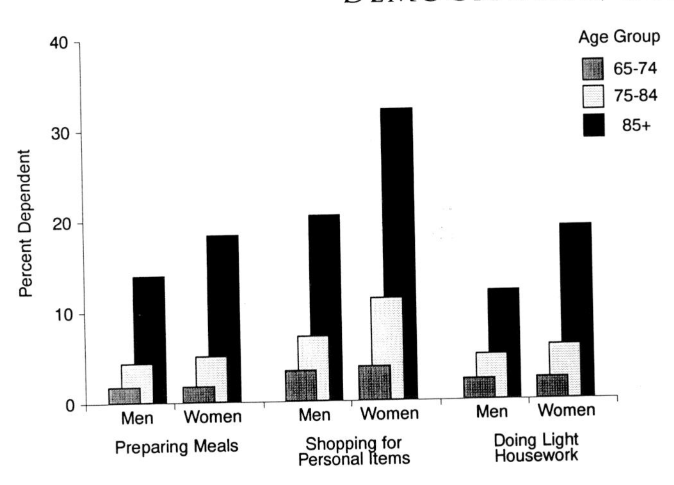

FIGURE 2-5 *Percentage of population dependent in three instrumental activities of daily living (IADL) (U.S., 1984).* Dependence is defined as having difficulty or being unable to perform a specific activity by oneself because of a health problem. Dependence increases with age in all three types of activity. At older ages (74–85+ years), women show a greater degree of dependence than men. (From Guralnik, J. M. and Lacroix, A. Z., *The Epidemiologic Study of the Elderly,* Wallace, R. B. and Woolson, R. F., Eds., Oxford University Press, New York, 1992, 159. With permission.)

than men (see above) live longer with disability. The cause of this sex difference is unknown. Probably, females have higher risk of a number of chronic degenerative conditions (e.g., osteoporosis, diabetes, arthritis) that interfere with those functions (e.g., walking, doing housework) necessary for independent living.

Evaluation of Geriatric Screening and Assessment Programs

Geriatric assessment traces its origins to the pioneering work of British geriatricians in the 1930s and 1940s. They demonstrated that end-stage, bed-bound, geriatric patients could be dramatically mobilized and many discharged to community settings following careful assessment and rehabilitation. Today, in this country, an increasing number of such assessment programs are available within a variety of health care structures, ranging from hospitals to nursing homes to out-patient clinics.

Two basic tactics for the study of human aging, using the tests discussed above, are the *cross-sectional study* and the *longitudinal study. Cross sectional methods* compare characteristics among different age groups at one time. Studies with this method can be conducted with relative accuracy and rapidity in a large group of people; for example, a population with a broad span of ages, perhaps birth to 90 years, is sectioned into narrow, age-defined subsets, and measurements are made in identical fashion in each group. This type of study has been used in gerontologic research for such programs as short-term testing of new drugs or regimens capable of influencing some aspects of the aging process. This method has also been used profitably in animal studies conducted in such species as rats and mice, which can be raised and maintained under special, but standardized conditions (e.g., diet, drugs, exercise).

Human populations, however, live under widely different environmental conditions, and the same measurements obtained in cohorts of individuals of the same age are subject to multifactorial influences only one of which is aging per se. Thus, the cross-sectional study, although less time consuming, is subject to a number of errors. For example, because of the secular trend in stature (that is, changes — in this case, an increase in stature — that occur in a population as a whole over time), the 20-year-old in 1980 is taller than was the 20-year-old in 1910, suggesting an aging-related decrease in height, the oldest people in comparison being born when body height was lower. Similarly, differential survivorship depends on the selected survival only of these individuals with a particular trait. To continue with the example of stature, tall, lean individuals may tend to live longer than the short and obese; if true, this would bias the outcome of the study.

The *longitudinal study* avoids some of these errors and therefore is the preferred method of many gerontologists. In the longitudinal study, the same individuals are examined at regular intervals throughout the lifespan so that the process of aging can be compared in a dynamic fashion, each individual being his/her own control between two or several ages. Longitudinal studies have been used with great success in the study of growth processes. In aging, however, this method has many obvious disadvantages, especially in the study of human populations; not only is the human lifespan relatively long, but in industrialized countries, human populations are extremely mobile and many of the components of the cohort are lost during the study.

A compromise approach combining cross-sectional and longitudinal studies is often used; in this type of study, initial data from cross-sectional studies are supplemented and corrected by data from longitudinal follow-ups. Another practical approach is to restrict the longitudinal survey to "critical" life periods (for example, 5 years before and 5 years after menopause or retirement).

In Vitro Measurements of Biologic Age

The complexities and difficulties of formulating indices of biologic age in humans and, eventually, other animals involve, as discussed above, an examination of the differential aging of organ systems, cells, and molecules and their changing responses to internal (e.g., hormonal) and external (e.g., environmental) demands. An alternate approach utilizes single cells removed from the body and cultured under *in vitro* conditions. Such studies establish standardized conditions of cell culture in cells derived from individuals of different chronologic ages and compare specific parameters of cell function (replication, metabolism).

One example is the systematic study of *human skin fibroblast cell cultures* established from skin biopsies of members of a human longitudinal study (the Baltimore Longitudinal Study).[68] *Cell doubling capacity* was taken as one index (from several) of cell reproductive capability; the initial culture flask when the cells had reached confluency (i.e., formed a monolayer of tightly packed cells covering the entire flask bottom) was designated as cell population doubling 1, and the subsequent flasks with the correspondingly subsequent doublings were given progressive numbers. In cultured fibroblasts, the number of doublings is about 50, and this number may be influenced by genetic and environmental factors.[69-71] Cell replication is slower and the number of doublings is reduced in cultures derived from old as compared to young donors. Similarly, other parameters of replicative capability showed a significant decline in the cultures from old donors.

Other examples of the usefulness of the *in vitro* approach to the study of human aging are presented in subsequent chapters throughout this book. Cultured cells and tissues can be utilized to draw correlations with *in vivo* physiologic measurements and can serve as models for the longitudinal analysis of cellular aging. Tissue culture models may also prove useful in mimicking aging pathologies characteristic of specific types of cellular aging (Chapter 3).

4 THE ELDERLY AS AN ASSET

This segment of society constitutes a group of *skilled, literate, often well-informed individuals* who are the products of full, active, and, hopefully, useful lives. They have coped with life for many years and have survived its many problems. Surely, such individuals care to be considered an asset.

One can find impressive examples of elderly people who are successful in public and political life, not only in early history (e.g., Roman senators) and many civilizations (e.g., Chinese elders), but also in current times.[72] Other examples of long and productive lives exist in the arts, humanities, and sciences. As old age is the final phase of growth and development, the longer the individual's life, the greater are the opportunities for the greatest progress. As written almost 2000 years ago, "Nor is age any bar to pursuits of every other kind...to the verge of extreme old age" (Cicero, ca. 44 BC). To live into old age, therefore, gives the greatest chance of fulfillment of potential, not only for those individuals with extraordinary talents but also for ordinary people whose contributions to society, minor as they may be, will be of benefit in the planning of the future of mankind.[73] As written by an active centenarian, "At my age, I stand, as it were, on a peak alone... But that very isolation gives me a less biased view of the vast panorama of human life which is spread before the eyes of a centenarian and...can see clearly the advance of mankind."[74]

Yet there is a tendency among the general public to concentrate on the *negative aspects of age*. We bemoan the stresses and strains placed on our social services as well as on the family finances and goodwill by the elderly, the victims of increased pathology.

Despite the remarkable lengthening of the lifespan during the 20th century, we dwell on the transitory nature of life and the rapid passage of time; from among many similar ones, one exerpt of Edward Young (*Night Thoughts*, 1742–1746, quoted in Sampson and Sampson[75]) — "Our birth is nothing but our death begun" — seems to embody and encourage this pessimistic obsession with aging. In our dislike and dread of aging, we overlook the positive contributions that can be made by the "third age" of life.

There are periods of history — generally associated with wars, natural disasters, or social upheavals — when old age was devalued and despised.[76-78] Today, sociodemographic pressures have renewed public policy and biomedical interest in the elderly. While some face difficult economic conditions, many of the elderly are relatively well-off and represent a new and strong market force. Every discipline and market gradually must modify its views and methods by taking into account characteristics and needs of the older population. Old age, which over the centuries has been essentially a private and family concern, has become a social problem so widespread to attract the attention and intervention of local and national governments. Historians, who so far have had little interest in the history of old age, are now closely following the fluctuations in the social and political role of the elderly. "Every society has the old people it deserves...Every type of socio-economic and cultural organization is responsible for the role and image of its old people."[77]

In a scholarly and compassionate treatise on old age, Simone de Beauvoir[48] rests the analogy between "the infant scarcely come to life and the old man scarcely alive" on the *dependency* shared by both on the young and powerful members of society who possess the vigor and fertility necessary for the survival of the species and the perpetuation of social organization. Thus, whether the elderly members of society are neglected as a group or are integrated into the social structure varies with respect to the values of the society as established by its adult members. We find wide variation in living conditions and active social participation of the elderly in rural versus urban environment, primitive versus technologically advanced societies, nomadic versus sedentary life-styles — all of which, in turn, show significant differences depending on the historic era and specific culture considered.[72] In the strongly organized society of Western civilization, it is expected that the contributions of the elderly, mainly *experience* and various degrees of technical skills, represent a useful commodity; yet, in the view of many, these qualities are often devalued. Practical skills can easily become redundant with technologic advances. What remains of value are the *organizational, advisory, and educational activities* in which the elderly excel. Knowledge of the past is a great asset in these activities, for it may be used to understand the present and foretell the future.

The "third age" should be viewed as a time of opportunity, a phase of life for reviewing one's goals and for making changes. "My remaining days, I may now consider a free gift," wrote Goethe at 81, and a year later, "Even at my great age, ideas come to me the pursuit and development of which would require another lifetime."[73] The special skills and talents of the elderly are valuable assets that can be used to the advantage of the remainder of society. It should be the duty and interest of the young and adult to assure that these talents and skills are fully utilized. There is a need to incorporate older people into meaningful societal roles and establish a clear progression of occupational steps that one ascends with age. Modern societies, despite the concern of many social scientists working in gerontology/geriatrics, have not been successful, so far, in making the necessary innovations that would provide adequately for the elderly. The "leisure town" communities constructed in the U.S. and other Western societies for retired persons of adequate income represent an entrepreneurial venture, more lucrative to business speculators than beneficial to the persons they purportedly serve. Another entrepreneurial venture is the nursing home, which is neither a hospital nor a "home", but has combined the negative aspects of both in confining and restricting the elderly.

In the view of Maddox,[79] the aged should not be isolated as a recipient group for purposes either of living or of policy making. Rather they present problems that are "likely to be the general problems of society, only more intensified." Not all elderly persons are handicapped by illness (as not all young and adult persons are in excellent health!), institutionalized, or dependent on public support. For this large segment of the aged population, the problems of adjusting to a "roleless role" may in large measure be solved by the aged themselves, particularly in a country like the U.S. in which private resources offer a viable adjunct to governmental social services agencies. The chief obstacle to such a self-help group among the aged is the fact that, given such qualifications as physical and mental health and economic independence, one refuses to view oneself as a member of the "subculture" of the elderly. Quite to the contrary, *the elderly have much to offer* and we should not hesitate to use them. We should be proud to learn from their previous experience, including mistakes. In many cases, the older the person, the more precious his/her knowledge and the greater the need to use it for the advantage of the entire society.

Unfortunately, today there is no system to care for the elderly that has sufficient imagination and flexibility to permit full utilization of the *manpower resources* they present. This is true even in countries in which the relative proportion of the aged is high. The services, facilities, and programs that do exist, even in our own technologically advanced society, assume a patronizing attitude that reflects the continuing underestimation of the elderly potential. The phrase "you can't teach an old dog new tricks" also needs to be critically revised. There is ample evidence that the elderly can master new subjects when they have the time and the enthusiasm. For the brain and mental function as for the muscle and physical activity the motto is "use it or lose it!" According to a Japanese proverb, "aging begins when we stop learning." The increasing number of educational associations and the success of the French and British "Universities of the Third Age", catering to the elderly, are proof that elderly are willing and able to learn.

Aging affects us all, personally and through our families. Although this book underlines the changes that occur with aging and the majority of these are of a decremental nature, we must not give way to negative attitudes of gloom and despondency. "I did affirm to my readers in my young years that I was the happiest man I ever met and I reaffirm it at the age ninety-one."[80]

The worldwide dramatic advances in science, technology, and social conditions foreshadow an era in which the elderly will comprise an increasing number of the population whose needs cannot be ignored. And, indeed, these advances have already significantly prolonged the average lifespan in prosperous as well as less economically advantaged countries. To meet the challenge of a growing elderly population, we must try to become more informed about the nature of the aging process and its physiologic and pathologic consequences. We must keep a positive and optimistic approach and maintain continued *high expectations* that can enable us to achieve the goal of growing old successfully.

REFERENCES

1. Hoover, S. L. and Siegel, J. A., International demographic trends and perspectives on ageing, *J. Cross-Cult. Gerontol.*, 1, 5, 1986.
2. Rabin, D. L., Waxing of the gray, waning of the green, in *America's Aging: Health in an Older Society,* Committee on an Aging Society, Institute on Medicine and National Research Council, National Academy Press, Washington, D.C., 1985, 28.
3. Brody, J. A., Brock, D. B., and Williams, T. F., Trends in the health of the elderly population, *Annu. Rev. Public Health,* 8, 211, 1987.
4. Coale, A. J. and Demeny, P., *Regional Model Life Tables and Stable Populations,* 2nd ed., Academic Press, New York, 1983.
5. Fries, J. F., Aging, natural death and the compression of morbidity, *N. Engl. J. Med.,* 303, 130, 1980.
6. Fries, J. F. and Crapo, L. M., *Vitality and Aging,* W.H. Freeman, San Francisco, 1981.
7. Schneider, E. L. and Brody, J. A., Aging, natural death and the compression of morbidity: another view, *N. Engl. J. Med.,* 309, 854, 1983.
8. Siegel, J. S. and Taeuber, C. M., Demographic perspectives on the long-lived society, *Daedalus,* 115, 77, 1986.
9. U.S. Bureau of the Census, *Demographic and Socioeconomic Aspects of Aging in the United States,* Current Population Reports, Series P-3, N 138, U.S. Government Printing Office, Washington D.C., 1984.
10. Lopez, A. D., Sex differentials in mortality, *WHO Chronicle,* 38, 217, 1984.
11. Seely, S., The gender gap: why do women live longer than men?, *Int. J. Cardiol.,* 29, 113, 1990.
12. Moore, T. J., *Lifespan, Who Lives Longer— and Why,* Simon and Schuster, New York, 1993.
13. Smith, D. W., Is greater female longevity a general finding among animals?, *Biol. Rev. Cambridge Philos. Soc.,* 64, 1, 1989.
14. Colsher, P. L., Epidemiologic studies of aging in historically understudied populations, in *The Epidemiologic Study of the Elderly,* Wallace, R. B. and Woolson, R. F., Eds., Oxford University Press, New York, 1992, 287.
15. Jackson, J. S., Chatters, L. M., and Taylor, R. J., Eds., *Aging in Black America,* Sage Publications, Newbury Park, CA, 1993.
16. Jackson, J. J. and Perry, C., Physical health conditions in middle aged and aged Blacks, in *Aging and Health: Perspectives on Gender, Ethnicity, and Class,* Markides, K. S., Ed., Sage Publications, Newbury Park, CA, 1989, 111.
17. Markides, K. S., Ed., *Aging and Health: Perspectives on Gender, Ethnicity, and Class,* Sage Publications, Newbury Park, CA, 1989.
18. Manton, K. G., Epidemiological, demographic and social correlates of disability among the elderly, *Milbank Q.,* 67, (Suppl. 2), 13, 1989.
19. Cohen, C., Teresi, J., Holmes, D., and Roth, E., Survival strategies of older homeless men, *Gerontologist,* 28, 58, 1990.
20. Overbo, B., Ollie Randall Symposium: Homeless elders: a growing national crisis, *Gerontologist,* 30, 120A, 1990.
21. Martin, M. A., The homeless elderly: no room at the end, in *The Vulnerable Aged,* Harel, Z., Ehrlich, P., and Hubbard, R., Eds., Springer Publishing Co., New York, 1990, 149.
22. Mooleson, T. I., Skeletal age and paleodemography, in *The Biology of Human Aging,* Bittles, A. H. and Collins, K. J., Eds., Cambridge University Press, Cambridge, 1986, 95.
23. *Directory and Yearbook,* Government of India, Official Publications, New Delhi, 1981, 264.
24. Hande, H. V., Proc. Eleventh Joint Conf. Central Councils of Health and Family Welfare, New Delhi, Sept. 2–5, 1985.
25. Olshansky, S. J., Carnes, B. A., and Cassel, C. K., In search of Methuselah: estimating the upper limits to human longevity, *Science,* 250, 634, 1990.
26. Smith, D. W., *Human Longevity,* Oxford University Press, New York, 1993.
27. Mazess, R. B. and Forman, S. H., Longevity and age exaggeration in Vilcabamba, Ecuador, *J. Gerontol.,* 34, 94, 1979.
28. Leaf, A., Long-lived populations: extreme old age, *J. Am. Geriatr. Soc.,* 30, 485, 1982.
29. Medvedev, Zh. A., Caucasus and Altay longevity: a biological or social problem?, *Gerontologist,* 14, 381, 1974.
30. Medvedev, Zh. A., Age structure of Soviet population in the Caucasus: facts and myths, in *The Biology of Human Aging,* Bittles, A. H. and Collins, K. J., Eds., Cambridge University Press, Cambridge, 1986, 181.
31. Bennett, N. G. and Garson, L. K., Extraordinary longevity in the Soviet Union: fact or artifact, *Gerontologist,* 26, 358, 1986.
32. Kirkwood, T. B. L., Comparative and evolutionary aspects of longevity, in *Handbook of the Biology of Aging,* 2nd ed., Finch, C. E. and Schneider, E. L., Eds., Van Nostrand Reinhold, New York, 1985, 27.

33. Kirkwood, T. B. L., Immortality of the germ-line versus disposability of the soma, in *Evolution of Longevity in Animals,* Woodhead, A. D. and Thompson, K. H., Eds., Plenum Press, New York, 1987, 209.
34. Comfort, A., *The Biology of Senescence,* 3rd ed., Elsevier, New York, 1979.
35. Finch, C. E. and Schneider, E. L., *Handbook of the Biology of Aging,* 2nd ed., Van Nostrand Reinhold, New York, 1985.
36. Strehler, B. L., *Time, Cells and Aging,* 2nd ed., Academic Press, New York, 1977.
37. Wolstenholme, G. E. W. and O'Connor, M., Eds., *The Lifespan of Animals,* Vol. 5, Ciba Found. Coll. on Ageing, Little, Brown, and Co., Boston, 1959.
38. Sacher, G. A., Relation of lifespan to brain weight and body weight in mammals, in *The Lifespan of Animals.,* Vol. 5, Wolstenholme, C. E. W. and O'Connor, M., Eds., Little, Brown, and Co., Boston, 1959, 115.
39. Hofman, M. A., Energy metabolism, brain size and longevity in mammals, *Q. Rev. Biol.,* 58, 495, 1983.
40. Salk, J., *The Survival of the Wisest,* Harper & Row, New York, 1973.
41. Finch, C. E., *Longevity, Senescence, and the Genome,* University of Chicago Press, Chicago, 1990, 248.
42. Sacher, G. A., Life table modification and life prolongation, in *Handbook of the Biology of Aging,* Finch, C. E. and Hayflick, K., Eds., Van Nostrand Reinhold, New York, 1977, 582.
43. Hart, R. W. and Setlow, R. B., Correlation between deoxyribonucleic acid excision-repair and life-span in a number of mammalian species, *Proc. Natl. Acad. Sci. U.S.A.,* 71, 2169, 1974.
44. Robertson, O. H. and Wexler, B. C., Prolongation of the lifespan of Kokanec salmon *(Oncorhynkus nerka Kennerlyi)* by castration before beginning of gonad development, *Proc. Natl. Acad. Sci. U.S.A.,* 47, 609, 1961.
45. Liu, R. K. and Walford, R. L., Observations of the lifespans of several species of annual fishes and of the world's smallest fishes, *Exp. Gerontol.,* 5, 241, 1970.
46. Segall, P. E. and Timiras, P. S., Patho-physiologic findings after chronic tryptophan deficiency in rats: a model for delayed growth and aging, *Mech. Ageing Dev.,* 5, 109, 1976.
47. Segall, P. E., Timiras, P. S., and Walton, J. R., Low tryptophan diets delay reproductive ageing, *Mech. Ageing Dev.,* 23, 245, 1983.
48. de Beauvoir, S., *The Coming of Age,* 1st American ed., translated by Patrick O'Brien, Putman, New York, 1972.
49. Rose, M. R., *Evolutionary Biology of Aging,* Oxford University Press, New York, 1991.
50. Rowe, J. W. and Kahn, R. L., Human aging: usual and successful, *Science,* 237, 143, 1987.
51. Martin, G. M., Interactions of aging and environmental agents: the gerontological perspective, in *Environmental Toxicity and the Aging Process,* Baker, S. R. and Rogul, M., Eds., Alan R. Liss, New York, 1987, 25.
52. Suzman, R. M., Willis, D. P., and Manton, K. G., Eds., *The Oldest Old,* Oxford University Press, New York, 1992.
53. Shock, N. W., Age changes in physiological functions in the total animal: the role of tissue loss, in *The Biology of Aging: A Symposium,* Strehler, B. L., Ebert, J. D., Glass, H. B., and Shock, N. W., Eds., American Institute of Biological Sciences, Washington, D.C., 1960, 258.
54. Cavalieri, T. A., Chopra, A., and Bryman, P. N., When outside the norm is normal: interpreting lab data in the aged, *Geriatrics,* 47, 66, 1992.
55. Timiras, P. S., Physiology of ageing: aspects of neuroendocrine regulation, in *Principles and Practice of Geriatric Medicine,* Pathy, M. S. J., Ed., John Wiley & Sons, New York, 1991, 31.
56. World Health Organization (WHO), *Epidemiological Studies on Social and Medical Conditions of the Elderly; Report on a Survey,* EURO Reports and Studies n.62, Copenhagen, 1982.
57. Besdine, R. W., Dementia and delirium, in *Geriatric Medicine,* Rowe, J. W. and Besdine, R. W., Eds., Little, Brown, Boston, 1988, 375.
58. Kane, R. A. and Kane, R. L., *Assessing the Elderly: a Practical Guide to Measurement,* Lexington Books, Lexington, MA, 1981.
59. Rubenstein, L. Z., Josephson, K. R., Nichol-Seamons, M., and Robbins, A. S., Comprehensive health screening of well elderly adults: an analysis of a community program, *J. Gerontol.,* 41, 342, 1986.
60. Rubenstein, L. V., Calkins, D. R., Greenfield, S., et al., Health status assessment for elderly patients, Report of the Society of General Internal Medicine Task Force on Health Assessment, *J. Am. Geriatr. Soc.,* 37, 562, 1989.
61. Lachs, M. S., Feinstein, A. R., Cooney, L. M., Jr., et al., A simple procedure for general screening for functional disability in the elderly, *Ann. Int. Med.,* 112, 699, 1990.
62. Colsher, P. L., Ethical issues in conducting surveys of the elderly, in *The Epidemiologic Study of the Elderly,* Wallace, R. B. and Woolson, R. F., Eds., Oxford University Press, New York, 1992, 287.
63. Hendriksen, C., Lund, E., and Stomgard, E., Consequences of assessment and intervention among elderly people: a three-year randomised controlled trial, *Br. Med. J.,* 289, 1522, 1984.
64. Annas, G. J. and Glantz, L. H., Rules for research in nursing homes, *N. Engl. J. Med.,* 315, 1157, 1986.
65. Harel, Z., Ehrlich, P., and Hubbard, R., Eds., *The Vulnerable Aged: People, Services, and Policies:* Springer-Verlag, New York, 1990.
66. Kaye, J. M., Lawton, P., and Kaye, D., Attitudes of elderly people about clinical research on aging, *Gerontologist,* 30, 100, 1990.
67. Guralnik, J. M. and Lacroix, A. Z., Assessing physical function in older populations, in *The Epidemiologic Study of the Elderly,* Wallace, R. B. and Woolson, R. F., Eds., Oxford University Press, New York, 1992, 159.
68. Schneider, E. L., A new approach to measuring human biological age: studies on human skin fibroblasts in tissue culture, in *Aging: A Challenge to Science and Society,* Danon, D., Shock, N. W., and Marois, M., Eds., Oxford University Press, Oxford, New York, 1981, 295.
69. Hayflick, L., The limited *in vitro* lifetime of human diploid cell strains, *Exp. Cell. Res.,* 37, 614, 1965.
70. Cristofalo, V. J. and Balin, A. K., Oxygen tension and vitamin E: Effects on cellular senescence in vitro, in *Aging: A Challenge to Science and Society,* Danon, D., Shock, N. W., and Marois, M., Eds., Oxford University Press, New York, 1981, 99.
71. Martin, G. M., Ogburn, C. E., and Sprague, C. A., Effects of age on cell division capacity, in *Aging: A Challenge to Science and Society,* Danon, D., Shock, N. W., and Marois, M., Eds., Oxford University Press, New York, 1981, 124.
72. Kertzer, D. I. and Keith, J., *Age and Anthropological Theory,* Cornell University Press, Ithaca, NY, 1984.
73. Comfort, A., *A Good Age,* Crown Publishers, New York, 1976.

74. Murray, M. A., *My First Hundred Years,* W. Kimber, London, 1963.
75. Sampson, A. and Sampson, S., *The Oxford Book of Ages,* Oxford University Press, New York, 1985.
76. Bois, J.-P., *Les Vieux,* Fayard, Paris, 1989.
77. Minois, G., *History of Old Age: from Antiquity to the Renaissance,* University of Chicago Press, Chicago, 1989.
78. Cole, T. R., *The Journey of Life: A Cultural History of Aging in America,* Cambridge University Press, Cambridge, 1992.
79. Maddox, G. L., Growing old: getting beyond the stereotypes, in *Foundations of Practical Gerontology,* Boyd, R. R. and Oakes, C. G., Eds., University of South Carolina Press, Columbia, 1969, 5.
80. Rubenstein, A., *My Many Years,* 1st ed., Knopf, New York, 1980.

I
GENERAL
PERSPECTIVES

AGING AND DISEASE

3

Paola S. Timiras

In this chapter, death-causing diseases and diseases associated specifically with aging will be considered first, in the light of current medical, political, economic, and ethical issues related to morbidity in the elderly. Second, the condition of "disuse" as a consequence of aging and disease will be discussed briefly, together with its impact on the rate of aging and the possible influence of increased physical and mental activity in delaying or alleviating the consequences of disuse and aging. Third, some examples will be presented in which disease has been used to study aging or as a tool for inducing aging (as illustrated by *in vivo* and *in vitro* models).

1 COMPLEXITY OF PATHOLOGY WITH AGING

1.1 Some Differences Between Aging and Disease

Aging, usually considered to be a normal phenomenon that occurs in all members of a population, is associated with *increased incidence and severity of diseases, accidents, and stress.* Incidence of disease is also referred to as *morbidity.* Deleterious factors, not lethal in themselves, may, from a very early age, gradually add to physiologic decline and predispose the individual to functional losses or to specific diseases in later life. Death from "pure" old age is rare; rather, it may result "prematurely" from increased multiple pathology superimposed on homeostatic insufficiency. The study of the physiopathology of aging is important in understanding the succession of events that finally culminate in functional failure and disease; it may also uncover valuable clues concerning the overall process of aging.

It is difficult to isolate the effects of aging alone from those consequent to (1) disease or (2) gradual degenerative changes that develop fully with the passage of time. Any demarcation among these effects can be only tentatively drawn at the present state of our knowledge. For example, it is reasonable to question whether the atheroma — the characteristic lesion of atherosclerosis — represents a degenerative process or a disease; in other words, whether it represents the consequences of age-related cellular and molecular changes in one or several of the arterial wall constituents, or the consequence of mechanical injury or of a virus. Dementia (contrary to the opinion of some pessimists!) is not a normal, unavoidable consequence of aging and should be investigated as any other disease process. Likewise, anemia is not a normal correlate of aging, and a cause for this condition should be investigated. Indeed, one of the tasks of the geriatrician is to be able to differentiate aging from disease and to treat both as independent but related entities.

It has long been accepted, and with reason, that "old age is almost always combined with or masked by morbid processes." Certainly, late life is a period of increasing and *multiple pathology* during which it becomes progressively more difficult to distinguish conceptually between "normal" aging, which is universal, and specific diseases, which affect people differently.

As already indicated, aging may be viewed as a process which is

Universal — shared by all living organisms
Intrinsic — independent of environmental factors
Progressive — continuous although, in some cases, slow-paced
Deleterious — likely to reduce functional competence
Irreversible — rarely treatable

In contrast, disease may be viewed as a process which is

Selective — varies with the species, tissue, organ, cell, and molecule
Intrinsic and extrinsic — may depend on both environmental and genetic factors
Discontinuous — may progress, regress, or be arrested
Occasionally deleterious — damage is often variable, reversible
Often treatable — with known etiopathology, cure may be available

The close association between aging and diseases affects the practical orientation of aging research.[1] One essentially pragmatic approach focuses on traditional research aimed at specific functional and clinical entities rather than at aging as a whole. Clarification of the etiopathogenesis and eventual treatment of a specific disease spotlights the search for the immediate improvement of health and quality of life of the elderly. While the hope of conquering aging may seem too theoretical and far removed, the study of a specific disease projects a much more concrete image and a task achievable within a limited time.

Another, more fundamental, approach is to study the aging process *per se.* This study responds to the reality that progress in biomedical research and conquest over many diseases represents only one of many factors responsible for today's longer lifespan and increasing percentage of the elderly in the human population, Both research goals have merits and should be pursued simultaneously.

1.2 Cellular Defense Mechanisms and Stress

If resistance to environmental stress declines with aging, then death can be attributed to any intervening accident or disease, however minor; in other words, the probability of dying from disease or injury would inevitably increase with the passage of time. In every organ, tissue, and in many cells, a time-dependent loss of structure, function, and chemistry proceeds slowly as the consequence of small, but accumulating, injuries (micro-insults).[1] For example, deterioration of skin elastin

0-8493-8979-8/94/$0.00+$.50

may result from bombardment by ultraviolet photons; degeneration of articular cartilage, from countless mechanical insults; opacification of the crystalline lens, from molecular injuries. Indeed the most significant consequences for aging may occur when these small injuries act at the molecular level, such as failure of DNA repair or progressive cross-linking of collagen, the major structural protein of the body.

In this context, cell death may depend on loss of efficiency of cellular defense mechanisms. These mechanisms may be the same or similar for cell death as for cell proliferation. Cell death may be as tightly regulated as cell proliferation and, accordingly, viewed more as a physiologic than a pathologic event. Cell defense mechanisms that may be important in aging include: (1) DNA repair; (2) detoxification of xenobiotics (e.g., drugs, pesticides, carcinogens); (3) antioxidants; and (4) heat shock proteins. Failure in any one of these mechanisms may not be sufficient to irreversibly induce lethal cell damage, but several systems failing simultaneously may well induce cell death.

Cell death may also be viewed as a safety mechanism to avoid cell senescence and cell transformation. Cells may respond to specific physiologic stimuli by actually programming themselves to die (apoptosis) (Chapter 5). This self-cell sacrifice (cell suicide) to protect the organism is illustrated by the strategy of the immune responses to kill virus-infected or transformed cells to eliminate the virus or cancer while sparing the entire organism (Chapter 7).

2 DEATH RATES THROUGHOUT THE LIFESPAN

Death represents the end point of aging. Therefore, information regarding the mechanisms underlying aging can be obtained from evaluation of the death rates, *mortality,* of different populations, *Death rate* (i.e., age-specific death rate) *is expressed as the number of deaths in a given age interval divided by the number of individuals in that age interval.* The death certificate on which death numbers and cause of death are based represents the primary source of information for mortality statistics. Thus, the validity of these statistics in providing fundamental information on aging and disease in older persons depends on the quality of the original death certificate.[2]

Overall, the pathology of the elderly is characterized by (1) multiple diseases and disabilities affecting the individual at the same time (hence the expression of "comorbidity") and (2) the chronic nature of such diseases and disabilities. Because of the multiple pathologies, it is often difficult to accurately report the cause of death and the completeness, and reliability of the certificate of death may often be difficult to validate. Despite these difficulties, death records are a valuable and ultimate source of epidemiologic information of the health status of older persons.

Age-specific death rate may also be interpreted as the probability of death within a given age interval and a specific population. According to U.S. data, the probability of death within the next 24 h or the next year is much greater at age 80 than at age 60. This increased probability of death with advancing age was reported for the first time in 1825 by the English actuary Benjamin Gompertz,[3] who suggested that human mortality was governed by an equation with two terms. The first accounted for the chance deaths that would occur at any age. The second, characteristic of the species, represented the exponential death increase with time. That is, *the older we get, the faster we get old.* Data acquired since the last century from both human and animal populations show a remarkable similarity among mortality curves for different populations, whether the total lifespan is measured in days, as in *Drosophila,* or in years, as in humans. Regardless of the species, the death rate, as illustrated in Figure 3-1, increases exponentially until the population is exhausted. The principle that the probability of dying increases with age is considered fundamental for all living organisms. However, it is important to emphasize that it may differ between males and females, vary with the race, and change through time.

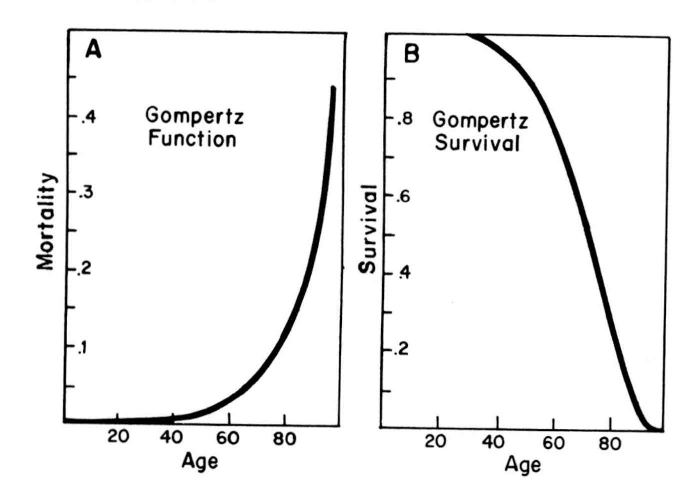

FIGURE 3-1 *Age-specific mortality rate and survival curves.* (A) Age-specific mortality rates for white males; (B) derived survival curve.

The reasons for the logarithmic increase in death rate with aging have not, so far, been adequately explained. It is particularly difficult to reconcile the physiologic decline, which progresses linearly with aging, with the increasing death rate, which progresses logarithmically. One suggested interpretation is that biologic systems have a certain built-in redundancy or *functional reserve.* A steady decline in function may proceed for many years without lowering the functional capacity below that required for homeostasis. Only when homeostatic needs can no longer be matched does failure of adaptive competence lead to disease and death.

Of interest is the fact that this logarithmic mortality continues even when severe public health problems are alleviated. For example, despite the progressive decline in infant and childhood mortality since the turn of the century, the tendency of the population to die during adulthood continued to double (in 1955) every 8.5 years. Under the current improved conditions of medical care, nutrition, and sanitation, the rise in the death rate occurs at a later age, but once the curve has begun to rise, the doubling time of 8.5 years remains essentially unchanged. Of further interest is the observation that the increase is logarithmic in all countries whether the tendency to die is great, as in developing countries in which mortality rate is typically high, or slight, as in developed countries that have experienced a rapid reduction in death rate.

3 DISEASES OF OLD AGE — CAUSES OF DEATH?

3.1 Causes of Death and Comorbidity

The relationship between the aging process and disease is also supported by demographic studies of diseases of old

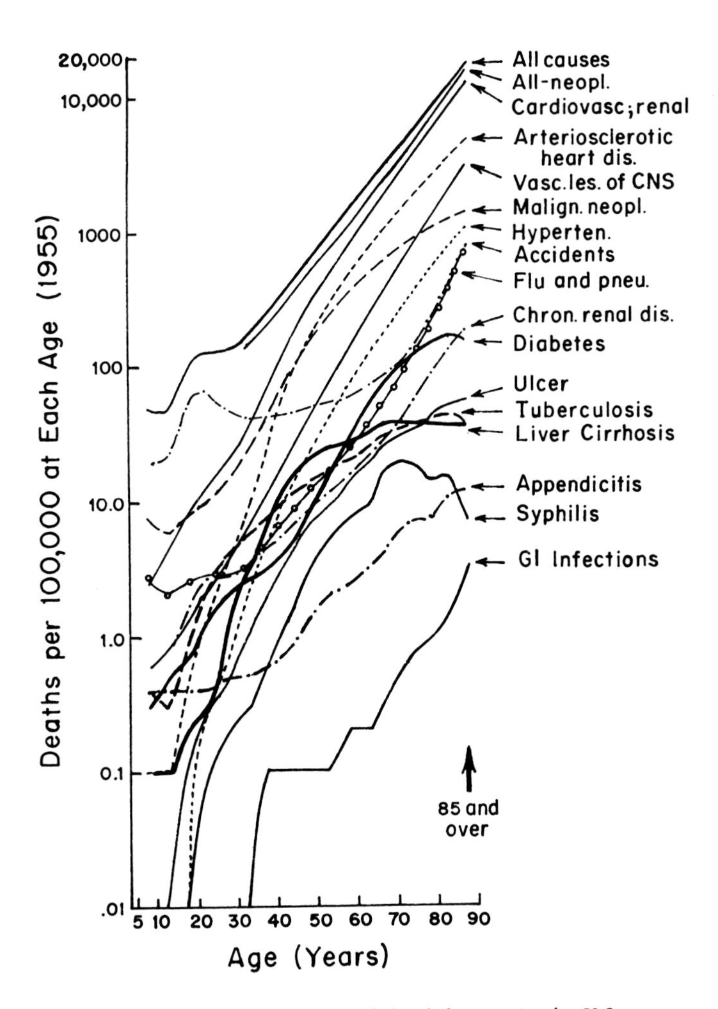

FIGURE 3-2 *Common causes of death by age in the U.S.*

age. Deaths from major causes in the U.S. population are obtained from Vital Statistics and can be plotted as a function of age. Those for 1955 were graphed by Kohn[4-6] and represent an often quoted source of information. While the patterns of death have varied considerably from the beginning of the century to now, values for 1955 are not too different from today's and, therefore, are presented here (Figure 3-2). In 1900, and the early years of this century, the three major causes of death in this country were tuberculosis, pneumonia, and diarrhea-enteritis. By 1949, and persisting to today, the three major causes of death are cardiovascular diseases, cancer (neoplasms), and accidents (including those consequent to violence). The graph of all causes of death is a straight line after maturity. This agrees with the suggestion that the probability of dying increases exponentially with aging, as already discussed.

The death rate due to accidents increases sharply in late adolescence and young adulthood; it then remains constant until about 70 years of age, when it starts to rise again. Aging-related diseases such as those of cardiovascular pathology or cancer do not equally affect all individuals, hence the heterogeneity of these diseases. In addition, with increasing age, individuals are affected by several diseases simultaneously. Functional disabilities and the *presence of concurrent health conditions, also known as "comorbidity"*, have adverse effects, or, at least, complicate diagnosis, treatment, and prognosis of many of the diseases of the elderly. Consequently, death may be due to a multiplicity of causes rather than a single one.

The major causes of death by age in the U.S. show a close relationship to cardiovascular pathology, primarily atherosclerosis (responsible for heart, kidney, and brain pathology) and hypertension. While, in Figure 3-2, neoplasms are shown as the first cause of death, they currently, in fact, represent the second cause. For example, in 1984, the annual mortality rate per 100,000 individuals in the U.S. was 362 deaths due to cardiovascular disease, 189 due to malignant neoplasms, and only 19 due to pneumonia and influenza. However, even mortality from respiratory diseases, although considerably reduced now as compared to 1900, increases with aging. This is particularly evident for tuberculosis which had almost disappeared but is now recurring especially in the old and those affected with AIDS (Chapters 7 and 18).

Comparing the incidence of deaths due to cardiovascular diseases and malignant neoplasm, we find that at the age of 50 to 55, neoplasms are responsible for 20 to 30% of deaths in males and 30% in females while, at the same time, cardiovascular diseases are responsible for 50% of deaths in males and 40% in females. After 80 years of age, however, cancer mortality declines to less than 10% and cardiovascular mortality increases to 60% or more for both sexes.

It has been claimed that the life expectancy of adults would be extended by 1 to 3 years if malignant neoplasms were cured, by 5 to 7 years if atherosclerosis were prevented, and by approximately 10 years if both diseases were abolished. It is true that the spectacular advances in medicine in the last 100 years have been responsible in part for the dramatic increase in the average lifespan, but further lengthening on this basis alone appears unlikely. Efforts, therefore, should be expended to elucidate the basic mechanisms, including the physiology of the aging process *per se*. An appreciation of these mechanisms is indispensable for a better understanding of the aging process. Without such an understanding, the etiopathology of the diseases of old age and their rational treatment cannot be achieved and the lifespan neither further improved nor prolonged.

Risk-Factor Reduction or Medical Care: Which Is Responsible for the Current Decline in Cardiovascular Disease Mortality?

As will be discussed in detail in Chapter 16, one of the major life-threatening cardiovascular diseases involves atherosclerosis of the coronary arteries, that is, the arteries that supply blood to the heart. This disease, known as coronary heart disease (CHD or coronary artery disease or ischemic heart disease), leads to reduction of blood flow to the heart with resulting angina pectoris, myocardial infarction, or sudden death; it represents the major single cause of disability and death in the U.S. CHD prevalence increases with age but is less in women than men until about 70 years of age, when it becomes the same in both sexes.

In the last 25 years, CHD prevalence has declined significantly in all sectors of the adult population. For persons aged 35 to 74 years, the rate of mortality for CHD has fallen by 30%;[7] this fall is all the more remarkable because it occurred after a sustained period of rising death rates from this cause dating back to at least 1940. The recent decline in the U.S. is also striking, for it is larger than that observed in any other country (although we must not forget that the CHD incidence is also among the highest!). This decline in CHD morbidity and mortality may be ascribed to our better understanding of the nature of the disease itself and, hence, the availability of more effective preventive and therapeutic interventions.

The most important factor in this decline appears to be the "improvement of the life-style", primarily, amelioration of the diet, increased physical exercise, and cessation of cigarette smoking. Clearly, we have been doing the "right things" with respect to improving the hygienic and dietary conditions of our life.[8] The second factor is therapeutic improvement, including better control of hypertension and diabetes mellitus, better medical care in hospitals, and more widespread use of coronary bypass surgery and other methods of coronary revascularization.

From the available data, improved medical care probably made some contribution to the decline in mortality from CHD, but the major source of the decline has been a downward trend in the incidence of the disease. The latter has been brought about by our better understanding of both the underlying effect of age, in general, and the factors that influence the aging of the cardiovascular system, in particular.

4 DISEASES OF OLD AGE — HOW THEY DIFFER

4.1 Multiple Pathology

Geriatrics may represent "a reaction against the belief that, after the age of 65, a patient is too old to be clinically interesting and therapeutically rewarding."[9] As previously discussed, the diminished physiologic competence of old people may lead to selective vulnerability and impaired capacity for recovery and rehabilitation. Yet, the manifest success of some preventive and therapeutic interventions in the elderly negates the viewpoint that little in terms of ameliorating health and quality of life can be done at a late age. Rather, the opposite appears to be true, that is, even in old age measures can be operative in improving health and quality of life.

As children are not just "little or young adults," the elderly are not just "old adults". They have many specific characteristics and concerns related to their physical and mental health. Of course, the wide range of individuality and variability makes many generalizations inappropriate; however, certain principles should be borne in mind when considering the clinical manifestations of disease in the elderly (Table 3-1).

One of the challenges of geriatrics arises from the multiplicity of problems confronting the elderly.[10] One cannot adequately treat disease without considering the psychologic, economic, and social situation of each individual (Table 3-2). This global, "holistic" view of the individual should be taken for all ages, of course, but it becomes crucial for the elderly, for whom loneliness, social instability, and often financial hardship have enormous impacts on health and well-being.

As mentioned previously and worth repeating, one of the main characteristics of the pathology of old age is *the comorbidity, that is, the multiplicity of the diseases simultaneously affecting the same individual.*[11-13] One recent study,[11] demonstrates that the prevalence of two or more diseases increases with age for both men and women. Included in this study are reports of arthritis, hypertension, cataracts, varicose veins, diabetes, cancer, osteoporosis or hip fracture, and stroke. Two or more conditions are reported by 45% of women aged 60 to 69 years, 61% of women aged 70 to 79 years, and 70% of women 80 years old and older. For men, the percentages reported for the same age groups are 35%, 47%, and 43%, respectively. As for functional disabilities, there appears to be a sex difference in comorbidity, with women being more severely affected than men (Chapter 2).

Autopsy often reveals numerous lesions involving so many organs of the body that it is difficult to know which one was responsible for death. The pathologist is often perplexed as to how the patient managed to live so long with such a "load" of diseases. This pathologic multiplicity is particularly evident in the 70- to 90-year group in contrast to the 60- to 70-year-olds who reveal a limited number of lesions at autopsy.

The multiple pathology of the elderly poses many problems in diagnosis and treatment. Particularly crucial becomes the need for *multiple therapy and the danger associated with "polypharmacy"* (Chapter 23). Due to their impaired homeostatic mechanisms, the elderly do not tolerate therapeutic mistakes as well as younger patients. Considerations of risk versus benefit, while important in all medical decisions, become crucial when dealing with elderly patients. Altered drug reactions and interactions to commonly used drugs should be considered potential dangers lest they lead, together with the decreased physiologic competence and the multiple pathology, to the most serious of complications of medical treatment of the elderly, iatrogenic disease, that is, medically induced disease (Chapter 23).

TABLE 3-1

General Characteristics of Disease in the Elderly

Symptoms
- Vague and subtle
- Atypical
- Unreported

Chronic versus acute
Multisystem disease
Altered response to treatment
Increased danger of iatrogenicity (medically induced morbidity and/or mortality)

TABLE 3-2

Holistic View of the Elderly

In geriatrics, it is necessary to
- Differentiate the aging process from disease
- Correlate physical state with psychosocial environment

Compression of Morbidity

Compression of morbidity is the term used by some epidemiologists to indicate *a shortening in the length of time between onset of disease and death.* Delay of the onset of functional limitation of ADLs (activities of daily living), IADLs (instrumental activities of daily living; Chapter 2), and of disease, even without further improvements in life expectancy, would greatly improve the quality of the later years of life.

According to these epidemiologists, such a compression of morbidity (resulting from changes in diet, exercise, and daily routines) would postpone the onset age of both the major fatal diseases (heart disease, cancer, and stroke) and of the degenerative diseases of old age (including Alzheimer's disease, osteoporosis, and sensory impairments).[14,15] According to these same epidemiologists, compression of morbidity is already an ongoing process and is expected to continue and lead to rectangularization of the survivorship curve (Chapter 2, Figure 2-3) and to an healthier old age: declining function and increasing pathology would be banished to the very last months or weeks of life.

This hypothesis is challenged by many epidemiologists. These opponents emphasize that the lengthening of the lifespan will result in an expansion (rather than compression) of morbidity.[16,17] They argue that the environmental and behavioral factors known to reduce the risks from fatal diseases do not change the onset and progression of most of the debilitating

TABLE 3-3

Common Fatal Diseases in Old Age

Disease	General Hospital (% Affected) Age 65–69	70–74	75–79	80+	Geriatric Unit (% Affected)	Age 80–89	90+
Cancer	29	27	27	24	Atherosclerosis	21	30
Cardiovascular	25	25	32	36	Myocardial infarct	19	10
Respiratory	14	12	13	10	Bronchopneumonia	17	25
Digestive	12	9	13	16	Cancer	10	7
Nervous system	11	9	8	6	Cerebral thrombosis	9	—
Renal tract	4	7	5	3	Chronic bronchitis	7	—
Others[a]	5	1	2	5	Others[a]	16	28

[a] Senile dementia frequent in old age. Due to chronicity, patients are placed in long-term care facilities.

diseases associated with old age. Reductions in old-age mortality could only extend the time during which disability and disease can be expressed. The trade-off becomes reduced mortality in the middle and older ages for an expansion of morbidity.[18]

4.2 Diseases Associated with Old Age

A number of diseases are more frequent in old individuals than in young and are responsible for their death. A list of common fatal lesions in old individuals compiled from hospital records in the U.S. and other developed countries shows atherosclerosis (including hypertension and myocardial and cerebral vascular accidents) and cancer to be the diseases most related to old age (Table 3-3). It should be noted that when the information is secured not only from general hospitals but specifically from geriatric units, the distribution of diseases is somewhat different, with atherosclerosis and dementia being the major causes of hospitalization and deaths, and cancer a lesser cause, especially after 80 years of age (Table 3-3).

It may be mentioned here that after 60 years of age, *hospitalization days* per person per year increase dramatically from 2 days to 2 weeks (Figure 3-3). Such an increase has a significant impact on the cost of medical care.

Of the many diseases that afflict the elderly, certain medical problems are clearly relegated to the older population while some overlap with those found in younger adults (Table 3-4). Diseases that are primarily limited to the elderly include osteoporosis, osteoarthritis, adenocarcinoma of the prostate, temporal arteritis, and polymyalgia rheumatica. Diseases associated with aging include adult-onset diabetes (non-insulin dependent), Alzheimer's and Parkinson's diseases, neoplasms, emphysema, and hypertension.

Some diseases in these first two groups will be discussed in the corresponding chapters as a comparison between "normal" aging versus disease. Examples of diseases associated with aging and with more clear-cut etiologies include septicemia, pneumonia, cirrhosis, nephritis, and diseases due to the advanced stage of atherosclerosis.

Another problem of the pathology of old age is the *atypical fashion* of disease presentation (Table 3-1). Sepsis without fever is common. Myocardial infarction may occur without chest pain and present either asymptomatically or with syncope or congestive heart failure. Even for such well-characterized diseases as thyrotoxicosis, appendicitis, peptic ulcer, and pneumonia, the senile patient often has atypical symptomatology. For example, pneumonia may present with a chief complaint of confusion but lack any history of cough.

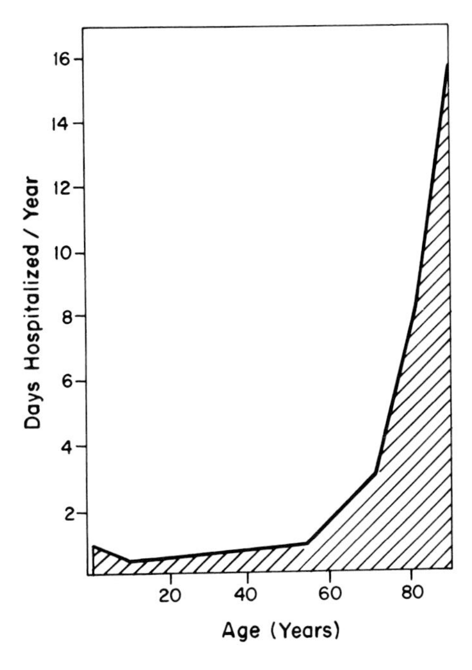

FIGURE 3-3 *Days spent in hospital according to age.* Number of days hospitalized per year according to age increases after 55 years of age and increases precipitously after age 70.

Characteristic of the elderly is that *diseases tend to be chronic and debilitating rather than acute and self-limiting;* symptomatology tends to be more *subtle and vague.* Thus, recognition and diagnosis of disease in the elderly require a high degree of alertness on the part of health care providers.

Self-Awareness of Disease in the Elderly

It has been claimed that the elderly complain excessively about poor health, aches and pains, and a variety of symptoms. In fact, considering their multiple pathology, they do not complain enough. Overall, they view their health status positively, as is the case of 75% of the 55 to 64 years group; it is only in 13% of the older age group (85+) that this optimistic view is replaced by a "poor" health outlook (Table 3-5).

In addition to the possibility of self-overevaluation, the reticence of old people to seek help from health care providers may be ascribed to several factors: (1) the often mistaken assumption that nothing can be done about the problem; (2) the fear of dreaded disease, i.e., cancer; (3) the lack of knowledge of what is normal aging; and therefore, (4) the failure to discriminate among a variety of deficits and disturbances, those due to "old age" from those due to disease.

To remedy some of these misconceptions, a vigorous educational campaign must be promoted. Such a campaign should include information on normal changes with age in various body functions, geared to the level of the particular audience and accompanied by brief descriptions of the diseases associated with aging, and should promote awareness of the many interventions possible to extend and improve the quality of life. Such a campaign should not be limited to geriatric centers; education of aging should be started in secondary schools in parallel with other programs (e.g., sex education) related to health and social issues.

TABLES 3-4

Diseases of the Elderly

Limited to aging
Osteoporosis
Osteoarthritis
Prostatic adenocarcinoma
Polymyaglia rheumatica
Temporal arteritis
Associated with aging
Known etiology
Septicemia
Pneumonia
Cirrhosis
Nephritis
Cerebrovascular disease
Myocardial infarction
Unknown etiology
Adult-onset diabetes (noninsulin dependent)
Neoplasm
Hypertension
Alzheimer's disease
Parkinson's disease
Emphysema

TABLE 3-5

Self-Assessed Health Status by Selected Characteristics

	Self-assessed health status (%)			
	Excellent or very good	Good	Fair	Poor
Age				
55–64	44.1	30.8	16.6	8.4
65–74	36.5	32.0	21.1	10.3
75–84	36.0	31.1	20.7	12.2
85+	35.0	28.6	23.2	13.2
Gender				
Male	39.4	30.3	19.1	11.2
Female	37.9	32.0	20.4	9.7
Family Income				
Under $15,000	30.2	30.5	24.7	14.6
$15,000 or more	46.9	31.8	15.2	6.1
Race				
White	39.5	31.8	19.1	9.6
Nonwhite	28.7	25.6	27.1	18.5
Residence				
Central city	38.0	31.1	20.6	10.3
Suburban	42.0	32.3	17.5	8.2
Rural	35.3	30.3	21.7	12.7

Note: Based on 1984 Supplement on Aging (SOA) to the 1984 National Health Interview Survey (NHIS). In addition to age, self-evaluation of health status depends on several variables including income, race, and geographic residence (affluent, white suburbanites have a more positive evaluation of health than low income, black residents of central cities or rural area. Gender differences are not evident despite the greater degree of disability in old women (Chapter 2).

4.3 Treatment Goals

The goal of treatment is likewise modified when compared to treatment for the young and adults. Frequently, a cure is not the main objective; rather, efforts should be shifted toward prevention and relief. Often the priority is to maximize the ability of the elderly to function. When cure is not possible, rehabilitation can help in some cases; in other cases, the provision of proper care can assist in preventing the development of further complicating illnesses. Quality of life versus prolongation of life becomes a significant issue that creates medicolegal and ethical dilemmas that are heatedly debated but remain still largely unresolved.

Too Old for Health Care? Setting or Not Setting Limits of Health Care for the Elderly?

The quality of life in later years and the cost to be paid to achieve it satisfactorily continues to be a subject of heated debate and controversy. In the 1980s, the concept was first publicly formulated that health care should be denied or "rationated" to older persons in the U.S.[19] In support of such a concept it was argued that "In the name of medical progress, we have carried out a relentless war against death and (health) decline, failing to ask ... if that will give us a better society. Neither a longer lifetime nor more life-extending technology is the way to that goal."[20,21]

This and similar statements represent a backlash against a stereotyped group of old individuals who are all the same, in need of assistance and ready to reap the benefits of a "welfare state for the elderly."[22] This negative view was compounded by the tremendous growth of federal (e.g., Medicare) spending for the elderly, by the rapidly increasing costs of medical care, and by the realization that not all elderly are poor and in need of public aid but, on the contrary, many represent a well-off elite.

Arguments against the "setting of limits", based exclusively on age include (1) the heterogeneity of the elderly, many of whom age "successfully" and require little or no public support; (2) the realization that denying access to high-technology medicine to the elderly may not substantially reduce overall health costs (primarily incurred in neonatal care) and save money, and (3) the difficulty of managing ethical choices and legal consequences. Setting limits would: "burden the elderly, undermine our (U.S.) democratic freedoms and would not guarantee any significant reductions in expenditures."[23]

Compression of morbidity (discussed above) would result in significant cost-reduction for the elderly and relieve the anxiety for the future of health care costs for all U.S. citizens. However, it cannot at present be considered realistically for the planning of health care policies.[24] Medical decisions should be based on a number of factors, including age, but not on age alone.[25] Life-saving interventions through advanced medical technologies, such as dialysis and organ transplantation, have been so far used rarely in individuals older than 65. Those in whom they have been used have profited from them, sometimes more than younger patients, and the interventions have proven to be cost effective.[26] The increase in the proportion of the elderly in the population and, within this group, that of the very old dictates that policies for health and medical care be reorganized to meet the changing demographic demands. It is to be hoped that their reorganization will follow those guidelines of medicine, law, ethics, public policy, religion, and economics that are acceptable to a civilized and caring society. Americans should decide for themselves just what is a "natural life course" and whether any of us is "too old" for health care. "Be careful! Your decisions about someone else's life might affect you sooner than you think!"[27]

5 DISUSE AND AGING

Many of the changes that accompany aging coincide closely with those associated with *physical inactivity and are generally grouped under the term of "disuse"*. Changes induced by bed rest as a consequence of disease may be superimposed on aging changes and further accelerate the aging process. Studies of long-term space travel have revealed that weightlessness in space also induces changes resembling those of aging and physical inactivity. The relationship between disuse (due to bed rest, insufficient exercise, or lack of gravity) and aging has some definite practical implications inasmuch as it permits us to envision that prevention and rehabilitation of the disuse phenomena may also ameliorate some of the deficits of old age.

5.1 Functional Changes with Disuse

This association of disuse and aging is apparent at all functional levels. A brief list of the functions affected is presented in Table 3-6, where we have also noted the eventual beneficial effects of physical exercise programs. Only a few examples will be discussed here. Information on physical assessment of the elderly has already been presented in Chapter 2 and more details will be presented in the chapters dedicated to aging of the respective systems.

TABLE 3-6

Alternations of Physiologic Parameters in Aging, Physical Inactivity, and Weightlessness (Possibly Responsive to Physical Activity)

Reduced	***Increased***
Maximum oxygen consumption (VO_2 max)	Systolic blood pressure and peripheral resistance
Resting and maximum cardiac output	Vestibular sensitivity
Stroke volume	Serum total cholesterol
Sense of balance	Urinary nitrogen and creatinine
Body water and sodium	Bone calcium
Blood cell mass	
Lean body mass	
Glucose tolerance test	***Variable***
Sympathetic activity and neurotransmission	Endocrine changes
Thermoregulation	Altered EEG
Immune responses	Altered sleep
	Changes in specific senses

At the top of the list is a decrease in maximum oxygen consumption (VO_2), which measures indirectly the ability of the organism to transport oxygen from the atmosphere to the tissues. This transport is significantly reduced with age[28,29] at the rate of about 1% per year[30,31] and with bed rest.[32,33] In both cases, a program of physical activity slows down the decline.[34-36] VO_2 max depends on cardiac output, which also decreases with age and bed rest, as a result of decreased stroke volume (Chapter 21). Simultaneously, blood pressure increases with age and weightlessness, probably due to increased peripheral circulatory resistance.

5.2 Beneficial Effects of Exercise

In all cases, both younger and older subjects are able to benefit from exercise. Studies of exercise and movement programs for elders (including nonagenarians) have shown significant increased lean body (muscle) mass, improvement in several joint movements, and subjective perception of improved mobility and well-being.[37-40] The capacity for improvement in muscle mass and power upon exercise is quite striking; for example, in old rats, treadmill running and ablation of one of two synergistic muscles (thereby increasing demand on the remaining muscle) rises by 45 to 75% muscle mass and force-generating capacity. The increase in old animals is comparable to that occurring in adult and young rats.[41] At all ages, "practice makes perfect". However, for the elderly more than for other age groups, activity programs must be tailored to the individual for optimal duration, frequency, and intensity as well as conditioning.

Although numerous investigators have endeavored to identify the best exercise regimen for each case, a consistent regimen has not been conclusively devised so far. Nevertheless, some generalizations have emerged that can be profitably applied to many elderly subjects. Depending on the health status of the subject, the exercise must be moderate and its duration, frequency, and intensity increased slowly; warmup and stretching exercises are recommended to guard against injuries.

To profit from the exercise program not only physically but also psychologically, it is recommended to choose interesting and enjoyable physical activities, preferably involving group participation. The latter can also alleviate loneliness, one of the most devastating problems of the elderly.

In relation to the potential psychologic and mental value of exercise, it should be noted that experiments in rats have shown that raising and maintaining the animals in an "enriched" environment with many stimuli increases brain weight and dendritic branching.[42] Likewise, an exercise program is capable of delaying the involutional changes seen with aging within the spinal cord and peripheral nerves.[43,44] Thus, the common aphorism "use it or lose it" may have significant implications for the well-being of the elderly.[35] Physical inactivity is clearly not the cause of aging nor does current evidence suggest that exercise arrests the aging process. Conversely, some investigators claim that strong exercise may stimulate the formation of free radicals and accelerate aging (Chapter 6). While it has not yet been possible to test the impact of exercise on longevity in humans, following a physically active life and other hygienic principles may allow us "to approach our true biogenetic potential for longevity."[35]

6 INDUCTION OF DISEASE AS A TOOL FOR THE STUDY OF AGING

A cause-effect relationship between disease and aging has been variously explored, depending on the specific hypotheses entertained by different investigators. Most often, disease has been used to accelerate the onset and course of aging processes and, thereby, to test therapeutic and prophylactic measures that might prevent or slow down both disease and aging. Other attempts have focused on "segmental" aging, that is, induction of aging in a selected organ, tissue, or cell type to mimic specific aspects of aging. Examples of these experimental approaches include genetic manipulation, increased accumulation of free radicals, inoculation of slow viruses, interference with nervous and endocrine functions, induction of wear-and-tear and stress, administration of mutagens and carcinogens, and so forth.

As for the study of normal aging (Chapter 2), the use of tissue or cells in culture has proven quite useful in inducing abnormal cellular aging and identifying causes, manifestations, and consequences of the abnormalities. A number of human diseases associated with premature or accelerated aging, the so-called progeria and progeria-like syndromes, and some *in vivo* and *in vitro* models of aging illustrate the respective usefulness and shortcomings of clinical and experimental approaches, both *in vivo* and *in vitro*.

6.1 Human Diseases Associated With Accelerated Aging: Progeria and Progeroid Syndromes

Sporadic cases of syndromes having multiple characteristics of premature (early onset) or accelerated (rapid progression) aging occur in humans. It is unclear how far any of these syndromes can be regarded as a genuine acceleration of timing mechanisms that determine senescence. They are apparently pleiotropic genetic defects, and when one of the major features is accelerated aging, they are designated as progeria. When accelerated aging is associated with other prevalent defects, they are called progeria-like or progeroid or "segmental" syndromes; one example is Down's syndrome.

Progeria syndromes do not quite duplicate all the pathophysiology of aging; each syndrome presents an acceleration of only some of the characteristics associated with normal aging. Progeria is described as being of two main types: infantile and adult. The infantile form (Hutchinson-Gilford syndrome) becomes apparent at a very early age and is associated with stunted growth, failure of sex maturation, and signs of aging, such as skin atrophy, hypertension, and severe atherosclerosis; death occurs in the twenties, usually consequent to coronary heart disease. The adult form (Werner's syndrome) resembles more closely the changes associated with aging, with respect to both the affected individual's physical appearance and the disease pattern. The onset of this premature aging syndrome occurs between the ages of 20 and 30 years, and death ensues a few years from the onset, usually due to a cardiovascular accident.

Similarities and Differences Between Werner's Syndrome and Aging

Major features of Werner's syndrome include: shortness of stature; senile appearance; cataracts and graying of the hair beginning at 20 to 30 years; skin changes (i.e., tautness, atrophy or thickening, ulceration) designated as scleroderma; joint deformities, soft tissue calcifications, and osteoporosis; atrophy of muscles and connective tissue; early cessation of menstruation; and increased incidence of neoplasms. Most of these features occur in aging as well, but at a later age and in different degrees.

Among some of the major differences between Werner's syndrome and aging are the type of inheritance — universal, multifactorial in aging and autosomal, recessive in Werner's; the high incidence of hypertension in aging, but not in Werner's; the presence of dementia and other degenerative disorders of the central nervous system in aging, but not in Werner's; the occurrence of soft tissue calcifications, uncommon in aging, but common in Werner's.

These differences are sufficient to justify the statement that Werner's syndrome is not merely a process of premature or accelerated aging. Rather, it should be viewed as a "caricature" of aging.[45] Both Werner's and aging may represent the result of generalized metabolic processes or aberrations thereof. Indeed, the overlap between the two entities is not surprising inasmuch as the various tissues of the human organisms have only a limited repertoire of reactions to genetic abnormalities and environmental insults.[46] Irrespective of similarities or differences, a study of the features of Werner's syndrome and aging will conceivably be useful in achieving an understanding of both.

Tissue culture studies of fibroblasts, in both infantile and adult syndromes, reveal a shortening of the cell-replicating ability which has been interpreted as supportive of accelerated aging.[46] The etiology of the syndromes remains obscure, but among the several causes proposed, neuro-endocrinologic dysfunction is supported by the stunted growth, failure of gonadal maturation, and diabetes, either singly or in combination.

Another example of a syndrome characterized by several symptoms including accelerated aging and premature death is Down's syndrome, or mongolism, which is due to trisomy at chromosome 21. The incidence of the syndrome is greatest among children born from mothers 40 years of age and older, and the genetic abnormality has therefore been related to aging processes involving the oocytes (Chapter 12). Although in 20 to 30% of cases the extra chromosome is contributed by the father, paternal age does not seem to have any significant effect on the incidence of the syndrome.[47] Individuals affected by Down's syndrome may present somatic malformations, but the major deficit is represented by severe mental disability. Affected subjects, who live to reach 30 years of age and more, present many signs of accelerated aging, including senile dementia of the Alzheimer type (SDAT) superimposed on the mental retardation[48] (Chapter 8). Animal models have also been proposed for the study of Down's syndrome. The mouse is the animal of choice, for it is possible to introduce trisomy of chromosome 16[49] and produce individuals with some phenotypic characteristics of Down's syndrome, such as cardiovascular defects, neurologic alterations, and retardation of brain maturation[50] as well as early aging.[51] Proposed as a good *in vitro* model to induce experimentally the premature aging of neurons is the transfer of the trisomy 21 from the fibroblast donor cells to neuroblastoma cells.[52]

6.2 Experimentally Induced Aging in Animals

Premature aging has been induced in a number of laboratory animals as well as in captive wild animals by various methods derived from the rationale of the different theories of aging, especially the neuroendocrinologic, immunologic, and free radical theories. One of the favorite methods is to expose the animal to stress, that is, to excessive environmental (including physical, emotional, and social) demands. Stress will activate or interfere with all three regulatory systems, i.e., will disturb neuroendocrine balance, alter immunologic competence and tolerance, and increase the production of free radicals (Chapter 6). Stress, produced by repeated disruption of conditioned responses in young adult dogs, leads to neurologic disturbances, EEG abnormalities resembling those of old animals, graying of the coat, muscular weakness, weight loss, and so forth.[53] Stress or injection of cortisol to mimic adrenocortical stimulation, in association with a high-lipid or high-cholesterol diet, accelerates atheroma formation in rabbits.[54] High doses of vitamin D given along with calcium to rats precipitate calcification of the skin, heart, blood vessels, and other tissues usually susceptible to calcium deposits in old age.[55] Other methods used often to induce premature aging in animals involve administration of the appropriate antigens to induce autoimmune responses against sequestered proteins. This technique is employed, for example, in the experimental induction of multiple sclerosis. Still other interventions utilize inoculation of animals with viruses or viral particles (e.g., the so-called prions) to mimic some of the degenerative diseases of old age. One model that resembles symptomatically and pathologically (e.g., accumulation of amyloid) human SDAT is the disease in sheep, so-called "scrapie", which is caused by a slow virus and can be reproduced in laboratory animals.[56 58] Attempts to isolate the virus or prions of SDAT and to induce the disease in laboratory animals are still unsuccessful (see below). Another condition caused by a retrovirus and often associated with senile dementia is the acquired immune deficiency syndrome (AIDS) (Chapter 7). As for SDAT, attempts to reproduce the neurologic symptoms (dementia) and neuropathology (e.g., brain atrophy, loss of neural cells, abnormal proteins) of AIDS in experimental animals (sheep, mice, primates) have so far been unsuccessful since only chimpanzees are infected with the virus, but the neural lesions are quite different from those of humans.[59]

As in the case of progeria, none of the animal models exhibits all the symptoms and pathology of old age; each model only partially mimics the aging process of humans. The aging process of humans is complex and its mechanisms are difficult to isolate; therefore, mammals with aging processes resembling those of the human are used for such studies. Humans and other primates take decades to age, and even the 2 to 3 years required in a short-lived mammal like the rat or mouse is a long, impractical period for experimentation and innovation.

A major obstacle, then, to a better understanding of the complicated and lengthy processes governing mammalian aging and age-related pathology is the paucity of useful model systems wherein events can be accelerated and, hopefully, simplified. Moreover, within a model system, markers reflecting adequate indices of some segment of the aging or pathologic process must be found.[60,61] In this respect, *in vitro* models involving only a single group of cells in rigorously controlled conditions can be and are being widely and profitably utilized to study cellular and molecular aging.

6.3 *In Vitro* Models for Aging

The most successful *in vitro* models are those utilized in the study of cancer, one of the major causes of death in the elderly, although not exclusively a disease of old age. Much cancer research is based on the reasonable premise that human cancer can be unraveled by studying carcinogenesis in animals. Indeed, cancer is a major age-related pathology, and the process of carcinogenesis may be studied both *in vivo* and *in vitro* because transformed (i.e., cancer) cells exhibit a set of known properties that can be readily selected for, assayed, and identified. During the last 80 years, since coal tar was shown to be carcinogenic to the skin of rabbits, many ways have been found to produce cancer in experimental animals.[62] The many shortcomings of these models soon became apparent, such as the difficulty in deciding on the best models for human cancers and creating a cancer cell in which to study the early "initiation" and subsequent "promotion" steps of the carcinogenic process. Thus, established *in vitro* models for carcinogenesis[63] and the related but more general process of mutagenesis[64] have been instrumental in our understanding of oncogenes, their activation, and the resultant cancer.

There are two essential differences between normal cells and cancer (tumor) cells; one is that the tumor cells have the capacity to divide forever — they are immortal and, thus, escape senescence; the other is that their growth is controlled abnormally. There is emerging recognition that the development of tumors may be due as much to loss of growth inhibition as to increased growth stimulation.[65] In order to understand carcinogenesis, both growth-inhibitory and -stimulatory processes must be clarified. Indeed, natural carcinogenesis is a multistage process, involving, as a first step, clonal expansion and, as a second step, transformation into cancer.

Evidence for the multistep nature of carcinogenesis can be found in epidemiologic studies, such as those that show a relationship between the incidence of lung cancer and the frequency and duration of smoking. Another example is cervical carcinoma. The first visible step in the production of the cancer of the uterine cervix is, commonly, the appearance of an expanding clone of abnormal cells (cervical dysplasia). The clone generally regresses and disappears, but sometimes it gives rise to a still more abnormal family of cells (carcinoma *in situ*), and it is among these that the fully invasive cancer arises. The sequence then involves: (1) the release or liberation of cells from the physical (cell-to-cell interactions) and chemical (growth inhibitory factors) restraints of an organized epithelium (the uterine epithelium); (2) a relaxation of the rules of differentiation, which would oppose continuing cell division; and (3) exemption from the rule of finite lifespan (which regulates "normal" cells; Chapter 2). It is a great simplification to study these different steps, so to speak, "telescoped" in cultured cells, where the first step — i.e., liberation from regional restraints *in situ* — has already been achieved.[66] Certain cell lines (e.g., the NIH line 3T3 mouse fibroblasts) are capable of taking up raw DNA and of being transformed to permanent tumorigenicity if the absorbed DNA is derived from animal or human tumors or tumorigenic cell lines. Since human DNA contains specific sequences not found in mouse DNA, it has been possible to isolate these human genes that are tumorigenic for 3T3 cells and, in part, presumably responsible for the cancers from which they were derived. The results of these experiments demonstrate dramatically that alteration of the genome is an essential step in the production of some cancers.[67]

Other factors involved in tumorigenicity include a variety of mutagens capable of cell mutation and oncogenes capable of producing transformation by activation of specific genes through rearrangement, mutation, or viral infection. A number of cellular oncogenes have been described, such as *ras* and *myc,* originating in retrovirus and capable of inducing abnormal growth or cancer transformation or both.[68,69]

Aging is a more general phenomenon than cancer and other pathologies, and we lack sufficiently specific markers to characterize the induction of aging processes in a model system. Because aging is complex, and different tissues may age and develop age-related pathology from multiple causes and at different times (Chapter 2), induction of true aging in a whole animal may actually prove to be impossible. The limited "segmental" progeroid pathology discussed above and observed in all the known human syndromes manifesting some signs of accelerated aging reinforces this point. The induction of a limited set of markers for aging or age-related pathology has become a more attainable goal.[60,61]

In Vivo and *In Vitro* Models for Senile Dementia of the Alzheimer Type

Current efforts to identify a model for senile dementia of the Alzheimer's type (abbreviated as SDAT or, the shorter form, AD for Alzheimer's Disease) represent one example of experimental studies intended to mimic at least some aspects of aging. As this degenerative disease of the elderly is rapidly increasing in incidence, it has received considerable attention in the last few years and has stimulated the need to create models for its study. Attempts have been made to reproduce this condition, characterized by marked cognitive and behavioral abnormalities and specific pathology (Chapter 8), utilizing *in vivo* (Table 3-7) and *in vitro* (Table 3-8) models.

TABLE 3-7

In Vivo Models for Senile Dementia Alzheimer Type (SDAT)

1. Spontaneous neuropathologic lesions
   - Senile plaques in brains of old dogs and monkeys
   - Abnormal filaments in old monkey brains
2. Experimentally induced neuropathologic SDAT-like lesions (cognitive impairment and behavioral changes)
   - Slow viruses (prions)
     - Senile plaques in selected mouse strains
     - Spongioform changes in sheep and other species
   - Aluminum
   - Microtubule disrupting agents (colchicine, taxol)
   - Accumulation of straight filaments
3. Surgical, electrical, or chemical lesions of SDAT-susceptible brain regions (cognitive impairment and behavioral changes)
   - Hippocampus
   - Basal nucleus of Meynert
4. Use of transgenic mice with transfer of β-amyloid gene
5. Injection of human amyloid cores into cortex and hippocampus of rats; intraventricular injection of excitotoxins and trauma in rats to induce β-amyloid immunoreactivity

TABLE 3-8

In Vitro Models for Senile Dementia Alzheimer Type (SDAT)

1. Spontaneously occurring paired helical filaments (PHF) in cultured cortical human cells (Alzheimer brain extracts not necessary for PHF induction)
2. Experimentally induced SDAT-like lesions by addition to cultured neurons of:
   - Aluminum and microtubule-disrupting agents
     - Neurofibrillary tangles of straight filaments in cultured neurons from humans, rabbits, rats, and mice
   - Excitatory amino acids, glutamate, and/or aspartate
     - PHF in human-cultured neurons
3. Addition to differentiated human teratocarcinoma neurons, individually or combined, of:
   - Aluminum, colchicine
     - Tangles of straight filaments
   - Doxorubicin
     - PHF reactive fibers and tangles
4. Transfection of human neuroblastoma cells with a mutation of β-amyloid protein precursor

The *in vivo* models include the following:

- Spontaneously occurring neuropathologic lesions resembling (but not exactly reproducing) AD in several animal species (e.g., monkeys, dogs, rodents).[70-73] Indeed, aged nonhuman primates develop pathologic brain alterations including memory deficits similar to those of AD, but these animals are too scarce and expensive to be used for routine research.
- Induced lesions resembling those of AD through (1) the inoculation of slow viruses,[56,74] (2) the administration of specific substances such as aluminum,[75] (3) the induction of autoimmune reactions,[76] or (4) lesioning-specific brain areas related to memory and cognitive functions[77-81] (Table 3-6).
- The use of transgenic mice as proposed independently (1991) by three groups of investigators. Models were created by transferring into the animals the gene for β-amyloid, a protein found to accumulate in the neuritic plaques and around the cerebral blood vessels of AD patients (Chapters 5 and 8). By inducing this protein in mice, it was claimed that mice would also show, in addition to amyloid deposits, neurofibrillary tangles and degenerating neurons (two other signs of AD). However, some of the published work was retracted by the investigators on the ground that they could not reproduce the pathology in additional mice and the work that was not retracted is still unconfirmed.[82] Clearly the usefulness of transgenic mice as an AD model, although originally promising, needs to be reevaluated.[83]
- A variation of this approach adopts the direct injection of amyloid cores isolated from AD brains into the cortex and hippocampus of rats.[84,85] The injected amyloid exerts neurotoxic effects and induces antigens similar to those found in AD brains; it is ingested by phagocytes which then migrate to the brain vessels and ventricles carrying the amyloid with them. Still another approach combines intraventricular injection of excitotoxins and amyloid injection with trauma to induce beta-amyloid immunoreactivity in phagocytes and possibly neurons.[86]

None of these models has yet proven quite satisfactory, especially with respect to reproducing the complete pathology of the disease (i.e., the presence, in discrete brain areas, of abundant neurofibrillary tangles of paired helical filaments [PHF], neurotic plaques, and vascular amyloid) nor the behavioral deficits (i.e., loss of all cognitive functions).

It may be possible, however, to model certain aspects of neuronal senescence *in vitro* by utilizing some specific markers, such as those of AD pathology. Indeed, segmental neuronal aging *in vitro* has been demonstrated in response to specific types of damage due to free radicals, hormones, or other toxins. In studies utilizing human tumor cell lines (neuroblastoma and neurogenic teratocarcinoma), several markers of AD can be induced in response to a number of agents, such as aluminum[87] and doxorubicin,[52,88] known to produce a marked axonopathy with neurofilamentous tangles, nuclear heterochromatization, and decreased transcription.[88,90]

Evidence that a particular form of familial AD may be due to a mutation in the gene encoding β-amyloid has led to transfection of human neuroblastoma cells with a mutation of the β-amyloid protein precursor. As a consequence, the production of the protein is markedly increased.[91,92] While animal models are important, the advantages (e.g., low cost, rapid utilization, easy manipulation of the cells and their environment) of using cultured cells to study AD pathogenesis should not be underestimated.

From these studies, normal neuronal aging and age-related neuronal pathology may be viewed as a response to specific types of mutations or to damage against the general background of responses of cells to stress. At the systemic physiologic level, many investigators have noted an apparent relationship between stress and aging;[93] hence the use of stress to induce aging in animals, as discussed above. However, despite some studies of generalized cellular response to stress such as heat shock proteins,[94-96] the cellular response has been sadly neglected by most gerontologists and pathologists. The production *in vitro* of AD-like lesions in response to selected types of damage argues for its production *in vivo* by a similar damage. This damage or stress response may be conceptualized in the context of the general cellular stress response; it is being applied to put AD pathology in a useful heuristic framework for further experimentation in AD material and in neuronal aging.[52,88]

REFERENCES

1. Johnson, H. A., Ed., *Relations Between Normal Aging and Disease,* Vol. 28, Aging, Raven Press, New York, 1985.
2. Havlik, R. J. and Rosenberg, H. M., The quality and application of death records of older persons, in *The Epidemiologic Study of the Elderly,* Wallace, R. B. and Woolson, R. F., Eds., Oxford University Press, New York, 1992, 262.
3. Gompertz, B., On the nature of the function expressive of the law of human mortality and a new mode of determining life contingencies, *Phil. Trans. R. Soc. London,* 2, 513, 1825.
4. Kohn, R. R., Human aging and disease, *J. Chronic Dis.,* 16, 5, 1963.
5. Kohn, R. R., *Principles of Mammalian Aging,* 2nd ed., Prentice-Hall, Englewood Cliffs, NJ, 1978.
6. Kohn, R. R., Aging and aging-related diseases: normal processes, in *Relations Between Normal Aging and Disease,* Vol. 28, Aging, Johnson, H. A., Ed., Raven Press, New York, 1985, 1.
7. Pell, S. and Fayerweather, W. E., Trends in the incidence of myocardial infarction and in associated mortality and morbidity in a large employed population 1957–1983, *N. Engl. J. Med.,* 312, 1005, 1985.
8. Stamler, J., Coronary heart disease: doing the "right things", *N. Engl. J. Med.,* 312, 1053, 1985.
9. Howell, T. H., *A Student's Guide to Geriatrics,* 2nd ed., Staples, London, 1970.
10. Fried, L. P. and Wallace, R. B., The complexity of chronic illness in the elderly: from clinic to community, in *The Epidemiologic Study of the Elderly,* Wallace, R. B. and Woolson, R. F., Eds., Oxford University Press, New York, 1992, 10.
11. Guralnik, J. M., LaCroix, A. Z., Everett, D. F., and Kovar, M. G., Aging in the eighties: the prevalence of comorbidity and its association with disability, *National Center for Health Statistics, Advance Data from Vital and Health Sciences,* No. 170, Hyattsville, MD, 1989.
12. Verbrugge, L. M., Lepkowski, J. M., and Imanaka, Y., Comorbidity and its impact on disability, *Milbank Q.,* 67, 450, 1989.
13. Satariano, W. A., Comorbidity and functional status in older women with breast cancer: implications for screening, treatment, and prognosis, *J. Gerontol.,* 47, 24, 1992.
14. Fries, J. F., Aging, natural death and the compression of morbidity, *N. Engl. J. Med.,* 303, 130, 1980.
15. Fries, J. F. and Caprol, L. M., *Vitality and Aging,* W. H. Freeman, San Francisco, 1981.
16. Schneider, E. L. and Brody, J. A., Aging, natural death and the compression of morbidity: another view, *N. Engl. J. Med.,* 309, 854, 1983.
17. Kaplan, G. A., Epidemiologic observations on the compression of morbidity; evidence from the Alameda County study, *J. Aging Health,* 3, 155, 1991.
18. Olshansky, S. J., Rudberg, M. A., Carnes, B. A., Cassel, C. K., and Brody, J. A., Trading off longer life for worsening health: the expansion of morbidity hypothesis, *J. Aging Health,* 3, 194, 1991.
19. Smeeding, T. M., Battin, M. P., Francis, L. P., and Landesman, B. M., Eds., *Should Medical Care Be Rationed by Age?,* Rowman and Littlefield, Totowa, NJ, 1987.
20. Callahan, D., *Setting Limits: Medical Goals in an Aging Society,* Simon and Schuster, New York, 1987.
21. Callahan, D., *What Kind of Life: The Limits of Medical Progress,* Simon and Schuster, New York, 1990.
22. Myles, J. F., *Old Age in the Welfare State: the Political Economy of Public Pensions,* University Press of Kansas, Lawrence, KS, 1989.
23. Barry, R. L. and Bradley, G. V., *Set No Limits: A Rebuttal to Daniel Callahan's Proposal to Limit Health Care for the Elderly,* University of Illinois Press, Urbana, IL, 1991.
24. Binkstock, R. H. and Post, S. G., *Too Old for Health Care?,* Johns Hopkins University Press, Baltimore, 1991.
25. Homer, P. and Holstein, M., Eds., *A Good Old Age? The Paradox of Setting Limits,* Simon and Schuster, New York, 1990.
26. Evans, R. W., Advanced medical technology and elderly people, in *Too Old for Health Care?,* Binkstock, R. H. and Post, S. G., Eds., Johns Hopkins University Press, Baltimore, 1991, 44.
27. Koop, C. E., Foreword, in *Too Old for Health Care?,* Binkstock, R. H. and Post, S. G., Eds., Johns Hopkins University Press, Baltimore, 1991, vii.
28. Astrand, I., Aerobic work capacity in men and women with special reference to age, *Acta Physiol. Scand.,* 49 (Suppl. 169), 92, 1960.
29. Shephard, R. J., World standards of cardiorespiratory performance, *Arch. Environ. Health,* 13, 664, 1966.
30. De Vries, H., Physiological effects of an exercise training regimen upon men aged 52 to 88, *J. Gerontol.,* 25, 325, 1970.
31. Dehn, M. M. and Bruce, R. A., Longitudinal variations in maximal oxygen intake with age and activity, *J. Appl. Physiol.,* 33, 805, 1972.
32. Saltin, B., Blomqvist, G., Mitchell, J. H., et al., Response to exercise after bed-rest and after training: a longitudinal study of adaptive changes in oxygen transport and composition, *Circulation,* 38 (Suppl. 5), VII1, 1968.
33. Birkhead, N. C., Haupt, G. J., and Issekutz, B., Jr., Circulatory and metabolic effects of different types of prolonged inactivity, *Am. J. Med. Sci.,* 247, 243, 1964.
34. Bortz, W. M., 2nd, Effects of exercise on aging — effect of aging on exercise, *J. Am. Geriatr. Soc.,* 28, 49, 1980.
35. Bortz, W. M., 2nd, Disuse and aging, *JAMA,* 248, 1203, 1982.
36. Bortz, W. M., 2nd, *We Live Too Short and Die Too Long,* Bantam Trade Paperback, New York, 1992.
37. Faulkner, J. A. and White, T. P., Adaptations of skeletal muscle to physical activity, in *Exercise, Fitness and Health,* Bouchard, C., Shephard, R. J., Stephens, T., Sutton, J. R., and McPherson, B., Eds., Human Kinetics Publishers, Champaign, IL, 1990.

38. Fiatarone, M. A. and Evans, W. J., Exercise in the oldest old, *Top. Geriatr. Rehabil.*, 5, 63, 1990.
39. Fiatarone, M. A., Marks, E. C., Ryan, N. D., Meredith, C. N., Lipsitz, L. A., and Evans, W. J., High-intensity strength training in nonagenarians. Effects on Skeletal muscle, *JAMA*, 263, 3029, 1990.
40. Blumenthal, J. A., Emery, C. F., Madden, D. J., Schniebolk, S., Riddle, M. W., Cobb, F. R., Higginbotham, M., and Coleman, R. E., Effects of exercise training on bone density in older men and women, *J. Am. Geriatr. Soc.*, 39, 1065, 1991.
41. White, T. P., personal communication, 1993.
42. Diamond, M. C., *Enriching Heredity: the Impact of the Environment on the Anatomy of the Brain,* Free Press, New York, 1988.
43. Vogt, C. and Vogt, O., Aging of nerve cells, *Nature,* 158, 304, 1946.
44. Retzlaff, E. and Fontaine, J., Functional and structural changes in motor neurons with age, in *Behavior, Aging and the Nervous System,* Charles C Thomas, Springfield, IL, 1965, 340.
45. Epstein, C. J., Werner's syndrome and aging: a reappraisal, in *Werner's Syndrome and Human Aging,* Salk, D., Fujiwara, Y., and Martin, G. M., Eds., Plenum Press, New York, 1985, 219.
46. Salk, D., Fujiwara, Y., and Martin, C. M., Eds., *Werner's Syndrome and Human Aging,* Plenum Press, New York, 1985.
47. Erickson, J. D., Paternal age and Down syndrome, *Am. J. Hum. Genet.*, 31, 489, 1979.
48. Heston, L. L., Alzheimer's dementia and Down's syndrome: genetic evidence suggesting an association, *Ann. N.Y. Acad. Sci. U.S.A.*, 396, 29, 1982.
49. Epstein C. J., Cox, D. R., and Epstein, L. B., Mouse trisomy 16: an animal model of human trisomy 21 (Down syndrome), *Ann. N.Y. Acad. Sci. U.S.A.*, 450, 157, 1985.
50. de la Cruz, F. F. and Oster-Granite, M. L., Neural bases of mental retardation, in *Handbook of Human Growth and Developmental Biology,* Meisami, E. and Timiras, P. S., Eds., CRC Press, Boca Raton, FL, 1988, 1.
51. Brooksbank, B. W. L. and Balazs, R., Development and aging of the brain in a common human aneuploidy-Down's syndrome, in *Handbook of Human Growth and Developmental Biology,* Meisami, E. and Timiras, P. S., Eds., CRC Press, Boca Raton, FL, 1988, 21.
52. Cole, G. M. and Timiras, P. S., Aging-related pathology in human neuroblastoma and teratocarcinoma cell lines, in *Model Systems of Development and Aging of the Nervous System,* Vernadakis, A., Privat, A., Lauder, J. M., Timiras, P. S., and Giacobini, E., Eds., Martinius Nijhoff, Boston, 1987, 453.
53. Frolkis, V. V., *Aging and Life-Prolonging Processes,* Springer-Verlag, New York, 1982.
54. Constantinides, P., *Experimental Atherosclerosis,* Elsevier, Amsterdam, 1965.
55. Selye, H., Gentile, G., and Prioreschi, P., Cutaneous molt induced by calciphylaxis in the rat, *Science,* 134, 1876, 1961.
56. Prusiner, S. B., Some speculations about prions, amyloid and Alzheimer's disease, *N. Engl. J. Med.*, 310, 661, 1984.
57. Carp, R. I., Merz, G. S., and Wisniewski, H. M., Transmission of unconventional slow virus diseases and the relevance to AD/SDAT transmission studies, in *Senile Dementia. Outlook for the Future,* Wertheimer, J. and Marois, M., Eds., Alan R. Liss, New York, 1984, 31.
58. Merz, P. A., Rohwer, R. G., Kascsak, R., Wisniewski, H. M., Somerville, R. A., Gibbs, C. Jr., and Gajdusek, D. C., Infection-specific particle from the unconventional slow virus diseases, *Science,* 225, 437, 1984.
59. Barnes, D. M., Neurosciences advance in basic and clinical realms, *Science,* 234, 1324, 1986.
60. Gibson, D. C., Adelman, R. C., and Finch, C. E., Development of the rodent as a model system of aging, DHEW publication, no. (NIH) 79-161, U.S. Government Printing Office, Washington, D.C., 1978.
61. Reff, M. E. and Schneider, E. L., *Biological Markers of Aging,* NIH publication no. 82-2221, U.S. Government Printing Office, Washington, D.C., 1982.
62. Cairns, J. and Logan, J., Step by step into carcinogenesis, *Nature,* 304, 582, 1983.
63. Kennedy, A. R., Fox, M., Murphy, G., and Little, J. B., Relationship between x-ray exposure and malignant transformation in C3H10T 1/2 cells, *Proc. Natl. Acad. Sci. U.S.A.*, 77, 7262, 1980.
64. Ames, B. N. and Gold, L. S., Endogenous mutagens and the causes of aging and cancer, *Mutat. Res.*, 250, 3, 1991.
65. Marx, J. L., The yin and yang of cell growth control, *Science,* 232, 1093, 1986.
66. Heldin, C. H. and Westermark, B., Growth factors: mechanism of action and relation to oncogenes, *Cell,* 37, 9, 1984.
67. Weinstein, I. B., Gattoni-Celli, S., Kirschmeier, P., Hsiao, W., Horowitz, A., and Jeffrey, A., Cellular targets and host genes in multistage carcinogenesis, *Fed. Proc.*, 43, 2287, 1984.
68. Marshall, C., Functions of *ras* oncogenes, *Nature,* 310, 448, 1984.
69. Robertson, M., Paradox and paradigm: the message and meaning of *myc, Nature,* 306, 733, 1983.
70. Ball, M. J., MacCregor, J., Fyfe, I. M., Rapoport, S. I., and London, E. D., Paucity of morphological changes in the brains of ageing beagle dogs: further evidence that Alzheimer lesions are unique for primate central nervous system, *Neurobiol. Ageing,* 4, 127, 1983.
71. Campbell, B. A., Krauter, E. E., and Wallace, J. E., Animal models and aging: sensory-motor and cognitive function in the aged rat, in *Psychobiology of Aging: Problems and Perspectives,* Stein, D. G., Ed., Elsevier, Amsterdam, 1980, 201.
72. Dean, R. L. and Bartus, R. T., Animal models of geriatric cognitive dysfunction: evidence for an important cholinergic involvement, in *Senile Dementia of the Alzheimer Type: Early Diagnosis, Neuropathology and Animal Models,* Traber, J. and Gispen, W. H., Eds., Springer-Verlag, Berlin, 1985, 269.
73. Zilles, K., Morphological studies on brain structures of the NZB mouse: an animal model for the aging human brain?, in *Senile Dementia of the Alzheimer Type: Early Diagnosis, Neuropathology and Animal Models,* Traber, J. and Gispen, W. H., Eds., Springer-Verlag, Berlin, 1985, 355.
74. Fraser, H. and McBride, P. A., Parallels and contrasts between scrapie and dementia of the Alzheimer type and ageing: strategies and problems for experiments involving life span studies, in *Senile Dementia of the Alzheimer Type: Early Diagnosis, Neuropathology and Animal Models,* Traber, J. and Gispen, W. H., Eds., Springer-Verlag, Berlin, 1985, 250.
75. Petit, T. L., Biederman, G. B., and McMullen, P. A., Neurofibrillary degeneration, dendritic dying back and learning memory deficits after aluminum administration: implications for brain aging, *Exp. Neurol.*, 67, 152, 1980.
76. Lal, H., Forster, M. J., and Nandy, K., Immunologic factors related to cognitive/behavioral dysfunctions in aging, in *Senile Dementia of the Alzheimer Type: Early Diagnosis, Neuropathology and Animal Models,* Traber, J. and Gispen, W. H., Eds., Springer-Verlag, Berlin, 1985, 343.
77. Arendash, G. W., Strong, P. N., and Mouton, P. R., Intracerebral transplantation of cholinergic neurons in a new animal model for Alzheimer's disease, in *Senile Dementia of the Alzheimer Type,* Hutton J. T. and Kenny, A. D., Eds., Alan R. Liss, New York, 1984, 351.

78. Azmitia, E. C, Perlow, M. J., Brennan, M. J., and Lauder, J. M., Fetal raphe and hippocampal transplants into adult and aged C57B1/6N mice: a preliminary immunocytochemical study, *Brain Res. Bull.,* 7, 703, 1981.

79. Gage, F. H., Bjorklund, R., Stenevi, U., Dunnett, S. B., and Kelly, P. A., Intrahippocampal septal grafts ameliorate learning impairments in aged rats, *Science,* 225, 533, 1984.

80. Pepeu, G., Casamenti, F., Bracco, L., Ladinsky, H., and Consolo, S., Lesions of the nucleus basalis in the rat: functional changes, in *Senile Dementia of the Alzheimer Type: Early Diagnosis, Neuropathology and Animal Models,* Traber, J. and Gispen, W. H., Eds., Springer-Verlag, Berlin, 1985, 305.

81. Spencer, D. G., Jr., Horvath, E., Luiten, P., Schuurman, T., and Traber, J., Novel approaches in the study of brain acetylcholine function: neuropharmacology, neuroanatomy, and behavior, in *Senile Dementia of the Alzheimer Type: Early Diagnosis, Neuropathology and Animal Models,* Traber, J. and Gispen, W. H., Eds., Springer-Verlag, Berlin, 1985, 325.

82. Quon, D., Wang, Y., Catalano, R., Scardina, J. M., Murakami, K., and Cordell, B., Formation of beta-amyloid protein deposits in brains of transgenic mice, *Nature,* 352, 239, 1991.

83. Marx, J., Major setback for Alzheimer's models, *Science,* 255, 1200, 1992.

84. Frautschy, S. A., Baird, A., and Cole, G. M., Effects of injected Alzheimer beta-amyloid cores in rat brain, *Proc. Natl. Acad. Sci. U.S.A.,* 88, 8362, 1991.

85. Frautschy, S. A., Cole, G. M., and Baird, A., Phagocytosis and deposition of vascular beta-amyloid in rat brains injected with Alzheimer beta-amyloid, *Am. J. Pathol.,* 140, 1389, 1992.

86. Frautschy, S. A., Cole, G. M., and Baird, A., The comparative effects of injury and amyloid injection on Alzheimer's antigens in the rat brain, in *Alzheimer's Disease: Advances in Clinical and Basic Research,* Corain, B., Iqbal, K., Nicolini, M., Winblad, B., Wisniewski, H., and Zatyta, P., Eds., John Wiley & Sons, New York, 1993, 50.

87. Mesco, E. R., Kachen, C., and Timiras, P. S., Effects of aluminum on tau proteins in human neuroblastoma cells, *Mol. Chem. Neuropathol.,* 14, 199, 1991.

88. Cole, G. M., Wu, K., and Timiras, P. S., A culture model for age-related human neurofibrillary pathology, *Int. J. Dev. Neurosci.,* 3, 23, 1985.

89. Cole, G. M. and Timiras, P. S., Lipid peroxidation and Alzheimer amyloid precursor processing *in vitro,* in *Phospholipid Research and the Nervous System,* Bazan, N. G., Horrocks, L. A., and Toffano, G., Eds., Fidia Res. Series, Vol. 17, Liviana Press, Padova, Italy, 1989, 115.

90. Argasinski, A., Sternberg, H., Fingado, B., and Huynh, P., Doxorubicin affects tau protein metabolism in human neuroblastoma cells, *Neurochem. Res.,* 14, 927, 1989.

91. Cai, X.-D., Golde, T. E., and Younkin, S. G., Release of excess myloid beta protein from a mutant amyloid beta protein precursor, *Science,* 259, 514, 1993.

92. Marx, J., Alzheimer's pathology begins to yield its secrets, *Science,* 259, 457, 1993.

93. Selye, H., Stress and aging, *J. Am. Geriatr. Soc.,* 18, 669, 1970.

94. Napolitano, E. W., Hutchinson, R. K., and Liem, H., Association of an endogenous heat shock protein (HSP70) with neuronal and non-neuronal cytoskeletal elements, *J. Cell Biol.,* 101, 30a, 1985.

95. Niedzwiecki, A. and Fleming, J. E., Changes in protein turnover after heat shock are related to accumulation of abnormal proteins in aging *Drosophila melanogaster, Mech. Ageing Dev.,* 52, 295, 1990.

96. Fargnoli, J., Blake, M. J., and Holbrook, N. J., In Vivo and in vitro studies on the heat shock response and aging, in *Molecular Biology of Aging,* Finch, C. E. and Johnson, T. E., Eds., Wiley-Liss, New York, 1990, 379.

I
GENERAL
PERSPECTIVES

THEORIES OF AGING

4

Ramesh Sharma

1 GENETIC/ENVIRONMENTAL INTERACTIONS IN AGING

The maximum lifespan potential is a constitutional feature of speciation to polygenic controls and to environmental influences. The enormous genetic heterogeneity that characterizes many species, particularly humans, and the complexity of environmental experiences create quantitative and qualitative variations of the senescent phenotype. Until now, no single theory has accounted for all phenotypes, although many scientists have tried to explain at least some of the major and most frequent aging phenomena. Almost all phenotypes result from an interaction between nature and nurture, and an integrating view of these interactions may help provide a more fundamental understanding of aging. Thus, an analysis of molecular, cellular, and systemic events may reveal a productive path for understanding the biology and pathology of aging. It is with this rationale in mind that the present chapter presents a comprehensive account of the major theories of aging categorized as molecular, cellular, and systemic.

Molecular theories propose that the lifespan of any species is governed by the genes interacting with environmental factors. Genetic information is stored in the genes (nucleotides segment of DNA), is transcribed to RNA, and is subsequently translated into proteins. These proteins, either structural or functional, govern the form and function of organisms. Aging may result from changes in DNA template activity, which regulates the formation of the final cellular products. It is believed that gene expressions are carefully regulated and that the proteins produced by gene activity are involved in multiple interacting processes.

A number of theories propose that changes in cellular proteins and other macromolecules occur as a function of age. These changes occur with the passage of time under the influence of environmental factors (e.g., nutrition and stress). They may be chemical and/or morphologic and involve enzymes, hormones, age pigments, free radicals, membrane permeability, macromolecule crosslinking, and changes in various cell organelles such as lysosomes and mitochondria.

Systemic theories ascribe aging of the entire organism to decrements in the function of a key system, such as the nervous, endocrine, or immune system. Such decrements could be genetically programmed, as are the early developmental phases of the lifespan, or be the consequence of environmental insults. Alterations in the key system will generate changes throughout the entire organism.

The various theories of aging are listed in Table 4-1. Some of these theories are presented here and some are discussed in chapters applicable thereto (Chapters 6, 7, and 11).

2 MOLECULAR THEORIES

These theories begin with the following concepts:

1. All individuals within a species have an almost similar length of life.
2. Individuals from different species have different lifespans.

For example, mayflies live only 1 day, houseflies 30 days, rats 3 years, dogs 12 years, horses 25 years, and humans 100 years.[1] It is presumed that there is some genetic program which determines the maximum lifespan for each species. Another argument for a genetic basis of aging is that the offspring of long-lived parents have a longer lifespan than those born from average-lived parents.[2] The average lifespan for females is generally longer than for males in most developed countries, like the U.S., Sweden, and Japan; this sex difference is also observed in other groups of animals (Chapter 2).[3]

An equally significant contribution to a genetic basis of aging is deduced from the duration of the three phases of the lifespan — developmental, reproductive, and senescent. In most animals, the reproductive phase occupies a very significant period in the lifespan, followed by a postreproductive phase. In mammals, the time taken to reach reproductive maturity is directly correlated with maximum lifespan. Humans and other long-lived mammals take a longer time to reach reproductive maturity than other animals and continue to live even after reproduction has ceased. Conversely, certain lower vertebrates (Pacific salmon, Atlantic eel, and lamprey) and invertebrates (octopus) die soon after their first reproduction, as if reproduction might involve depletion of certain essential factors necessary for maintenance of later life (Chapter 2).

The expression of the genetic program that regulates the lifespan may be altered by various environmental factors. Evidence for genetically programmed lifespan has been reported in a colonial protochordate *Botryllus schlosseri*.[4] Such animals display a parent colony-specific timing of mortality.

Molecular theories of aging discussed here include codon restriction, somatic mutation, error theory, and gene regulation theory, as well as antagonistic pleiotropy, dysdifferentiation, and soma disposal hypotheses (Table 4-2).

2.1 Codon Restriction

All the genetic information stored in DNA directs the structure and function of the organism, although only part of the total DNA information is utilized by the cell at a given time. The information is transferred from DNA to messenger RNA (mRNA) by the process of transcription. The functional mRNA in eukaryotic cells is derived by excision of intervening sequences (introns) and splicing. This mRNA is then translated into protein. *The codon restriction theory of aging is based on*

0-8493-8979-8/94/$0.00+$.50

TABLE 4-1

Major Theories of Aging

Aging due to external causes	Lifespan indefinite were it not for environmental insults such as Foods/toxins Bacteria/viruses Radiation Pollutants	
Aging due to internal causes	Lifespan genetically determined for a finite period; genetic expression modulated through specific programs leading to Neuroendocrine theory Immunologic theory	
Aging due to cellular and molecular causes	Both internal and external causes may act at one or more cellular levels and/or specific molecules to produce	
	In membranes:	Changes in fluidity, permeability, transport Organelle biogenesis and intracellular molecular movements
	In cytoplasm:	Wear and tear Free radical accumulation Cross-linking Lipofuscin
	In nucleus:	Codon restriction DNA damage and DNA repair failure RNA catastrophe errors Mutations Gene regulation Antagonistic pleiotropy Dysdifferentiation Disposable soma

the hypothesis that the fidelity or accuracy of translation, which depends on the cell's ability to decode the triple codons (three bases) in mRNA molecules, is impaired with aging.[5] Accurate readings of codons are done by two main biomolecules: transfer RNAs (tRNAs) and aminoacyl-tRNA synthetases. Any changes in these tRNAs and aminoacyl-tRNA synthetases may alter the rate of translation.

There is experimental evidence for quantitative changes in the tRNAs and synthetases during development and aging. Ilan and Patel[6] have reported alterations in $tRNA^{tyr}$, $tRNA^{leu}$, and corresponding synthetases during the developmental period of the insect *Tenebrio molitor.*[7] These quantitative alterations also occur in the isoacceptors of $tRNA^{arg}$ and $tRNA^{tyr}$ during aging of the Free living nematode *Turbatrix aceti.* Support for this theory has come from the findings of Hosbach and Kubli[8] who demonstrated that tRNA isolated from 35-day-old *Drosophila melanogaster* cannot be aminoacylated as efficiently as that of 5-day-old flies. The efficiency of the aminoacylating ability of some synthetases of old flies is only 50% that of the young flies. The fetal rat liver contains six isoacceptors for $tRNA^{tyr}$ compared to the adult which has only three.[9] A lesser aminoacylation has been reported in hepatic parenchymal cells of old rats.[10]

Changes in tRNAs and aminoacyl-tRNA synthetases with aging occur in plant systems as well. Young and old tissues of soybean cotyledons differ from each other in the kinds of completely chargeable tRNA that are present. Moreover, extracts of old tissue are not only deficient in certain aminoacylating abilities but possess factors which inhibit the charging of some tRNAs by extracts of young cotyledons.[11] Gene sequencing of rabbit β-globin reveals a highly restricted use of the synonymous codons for various amino acids, only 39 of the 61 usable codons are employed in the framing of the message.[12] Comparison of the isoaccepting species of $tRNA^{lys}$ from early and late human fibroblasts shows a smaller proportion of these species in senescent cells than in those from early passage cultures.[13]

As a result of differentiation, cells would lose their ability to translate genetic information. Despite much supportive evidence, this theory, based on the view that sequential changes in the tRNAs and aminoacyl-tRNA synthetases during lifespan may lead to the aging of an organism, needs further validation. It is difficult to explain the basic cause(s) for the alterations with aging in these message-reading molecules and the implications of such changes in the aging phenomena.

TABLE 4-2

Molecular Theories of Aging

Codon restriction	Fidelity/accuracy of mRNA message translation is impaired with aging due to cell inability to decode the triple codons (bases) in mRNA molecules
Somatic mutation	Exposure to radiation shortens lifespan due to increased incidence of mutations and loss of functional genes
Error catastrophe	Errors in information transfer due to alterations in RNA polymerase and tRNA synthetase may increase exponentially with age resulting in increased production of abnormal proteins
Gene regulation	Changes in expression of genes regulating both development and aging
Antagonistic pleiotropy	Genes beneficial during development and deleterious at later ages
Dysdifferentiation	Gradual accumulation of random molecular damages impair regulation of gene expression
Disposable soma	Preferential allocation of energy resources for reproductive cells to the detriment of maintenance and survival of somatic cells

2.2 Somatic Mutation

Alteration in the structure of DNA molecules alters the genetic message and results in differences in protein structures which lead to physiologic deficits. This proposed theory was based on the report that rats exposed to limited *irradiation* died at a younger age than nonirradiated controls.[14] These considerations were extended to humans[15,16] and included a higher incidence of neoplasia in irradiated individuals, suggesting that irradiation accelerates the aging process. According to this theory, exposure to radiation damages DNA and subsequently induces mutations which, in turn, lead to progressive loss of genes in postmitotic cells throughout the lifespan. The increased rate of mutations and loss of functional genes decrease the rate of production of functional proteins and cause cell death at a critical level.

Support for this theory was provided by the observation that increased exposure to X-rays shortens life expectancy and increases chromosomal aberrations in a dosage-dependent way.[17] Older animals have a greater number of chromosomal abnormalities than younger and, in short-lived mice, the rate of development of abnormalities is more rapid than in long-lived ones. These data suggest that natural radiation also affects the aging process. Martin and associates[18] reported a fivefold higher frequency of chromosomal aberrations in primary cultures of kidney from 40-month-old mice compared to young animals. A general consensus has emerged that the frequency of chromosomal aberrations increases greatly with age. In young non-cigarette smoking adults, the frequencies of aneuploidy, breakage, and structural chromosomal rearrangements are six times less than they are in 60-year-old individuals. However, contrasting evidence negates a causative role of somatic mutation in aging.[19]

In some species, such as humans, the sex chromosomes of females are similar (XX) but those of males are different (XY), while in other species the reverse is true. If radiation is a cause of aging, then one might expect a longer life for individuals with identical sex chromosomes. In most species, the females generally live longer than males, irrespective of the chromosomal composition. Another example of the lack of influence of sex chromosomes is the wasp *Habrobracon,* in which males have either two sex chromosomes (diploid) or one (haploid). If both types of males are exposed to X-rays, the haploid males should die earlier than the diploid, but this is not the case: both males have similar lifespans despite the greater resistance of the diploid male to ionizing radiation because of the larger number of repairable chromosomes.[20]

Chemical substances which alter DNA structure have no effect on lifespan.[21] Exposure of human fetal lung fibroblast to colchicine (an alkaloid capable of inducing polyploidy, i.e., higher than normal number of chromosomes), produces 60% tetraploid cells (with four sets of chromosomes) which continue to divide and have growth rates and lifespans similar to diploid cells (two sets of chromosomes).[22] Diploid as well as tetraploid human skin fibroblasts likewise have similar lifespans.[23] The somatic mutation theory is further weakened by the results from studies on the effects of low-dose ionizing radiation on the lifespan of human fibroblasts *in vitro.* Irradiation of early embryonic as well as postnatal cells may shorten, prolong, or have no effect on doubling potential and lifespan.[24,25]

Somatic mutations are no longer regarded as a probable cause of aging because the rate at which they occur in the absence of ionizing radiation is too low to account for overall age changes.[26] Furthermore, this theory does not give a clear picture of the mutation load in different organs and tissues and also of the kinetics of mutation accumulation.[27] Mutation accumulation has been suggested as a consequence rather than a cause of aging.[27]

Most cells have mechanisms for the repair of damaged DNA molecules,[28,29] and there is little evidence that DNA repair mechanisms decline in senescent animals;[30,31] rather, these repair mechanisms appear more effective in long-lived species as compared to short-lived.[28] The species-specific differences in the lifespan of animals might be attributed to the ability of animals to tolerate DNA damage rather than to repair. The presence of multiple copies of the same message coded within the DNA would offer protection to DNA damage.

The number of repetitive genes for the major rRNA is 5 to 10 in bacteria, 100 to 130 in *Drosophila,* and 250 to 600 in vertebrates,[32] suggesting a positive correlation between the number of repetitive genes and the lifespan of the species. Cutler[33] reported that the average redundancy of the transcribing mRNA in the brain was greater in humans than in cows and greater in cows than in mice. The higher the redundancy of the transcribing mRNA the longer is the lifespan. Based on available experimental evidence, radiation does not seem to play a major role in accelerating the aging process or in causing aging.

2.3 Error Theory

The form and function of organisms are determined by specific structural and functional proteins. Certain proteins such as RNA polymerase and tRNA synthetases are involved in the synthesis of other proteins. Medvedev[34] first proposed that *errors in information transfer from DNA to proteins may be responsible for cellular aging.* This concept was extended in a search for errors in transcription and translation processes, which may lead to accumulation of proteins and cause aging.[35,36] It was further argued that production of functional proteins such as enzymes depends not only on the genetic information stored in DNA but also on the protein synthetic machinery. Inaccuracy may occur both in protein and DNA synthesis.[37] The initial error in proteins may be low, but errors may increase exponentially as a function of age and lead to error catastrophe and cell death.

Evidence for the error theory is based on error induced experimentally in fruit flies by feeding them amino acid analogs.[38] Much of the support of the theory came from the work of Holliday and Tarrant[39] who reported an increased accumulation of heat-labile glucose-6-phosphate dehydrogenase in old fibroblasts. This heat-labile enzyme in old fibroblasts also showed an altered substrate specificity suggesting the possibility of errors. Using immunological techniques, Lamb[40] found an age-dependent increase in the proportion of inactive lactate dehydrogenase (LDH) in human fibroblasts, isocitrate lyase in nematodes, and aldolase in mouse liver. Functionally altered enzymes are known to accumulate in various animal tissues with age,[41,42] and the consequent decrease in the functional activity of tissues with age would be due to accumulation of such altered proteins (Chapter 5).

Some experimental evidence is not consistent with the proposal that errors in protein cause aging. Kanungo and Gandhi[43] could not detect any age-related differences in liver malate dehydrogenase (MDH) by immunologic techniques. Kinetic properties (K_m and K_i) and electrophoretic mobilities of rat liver cytosolic alanine aminotransferase[44] and aspartate aminotransferase[45] do not reveal age-related differences. Studies of cytosolic superoxide dismutase (SOD) from livers, brains, and hearts of rats and mice have not revealed age differences in antigenicity, K_m and K_i, and electrophoretic mobility.[46]

The fidelity of protein synthesis *in vitro* remains unchanged in human diploid skin fibroblasts as a function of age.[47] Although significant quantitative changes occur in aging *Drosophila* mitochondrial proteins, there is no change in the molecular weight or isoelectric point of these proteins.[48] The fidelity of mitochondrial proteins seems to be preserved throughout the lifespan of *Drosophila*. Thus, there is evidence to believe that errors in the fidelity of protein synthetic machinery do not occur with age and, therefore, cannot be responsible for aging. Nevertheless, the findings of altered conformation in some proteins by oxidation may account for physiological impairment during aging of animals (Chapter 6).[48]

2.4 Gene Regulation Theory

According to this theory,[49] *senescence results from changes in the expression of genes after reproductive maturity is reached.* It is based on the presumption that senescence would follow a pattern similar to that of differentiation and growth i.e., a sequential activation and repression of certain genes which are unique to these phases. Sequential activation and repression of genes have been reported for various chains of hemoglobin during the gestational period in humans.[50] Hemoglobin, a tetramer metallo-protein, consists of α_2 ε_2 chains in the fetus at 1 to 2 months of gestation. In the later phase of gestation the a chain remains the same and the ε chain is replaced by the γ chain. Just before birth, the γ chain is further replaced by the β chain which gives rise to adult hemoglobin $\alpha_2\beta_2$. The synthesis of these chains is governed by different sets of genes which are sequentially activated and repressed during the development of the human fetus.

Another example of gene activation and repression is the differential expression of LDH isoenzymes in mice during embryonic development.[51] The proportion of M4-LDH is significantly lower in heart and skeletal muscle of old compared to young rats (Chapter 5).[52] Studies on rat liver cytosolic alanine aminotransferase (cAAT) have shown that the gene for A subtype is more active in the early period of the lifespan and subsequently repressed, while B subtype gene is activated in old age.[44] The sequential activation and repression of genes would not be restricted to development,[53] but would extend into adulthood and aging.[54,55]

The genes responsible for the synthesis of various enzymes do not appear to undergo any change in their basic sequences during the lifespan. Rather, the observed changes in levels of enzymes may be due to the alterations in the template activity of corresponding genes induced by various extrinsic and intrinsic factors (Chapter 5). For example, the level and inducibility of many enzymes by hormones change in different tissues as a function of age without any possibility of error incorporation into these molecules.[56] Modulating actions may either appear or disappear and/or their levels may change at different phases of the lifespan.[54] The products or by-products of the genes responsible for differentiation and growth, on reaching critical levels, stimulate certain unique genes responsible for the reproductive phase. However, as a result of continued reproduction, certain factors may be depleted and/or they may not be replenished as fast as they disappear. Such factors may be of crucial importance for keeping certain genes expressed or repressed. They may also activate some undesirable genes which may be responsible for a gradual decline in reproductive rate (e. g., number of offsprings) with age. This theory also predicts that should the organism be able to replenish the factors that become depleted due to continued reproduction, the reproductive period and lifespan would be lengthened. It is supported by the data on the lifespan of mammals, particularly the reproductive phase, which has continuously lengthened with the progress of evolution.[57]

The lifespan of a species may be divided into three phases:

1. Developmental
2. Reproductive
3. Senescent

Each phase has a characteristic

1. Duration
2. Rate, i.e., velocity
3. Sequential timetable of events
4. Regulatory mechanisms

Initiation and duration of developmental and reproductive phases depend on a unique set of genes that are sequentially activated and repressed.

Human genetic diseases like *progeria* and *progeroid syndromes* are in agreement with this sequence.[58] Progeria is caused by the mutation of an autosomal gene. In this case, the newborn child appears normal and grows normally up to about 6 years, when signs of aging (e.g., atherosclerosis, accumulation of lipofuscin, graying of hairs) do appear. Fibroblasts taken from a 10-year-old progeria patient do not undergo as many population doublings as those of a normal child of the same age. It appears that some genes responsible for normal development are altered to induce this condition. Perhaps the production of essential factors necessary for development and growth is prevented by this mutation. The reproductive phase is not initiated due to lack of switching on one of the necessary genes during the later phases of development and growth. The lifespan is shortened following expression of the mutated gene (Chapters 3 and 5).

Another example is the sudden death of the female *octopus* which lays eggs only once, broods them, reduces food intake, and dies soon after the hatching of the young ones.[59] Removal of the female's paired optic glands after spawning prevents brooding, and the octopus continues to eat and to grow and its longevity increases. It is apparent that some factors produced by the optic gland are essential for brooding and cessation of feeding followed by senescence and death. Egg laying may deplete these factors, which may in turn cause the optic gland to produce a hormone that causes behavioral change.

A similar phenomenon of semelparous (pertaining to organisms reproducing only once) degeneration and death is observed in *Pacific salmon*[60] and in *marsupial mice*.[61] In all three cases, there is marked evidence for hormone-dependent degeneration and death. Each species has a unique set of genes for development and reproduction. The sequential activation or repression determines the duration of development and the onset of reproduction and is governed by the proper balance of various factors that are essential for maintenance of the reproductive phase. No unique gene would be responsible for aging; rather, aging would merely be a consequence of the organism's attaining reproductive maturity, irrespective of whether or not it reproduced.

Evidence of semelparous degeneration and death in certain species also provides support for developmentally programmed aging.[1,55] It assumes that aging is controlled in ways similar to those that operate during development. These events are primarily controlled by hormones, which are produced or depleted during and after reproduction. Hormones would play a significant role not only in development but also in regulating the aging of an organism. Developmentally programmed aging,[55] in spite of supportive evidence, needs further testing of genetic approaches to both development and aging in the same organism. Discoveries of homeotic genes[62,63] controlling development of *Drosophila* have provided a way to investigate the role of such related genes in regulating aging of an organism.

Sequential activation and repression of genes for certain senescence marker proteins have been reported.[64] Most striking is the reversible expression of liver senescence marker protein 2 (SMP-2) gene as a function of age. It is expressed maximally during both prepuberty and senescence when liver is insensitive to androgen because of lack of functional androgen receptors, and mRNA expression drops significantly in the postpubertal adult male rat when liver is most responsive to androgen. SMP-2 gene is reported to be an androgen-repressible gene as its high level of expression is maintained in young adult females. These findings are consistent with the concept of hormone-dependent sequential regulation of genes during aging.

Studies of long-lived organisms compared to normal-lived ones may reveal which factors the normal organisms are missing.[65] By selective breeding, longer-lived *Drosophilae* have been generated[65,66] that produce a remarkably active form of SOD, an antioxidant enzyme, indicating that they contain a variant of the normal enzyme encoding gene. An active form of SOD neutralizes superoxide ($O_2^{\cdot}$) more effectively leading to a longer lifespan of the flies. Normal fruit flies age more quickly because their free radical defenses are not as effective as those of specifically bred flies.

Another example of genetic clues comes from the search for a gene in the soil nematode *(Caenorhabditis elegans)* that is differentially expressed in the long-lived and normal organism.[67] Mutation of a single gene called age-1 can increase the average lifespan of *C. elegans* by about 70%. These mutant worms also produced enhanced levels of SOD and catalase. Johnson[67] postulated that mutation of age-1 may deplete its protein product, which might have suppressed SOD and catalase gene activity in normal-lived worms.

In a similar approach, Jazwinski[68] identified several genes that prolong the life of brewer's yeast *(Saccharomyces cerevisiae)*. The best studied of these is LAG-1 (longevity assurance gene-1) which is more active in young than in old cells. LAG-1 activity in older cells extends the lifespan of yeast by one third. Strikingly, aged yeast cells harboring the extra-active gene do not become immortal. They simply remain youthful for a longer period of time. The LAG-1 gene product has not yet been functionally characterized. Attempts are being made to isolate such longevity assurance gene(s) from human cells and to correlate their presence with the human lifespan.[65] These examples of longevity-controlling genes reflect a unique feature of gene regulation during aging. Although, the function of such genes remains obscure, the discovery of induced antioxidant enzymes seems to have the potential for affecting longevity in both fruit flies and *C. elegans*.

3 ANTAGONISTIC PLEIOTROPY, DYSDIFFERENTIATION, AND DISPOSABLE SOMA HYPOTHESES

3.1 Antagonistic Pleiotropy

A long-held theory of the cause of senescence is *the declining force of natural selection as a function of age of adult somatic cells*.[69,70] Natural selection that favors genes with early beneficial effects leads to deleterious effects later on. Certain *genes confer survival advantages early in life and cause harmful physiological effects in later stages of lifespan*,[69,71] a phenomenon termed negative or antagonistic pleiotropy. The evolutionary view of aging is supported by mathematics and interesting experimental data. Friedman and Johnson[72,73] have isolated and characterized a mutant allele, age-1, from *C. elegans* which increases maximum lifespan by 60% at 20°C. At the same time, it decreases fertility by 75% in self-fertilizing hermaphrodites.[67] Genes that specify instructions for synthesizing reproductive hormones also serve as examples of antagonistic pleiotropy.[74] Long-term exposure of the estrogen used to enhance fertility increases the risk of breast cancer in aged women. A variety of other cases have been noted. Hypothalamus and pituitary gland control ovarian function and also contribute to aging of the ovary in rodents. At the same time, ovarian signals appear to promote aging of the hypothalamus and pituitary. Age-dependent cytotoxic effect of glucocorticoids on hippocampal and hypothalamic neurons also correlate with antagonistic pleiotropy.[75] However, there is some evidence that is inconsistent with this hypothesis.[76]

3.2 Dysdifferentiation

Gradual accumulation of random molecular damages impairs the normal regulation of gene activity, potentially triggering a cascade of injurious consequences. Cutler[77] has called this process dysdifferentiation. Dysregulation of genes may provide a mechanism that links the antagonistic pleiotropy and disposable soma hypotheses into a unified concept of aging.[78] It is evident that genes are carefully regulated and that the proteins produced by gene activity are involved in multiple, often interacting, processes. Aging may occur when the normal repair and maintenance functions of cells become dysregulated and gradually lead to impaired physiologic functions.

Aberrant expression of genes may play a role in the aging process.[79] Evidence for this came from the discovery of increased levels of globin RNA in the liver and brain of older mice.[80] Loss of epigenetic control has been proposed as a major causal factor for reactivation of genes in aged.[79] DNA methylation at C≡G bases relates to the inactivation of gene expression and has been shown to decrease with age in several systems,[81,82] leading to the reactivation of genes. The extent of demethylation correlates with longevity in two different mouse species.[81] Furthermore, treatment of cultured cells with 5-azacytidine, an inhibitor of methylase activity, results in a decreased doubling potential.[83]

In an attempt to test the universality of the loss of epigenetic control and aberrant gene expression,[79,80] Slagboom and Vijg[84] examined the age-related expression of a number of genes. It appears that the expression of many genes decreases, the expression of some increases, and others do not change in different systems with advancing age of animals.[84,85] In systematic studies to test the aberrant gene expression phenomena, Sato and his associates[86] measured the steady state level of mRNA for five tissue-specific genes (myelin base protein, atrial natriuretic factor, albumin, κ immunoglobin, and keratin gene expressed in brain, heart, liver, spleen B lymphocytes, and skin, respectively) and found no aberrant expression of these genes with age. Another molecular abnormality with aging is telomere shortening. Telomeres are the tail portion of the chromosomes, which they help to stabilize during cell division. As telomeres shrink and are shed with each cell division, their length gives some indication of the number of divisions undergone and still to occur. In cancer cells, in the presence of the enzyme telomerase, telomeres are not shortened but continue to replace lost sequences; in aging cells, telomerase activity may be reduced, telomeres may be shortened, and cell division may be curtailed.[87] Overall, these studies indicate that aberrant gene products, telomere shortening, or other changes in gene expression are not necessarily a typical feature of the aging process *in vivo*.[86]

3.3 Disposable Soma Hypothesis

This hypothesis suggests that aging has evolved as a by-product of optimization of the allocation of energy and resources for the various works performed by the organism. It assumes that *energy resources are better spent for maintenance of reproductive cells, responsible for species survival.* The minimum required for maintenance, repair, and survival of somatic cells is not cost-effective; it is too expensive in terms of energy and the nonreproductive cells of the body are consequently expendable.[88] Aging then would result from a progressive accumulation of somatic defects and damage. Maintenance and repair include the prevention and removal of DNA damage, accuracy in macromolecular synthesis, and degradation of defective proteins. Lifespan of different species depends on the differential level of somatic maintenance and repair. Long-lived species in general have a greater level of maintenance and repair systems compared to short-lived species.

The disposable soma hypothesis *balances the maintenance and repair of somatic cells on one side and the reproduction and fertility on the other.* If more energy is used for maintenance of soma, less will be available for reproduction, and *vice versa*. This hypothesis treats *senescence as a price paid for sexual reproduction*.[89] There exists a direct correlation between the time taken to reach reproductive maturity and the species lifespan (Chapter 2). Experimental animals whose reproductive age is delayed tend to live longer than normal animals.[90]

Semelparous reproductive degeneration and death in certain species also provide support for this hypothesis. Animals with exhaustive reproductive activity seem to expend much more energy than is allocated for this purpose; they are left with little to maintain and repair the somatic cells and hence die soon after their single reproduction. This cost-effectiveness theory also draws support from the free radical theory of aging (Chapter 6).[91] The maintenance and repair of free radical damage to various structural and functional biomolecules play an important role in determining the aging of an organism.[91] Accumulation of age-dependent advanced glycosylation end products (AGEs) occurs because of failure of prevention and repair systems for such damages.[92]

TABLE 4-3

Cellular Theories of Aging

Wear and tear	Intrinsic (e.g., oxidative processes) and extrinsic (e.g., ambient temperature) influence the lifespan
Free radical accumulation	Free radicals formed by oxidative reaction accumulate and damage membrane, cytoplasm, and nucleus (also discussed in Chapter 6)
Age pigments	Accumulation of lipofuscin (fluorescent age pigment) causes several pathophysiological complications and is inversely correlated with aging (also discussed in Chapter 6)
Cross-linking theory	Cross-linkages among molecules develop with aging and alter chemical/physical properties of cell molecules

4 CELLULAR THEORIES

These theories relate to changes that occur in structural and functional elements of cells with the passage of time (Table 4-3). They also concern the biomolecules after their synthesis is over, suggesting that these changes impair the effectiveness of molecules as a function of age.

4.1 Wear and Tear

The idea of wear and tear compares living organisms with machines,[16] i.e., with repeated use, parts wear out and become defective, and the machinery finally fails to function. This comparison is not entirely appropriate: organisms have a mechanism by which they can repair their damages, whereas machines do not.

The premise of this theory originates from the observation that the *lifespan of poikilotherms is shortened by increasing the environmental temperature and prolonged by decreasing it;* rates of chemical reactions increased with increasing temperatures, and the reverse is true for low temperatures (Chapter 2). This phenomenon has been reported for fruit flies[93] and rotifers.[94] Even a slower rate of aging has been reported for rat tail tendon collagen, an extracellular protein, at low temperature.[95]

On the other hand, an increase in metabolic rate may shorten the lifespan by accelerating wear and tear. The lifespans of different animal species are inversely proportional to basal metabolic rate.[96] Basal oxygen consumption rates of short-lived animals, such as rats and mice, are much higher than those of long-lived animals, such as elephants and men. However, within the same species, it is difficult to correlate individual differences in lifespan with the metabolic rate.

4.2 Age Pigments

Accumulation of lipofuscin or age pigment is the most prominent age-associated change present in a variety of cell types of many organisms. It is deposited predominantly in nondividing cells such as neurons and cardiac myocytes as a function of age.

Lipofuscin accumulation has been observed in the cortex and hippocampus of man, Rhesus monkey, and rat as one of the common morphological features of aging and has been correlated with the loss of neurons in old age (Chapters 5 and 8). Lipofuscin is also deposited in dividing cells of liver, adrenal cortex, and testes.

Lipofuscin accumulation causes the loss of cytoplasmic mass, mitochondrial number, rough endoplasmic reticulum, and is associated with vacuolization of cytoplasm. Indeed, lipofuscin accumulation may represent a basic feature of cellular aging (Chapters 5, 6, and 8).

TABLE 4-4

System Level Theories of Aging

Neuroendocrine	Control of homeostasis by neural and endocrine signals becomes disorganized with aging; physiologic performance declines while pathologic responses to stress increase in number and severity (also discussed in Chapter 11)
Immunologic	Immune system reduces its defenses against antigens and loses the capacity to recognize self, resulting in increasing incidence of infections and autoimmune diseases (also discussed in Chapter 7)

4.3 Crosslinking Theory

With passage of time, many biological macromolecules develop cross-linkages between identical or different molecules. These linkages alter the physical and chemical properties of the molecules.[97] Major support for this theory was provided by the studies of the extracellular fibrous protein collagen by Verzar.[98] Collagen is synthesized in all cell types, particularly connective tissue cells, and is deposited extracellularly in all tissues. The structural unit of collagen is tropocollagen; units are packed together side by side and stabilized by chemical cross-links between the chains (Chapter 22). The mode of packing creates periodic striations in the structure of collagen fibers. The number of striations in rat tail-tendon collagen and its thermal stability increase with age, while its solubility decreases due to the increased cross-linkages.[99]

Cross-linking agents with charged groups are produced during normal metabolism. Such ionized groups are replaced in early life by normal metabolic processes but accumulate in larger amounts in old age.[99] The groups react irreversibly with macromolecules such as DNA and proteins, inactivating them and thus reducing their functional competence.[98,100]

Increased cross-linking of aged collagen has been correlated with an increased rigidity of the cell membrane, a probable cause of the decreased potassium conductance of the membrane.[101] The higher intracellular potassium would, in turn, increase the intracellular ionic strength and lead to a decreased rate of transcription by chromatin and a decreased rate of protein synthesis.

The free radical hypothesis of aging (Chapter 6) suggests that the important causative agent of aging is the active oxygen species, which produces more damaging effects in compact structures such as cellular membranes than in diluted systems like cytosol.[102] The probability of cross-linking is enhanced in closely packed molecules, making the membranes the most likely targets to be damaged. Cross-linkages are present not only in extracellular collagen but also in intracellular proteins (enzymes) and in nucleic acids (DNA). The decrease in the extractability of chromosomal proteins from chromatin may be attributable to their increased cross-linking with DNA.[103]

Cross-links are produced not only by charged groups but also by some inert molecules such as glucose.[91] Nonenzymatic glycosylation, a chemical attachment of glucose to proteins and nucleic acids, has been implicated in the production of cross-links in these macromolecules.[91] Extensive cross-linking of proteins may contribute to the stiffening and loss of elasticity characteristic of aging tissues.

Even nonenzymatic addition of glucose to nucleic acids may gradually damage DNA. The reaction between glucose and proteins is known as the Maillard or browning reaction. It begins when an aldehyde group (–CHO) of glucose combines with an amino group ($-NH_2$), a Schiff base. This combination is unstable and quickly converts to a substance known as Amadori products. In long-lived proteins, these Amadori products slowly dehydrate and rearrange irreversibly into structures called advanced glycosylation end products (AGEs). Many of these AGEs are also able to cross-link adjacent proteins. It has been suggested that nonenzymatic glycosylation of lens crystallins may contribute to cataract formation in aging individuals. Cross-links generated in proteins and nucleic acids by nonenzymatic glycosylation may contribute to age-related declines in the functioning of cells and tissues.

5 SYSTEM LEVEL THEORIES

Major systemic theory includes the neuroendocrine and immunologic theories (Table 4-4). *For the immunologic theory, the reader is referred to Chapter 7.* Some aspects of the neuroendocrine theory are also discussed in Chapter 11. Because of the interrelation of the immune and neuroendocrine systems, a section including discussion of this interrelation with aging is presented here.

5.1 Neuroendocrine Control Theory

The overall performance of an animal is closely related to the efficacy of a variety of control mechanisms that regulate the interaction between different organs and tissues.[104] *The effectiveness of homeostatic adjustments declines with aging* and leads to consequent failure of adaptive mechanisms, aging, and death.[105] Adaptation to external and/or internal stress depends on *control mechanisms orchestrated by the combined interplay of the nervous and endocrine systems.* The activity of several peripheral endocrine glands, such as thyroid, adrenal, and gonads, is controlled directly by the pituitary gland and indirectly by higher nervous centers, mainly the hypothalamus, which signal the pituitary. For efficient adaptation, nervous and endocrine signals must be synchronized and be responsive to the needs of the many functions they regulate.[106-108] However, with aging, some of the efficiency of the hypothalamo–pituitary interaction is lost or altered, leading to decreased function and increased pathology of most organs and tissue systems.[109,110]

Hormones secreted by the hypothalamus–pituitary–endocrine axis are necessary for the proper functioning of almost every cell in the body. This axis is controlled by a complex mechanism that includes interactions with neurotransmitters of the brain, hormones produced by different endocrine glands, and nutrients from the small intestine and liver. The neuroendocrine theory views aging as part of a lifespan program regulated by neural and hormonal signals. The program unfolds from fertilization through birth, childhood, adulthood, and finally old age and death; command neurons in higher brain centers act as "pacemakers" that regulate the "biological clock" governing development and aging. With the passage of time, aging changes may result from programmed deterioration or cessation of the programming that regulates homeostasis.[111] In either case, aging would be manifested through a slowing down or imbalance in the activity of the pacemaker neurons with consequent neurotransmitter and hormonal alterations and their repercussion on neural, muscular, and secretory functions. Such functional decrements are exemplified by involution of reproductive organs, loss of fertility, diminished muscular strength, lesser ability to recover from stress, and impairment of cardiovascular and respiratory activity.

5.2 Endocrine–Immunologic Relationship

Hormonal and neural influences on the immune system have long been known; for example, the involutionary action of glucocorticoids on the thymus and lymphatic system has been used for the treatment of allergies and in organ transplants. In addition, the nervous system may regulate some aspects of the immune response. Thus, *neuroendocrine and immunologic theories of aging may converge, or changes in their function may articulate with each other to lead to age-related decline in several bodily functions.*[112]

A neuroendocrine-immunomodulation of thymic aging has been demonstrated in experiments in which tumor (GH3) pituitary adenoma cells, which secrete both growth hormone and prolactin, can reconstitute thymic structure and improve T-cell production and function when implanted in old rats.[113] The possibility of thymic rehabilitation in old age suggests that lymphoid cells in aged animals are not inherently defective, but given the proper stimulus can return to normal function. It will be of immense importance to reactivate the aging thymus, even replacing old T cells with the young ones to make the body's immune system functional for a longer period of time. In addition to cell replacement therapy, one might expect a log of potential of somatic gene therapy in the prevention of immunosenescence.[114,115] These age-related decrements of the body's defense and/or of neuroendocrine systems can be enhanced by using somatic gene therapy, once it becomes fully operative. It can be of great importance to several other age-related disorders such as Parkinson's and Alzheimer's diseases.

5.3 Perspectives on Aging

In spite of tremendous progress in the field of aging research, the mechanism of this process remains elusive. None of the theories proposed explains all the cause(s) for aging; rather, aging seems to be a multifactorial process. Growing evidence suggests that a multitude of parallel and often interacting processes govern the aging of an organism, many of them controlled jointly by a combination of genetic and environmental factors. Although at present it is difficult to advocate a coherent theory of the cause of aging, progress achieved so far and continuing efforts by numerous scientists let us hope that an answer will be forthcoming in the near future.

REFERENCES

1. Comfort, A., *The Biology of Senescence,* 3rd ed., Elsevier, New York, 1979.
2. Dublin, L. I., *Length of Life: A Study of the Life Table,* Ronald Press, New York, 1949.
3. Rockstein, M., *Theoretical Aspects of Aging,* Academic Press, New York, 1974.
4. Rinkevich, B., Lauzon, R. J., Brown, B. W. M., and Weissman, I. L., Evidence for a programmed lifespan in a colonial protochordate, *Proc. Natl. Acad. Sci. U.S.A.,* 89, 3546, 1992.
5. Strehler, B. L., *Time Cells and Aging,* 2nd ed., Academic Press, New York, 1977.
6. Ilan, J. and Patel, N., Mechanism of Gene expression in *Tenebrio molitor, J. Biol. Chem.,* 245, 1275, 1970.
7. Reitz, M. S. and Sanadi, D. R., An aspect of translational control of protein synthesis in aging: changes in the isoaccepting forms of tRNA in *Turbatrix aceti, Exp. Gerontol.,* 7, 119, 1972.
8. Hosbach, M. A. and Kubli, E., Transfer RNA in aging Drosophila: extent of aminoacylation, *Mech. Ageing Dev.,* 10, 131, 1979.
9. Yang, W. K., Isoaccepting transfer RNAs in mammalian differentiated cells and tumor tissues, *Cancer Res.,* 31, 639, 1971.
10. Mays, L. L., Lawrence, A. E., Ho, R. W., and Ackley, S., Age related changes in function of transfer ribonucleic acid of rat livers, *Fed. Proc.,* 38, 1984, 1979.
11. Bick, M. D. and Strehler, B. L., Leucyl- transfer RNA synthetase activity in old cotyledons: evidence on repressor accumulation, *Mech. Ageing Dev.,* 1, 33, 1972.
12. Efstratiadis, A., Kafatos, F. C., and Maniatis, T., The primary structure of rabbit β-globin mRNA as determined from cloned DNA, *Cell,* 10, 571, 1977.
13. Agris, P. F., Boak, A., Basler, J. W., Voorn, C. V., Smith, C., and Reichlin, M., Analysis of cellular senescence through detection and assessment of RNAs and proteins important to gene expression: transfer RNAs and autoimmune antigens, in *Werners Syndrome and Human Aging,* Salk, D., Fujiwara, Y., and Martin, G., Eds., Plenum Press, New York, 1985.
14. Szilard, L., On the nature of the aging process, *Proc. Natl. Acad. Sci. U.S.A.,* 45, 30, 1959.
15. Failla, G., The aging process and somatic mutations, in *The Biology of Aging,* Strehler, B. L., Ed., American Institute of Biological Sciences, Washington, D.C., 1960.
16. Sacher, G. A., Life table modification and life prolongation, in *Handbook of the Biology of Aging,* Finch, C. E. and Hayflick, L., Eds., Van Nostrand Reinhold, New York, 1977.
17. Curtis, H. J., Cellular processes involved in aging, *Fed. Proc.,* 23, 662, 1964.
18. Martin, G. M., Smith, A. C., Ketterer, D. J., Ogburn, C. E., and Disteche, C. M., Increased chromosomal aberrations in first metaphases of cells isolated from the kidneys of aged mice, *Isr. J. Sci.,* 21, 296, 1985.
19. Evans, H. J., Cytogenetics: overview, in *Mutation and the Environments, part B: Metabolism, Testing Methods and Chromosomes,* Mendelson, M. L. and Albertini, R. J., Eds., Wiley-Liss, New York, 1990.
20. Clark, A. M. and Rubin, M. A., The modification by X-irradiation of the lifespan of haploids and diploids of the wasp, Habrobracon sp., *Radiat. Res.,* 15, 244, 1961.
21. Curtis, H. J., *Biological Mechanisms of Aging,* Charles C Thomas, Springfield, IL, 1966.
22. Thompson, K. V. A. and Holliday, R., The longevity of diploid and polyploid human fibroblasts. Evidence against the somatic mutation theory of cellular aging, *Exp. Cell Res.,* 112, 281, 1978.
23. Hoehn, H., Bryant, E. M., Johnston, P. H., Norwood, T. H., and Martin, G. M., Non-selective isolation, stability and longevity of hybrids between normal human somatic cells, *Nature,* 258, 608, 1975.
24. Macieira-Coelho, A., Diatloff, C., Billard, M., Fertil, B., Malaise, E., and Fries, D., Effects of low dose rate irradiation on the division potential of cells in vitro. IV. Embryonic and adult human lung fibroblast-like cells, *J. Cell. Physiol.,* 95, 235, 1978.
25. Azzarone, B., Diatloff-Zito, C., Billard, C., and Macieira-Coelho, A., Effect of low dose rate irradiation on the division potential of cells in vitro. VII. Human fibroblasts from young and adult donors, *In Vitro,* 16, 634, 1980.

26. Maynard-Smith, J., Theories of aging, in *Topics in Biology of Aging,* Krohn, P. L., Ed., Interscience, New York, 1966.
27. Vijg, J. and Gossen, J. A., Somatic mutations and cellular aging, *Comp. Biochem. Physiol.,* 104, 429, 1993.
28. Hart, R. W. and Setlow, R. B., Correlation between deoxyribonucleic acid excision-repair and life-span in a number of mammalian species, *Proc. Natl. Acad. Sci. U.S.A.,* 71, 2169, 1974.
29. Wheeler, K. T. and Lett, J. T., On the possibility that DNA repair is related to age in non-dividing cells, *Proc. Natl. Acad. Sci. U.S.A.,* 71, 1862, 1974.
30. Tice, R. R., Aging and DNA repair capability, in *The Genetics of Aging,* Schneider, E. L., Ed., Plenum Press, New York, 1978.
31. Hanawalt, P. C., On the role of DNA damage and repair processes in aging: evidence for and against, in *Modern Biological Theories of Aging,* Warner, H. R., Ed., Raven Press, New York, 1987.
32. Medvedev, Z. A., Repetition of molecular-genetic information as a possible factor in evolutionary changes life span, *Exp. Gerontol.,* 7, 227, 1972.
33. Cutler, R. G., Redundancy of information content in the genome of mammalian species as a protective mechanism determining aging rate, *Mech. Ageing Dev.,* 2, 381, 1973.
34. Medvedev, Z. A., The molecular processes of aging, *Sowjet-wiss Naturwiss, Beitr.,* 12, 1273, 1961.
35. Orgel, L. A., The maintenance of the accuracy of protein synthesis and its relevance to aging, *Proc. Natl. Acad. Sci. U.S.A.,* 49, 517, 1963.
36. Medvedev, Z. A., The nucleic acids in development and aging, in *Advances in Gerontological Research,* Strehler, B. L., Ed., Vol. 1, Academic Press, New York, 1964.
37. Orgel, L. E., Ageing of clones of mammalian cells, *Nature,* 243, 441, 1973.
38. Harrison, B. J. and Holliday, R., Senescence and the fidelity of protein synthesis in Drosophila, *Nature,* 213, 990, 1967.
39. Holliday, R. and Tarrant, G. M., Altered enzymes in aging human fibroblasts, *Nature,* 238, 26, 1972.
40. Lamb, M. J., *Biology of Aging,* John Wiley & Sons, New York, 1977.
41. Gershon, D., Current status of age-related enzymes: alternative mechanisms, *Mech. Ageing Dev.,* 9, 189, 1979.
42. Rothstein, M., The formation of altered enzymes in aging animals, *Mech. Ageing Dev.,* 9, 197, 1979.
43. Kanungo, M. S. and Gandhi, B. S., Induction of malate dehydrogenase isoenzymes in livers of young and old rats, *Proc. Natl. Acad. Sci. U.S.A.,* 69, 2035, 1972.
44. Patnaik, S. K. and Kanungo, M. S., Soluble alanine aminotransferase of the liver of rats of various ages: induction, characterization and change in patterns, *Indian J. Biochem. Biophys.,* 13, 117, 1976.
45. Sharma, R. and Patnaik, S. K., Properties of liver cytoplasmic aspartate aminotransferase of rats of various ages, *Biochem. Int.,* 5, 561, 1982.
46. Reiss, V. and Gershon, D., Comparison of cytoplasmic superoxide dismutase in liver, heart, and brain of aging rats and mice, *Biochem. Biophys. Res. Commun.,* 73, 255, 1976.
47. Goldstein, S., Wojtyk, R. I., Harley, C. B., Pollard, J. W. Chamberlain, J. W., and Stanners, C. P., Protein synthetic fidelity in aging human fibroblasts, in *Werners Syndrome and Human Aging,* Salk, D., Fujiwara, Y., and Martin, G. M., Eds., Plenum Press, New York, 1985.
48. Fleming, J. E., Melnikoff, P S., Latter, G. T., Chandra, D., and Bensch, K. G., Age-dependent changes in the expression of Drosophila mitochondrial proteins, *Mech. Ageing Dev.,* 34, 63, 1986.
49. Kanungo, M. S., A model for ageing, *J. Theor. Biol.,* 53, 253, 1975.
50. Zuckerkandl, E., The evolution of hemoglobin, *Sci. Am.,* 212, 110, 1965.
51. Markert, C. L. and Ursprung, H., The ontogeny of isoenzyme patterns of lactate dehydrogenase in the mouse, *Dev. Biol.,* 5, 363, 1962.
52. Singh, S. N. and Kanungo, M. S., Alterations in lactate dehydrogenase of the brain, heart, skeletal muscle, and liver of rats of various ages, *J. Biol. Chem.,* 243, 4526, 1968.
53. Caplan, A. I. and Ordahl, C. P., Irreversible gene expression model for control of development, *Science,* 201, 120, 1978.
54. Kanungo, M. S., *Biochemistry of Aging,* Academic Press, London, 1980.
55. Russell, R. L., Evidence for and against the theory of developmentally programmed aging, in *Modern Biological Theories of Aging,* Warner, H. R., Ed., Raven Press, New York, 1987.
56. Sharma, R., Enzymatic changes during aging, in *Physiological Basis of Aging and Geriatrics,* Timiras, P. S., Ed., Macmillan Press, New York, 1988.
57. Cutler, R. G., Evolution of human longevity and the genetic complexity governing aging rate, *Proc. Natl. Acad. Sci. U.S.A.,* 72, 4664, 1975.
58. Brown, W. T., Genetics of human aging, in *Review of Biological Research in Aging,* Vol. 2, Rothstein, M., Ed., Alan R. Liss, New York, 1985.
59. Wodinsky, J., Hormonal inhibition of feeding and death in octopus: control by optic gland secretion, *Science,* 198, 948. 1977.
60. Robertson, O. H., Prolongation of the lifespan of kokanee salmon by castration before beginning of gonad development, *Proc. Natl. Acad. Sci. U.S.A.,* 47, 609, 1961.
61. Diamond, J. M., Big-bang reproduction and ageing in male marsupial mice, *Nature,* 298, 115, 1982.
62. Scott, M. R. and Carroll, S. B., The segmentation and homeotic gene network in early Drosophila development, *Cell,* 51, 689, 1987.
63. Ingham, P. W., The molecular genetics of embryonic pattern formation in Drosophila, *Nature,* 335, 25, 1988.
64. Chatterjee, B. and Roy, A. K., Changes in hepatic androgen sensitivity and gene expression during aging, *J. Steroid Biochem. Mol. Biol.,* 37, 437, 1990.
65. Rusting, R. L., Why do we age, *Sci. Am.,* 267, 131, 1992.
66. Rose, M. R. and Graves, J. L., Jr., Evolution of aging, in *Review of Biological Research in Aging,* Vol. 4, Rothstein, M., Ed., Alan R. Liss, New York, 1990.
67. Johnson, T. E. and Hutchinson, E. W., Aging in *Caenorhabditis elegans:* update 1988, in *Review of Biological Research in Aging,* Vol. 4, Rothstein, M., Ed., Alan R. Liss, New York, 1990.
68. Chen, J. B., Egilmez, N. K., Jazwinski, S. M., Differential gene expression during aging of the yeast *Saccharomyces cerevisiae, Fed. Am. Soc. Exp. Biol.,* 3, A570, 1989.
69. Williams, G. C., Pleiotropy, natural selection, and the evolution of senescence, *Evolution,* 11, 398, 1957.
70. Charlesworth, B., *Evolution in Age-structured Population,* Cambridge University Press, London, 1980.
71. Rose, M. R., Life history evolution with antagonistic pleiotropy and overlapping generations, *Theor. Popul. Biol.,* 28, 342, 1985.
72. Friedman, D. B. and Johnson, T. E., A mutation in the age-1 gene in *Caenorhabditis elegans* lengthens life and reduces hermaphrodite fertility, *Genetics,* 118, 75, 1988.
73. Friedman, D. B. and Johnson, T. E., Three mutants that extend both mean and maximum lifespan of the nematode, *C. elegans,* define the age-1 gene, *J. Gerontol.,* 43, B102, 1988.

74. Finch, C. E., *Longevity, Senescence, and the Genome,* University of Chicago Press, Chicago, 1990.
75. Sapolsky, R. M., Krey, L. C., and McEwen, B. S., The neuroendocrinology of stress and aging: the glucocorticoid cascade hypothesis, *Endocr. Rev.,* 7, 284, 1986.
76. Le Bourg, E., Lints, F. A., Delince, J., and Lints, C. V., Reproductive fitness and longevity in *Drosophila melanogaster, Exp. Gerontol.,* 23, 491, 1988.
77. Cutler, R. G., The dysdifferentiative hypothesis of mammalian aging and longevity, in *The Aging Brain: Cellular and Molecular Mechanisms of Aging in the Nervous System,* Giacobini, E., et al., Eds., Raven Press, New York, 1982.
78. Olshansky, S. J., Carnes, B. A., and Cassel, C. K., The aging of the human species, *Sci. Am.,* 268, 46, 1993.
79. Holliday, R., The inheritance of epigenetic defects, *Science,* 238, 163, 1987.
80. Wareham, K. A., Lyon, M. F., Glenister, P. H., and Williams, E. D., Age-related reactivation of an X-linked gene, *Nature,* 327, 725, 1987.
81. Singhal, R. P., Mays-Hoopes, L. L., and Eichhorn, G. L., DNA methylation in aging of mice, *Mech. Ageing Dev.,* 41, 199, 1987.
82. Holliday, R., Strong effects of 5-azacytidine on the in vitro lifespan of human diploid fibroblasts, *Cell Res.,* 166, 543, 1986.
83. Fairweather, D. S., Fox, M., and Margison, G. P., The in vitro lifespan of MRC-5 cells is shortened by 5-azacytidine induced demethylation, *Cell Res.,* 168, 153, 1987.
84. Slagboom, P. E. and Vijg, J., Genetic instability and aging: theories, facts and future perspectives, *Genome,* 31, 373, 1989.
85. Thakur, M. K., Oka, T., and Natori, Y., Gene expression and aging, *Mech. Ageing Dev.,* 66, 283, 1993.
86. Sato, A. I., Schneider, E. L., and Danner, D. B., Aberrant gene expression and aging: examination of tissue-specific mRNAs in young and old rats, *Mech. Ageing Dev.,* 54, 1, 1990.
87. Levy, M. Z., Allsopp, R. C., Futcher, A. B., Greider, C. W., and Harley, C. B., Telomere end-replication problem and cell aging, *J. Mol. Biol.,* 225, 951, 1992.
88. Kirkwood, T. B. L., and Holliday, R., Aging as a consequence of natural selection, in *The Biology of Human Aging,* Bittles, A. H. and Collins, K. J., Eds., Cambridge University Press, Cambridge, 1986, 1.
89. Kirkwood, T. B. L., Repair and its evolution: survival versus reproduction, in *Physiological Ecology. An Evolutionary Approach to Resources Use,* Townsend, C. R. and Calow, P., Eds., Blackwell Scientific Publications, Oxford, 1981, 165.
90. Segall, P. E., Timiras, P. S., and Walton, J. R., Low tryptophan diets delay reproductive aging, *Mech. Ageing Dev.,* 23, 245, 1983.
91. Harman, D., The aging process: major risk factor for disease and death, *Proc. Natl. Acad. Sci. U.S.A.,* 88, 5360, 1991.
92. Cerami, A., Vlassara, H., and Brownlee, M., Glucose and aging *Sci. Am.,* 256, 90, 1987.
93. Strehler, B. L., Further studies on the thermally induced aging of *Drosophila melanogaster, J. Gerontol., 17, 347, 1962.*
94. Fanestil, D. D. and Barrows, C. H., Jr., Aging in the rotifer, *J. Gerontol.,* 20, 462, 1965.
95. Everitt, A. V., Porter, B. D., and Steele, M., Dietary, caging and temperature factors in the ageing of collagen fibers in rat tail tendon, *Gerontology,* 27, 37, 1981.
96. Sohal, R. S., Metabolic rate and lifespan, in *Interdisciplinary Topics in Gerontology,* Vol 9, Cutler, R. G., Ed., Karger, Basel, 1976.
97. Bjorksten, J., The crosslinkage therory of aging, *J. Am. Geriatr. Soc.,* 16, 408, 1968.
98. Verzar, F., *Lectures on Experimental Gerontology,* Charles C Thomas, Springfield, IL, 1963.
99. Verzar, F., Aging of the collagen fiber, *Int. Rev. Connect. Tissue Res.,* 2, 245, 1964.
100. Kohn, R. R., *Principles of Mammalian Aging,* 2nd ed., Prentice Hall, Englewood Cliffs, NJ, 1978.
101. Nagy I. Zs., A membrane hypothesis of aging, *J. Theor. Biol,* 75, 189, 1978.
102. Nagy I. Zs., Cutler, R. G., and Semsei, I., Dysdifferentiation hypothesis of aging and cancer: a comparison with the membrane hypothesis of aging, *Ann. N. Y. Acad. Sci.,* 521, 215, 1988.
103. Hahn, H. P. V., The regulation of protein synthesis in the ageing cell, *Exp. Gerontol.,* 5, 323, 1970.
104. Shock, N. W., Systems physiology and aging: introduction, *Fed. Proc.,* 38, 161, 1979.
105. Frolkis, V. V., *Aging and life-Prolonging Processes,* Springer-Verlag, New York, 1982.
106. Timiras, P. S., Neuroendocrinology of aging: retrospective current and prospective views, in *Neuroendocrinology of Aging,* Meites, J., Ed., Plenum Press, New York, 1983, 5.
107. Timiras, P. S., Physiology of ageing: aspects of neuroendocrine regulation, in *Principles and Practice of Geriatric Medicine,* Pathy, M. S. J., Ed., John Wiley & Sons, New York, 1991, 31.
108. Sharma, R. and Timiras, P. S., Glucocorticoid receptors, stress and aging, *Interdiscipl. Topics Gerontol,* 24, 98, 1988.
109. Everitt, A. V., and Walton, J. R., Regulation of aging along the hypothalamo-pituitary-endocrine axis, *Interdiscipl. Topics Gerontol,* 24, 1, 1988.
110. Meites, J., Effects of aging on the hypothalamo-pituitary axis, *Review of Biological Research in Aging,* Vol. 4, Rothstein, M. Ed., Alan Liss, New York, 1990, 253.
111. Walker, R. F. and Timiras, P. S., Pacemaker insufficiency and the onset of aging, in *Cellular Pacemakers,* Vol. 2, Carpenter, D. O., Ed., John Wiley & Sons, New York, 1982, 345.
112. Schmoll, H.-J., Tewes, U., and Plotnikoff, N. P., Eds., *Psychoneuroimmunology: Interactions Between Brain, Nervous System, Behavior, Endocrine and Immune System,* Hogrefe and Huber, Lewiston, NY, 1992.
113. Kelley, K. W., Brief, S., Westley, H. J., Navakofski, J., Bechtel, P. J., Simon, J., and Walker, E. B., Gh_3 pituitary adenoma cells can reverse thyrmic aging in rats, *Proc. Natl. Acad. Sci. U.S.A.,* 83, 5663, 1986.
114. Currie, M. S., Immunosenescence, *Comp. Ther.* 18, 26, 1992.
115. Patel. P. I., Identification of disease genes and somatic gene therapy: an overview and prospects for the aged, *J. Gerontol.,* 48, B80, 1993.

II
MOLECULAR AND CELLULAR AGING

5 DEGENERATIVE CHANGES IN CELLS AND CELL DEATH

Paola S. Timiras

The study of aging at the cellular level is concerned with the basic functions of molecules within the cells and of the cells within the tissues. Implicit in this study is the objective of discovering universal principles capable of explaining aging. Cellular and molecular theories of aging hold that intrinsic processes occurring in the cell (eventually in combination with external factors) are responsible for aging; these theories are discussed in Chapters 4, 6, and 7.

This chapter is concerned with cellular aging processes leading to cellular "degeneration" (i.e., progressive deterioration with impaired structure and diminished function) and, ultimately, to cellular death. Age-related enzymatic changes are summarized and related to other aspects of cellular aging. Given our interest in humans, focus will be on complex organisms such as mammals. Major age-related changes in cells *in vivo* and *in vitro* will be presented. The usefulness of studying cultured cells, mentioned in previous chapters, will be further illustrated by selected examples here. With current technologic progress, organs, tissues, and cells can be transplanted from one individual to another to replace lost or inefficient function. The benefits of transplantation technology for the elderly and its implications for our understanding of aging will be reviewed briefly.

1 VARIABILITY OF CELLULAR AND MOLECULAR AGING

In complex organisms such as humans, organs are formed by a variety of tissues, and tissues by a variety of cellular and extracellular elements, each aging at its own rate. Variation is a characteristic of biological systems that increases with aging.[1,2] At any given age and in any given tissue or organ, *the degree of impairment varies with the cell type*. In the nervous tissue, neurons do not divide after birth and show more age-related changes than glial cells which continue to replicate throughout life. Within the cell, molecules and organelles age at different rates, and even a slight age-related impairment may have crucial functional and biochemical repercussions on the entire cell.

The relationship between the magnitudes of the "error" or "damage" or "alteration" rates and their effects on cell growth, metabolism, and aging is still quite unclear. The picturesque words "error catastrophe" (Chapter 4) or "degenerative changes" imply an obvious state of disintegration,[3] but even quite modest increases in error or damage levels may have a severe effect in altering function and impairing cell survival. The consequent attempts to repair the damage, such as an increase in the rate of scavenging mechanisms, may further reduce cell efficiency, increase cell pathology, and lead to cell death.

Although each cell type follows its own timetable of aging, some processes are shared by all cells or at least a large group of cells. A number of these common features will be summarized in this chapter with the caution, however, that it is often impossible to separate age-related cellular changes from pathologic cellular changes associated with the diseases of old age. The ways in which cells, tissues, and organs, as well as whole organisms, respond to the innumerable noxious agents to which they are continually exposed are limited; consequently, the *repertoire of structural, functional, and biochemical alterations is restricted to the types of responses of which the cell is capable,* irrespective of the myriad causes related and unrelated to aging and/or disease. The older individual will inevitably have been exposed to more diseases than the younger, and changes due to physiologic aging may be expected to be compounded with those consequent to pathologic processes.

Given these limitations, attempts are being made to identify in the aging process certain basic criteria that could be meaningfully applicable at different levels of biologic organization. Thus, as already mentioned in Chapter 3, changes with aging may be qualified as universal, intrinsic, progressive, and deleterious.[4] A more clinically oriented listing categorizes age-related functional impairments in the so-called dreaded five "I's" of geriatrics: Instability, Immobility, Incontinence, Impaired cognition, and Iatrogenic diseases.[5] Causes and consequences of these conditions are discussed in the corresponding chapters. These and similar classifications, although still controversial, offer some guidelines for the selection of specific characteristics of cellular aging as discussed below.

Irrespective of their cause, cellular and molecular changes often represent the final common pathway(s) of all degenerative alterations responsible for impairment and failure of cell function that are the consequence of cell injury in general. Severe alterations involved in the final stages of cell degeneration and resulting in cell death are far easier to recognize than changes that, especially at their onset, are often too subtle to be analyzed by techniques presently available and, in many cases, only become manifest at advanced stages of aging.

Only a few specific cellular changes can be ascribed to aging *per se*. Exceptions are perhaps represented by (1) the intracellular accumulation of the so-called "age pigments," and (2) the demonstration of the limited proliferative capability of cells cultured *in vitro,* the latter utilized as a model of cellular aging.

2 CELLULAR AGING

Cellular Organization and Function

Cells have undergone considerable specialization among higher animals, and no human cell can be called typical of all cells in the body. However, a composite model shows that most cells have a membrane, a cytoplasm containing various organelles, and a nucleus (Figure 5-1).

The *membrane* or *plasma membrane* surrounding the cell is approximately 100 Å thick and, like all membranes, it contains proteins and lipids. Other membranes are located inside the cell where they form independent

0-8493-8979-8/94/$0.00+$.50

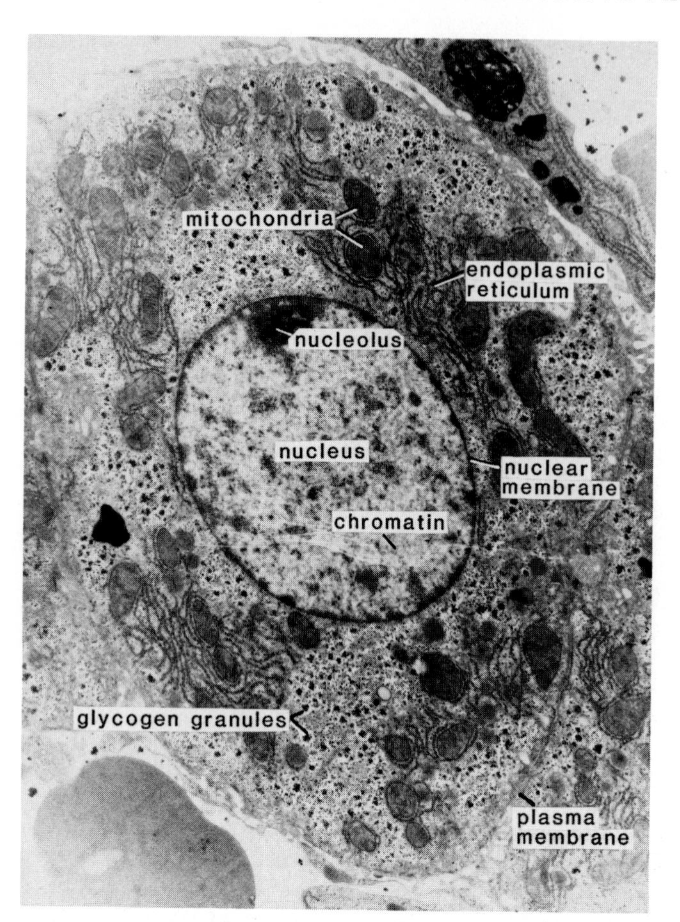

FIGURE 5-1 *Transmission electron micrograph of a mouse hepatocyte.* Overview of the major cell organelles. Magnification × 3675. (Courtesy of Dr. M. Schliwa.)

systems or surround specific organelles (see below). The ratio of the membrane's two major constituents, however, varies enormously: protein represents 76% of the inner mitochondrial membrane but only 18% of myelin. Likewise, the composition of lipids varies greatly among different membranes. The major structural unit is the phospholipid bilayer with the hydrophobic fatty acid chains forming the interior of the membrane and the polar heads facing the surrounding watery surface. At physiologic temperatures, lipid (hydrocarbon) chains are relatively fluid and the bulk of lipids is in a liquid-crystalline state. This state, characterized by relatively high fluidity, is to be distinguished from a gel phase in which lipid chains are rigid and closely packed with low fluidity. Carbohydrates, too, are important constituents of many membranes.

The membrane is semipermeable and capable of active transport; that is, it allows some substances to pass through (e.g., potassium ions) but excludes others (e.g., sodium ions). Large or lipid-insoluble molecules can gain entry to the interior of the cell by passing through holes or pores by a process of "facilitated diffusion" due to the presence of membrane proteins that act as "carriers" and "channels". Substances can also bind to "receptor" sites on the surface membrane, be internalized, and stimulate cell activity. Thus, the cell surface contains a number of molecular structures — cell-recognition sites or receptors — that sense or recognize external signals.

Other membranes that develop around the nucleus and the organelles within the cell have been given the name of *endoplasmic reticulum* (ER). These internal membranes are continuous with the plasma and nuclear membranes, furnishing a vastly increased and deeply invaginated surface area for exchanges within the fluid in which the cell is bathed. There are two types of ER: "rough", studded with ribosomes, and "smooth", without ribosomes. The rough ER participates in the synthesis of certain membrane and organelle proteins as well as the proteins to be secreted by the cell. The smooth ER is the site of the synthesis and catabolism of fatty acids and phospholipids.

The *Golgi apparatus,* a canalicular structure located in the cytoplasm near the nucleus, is the factory for the production or, more accurately, for the "packaging" of protein to be secreted by the cell and necessary to the production of membranes. The Golgi bodies are prominent in nervous and secretory cells. They are involved in the transport of secretory products from the inside of the cell to the outside by the process of "exocytosis".

Within the cytoplasm, the most prominent organelles are the *mitochondria*. These are the "power plants" adapted to extract energy from the chemical bonds in the nutrients of the cell by oxidation. From each of these power plants, the yield of energy is made available to the energy-consuming processes of the cell, neatly stored in the phosphate bonds of the compound adenosine triphosphate (ATP). Each mitochondrion is a sausage-shaped membranous structure about 150,000 Å long and 5,000 Å in diameter, containing shelves or cristae. Stacked on these shelves and studding the exterior surface of the mitochondria are packets of enzymes. These enzymes are involved in the transfer of electrons in biologic oxidations and in the synthesis of ATP through oxidative phosphorylation.

The *lysosomes* are membrane-surrounded spherical vesicles in the cytoplasm. They contain a variety of degrading enzymes, such as acid hydrolases, which hydrolyze biopolymers (proteins, carbohydrates, and lipids) that are no longer needed. They also contain ribonucleases and deoxyribonucleases, that degrade RNA and DNA.

Ribosomes are granules attached to the ER but may also be present in the cell as free ribosomes. These are about 150 Å in diameter and are made of one large and one small subunit. Sometimes three to five ribosomes are clumped together forming polyribosomes (polysomes). Ribosomes contain mostly ribonucleic acid, RNA, and proteins. They are the site of protein synthesis. Free ribosomes are involved in the synthesis of cytoplasmic proteins, whereas ER-bound ribosomes are involved in the synthesis of secretory and membrane/organelle bound proteins. These proteins are synthesized on the ribosomes using messenger RNAs transcribed from the DNA.

Different kinds of microtubules and microfilaments together form a flexible framework, the *cytoskeleton*. These protein structures give shape and form to the cell, are essential for cell motility and contractility, and facilitate intracellular communication or segregation. A third element of the cytoskeleton intracellular communication or segregation is *the microtabecular network,* originally designated as the ground substance and now visualized as a network.

Another type of membrane-surrounded organelle in the cytoplasm is the *peroxisome* or microbody. Segregated within this structure are enzymes that form and use hydrogen peroxide (hence the name). Hydrogen peroxide, extremely toxic to the cell, is hydrolyzed by the enzyme catalase, which protects the cell from self-destruction by the action of peroxides (Chapter 6).

A *nucleus,* present in all cells that divide, contains DNA responsible for the replication not only of proteins but also of the nucleic acids DNA and RNA. When the cell is in the resting stage, i.e., in the process of growth between divisions, the chromatin (consisting of DNA, RNA, and specialized proteins) is diffusely distributed in the nucleus. The DNA thus makes maximum surface contact with other material in the nucleus from which it presumably pieces together the molecules of RNA and replicates itself. In preparation for cell division (by mitosis for somatic cells, or by meiosis for the germ cells), the chromatin coils up tightly to form the *chromosomes*. The chromosomes, containing the *genes,* are the structures that carry a complete blueprint for all the heritable characteristics of the individual; a fixed number (diploid number) are present in each somatic cell to be distributed equally to each daughter cell, except in the germ cells, where the number of chromosomes is halved (haploid number). The nucleus is surrounded by a nuclear membrane with pores. Inside the nucleus is the *nucleolus,* rich in RNA, which is the site for the assembly of ribosomes.

2.1 Aging Changes in Membranes

Microvilli, identifiable on the *cell surface* by scanning microscopy, are a common cellular feature. Their number increases with advancing age in specific cell types *in vivo* (e.g., podocytes of renal glomerulus) and in fibroblasts in later population doublings *in vitro*.[6-8] Such an increase is reduced by dietary restriction, a procedure that increases the lifespan of laboratory animals (Chapter 24).[9-11] Cell surfaces are involved in several important functions such as cell-to-cell interaction, external signal reception, cell adhesion and motility, and maintenance of cell shape. Microvilli represent an extension of the cell surface, especially important in increasing membrane transport and as a reserve for cell movement and division. Their increase with aging may be viewed as a compensatory attempt to correct, in part, decrements in cellular intercommunication;[12] disrupted communication (as has been described in hepatocytes[13]) would lead to a diminished ability to maintain tissue, organ, and organism homeostasis, thereby precipitating aging, pathology, and death.

Age-related changes concern both *major membrane constituents* — lipids and proteins — as well as certain cytoskeletal elements that are membrane bound. So far, emphasis in membrane research has been directed to the study of the lipid bilayer membrane fluidity and the effects of aging thereon. According to the "fluid mosaic model", membrane components are able to diffuse laterally in the plane of the membrane. This diffusion offers a potential mechanism for directed changes in the topography of specific membrane components, with proteins generally diffusing to a greater degree than lipids. Membrane fluidity is influenced by several factors, the most important being temperature, fatty acid unsaturation, cholesterol, and protein content. With aging, membrane lipids are susceptible to peroxidation, and membrane fluidity is markedly altered by free radical accumulation (Chapter 6).

Aging-associated changes in membrane composition, especially lipid components, vary depending on cell and membrane type, animal species, diet, exercise, hormones, antigens, mitogens, and so forth. For example, in myocardial mitochondrial membranes, although several phospholipids remain unchanged, cholesterol and cardiolipin are significantly increased in old (33 months) rats as compared to young (4 months), and variable changes in fatty acid composition lead to decreased unsaturation index, suggestive of decreased fluidity.[14] Changes in *membrane fluidity* may induce alterations in cell responses involving excitability, transport, and receptor binding, thereby altering responsiveness to endogenous and exogenous stimulation. As membrane receptors and proteins (and glycoproteins) move along the membrane lipid bilayer, they bind to ligands forming receptor-ligand complexes that aggregate in clusters, patches, or caps. The capacity for the formation and motility of receptor-ligand complexes seems to be diminished in aged individuals and in some forms of accelerated aging such as Down's syndrome.[15]

Membrane-Associated Cytoskeletal Structures

The affinity or relationship between plasma membrane and cytoskeleton appears to play an important role in the mobility and redistribution of receptor-ligand complexes. For example, in the process of receptor-ligand aggregations, cross-linked surface immunoglobulin becomes attached to actin, and this process is significantly lower in the aged (85 years and older). Another example is the implication of tubulin, the microtubule protein subunit, and of "tau", one of the microtubule-associated proteins, in some aspects of the pathology of the aged brain. Tau proteins have been implicated in the pathology of senile dementia of the Alzheimer type (Chapter 9). These proteins appear associated with the formation of neurofibrillary tangles, where they are a major constituent of the paired helical filament (Chapter 8). Extensive phosphorylation of tau proteins favors cross-linking of tubulin to actin, whereas dephosphorylation inhibits it. In Alzheimer's disease, tau protein alterations may result in tau-ubiquitin complexes probably involved in the formation of neurofibrillary tangles.[16-18] Alterations of adult tau may induce the expression of the fetal form of tau in an attempt to regenerate the altered adult protein.[19,20] Impairment of the microtubule system, particularly tubulin, may occur as a consequence of genetic or epigenetic factors in neurons and nonneuronal cells and may serve as the basis for a unifying hypothesis of cell degenerative processes such as those occurring in Alzheimer's disease (Chapter 9).[21]

Cytoskeletal elements, by conveying information from cell membrane to intracellular components,[22] play an important role in regulating several cell functions, including response of lymphoid cells to mitogens, chemotactic responsiveness of neutrofils, lipofuscin lysosomal content, and cell division.[23] Alteration of these functions with aging suggests that disorganization of cytoskeletal structures may be a common denominator of some aging phenomena.

Other membrane alterations with aging include lipofuscin accumulation and enzyme alterations (see below), peroxidation of intracellular membranes (Chapter 6), physicochemical changes (leading to increased rigidity, decreased ion conductance), and reduced binding of several specific antigens such as histocompatibility antigens (Chapter 7). These antigens are encoded by the major histocompatibility complex (MHC) genes; their receptors are transmembrane cell surface glycoproteins which are some of the best studied surface markers of cells. Quantitative and qualitative receptor changes may be responsible for the increase in autoimmune diseases and other alterations of the immune system with aging. MHC alleles influence maximum lifespan, DNA repair, free radical scavengers, and some enzyme activity (Chapters 4, 6, and 7).

2.2 Aging Changes in the Cytoplasm

Data on age-related changes in the ER, the site of protein and steroid synthesis (rough ER) and lipid metabolism (smooth ER), are conflicting. Such changes are important inasmuch as survival depends on the fidelity or accuracy of protein synthesis,[24] and several theories of aging underline the failure of accuracy in molecular processes (Chapter 4). Biochemical studies show no or minimal ER change with aging. In the light microscope, RNA-associated structures represented by the "Nissl bodies" decrease with aging in several cell types, especially those, such as neurons, that do not divide after birth and persist throughout the lifespan.

The number of *mitochondria* is reduced with aging in neurons, presynaptic terminals of neuromuscular junction, liver and heart cells, and *in vitro* human fibroblasts. Mitochondrial size, on the other hand, is increased in some cells; this enlargement or "swelling" occurs not only in aging cells but also in pathologic and inadequately fixed (for histology) cells. The latter is particularly true of the nervous system, where mitochondria often show vacuolations and the accumulation of "unusual" substances *in vivo* and *in vitro*.[25] Mitochondrial inclusions and changes in shape may result from altered metabolism, such as uncoupled respiration, and/or artifacts, such as hypoxia during fixation. Because of damage induced by free radicals originating as by-products of oxygen reduction during respiration (Chapter 6), disorganization of mitochondria has been considered the "Achilles' heel" of the aging cell.[26,27] In the brain, the magnitude of the shrinkage during fixation is more pronounced in tissues from young than old brains, probably because of the higher water content of the former.[28] Tissue shrinkage would, undoubtedly, influence cell density as well as distribution and structure of organelles within the cell.

Mitochondrial Mutations and Degenerative Diseases

The genetics of oxidative mitochondrial process show that mitochondrial DNA (mtDNA) is predominantly inherited from the mother (only 0.1% of mtRNA is provided by the sperm). Mutations of mtDNA have been implicated in the etiology of a number of degenerative diseases with apparently late onset and associated with defects of oxidative phosphorylation (e.g., Alzheimer's disease, Parkinson's disease).[29] It is suggested that mutations may occur early in development and induce cumulative cell damage and loss (e.g., by free radical accumulation) which become manifest in old age (Chapter 6). Prevention of mtDNA damage in the mother may represent an equally or more efficient intervention than prevention of mtDNA damage in the elderly for reducing the risk of age-related degenerative diseases.[29]

It is difficult to measure the volume of *Golgi complex* even with new techniques because it is not contained within a distinct boundary. It appears, however, to be increased in volume in some aging cells, such as rat cerebral cortical and vestibular neurons and fibroblasts.

2.3 Aging Changes in the Nucleus

Aging affects proliferative homeostasis[11] and proliferative capacity of the cells[30,31] under normal conditions and after mitogenic stimulation.[32,33] These changes have been correlated with a number of structural alterations in the nucleus with aging as identified by electron microscopy *in vivo* and *in vitro*. They include: irregularities of shape and invagination, interpreted as consequent to reduced blood supply and cell metabolism; chromatin condensation, suggesting lowered functional activity; presence of inclusion bodies, particularly evident in pineal cells and neurons, perhaps related to alterations of intranuclear microtubules and microfilaments; and increased number of nucleoli *in vivo* but decreased *in vitro*.

Changes in DNA, RNA systems, mutations, errors, and repair mechanisms have been discussed in Chapter 4 together with their possible role in the aging process.

3 ENZYMATIC CHANGES DURING AGING*

3.1 Enzyme Levels During Aging

Changes in enzymatic levels and activities were among the first biochemical observations demonstrated to occur during aging and have been extensively reviewed.[34] The collected information is hard to integrate because of the variable and often contradictory nature of the reported data. Nonetheless, studies on enzymes have stimulated interest in aging research and have contributed considerably to the field.

Enzymes and Enzymatic Activity

Enzymes are specific proteins that

- Catalyze chemical reactions in biological systems
- Have enormous catalytic power
- Are highly specific with regard to both the reaction catalyzed and the choice of reactants, called substrates

Concentrations of enzymes and their rate of synthesis are under genetic controls and are influenced by small molecules such as hormones, substrates, and products of the metabolic pathways. Some enzymes are synthesized in an inactive precursor form and are activated at a suitable physiologic milieu. Another controlling mechanism is the covalent insertion of small groups on the enzyme.

The extent to which an enzyme can increase the rate of a reaction depends on the activity of the enzyme, which is a function of (1) the amount of active enzyme available, (2) the concentration of substrates and the presence of cofactors, inhibitors, and activators, and (3) the innate property of the enzyme itself.

Changes in the levels or properties of enzymes may alter the functional activity of an organism. Since enzymes are responsible for specific functions, the various phases of the lifespan (differentiation, development, reproductive maturity, aging) may depend on the activity of specific enzymes.

Depending on the animal species, *the activities of several enzymes decrease, increase, or remain unchanged with aging*. Even within each class, the activities of most enzymes do not follow a specific age-related pattern; indeed, few single enzymes consistently change activity with aging. However, alterations in the activities of various enzymes may affect the functional ability of the aging organism.

The metabolic status of animals is coordinated by a network of enzymes at different phases of the lifespan. Activities of certain lipogenic and gluconeogenic enzymes in the rat liver change at different ages. Lipogenic activity is diminished because of an age-dependent decline in nicotinamide-adenine dinucleotide (NADP)-malic enzyme and ATP-citrate lyase activity.[35] Studies on oxaloacetate metabolism in the liver and brain of male Wistar rats show that activities of cytoplasmic aspartate aminotransferase and malate dehydrogenase (gluconeogenic enzymes) are higher in older rats.[34] This increase may be correlated with the greater involvement of the enzymes in the conversion of the oxaloacetate pool for gluconeogenesis in older rats.

The activity of poly(ADP-ribose) polymerase (an enzyme involved in DNA repair) is higher in long-lived mammals. Such increased activity may contribute to the efficient maintenance of the genome integrity and stability over a longer lifespan.[36] Long-lived flies *(Drosophila melanogaster)* express an unusually active form of the antioxidant enzyme superoxide dismutase; this higher activity helps to neutralize superoxide quite efficiently, thereby supporting the longer life of these flies.[37] Activities of the antioxidant enzymes, superoxide dismutase and catalase, decrease significantly in the liver of old rats, but can be returned to normal by dietary restriction (which also raises the levels of mRNAs coding for these enzymes).[38] Indeed, the observed decrease or increase in enzyme levels may be correlated with the decrease or increase in the template activity of the corresponding genes and/or their internal regulation, depending on the changing metabolic needs with advancing age.

---
* This section (pp. 50–52) was contributed by Dr. Ramesh Sharma.

3.2 Changes in Isoenzyme Composition

During development, many proteins change from the fetal to the adult form. Differentiation and development depend on the sequential activation and repression of genes.[39] This dependence is best exemplified during gestation in humans by the shift of hemoglobin from the embryonic to the fetal and finally to the adult form.

Isoenzymes are functionally related proteins present in more than one molecular form within the same individual and species. The best example is lactate dehydrogenase (LDH), a tetramer of two different types of subunits designated M (the predominant form in skeletal muscle) and H (the predominant form in heart). These subunits are controlled by two separate genes. Several LDH isoenzymes are formed by combinations of M and H subunits (M4, M3H, M2H2, MH3, and H4). Each isoenzyme is characteristic of a specific tissue or cell population and is subject to different regulatory signals. Aerobic and gluconeogenic tissues contain mostly the H form, which is primarily concerned with the conversion of lactate to pyruvate. M-type LDH predominates in anaerobic and glycolytic tissues, where it converts pyruvate to lactate.[40] LDH isoenzymic composition is not only tissue specific, but it also changes in the same tissue with development.[41] The greater proportion of M4 isoenzyme is present in the developing embryo of mammals, as their metabolism is mostly anaerobic. A shift towards H4 occurs as development proceeds and the developing organism becomes increasingly dependent on aerobic metabolism. In the chick embryo which develops in an aerobic environment, the H4 isoenzyme predominates and shifts to M4-type isoenzyme during the later stages of development.

These changes in isoenzyme composition are not restricted to development, but extend into adulthood and old age.[42] The proportion of M4-LDH is considerably lower in the heart, skeletal muscle and brain of old as compared to young rats, with a concomitant increase of H4-LDH in old age. This shift in LDH isoenzymes has been correlated with changing metabolic functions with advancing age: decreased M4-LDH may result in a decreased ability of tissues from old animals to cope with anaerobic conditions.[43]

The isoenzyme pattern of isocitrate lyase is markedly altered in old as compared to young nematodes *(Turbatrix aceti)*.[44] Studies on cytoplasmic alanine aminotransferase (c-AAT) in rat liver reveal that the phenomena of sequential changes do extend to old age.[45] Liver from young (5 weeks of age) rats has the A-type isoenzyme, the adult (52 weeks) has both the A and B types, while the old (100 weeks) has only the B isotype. Both subunits are under the control of two separate genes which are sequentially activated and repressed at different ages.[46]

The mono-, di-, and tetrameric forms of the enzyme glucose-6-phosphate dehydrogenase differ quantitatively in the liver of young as compared to old rats.[47] Selective enhancement of isozymic variants of β-D-glucosidase and acid phosphatase has also been reported in aging *Caenorhabditis elegans*.[48] Overall, these isoenzymes shifts may be explained by regulatory changes in the activity of their corresponding genes consequent to endogenous programmed signals.

3.3 Enzyme Induction During Aging

Induction of enzymes is caused by "inducers" or "effectors", which may be either the substrate, a metabolite, or an exogenous factor. Induction is an adaptive process. The ability to initiate adaptive changes in the activity of many enzymes is impaired with old age.[49] The induction of enzymes by hormones during aging has been extensively reviewed.[34] The magnitude of induction of many enzymes decreases, increases, or remains unchanged with the increasing age of the animal.

The expression of age-related adaptive changes in enzyme induction has been categorized into four general patterns of response.[49] With aging, and depending on a number of variables (animal species, strain, sex, and physiologic state, environmental conditions), the response (1) may have an altered adaptive latent period (or initiation time) following the stimulus, without affecting the magnitude of the induction, (2) may decrease or increase in the magnitude of induction without changes in latency, (3) may show alterations in both magnitude and latency, and (4) may fail to show any changes in the induction pattern.

Despite the great variability of responses, an overall picture shows a decrease in the magnitude of the adaptive response with an increase in latency. Identical stimuli elicit weaker responses in older animals, although the reverse is seen in a few cases.[50] Among inducers, hormones are important in the maintenance of overall adaptive responses of enzymes during the lifespan. With respect to hormonal actions, the degree of tissue responsiveness would be directly proportional to the amount of hormone bound to its specific receptor.[51] Alterations in responsiveness may depend on changes in the number of receptors and in the physicochemical properties of receptor molecules[52-54] and/or transacting factors involved in hormonal regulation of gene expression.[55-57]

3.4 Changes in the Kinetic Properties of Enzymes

The hypothesis that proteins are altered during aging is discussed in Chapters 4 and 6. The proposed concept assumes that faulty proteins are formed with aging. The hypothesis creditably stimulated a considerable amount of research both theoretical and practical to confirm or negate the production with aging of error-containing proteins and their possible causative role in aging.[58-61] Search for errors was vigorously pursued by measuring in great detail the kinetic properties, molecular weight, electrophoretic mobility, heat stability, and antigenicity of many enzymes at progressive ages and in several animal species.

Despite some differences in the molecular properties of a few enzymes, the overwhelming evidence negates the idea that errors in proteins may cause aging. Significant differences in young as compared to old enzymes have not been convincingly demonstrated. If errors were made in transcription by RNA polymerase or during translation by tRNA synthetase, then all proteins would contain errors, which in fact does not occur in old animals. Some enzymatic alterations reported in old animals may simply be due to a change in the shape of the molecules with no covalent modifications.[61] For example, aldolase A from rabbit muscle differs in conformation between young and old animals.[62] Comparison of the properties of glyceraldehyde-3-phosphate dehydrogenase (GPDH) and phosphoglycerate kinase (PGK) in tissues from young and old animals indicates lower specific activity of the enzymes and greater heat sensitivity, without differences in amino acid composition, –SH groups, UV spectra, and sedimentation coefficient.

Even in the absence of errors in the protein synthetic machinery, it is after completion of synthesis that the altered enzymes may be produced by altered protein folding. Oxidative damage is one of the possible mechanisms capable of altering proteins and enzymes during aging (Chapter 6). The apparent changes in the activity of enzymes and the shift in isoenzyme patterns may be viewed as expression of adaptive responses, coordinating with intrinsic and extrinsic factors responsible for the hormonal and metabolic status of the organism at progressive ages.

4 CELL INJURY AND CELL DEATH

The more our knowledge of cell complexity increases, the more we become aware of the difficulty of distinguishing between a normal and an abnormal cell and, even more so, between a normal adult and an aging cell. Formerly, it was considered an easy task to identify a living amoeba: it withdrew from a noxious stimulus and, if it failed to withdraw, it was either sick or dead. Today, with rapid advances in cell and molecular biology, the line between health and injury, even in a simple organism such as the unicellular amoeba, is being more finely drawn. Slight swelling of the mitochondria or minute increases in intracellular sodium levels have now become criteria for assessing injury or aging of a cell. Changes in cellular morphology represent only one aspect of cellular aging. Biochemical and functional changes consistently accompany, precede, or cause changes in structure.

The view of aging as a progressive alteration of structure and an impairment of function with time makes it extremely difficult to differentiate cellular changes due to age from those due to injury. In both cases, morphologic changes underlie the functional disturbances: if the injury is severe enough and the aging process sufficiently advanced, they will lead to cellular death. This section considers briefly certain types of morphologic alterations and cell death. Major causes of cellular aging and death have already been discussed in Chapter 4 with the theories of aging.

4.1 Cell Degeneration

Cellular degeneration has already been defined as reflecting altered structure and impaired function. It may be (1) reversible or (2) irreversible and lead to cell death (when aging or injury are applied over a longer period of time or to a more intense degree). In this sense, cellular degeneration and cellular death merely reflect two levels of severity of cell damage, the former compatible with recovery and the latter resulting in death. Cell degeneration is often accompanied by infiltration, that is, entrance (and accumulation) into the cell of normal or abnormal substances usually kept outside. Cellular degenerations and infiltrations have classically been subdivided into specific morphologic patterns based on damage location or nature of the metabolite that accumulates within the cell. Some of the principal changes of this type are listed here; for more information on this subject, the reader is referred to specialized textbooks of pathology.

4.2 Cloudy Swelling/Dehydration

Changes in cell membrane transport and in cell metabolism result in abnormal movements of water and solutes in and out of the cell. Water accumulation in the cell is referred to as "edema" or "swelling" and, when associated with high solute content, as "cloudy swelling". The opposite condition, i.e., water loss, is referred to as "dehydration".

Intracellular water accumulates in the ER or the mitochondria, probably due to impairment of respiratory energy-releasing mechanisms. The loss of this energy would inhibit continued excretion of sodium and water. Whether sodium is primarily affected and increased intracellular water is secondary or whether changes in membrane permeability directly promote the passage of water remains to be established. Edema is often preceded by a decrease in protein synthesis, but the relationship between this decrease and edema of mitochondria and ER is not clear. A more severe form of intracellular edema, "hydropic" or "vacuolar" degeneration, is produced in the same way as cloudy swelling but implies a greater accumulation of water, often in vacuoles and throughout the cytoplasm.

A certain degree of dehydration occurs usually with age, as manifested in the shrunken appearance of skin in general and the "dried-up look" associated with senility. Water content decreases during growth, from a total body water value of approximately 80% at birth to about 50% in the adult; however, further decrease of body water with advancing years occurs at a very slow rate. Indeed, some investigators claim that dehydration *per se* is not a characteristic of aging and that changes in body water with age involve a redistribution of water between extracellular and intracellular compartments rather than a net loss. Such a redistribution may be consequent to a number of factors related either to the cell population of the tissue considered (e.g., loss of cells, redistribution of cell population, replacement of specific cells with fibroblasts) or to shifts in water between intracellular and extracellular spaces (e.g., alterations in membrane permeability, ion distribution, energy required for transport). For example, in skeletal muscle of old rats, the extracellular fluid is nearly double that found in young and mature animals, whereas intracellular water and potassium remain relatively constant. In heart, brain, and liver tissue of rats, little change occurs in the amount and composition of either extracellular or intracellular compartments. Conversely, chemical analysis of the kidney in both humans and rats indicates an increase in extracellular water and a considerable decrease in cell mass with age (Chapter 21). The decrease of protoplasmic units (e.g., loss of muscle fibers) would show a positive correlation with an increase in the proportion of extracellular water to total body water and a negative correlation with the quantity of intracellular constituents.

4.3 Fat/Glycogen Changes

With middle and advanced age, whole-body adipose (fatty) tissue has the tendency to accumulate in many mammals including humans.[63] In addition to overall increased body fat, abnormal accumulation of fat occurs within cells (other than those of the adipose tissue), *Fatty change* (also referred to as fatty degeneration, fatty infiltration, fatty metamorphosis) represents a common response to a derangement of cellular metabolism. It is indicative of severe cellular dysfunction and, although in itself it is a reversible alteration, it is often the harbinger of cell death. Examples of fatty change can be found in the lipid (fatty) loading of smooth muscle "foam" cells in the arterial wall and in the lipid accumulation in the atheroma, both characteristic lesions of atherosclerosis (Chapters 16 and 17).

Glycogen is normally present in all cells and is particularly abundant in muscle and liver cells. In certain conditions, increased amounts of glycogen accumulate in microscopically visible droplets (vacuoles, after histologic fixation) within the cytoplasm or nuclei of the affected cells. This accumulation is considered to represent an imbalance between glycogen synthesis and catabolism. It is found chiefly in two disorders: diabetes mellitus and glycogen-storage diseases (e.g., von Gierke's disease). To what extent diabetes mellitus can be considered a disease of aging is discussed elsewhere (Chapter 15). In the glycogen-storage diseases, the defect is one of mobilization, usually consequent to the inherited deficiency of an enzyme necessary for normal carbohydrate metabolism.

4.4 Age Pigments or Lipofuscin

The "age pigments", so called because of their accumulation in the cytoplasm of the cells with the passage of time (and increasing age), have been designated in the past by a variety of names, among which the most common is "lipofuscin" (*lipo* from the Greek, fat; *fuscin* from the Latin, dusky). They are ubiquitous as they are found in all animal organisms from Protozoa to Chordata, including humans.[64] Indeed, the gradual accumulation of these pigments with age, first detected in nerve cells, was one of their earliest recognized aging-related characteristics.

At the light microscope level, the pigment is seen as yellow to brown round or oblong granules, either diffused throughout the cytoplasm, or in perinuclear clusters or polar aggregates. The latter distribution, found in older tissues, probably represents the more advanced form.

Lipofuscin is a by-product of lipid peroxidation, with both a lipid and a protein component, and is formed during the intracellular process of autophagia (self-digestion; Chapter 6). Lipofuscin accumulation with aging may be consequent to a number of factors including: (1) deficiency of lysosomal enzymes to hydrolyze lipids and other slowly degradable materials; (2) increase in the amounts of slowly degradable materials such as antibody-antigen complexes; (3) extrinsic factors such as diet, drugs, hypoxia, vitamin E deficiency, cirrhosis, and arterial plaques. Some of these factors may induce cell injury and thereby require activation of the lysosomal system (which may be unable to cope as a consequence of age-related reduced efficiency); and (4) genetic factors regulating lysosomal enzyme formation and activity.

Lipofuscin accumulates in some tissues but not all. For example, the zona reticularis of the adrenal gland of the senile rat has five time more pigment than any other zone in that structure; in the rat ovary, the pigment is localized exclusively in the perifollicular macrophages, not the ovarian follicles.[65] Lipofuscin accumulates preferentially in the muscles of locomotion as opposed to those that maintain posture.[66] Given such cellular and regional differences within single structures, and the fact that some tissues do not accumulate pigment at all, it may be opportune to question the assumption that age pigments (1) are an homogeneous group of substances, and (2) that their accumulation in different cells of the body results from the same processes in every case.

The rate of lipofuscin deposition is consistent enough to render it a reliable index of chronologic age. Lipofuscin is scant or absent until age 10 in heart muscle and accumulates thereafter at the rate of 0.3% of the total heart volume per decade, or 0.6% of the intracellular volume per decade, independent of cardiac pathology or the presence or absence of cardiac failure. Thus, the heart of a person living to be 90 years of age would be expected to have 6 to 7% of its intracellular volume occupied by lipofuscin.[67]

Properties of Lipofuscin Granules

One of the most consistent properties of lipofuscin is its autofluorescence in response to stimulation by visible or near-ultraviolet light. When viewed in the fluorescent microscope, it appears as a yellow, yellowish-green, and orange mass. Despite inherent differences in the composition of the granules from different specimens, the relative distribution of its three main constituents is 19 to 51% lipid, 30 to 50% protein, and 9 to 30% acid hydrolysis-resistant residue.

Three fourths of the lipid component are phospholipids.[68] Other lipids identified are cholesterol and its derivatives, triglycerides, cephalins, lecithin, spingomyelin, and traces of fatty acids, sulfatides, and gangliosides.

The composition of the protein fraction after acid hydrolysis is essentially similar for lipofuscin as for other structural proteins of various tissues; it shows a high lipid-binding capacity[69] with a range of amino acid values[70,71] and a number of enzymes with few differences in amount and activity compared to the enzymes from intact tissues.

The significance of lipofuscin accumulation for cellular function is still uncertain. The fact that it may be influenced by extrinsic factors has been interpreted to indicate that "rare and irreparable metabolic accidents beyond the genetic design gradually result in the accumulation of insoluble, non functional, or noxious by-products of metabolism." This view of lipofuscin resulting from and leading to metabolic errors (incorporated in the so-called "clinker theory") was modified to include a genetic omission rendering nondividing, nonreplaceable cells susceptible to slow deteriorative changes (stochastic senescence).[4]

Yet, the evidence against generalized impairment of function in nondividing cells has been rather convincing. Human, rat, and mouse myocardia do not lose the ability to hypertrophy throughout adulthood despite increased deposition of age pigments. Although pigment accumulation in neural cells is sufficient to displace the cell nucleus (Chapter 8), it is not known whether this is an excessive accumulation of lipofuscin or whether the cell has degenerated from other causes and pigment masses have formed as a result of such deterioration. It is possible that the aggregations of lipofuscin characteristic of advanced age decrease cellular plasticity and, thus, interfere with the ability of the cell to maintain itself. The question of whether the changes associated with the presence of lipofuscin are deleterious, and whether they are causative or symptomatic of aging remains unanswered.

4.5 Amyloid and Amyloidoses

Amyloid is a generic term for proteinaceous fibrillar deposits with (1) the properties of yellow-to-green birefringence in crossed polaroid illumination after staining with Congo red, (2) a β-pleated sheet structure of the fibrils, and (3) insolubility and resistance to enzymatic digestion.[72,73] *Amyloidosis includes a number of diseases characterized by the extracellular accumulation of these insoluble fibrillar proteins.*

Currently, amyloidoses are classified on the basis of the biochemical composition of amyloid subunit protein[74] in

1. *Primary systemic amyloidosis* (AL), as in myeloma; amyloid is produced by proteolytic cleavage of the N-terminal variable region of immunoglobulin (Ig) light chain in phagocytic cells
2. *Secondary amyloidoses* (AA) occurring in chronic inflammation and characterized by accumulation of amyloid A protein[75]
3. *Familial amyloidoses* (AF) in which prealbumin is the major accumulating protein.[76]

Several other proteins have now been identified as components of amyloid complexes: atrial natriuretic peptide (Chapter 21) and prealbumin in senile cardiac amyloidoses; A4 and A-β proteins are found in cerebral vessels, plaques, and neurofibrillary tangles of patients with Down's syndrome and Alzheimer's disease.[77,78]

Senile cardiac amyloidosis is present in 65% of hearts from persons 90 years of age and older. Focal deposition of this amyloid appears to have little functional consequences, but extensive distribution has been associated with congestive heart failure and fibrillation. Other forms are found in senile lungs, liver, kidneys, and in general in senescent or injured tissues. Controversy still exists regarding its origin and the mechanisms responsible for its deposition,[79] although a deficient function of lysosomes may be implicated.[80]

4.6 Cell Death, Necrosis, and Apoptosis

All "normal" cells have a limited lifespan; death is the ultimate step in the aging process. Cells may reach this endpoint prematurely if they encounter a hostile environment. In contrast, "abnormal" cells, either tumor derived or virally or carcinogen-treated, proliferate indefinitely, i.e., are "immortal". Thus, identification of factors causing cell death may provide useful insights on the aging process itself.[81,82]

The terms "cell death" and "necrosis" are often used interchangeably, although they refer to quite different phenomena. As with the whole organism, *cells die when their vital functions cease.* All cells engage in four basic major activities: (1) the transformation of energy needed for vital processes, (2) synthesis of proteins (both enzymatic and structural), (3) maintenance of the chemical and osmotic homeostasis of the cell, and (4) reproduction. Sometimes one particular vital activity ceases before the others, e.g., the heart may continue to beat for a short time after respiration has ceased. When cellular respiratory mechanisms are destroyed, synthetic activities may continue very briefly. This dissociation of vital functions at the cellular level makes it very difficult to be certain of the precise moment of cell death.

Necrosis refers to distinct morphologic changes that permit recognition of the fact that the cell has died. Such changes occur only gradually; sudden cellular death, as occurs, for example, when cells are fixed in formalin for histologic examination, is associated with few alterations in cell morphology. Thus, the individual who dies suddenly of myocardial infarction may show little morphologic alterations. Had this person survived 24 hours or longer, enzymatic changes would have ensued and eventually permitted recognition of the necrotic myocardium. The structural alterations of necrosis are biochemically induced by enzymes of intra- or extracellular origin.

The ultimate causes of cell death must involve a derangement of vital functions that are initiated at the molecular level. DNA is one of the major targets of several damaging agents, and DNA damage has long been considered a major cause of cell death. While DNA damage may be repaired (Chapter 4), the consequences of DNA breaks on cell function may lead to cell pathology and death. In this manner, cell death may be programmed to achieve a specific function, that is, cells may be active in planning their own death. This type of *"programmed cell death"* or *"apoptosis"* is involved in many fundamental biologic processes including differentiation, embryogenesis, and metamorphosis as well as death.[83] The mechanisms responsible for allowing specific cells to undergo apoptosis are unknown. Target cells are varied; the most often studied are hormonal target cells, neural and immune cells. Biochemical events leading to death include chromatin condensation resulting in pycnosis and increased nuclease activity resulting in the generation of DNA fragments and DNA degradation.[83-86]

5 TRANSPLANTATION STUDIES IN AGING

Surgical techniques and follow-up therapy of recipient patients have currently progressed to the extent that major restrictions to transplantation intervention are the lack or scarcity of available organs, rather than technological limitations.[87] Tissue and organ transplantation in the elderly raise several questions of a biological and ethical nature.

With respect to biological questions, one of the most important considerations is the effectiveness of the transplantation. As discussed in Chapter 7, tolerance to foreign materials is decreased in the elderly; therefore, tissue rejection is less likely to occur. Indeed, transplantation interventions have been used successfully in the elderly.[88] However, due to the high cost of the intervention and especially the scarcity of the organs available for transplantation, it has been proposed to restrict this type of intervention to the young and adult, thereby setting priority limits with a clear disadvantage for the aged (Chapter 3).

Some of the first experiments in which transplantation usefulness was tested in aged animals were conducted in the 1950s; adrenocortical tissue from middle-aged rats (donors) was transplanted into young recipients, and these subsequently lived a little longer than control lifespans, with normal growth rates and reproductive performances.[89,90] Although the lifespan of the recipient rats was prolonged slightly, it remained short of the maximal rat longevities obtained with other types of interventions, such as dietary restriction (Chapter 24). Since then, many other experiments have confirmed that a large number of tissues — adrenal cortex, skin, pituitary, mammary epithelium, kidneys, blood stem cells — are capable of functioning normally well beyond donor lifespans when transplanted from an old donor to a young recipient.[91] In view of the scarcity of organs for transplantation, the possibility exists for collecting these organs after the death of older donors, provided techniques for harvesting and preserving the organs be perfected. Current progress in perfusion fluids, cryopreservation, and distribution techniques[87] suggests that the elderly may soon benefit from

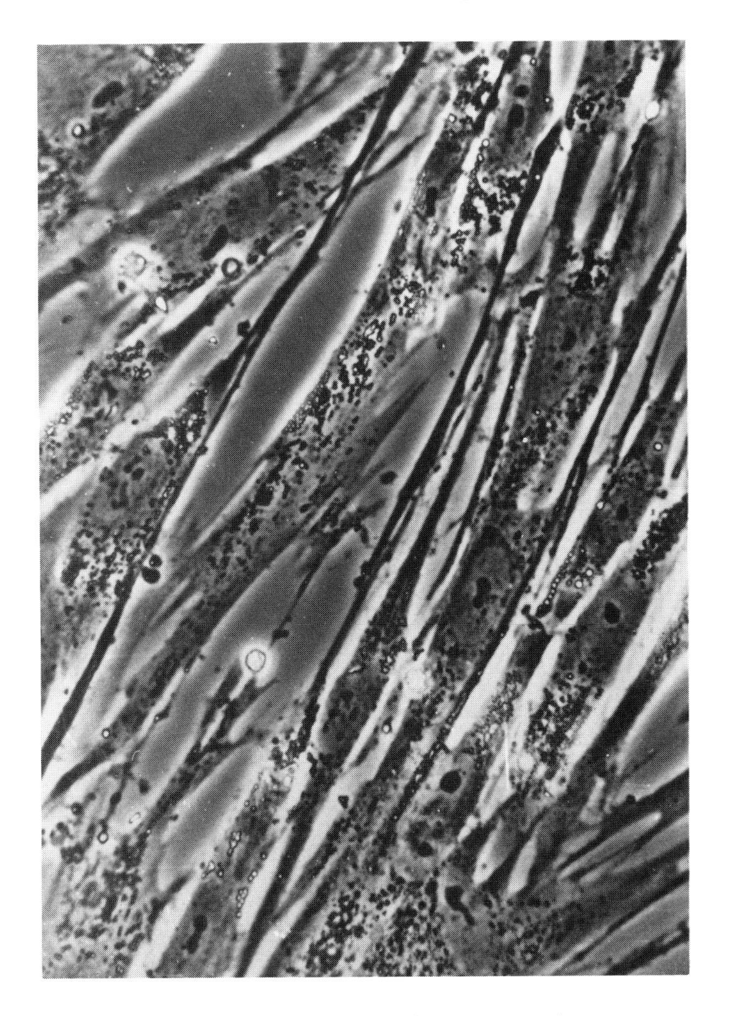

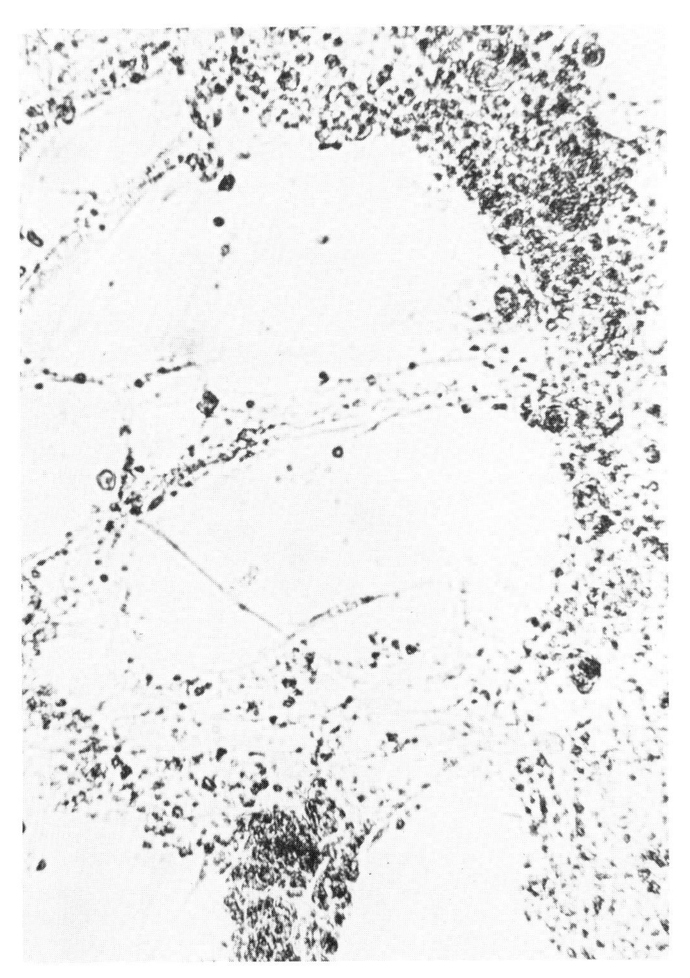

FIGURE 5-2 *Aging of human fibroblast* in vitro. Phase contrast micrographs of the human diploid cell strain WI-38: (A) during active cell division; (B) at the end of its *in vitro* lifetime (after 50 doublings in cell culture). (Courtesy of Dr. L. Hayflick.)

available transplantations by contributing to these interventions both as recipients and donors.

Transplantations have also served to validate or deny some theories of aging. For example, theories predicting that all tissues, and cells, age intrinsically must be modified to recognize that, in many cases, tissues do not age as rapidly as the animal from which they come. A major consideration in assessing the value of transplanted tissues is the ability to establish some functional criteria for the identification of the transplanted tissues and their functional status. These may be at the organismic or tissue level, such as the color of hair produced by skin, the genotype of offspring produced by an ovary, the hemoglobin type produced by erythrocytes. Other methods may utilize molecular techniques, such as messenger RNA, DNA clones, or embryonic nuclear transplantation. Utilizing the latter, for example, nuclei from embryonic cells were transferred into fertilized eggs, and normal mice were born with genetic markers characteristic of the transplanted nuclei.[92] These results were obtained from embryonic nuclei, and it must yet be tested whether such products could also arise utilizing older or aged nuclei. Another approach which shows promise is the use of cells cultured *in vitro* for use in *in vivo* transplants.

6 AGING UNDER GLASS: THE HUMAN FIBROBLAST

Tissue culture methods, "aging under glass", have potential value in transplantation but are also a valuable means of investigating developmental and aging phenomena. By these techniques, control of the milieu of the growing tissue or cell permits easier distinction of extrinsic and intrinsic factors in aging.[93] Several experiments have shown that somatic cell strains have a finite doubling potential *in vitro* and undergo aging and death even under optimal conditions of culture.[94] Normal human fibroblasts, spindle-shaped cells most capable of sustained proliferation, derived from several embryonic organs undergo a finite number of population doublings (approximately 50 doublings) after which the cultured cells degenerate and die. Once established in culture, the fibroblasts grow and multiply through successive passages: each time a culture has doubled in size, it is divided into two subcultures, each of which is allowed to double in size again; this process is repeated through a specified number of so-called doublings. Finally, the proliferative activity declines, the time required for the cells to double in number becomes longer, and this increased growth latency leads to complete failure of cell replication (Figure 5-2). This phenomenon is not caused by any deficiency or toxicity of the culture medium or methodological artifact but has been interpreted as a manifestation of intrinsic cellular senescence.[95]

6.1 Immortality versus Mortality of Cells

According to current views, most normal animal cells have a limited proliferative capacity when grown *in vitro*. The early (1912) experiments of Carrel[96] held that isolated cells maintained in culture were capable of unlimited proliferation —a view contradicted by current findings. The "immortality" of the fibroblasts claimed by Carrel in his studies of the heart of the chick embryo is now viewed as the result of a technical artifact. Two fundamental types of cells are now identified *in vitro*; according to the Tissue Culture Association, cell types can be distinguished according to their growth potential by the terms of "finite" or "continuous" used as prefixes to the general term of "cell line", which denotes a culture of multiple lineages arising from the subcultivation of a primary culture. Cells growing *in vitro* usually demonstrate a finite proliferative capacity; those with unlimited capacity for division are inevitably abnormal in one or more properties and most often resemble cancer cells.[97] The spontaneous occurrence of a transformed cell line with cells of unlimited growth *in vitro* is a rare event regardless of the species studied. The question of the mechanism(s) that limit the growth potential of fibroblasts remains unanswered as well as that of the factors that stimulate growth potential. Molecular techniques utilizing viruses to study proliferative mechanisms may provide the needed information. For example, the temperature-sensitive strain of the tumor virus SV-40 can reinitiate DNA synthesis in senescent cells.[98]

It is not yet clear whether the limited *in vitro* proliferative potential of cultured cells is determined by cell division rather than by calendar time.[99] Irrespective of its nature, the proliferative behavior and growth potential of cultured fibroblasts is influenced by a number of variables: medium composition (e.g., serum batch, nutrients), hormones (e.g., extension of cell lifespan by hydrocortisone), donor genotype (e.g., smaller number of doublings in some progeroid syndrome), and tissue of origin (e.g., fibroblasts from lungs divide more rapidly than those from skin).

6.2 Relevance to Cellular Aging

Several studies have shown that the doubling capacity of cultured fibroblasts does vary with the age of the individual from whom the cells were taken, even though the cells appear biologically identical. For example, cells from lung tissue of adult humans reach the phase of declining mitotic activity and increasing degenerative changes after only 20 doublings, whereas cells from embryonic lung tissue undergo approximately 50 doublings before failure to proliferate.[100] A number of experimental approaches have corroborated the concept that the proliferative activity of cells is finite. Thus, in humans, cells taken from several tissues of normal individuals of different ages and of individuals affected by specific progeroid syndromes show that the replicative lifespan (as measured by the number of doublings of cultured human somatic cells) is:

1. A function of the tissue of origin, e.g., fibroblasts from the skin showed the greatest (30 to 45) number of doublings and the bone the smallest (10), with the muscle being intermediate (29 to 30 doublings)
2. Inversely related to the age of the donor, e.g., the number of doublings declines from approximately 45 from birth to 10 years of age to 30 by 80 to 90 years of age
3. A function of the genotype of the donor, e.g., the number of cell doublings decreases to approximately 10 in Werner's syndrome (Chapter 3), compared with the 40 doublings from normal individuals of the same age

In other types of progeria, the number of doublings is similarly reduced. However, in Down's syndrome, characterized also by accelerated segmental aging (e.g., degenerative brain changes; Chapters 6 and 9), and in diabetes mellitus, characterized by altered glucose transport (Chapter 15), contradictory results seem to support or negate a reduced number of doublings.

If the capacity for cell division is taken as an index of cell function, then its loss can be extrapolated to indicate functional impairment with aging, *In vitro* aging is corroborated by numerous morphologic and biochemical alterations in human fibroblasts at or near the end of their proliferative lifespan; these have been extensively reviewed.[94,101] Of particular significance are (1) increasing heterogeneity of cell cycle activity, (2) increased number of lysosomes associated with the formation of autophagic vacuoles and residual bodies, (3) irregular lobation of nuclei, (4) swollen mitochondria, particularly the cristae, and (5) abnormalities in the assembly of gap junctions.

Current studies also suggest abnormal responses to external stimuli such as growth factors (e.g., insulin, mitogenic peptides). This altered responsiveness has been ascribed to a reduction of binding sites perhaps due to plasma membrane/cell surface abnormalities. For example, a membrane-associated glycoprotein, fibronectin, appears to be significantly reduced in the senescent fetal lung cultures (Chapter 18). Altered cell surface molecules may be of greater significance in aging *in vitro* and, eventually by extrapolation *in vivo,* than the decrements in proliferative capacity.

REFERENCES

1. Economos, A. C., Miquel, J., Ballard, R. C., and Johnson, J. E., Jr., Variation: principles and applications in the study of cell structure and aging, in *Aging and Cell Structure,* Vol. 1, Johnson, J. E., Jr., Ed., Plenum Press, New York, 1981, 187.
2. Phelan, J. P., Genetic variability and rodent models of human aging, *Exp. Gerontol.,* 27, 147, 1992.
3. Rothstein, M., Altered proteins, errors and aging, in *Protein Metabolism in Aging,* Segal, H. L., Rothstein, M., and Bergamini, E., Eds., Wiley-Liss, New York, 1990, 3.
4. Strehler, B. L., *Time, Cells, and Aging,* 2nd ed., Academic Press, New York, 1977.
5. Feigenbaum, L., Geriatric medicine and the elderly patient, in *Current Medical Diagnosis and Treatment,* Schroeder, S., Krupp, M. A., and Tierney, L. M., Jr., Eds., Lange, Los Altos, CA, 1988, 17.
6. Bolton, W. K. and Sturgill, B. C., Ultrastructure of the aging kidney, in *Aging and Cell Structure,* Vol. 1, Johnson, J. E., Jr., Ed., Plenum Press, New York, 1981, 215.
7. Johnson, J. E., Jr., Fine structure of IMR-90 cells in culture as examined by scanning and transmission electron microscopy, *Mech. Ageing Dev.,* 10, 405, 1979.
8. Kelley, R. O. and Vogel, K. G., The aging cell surface: structural and biochemical alterations associated with progressive subcultivation of human diploid fibroblasts, in *Aging and Cell Structure,* Vol. 2, Johnson, J. E., Jr., Ed., Plenum Press, New York, 1984, 1.
9. Johnson, J. E., Jr. and Barrows, C. H., Jr., Effects of age and dietary restriction on the kidney glomeruli of mice: observations by scanning electron microscopy, *Anat. Rec.,* 196, 145, 1980.

10. Masoro, E. J., Nutrition as a modulator of aging and age-associated problems, in *Protein Metabolism in Aging,* Segal, H. L., Rothstein, M., and Bergamini, E., Eds., Wiley-Liss, New York, 1990, 257.
11. Masoro, E.J., Aging and proliferative homeostasis, modulation by food restriction in rodents, *Lab. Anim. Sci.,* 42, 132, 1992.
12. Johnson, J. E., Jr., *In vivo* and *in vitro* comparisons of age-related fine structural changes in cell components, in *Aging and Cell Structure,* Vol. 2, Johnson, J. E., Jr., Ed., Plenum Press, New York, 1984, 37.
13. Doyle, D., Diamond, M., and Petell, J., Turnover of plasma membrane proteins in hepatocytes, in *Protein Metabolism and Aging,* Segal, H. L., Rothstein, M., and Bergamini, E., Eds., Wiley-Liss, New York, 1990, 79.
14. Lewin, M. D. and Timiras, P. S., Lipid changes with aging in cardiac mitochondrial membranes, *Mech. Ageing Dev.,* 24, 343, 1984.
15. Naeim, F. and Walford, R. L., Disturbance of redistribution of surface membrane receptors on peripheral mononuclear cells of patients with Down's syndrome and of aged individuals, *J. Gerontol.,* 35, 650, 1980.
16. Cole, G. M. and Timiras, P. S., Ubiquitin-protein conjugates in Alzheimer's lesions, *Neurosci. Lett.,* 79, 207, 1987.
17. Haas, A. L., Role of ubiquitin in protein degradation, in *Protein Metabolism in Aging,* Segal, H. L., Rothstein, M., and Bergamini, E., Eds., Wiley-Liss, New York, 1990, 135.
18. Mesco, E. R. and Timiras, P. S., Tau-ubiquitin protein conjugates in a human cell line, *Mech. Ageing Dev.,* 61, 1, 1991.
19. Kosik, K. S., Joachim, C. L., and Selkoe, D. J., Microtubule-associated protein tau is a major antigenic component of paired helical filaments in Alzheimer's disease, *Proc. Natl. Acad. Sci. U.S.A.,* 83, 4044, 1986.
20. Miller, F. D. and Geddes, J. W., Increased expression of the major α-tubulin mRNA, Tα1, during neuronal regeneration, sprouting, and in Alzheimer's disease, in *Progress in Brain Research,* Vol. 86, Coleman, P., Higgins, G., and Phelps, C., Eds., Elsevier, Amsterdam, 1990, 321.
21. Matsuyama, S. S. and Jarvik, L. F., Hypothesis: microtubules, a key to Alzheimer disease, *Proc. Natl. Acad. Sci. U.S.A.,* 86, 8152, 1989.
22. Puck, T. T., Cyclic AMP, the microtubule-microfilament system and cancer, *Proc. Natl. Acad. Sci. U.S.A.,* 74, 4491, 1977.
23. Naeim, F. and Walford, R. L., Aging and cell membrane complexes. The lipid bilayer, integral proteins and cytoskeleton, in *Handbook of The Biology of Aging,* Finch, C. E. and Schneider, E. L., Eds., Van Nostrand Reinhold, New York, 1985, 272.
24. Kirkwood, T. B. L., Rosenberger, R. F., and Galas, D. J., *Accuracy in Molecular Processes: Its Control and Relevance to Living Systems,* Chapman and Hall, London, 1986.
25. Spoerri, P. E., Mitochondrial alterations in ageing mouse neuroblastoma cells in culture, in *Cellular Ageing,* Sauer, H. W., Ed., S. Karger, Basel, 1984, 210.
26. Fleming, J. E., Miquel, J., and Bensch, K. G., Age dependent changes in mitochondria, in *Molecular Biology of Aging,* Woodhead, A. V., Blackett, A. D., and Hollaender, A., Eds., Basic Life Sciences, Vol. 35, Plenum Press, New York, 1985, 143.
27. Miquel, J., Economos, A. C., and Johnson, J. E., Jr., A system analysis — thermodynamic view of cellular and organismic aging, in *Aging and Cell Function,* Johnson, J. E., Jr., Ed., Plenum Press, New York, 1984, 247.
28. Haug, H., Are neurons of the human cerebral cortex really lost during aging? A morphometric examination, in *Senile Dementia of the Alzheimer's Type: Early Diagnosis, Neuropathology and Animal Models,* Traber, Y. and Gispen, W. H., Eds., Springer-Verlag, New York, 1985, 150.
29. Wallace, D. C., Mitochondrial genetics, a paradigm for aging and degenerative diseases?, *Science,* 256, 628, 1992.
30. Hayflick, L., Theories of biological aging, *Exp. Gerontol.,* 20, 145, 1985.
31. Finch, C. E., Mechanisms in senescence: some thoughts in April 1990, *Exp. Gerontol.,* 27, 7, 1992.
32. Bradshaw, R. A. and Prentis, S., Eds.,*Oncogenes and Growth Factors,* Elsevier, Amsterdam, 1987.
33. Powers, D. C., Morley, J. E., and Flood, J. F., Age-related changes in LFA-1 expression, cell adhesion, and PHA-induced proliferation by lymphocytes from senescence accelerated mouse (SAM-P/8 and SAM-R/1) substrains, *Cell. Immunol.,* 141, 444, 1992.
34. Sharma, R., Enzymatic changes during aging, in *Physiological Basis of Aging and Geriatrics,* Timiras, P. S., Ed., Macmillan, New York, 1988, 75.
35. Vitorica, J., Satrustegui, J., and Machado, A., Metabolic implications of aging: changes in activities of key lipogenic and gluconeogenic enzymes in the aged rat liver, *Enzyme,* 26, 144, 1981.
36. Grube, K. and Burkle, A., Poly (ADP-ribose) polymerase activity in mononuclear leukocytes of 13 mammalian species and correlates with species-specific lifespan, *Proc. Natl. Acad. Sci. U.S.A.,* 89, 11759, 1992.
37. Rusting, R. L., Why do we age?, *Sci. Am.,* 267, 131, 1992.
38. Rao, G., Xia, E., Nadakavukaren, M. J., and Richardson, A., Effect of dietary restriction on the age-dependent changes in the expression of antioxidant enzymes in rat liver, *J. Nutr.,* 20, 602, 1990.
39. Davidson, E. H. and Britten, R. J., Regulation of gene expression: possible role of repetitive sequences, *Science,* 204, 1052, 1979.
40. Markert, C. L. and Ursprung, H., The ontogeny of isoenzyme patterns of lactate dehydrogenase in mouse, *Dev. Biol.,* 5, 363, 1962.
41. Markert, C. L. and Moller, F., Multiple forms of enzymes: tissue, ontogeny, and species-specific patterns, *Proc. Natl. Acad. Sci. U.S.A.,* 45, 753, 1959.
42. Kanungo, M. S., *Biochemistry of Aging,* Academic Press, New York, 1980.
43. Singh, S. N. and Kanungo, M. S., Alterations in lactate dehydrogenase of the brain, heart, skeletal muscle, and liver of rats at various sizes, *J. Biol. Chem.,* 243, 4526, 1968.
44. Reiss, U. and Rothstein, M., Age-related changes in isocitrate lyase from the free living nematode, *Turbatrix aceti, J. Biol. Chem.,* 250, 826, 1975.
45. Patnaik, S. K. and Kanungo, M. S., Soluble alanine aminotransferase of the liver of rats of various ages: induction, characterization, and changes in patterns, *Indian J. Biochem. Biophys.,* 13, 117, 1976.
46. Chen, S. H. and Giblett, E. R., Polymorphism of soluble glucatimic-pyruvic transaminase: a new genetic marker in man, *Science,* 173, 148, 1971.
47. Wang, R. K., and Mays, L. L.,Opposite changes in rat liver glucose-6-phosphate dehydrogenase during aging in Sprague-Dawley and Fischer 344 male rats, *Exp. Gerontol.,* 12, 117, 1977.
48. Russell, R. L., Evidence for and against the theory of developmentally programmed aging, in *Modern Biological Theories of Aging,* Warner, H. R. et al., Eds., Raven Press, New York, 1987, 35.
49. Obenrader, M., Sartin, J. L., and Adelman, R. C., Enzyme adaptation during aging, in *Handbook of Biochemistry of Aging,* Florini, J. R., Ed., CRC Press, Boca Raton, FL, 1981, 263.

50. Gesseck, D., Endocrine mechanisms and aging, in *Advances in Gerontological Research,* Strehler, B. L., Ed., Academic Press, New York, 1972, 105.
51. Kalimi, M., Hubbard, J., and Gupta, S., Modulation of glucocorticoid receptor from development to aging, *Ann. N.Y. Acad. Sci.,* 521, 149, 1988.
52. Sharma, R. and Timiras, P. S., Regulatory changes in glucocorticoid receptors in the skeletal muscle of immature and mature male rats, *Mech. Ageing Dev.,* 37, 249, 1987.
53. Sharma, R. and Timiras, P. S., Regulation of glucocorticoid receptors in the kidney of immature and mature male rats, *Int. J. Biochem.,* 20, 141, 1988.
54. Roth, G. S., Mechanism of altered hormone and neurotransmitter action during aging: the role of impaired calcium mobilization, *Ann. N.Y. Acad. Sci.,* 521, 170, 1988.
55. Yamamoto, K. R., Steroid receptor regulated transcription of specific genes and gene networks, *Annu. Rev. Genet.,* 19, 209, 1985.
56. Ptashne, M. and Gann, A. A. F., Activators and targets, *Nature,* 346, 329, 1990.
57. Sharma, R., Glucocorticoid actions and biomodulators: an integrated biological control, *Indian J. Biochem. Biophys.,* 28, 159, 1991.
58. Gershon, H. and Gershon, D., Inactive enzyme molecules in aging mice: liver aldolase, *Proc. Natl. Acad. Sci. U.S.A.,* 70, 909, 1973.
59. Gallant, J. and Palmer, L., Error propagation in viable cells, *Mech. Ageing Dev.,* 10, 27, 1979.
60. Kirkwood, T. B. L., Error propagation in intracellular information transfer, *J. Theoret. Biol.,* 82, 363, 1980.
61. Rothstein, M., *Biochemical Approaches to Aging,* Academic Press, New York, 1982.
62. Demchenko, A. P. and Orlovska, N. N., Age-dependent changes of protein structure. II. Conformational differences of aldolase of young and old rabbits, *Exp. Gerontol.,* 15, 619, 1980.
63. Shimokata, H., Andres, R., Coon, P. J., Elahi, D., Muller, D. C., and Tobin, J. D., Studies in the distribution of body fat. II. Longitudinal effects of change in weight, *Int. J. Obesity,* 13, 455, 1989.
64. Sohal, R.S., Ed., *Age Pigments,* Elsevier/North-Holland, Amsterdam, 1981.
65. Reichel, W., Lipofuscin pigment accumulation and distribution in five rat organs as a function of age, *J. Gerontol.,* 23, 145, 1968.
66. Kny, W., Uber die Verteilung des Lipofuscins in der Skeletmuskulatur in ihrer Beziehung zur Funktion, *Virchows Arch. Pathol. Anat.,* 299, 468, 1937.
67. Strehler, B. L., Mark, D. D., Mildvan, A. S., and Gee, M. V., Rate and magnitude of age pigment accumulation in the human myocardium, *J. Gerontol.,* 14, 430, 1959.
68. Elleder, M., Chemical characterization of age pigments, in *Age Pigments,* Sohal, R. S., Ed., Elsevier/North Holland, Amsterdam, 1981, 204.
69. Packer, L., Deamer, D. W., and Health, R. L., Regulation and deterioration of structure in membranes. VI. Age pigments, in *Advances in Gerontological Research,* Strehler, B. L., Ed., Vol. 2, Academic Press, New York, 1967, 77.
70. Hendley, D. D., Mildvan, A. S., Reporter, M. C., and Strehler, B. L., The properties of isolated human cardiac age pigment. I. Preparation and physical properties, *J. Gerontol.,* 18, 144, 1963.
71. Hendley, D. D., Mildvan, A. S., Reporter, M. C., and Strehler, B. L., The properties of isolated human cardiac age pigment. II. Chemical and enzymatic properties, *J. Gerontol.,* 18, 250, 1963.
72. Glenner, G. G., Amyloid deposits and amyloidosis. The beta fibrilloses (first of two parts), *N. Engl. J. Med.,* 302, 1283, 1980.
73. Glenner, G. G., Amyloid deposits and amyloidosis. The beta fibrilloses (second of two parts), *N. Engl. J. Med.,* 302, 1333, 1980.
74. Gertz, M. A. and Kyle, R. A., Primary systemic amyloidosis, a diagnostic primer, *Mayo Clin. Proc.,* 64, 1505, 1989.
75. Glenner, G. G., Ein, D., Eanes, E. D., Bladen, H. A., Terry, W., and Page, D. L., Creation of "amyloid" fibrils from Bence Jones proteins *in vitro, Science,* 174, 712, 1971.
76. Ghiso, J., Pons-Estel, B., and Frangione, B., Hereditary cerebral amyloid angiopathy: the amyloid fibrils contain a protein which is a variant of cystatin C, an inhibitor of lysosomal cysteine proteases, *Biochem. Biophys. Res. Commun.,* 136, 548, 1986.
77. Glenner, G. G. and Wong, C. W., Alzheimer's disease: initial report of the purification and characterization of a novel cerebrovascular amyloid protein, *Biochem. Biophys. Res. Commun.,* 120, 885, 1984.
78. Glenner, G. G. and Wong, C. W., Alzheimer's disease and Down's syndrome: sharing of a unique cerebrovascular amyloid fibril protein, *Biochem. Biophys. Res. Commun.,* 122, 1131, 1984.
79. Gorevic, P. D., Elias, J., and Peress, N., Amyloid, immunopathology and aging, in *Molecular Biology of Aging,* Woodhead, A. D., Blackett, A. D., and Hollaender, A., Eds., Basic Life Sciences, Vol. 35, Plenum Press, New York, 1985, 397.
80. Golde, T. E., Estus, S., Younkin, L. H., Selkoe, D. J., and Younkin, S. G., Processing of the amyloid protein precursor to potentially amyloidogenic derivatives, *Science,* 255, 728, 1992.
81. Bowen, I. D. and Lockshin, R. A., Eds., *Cell Death in Biology and Pathology,* Chapman and Hall, London, 1981.
82. Davies, I. and Sigee, D. C., Eds., *Cell Ageing and Cell Death,* Cambridge University Press, Cambridge, 1984.
83. Kerr, J. F. R., Searle, J., Harmon, B. V., and Bishop, C. J., Apoptosis, in *Perspectives on Mammalian Cell Death,* Potten, C. S., Ed., Oxford University Press, New York, 1987, 93.
84. Wyllie, A. H., Apoptosis, cell death in tissue regulation, *J. Pathol.,* 153, 313, 1987.
85. Gaido, M. L., Schwartzman, R. A., Caron, L.-A. M., and Cidlowski, J. A., Glucocorticoids and cell death: biochemical mechanisms, in *Molecular Biology of Aging,* Finch, C. E. and Johnson, T. E., Eds., Wiley-Liss, New York, 1990, 299.
86. Franceschi, C., Cossarizza, A., Monti, D., Comaschi, V., Farruggia, G., Sartor, G., and Masotti, L., Cellular defence mechanisms, cell death and aging: an integrated view, in *Protein Metabolism in Aging,* Segal, H. L., Rothstein, M., and Bergamini, E., Eds., Wiley-Liss, New York, 1990, 395.
87. Murray, J. E., Human organ transplantation: background and consequences, *Science,* 256, 1411, 1992.
88. Evans, R. W., Advanced medical technology and elderly people, in *Too Old for Health Care?,* Binstock, R. H. and Post, S. G., Eds., The Johns Hopkins University Press, Baltimore, 1991, 44.
89. Geiringer, E., Homotransplantation as a method of gerontological research, *J. Gerontol.,* 9, 142, 1954.
90. Geiringer, E., Young rats with adult adrenal transplants, a physiologic study, *J. Gerontol.,* 11, 8, 1956.
91. Harrison, D. E., Cell and tissue transplantation: a means of studying the aging process, in *Handbook of The Biology of Aging,* Vol.2, Finch, C. E. and Schneider, E. L., Eds., Van Nostrand Reinhold, New York, 1985, 322.
92. Illmensee, K. and Hoppe, P. C., Nuclear transplantation in *Mus musculus:* developmental potential of nuclei from preimplantation embryos, *Cell,* 23, 9, 1981.
93. Norwood, T. H. and Smith, J. R., The cultured fibroblast-like cell as a model for the study of aging, in *Handbook of The Biology of Aging,* Vol. 2, Finch, C. E. and Schneider, E. L., Eds., Van Nostrand Reinhold, New York, 1985, 291.
94. Hayflick, L., and Moorhead, P. S., The serial cultivation of human diploid cell strains, *Exp. Cell Res.,* 25, 585, 1961.

95. Pool, T. B. and Metter, J. D., New concepts in regulation of the lifespan of human diploid fibroblasts *in vitro*, in *Aging and Cell Structure*, Vol.2, Johnson, J. E., Jr., Ed., Plenum Press, New York, 1984, 89.

96. Carrel, A., On the permanent life of tissues outside of the organism, *J. Exp. Med.*, 15, 516, 1912.

97. Rozengurt, E., Early signs in the mitogenic response, *Science*, 234, 161, 1986.

98. Gorman, S. D. and Cristofalo, V. J., Evidence that senescent WI-38 cells are blocked in late G1, *In Vitro*, 20, 281, 1984.

99. Roberts, T. W. and Smith, J. R., The proliferative potential of chick embryo fibroblasts: population doubling vs. time in culture, *Cell Biol. Int. Rep.*, 4, 1057, 1980.

100. Martin, G. M., Sprague, C. A., and Epstein, C. J., Replicative lifespan of cultivated human cells: effects of donor's age, tissue, and genotype, *Lab. Invest.*, 23, 86, 1970.

101. Cristofalo, V. J., Doggett, D. L., Brooks-Frederick, K. M., Cianciarulo, F. L., Gerhard, G. S., and Phillips, P. D., The regulation of cell senescence, in *Protein Metabolism and Aging*, Segal, H. L., Rothstein, M., and Bergamini, E., Eds., Wiley-Liss, New York, 1990, 195.

II
MOLECULAR AND
CELLULAR AGING

6 OXIDANTS AND ANTIOXIDANTS IN AGING

Rolf J. Mehlhorn

Many theories have been put forward to explain aging as a consequence of some specific cumulative damage process. Biological organisms have the capacity to minimize adverse chemical reactions and, for the few undesired reactions that do occur, to repair the damage or to compensate for its effects (e.g., by replacing an irreversibly damaged cell with a new cell). Hence, the various formulations of *cumulative damage theories must contend with the question of why repair is incomplete or why compensatory processes fail.* One important consideration is that no repair system or compensatory mechanism can be absolutely efficient, particularly if the damage is so random that macromolecules are altered in virtually limitless variety. A wear-and-tear theory, which embodies the characteristic of unpredictable molecular alterations, is the free radical theory of aging. It was proposed by Harman in 1956 as follows: *aging results from the deleterious effects of free radicals produced in the course of cellular metabolism.*[1] Growing evidence supports an involvement of free radicals in life-shortening diseases, although efforts to use predictions of the free radical theory to achieve increases in maximum lifespans have met with limited success at best.[2] This chapter will critically assess the current status of the available evidence concerning free radical involvement in aging.

1 CHEMISTRY AND BIOCHEMISTRY

1.1 Free Radical Chemistry

Free radicals are molecules, rarely occurring in nature, that contain one or more "unpaired" electrons, sometimes also referred to as "free spins", in their outer shells. A free radical molecule containing a single unpaired electron is frequently denoted by R·, where the dot refers to the unpaired electron, e.g., the hydroxyl radical OH·.

Free radicals rarely arise naturally because of the familiar chemical principle that valence electrons in atoms form chemical bonds, each of which consists of an electron pair. Energy must be supplied to break chemical bonds, and usually the resulting molecular fragments are free radicals. Once formed, these free radical fragments will combine rapidly with other free radical fragments, or with each other, to form nonradical products:

$$R\cdot + R'\cdot \rightarrow R{:}R'$$

Hence, even if sufficient energy is available to break bonds, the proportion of free radicals among nonradicals is usually very low. Examples of reactive free radicals include the following:

- The very reactive hydroxyl radical (OH·), which is produced when ionizing radiation passes through water
- The moderately reactive thiyl radical (RS·), which is formed when reactive radicals like the hydroxyl radical react with sulfhydryl groups, e.g., cysteine side chains in proteins
- The weakly reactive nitric oxide (NO·) radical, which is found in polluted urban air

Free radicals can pose a considerable hazard to biological systems because of their unique chemistry, which distinguishes free radicals from other toxic agents. The most damaging free radicals exhibit some or all of the following reaction patterns:

- Attack other molecules indiscriminately
- Produce oxygen-consuming chain reactions, such that a single free radical effectively damages a large number of other molecules
- Cause fragmentation or cross-linking of molecules, including vital macromolecules like DNA and critically important enzymes

Because of the random reactions that can occur, some of the products of free radical chemistry are completely foreign to the repair or turnover enzymes of the cell. For example, when two proteins become cross-linked, the product of this reaction may be resistant to attack by proteolytic enzymes. Hence, the usual metabolic turnover processes may fail to degrade such cross-linked molecules to provide amino acids for the synthesis of new biomolecules. Cross-linked products can accumulate progressively in cells, as in the example of age pigments, which increase with age in the cells of animals (Chapter 5).

1.2 Oxygen

Ordinary oxygen, also referred to as triplet oxygen, has two electrons with unpaired spins. Thus, it can be thought of as a biradical. Oxygen is an example of a radical that is normally quite stable. However, oxygen can be extremely reactive in the presence of other free radicals, or in collisions with molecules that are chemically very reducing.

When an oxygen molecule collides with another free radical, the favorable energetics of electron pairing imply that a new addition product, called a peroxyl radical, will frequently be formed. If oxygen reacts with some nonradical molecule that has a sufficiently reducing chemical potential, a one-electron transfer to oxygen can occur, resuting in O_2^-, the superoxide radical. In such an electron transfer reaction, the energy that is required to break one bond is recovered in the formation of another bond. An example of this reaction in the laboratory is the formation of superoxide radicals when molecular oxygen reacts with photochemically reduced flavins. Fortunately, oxygen is inert towards most nonradical molecules so it cannot initiate free radical chain reactions. When oxygen reacts with an organic radical whose unpaired electron is associated with a carbon atom, the resulting peroxyl radical molecules are moderately reactive and will remove hydrogen atoms from a variety of other molecules, leading to chain reactions and an accumulation of hydroperoxides. Such free radical chain reactions are familiar to food chemists, occurring readily in fats to cause rancidity. The facile

0-8493-8979-8/94/$0.00+$.50

occurrence of analogous free radical chain reactions in biological membranes, referred to as lipid peroxidation, is a major oxidative damage mechanism in cells.

The chemical products of most of the free radical reactions involving oxygen are so reactive themselves as to have earned many of them the designation "active oxygen" molecules. Several active oxygen species are of interest in biological damage processes. These include the free radicals:

- Hydroxyl radical, OH·
- Superoxide radical, O_2^- and the two nonradicals
- Singlet oxygen, $^1O_2$
- Hydrogen peroxide, H_2O_2

The most reactive is the hydroxyl radical — virtually no biological molecule is immune to its attack. Hence, organisms have evolved elaborate defenses to ensure that the hydroxyl radical arises as infrequently as possible. Singlet oxygen, which may arise during peroxyl radical reactions, is also reactive, although not as reactive as the hydroxyl radical.

Hydrogen peroxide is of interest as a potential damaging species because, in the presence of freely dissolved or loosely bound iron or copper and mild reducing agents like ascorbic acid, it can decompose to produce hydroxyl radicals. Because of the destructiveness of hydroxyl radicals, hydrogen peroxide decomposition could be a major source of biological damage under conditions of iron or copper release from protein binding sites. Hydrogen peroxide can also interact with heme proteins to produce highly oxidizing products (ferryl species) that can initiate free radical reactions.

The superoxide radical is not very reactive chemically, but its lack of reactivity may be offset by the abundance of this radical species in aerobic biological environments. It may exert its damaging effects by serving as a reductant for ferric iron, particularly when iron is sequestered within proteins, thus releasing the ferrous iron into aqueous solution. Superoxide radicals react with each other or with appropriate reducing agents to produce hydrogen peroxide. The fact that no organism can survive life in air without the superoxide dismutase protective enzyme (or high concentrations of free manganese ions in one exceptional microorganism) provides compelling evidence that superoxide poses a great threat to aerobic cells.

While endogenous active oxygen molecules may play a role in unavoidable aging, avoidable exogenous agents that either are free radicals or can become free radicals after metabolic activation are linked to life-shortening diseases and may accelerate the basic aging process. Cigarette smoke is one of the best characterized of these avoidable agents.

1.3 Cigarette Smoke

Free radicals are known to be present in cigarette smoke and to react with a variety of biological molecules.[3] Among free radicals in cigarette smoke are remarkably persistent organic free radicals in the "tar phase", which can be trapped as a dark brown material on glass wool and extracted into ethanol (Figure 6-1). The persistence of these free radicals in ethanol suggests that they would also persist significantly in tissue fluids, possibly exerting adverse effects far from the lung. Another type of free radical is found in filtered cigarette smoke, which is designated as "gas phase" cigarette smoke. For example, gas phase cigarette smoke contains 300 to 500 ppm nitric oxide (NO·), a weakly reactive free radical and a potent vasodilator.[4] NO· is converted to the nitrogen dioxide radical (NO_2·) in the presence of oxygen.[3] NO_2· is more reactive than is NO·, e.g., with unsaturated lipids.[5] Cigarette smoke is known to contain hydroperoxides,[6] to generate hydrogen peroxide in aqueous solution,[7] and to stimulate peroxide production by leukocytes. Hydroperoxides degrade hemoglobin *in vitro* to products, including loosely bound iron, that catalyze free radical reactions.[8-10] High iron levels have been detected in smokers' lungs.[11]

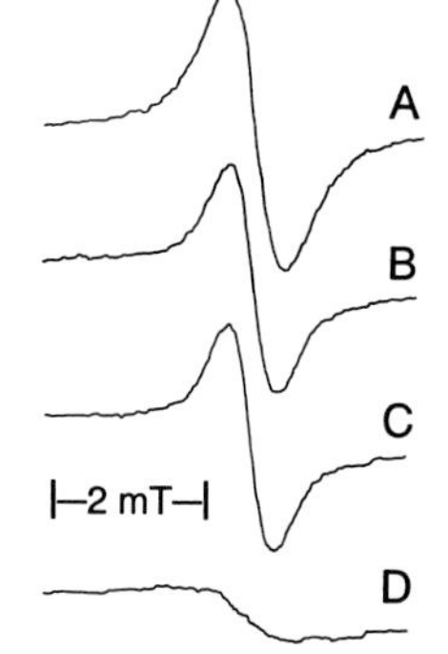

FIGURE 6-1 *Free radicals in cigarette smoke trapped on glass wool and extracted into ethanol.* Electron spin resonance (ESR) spectra of cigarette tar on (A) glass wool, (B) ethanol extract 5 min after extraction, (C) ethanol extract 10 min after extraction, and (D) ethanol extract 50 min after extraction.

Cigarette smoke is also known to contain powerful metal chelators, e.g., catechol, which would be expected to stabilize hemoglobin decomposition products and to bind iron that may be released during hemoglobin decomposition.[8] Metal binding to quinols can be particularly destructive to cells when the quinol-metal complex has a high affinity for DNA.[12] Intercalation of iron complexes into DNA can induce base modifications and strand breaks. This is a particularly important damage mechanism because it may not be responsive to antioxidants. In principle, transition metal ion complexes can catalyze free radical reactions at many other sites, e.g., in lipoproteins. Alpha-1-proteinase, whose inactivation is generally thought to play a role in the development of emphysema (Chapter 8), can be protected by antioxidants and reducing agents,[13] implicating free radicals in its oxidation.

1.4 Antioxidants

Biological systems have evolved a multiplicity of defenses against oxidative attack. Indeed, the discovery of one defensive substance, the enzyme superoxide dismutase (SOD), has played a pivotal role in the explosive expansion of our understanding of oxidative damage in recent years. We now know that a considerable variety of proteins and smaller molecules are involved in protecting the cell against the threat of free radical reactions. In addition, considerable energy expenditures are made by cells to repair such damage when the protective systems fail to intercept all of the endogenous and exogenous free radical initiators. In excess of 40 enzymes are constantly repairing damaged DNA, and both lipids and proteins are being turned over continuously. Thus, it seems that organisms pay dearly for the privilege of aerobic life.

Superoxide Dismutases (SODs)

At least three different types of SODs are now known. In mammals, a copper-zinc enzyme is found in cells, excluding the intramitochondrial matrix which contains a manganese-containing enzyme. A number of microbes have a third type of SOD that contains iron at its active center. Concentrations of SODs in cells are relatively high, e.g., in spinach chloroplasts, 0.24% of the total soluble protein is SOD.

Catalase
This enzyme decomposes hydrogen peroxide and oxidizes low molecular weight alcohols and other reduced organic molecules using H_2O_2 as a hydrogen acceptor. Catalase is located exclusively in the peroxisomes of most cells. It is also found in red cells. Some humans lack this enzyme altogether, yet seem to suffer no adverse effects under normal circumstances, possibly because of the existence of another peroxide-decomposing enzyme (see below).

Glutathione Peroxidase
Glutathione peroxidase and glutathione transferases also act on hydrogen peroxide but can use organic hydroperoxides as electron acceptors. The most active form of these enzymes contains selenium and is found throughout the cell except peroxisomes. The glutathione-S-transferases react with organic hydroperoxides. Some 10% of the total soluble liver protein consists of these enzymes.

Vitamin E
Also referred to as α-tocopherol, vitamin E is a lipid-soluble vitamin that acts principally as a phenolic antioxidant by donating a hydrogen atom to lipid peroxyl radicals. The resulting vitamin E radical is relatively unreactive and hence free radical chains that could arise in biological membranes are terminated highly effectively by this antioxidant. A number of reducing agents, including vitamin C, can react with the vitamin E radical to regenerate vitamin E. Therefore, vitamin E usually acts as a catalytic protective agent that transfers reducing power derived from cellular metabolism to membrane free radicals. If the concentration of lipid radicals is too high, the reduction of vitamin E radicals cannot keep pace with vitamin E oxidation. Under these conditions, the vitamin E radical is irreversibly destroyed.

Vitamin C
Also known as ascorbic acid, this widely used nutritional supplement is an effective aqueous antioxidant which can terminate free radical chain reactions by donating hydrogen atoms to a variety of free radicals. Ascorbate may play a role in protecting membranes by donating hydrogen to vitamin E radicals, thus regenerating the lipid antioxidant. Ascorbate can act as a free radical initiator in the presence of "free" iron or copper, however. (Note that iron and copper would usually be bound to small molecules or macromolecular surfaces in most cells, and the term "free" is here used to distinguish such bound metal ions from enzyme-bound forms that are usually not accessible to peroxides.)

Carotenes
Beta carotene is a potent quencher of singlet oxygen which acts by an energy transfer mechanism whereby the singlet oxygen molecule is converted to the relatively harmless triplet oxygen molecule, while the carotene molecule is promoted to its triplet state. Subsequently, the excited carotene releases its energy harmlessly as heat. In this reaction carotene acts catalytically, without being consumed in the reaction. Epidemiologically, carotenes, as are vitamin E and selenium, are correlated with good health, including resistance to cancer.

Uric Acid
A striking difference between the higher apes and other animals is their relatively long lifespan. This longevity and the attribute of a large brain, which accounts for a large fraction of the total oxygen consumption, may be partially the result of retention of the antioxidant uric acid, which is a product of purine metabolism. Two evolutionary alterations have led to high tissue concentrations of urate: loss of peroxisomal urate oxidase and active reabsorption of urate from the kidneys. In model experiments, urate is an effective antioxidant and singlet oxygen scavenger.

TABLE 6-1
Subcellular Sources of Hydrogen Peroxide

Organelle	Function	H_2O_2 Production (% of total)
Mitochondria	Electron transport	15
Peroxisomes	Divalent oxidations	35
Endoplasmic reticulum	Mixed function oxidations	45
Cytosol	Xanthine oxidation	5

Reprinted from Chance et al., *Physiol. Rev.*, 59, 527, 1979. With permission.

2 CELLS, TISSUES, AND ORGANISMS

2.1 Free Radicals in Aerobic Cells

Free radicals are produced in aerobic cells during normal oxidative metabolism.[14] Elaborate defense systems wage a constant battle against active oxygen species in all aerobic organisms. A measure of the threat posed by oxygen radicals is that the antioxidant enzymes comprise a significant fraction of the total protein content of aerobic cells.

Most oxygen consumption in mammals occurs in their *mitochondria*. Although it is known that the mitochondrial respiratory chain operates as an array of enzymes that perform electron transfer reactions very similar to those occurring in free radical reactions, the structures of these enzymes ensure that no uncontrolled chemical reactions occur. In particular, no release of active oxygen molecules occurs from the site of oxygen binding — cytochrome oxidase. However, despite the tightly controlled flow of reducing equivalents through the respiratory chain, some leakage of reductants from mitochondria does occur at sites other than cytochrome oxidase, causing release of superoxide radicals into the cell. In addition, other organelles and some soluble enzymes appear to cause both univalent and divalent oxygen reduction. Of possible significance as a risk factor in fat consumption is beta oxidation of fatty acids in peroxisomes, which generates hydrogen peroxide. Estimates of hydrogen peroxide production from isolated subcellular fractions are set forth in Table 6-1, demonstrating that a variety of potential sources of active oxygen exist in most cells.

Cells of the immune system destroy invasive microorganisms (Chapter 7) by mechanisms involving active oxygen molecules. The destructive molecular species produced by immune system cells may sometimes contribute a substantial fraction of an animal's total oxidative damage burden. Of particular interest is the potential posed by these cells for amplifying the damage produced by free radicals. Increasing evidence shows that oxidative damage, particularly lipid peroxidation, generates chemotactic factors that attract leukocytes to an area of high free radical activity and that the resultant inflammatory process can lead to

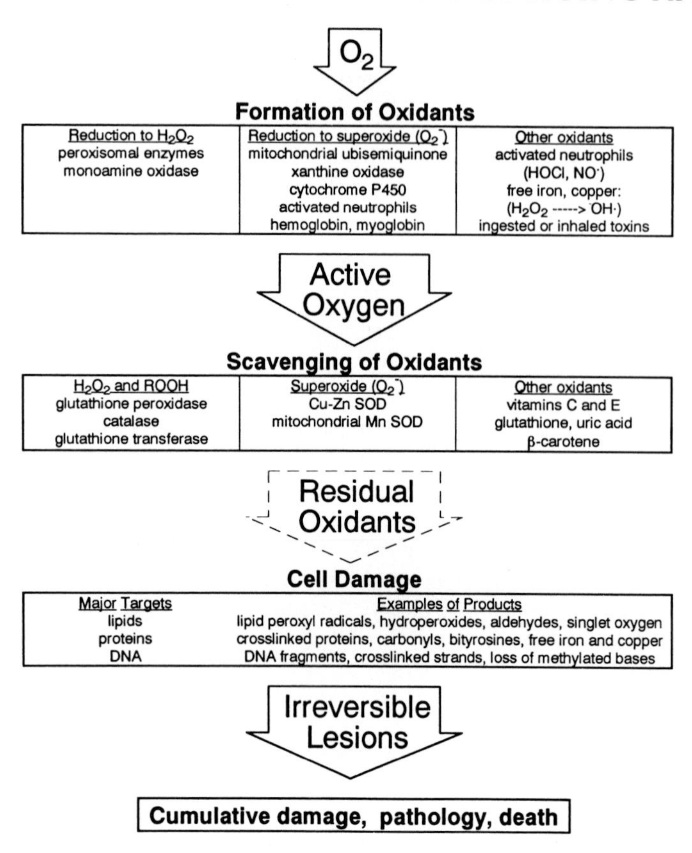

FIGURE 6-2 *Scheme of how free radicals and antioxidants may affect aging.*

increases of free radical damage. Such processes are probably important factors in pulmonary damage associated with emphysema (Chapter 8) and pollutant exposure and in atherosclerosis.

It could be argued that the protective systems, as they have evolved in the higher species, are so effective that the accumulation of damage is negligible, and that aging arises from causes other than wear and tear. The gradual progress of the changes that occur with aging implies that highly sensitive assays of damage, coupled to sophisticated data analysis, will be required to demonstrate a net accumulation of free radical damage, consistent with a progressive and irreversible decline of viability.

A possible involvement of free radicals in aging is outlined schematically in Figure 6-2. This scheme identifies the major sources of superoxide radicals and other oxidants in the cell, protective systems that remove these oxidants and their major cellular targets. The fundamental precept of the free radical theory of aging is that some reaction targets are not repaired and damage gradually accumulates, thus compromising cellular integrity. While oxidized lipids can be replaced, and most protein and DNA damage can be recognized or repaired, some types of DNA damage, such as loss of methylated bases, cannot always be repaired and will accumulate with age.

2.2 Cellular Targets of Free Radicals

Membranes

Most cell membranes contain polyunsaturated *fatty acids* that are highly susceptible to free radical oxidation. As long as vitamin E is present, the extent of oxidation is controlled because the vitamin prevents the occurrence of chain reactions. Depletion of vitamin E can lead to considerable membrane oxidation, whose consequences can include major alterations of membrane structure, a release of lipid oxidation products, including cytotoxic molecules like hydroxynonenal, and an increasing permeability to ions. Lipid radicals react avidly with oxygen, and membrane free radical chain reactions can potentially create an oxygen deficit with a concomitant loss of oxidative phosphorylation. The reduction of hydroperoxides by glutathione peroxidase consumes cellular reducing power and can also contribute to energy depletion. Thus, it is not surprising that free radical damage is often associated with impaired bioenergetics.

Proteins

Some *amino acids* are readily oxidized. Cysteine reacts with many free radicals and even with hydroperoxides to form products that react with thiols to produce disulfides. Disulfides can subsequently be reduced again to repair the lesion; indeed, the formation of protein mixed disulfides with glutathione may serve a protective function. Oxidation of methionine produces the sulfoxide, which can be repaired by specific reductases. Many amino acids may be irreversibly oxidized by free radicals that are much less reactive than the hydroxyl radical. The reactive amino acids include histidine, tryptophan, and tyrosine. Tyrosine oxidation can lead to bityrosine, a strongly fluorescent, free radical damage marker that is implicated in irreversible protein cross-linking.

Heme proteins, including *hemoglobin and myoglobin,* react with hydroperoxides to form highly oxidizing species that can produce amino acid radicals within the heme protein or oxidize nearby molecules, e.g., initiate free radical chain reactions in membranes. Activation of hydroperoxides by heme proteins can also exert its damaging effects more indirectly, by inducing a release of iron from oxidatively cleaved porphyrins. The released iron can subsequently catalyze hydroxyl radical production or other reactions at iron binding sites. Among the site-specific reactions of protein-bound iron, the oxidations of histidine, proline, and arginine side chains have been cited.[15]

Although most oxidized proteins are either repaired or degraded and replaced by new ones, these processes are not perfect and some altered proteins may persist for considerable periods, particularly in energy-deficient cells. A dramatic example of a persistent altered protein is seen in brain myelin protein. Racemized amino acids accumulate in this protein throughout life, indicating that these altered proteins are not turned over significantly (Chapters 8 and 9).

An accumulation of oxidatively modified nonfunctional or dysfunctional proteins, which are not recognized or are turned over at an inadequate rate, appears to play an important role in aging.[15] The observation that catalytic activity per unit of protein antigen decreases with age indicates that the persistence of altered proteins in cells is age dependent. This may be due, in part, to a decreased activity of proteases involved in the degradation of altered enzymes.

Nucleic Acids

Base methylation plays an important regulatory role in gene expression and in the differentiation that characterizes multicellular organisms. Dedifferentiation and a loss of methylated bases occur with age, and the resulting organismal dysfunctions may explain some of the phenomenology of the aging process.[16] Free radical oxidants hydroxylate and otherwise modify DNA bases at high rates.[17] The repair of oxidant-modified methylated bases may lead to a loss of normally methylated bases.[18] Therefore, free radical damage could play an important role in age-dependent base demethylation and altered gene expression.

The effects of altered base methylation may exert deleterious feedback effects on oxidative stress. Free radical-mediated base demethylation is likely to be a random process, resulting in a random loss of differentiated cell functions and a loss of homeostasis. It is quite conceivable that imbalances in free radical production and protective systems will arise during random dedifferentiation, leading to subpopulations of vulnerable and oxidant-resistant cells. Increased susceptibility to peroxide damage in dedifferentiated cells seems especially plausible. This follows from the critical importance of the glutathione peroxidase system in protecting cells from powerful peroxide-derived free radicals. This system requires reducing equivalents from a complex array of enzymes, coenzymes, and substrates linked to cellular metabolism; the multiplicity of its constituents renders this system especially sensitive to the adverse effects of randomly altered gene expression.

2.3 Genetic Disorders

In principle, an analysis of genetic disorders, involving defects in antioxidant defenses, could resolve whether the free radical theory of aging is valid. Unfortunately, although several examples of human mutations in one or more of the protective enzymes or antioxidant vitamins are known, including individuals with reduced levels of catalase, glutathione (GSH), or GSH peroxidase, or defective vitamin E absorption, none of these show signs of accelerated aging that can be clearly distinguished from pathology.[2]

Conversely, several genetic syndromes that exhibit some features of accelerated aging, the so-called "segmental progeroid syndromes", exhibit damage that could be consistent with increased free radical damage (see also Chapter 3). These include Down's syndrome, Ataxia Telangiectasia, Cockayne's syndrome, and, possibly, Werner's syndrome, which exhibits genetic instability yet has normal levels of SOD, GSH peroxidase, and radiation-induced repair. The molecular basis of these "geromimetic" diseases may involve accelerated rates of chromosome breakage, and the diseases frequently exhibit radiation sensitivity that suggests a high susceptibility to free radical damage.

Fanconi's anemia also shows characteristics of oxygen radical damage, but the pathology is so severe that death occurs in infancy or early childhood, and the only sign of an age-related effect is increased malignancy. Devastating genetic deficiencies of this type do not permit normal development and, frequently, disease-linked premature death precludes distinguishing symptoms of accelerated aging from pathology.

Down's syndrome is characterized by a 50% elevation of the copper–zinc SOD (Cu/Zn-SOD) above normal levels. This increased protection against superoxide radicals fails to confer anti-aging benefits; indeed, some aspects of age-related pathology, notably senile dementia, are accelerated. There are conflicting reports concerning altered brain lipofuscin accumulation, an *in vivo* index of lipid peroxidation (Chapter 5). *In vitro* lipid peroxidation appears to be accelerated. In patients with trisomy 21, red blood cells are abnormally sensitive to lysis in the presence of paraquat, a molecule that causes an increase in cellular superoxide production. Fibroblasts from trisomy 21 sufferers exhibit enhanced lipid peroxidation. The apparent increase in free radical damage under conditions of elevated Cu/Zn-SOD levels can be explained by a concurrent decrease in manganese SOD (Mn-SOD) levels in some tissues. In primates, Mn-SOD is not found exclusively in mitochondria, but also occurs in nuclear and other cellular compartments. In some tissues, Mn-SOD may account for more than 50% of the total SOD content. Hence, an increase in Mn-SOD in Down's syndrome might readily compensate for a decrease in the copper-zinc enzyme, but it is not yet clear which tissues undergo such compensations. In patients with monosomy 21 ("21q- or anti-Down's syndrome"), cells possess only 50% of normal Cu/Zn-SOD but normal Mn-SOD. While these patients suffer from developmental abnormalities and poor survival, there are no obvious signs of accelerated aging.

The difficulty of dissociating disease from aging has been a major obstacle in exploiting genetic analyses to resolve aging mechanisms. Nevertheless, the available genetic data suggest that biological concentrations of catalase, vitamin E, glutathione, and glutathione peroxidase are probably not critical determinants of rates of aging, while levels of superoxide dismutase and the overall oxygen radical defensive capacity (viz., whole-body radiation) may be important factors. Thus, the evidence can presently offer only modest support for the free radical theory of aging.

2.4 Lifespans and Metabolic Rate

To a first approximation, the specific metabolic rate of homeothermic species is inversely proportional to the maximum lifespan (Chapter 2).[8,19,20] This correlation can also be expressed as follows: *the total oxygen (or energy) consumption per unit weight is the same for all animals when they attain their maximum lifespans.* If one assumes that free radical production is proportional to the rate of aerobic metabolism, then the association between metabolic rates and lifespans can be rationalized in terms of the free radical theory of aging. More direct support for a role of free radicals in determining lifespans has been sought by attempting to correlate deviations from expected lifespans (as predicted from metabolic rates) with varying antioxidant protection. Species that live longer than predicted from their metabolic rates, notably the primates, have higher concentrations of SOD and some other antioxidants, at least in some of their tissues. The correlation of SOD with the maximum attainable lifespans of a group of animals, where the latter is effectively expressed as the lifetime energy potential per unit tissue weight (LEP), is shown in Figure 6-3.

However, while a direct correlation between superoxide dismutase and species lifespan has been found in a number of tissues, other antioxidant factors showed no such correlation. Cutler and his collaborators have found that, in addition to SOD, serum carotenoids and vitamin E appear to correlate positively with longevity, while vitamins A and C are relatively uncorrelated. Glutathione, glutathione peroxidase, and glutathione-S-transferase seem to be negatively correlated with lifespans. The same group of investigators has also reported that the susceptibility of whole brain homogenates to autoxidation is an inverse function of species lifespan.

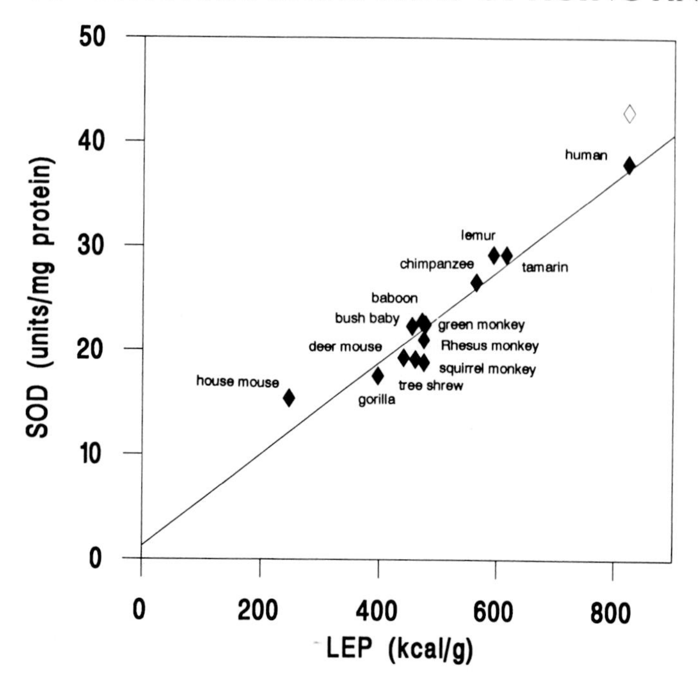

FIGURE 6-3 *Lifetime energy potential (LEP) or consumption of different animal species as a function of their liver SOD activities.* The open symbol shows human liver SOD activity, corrected for enzyme degradation during tissue processing. (Adapted from Cutler, *Gerontology,* 29, 113, 1983. With permission.)

Correlative evidence for an association between metabolic rate and free radical processes has been obtained by direct detection of tissue free radicals with electron spin resonance (ESR) spectroscopy. ESR signals have magnitudes that correlate directly with metabolic rate and inversely with lifespan. Lipofuscin particles, which are products of free radical-mediated oxidative polymerization, sometimes contain ESR signals characteristic of metal-ion complexes. Some lipofuscin particles isolated from the human brain contain both copper and iron and exhibit free radical signals that increase progressively with age. Thus, cumulative free radical damage may occur with age, but further basic studies are required to establish the relationship of ESR signals to metabolically generated free radicals.

Studies of free radical production by isolated mitochondria suggest that altered free radical production may be a more significant factor in aging than impaired antioxidant protection. Sohal and his collaborators have shown that hydrogen peroxide production by mitochondria correlates with age and species lifespan and that antioxidant enzymes do not exhibit consistent correlations. They suggest that altered free radical production is a key determinant of maximum attainable lifespan and that variations in antioxidant enzymes exert relatively little influence.[21-23]

2.5 Lifespan Modification

Ideally, one would like to test the free radical theory definitively by specifically manipulating the rate of free radical damage and demonstrating a corresponding effect on maximum lifespans. In principle, free radical damage is a function of both the rate of radical formation and the interception of these radicals by protective factors, either of which may be amenable to experimental control.

Effects of Ionizing Radiation and Chemical Mutagens

A straightforward example of free radical-mediated damage is that of radiation injury (see Chapter 4). However, there is no clear evidence that a slight increase in free radicals, imposed by low level radiation, shortens lifespan. Rather, chronic radiation exposure causes a variety of cancers and other diseases. Similarly, chronic exposure to mutagens (nitrogen mustards) does not appear to shorten lifespans, once the effects of increased disease incidence are taken into account. The implications of these studies for the free radical theory of aging are difficult to interpret, in part because of a lack of a good aging parameter, which would facilitate discerning effects of disease from those of aging.

Effects of Cigarette Smoking

Although it has not been established whether cigarette smoking affects maximum lifespan, its positive correlation with major life-shortening diseases, including cardiovascular disease and cancer, implies that cigarette smoking reduces mean life expectancy. Skin wrinkling, the most familiar marker of aging in humans, is accelerated in white smokers by cigarette consumption and, independently of smoking, by sunlight exposure.[24,25] The effects of smoking and sunlight exposure on skin wrinkling are additive. Rodents exposed to cigarette smoke exhibit increased levels of lipid peroxidation products, prompting the suggestion that smoking accelerates the aging process.[26] In humans, the body mass index (weight divided by the square of the height) and the distribution of body fat (waist circumference divided by hip circumference) decrease with smoking and with increasing age,[27,28] consistent with the view that smoking accelerates some aspects of the aging process.

Caloric Restriction

The most dramatic lifespan extensions that have been achieved with dietary manipulations are increases in maximum lifespans of up to 50% that have been achieved by caloric restriction. The efficacy of caloric restriction in prolonging lifespans (Figure 6-4) might suggest, at first sight, that there is a direct relationship between caloric intake and metabolic rate, and, hence, that such restriction lowers the rate of free radical production and damage.[29] However, caloric restriction leads to lowered weight, and this must be taken into account. In studies with rats, when the effects of food restriction were analyzed in terms of both decreased body weights and metabolic rates, it was found that the specific metabolic rates of the smaller food-restricted animals were in fact greater than those of *ad libitum*–fed animals. The food-restricted animals lived longer than the controls, contrary to the expectation that specific metabolic rate would be inversely related to the aging rate. Thus, at first sight, the data on caloric restriction seem to contradict the free radical theory. Further progress in achieving a meaningful interpretation of the effects of caloric restriction may hinge upon being able to relate metabolic rates to free radical fluxes.

Effects of Antioxidants

Studies with dietary antioxidants have shown that while the mean lifespan is increased, there is no significant increase in the maximum lifespan, as would be expected if the aging rate had been altered.[2] Side effects of antioxidants must also be

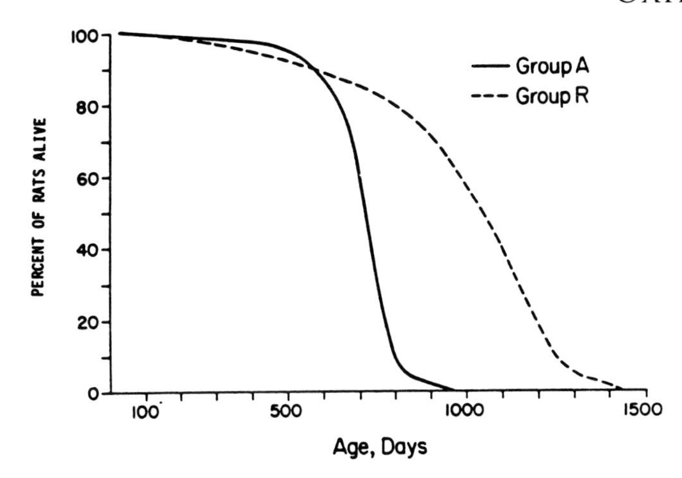

FIGURE 6-4 *Survival curves for* ad libitum*-fed* (group A) *and food-restricted* (group B) *SPF Fischer 344 Rats.* (Reprinted from Yu et al., *J. Gerontol.*, 37, 130, 1982. With permission.)

considered, e.g., possible caloric restriction effects associated with foul-smelling food additives like 2-mercaptoethylamine. Another potential problem with dietary antioxidants in whole animals is a possible lack of specificity. For example, antioxidants are claimed to enhance immune function, and some quinone antioxidants like 2(3)-*tert*-butyl-4-hydroxy-anisole (BHA) induce the microsomal DT diaphorase and UDP-glucuronyl transferase systems and hence may accelerate the removal of potentially mutagenic chemicals that could affect aging by nonradical mechanisms. Two other synthetic antioxidants, BHT and ethoxyquin, have been shown to induce hepatic enzymes. Thus, increased lifespan that correlates with antioxidant feeding may be due to effects other than free radical quenching.

The accumulation of lipofuscin is increased with vitamin E deficiency and decreased with dietary antioxidants. However, there is as yet no convincing evidence that lipofuscin accumulation correlates with age-related cell loss. On the contrary, among the brain stem nuclei, the inferior olive has high lipofuscin levels and shows no age-related cell loss. Moreover, there are examples of cells that accumulate lipofuscin but do not exhibit a decline in function. This was illustrated by cell sorting experiments with cultured fibroblasts, which separated populations of cells on the basis of fluorescence intensity, and which showed that fluorescent cells suffered no loss of proliferative potential.

Thus, the evidence for antioxidant-mediated lifespan extension in vertebrates does not argue persuasively for a causal role of free radicals in aging. This fact is generally acknowledged by advocates of the free radical theory, and they offer two general explanations that could sustain the basic concept:

1. The principal site of damage (e.g., mitochondrial DNA) is not protected by exogenous antioxidants
2. The endogenous defense system is regulated to maintain a fixed overall level of protection so that simple attempts to increase protection by some dietary manipulation of one or a few antioxidants fail because of compensatory decreases in endogenous defenses ("compensatory downregulation").

It has been suggested that mitochondria could play a role as pacemakers. This hypothesis could explain the failure of some antioxidants to delay aging. Excessive concentrations of hydrophobic phenolic antioxidants could react with mitochondrial electron transport components to impair mitochondrial metabolism, which may explain the futility of extending lifespans with antioxidants of this type.

Another explanation for the failure of antioxidants to confer substantial lifespan extension — "compensatory downregulation of endogenous antioxidant defenses" — argues that when increased concentrations of exogenous antioxidants reduce the flux of radicals, the cell decreases its production of other defensive enzymes. However, the fact that lipofuscin accumulation can be reduced by feeding antioxidants in the absence of lifespan extension, and that tumor incidence can be reduced by feeding exogenous antioxidants, suggests that antioxidants can effectively reduce the rate of free radical damage in some tissues without undue compensatory downregulation.

In Vitro *"Aging Models"*

The doubling potential of cultured mammalian cells, like fibroblasts, has been perceived as a meaningful *in vitro* aging model because it correlates with species lifespan and with the age of the cell donor and is reduced by life-shortening genetic disorders in humans (Chapter 5). To explore the possibility that the finite doubling potential of the cells is due to free radical damage, a number of investigators have varied the oxygen tension during cell growth and studied how this affected the doubling limit. No increase in the maximum number of doublings could be achieved by lowering the oxygen concentration below ambient levels, although increased oxygen concentrations curtailed both growth rates and the doubling number.[30]

Cultured human fibroblasts can be transformed by treatment with SV40 virus to remove the limited growth potential, i.e., the transformed cells become "immortal". The transformed cells are aneuploid, genetically unstable, and, often, DNA repair-deficient. When these transformed cells are exposed to higher-than-ambient oxygen concentrations, they too are slowed in their growth rate. This oxygen-induced decline in growth rate may be due to an inhibition of DNA replication, suggesting that the effect of high oxygen concentrations on growth rates is not necessarily related to free radical formation. In normal fibroblasts, oxygen also modulates growth and can affect the process of differentiation. A stimulating effect of low oxygen tension on fibroblast growth rate may serve a regulatory role in wound healing. These multiple effects of oxygen in fibroblasts underscore the ambiguities associated with interpreting experiments with the replicative limit model in terms of free radical damage alone.

Adrenocortical cells in long-term culture, grown on certain batches of 10% fetal bovine serum, are essentially completely deficient in both vitamin E and selenium when not supplemented with these nutrients.[31] This has set the stage for analyzing effects of these antioxidants in determining the *in vitro* "lifespans" of these cells. Supplementation with either vitamin E or selenium protects the cells from the toxic effects of added cumene hydroperoxide, demonstrating the efficacy of these nutrients in ameliorating the effects of oxidative damage. However, the long-term growth of the cells is nearly identical in the absence or presence of these antioxidants. These data suggest that either (1) oxidative damage is not a significant determinant of the doubling potential in adrenocortical cells, or (2) rapid cell growth under optimal conditions enables cells to outpace the damaging effects of free radical chemistry such that damage only becomes apparent with cumulative radical damage when metabolically active cells have a limited capacity to divide.

The validity of the replicative limit model as an aging model is unclear;[32] hence, negative evidence obtained with it should not necessarily be taken as an indictment of the free radical theory of aging. The central difficulty with the proliferative limit model lies in its questionable significance for *in vivo* aging. There is no solid evidence for a proliferative limit occurring during normal *in vivo* aging, and it seems likely that functional losses which precede the *in vitro* proliferative limit are responsible for aging.

Exercise

A promising experimental strategy for manipulating metabolic rates is exercise. The analysis of exercise effects requires that both the intensity of exercise and prior training be taken into account. Increased radical damage appears to occur with intensive exercise in untrained animals, and positive adaptations, conferring increased resistance to oxidative damage, are observed in endurance-trained animals.

Studies with intense physical activity in untrained rats have suggested an increase in free radical damage, e.g., increased lipid peroxidation products in liver and skeletal muscle homogenates and in mitochondrial fractions. Experiments with rats have shown that plasma glutathione disulfide increases significantly with intensive exercise and that vitamin E deficiency is associated with a marked decrease in endurance capacity, suggesting that lipid peroxidation may play a role in muscle fatigue upon exhaustion.

Rodents that have been exercise trained throughout life live longer than their sedentary counterparts, consistent with a positive effect of exercise training on mean lifespans. Although the mechanism of the lifespan extension is unresolved, it may involve caloric restriction, since exercised animals accumulate less body fat than sedentary animals. Endurance training appears to decrease the susceptibility of skeletal muscles to lipid peroxidation, consistent with an adaptive response of protective systems. Endurance training raises the activity of the enzyme glutathione reductase in both skeletal and cardiac muscles. This enzyme serves an antioxidant function by maintaining glutathione in its reduced state, thus enabling glutathione peroxidase to perform its function of hydroperoxide reduction. Exercise training markedly retards the appearance of high plasma glutathione disulfide concentrations during intensive exercise, supporting the notion that exercise training increases the capacity of an animal to maintain highly reduced glutathione levels for long periods of intensive activity.

2.6 Susceptible Tissues, Cells, and Organelles

Because the cell populations in different tissues are likely to vary in their redundancy, regenerative capacity, free radical production rates, antioxidant defenses, and DNA repair rates, some tissues may be particularly sensitive to free radical damage, and these could determine aging rates in the organism.

Neuronal tissue

Metabolically active, nondividing cells like the neurons could well be the most susceptible targets of oxidative damage. Indeed, there is a loss of neurons with age, but such loss appears to be species specific and to be a function of the tissue location. The surprising durability of neuronal tissue despite its large oxygen utilization appears, at first sight, to be incompatible with free radical involvement in age-related functional declines. However, a large portion of the oxygen utilization is in axons and terminals that are physically removed from sensitive and nonrenewable targets like the nuclear genome. Moreover, neuronal mitochondria do not appear to produce superoxide radicals at a site that is a major source of these radicals in other cells. Finally, considerable cell redundancy may prevent functional deficits from becoming clinically significant. As a rule, more than 40% cell loss is required in a neuronal system before functional loss in the central nervous system is apparent.

Mitochondria

Mitochondrial dysfunction and depletion occurs in aging postmitotic cells and may prove to be a major factor in aging. Miquel and his colleagues have postulated that mitochondria are the "Achilles' heel" of postmitotic cells (Chapter 5). Mitochondrial DNA (1) is not protected by a histone coat, (2) is in close proximity to sites of oxygen radical production and lipid peroxidation, and (3) relies on a possibly inadequate DNA-repair system.[33] However, because mitochondria are dependent on the nuclear genome for many of their proteins and are sensitive to the cellular environment (ions, substrates) and even hormonal stimulation (viz., thyroid hormone effects on mitochondrial replication), it is difficult to unambiguously dissociate intrinsic mitochondrial damage from extrinsic factors.

The high susceptibility of mitochondrial DNA (mtDNA) to oxidant damage is indicated by high levels of 8-hydroxydeoxyguanosine[34,35] and by an increasing number of mtDNA deletions with age.[36,37] Altered protein synthesis due to mtDNA damage may play a role in age-dependent mitochondrial dysfunction.[38,39] Among the four electron transport complexes of the respiratory chain, the activity of complex I declines most rapidly with age.[38,40] Apart from effects on mitochondrial bioenergetics, more indirect damage scenarios, possibly involving an incorporation of mtDNA fragments into the nuclear genome, may provide a link between mitochondrial free radicals and aging.[41]

2.7 Systemic Effects

One of the most obvious characteristics of aging is the gradual decline of physiologic integrity, including the progressive impairment of neurological, immunological, humoral, and metabolic function. Evidence for free radical involvement in all of these debilitating processes is growing. However, organelle and cellular redundancy and renewal may be able to compensate for the damage sufficiently to sustain function. Thus, the epithelia lining the digestive tract and skin, the blood cells, and the liver may be extensively damaged and yet regenerate on a regular basis. In these renewing cell populations, the principal danger of damage appears to be neoplastic transformation.

Because aging is characterized by an increased incidence of infectious diseases, one might expect immunosenescence to play a crucial role in physiological aging (Chapter 7). Declines in *immune function* begin relatively early and appear to be heavily dependent on thymic involution and selective T-cell aging. Immunological declines can be, at least partially, reversed by thymic grafts or thymic hormone therapy, but complete restoration requires an additional young bone marrow graft.

Evidence for free radical involvement in immunosenescence is based on selective vulnerability of the immune system to radiation and other free radical-generating agents. T cells, which age more rapidly than B cells, are reported to be more vulnerable to oxygen radicals and to accumulate more lipofuscin; treatment of aging mice with 2-mercaptoethanol delays the accumulation of T-cell lipofuscin and the decline of immune function with age, and increases the mean lifespan.

The thymus may also be selectively vulnerable to free radical damage. The first age-related loss in size (thymic involution) can be ascribed to the loss of the most radiosensitive (cortical) lymphocytes. The medullary epithelial cells that secrete thymic hormones are heterogenous and the early loss is again that of the most radiosensitive cells, which are active metabolically, require vitamin C for their secretory activity, and appear to accumulate intrinsically autofluorescent substances (age pigments). Although there are probably developmentally programmed cellular and hormonal controls of thymic involution, significant oxidative damage seems to be involved as well.

Many antioxidants are immune stimulants and will enhance immune function both *in vivo* and *in vitro* with both young and old cells. Part of this effect appears to be due to the maintenance of reduced sulfhydryl groups. Although antioxidant treatment can boost immune function at any age, it does not markedly reduce the rate of basic aging of the immune system.

2.8 Disease

Balin has emphasized the importance of separating the effects of specific life-extending treatments that may reduce the incidence of disease, and hence most likely affect only mean lifespans from those that affect basic aging rates.[42] As reviewed in the preceding sections, free radicals do not seem to determine maximum lifespan. However, mean lifespan is well correlated with the incidence of disease, and many diseases that shorten mean lifespans are aggravated by free radical processes (see, for example, Pryor[43]). Some of the support for free radical involvement in disease is outlined in the following sections.

Cancer

The process whereby a normal cell becomes transformed, i.e., whereby it assumes the uncontrolled growth characteristics of a cancer cell, seems to involve an alteration of the DNA, as shown by the fact that many chemicals that cause cancer in animals also cause mutations in microorganisms in the well-known Ames Salmonella revertant test. This test is often used as a predictor of risk for chemicals that are suspected to have carcinogenic potential.

The relationship between aging and cancer is complex; for example, neuroblastomas, many leukemias, and hormone-dependent tumors have radically different age patterns than most sarcomas and many carcinomas that rise with the second and fourth power of age, respectively. The incidence of most tumors, in a large variety of species, rises dramatically with age. This increase in cancer incidence with age may be attributed to several factors:[44]

1. Long-term carcinogen exposure increases the risk of initiation, e.g., lung cancer incidence reflects duration of smoking rather than chronological age.
2. The prolonged period required for one malignant cell to multiply and develop into a detectable tumor.
3. Aging itself increases the risk because of
   - a possible reduction in natural killer or other immune surveillance function
   - an increase in the activation of procarcinogens
   - other factors including demethylation-induced epigenetic instability and free radical–induced DNA damage

The pathogenesis of neoplasia has been thought to be a two-stage process:

1. Initiation, produced by mutagenesis, abnormal differentiation, or viral infection
2. Promotion, characterized by a prolonged latency allowing for the multiplication and evolution of initiated cells into a tumor (Chapter 3)

The process involves many variables, including tissue type, hormonal influence, proliferative rates, DNA repair capacity, environmental carcinogen exposure, viral infection, immune surveillance, and genotype (see also Chapter 3). The proposal that cancer is the result of a two-stage process has been challenged recently by Ames and others. Ames has suggested that cell division greatly increases the potential for nonrepairable DNA damage and cancer.[45] He has postulated that many carcinogenic agents exert their effects indirectly by killing cells, which stimulates the growth of new cells and concomitant DNA damage.

Work has progressed rapidly on the relationship between carcinogens and mutagens (i.e., agents that cause cancer and mutations, respectively). Point mutations are implicated in the activation of *ras* oncogenes in human tumors, for example. However, for other oncogenes — for example, in the *myc* family (e.g., Burkitt's lymphoma) — there is a chromosomal translocation of the oncogene to an enhancer-amplified site, presumably activating oncogene expression. Epidemiological evidence suggests that the point mutations may stem from a variety of environmental and intrinsic mutagens, including free radicals.

A number of potent chemical carcinogens, like benzo[a]pyrene, form free radical derivatives in simple chemical systems and in isolated membranes of the endoplasmic reticulum. Some molecules are also converted to quinones by the cytochrome P450 system, and these quinones can be substantially more toxic to cells (e.g., cultured fibroblasts) than their precursors. The mechanism of the quinone toxicity may be a catalytic reaction whereby the quinones undergo redox cycling with cellular electron donors and with molecular oxygen to produce superoxide radicals. For example, the quinones that result from the microsomal oxidation of benzo[a]pyrene are readily reduced by electron donors like glutathione, NADH, and NADPH and by the DT diaphorase enzyme. As soon as the reduced quinol is formed, it reacts with molecular oxygen in one-electron transfer steps to form superoxide radicals. In the process of redox cycling, some semiquinone radical species can also react directly with cellular macromolecules, including DNA.

Gross genetic rearrangements, transpositions, and amplification appear to play a major role in the promotion process of carcinogenesis. It is quite probable that free radical damage is also involved in these processes. Since oxygen-derived radicals are capable of inducing mutations, and since they appear to have the capacity to cause chromosomal strand breakage, they should be able to participate in the gene rearrangements necessary to link a protooncogene with an enhancer element.

Current evidence that oxygen radicals are involved in carcinogenesis includes the following:

1. Several genetic disorders with a high incidence of both chromosomal rearrangement and neoplasia, e.g., Fanconi's and Bloom's syndromes, appear to involve abnormally high oxygen radical-mediated damage. Unlike other syndromes predisposed to malignancy, Fanconi's anemia and Bloom's syndrome appear to have normal DNA-repair capacity, but increased sensitivity to oxygen radical-generated damage.
2. Epidemiological evidence suggests that dietary or endogenous factors rather than industrial products or environmental agents are responsible for the majority of human cancers.[46,47] An involvement of lipid-peroxidation-linked oxygen radicals is supported by the implication of fat as a risk factor for many types of tumors and the inverse correlation of risk with dietary antioxidant intake.[48] Other epidemiological data show that high dietary intake of quinones or quinone precursors, which are capable of promoting free radical reactions, is attended by increased cancer incidence. Ionizing radiation also induces cancer in a dose-dependent manner. When cigarette smoke is trapped on glass wool, a large ESR signal can be observed and it seems quite plausible that this free radical component of cigarette smoke is a causative factor in its well-established link with cancer.
3. Oxygen-derived radicals are implicated in tumor promotion. Phorbol esters, the classical tumor promoters, produce extensive chromosomal damage and stimulate polymorphonuclear leukocytes to produce an abundance of superoxide and peroxides. Antipromoters such as retinoids and protease inhibitors inhibit oxygen radical production. DNA strand breakage correlates directly with promotional efficiency, and antipromotional ability correlates with inhibition of strand breaks. Promoter-induced chromosomal damage is inhibited by superoxide dismutase, but, in some systems (lymphocytes), it is also inhibited by indomethacin and other inhibitors of prostaglandin synthetase that can produce unstable lipid peroxides and attendant oxygen radicals.

Phorbol esters have several other effects on cellular metabolism; in particular, they rapidly activate the protein kinase C pathway, a principal control point in the regulation of proliferation, and rapidly and reversibly induce many of the properties of the transformed phenotype. Hence, phorbol esters have a rapid and reversible short-term effect on cells that stimulate cell proliferation. They also generate free radical-mediated chromosomal damage, which may allow for an emergence of clones with activated oncogenes and a permanently malignant phenotype.

Phorbol esters are not the only promoters capable of producing oxygen radicals. In a recent review, Ames has assembled evidence that most promoters (e.g., phorbol esters, fat, TCDD, lead, cadmium, wounded tissue, asbestos, anthralin, peroxides, mezerein, telociden B, phenobarbitol, radiation, nitroso and nitro compounds, hydrazines, and polycyclic hydrocarbons) are also capable of generating oxygen radicals.[48] While all of these agents have a multiplicity of other effects and there is, at present, no quantitative correlation between the efficiency of promotion and oxygen radical production, the qualitative correlation is striking. The evidence for a causal relation between oxygen radical production and promotion is also supported by the efficacy of antioxidants as anticarcinogens.[48] In particular, vitamin E, selenium, glutathione, ascorbic acid, and beta carotene have proven effective in reducing tumor incidence.

Parkinson's Disease

Parkinson's disease is a strongly age-related pathology marked by a selective and progressive loss of pigmented catecholaminergic, particularly dopaminergic, neurons of the *substantia nigra* (see also Chapter 8). This loss, which may well be due to a selective vulnerability to oxidative damage, averages about 80% in autopsied patients, compared to only about a 40% loss seen in the oldest normal individuals. The current epidemiological data suggest that Parkinson's disease is caused by some environmental agent(s), and reasonable candidates would be those producing excessive fluxes of oxygen radicals or other active oxygen species. Several investigators have speculated about possible oxidative mechanisms in the development of Parkinson's disease, including redox cycling of quinones derived from catecholamines and a role for hydrogen peroxide derived from monoamine oxidase.

Glutathione concentrations are low in human *substantia nigra* and almost absent from nigral tissue of patients who have succumbed to Parkinson's disease. Catalase and glutathione peroxidase have also been found to be abnormally low in patients with this disease. It is conceivable, however, that these reductions are the result of neuronal loss rather than decreases in cell-specific enzyme activities.

It has been claimed that brains of patients with Parkinson's disease contain abnormally high concentrations of iron. Since iron is normally sequestered in highly stable storage and binding proteins, it is conceivable that "free" iron may be present in these brains, which could catalyze hydroxyl radical formation in a Fenton-type of reaction.

An environmental neurotoxin like *N*-methyl-4-phenyl-1,2,3,6-tetrahydropyridine (MPTP), which selectively kills dopaminergic neurons, could displace dopamine from its storage vesicles (which are at acidic pH and therefore provide some protection against autoxidation) and thus drive an increased catecholamine autoxidation and an increased peroxide release from monoamine oxidase.

Thus, the evidence for free radical involvement in Parkinson's disease is suggestive, but further experiments are necessary before an unequivocal resolution of the mechanisms involved can be offered.

Autoimmune Disease

Because tolerance to self appears to require an active thymic role in the production of T-suppressor cells as well as in the deletion of self-reactive clones, one might expect thymic involution and age-related immune dysregulation to result in an age-related increase in autoimmune disease. Further, the emergence of altered self-antigens through persistent viral infection, post-translational modifications, somatic mutation, or even postmaturational development of newly expressed genes could also bring about an increased number of autoimmune reactions with age (Chapter 7).

Autoimmune phenomena like autoantibodies, glomerulonephritis, periarteritis, and probably some classes of senile amyloid increase with age. A major role of autoimmune pathology is played in the aging of some rodents, but not all strains and species seem to be affected. Hence, while some workers have hypothesized that autoimmunity is the major aging process and likened senescence to a chronic graft-versus-host reaction, the evidence is not consistent. In humans, most known or suspected autoimmune diseases, including rheumatoid arthritis, have a peak incidence in middle age, and it is difficult to assess the significance of autoimmune phenomena in human aging. Among rodents, in strains that are clearly autoimmune susceptible, such as NZB mice, antioxidant feeding has produced a delay in disease onset and lifespan extension. If autoimmunity is a major aging process and not a secondary pathology, then antioxidant feeding may be said to delay the accelerated aging that has been reported in NZB mice.

Several other diseases seem to involve free radicals in their etiology, including atherosclerosis, emphysema, arthritis, cirrhosis, and diabetes (Chapters 15 to 18, 20, and 21).[43]

3 INTERVENTIVE STRATEGIES

3.1 Genetic Approaches towards Altering Free Radical Damage

Genetic manipulation is one promising approach towards understanding how an altered expression of endogenous oxidative defenses affects aging. However, studies of antioxidant enzymes have been plagued by problems associated with species and tissue variability, compounded by occasional conflicting reports from different laboratories. For example, in the rat brain, mitochondrial SOD is unchanged or increased with age, while CuSOD is unchanged or decreased. Rat brain cytosolic glutathione peroxidase is unchanged or decreased, mitochondrial glutathione peroxidase and reductase are increased, and catalase is unchanged with age. Because of the conflicting data and lack of correspondingly reliable information about metabolic rates and free radical production, the interpretation of data on protective enzymes remains obscure and further studies are needed to assess the significance of observed enzyme level changes.

3.2 Lifespan Extension Strategies Based on Free Radical Concepts

Even if free radicals are implicated only in life-shortening diseases and do not make a contribution to the basic aging process, the general problems associated with diseases of the aged are so great that intervention in the free radical-mediated damage that is implicated in these diseases is highly worthwhile.[49] This section addresses some of the interventive measures that might be considered to reduce the incidence of free radical potentiated diseases. Refer to Table 6-2.

There is a growing awareness of the hazards of exogenous free radicals. These include cigarettes, oxidized fat, moldy nuts, pickled vegetables, burned meats, combustion products, and asbestos. Avoidance of these agents is likely to reduce the risk of the free radical-mediated diseases.

It is also feasible to reduce the impact of exogenous or metabolically generated radicals on cellular targets by ingesting antioxidants. Vitamin E, beta carotene, and BHT can protect against active oxygen species in the fatty tissues and in cell membranes. They are particularly effective in protecting against the harmful effects of certain drugs used in chemotherapy, which kill cancer cells by free radical mechanisms.

TABLE 6-2

Potential Strategies, Based on Free-Radical Causality Assumption, for Intervening in the Aging Process

Decrease free-radical stress	Increase protection
Stress	Antioxidant factor
Cigarettes	Vitamin C
High fat diet	Vitamin E
Air pollutants (ozone, nitrogen oxides, asbestos)	Selenium
	Beta carotene
Water pollutants (halogenated solvents, chromate)	Transition metal control
	Zinc
	Nutrients
Ultraviolet light (including sunlight)	Acetyl carnitine
Immune Effects	
Reduce infections	
Speed wound healing	
Control metabolism	
Caloric restriction	

Aqueous chemical reactions may be a more important free radical damage route than fatty tissue reactions. Unfortunately, there are no facile means to enhance the protection afforded by the water-soluble enzymes catalase and SOD. However, dietary selenium can be used to enhance glutathione peroxidase activity, although toxic overdoses are a concern. One can also consider vitamin C supplements, but this vitamin has the potential of acting as a prooxidant rather than an antioxidant in the presence of free-transition metal ions. Uric acid is a water-soluble antioxidant factor that may confer benefits against singlet oxygen and may protect against the ravages of superoxide produced by xanthine oxidase. There may also be value in cysteine supplementation to increase the biosynthesis of glutathione.

Since the immune system appears to do much of its work by exploiting the toxic properties of free radicals, another line of protection is to minimize activation of such free radical reactions. This can be achieved by avoiding infections, treating wounds promptly and effectively, using drugs to speed postoperative recovery, and using antiinflammatory agents when appropriate.

Because of the possibility that transition metals may contribute substantially to free radical damage, it would be worthwhile to pursue strategies for minimizing their reactions. One suggestion that has been made is to use zinc supplementation as a strategy for displacing iron from binding sites that might otherwise promote free radical reactions.[49]

4 SUMMARY

The validity of the free radical theory of aging remains to be resolved. A major difficulty with the theory is the failure of antioxidant supplementation to significantly extend the maximum lifespans of mammals. This failure is particularly troubling because the administration of antioxidants has clear-cut effects in reducing the extent of lipofuscin accumulation, an indication that the antioxidants do protect some cellular targets susceptible to peroxidation. Antioxidant administration can also confer anticarcinogenic and other health benefits, which affect mean, but apparently not maximum, lifespans. On the other hand, recent evidence on caloric restriction, the only known strategy for dramatic lifespan extension, indicates that at least some subcellular fractions prepared from calorically restricted rats generate fewer free radicals and are endowed with enhanced antioxidant protection relative to fractions from control animals. Recent work on mitochondria is consistent with radical-mediated damage of mitochondrial DNA and an age-dependent dysfunction that correlates with increased free radical production and decreased antioxidant capacity. These mitochondrial data generally seem consistent with the free radical theory of aging.

ACKNOWLEDGMENTS

This work was supported by funds provided by the Cigarette and Tobacco Surtax Fund of the State of California through the Tobacco-Related Disease Research Program of the University of California (Grant RT 88).

REFERENCES

1. Harman, D., Aging: a theory based on free radical and radiation chemistry, *J. Gerontol.*, 11, 298, 1956.
2. Mehlhorn, R. J. and Cole, G., The free radicals theory of aging: a critical review, *Adv. Free Radical Biol. Med.*, 1, 165, 1985.
3. Church, D. F. and Pryor, W. A., Free radical chemistry of cigarette smoke and its toxicological implications, *Env. Health Perspect.*, 64, 111, 1985.
4. Palmer, R. M. J., Ferrige, A. G., and Moncada, S., Nitric oxide release accounts for the biological activity of endothelium-derived relaxing factor, *Nature*, 327, 524, 1987.
5. Estefan, R. M., Gause, E. M., and Rowlands, J. R., Electron spin resonance and optical studies of the interaction between NO_2 and unsaturated lipid components, *Environ. Health Res.*, 3, 62, 1970.
6. Nakayama, T., Kodama, M., and Nagata, C., Generation of hydrogen peroxide and superoxide anion radical from cigarette smoke, *Gann*, 75, 95, 1984.
7. Cosgrove, J. P., Borish, E. T., Church, D. F., and Pryor, W. A., The metal-mediated formation of hydroxyl radical by aqueous extracts of cigarette tar, *Biochem. Biophys. Res. Commun.*, 132, 390, 1985.
8. Puppo, A. and Halliwell, B., Formation of hydroxyl radicals from hydrogen peroxide in the presence of iron. Is haemoglobin a biological Fenton reagent?, *Biochem. J.*, 249, 185, 1988.
9. Winterbourn, C. C., Free radical production and oxidative reactions of hemoglobin, *Environ. Health Perspect.*, 64, 321, 1985.
10. Davies, M. J., Detection of peroxyl and alkoxyl radicals produced by reaction of hydroperoxides with heme-proteins by electron spin resonance spectroscopy, *Biochem. Biophys. Acta*, 964, 28, 1988.
11. Qian, M. W. and Eaton, J. W., Tobacco-borne siderophoric activity, *Arch. Biochem. Biophys.*, 275, 280, 1989.
12. Antholine, W. E., Kalyanaraman, B., and Petering, D. H., ESR of copper and iron complexes with antitumor and cytotoxic properties, *Environ. Health Perspect.*, 64, 19, 1985.
13. Pryor, W. A., Dooley, M. M., and Church, D. F., The inactivation of alpha-1-proteinase inhibitor by gas-phase cigarette smoke: protection by antioxidants and reducing species, *Chem. Biol. Interact.*, 57, 271, 1986.
14. Chance, B., Sies, H., and Boveris, A., Hydrogen peroxide metabolism in mammalian organs, *Physiol. Rev.*, 59, 527, 1979.
15. Stadtman, E. R., Protein modification in aging, *J. Gerontol.*, 43, B112, 1988.
16. Holliday, R., The inheritance of epigenetic defects, *Science*, 238, 163, 1987.
17. Fraga, C. G., Shigenaga, M. K., Park, J.-W., Degan, P., and Ames, B. N., Oxidative damage to DNA during aging: 8-hydroxy-2′-deoxyguanosine in rat organ DNA and urine, *Proc. Natl. Acad. Sci. U.S.A.*, 87, 4533, 1990.
18. Cannon, S. V., Cummings, A., and Teebor, G. W., 5-Hydroxymethylcytosine DNA glycosylase activity in mammalian tissue, *Biochem. Biophys. Res. Commun.*, 151, 1173, 1988.
19. Tolmasoff, J. M., Ono, T., and Cutler, R. G., Superoxide dismutase: correlation with life-span and specific metabolic rate, *Proc. Natl. Acad. Sci. U.S.A.*, 77, 2777, 1980.
20. Cutler, R. G., Superoxide dismutase, longevity and specific metabolic rate: a reply, *Gerontology*, 29, 113, 1983.
21. Sohal, R. S., Svensson, I., and Brunk, U. T., Hydrogen peroxide production by liver mitochondria in different species, *Mech. Ageing Dev.*, 53, 209, 1990.
22. Sohal, R. S., Arnold, L. A., and Sohal, B. H., Age-related changes in antioxidant enzymes and prooxidant generation in tissues of the rat with special reference to parameters in two insect species, *Free Rad. Biol. Med.*, 9, 495, 1990.
23. Sohal, R. S. and Brunk, U. T., Mitochondrial production of prooxidants and cellular senescence, *Mutat. Res.*, 275, 295, 1992.
24. Kadunce, D. P., Burr, R., Gress, R., Kanner, R., Lyon, J. L., and Zone, J. J., Cigarette smoking: risk factor for premature facial wrinkling, *Ann. Intern. Med.*, 114, 840, 1991.
25. Grady, D. and Ernster, V., Does cigarette smoking make you ugly and old?, *Am. J. Epidemiol.*, 135, 839, 1992.
26. Boross, M., Penzes, L., Izsak, J., Rajczy, K., and Beregi, E., Effect of smoking on different biological parameters in aging mice, *Z. Gerontol.*, 24, 76, 1991.
27. Barrett-Connor, E. and Khaw, K.-T., Cigarette smoking and increased central adiposity, *Ann. Intern. Med.*, 111, 783, 1989.
28. Troisi, R. J., Heinhold, J. W., Vokonas, P. S., and Weiss, S. T., Cigarette smoking, dietary intake, and physical activity: effects on body fat distribution — the Normative Aging Study, *Am. J. Clin. Nutr.*, 53, 1104, 1991.
29. Yu, B. P., Masoro, E. J., Murata, I., Bertrand, H. A., and Lynd, F. T., Life span study of SPF Fischer 344 male rats fed *ad libitum* or restricted diets: longevity, growth, lean body mass and disease, *J. Gerontol.*, 37, 130, 1982.
30. Balin, A. K., Goodman, D. B. P., Rasmussen, H., and Cristofalo, V. J., The effect of oxygen and vitamin E on the lifespan of human diploid cells *in vitro*, *J. Cell Biol.*, 74, 58, 1977.
31. Hornsby, P. J., Aldern, K. A., and Harris, S. E., Adrenocortical cultures as model systems for investigating cellular aging, in *Free Radicals in Molecular Biology, Aging, and Disease*, Armstrong, D., Sohal, R. S., Cutler, R. G., and Slater, T. S., Eds., Raven Press, New York, 1984, 203–222.
32. Bell, E., Cells may have escaped mortality, the price paid for fitness by multicellular organisms, in *Comparative Pathobiology of Major Age-related Diseases: Current Status and Research Frontiers*, Scarpelli, D. G. and Migaki, G., Eds., Alan R. Liss, New York, 1984, 69-87.
33. Fleming, J. E., Miquel, J., Cottrell, S. F., Yengoyan, L. S., and Economos, A. C., Is cell aging caused by respiration-dependent injury to the mitochondrial genome?, *Gerontology*, 28, 44, 1982.
34. Hayakawa, M., Hattori, K., Sugiyama, S., and Ozawa, T., Age-associated oxygen damage and mutations in mitochondrial DNA in human hearts, *Biochem. Biophys. Res. Commun.*, 189, 979, 1992.

35. Richter, C., Reactive oxygen and DNA damage in mitochondria, *Mutat. Res.,* 275, 249, 1992.
36. Cooper, J. M., Mann, V. M., and Schapira, A. H., Analyses of mitochondrial respiratory chain function and mitochondrial DNA deletion in human skeletal muscle: effect of ageing, *J. Neurol. Sci.,* 113, 91, 1992.
37. Katayama, M., Tanaka, M., Yamamoto, H., Ohbayashi, T., Nimura, Y., and Ozawa, T., Deleted mitochondrial DNA in the skeletal muscle of aged individuals, *Biochem. Int.,* 25, 47, 1991.
38. Cooper, J. M., Mann, V. M., Krige, D., and Schapira, A. H., Human mitochondrial complex I dysfunction, *Biochim. Biophys. Acta,* 1101, 198, 1992.
39. Munscher, C., Rieger, T., Muller, H. J., and Kadenbach, B., The point mutation of mitochondrial DNA characteristic for MERRF disease is found also in healthy people of different ages, *FEBS. Lett.,* 317, 27, 1993.
40. Torii, K., Sugiyama, S., Takagi, K., Satake, T., and Ozawa, T., Age-related decrease in respiratory muscle mitochondrial function in rats, *Am. J. Respir. Cell Mol. Biol.,* 6, 88, 1992.
41. Wei, Y. H., Mitochondrial DNA alterations as ageing-associated molecular events, *Mutat. Res.,* 275, 145, 1992.
42. Balin, A. K., Testing the free radical theory of aging, in *Testing the Theories of Aging,* Adelman, R. C. and Roth, G. S., Eds., CRC Press, Boca Raton, FL, 1982, 137–182.
43. Pryor, W. A., The free radical theory of aging revisited: a critique and a suggested disease-specific theory, in *Modern Biological Theories of Aging,* Butler, R. N., Sprott, R. L., Schneider, E. L., and Warner, H. R., Eds., Raven Press, New York, 1986.
44. Ebbesen, P., Cancer and normal aging, *Mech. Ageing Dev.,* 25, 269, 1984.
45. Ames, B. N. and Gold, L. S., Chemical carcinogens: too many rodent carcinogens, *Proc. Natl. Acad. Sci. U.S.A.,* 87, 7772, 1990.
46. Doll, R. and Peto, R., The causes of cancer-quantifiable estimates of avoidable risks of cancer in the United States today, *J. Natl. Cancer Inst.,* 66, 1190, 1981.
47. Totter, J. R., Spontaneous cancer and it possible relationship to oxygen metabolism, *Proc. Natl. Acad. Sci. U.S.A.,* 77, 1763, 1980.
48. Ames, B. N., Dietary carcinogens and anticarcinogens, *Science,* 221, 1256, 1984.
49. Willson, R. L., Iron, zinc, free radicals and oxygen in tissue disorders and cancer control, in *Iron Metabolism: CIBA Foundation Symposium 65,* 331, 1978.

II
MOLECULAR AND CELLULAR AGING

7 AGING OF THE IMMUNE SYSTEM

Hal Sternberg

Perhaps the most dramatic and consequence-bearing age-related phenomenon is the decline in immunologic function with old age. Immune dysfunction, which is known to accompany aging, increases susceptibility to a number of disabling diseases having different etiologies. Despite great progress in pharmacologic and medical treatments, infectious diseases, such as pneumonia and influenza,[1] rise exponentially after the age of 25, along with the increased incidence of cancer[2] and autoimmune disease.[3]

Immune function is dependent on a great number of different factors, such as major histocompatibility genes, hormonal, psychologic, and nutritional status, age, and prior history of antigenic exposure. As a result of these many variables, contradictory data exist regarding the aging of this function in humans. However, much has been learned concerning the mechanism of immune reactions at a cellular and molecular level,[4-6] and the recent progress on immunologic aging is the focus of this chapter.

Understanding the mechanisms of immune function and the influence of aging on immune function is not a simple matter. Immunologic reactions are often highly complex (Table 7-1) and require the participation of numerous humoral factors, cell types, tissues, and organs; therefore, it is difficult to present a simple, comprehensive overview of the field. The present chapter will discuss age-related changes of the various components of the immune system. We begin with some background information concerning the immune system and an overview of the significance of age-related immune dysfunction. Thereafter, we will discuss age-related changes of the components of the immune system (from the thymus and thymic factors, to stem cells, lymphocytes, phagocytic cells, and natural killer cells), concluding with a brief summary and the possible reversibility of immunologic senescence.

1 SPECTRUM OF VITAL PROPERTIES AND ACTIVITIES

The immune system has the enormous task of recognizing that which is self from that which is nonself (or foreign). It must be diverse enough to defend against an almost unlimited variety of pathogenic microbes (viruses and bacteria) and sensitive enough to distinguish transformed cells (tumor cells) from normal cells. The specificity of the immune system toward potentially harmful foreign substances is indeed remarkable. This specificity guards against autoimmune reactions and allows for the detection of subtle changes that occur during tumorigenesis. Age-related changes in imm-une function may have profound effects on susceptibility toward a variety of diseases of different etiologies, as well as influencing the ability to accept allogeneic transplants. Because breakdown in immune function with aging leading to increased autoantibody production and tissue degeneration could account for much age-related pathology, gerontologists have proposed that immune system failure may cause human aging,[7] a proposal that has generated a great deal of interest and research in this area. Evidence does support the concept that aging of the immune system is rate limiting on the human lifespan but cannot account for all manifestations of aging (collagen cross-linking, limited proliferative abilities of dividing cells, the accumulation of lipofuscin).

1.1 Complexity of the Immune System and Immunologic Senescence

Numerous components interact and participate in eliciting immune reactions (Table 7-2 and Figure 7-1). In addition to the organs and tissues listed in Table 7-2, many different cell types and secretory factors are involved. The cell types include B lymphocytes, T lymphocytes, monocytes, granulocytes, and natural killer cells, all of which are derived through bone marrow stem cell differentiation. The secretory factors include thymic factors and lymphokines (factors released by activated leukocytes that, in turn, stimulate other leukocytes), which, too, are critical in immune reactions.

The components that are involved in a particular reaction are not only dependent on the immunogen but, as mentioned above, are also dependent on histocompatibility genes and prior antigenic exposure. Another variable reflects the differences in methodology used in the experimental approaches, which may be responsible for differences of interpretation.

Not all components of the immune system may be equally affected by aging; a change in one component may not influence all immunologic reactions. For example, an age-related deficit in T-killer-cell activity may not influence the removal of an antigen that is normally destroyed by complement-mediated lysis.

1.2 Immunologic Defense

Immunologic reactions can be triggered by an almost limitless number of substances called immunogens. These immunogens may or may not be associated with the surfaces of microorganisms. Among the substances that may be immunogenic and elicit an immune response are proteins, carbohydrates, lipids, or any combination of these. Immunogens that can induce antibody production and specifically bind to these antibodies are called antigens.

Generally accepted schemes of an immune reaction toward a viral and bacterial infection are shown in Figure 7-1. In section A, the bacteria (along with its surface antigens) is engulfed by a macrophage. The macrophage processes the bacteria and presents a portion of a bacterial surface antigen on the exterior of the macrophages cell membrane. The antigen is associated with a membrane glycoprotein known as a major histocompatibility complex class II molecule (MHC II). The MHC II-antigen complex is recognized by a specific pre-T-helper cell surface receptor (TCR) and the MHC II molecule interacts with a CD4 receptor (CD = cluster of differentiation) on the pre-T-helper cell. During the maturation of T cells in the thymus, a cell's fate to possess either CD4

0-8493-8979-8/94/$0.00+$.50

TABLE 7-1

Complexity of the Immune System and Immunologic Senescence

Components are multiple:
- Organs and tissues
- Cells
- Secretory factors

Differential reactions depend on:
- Immunogen (composition, route, dose, half-life)
- Histocompatibility genes
- Prior humoral, psychologic, and nutritional state
- Antigenic history
- Age

Immunologic senescence does not equally affect all components and activities of the immune system

Differences in experimental approaches and subjective interpretation of data often lead to contradictory conclusions

receptor or CD8 receptor is made. T cells that possess the CD4 receptor become T-helper cells, while those possessing CD8 become T-killer cells. The CD4 receptor interacts only with MHC II surface molecules, while CD8 receptors will only interact with MHC I surface molecules.

The double interaction (MHC II-antigen complex with both TCR and CD4) causes the macrophage to secrete a cytokine interleukin I, which results in the maturation and multiplication of the T-helper cell. The T-helper cell then interacts with a B cell that has previously engulfed the bacteria. Like the macrophage, the B cell has processed and presented a portion of the bacteria on the B-cell membrane associated to MHC II. Upon the interaction of the T-helper cell receptor and CD4 receptor with the MHC II-antigen complex, the helper cell secretes cytokines. This causes the B cell to mature and divide. The result is a B-memory cell and a plasma cell. The plasma cells secrete large amounts of specific antibodies, which complex with the bacteria's surface antigens. The antibody antigen complexes are destroyed by complement-mediated lysis and removed by macrophages (Table 7-3).

In section B of Figure 7-1, a normal cell becomes infected with a virus. The normal cell presents a portion of the viral coat protein antigen on its surface associated to a glycoprotein known as a major histocompatibility class I molecule (MHC I). This antigen-MHC I complex on the viral infected cell is recognized by a pre-T-killer cell. Upon the interaction of the T-cell receptor and CD8 antigen with the MHC I-antigen complex, the pre-T-killer cell is stimulated to mature and divide. It returns to the infected cell, and upon interacting with the MHC I-antigen complex, it releases factors that destroy the abnormal cell.

TABLE 7-2

Major Structures of the Immune System

Lymph nodes: Gland-like structures, arranged in groups, interspersed throughout the lymphatic circulation. They consist of a fibrillar network where lymphocytes are organized, mature, and interact. They serve as sites where antigens are trapped and destroyed. Major lymphoid organs include adenoids, appendix, Peyer's patches, spleen, and tonsils.

Bone marrow: Meshwork of connective tissue and stem cells contained within bone cavities. Stem cells are multipotent and differentiate into leukocytes and reticulocytes.

Lymphatic vessels: Thin-walled vessels that direct the flow of lymph in a particular direction using valves. The lymph is pumped, upon muscle contraction and osmotic pressure, through the lymphatic system, into the lymphatic duct, and then into the large subclavian veins.

Reticuloendothelial system: Phagocytic cells contained in reticular tissues located in lymph nodes and liver.

Spleen: A relatively large lymphoid organ situated in the upper quadrant of the abdominal cavity. Like the thymus, it has a cortex and medulla. The cortex contains densely packed lymphocytes and germinal centers. The medulla has a variety of leukocytes and encapsulates the cortex (unlike other lymphoid tissues). Some regions are particularly rich in B lymphocytes and appear important for B-lymphocyte storage and activation.

Thymus: A lymphoid organ under the sternum that consists of a network of epithelial cells that secrete various polypeptide factors important for the maturation of thymocytes (which are also contained within the thymus) into T cells and immune function in general.

TABLE 7-3

Possible Fates of Antibody-Coated Antigen

Phagocytic engulfment by neutrophils
Phagocytic engulfment by macrophages
Complement-mediated lysis

Major Histocompatibility Complex

Many of the mechanisms of immune function are encoded by a large gene cluster known as the *major histocompatibility complex* (MHC). This "supergene" codes for a number of different products that play an important role in immunologic defense, such as complement levels, lymphocyte glycoprotein surface receptors, and lymphocyte ontogeny.[8] In humans, the loci of the MHC are called human leukocyte antigens (HLA). The HLA genes are in chromosome 6 of humans, while in mice the MHC are in loci designated H-2, within chromosome 17.

Susceptibility to disease is related in part to HLA alleles and resulting antigens.[9] Certain HLA alleles are correlated with a higher incidence of different diseases. For example, the HLA composition is an important component of susceptibility to such age-related diseases as rheumatoid arthritis, Graves' disease, multiple sclerosis, and diabetes, to name only a few.

The MHC may influence longevity. The percentage of women with the HLA-B8 allele declines with increasing age,[10] indicating that individuals with that allele have a shorter life expectancy than those having a different allele at that locus. A decreasing proportion with increasing age of HLA homozygous individuals suggests that HLA heterozygosity may be important for longer survival, particularly at the HLA-B8 locus. While it is clear

FIGURE 7-1 (facing page)

(A) *A bacterium is engulfed by a macrophage (top).* The macrophage presents a bacterial antigen on its membrane. A pre-T-helper receptor interacts with the antigen. The macrophage secretes IL-I. The pre-T-helper cell divides and matures. At top right, a bacterium is engulfed by a B cell which presents a bacterial antigen on its membrane. *An activated T-helper cell interacts with the antigen and secretes cytokines (bottom).* The B cell divides and matures into antibody-secreting plasma cells and B-memory cells. The antibodies complex with the bacterial antigens. The complexes are removed by complement-mediated lysis and macrophage engulfment. (B) *A normal cell becomes infected with a virus.* A portion of the viral coat is presented on the membrane of the now abnormal cell. A pre-T-killer cell recognizes the antigen on the abnormal cell. The T cell is stimulated to mature and divide. The activated T-killer cells interact with the abnormal cell and release destructive factors.

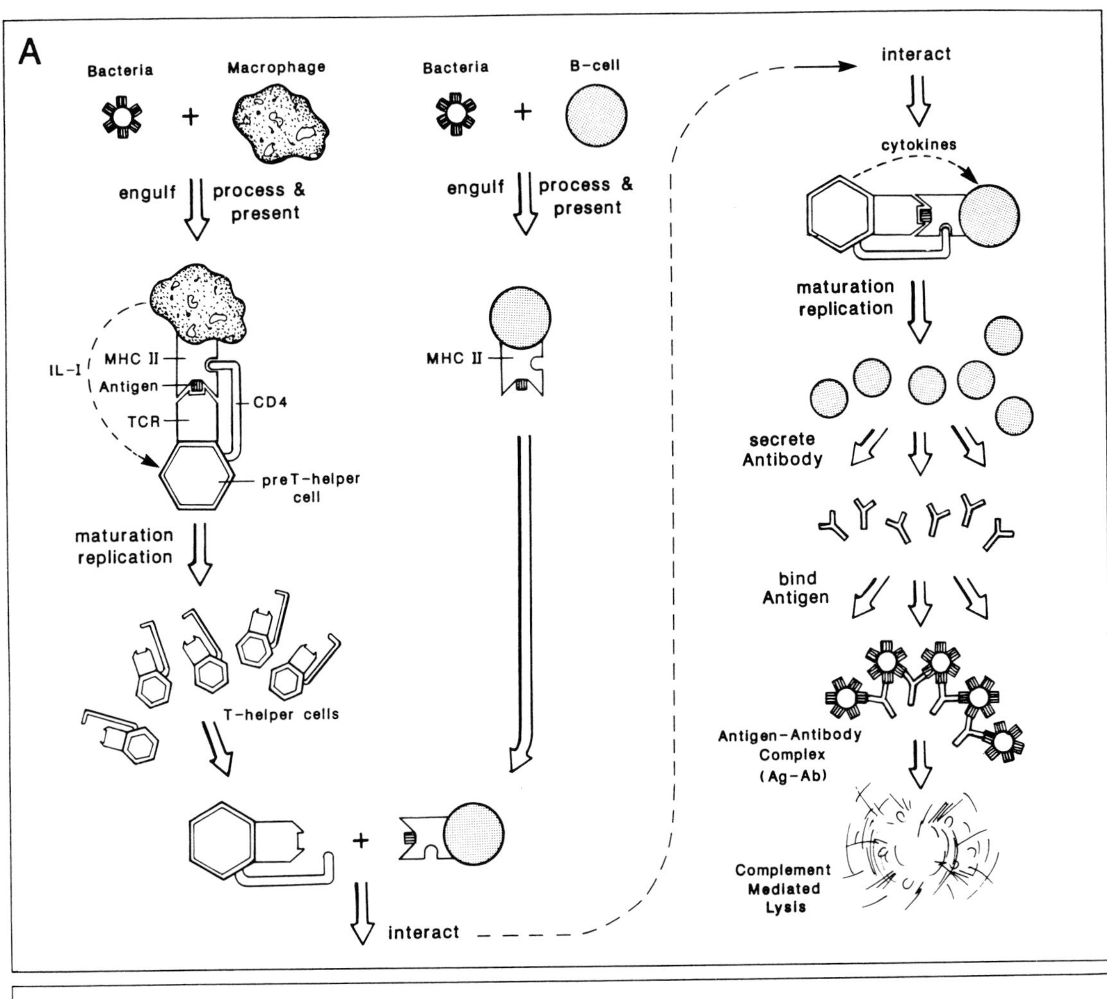
A
Bacteria
Macrophage
engulf
process &
present
IL-I
MHC II
Antigen
TCR
CD4
preT-helper
cell
maturation
replication
T-helper cells
Bacteria
B-cell
engulf
process &
present
MHC II
interact
interact
cytokines
maturation
replication
secrete
Antibody
bind
Antigen
Antigen-Antibody
Complex
(Ag-Ab)
Complement
Mediated
Lysis

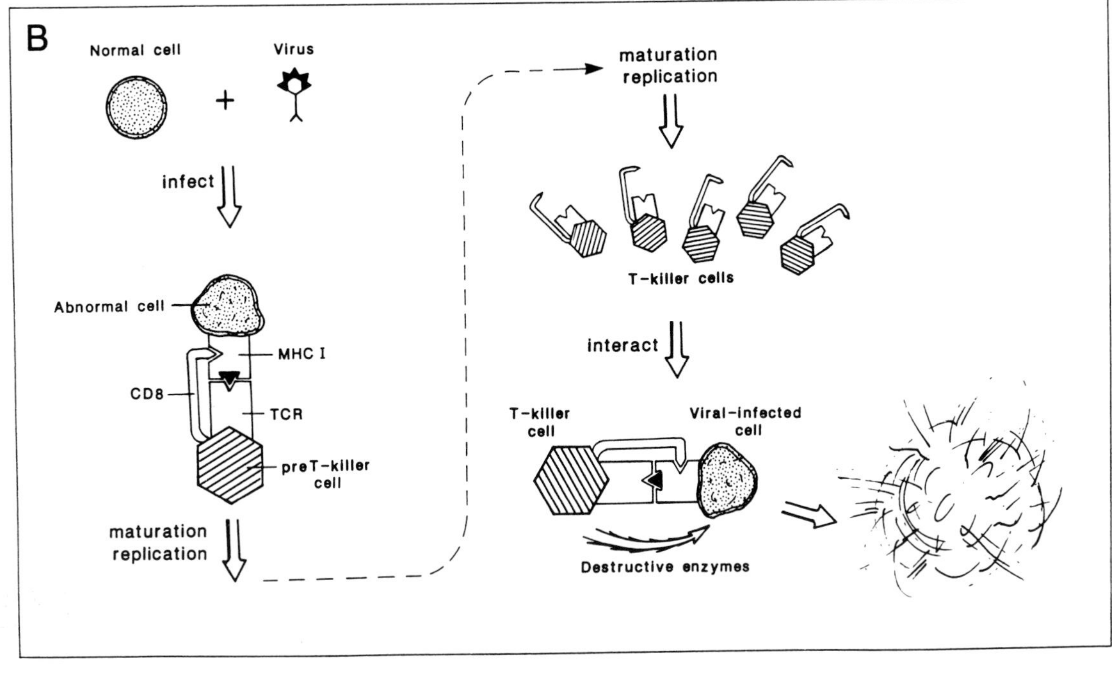
B
Normal cell
Virus
infect
Abnormal cell
MHC I
CD8
TCR
preT-killer
cell
maturation
replication
maturation
replication
T-killer cells
interact
T-killer
cell
Viral-infected
cell
Destructive enzymes

that HLA genotype influences predisposition to age-related disease, how it does so is not clear. However, knowing the histocompatibility genes (haplotypes) and genetic predisposition for certain disorders enables taking special precautions to prevent them.

1.3 Infectious Disease: Defense Against Microbes

Microorganisms reside in almost every corner of the environment, water, soil, and air. While some microbes are symbiotic with humans, others are pathogenic. To eliminate invading infectious agents there are *three lines of defense*. The first line is a physical barrier, such as the skin and mucous membranes lining the orifices. The second line of defense entails phagocytic digestion of invading agents by macrophages within various tissues. The third line of defense involves a humoral response toward the infectious agent. More specifically, the agent is destroyed by an immunologic reaction involving the binding of highly specific antibodies (that are secreted by mature B cells, plasma cells) to the antigenic components of the infectious agent.

Infectious diseases occur more frequently and are of greater consequence in the elderly.[11] Some common infectious diseases include pneumonia, influenza, tuberculosis, meningitis, and urinary tract infections. The age-related rise in susceptibility to infections may be attributed to a decline in antibody production and/or decline in cell-mediated immunity.

With aging in humans, the level of circulating antibodies begins to decline after sexual maturation. Perhaps the decline in serum levels is related to a decline in antibody responsiveness following antigen exposure. In fact, 10 times more antigen is required in old animals to induce maximal antibody responsiveness toward a primary antigen exposure (first time, in a lifetime, when exposed to a particular antigen).[12] While data suggest that older animals have a lowered primary antibody response toward certain antigens, a substantial decline in the secondary antibody response (i.e., previously exposed) is less apparent.

With aging, immunologic protection against some antigenic agents changes; for example, certain antigens may be lost from the repertoire of antigens to which the immune system can elicit an appropriate antibody response. Although much of the research regarding antibody responsiveness toward antigens has been done using mice and dogs, similar changes in immune function may be a contributing factor toward the increase in susceptibility toward infections in old humans.

1.4 Cell-Mediated Immunity (Delayed-Type Hypersensitivity Reactions)

Delayed-type hypersensitivity (DTH) reactions have also been assessed as a function of aging. DTH reactions are triggered by an antigen that induces a cell-mediated response involving both T cell and monocyte participation. More specifically, DTH reactions involve an interaction between the antigen and a particular T-helper lymphocyte subset. The activated T-helper cells release cytokines that both attract and stimulate macrophages and other leukocytes, resulting in localized tissue inflammation and nonspecific destruction of the antigen.

A variety of immunogens such as tuberculosis, streptokinase, and trichophyton have been used to demonstrate that DTH reactions are depressed with aging. Although some DTH reactions become depressed with aging,[13] others do not.[14] Therefore, the age-related decline in the ability of the immune system to perform certain delayed-type hypersensitivity reactions may result in heightened susceptibility toward only certain infectious diseases.

TABLE 7-4

Autoimmune Disorders

Name of Disorder	Target Tissue
Addison's disease	Adrenal gland
Agranulocytosis	Polymorphonuclear leukocytes
Allergic encephalitis	Brain
Episodic lymphopenia	T lymphocytes
Goodpasture's syndrome	Basement membranes (kidney glomerular and lung aveolar)
Graves' disease	Thyroid gland (hyperactive)
Hashimoto thyroiditis	Thyroid gland (hypoactive)
Hemolytic anemia	Erythrocytes
Multiple sclerosis	Central and peripheral nervous system
Myasthenia gravis	Diaphragm muscle
Pemphigus vulgaris	Skin
Rheumatic fever	Heart muscle
Rheumatoid arthritis	Synovial membranes
Systemic lupus erythematosus	Multisystem (cellular constituents)
Thrombocytopenic purpura	Platelets

1.5 Autoimmune Disease

Autoimmune diseases are characterized by the production of antibodies and/or cytotoxic T lymphocytes toward self-antigens leading to dysfunction of the tissues attacked. A breakdown in the immune system's ability to distinguish self from nonself can perceivably disrupt the function of every cell, tissue, and organ in the body. Not unexpectedly, there is a growing number of different diseases with etiology related to an autoimmune reaction. Examples of autoimmune disorders that have been observed to be related to aging of humans are rheumatoid arthritis, autoimmune thyroiditis, lupus, and chronic hepatitis. Some of the common human autoimmune disorders where antibodies to self-antigens have been detected are listed in Table 7-4.

The risk of contracting certain autoimmune disorders has been correlated to histocompatibility genes. Thus, some individuals have a genetic predisposition to acquiring a variety of autoimmune disorders with advancing age. Consistent with the theory that there is an age-related loss in the ability of the immune system to distinguish self from nonself, and the rise in autoimmune disorders, the levels of autoantibodies increase with aging in both mice and humans.[15,16] The detection of certain potentially autoreactive lymphocytes is thought to occur in the thymus. Thymic involution which accompanies aging may result in the escape of the autoreactive lymphocytes which would otherwise undergo cell death. The process by which certain useless or potentially harmful T lymphocytes are destroyed in the thymus is called clonal deletion.[17] The age-related disruption of clonal deletion may result in autoreactive lymphocytes which contribute to the higher incidence of autoimmune disease.

Tissue Grafts and Organ Transplants (Tolerance)

Tissue and organ transplantation has become an increasingly popular therapeutic technique in the past decade because of advances in surgical techniques and equipment, as well as improved pharmacologic protocols. Kidneys, liver, corneas, pancreas, and hearts are examples of organs that are commonly transplanted from allogeneic (genetically nonidentical individual of the same species) donors. One major difficulty of surgical transplantation as a therapeutic treatment concerns immunologic rejection by the recipient of the allogeneic tissue. The more dissimilar the major histocompatibility antigens of grafted tissue are from the recipient, the greater the rejection. Therefore, in order to lessen the magnitude of rejection, surgeons attempt to match the histocompatibility antigens of the tissue donor to that of the recipient. A number of immunosuppressive drugs, such as cyclophosphamide, fluorouracil, methotrexate, corticosteroids, cyclosporine, and FK506, improve the tolerance of allogeneic grafts. Unfortunately, many of these drugs have serious side effects. It has been reported that a factor present in early fetal calf thymus or serum can help induce immunologic tolerance.[18] This work has not yet been repeated, but it seems increasingly promising that pharmacologic agents and regimens will soon become available that have limited side effects and allow tolerance of allogeneic grafts. It is generally found that with aging there is a decline in the immunologic rejection of tissue grafts and consequently greater tolerance. Naturally, this is of great benefit to elderly individuals who can better tolerate grafted tissues and organs. Hence, not all age-related changes in immune function are necessarily deleterious, and some changes may be advantageous, such as reduced graft rejection.

1.6 Tumors and Cancer

An important function of the immune system is its role in preventing tumorigenic growth. Often associated with the development of neoplastic tissue are new (foreign) surface antigens that can potentially be distinguished and destroyed by the immune system. Immunologic reactions against tumor cells have been observed under experimental conditions. The components of the immune system implicated as playing a role in destroying tumor cells are activated macrophages, antibodies, complement, cytotoxic T cells, and natural killer cells. Despite the number of components and mechanisms by which tumor cells may be destroyed, some tumor cells may elude the immune system. One explanation for this is that some tumors may not express new, highly immunogenic, tumor-associated surface antigens. Thus, there is always a finite probability of contracting cancer even while immunologic competence is high.

With aging, there is a dramatic increase in the incidence of tumorigenesis,[2] possibly due to age-related changes in some or all of the components of the immune system that are involved in destroying tumor cells. Some common cancers whose incidence is known to increase with aging include: breast, lung, prostate, stomach, pancreas, and leukemia. Current cancer therapy involves various combinations of surgical removal, X-ray exposure, and chemotherapy. Many cancers can be eliminated by such regimes, but the side effects are often quite devastating. It is hoped that an understanding of immunologic senescence will provide an insight into cancer prevention and treatment.

1.7 Lifespan

Immunologic senescence has broad and profound manifestations and may influence the following:

- Predisposition to various disorders of different etiology
- Ability to be treated with allografts
- Ability to distinguish self from nonself, causing the potential attack and disruption of nearly every cell

Although it is well accepted that immunologic senescence is mostly responsible for the age-related increase in life-threatening diseases such as autoimmune disease, infectious disease, and cancer, it is not clear what role it may play in other disorders such as Alzheimer's disease and cardiovascular disease, whose incidence also rises with aging. If aging of the immune system is a process that limits longevity, then delaying immunologic senescence would lengthen the lifespan. For example, treating mice with a substance (anti-I-J antibody) to prevent the normal age-related enhancement of T-suppressor cell activity and number significantly extended lifespan.[19] On the surface of T-suppressor cells are antigens that are encoded for by the I-J region and the H-2 gene complex in mice. In an attempt to reduce the age-related enhancement of T-suppressor cell number, mice were treated on a weekly basis, from 18 months to death, with an anti-IJ antibody. This immunotherapy decreased T-suppressor activity and resulted in increased lifespan. An extension of lifespan was also observed upon treating mice with human dialyzable leukocyte extract, reported to enhance immune function. Although a slight increase in lifespan was observed, it was not as great as that found using the anti-I-J antibody. Another example showed that treating adult mice, rats, or pigs with a thymic extract (three times per week for 2.5 months), prepared from thymus tissue obtained during very early fetal development, dramatically lengthened the maximum lifespan of each species examined. In addition, the thymic extract treatment also prolonged the survival time of allogeneic and xenogeneic skin grafts[18] and reversed some age-related changes in thymic morphology. Finally, lifespan (of mice) has also been correlated to certain histocompatibility genes[20] and has been extended by splenectomy[21] and calorie restriction, a means of delaying immunologic senescence.[22] In summary, evidence, not yet conclusive, indicates immunologic senescence affects lifespan. Because age-related immune dysfunction may be an important phenomenon limiting human lifespan, discerning which components of the immune system are influenced by aging is of great interest. Identification of these components that contribute to immunologic senescence may provide insight on how to intervene and prevent its accompanying disorders.

2 THYMUS AND THYMOSINS

The thymus is a lymphoid organ that is composed of epithelial cells, thymocytes, fat, and connective tissue. The organ is fully developed by the third month of gestation in humans. It reaches its maximum weight-to-body-weight ratio during neonatal life but continues to grow until puberty.

While histologic studies have shown that spleen and lymph nodes undergo relatively subtle changes with aging (fewer germinal centers), the thymus undergoes gross morphologic changes.[23,24] After puberty, the thymus begins to involute, and thymic weight declines dramatically.[25,26] The percentage of fat and connective tissue within the cortex and medulla of the thymus (which contains thymocytes and epithelial cells) increases. Before puberty about 20% of the medullary and cortical tissue is connective tissue and fat, but, by age 40, about 80% of the tissue is composed of fat and connective tissue, with a corresponding decrease in thymocytes and epithelial cells. This intriguing, genetically programmed involutional event is thought to have a causal role in immunologic senescence.

Accompanying the morphologic changes of the thymus are changes in its biochemistry and physiologic function. The thymus has a network of epithelial cells that produce and secrete several polypeptide factors. Some of the better characterized thymic factors that are produced by thymic epithelial cells include thymosin a-1 thymic humoral factor (THF), thymosin fraction-5 (TF-5), thymulin (facteur thymique serique, FTS), thymopoietin, thymic factor-X (TFX), and thymostimulin.

With aging, the number of cortical epithelial cells of the thymus remains unchanged, while the number of medullary epithelial cells diminishes.[27] Consistent with this, the levels of thymosin a-1 progressively decline after age 20.[28] Similar decreases have been found in human blood for other thymic-derived factors such as FTS and thymopoietin.[29]

The thymic polypeptide factors appear to affect a variety of physiologic events. First, they are noted for their important role in regulating the maturation of pre-T lymphocytes (thymocytes), as monitored by the appearance of certain T-cell surface markers. In addition, the levels of thymic factors may regulate the production of lymphokines. Both TF-5 and thymosin a-1 affect the *in vivo* production (in rodents) of both migration inhibitory factor (which blocks the migration of macrophages in culture) and interferons (class of proteins secreted by natural killer cells). TF-5 also influences interleukin-2 levels and colony-stimulating factor (which is produced in the spleen and whose levels do not decline with aging). Other possible physiologic roles of thymic factors are being investigated. For example, thymosins may influence the levels of endocrine hormones such as ACTH and LH, thereby affecting systems other than the immune system.

The age-related decline in the levels of thymic factors may contribute significantly to immune dysfunction.[30,31] Administration of factors such as FTS, thymostimulin, thymopoietin, and thymosin a-1 reduces various aspects of age-related immune dysfunction. Also, when the thymus of old mice is replaced with the thymus of neonatal mice, the decline in the levels of brain-reactive autoantibodies suggests that thymic involution may aggravate age-related autoimmune disease and tissue degeneration.[15]

Accumulating evidence that immunologic dysfunction is at least in part promoted by thymic involution and by the declining levels of thymic-derived factors[29,32] has prompted experimentation in human subjects. The thymic factors TF-5 and thymosin a-1 have been utilized in human clinical situations, and preliminary studies have revealed that they may be of some benefit. At present, a great scientific effort is directed to obtain large amounts of the various thymic factors, using recombinant DNA technology; thus, the physiologic roles of these factors may be more thoroughly studied, in the hope they may be useful in treating disorders.

3 AGING OF CELLULAR CONSTITUENTS

3.1 Stem Cells

Stem cells comprise a multipotent dividing population residing primarily in bone marrow. They differentiate into erythrocytes and a variety of leukocytes such as T and B lymphocytes, monocytes, granulocytes, and so forth, which participate in immunologic reactions. Age-related changes in stem cell replication and/or differentiation would directly affect the population of the cells that can partake in immunologic defense.

Proliferative Abilities

There is evidence both in support of and against a change in stem cell proliferative abilities with aging. Stem cells have a limited capacity to divide, but this intrinsic capacity may not be significantly affected by aging.[33-35] In perhaps one of the most conclusive studies, mouse bone marrow cells of chromosomally marked young and old donors were serially transplanted together into young irradiated recipients.[33] After five such serial transplantations of the marked cells into recipients, the stem cells of both the young and old donors ceased to divide. Prior to the fifth serial transplant, the stem cells from both the young and old donors continued to respond identically to stimulation by mitogens, phytohemagglutinin (PHA), or lipopolysaccharide. This study indicates that aging does not affect the capacity of stem cells to replicate, but does suggest that stem cells have a limited intrinsic potential to divide. About five serial transplants were required to observe the limited proliferative potentials of these cells and thus would not be of consequence in a single, normal lifespan.

On the other hand, evidence supports an age-related change in stem cell proliferation kinetics.[36] Using the bromodeoxyuridine differential chromatid staining technique for *in vivo* analysis of cellular replication, it was found that the replication rates of stem cells within the bone marrow of old (24 month) rats were about 25% slower than those of younger (12 month) rats. Similar results were obtained using either male or female Wistar or Fisher rats. In addition, preliminary experimentation indicates that same is true for mice, suggesting that this slowing down of cellular replication rates with aging is not sex, strain, or species specific and perhaps is a general phenomenon prevalent in all senescent mammals.

Bone marrow stem cells are by no means a homogeneous population of cells. In the study discussed above,[36] it was not established whether particular subsets of stem cells were affected by aging while others were not. For example, it is quite plausible that only the replication rates of the erythrocyte lineage are affected by aging while all leukocyte lineages are unperturbed. One cannot conclude that immune function would necessarily be affected by slower average replication rates of the heterogeneous population of stem cells. Nevertheless, the study does indicate that at least certain lineages of stem cell differentiation must be affected by aging, and stem cells in general are not beyond being affected by age despite the fact they may have the potential to proliferate normally (when transplanted from old into young serially) four times greater than the duration of a normal lifespan. Perhaps this age-related decline in replication rates is a measure of an extrinsic environmentally induced phenomenon that is partly or wholly reversible. This is not inconsistent with Harrison's studies,[37] which suggest that stem cells do not undergo changes in a single lifespan. While bone marrow stem cells may divide more slowly in older animals owing to extrinsic (environmental) factors, their intrinsic capacity or potential to divide (i.e., when exposed to an environment such as that found in young animals) is not appreciably affected by aging.

Lifespan and Turnover Rate of Differentiated Stem Cells

Differentiated stem cells (lymphocytes, granulocytes, erythrocytes, and monocytes) have varying lifespans before they are removed from the circulation by macrophages. Cell types having the longer lifespan, and consequently the slower turnover rate, show the greatest age-related decline in function. A possible regulatory mechanism controlling the lifespan of differentiated stem cells may be the slow progressive appearance of a membrane component that is recognized by an IgG autoantibody. Such a mechanism has been demonstrated for the removal of red blood cells.[32] The attachment of enough IgG to the specific membrane glycoprotein (62 kDa) signals the engulfment and removal of "aged" red blood cells (RBC).

It is still unclear whether a similar mechanism regulates the lifespan and turnover rate of other differentiated stem cells. However, it seems likely that certain progressive changes in membrane composition will signal removal of other differentiated stem cells as well. Although still untested, it is possible that the age-related decline in cellular replication rates of stem cells is due to a decline in the progressive appearance of membrane signals, leading to slower turnover rates. It would be interesting to determine whether the progressive appearance of the RBC glycoprotein is influenced by organismic aging.

While little is known about the turnover of differentiated stem cells in general, further understanding of the mechanism of RBC removal and turnover may elucidate the mechanism controlling the removal and turnover of other differentiated stem cells (Chapter 18).

Effect of Aging on the Role of Stem Cells in Immunologic Reactions

There is evidence that the aging thymus induces changes in the stem cells that result in lowered antibody production toward antigen. However, these changes are reversible.[37,38] Antibody responsiveness toward the antigen sheep red blood cells declines with aging, but this decline is reversible. Indeed, stem cells from old donors, transplanted into young irradiated recipients, do not show a decline of antibody responsiveness with aging. However, if the young irradiated recipients are also thymectomized, stem cells transplanted from old animals show a decline in antibody production, similar to that of old. This suggests that the thymus of young animals is important in reversing the age-related changes that have occurred in stem cells from old animals, and that thymic involution may play a major role in immunologic senescence.

Stem Cell Differentiation Pattern

It is becoming clear that, with aging, not only do stem cell replication rates decline, but regulation of stem cell differentiation is altered.[39-41] Although these changes may be extrinsically induced and reversible,[39] they may be of serious consequence to an aging organism.

Stem cell differentiation is regulated by a variety of thymus-derived factors, such as thymosin and thymic factor-X, as well as the lymphokine interleukin-3 and colony-stimulating factor. The serum levels of those polypeptide factors decline with aging.[30] These decreasing hormonal factors also influence the pattern of bone marrow stem cell differentiation. Hormonally induced changes in stem cell differentiation patterns are believed to cause shifts in the numbers of certain progenitor cells and in the recognition repertoires of lymphocytes.[39,40] Such shifts would account for much of the data concerning immunologic senescence and are consistent with observations that the immunologic response of older individuals toward certain antigens differs from that of younger individuals. Though some immunologic reactions are not influenced by shifts in the pattern of bone marrow stem cell differentiation, others are, with a consequent increased susceptibility to cancer, autoimmune disease, and infectious disease.

3.2 T Lymphocytes

T lymphocytes are a heterogeneous class of cells presently classified on the basis of their activities and surface membrane components (antigens). The three general classes of T lymphocytes are T-killer (Tk), T-helper (Th), and T-suppressor (Ts) cells. Tk lymphocytes (also called cytolytic or cytotoxic T lymphocytes) are effector cells that possess cytolytic activities. Ts lymphocytes are regulator cells that are thought to suppress some immune reactions.[42] However, there has been some controversy whether the cells identified having the Ts phenotype really have suppressor function *in vivo*. Helper lymphocytes are regulator cells that can stimulate B-cell activities.

The proliferation and differentiation of T-cell progenitors (thymocytes) into mature T lymphocytes are dependent in part on various thymus-derived factors, such as thymosin and thymopoietin. With this in mind, one might expect that age-related involution of the thymus and the decline in the production of thymic factors[30] would influence T-lymphocyte differentiation and activities. Consistent with this, numerous investigators have demonstrated age-related changes regarding the numbers and activities of various T-lymphocyte subpopulations.

Changes in Responsiveness to Mitogens with Aging

Certain populations of T lymphocytes can be stimulated to divide in culture upon exposure to various mitogens. Numerous studies have demonstrated that mitogenic stimulation of T cells is influenced by aging. For example, mitogenic stimulation by phytohemagglutinin (PHA) of cells from rodents[43,44] or humans[45,46] declines with increasing donor age. Similarly, studies with the mitogen, concanavalin A (con A), which possibly stimulates a different population of T cells than PHA, show a decline in proliferation of cells from rodents[43] and humans[45] with increasing donor age, a decline also induced by other mitogens (tetradecanol-phorbolacetate [TPA] and pokeweed), including mitogenic monoclonal antibodies. While the decline in mitogenic responsiveness of T lymphocytes has been well demonstrated, the causes, direction, and magnitude of the age-related changes are still obscure. The decrease with aging of interleukin-2 (IL-2),[47,48] a T-cell growth factor, may contribute to the decline in T-lymphocyte proliferation, perhaps because of a decreased responsiveness of T cells to IL-2 or a decrease in the proportion of responsive T cells.[48]

Interestingly, mitogenic responsiveness of T lymphocytes from an involuted thymus (of older humans) is greater than that observed from a younger thymus.[49] The age-related enhancement of mitogenic responsiveness of thymic T cells may be related to the increase in the proportion of OKT3-positive T lymphocytes which have the T3-membrane antigen. OKT3-positive cells are identified on the basis of their possessing a surface receptor which binds specifically to a monoclonal antibody called OKT3. Cells with the T3-membrane antigen are known to be more sensitive to mitogens, so an increase in prevalence of this subpopulation may account for the overall increase in mitogenic responsiveness of thymic T cells with aging. However, the unchanged or decreased OKT3+ subpopulation in blood may account for the age-related decline in mitogenic activity of peripheral blood T lymphocytes.[50]

Although one might predict that the age-related decline in thymic hormones, which have activities promoting T-cell growth and maturation, would lead to an increase in undifferentiated T cells,[51] no convincing evidence supports such a claim (as monitored by the prevalence of certain membrane antigens whose appearance is dependent on T-cell maturation).[50] This adds some confusion regarding the role (if any) of thymic involution on T-cell proliferation and maturation. Perhaps other factors may compensate for the loss of thymic hormones with age. Indeed some cytokines such as IL-3, IL-4, and maybe IL-5 may rise with advancing age.[52]

Cytotoxic Activity (T-Killer Cells)

The cytotoxic activity of T lymphocytes appears to be depressed with aging.[39,53,54] In particular, the T lymphocytes of old mice display a reduced activity against allogeneic cells.[53] This decline may be due to an age-related decrease in the numbers of alloantigen-specific T cells possessing cytotoxic activity. Aged mice were found to have one tenth the number of alloantigen-specific precursor cells compared to young,[54] which may account for the lowered cytotoxic activity against allogeneic cells.

Contrary reports state that the cytotoxic response toward allogeneic cells (spleen or tumor) is not affected by aging.[55] Apparently, the cytotoxic response against alloantigens is dependent on the experimental method employed.

In addition to the reported decline in alloreactivity with aging, the heterogeneity of alloantigens recognized by cytotoxic T cells declines as well,[39] probably because of the reduced prevalence of certain subpopulations of T cells that have receptors for alloantigens. Such changes in T-cell recognition repertoire are thought to arise from changes in the regulation of differentiation of bone marrow stem cells.[56] The age-related change in stem cell differentiation pattern may result in a decline in both the heterogeneity of alloantigen-specific cytotoxic T cells and the reactivity of T cells toward alloantigens. Moreover, it may be a major reason for the increased tolerance of elderly individuals for allogeneic tissue and organ grafts. There is also an age-related decline in cytolytic activity toward both syngeneic tumor cells and parental strain spleen cells, which agrees with the known rise in cancer incidence. In conclusion, fundamental changes in the pattern of stem cell differentiation may largely contribute toward the age-related decline in T-killer cell activity, resulting in greater tolerance to allogeneic tissue transplants and higher tumor incidence.

Suppressor Activity of T Lymphocytes (T-Suppressor Cells)

Suppressor activity of T lymphocytes appears to rise with aging in mice[16,57] but may or may not rise in humans. Heightened T-suppressor cell activity due to an increased proportion of T-suppressor cells may contribute to the age-related decline in immunologic reactivity toward alloantigens[58] and may also account for the increase in tolerance of allografts and tumor growth with aging. In addition, increased suppressor cell activity may contribute to the age-related decline in T-cell–mediated DTH reactions. DTH reactions are often elicited in response to infectious agents, and a stronger suppression of such immunologic reactions may increase the predisposition toward infections.

Suppressor cells are identified on the basis of their activity and of their membrane markers, such as I-J and Ly23 in mice (humans possess similar markers, i.e., T5). The proportion of cells having these markers increases with aging.[59] Treatment of mice with anti-I-J alloantiserum enhances resistance to syngeneic tumor growth, suggesting that the high incidence of tumors with aging may be due in part to heightened suppressor activities.

Most studies of age-related increases in suppressor cell activity were performed using mice. Humans are exempt from a similar age-related change; in fact, suppressor activity has been observed to decline in humans.[60] Consistent with this, the proliferative response of splenic suppressor cells toward certain allogeneic H-2 antigens is lowered in mice as well.[61] Apparently, the effect of aging on T-suppressor activity is complex. Numerous fundamental changes in the subset proportions and attributes of T-suppressor cells arise with aging.

Suppressor T cells, stimulated by antigen in old mice, are effective in inducing tolerance only when transplanted in old syngeneic recipients, but not young recipients.[62] Similarly, suppressor cells that become stimulated in young mice are more active in inducing tolerance when transplanted into young syngeneic recipients. Perhaps a coordinated change in the pattern of differentiation of stem cells results in different subsets of suppressor cells that are active in immune reactions with aging.

Helper Activity of the T Lymphocytes (T-Helper Cells)

T-helper cells are a class of lymphocytes that help enhance the activity of either T-killer cells or B cells toward antigen. T-helper cells release (1) IL-2, which stimulates T-killer cell activity, or (2) other cytokines that stimulate B cells.

Several studies report an age-related decline in T-helper cell activity.[63,64] Such a decline may be related to the observed decrease in the number of T-helper cells with aging[65] but can be restored by *in vivo* administration of thymosin a-1. Thymosin administration may induce the maturation and proliferation of pre-T-helper cells, suggesting that the age-related decline in T-helper activity is due to the age-associated thymic involution and decline in thymic hormones such as thymosin a-1. Moreover, the decline in T-helper cell regulator activity may promote immune dysfunction by diminished activation of T-killer cells and antibody-producing B cells.

AIDS

T-helper cells appear to be implicated in the acquired immunodeficiency syndrome (AIDS), a disease that is on the rise. This disease has been found to be caused by a retrovirus (containing RNA as the genetic material) called HIV. The virus, which can be transferred by blood or seminal fluids, infects certain T-helper lymphocytes that possess a surface antigen designated CD-4. Individuals diagnosed as having AIDS have defects in various aspects of cell-mediated immunity, including immunosurveillance of tumorigenic tissue. The disease results in death due to infections or cancer (Kaposi's sarcoma). The susceptibility to AIDS does not appear to be related to age, sex, or race, but is correlated highly with sexual behavior, intravenous drug use, and frequency of blood transfusions. While a cure is being searched, prevention of the disease is already being implemented. Careful blood donor screening, the elimination of sharing needles for intravenous injections, and careful selection of sexual partners may eradicate this dreadful disease.

3.3 Phagocytic Cells: Macrophages

Macrophages are derived from stem cells along the monocyte lineage. Monocytes are smaller than macrophages, circulate in the blood, infiltrate various tissues, and differentiate into mature macrophages that can play a variety of roles in immunologic responses. These include: nonspecific phagocytosis of an antigen, specific phagocytosis of an antibody-tagged antigen, and degradation of the contents of the phagocytic vesicle by lysosomal enzymes after fusion with lysosomes. The macrophage, which characteristically can adhere tightly to glass surfaces, appears to be the first leukocyte to recognize and bind to antigen and to present the antigenic epitope to T helpers (in association with certain membrane attached class II, HLA gene products). In addition, macrophages can secrete cytokines, such as IL-1, that modulate lymphocyte activity and immunologic responsiveness. To maximize T- and B-cell participation in immune responses, it is necessary to expose these lymphocytes to antigen-pulsed macrophages. Macrophages exposed to antigen are very powerful enhancers of T- and B-cell proliferation and activity. However, not all macrophages are identical; some bind antigen, present it to T-helper cells, and are powerful inducers of T-lymphocyte proliferation, while others do not.

Lysosomal and Phagocytic Activity

The activity of lysosomal enzymes, such as β-glucuronidase, acid phosphatase, and cathepsin-D, becomes significantly elevated with aging.[66] Since lysosomal degradation of antigen and antigen-associated material follows phagocytosis, a rise in phagocytic activity might be expected. In contrast, phagocytic activity remains unchanged with aging or may be slightly depressed.[67,68] For instance, the phagocytic activity of human polymorphonuclear cells, granulocytes, and monocytes declines with aging.[67] It is unclear why lysosomal enzymes may become more prevalent with age, while phagocytic activity declines.

It is possible that the phagocytic cells are not turned over as quickly with aging and that the average age of these leukocytes increases with aging, as has been suggested for B cells.[69,70] Consistent with this, the membrane components and fluidity of phagocytes appear to be affected by aging,[71] as might be expected if turnover rates were slower.[32] Older phagocytic cells may accumulate lysosomal enzymes and have slightly lowered phagocytic activity.

Although age-related changes in lysosomal enzymes and phagocytic activity do appear relatively small, they may contribute to an increased incidence of infectious disease.

Antigen Presentation

Antigen presentation by macrophages can be monitored by observing the effects on respondent lymphocytes. The amount of antigen necessary to evoke a vigorous immune response may be lowered several hundred times if antigen-presenting accessory cells such as macrophages are employed. The "Ia" molecule is a surface protein on antigen-presenting macrophages (coded for by the I region of the H-2 complex in mice), which interacts with antigen. The Ia-antigen complex is recognized by and stimulates T-helper cells, which in turn will activate T- and B-effector cells. The proportion of the Ia-positive subpopulation of macrophages appears to rise with aging.[72] Because this subpopulation is important for antigen presentation, increased antigen presentation activity by macrophages is expected with aging, as shown by some[73] but not by all.[68] Further, this increase may be specific for only certain antigens.

The changes in macrophage subpopulation proportions that may be responsible for a slight enhancement in macrophage activity might improve protection against certain infectious agents. For example, protection against *Listeria monocytogenes* is heightened in aging mice, local and systemic infections are better prevented by nonspecifically activated macrophages of older mice, and peritoneal macrophages from older mice display improved suppression of bacteria grown *in vitro*.[73]

In conclusion, evidence suggests that under some circumstances, immunologic protection provided by macrophages may be superior in older than younger subjects, but inferior in others. The possible influence of antigen-presenting macrophages on the recognition repertoires of B cells[40] suggests that age-related changes are highly selective, coordinated, and perhaps quite purposeful.

3.4 Natural Killer Cells

Natural killer (Nk) cells are derived from bone marrow stem cells and are considered to be a subpopulation of lymphocytes. They have a number of morphologic and biochemical features in common with T lymphocytes (e.g., the presence of certain surface antigens and responsiveness to cytokines), suggesting that Nk cells may arise from a T-lymphocyte lineage.

Nk cells possess cytolytic activity against tumorigenic cells, as well as virus-infected cells without previous sensitization (unlike T-killer cells). They are important for immunologic surveillance of neoplastic tissue, they mediate antibody responses, and they also affect hematopoiesis.

Comparison of the number of Nk precursor cells in the spleens of old and young mice[74,75] shows fewer precursor cells in old mouse spleens and, in fact, as few as one half that of young.[74] The decreased number of Nk precursor cells with aging may be important in the age-related rise in tumorigenesis.

Does aging affect the cytolytic activity of Nk cells in humans? Some investigators could find no change in the ability of human Nk cells to lyse a variety of tumor cell lines *(in vitro)*,[76] while others, using mice, have detected a depression of Nk activity with aging.[74,75,77] Using IL-2, the cytolytic activity of Nk cells from spleens of older mice can be increased to a greater extent than that of young.[75] Because the spleens of old mice contain fewer cytokine-respondent Nk cells, the total Nk activity is still less than that of young, a deficit perhaps dependent on age-related changes in bone marrow effector cells.[74]

The activity of Nk cells can be modulated not only by cytokines, but also by other environmental conditions, such as the lipoprotein composition of blood. The presence of certain lipoproteins can influence the binding of Nk cells to the target tissue *in vitro;* some lipoproteins depress the binding of Nk cells from aged donors more than from young.[78] Although target binding of Nk cells is affected by lipoproteins, lytic activity of attached Nk cells is not. Thus, blood lipoproteins

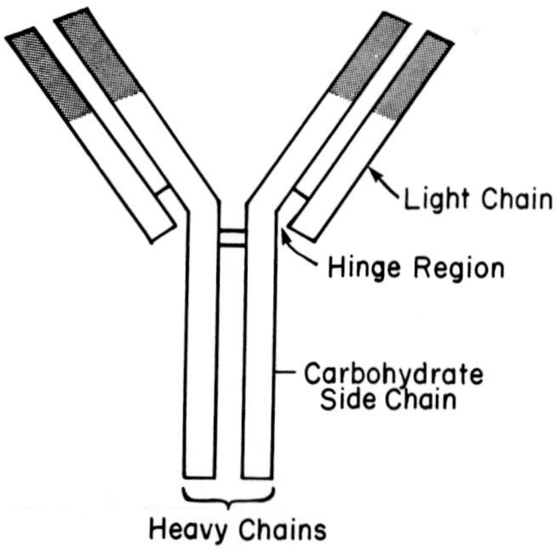

FIGURE 7-2 *Two identical heavy polypeptide chains and two identical light chains are associated by disulfide bonds (as illustrated).* The five major types of Ig heavy chains distinguish the five major classes of Ig molecules (IgA, D, E, G, M). Although the various types of heavy chains are structurally similar, they are not identical, which accounts for their functional differences. IgA, found in exocrine secretions, more stable to aberrant conditions; IgD, receptor on B-cell membranes; IgE, binds mast cells, contributes to allergic reactions; IgM, a multivalent pentamer, secreted in a primary immune response, can stimulate complement-mediated lysis and macrophage engulfment; IgG, the most abundant Ig in blood, stimulates complement-mediated lysis and macrophage engulfment. The antigen binding specificity of all immunoglobulins is contained in the shaded region (as depicted), which is near the amino terminus and has a highly variable amino acid sequence.

may contribute to the age-related decline in Nk cells activity and an enhanced risk of tumors.

4 B LYMPHOCYTES AND ANTIBODY PRODUCTION

B cells, which are derived from bone marrow stem cells, secrete specific antibodies (Figure 7-2). The antibody-antigen complexes that result are destroyed by complement-mediated lysis and/or engulfment by macrophages or neutrophils. Both the class of Ig associated with the antigen and its binding pattern dictate the mechanism of removal.

B-cell antibody production is dependent on prior antigenic history, duration, and dose of antigen exposure, and, usually, T-cell participation. The great diversity of specific antibodies secreted in response to a wide variety of antigens is generated in B lymphocytes through chromosomal rearrangements among genes that code for antibody structure.[79] Available evidence suggests a general decline in the ability of B cells to produce and secrete specific antibodies with aging.[80-82] After serial treatment with a T-cell-independent antigen (pneumococcal polysaccharide type III), B-cell antibody production was lower in old mice than in young.[80] The antigen used induces the production of the IgM class of Ig, the predominant Ig secreted in older animals. Usually IgM is secreted during a primary response; the more mature B cells secrete IgG. The age-related decline in antibody production to this T-cell-independent antigen may be due to the B-cell subset failing to differentiate into mature IgG-producing cells.[70] Not only does antibody production decline upon stimulation with T-cell-independent antigens, but it also declines in response to T-cell-dependent antigens.[82]

Some contrasting evidence suggests B-cell antibody production and subpopulation numbers do no change with aging. At the clonal level (using limiting dilution analysis) the number of mouse splenic B cells, respondent to a particular viral antigen (PR 8 influenza virus) is unchanged with aging.[82] However, overall antibody responses by B cells decline with aging.[39,82,83] Such a decline may be ascribed to a shift in the subpopulations of B lymphocytes (changing their recognition repertoires) in parallel with changes in the pattern of stem cell differentiation.

Consistent with this, limiting dilution analysis has revealed an age-related decline in the population of certain antigen-specific B lymphocytes as well as a decline in the population of B lymphocytes that are respondent to the mitogen lipopolysaccharide.[40] Like macrophages, B cells have the ability to process and present antigen to T-helper cells. It is not yet clear how this function may be influenced by aging.

In summary, only certain subpopulations of B cells are influenced by aging. Thus, the administration of some antigens results in a lowered antibody response.

Mitogenic Responsiveness

B-cell mitogenic responsiveness is dependent on various factors, such as the type of mitogen utilized, the animal species or strain, and the experimental approach of the investigators. Hence, investigators reach diverging conclusions: some investigators have detected a decline in B-cell mitogenic responsiveness toward plant lectins,[83-85] others have not.[86]

Interestingly, restoration of the age-related decline in B-cell proliferative responsiveness toward activated monocytes was possible by increasing the levels of IL-1 and/or IL-2 to which lymphocytes were exposed,[83] not without considerable variation from individual to individual, however.

A possible mechanism for changes in B-cell activity may involve an age-related shift in the proportion of B-cell subpopulations, arising during shifts in bone marrow stem cell differentiation, that are responsive to certain mitogens.[41,82]

4.1 B-Cell Number and Population Size

B lymphocytes are a heterogeneous class of cells that differ in their maturational state, surface components, and mechanism by which their activity is regulated. An age-related change in B-cell maturation and/or subpopulation proportions may influence the humoral response and susceptibility to disease.

Despite reports of lowered B-cell mitogenic responsiveness and maturation with aging, the bone marrow of older mice possesses greater numbers of mature B cells than that of young controls.[69] On the other hand, the number of circulating B cells does not appear to change with age, while the ratio of B cells to T cells increases owing to a decline in T-cell number.

How can the greater numbers of mature B cells in bone marrow be reconciled with reports of lowered maturational development of B cells with aging?[70] Perhaps turnover rates of mature B cells decline with aging. If fewer cells are removed from the circulation, and if proliferative rates are only moderately slowed with aging, B cells may accumulate in bone marrow. Consistent with this view are reports of fewer germinal centers in lymphatic follicles with aging, as one might expect if there is a reduced need for new cells due to slower turnover rates. It has been observed that cell types having longer lifespans and slower turnover rates demonstrate the greatest age-related decline in function.

5 IS AGE-RELATED IMMUNE DYSFUNCTION REVERSIBLE?

The components and activities of the immune system are selectively affected by aging (Table 7-5). This selectivity leads to the questions of which and to what extent these changes are reversible. Considerable evidence supports the proposi-

TABLE 7-5

Immunologic Senescence

Secretory factors		
Thymus		
Thymosin a-1	→	Decline
Thymulin		Secretory factors
Thymopoietin		
Thymic humoral factor		
Stem Cell Differentiation		
Influenced by:		
Bone marrow		
Thymosin	→	Decline
Colony-stimulating factor		T-helper cell proportion
Interleukin-3		Alloantigen-specific Tk
		Natural killer cell number
		Increase
		Ts cell proportion
		B-cell/T-cell ratio
		Shift
		B-cell characteristics
		Antibody production
Consequences		Increase
		Tissue graft tolerance
		Cancer incidence
		Autoimmune disease
		Infectious disease

tion that age-related immune dysfunction can be at least partially reversed. Thymic factors such as thymopoietin,[87] thymic fraction-5,[88] and thymosin a-1,[28] as well as cytokines IL-1, IL-2, and interferon-g,[30,75,83] can restore various parameters of age-related immune dysfunction. Other treatments, e.g., 2-mercaptoethanol, levamisole, and synthetic polynucleotides, may also delay immune senescence. This provides great optimism for the potential of delaying age-related immune dysfunction in humans.

On the other hand, there are also data suggesting that delaying human immunologic senescence is not a simple matter. The involution of the thymus and subsequent decrease in thymic hormone levels are well accepted factors in immunologic aging. Yet, the implantation of a thymus from a young mouse into an older thymectomized, syngeneic mouse does not provide prolonged improvement in immune function, as might have been predicted.[89] Similarly, a close correlation is lacking between thymic involution and a reduction in serum thymic hormones. Such observations indicate that immunologic aging is not solely dependent on thymic involution, and that multiple independent events regulate immune senescence.

An explanation for the relative ineffectiveness of thymic transplants is that the signal that might trigger thymic involution is still active and reinitiates thymic involution of the transplant. It is likely that such a signal is derived from the nervous system, and thymic involution may be a result of specific neuronal cell loss and alterations. Neural signals regulate development and aging of specific systems such as the reproductive system, suggesting that such a mechanism may induce aging of the immune system as well.[90]

Other data emphasize the difficulty of reversing or delaying human immune senescence. For example, old mice implanted with young bone marrow cells did not undergo improvement of immunologic function for extended durations.[89] The internal environment within older animals may be "hostile" to young cells so that the young cells rapidly acquire the properties of the old. Only after the environmental factors capable of inducing in the older animals a youthful environment are identified will implantations of young thymic tissue or bone marrow be effective in substantially delaying immune senescence.

There still remains a great deal to be learned concerning the mechanisms of immunologic aging, and while it may not be currently possible to delay, prevent, or reverse human immunologic senescence, continuing advances in immunology, and immunologic aging in particular, will guide us toward successful interventions in improving and prolonging life.

REFERENCES

1. Galpin, J., Immunity and microbial diseases, in *Handbook of Immunology in Aging,* Kay, M. M. B. and Makinodan, T., Eds., CRC Press, Boca Raton, FL, 1981, 141.
2. Ershler, W. B., The influence of aging and immune system on cancer incidence and progression, *J. Gerontol. Biol. Sci.,* 48, B3, 1993.
3. Teller, M. N., Interrelationships among aging, immunity and cancer, in *Tolerance, Autoimmunity and Aging,* Sigel, M. M. and Good, R. A., Eds., Charles C Thomas, Springfield, IL, 1972, 18.
4. Eisen, H. N., *Immunology,* 2nd ed., Harper & Row, Philadelphia, 1980.
5. Hood, L. E., Weissman, I. L., Wood, W. B., and Wilson, J. H., *Immunology,* 2nd ed., Benjamin/Cummings, Menlo Park, CA, 1984.
6. Roitt, I. M. and Delves, P. J., *Encyclopedia of Immunology,* Academic Press, San Diego, CA, 1992.
7. Walford, R. L., *The Immunological Theory of Aging,* Munksgaard, Copenhagen, 1969.
8. Albert, E. and Gotze, D., The major histocompatibility system in man, in *The Major Histocompatibility System in Man and Animals,* Gotze, D., Ed., Springer-Verlag, New York, 1977, 7.
9. Dausset, J., HLA complex in human biology in the light of associations with disease, *Transplant Proc.,* 9, 523, 1977.
10. Greenberg, L. J. and Yunis, E. J., Genetic control of autoimmune disease and immune responsiveness and the relationship to aging, in *Genetic Effects on Aging,* Bergsma, D. and Harrison, D. E., Eds., Alan R. Liss, New York, 1978, 249.
11. Garibaldi, R. R. and Nurse, B. A., Infections in the elderly, *Am. J. Med.,* 81, 53, 1986.
12. Nordin, A. A. and Makinodan, T., Humoral immunity in aging, *Fed. Proc.,* 33, 2033, 1974.
13. Gardner, I. D. and Remington, J. S., Age-related decline in resistance of mice to infection with intracellular pathogens, *Infect. Immunol.,* 16, 593, 1977.
14. Lvik, M. and North, R. J., Effect of aging on antimicrobial immunity: old mice display a normal capacity for generating protective T-cells and immunologic memory in response to infection with *Listeria monocytogenes, J. Immunol.,* 135, 3479, 1985.
15. Nandy, K. and Bennett, M., Immune manipulations and brain reactive antibody formation in aging mice, *Mech. Ageing Dev.,* 22, 3, 1983.
16. Goidl, E. A., Choy, J. W., Gibbons, J. J., Weksler, M. B., Thorbecke, G. J., and Siskind, G. W., Production of auto-antiidiotypic antibody during the normal immune response. VII. Analysis of the cellular basis for the increased auto-antiidiotype antibody production by aged mice, *J. Exp. Med.,* 157, 1635, 1983.

17. von Boehmer, H. and Kisielow, P., How the immune system learns about self, *Sci. Am.*, October, 74, 1991.
18. Czaplicki, J., Blonska, B., Klementys, A., Klementys, K., Kaluzewski, B., and Pawlikowski, M., The effects of blood sera from calf fetuses in various periods of fetal life on skin allograft survival time in mice and on human lymphocyte proliferation in vitro, *Thymus,* 3, 17, 1981.
19. Liu, J. J., Segre, D., Gelberg, H. B., Fudenberg, H. H., Tsang, K. Y., Khansari, N., Waltenbaugh, C. R., and Segre, M., Effects of long-term treatment of mice with anti-I-J monoclonal antibody and dialyzable leukocyte extract on immune function and lifespan, *Mech. Ageing Dev.,* 27, 359, 1984.
20. Smith, G. S. and Walford, R. L., Influence of the main histocompatibility complex in aging in mice, *Nature,* 46, 727, 1977.
21. Albright, J. F., Makinodan, T., and Deitchman, J. W., Presence of life-shortening factors in spleens of aged mice of long lifespan and extension of life expectancy by splenectomy, *Exp. Gerontol.,* 4, 267, 1969.
22. Weindruch, R. H., Kristie J. A., Naeim, H., Mullen, B. G., and Walford, R. L., Influence on weaning-initiated dietary restriction on responses to T-cell mitogens and on splenic T cell levels in a long-lived F1-hybrid mouse strain, *Exp. Gerontol.,* 17, 49, 1982.
23. Makinodan, T., Immunity and aging, in *Handbook of the Biology of Aging,* Finch, C. E. and Hayflick, L., Eds., Van Nostrand Reinhold, New York, 1977, 379.
24. Kay, M. M. B. and Baker, L. S., Cell changes associated with declining immune function, in *Physiology and Cell Biology of Aging,* Vol. 8, Cherkin, A., Finch, C. E., Kharasch, N., Makinodan, T., Scott, F. L., and Strehler, B. S., Eds., Raven Press, New York, 1979, 27.
25. Boyd, E., The weight of the thymus gland in health and disease, *Am. J. Dis. Child.,* 43, 1162, 1932.
26. Good, R. A. and Gabrielson, A. B., Eds., *Thymus in Immunobiology,* Hoeber-Harper, New York, 1964.
27. Haynes, B. F., Shimizu, K., and Eisenbarth, G. S., Identification of human and rodent thymic epithelium using tetanus toxin and monoclonal antibody A2B5, *J. Clin. Invest.,* 71, 9, 1983.
28. McClure, J. E., Lameris, N., Wara, D. W., and Goldstein, A. L., Immunochemical studies on thymosin: radioimmunoassay of thymosin alpha-1, *J. Immunol.,* 128, 368, 1982.
29. Lewis, V. M., Twomey, J. J., Bealmar, P., Goldstein, G., and Good, R. A., Age, thymic involution, and circulating thymic hormone activity, *J. Clin. Endocrinol. Metab.,* 47, 145, 1978.
30. Zatz, M. M. and Goldstein, A. L., Thymosins, lymphokines and the immunology of aging, *Gerontology,* 31, 263, 1985.
31. Goldstein, A. L. and Zatz, M. M., Thymosin and aging, in *Immunological Aspects of Aging,* Segre, D. and Smith, L., Eds., Dekker Publishing Co., New York, 1981, 371.
32. Kay, M. M. B., Immune system: expression and regulation of cellular aging, in *Thresholds in Aging,* Bergener, M., Ermini, M., and Stahelin, H. B., Eds., Academic Press, London, 1985, 59.
33. Harrison, D. E., Astle, C. M., and Delaittre, J. A., Loss of proliferative capacity in immunohemopoietic stem cells caused by serial transplantation rather than ageing, *J. Exp. Med.,* 147, 1526, 1978.
34. Curtis, H. J. and Tiley, J., The lifespan of dividing mammalian cells *in vivo, J. Gerontol.,* 26, 1, 1971.
35. Ogden, D. A. and Micklem, H. S., The fate of serially transplanted bone marrow cell populations from young and old donors, *Transplantation,* 22, 287, 1976.
36. Schneider, E. L., Sternberg, H., Tice, R., Senula, G., Kram, D., Smith, G., and Bynum, G., Cellular replication and aging, *Mech. Ageing Dev.,* 9, 313, 1979.
37. Harrison, D. E., Astle, C. M., and Doubleday, J. W., Stem cell lines from old immunodeficient donors give normal responses in young recipients, *J. Immunol.,* 118, 1223, 1977.
38. Farrar, J. J., Loughman, B. E., and Nordin, A. A., Lymphopoietic potential of bone marrow stem cells from aged mice: comparison of the cellular constituents of bone marrow from young and aged mice, *J. Immunol.,* 112, 1244, 1974.
39. Gorczynski, R. M., Kennedy, M., and MacRae, S., Alteration in lymphocyte recognition repertoire during aging. II. Changes in expressed T-cell receptor repertoire in aged mice and the persistence of that change after transplantation to a new differentiative environment, *Cell. Immunol.,* 75, 226, 1983.
40. Gorczynski, R. M., Chang, M . P., Kennedy, M., MacRae, S., Benting, K., and Price, G. B., Alterations in lymphocyte recognition repertoire during aging. I. Analysis of changes in immune response potential of B-lymphocytes from non-immunized aged mice, and the role of accessary cells in the expression of that potential, *Immunopharmacology,* 7, 179, 1984.
41. Manheimer, A. J., Victor-Kobrin, C., Stein, K. E., and Bona, C. A., Antimmunoglobulin antibodies. V. Age-dependent variation of clones stimulated by polysaccharide TI-2 antigens in 129 and MRL mice spontaneously producing anti-gamma-globulin antibodies, *J. Immunol.,* 133, 562, 1984.
42. Webb, D. R. and Devens, B. H., Suppressor T lymphocytes, in *Encyclopedia of Immunology,* Roitt, I. M. and Delves, P. J., Eds., Academic Press, San Diego, 1992, 1409.
43. Kaplan, P. J. and Garvey, J. S., Age-related changes in responsiveness of various rat tissue lymphocytes to mitogens, *Immunol. Lett.,* 3, 357, 1981.
44. Averill, L. E. and Wolf, N. S., The decline in murine splenic PHA and LPS responsiveness with age is primarily due to an intrinsic mechanism, *J. Immunol.,* 134, 3859, 1985.
45. Gillis, S., Kozak, R., Durante, M., and Weksler, M. E., Immunological studies of aging: decreased production of T-cell growth factor by lymphocytes from aged humans, *J. Clin. Invest.,* 67, 937, 1981.
46. Schwab, R., Hausman, P. B., Rinnooy-Kan, E., and Weksler, M. E., Immunological studies of ageing. X. Impaired T lymphocytes and normal monocyte response from elderly humans to the mitogenic antibodies OKT3 and Leu4, *Immunology,* 55, 677, 1985.
47. Thoman, M. L. and Weigle, W. O., Cell mediated immunity in aged mice: an underlying lesion in IL-2 synthesis, *J. Immunol.,* 128, 2358, 1982.
48. Miller, R. A., Age-associated decline in precursor frequency for different T-cell mediated reactions, with preservation of helper or cytotoxic effect per precursor cell, *J. Immunol.,* 132, 63, 1984.
49. Baroni, C. D., Valtieri, M., Stoppacciaro, A., Ruco, L. P., Uccini, S., and Ricci, C., The human thymus in ageing: histologic involution paralleled by increased mitogen response and by enrichment of OKT3+ lymphocytes, *Immunology,* 50, 519, 1983.
50. Jensen, T. L., Hallgren, H. M., Yasmineh, W. G., and O'Leary, J. J., Do immature T cells accumulate in advanced age?, *Mech. Ageing Dev.,* 33, 237, 1986.
51. O'Leary, J. J., Jackola, D. R., Hallgren, H. M., Abbasnezhad, M., and Yasmineh, W., Evidence for a less differentiated subpopulation of lymphocytes in advanced age, *Mech. Ageing Dev.,* 21, 109, 1983.
52. Cinadar, B., Aging and the immune system, in *Encyclopedia of Immunology,* Roitt, I. M. and Delves, P. J., Eds., Academic Press, San Diego, 1992, 45.
53. Zharhary, D. and Klinman, N. R., Antigen responsiveness of the mature and generative B cell populations of aged mice, *J. Exp. Med.,* 157, 1300, 1983.

54. Nordin, A. A. and Collins, G. D., Limiting dilution analysis of alloreactive cytotoxic precursor cells in aging mice, *J. Immunol.*, 131, 2215, 1983.
55. Fitzgerald, P. A. and Bennett, M., Aging of natural and acquired immunity of mice. II. Decreased T cell responses to syngeneic tumor cells and parental-strain spleen cells, *Cancer Invest.*, 1, 139, 1983.
56. Gorczynski, R. M., Kennedy, M., and MacRae, S., Altered lymphocyte recognition repertoire during aging. III. Changes in MHC restriction patterns in parental T-lymphocytes and diminution in T-suppressor function, *Immunology*, 52, 611, 1984.
57. Liu, J. J., Segre, M., and Segre, D., Changes in suppressor, helper and B-cell functions in aging mice, *Cell Immunol.*, 66, 372, 1982.
58. Gupta, S. and Good, R. A., Subpopulations of human T lymphocytes. X. Alterations in T, B, third population cells and T-cells with receptors for immunoglobulin M(Tu) or G(Ty) in ageing humans, *J. Immunol.*, 122, 1214, 1978.
59. Perry, L. L., McClusky, R. T., Benaceraff, B., and Green, M. I., Enhanced syngeneic tumor destruction by *in vivo* inhibition of suppressor cells using anti-I.J. alloantiserum, *Am. J. Pathol.*, 92, 491, 1978.
60. Ceuppens, J. L. and Goodwin, J. S., Regulation of immunoglobulin production in pokeweed mitogen-stimulated cultures of lymphocytes from young and old adults, *J. Immunol.*, 128, 2429, 1982.
61. Gottesman, S. R., Walford, R. L., and Thorbecke, G. J., Proliferative and cytotoxic immune functions in aging mice. II. Decreased generation of specific suppressor T-cells in alloreactive cultures, *J. Immunol.*, 133, 1782, 1984.
62. Hausman, P. B., Goidl, E. A., Siskind G. W., and Weksler, M. E., Immunological studies of aging. XI. Age related changes in idiotype repertoire of suppressor T cells stimulated during tolerance induction, *J. Immunol.*, 134, 3802, 1985.
63. Callard, R. E. and Basten, A., Immune function in aged mice. IV. Loss of T-cell and B-cell function in thymus dependent antibody responses, *Eur. J. Immunol.*, 8, 552, 1978.
64. Frasca, D., Garavini, M., and Doria, G., Recovery of T cell functions in aged mice injected with synthetic thymosin-alpha-1, *Cell. Immunol.*, 72, 384, 1982.
65. Doria, G., D'Agostaro, G., and Garavini, M., Age-dependent changes in B-cell reactivity and T cell-T cell interaction in the *in vitro* antibody response, *Cell. Immunol.*, 53, 195, 1980.
66. Tyan, M. L., Marrow stem cells during development and aging, in *Handbook of Immunology in Aging*, Kay, B. and Makinodan, T., Eds., CRC Press, Boca Raton, FL, 1981, 87.
67. Bongrand, P., Bartolin, R., Bouvenot, G., Arnaud, C., Delboy, C., and Depieds, R., Effect of age on different receptors and functions of phagocytic cells, *J. Clin. Lab. Immunol.*, 15, 45, 1984.
68. Vetvicka, V., Teaskalova-Hogenova, H., and Pospisil, M., Impaired antigen presenting function of macrophages from aged mice, *Immunol. Invest.*, 14, 105, 1985.
69. Tyan, M. L., Age-related decrease in mouse T-cell progenitors, *J. Immunol.*, 118, 846, 1977.
70. Mosier, D. E., Mond, Y. T., and Goldings, E. A., The ontogeny of thymic independent antibody responses in vitro in normal mice and mice with an x-linked B-cell defect, *J. Immunol.*, 119, 1874, 1977.
71. Vetvicka, V., Fornusek, L., Tlaskalova, H., Sima, P., and Pospisil, M., Some properties and functions of macrophages during ontogeny, *Dev. Comp. Immunol.*, 7, 747, 1983.
72. Doria, G., Frasca, D., and Adorini, L., Immunoregulation of antibody response in aging, in *Progress in Immunology V*, Yamamura, Y. and Tada, T., Eds., Academic Press, Tokyo, 1549, 1983.
73. Matsumoto, T., Miake, S., Mistuyama, M., Takeya, K., and Nomoto, K., Enhanced resistance to Listeria monocytogenes due to nonspecifically activated macrophages in aged mice, *J. Clin. Lab. Immunol.*, 8, 51, 1982.
74. Fitzgerald, P. A. and Bennett, M., Aging of natural and acquired immunity of mice. I. Decreased natural killer cell function and hybrid resistance, *Cancer Invest.*, 1, 15, 1983.
75. Saxena, Q. B. and Adler, W. H., Interleukin-2-induced activation of natural killer activity in spleen cells from old and young mice, *Immunology*, 51, 719, 1984.
76. Adler, W. H. and Nagel, J. E., Studies of immune function in a human population, in *Immunological Aspects of Aging*, Segre, D. and Smith, L., Eds., Marcel Dekker, New York, 1981, 295.
77. Albright, J. W. and Albright, J. F., Age-associated impairment of natural killer activity, *Proc. Natl. Acad Sci. U.S.A.*, 80, 6371, 1983.
78. Antonaci, S., Jirillo, E., Ventura, M. I., Garofalo, A. R., and Bonomo, L., Lipoprotein-induced inhibition of plaque-forming cells generation and natural killer cell frequency in aged donors, *Ann. Immunol. (Paris)*, 135e, 241, 1984.
79. Tonegawa, S., The molecules of the immune system, *Sci. Am.*, 253, 122, 1985.
80. Mason-Smith, A., The effects of age on the immune response to type III pneumococcal polysaccharide (SIII) and bacterial lipopolysaccharide (LPS) in BALB/c, SJL/I and C3H mice, *J. Immunol.*, 116, 469, 1976.
81. Zharhary, D., Segev, Y., and Gershon, H. E., T-cell cytotoxicity and ageing: differing cause of reduced response in individual mice, *Mech. Ageing Dev.*, 25, 129, 1984.
82. Zharhary, D. and Klinman, N. R., B-cell repertoire diversity to PR8 influenza virus does not decrease with age, *J. Immunol.*, 133, 2285, 1984.
83. Whistler, R. L. and Newhouse, Y. G., Immunosenescence of the human B-cell system: impaired activation proliferation in response to autologous monocytes pulsed with *Staph.* protein A and the effects of interleukins 1 and 2 compared to interferon, *Lymphokine Res.*, 4, 331, 1985.
84. Price, G. B. and Makinodan, T., Immunologic deficiencies in senescence. I. Characterization of intrinsic deficiencies, *J. Immunol.*, 108, 403, 1972.
85. Hollingsworth, M. A. and Evans, D. L., Changes in spleen cell proliferative responses and resistance to syngeneic tumor challenge in aging NBR rats, *Mech. Ageing Dev.*, 22, 321, 1983.
86. Kay, M. M. B., An overview of immune aging, *Mech. Ageing Dev.*, 9, 39, 1979.
87. Weksler, M. E., Innes, J. B., and Goldstein, G. J., Immunological studies of aging. IV. The contribution of thymic involution to the immune deficiencies of aging mice and reversal with thymopoietin, *Exp. Med.*, 148, 996, 1978.
88. Goldstein, A. L., Low, T. L. K., Hall, N., Naylor, P. H., and Zatz, M. M., Thymosin: can it retard aging by boosting the immune capacity?, in *Intervention in the Aging Process. Part A: Quantitation, Epidemiology and Clinical Research*, Regelson, W. and Sinex, F. M., Eds., Alan R. Liss, New York, 1983, 169.
89. Hirokawa, K., Albright, J. W., and Makinodan, T., Restoration of impaired immune functions in aging animals. I. Effect of syngeneic thymus and bone marrow grafts, *Clin. Immunol. Immunopathol.*, 5, 371, 1976.
90. Barnes, D. M., Neuroimmunology sits on broad research base, *Science*, 237, 1568, 1987.

III
SYSTEMIC AND
ORGANISMIC
AGING

8 AGING OF THE NERVOUS SYSTEM: STRUCTURAL AND BIOCHEMICAL CHANGES

Paola S. Timiras

The human nervous system — its development, adult state and aging, and alterations with disease — represents one of the most challenging and intriguing studies of our time. Despite increasing and rapid advances, we still have few direct answers to our many questions concerning the activities of the nervous system. Studies of the aging nervous system provide valuable information both on aging processes and nervous functions. From comparison of the aging brain, with and without neurologic and psychiatric diseases of old age, emerges specific differences in the morphologic, biochemical, and functional nature of the normal and diseased states. Normal aging of the nervous system, particularly of the central nervous system (CNS), will be emphasized here with references to age-related disorders. Locations, mechanisms, and consequences of changes with aging are summarized in Table 8-1. They are multiple and involve all organizational levels (Figure 8-1):

- *At the structural level,* aging affects neurons and glial cells, their number and branching, as well as the synapse, a key site for the reception and transmission of the nervous impulse.
- *At the chemical level,* aging influences the metabolism, turnover, and receptors of several neurotransmitters, substances carrying or transmitting the neural signal at the synapse as well as cellular metabolic processes. Aging also affects the membranes, both plasma and intracellular membranes, where lipid peroxidation and protein cross-linking accumulate, perhaps due to the continuing action of free radicals (Chapter 6). Biochemical alterations of neural membranes impair excitability, fluidity, and transport. Equally important are nuclear alterations in which DNA damage (and failure of DNA repair) may be reflected in disturbed protein synthesis and accumulation of abnormal proteins.
- *At the functional level,* aging influences electrical activity, motor and sensory, and cognitive and affective processes. Consequences of aging result in disturbances within the nervous system and in the organism as a whole. The key role of the nervous system in regulating all body functions implies that any impairment of its function will impact on the aging of the entire organism, particularly the efficacy of homeostatic adjustments to the environment.

1 STRUCTURAL CHANGES

1.1 Brain Weight

Brain weight decreases with aging in humans by about 6 to 11% (Figure 8-2), but some claim absence of changes in mentally normal older individuals. This contrasts with the reduction reported in many, but not all, demented individuals (Figure 8-3).[1-3] *Brain volume,* as measured by ventricular and sulcal dimensions at autopsy or by imaging techniques in the living, remains essentially unchanged in normal aged individuals. Most demented patients exhibit markedly enlarged ventricles with reduced cortical volume (Figure 8-4). Experimental animals vary in this regard: rats do not show any changes with aging, whereas dogs and monkeys demonstrate cerebral atrophy.

1.2 Neuronal Cell Loss

Because of the heterogeneity of nervous structures, aging affects the various portions of the nervous system differentially. One striking example of this diversity involves neuronal loss. In the normally aging human brain (i.e., without overt vascular, neurologic, or psychiatric disorders) *loss of neurons is limited to discrete areas and shows considerable individual variability* (Table 8-2).[4,5] In some areas of the cerebral cortex and in the cerebellum, the number of neurons decreases only moderately with old age; whereas in others, loss may be severe.

In the locus ceruleus (a major catecholaminergic structure, see below), cell number decreases by 40% after approximately 60 years of age in normal individuals, but the decrease is more severe in individuals with Alzheimer's disease (AD).[6] Loss of neurons in this area and in the substantia nigra have been related to a number of functional decrements (affecting sleep, locomotion, homeostasis) and specific neurologic disorders, such as Parkinson's disease (see below).

Another area which may undergo significant neuronal loss with aging is the nucleus basalis of Meynert which provides the primary cholinergic input to the cerebral cortex. Neuronal loss is small or absent in normal aging but is quite severe (75 to 80%) in AD patients. Given this loss, alterations of cholinergic transmission have been causally related to AD and restoration of cholinergic transmission suggested as a rational approach to the treatment for the disease[7,8] (Chapter 9). In the spinal cord, the number of motor neurons remains unchanged until approximately 60 years of age, after which it may decrease by as much as 50% at very advanced ages.

TABLE 8-1

Changes with Aging in the Nervous System

Locations
Regional selectivity
Neuronal loss/gliosis
Reduced dendrites and dendritic spines
Synaptic susceptibility
Vascular lesions
Mechanisms
Neurotransmitter imbalance
Membrane alterations
Metabolic disturbances
Intra/intercellular degeneration
Consequences
Sensory and motor decrements
Sleep alterations and EEG changes
Memory impairment
Increased neurologic and psychiatric pathology
Impaired homeostasis

0-8493-8979-8/94/$0.00+$.50

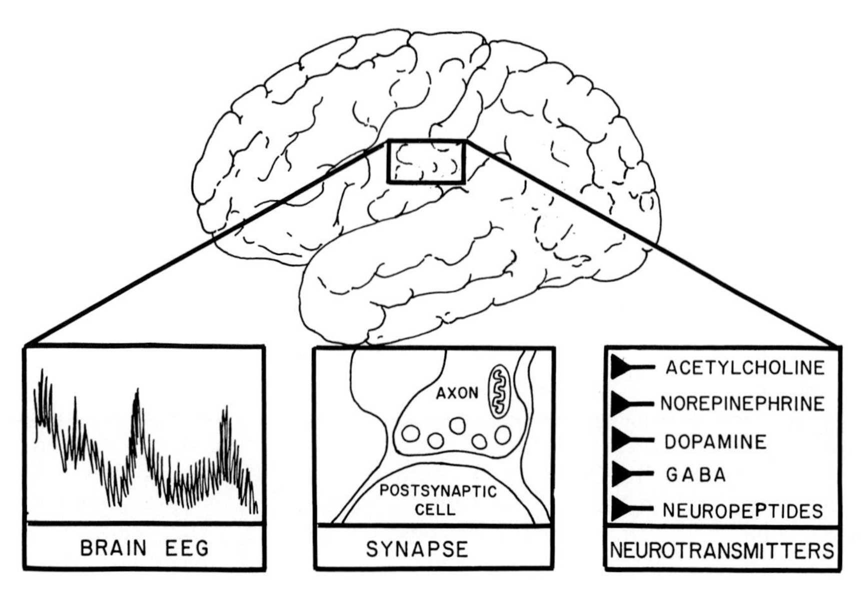

FIGURE 8-1 *Diagram of electrophysiological (brain EEG), functional (synapse), and chemical (neurotransmitters) sites for changes in the brain with aging.* (Drawing by S. Oklund.)

The number of glial cells increases with aging in most areas, and this increase, or *gliosis,* may represent *a compensatory response* not only to the limited neuronal loss but also to neuronal impairment.[9] Indeed, with aging, *neurons undergo degenerative changes* manifested by:

- Decreased number of dendrites and dendritic spines
- Intra- and intercellular accumulation of abnormal substances such as lipofuscin, melanin, neurofibrillary, and amyloid proteins
- Disturbances of synaptic function and neurotransmission
- Synaptic alterations ascribed to dendritic defects or losses
- Alterations at the pre- and post-synaptic structures of neurotransmitter synthesis, release or binding

Neuronal number *in experimental animals* decreases slightly with aging, depending on the area considered, the type of cell, and the animal species. For example, in the cerebral cortex of rats, the number of neurons remains unchanged with aging.[10] The glial cells — astrocytes and oligodendrocytes — increase in number while microglia are unchanged.[10] Reduced numbers of neurons have been reported in discrete brain areas of aged monkeys, guinea pigs, rats, and mice,[11] but when these animals are studied throughout the lifespan, a major reduction in cell number often occurs at young rather than old ages.[12,13]

Another compartment affected by aging is the cerebral vascular system which undergoes the same atherosclerotic changes that involve the entire arterial vasculature; these atherosclerotic lesions commence at an early age, progress with age, and profoundly influence the neural tissue, which has a critical dependence on oxygen and glucose for normal metabolism and function.

Cells of the Nervous System

Cells of the nervous system include neurons, glial, and endothelial cells. Neurons have long been regarded as the primary functional cells of the nervous system although glial cells are five times more numerous. Neural structure varies with respect to the cell body and branches. The latter include the dendrites, which extend from the cell body and arborize, and a longer extension, the axon. Communication between cells is through synapses located in specific structures, spines, and knobs, and composed of the presynaptic membrane, a cleft, and a postsynaptic membrane. Present in the spines or knobs are vesicles or granules containing the synaptic transmitter synthesized in the cell. The axons may be surrounded by a specialized membrane, myelin, or may be unmyelinated. The myelinated fibers transmit the nerve impulse more rapidly than do the nonmyelinated ones.

Glial cells comprise (1) astrocytes, which have end feet that surround blood vessels in the CNS; (2) oligodendrocytes involved in myelin formation; and (3) microglia, scavenger cells, which are part of the immune system. While the astrocytes have independent functions, they also interact to function as a neuronal-glial unit in such vital actions as memory, metabolism, and neural transmission.

Endothelial cells of cerebral capillaries form tight junctions that do not permit the passage of substances which pass through the junctions between endothelial cells in other tissues. In addition, cerebral capillaries are surrounded by the end feet of astrocytes. Both these special features of cerebral capillaries hinder the exchanges across capillary walls between plasma and brain interstitial fluid and represent the anatomical basis for the concept of a functional blood-brain barrier; this so-called barrier essentially protects the brain from entering endogenous metabolites and exogenous toxins and prevents drugs in the blood and from neurotransmitters escaping into the general circulation.

Overall, neuronal loss with aging in humans and experimental animals seems to be restricted to specific CNS structures and to be associated in animals, and possibly in human brains, with an increase in glial cell number. The relatively mild cell loss reported in normal aging contrasts with the severe cell death reported in several diseases of the nervous system associated with aging, such as senile dementia of the Alzheimer type and Parkinson's disease, or genetic abnormalities, such as Down's syndrome with impaired mental function and precocious aging.

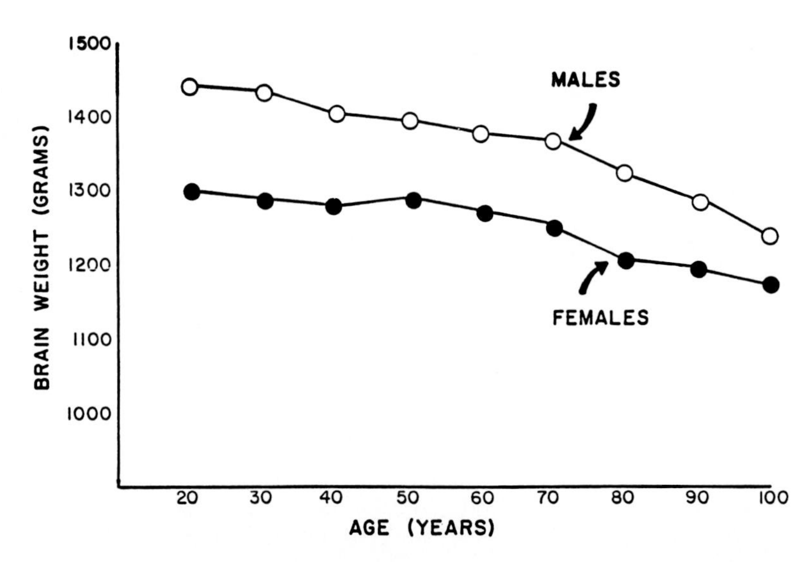

FIGURE 8-2 *Changes in brain weight with aging in human males and females.*

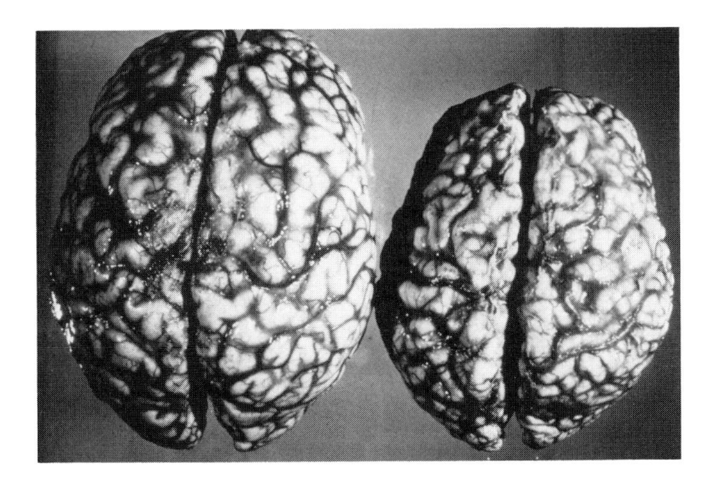

FIGURE 8-3 *Differences in size of normal (left) and Alzheimer's (right) brain.* (Courtesy of Dr. L. S. Forno.)

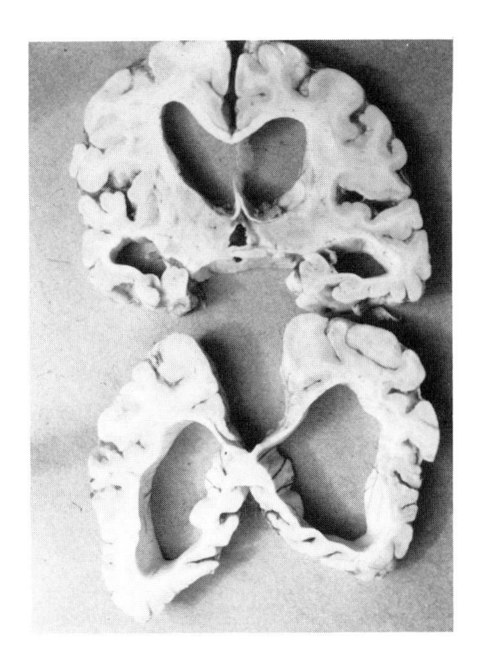

FIGURE 8-4 *Enlargement of the cerebral ventricles with severe cortical atrophy in a patient with Alzheimer's dementia.* Coronal sections at the frontal (top) and parieto-occipital (bottom) levels. (Courtesy of Dr. L. S. Forno.)

1.3 Dendritic Loss with Aging

The number of dendrites and dendritic spines is reduced in normal aging. For example, the cortex of a young individual shows large pyramidal cells with abundant dendrites and dendritic spines. A corresponding zone, in an old individual, shows quite a striking loss of dendrites and spines. Dendrites function as receptor membranes of the neurons and represent sites of excitatory and inhibitory activity. Dendritic spines, tiny and numerous on each dendrite, amplify such activity and isolate increases in synaptic calcium transport that may serve for induction of information storage.[14] Hence, loss of dendrites and spines will result in neuronal isolation and failure of interneuronal communication.

Dendrites undergo a certain degree of renewal continuously; therefore, it has been suggested that the *denudation of the neurons,* that is, the reduced number of dendrites, may not be a true loss, but, rather, a slowing down of the renewal process. When the loss of dendrites is viewed in a network of neurons, then the consequences of diminished connectivity become apparent.[15,16] In normal aging, with continued environmental stimulation dendritic loss may be minimal, absent, or even supplemented with a degree of dendritic outgrowth.[12,13] In AD, and more so in presenile dementia (Chapter 9), the dendritic loss is severe and progressive (Figure 8-5). With reduced dendrites, synapses are lost, neurotransmission is altered, and communication within and without the nervous system is impaired.

1.4 Synaptic Changes with Aging

The decreased number of *synapses* in discrete areas of the aging brain may coincide with the corresponding loss of dendrites and dendritic spines. Alterations in synaptic components — membrane, vesicles, and granules — have been variably reported. The "reactive synaptogenesis" or axonal sprouting that follows the loss of a neuron is not entirely lost, but it is slowed down with aging.[17] Reactive synaptogenesis represents

TABLE 8-2

Neuronal Loss in Selected Regions of the Human Brain During Normal Aging and in the Alzheimer's Disease (AD)[a]

Brain region	Age range (years)	Percentage decrease Normal aging	AD[b]
Frontal lobe			
Precentral gyrus	18–95	22–44	19–37
Association areas	19–95	15–51	22–40
Anterior cingulate gyrus	68–95	15–16	22
Temporal lobe			
Superior gyrus	18–95	34–57	33–53
Middle gyrus	19–97	23–50	25–67
Inferior gyrus	18–95	18–36	11–63
Parietal lobe			
Inferior parietal cortex	19–NA	15	9–37
Occipital lobe			
Area 17	18–95	31–54	17–22
Area 18	19–NA	14	14–30
Hippocampal cortex	45–90	27–80	47
Cerebellum			
Purkinje cells	60–100	25	NA
Subcortical areas			
Locus ceruleus	60–80	43	60–70
Substantia nigra	20–65	35	35–40
Nucleus basalis of Meynert	50–66	Unchanged?	75–80

Note: NA, not available.

[a] Data for this table were adapted from various authors[4,5] and several observations were combined for brevity. With normal aging, loss in neuron numbers varies significantly with the brain region considered. This variability has been ascribed to intrinsic, area-specific timetables of aging as well as secular trends in aging and differences in the methodology used to count neuron numbers. In AD, neural loss is increased in some but not in all areas with considerable overlap between AD and normal aging groups.

[b] Control (normal-aging) and AD values were matched for age. AD brains were originally[4] divided in juvenile and senile forms. They are combined here for brevity. It is to be noted that in most cases excess neuron loss was greater in juvenile AD.

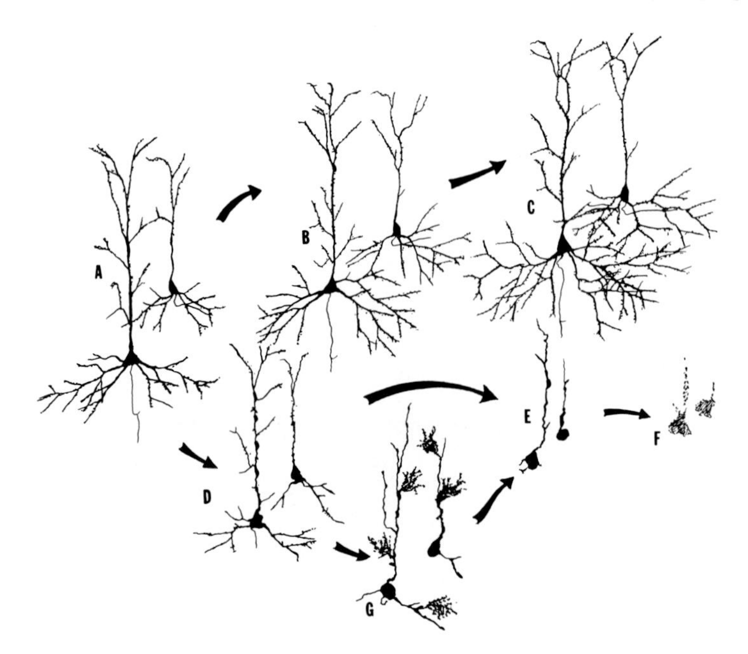

FIGURE 8-5 *Semi-schematized drawing summarizing the changes that may occur in pyramidal neurons of the aging human cerebral cortex.* **Sequence A,B,C** follows the changes which may occur in the normal aging cortex under the effects of continued physical and/or cognitive "challenge" and in the probable presence of continuing small amounts of neuronal loss. Increasing dendritic growth, especially in the peripheral portions of the basilar dendrites, reflects dendritic response to optimal cortical "loading", plus presumed supplementary growth to fill neurophil space left by dendrite systems of those neurons which die. **Sequence A,D,E,F** epitomizes the progressive degenerative changes which characterize senile dementia of Alzheimer type (AD), and includes progressive loss of dendritic spines and dendritic branches culminating in death of the cell. Basilar dendrite loss precedes loss of the apical shaft. **Sequence A,D,G,E,F** represents a unique pattern of deterioration of the dendrite tree found only in the familial type of presenile dementia of Alzheimer (Chapter 9). During the period of dendritic degeneration, bursts of spine-rich dendrites in clusters appear along the surface of the dying dendrite shafts. Such changes have been seen in neocortex, archicortex (hippocampus), and cerebellar cortex. The cause and/or mechanisms of these eruptive attempts at regrowth, even as the neurons are in a degenerative phase, are unknown. (Courtesy of Dr. A. B. Scheibel.)

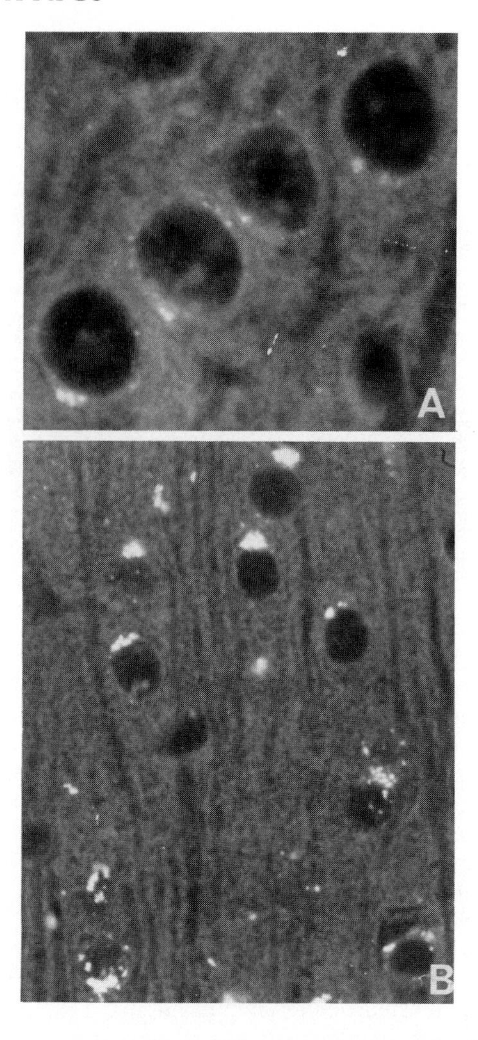

FIGURE 8-6 *Photomicrograph of unstained sections under blue light fluorescence illustrating lipofuscin pigment in neuron somata in cerebral cortex of male Long–Evans rats.*[18] (A) Young adults (109 to 113 days), magnification 32×; (B) aged rats (763 to 972 days), magnification 14×.

a compensatory reaction to neuronal loss or damage and is characterized by an increase in the number of synaptic contacts provided by the nearby neurons. Such a compensation, although less efficient, would persist in the old brain.

Even though generally quite moderate, alterations in the *axon* may contribute to the disruption of neural circuitry with aging. In the aging CNS, demyelination, axonal swelling, and changes in the number of neurofilaments and neurotubules show considerable variability. Changes in the neural cytoskeleton occurring with aging and repercussions therefrom on neuronal function are discussed in relation to disruption of neurofilaments (as in neurofibrillary tangles, particularly evident in some types of dementia) (Chapters 5 and 9).

1.5 Accumulation of Lipofuscin

Another aspect of neural aging is the accumulation of the age pigment, lipofuscin. Lipofuscin, the by-product of autophagia (self-digestion) and lipid peroxidation with both a protein and lipid component (Chapters 5 and 6), accumulates with aging in CNS cells, both neurons and glial cells, where it follows a regional distribution. Lipofuscin can be visualized as autofluorescent material *in vitro* or *in vivo* (Figure 8-6) as well as with the electron microscope as dark granules, either scattered in the cytoplasm, or clustered around the nucleus (Figures 8-7 and 8-8).[18] The functional significance of lipofuscin is unclear in neural cells as in other cells where it accumulates. The claims that lipofuscin accumulation interferes with intracellular function, or that antioxidants, by reducing brain lipofuscin, may lead to improved behavior have not been substantiated.

1.6 Neurofibrillary Tangles and Neuritic Plaques

Neurofibrillary tangles consist of *intracellular* tangled masses of fibrous elements, often flame-shaped bundles, coursing the entire cell body (Figure 8-9). They are present in normal aging, frequently found in the hippocampus, but their accumulation in the cortex and other brain areas is one of the pathologic hallmarks of the senile dementia of the Alzheimer type (AD).[19] In AD, tangles are in greatest number in the associative areas of the cortex, and their distribution is consistent with a primary toxin/infection spread through the

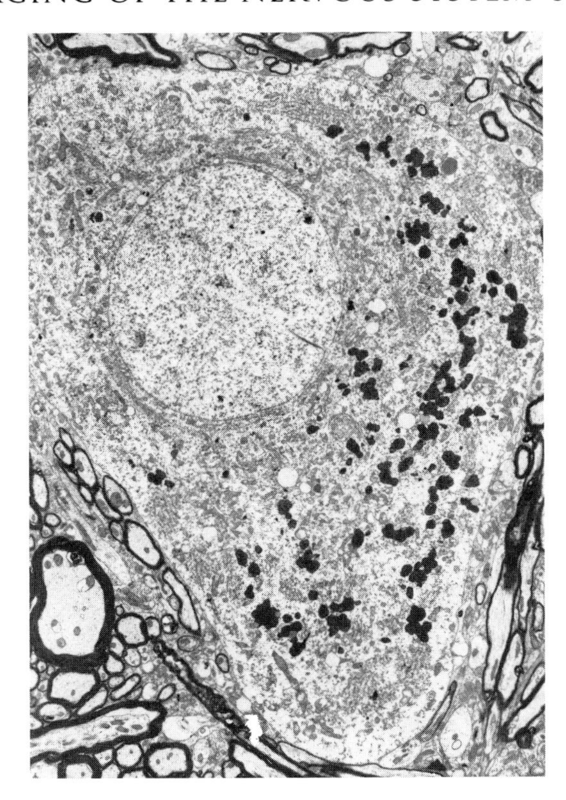

FIGURE 8-7 *A neuron from the cerebral cortex of a 605-day-old male Long–Evans rat.*[18] There are numerous dense bodies (lipofuscin) irregularly distributed in the perikaryon. Magnification 1440×.

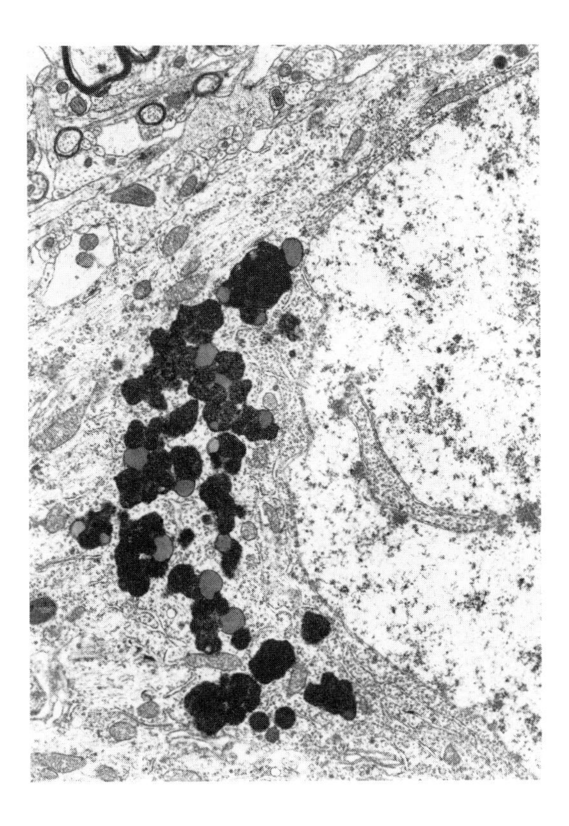

FIGURE 8-8 *A portion of a neuron from the cerebral cortex of a 605-day-old male Long–Evans rat.*[18] The lipofuscin granules are clustered at one pole of the nucleus. Magnification 6000×.

olfactory system.[20] Their density has also been correlated with the severity of the dementia, of the cell loss, and of the cholinergic deficits.[21]

Under the electron microscope, each fiber consists of a pair of filaments wound around each other to form a *paired helical filament or PHF.*[22] Chemically, PHF are highly insoluble and their precise protein composition remains unknown. However, they react with monoclonal antibodies specific for neurofilament proteins or for microtubule-associated proteins. Since both neurofilaments and microtubules are normal constituents of the cytoskeleton, the immunologic reactivity of PHF supports the hypothesis that several cytoskeletal proteins are linked in the formation of these abnormal, helical filaments (Chapter 5). Which proteins make up the core twisted filament and whether novel abnormal proteins are present are still unknown.

The *neuritic plaques* are situated *extracellularly*. Extremely rare in normal aging, they are abundant in AD where they are found in the frontal, temporal, and occipital cortex and the hippocampus. Their distribution is similar to that of the neurofibrillary tangles, and the two structures are often found in close proximity (Figure 8-10). Typically, the plaque consists of a central core of amyloid surrounded by coarse, silver-stained fibers represented at the electron microscope by many distrophic neurites (axon segments) with degenerated mitochondria, synaptic complexes, and dense bodies (Figure 8-10). Glial and microglial cells react to the plaque and accumulate about the abnormal nerve cell processes. One view is that the development of the plaque proceeds first from abnormalities of the neural processes to the deposition of amyloid and, only later, to reactive proliferation. Another view is that the primary alteration arises at the level of the brain capillaries and that amyloid or a precursor substance of blood origin leaks from the capillaries and diffuses into the cerebral tissue where it accumulates, leading to destruction of neurons and glial cells and plaque formation.[23] The possible role of amyloid deposition in AD is discussed in the next chapter, with the dementias (Chapter 9).

1.7 Circulatory Changes in Aging

Vascular changes accompany morphologic alterations. They include atherosclerotic lesions common throughout the arterial vascular system. One of the consequences of *atherosclerosis* is the production of infarcts — areas of dying tissue and scar formation following interruption of circulation due to obstruction or rupture of blood vessels. The occurrence of *multiple infarcts* leads to progressive destruction of brain tissue which may be responsible for a form of senile dementia, properly called "multiple-infarct senile dementia" contrasted with Alzheimer's dementia (Chapter 9). The cerebral capillaries are much more permeable at birth than in adulthood and the *blood-brain barrier* develops during the early years of life. This barrier is a functional concept based on specific mechanisms of exchange across the cerebral capillaries different from all other vascular beds. This peculiar type of exchange permits only a few substances to enter the nervous tissue — hence the term "barrier" — and, thus, serves a protective function (but it also limits the effectiveness of several medications which cannot cross the barrier and pass from the general circulation in the brain).

The efficiency of the blood-brain barrier increases with age during early development. With aging and disease, the blood-brain barrier may again become permeable at least to

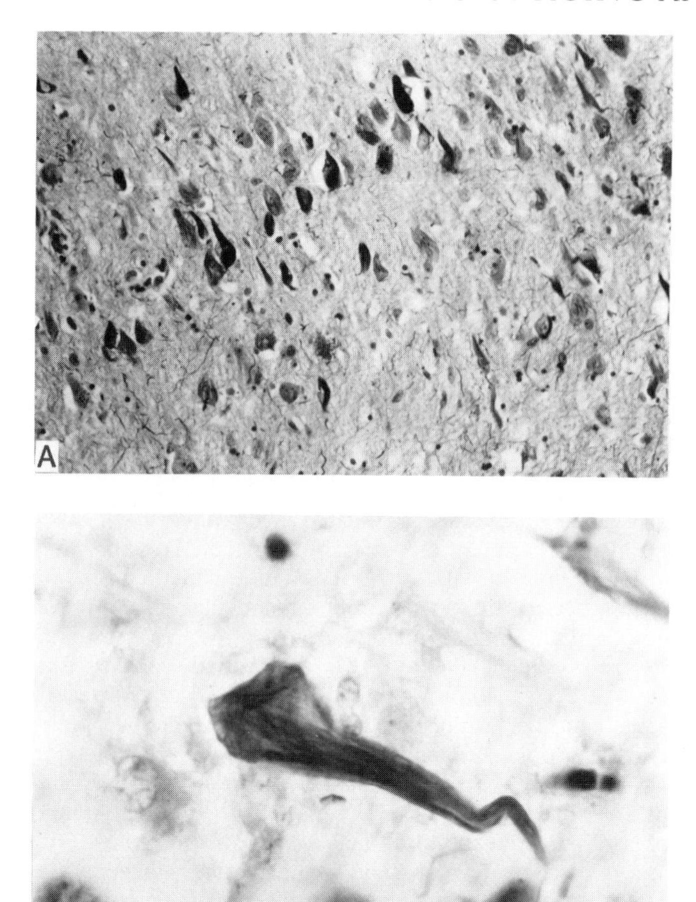

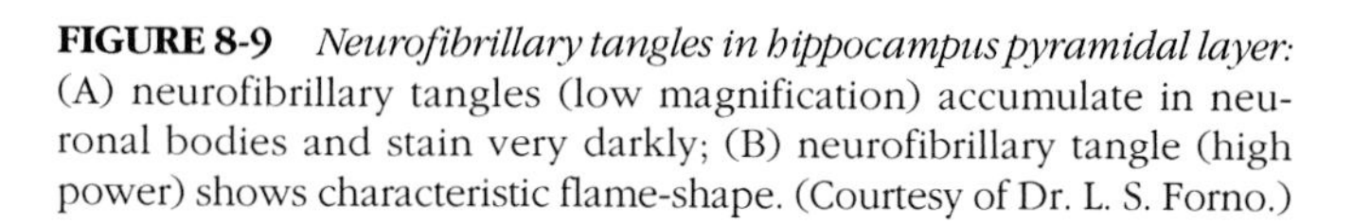

FIGURE 8-9 *Neurofibrillary tangles in hippocampus pyramidal layer:* (A) neurofibrillary tangles (low magnification) accumulate in neuronal bodies and stain very darkly; (B) neurofibrillary tangle (high power) shows characteristic flame-shape. (Courtesy of Dr. L. S. Forno.)

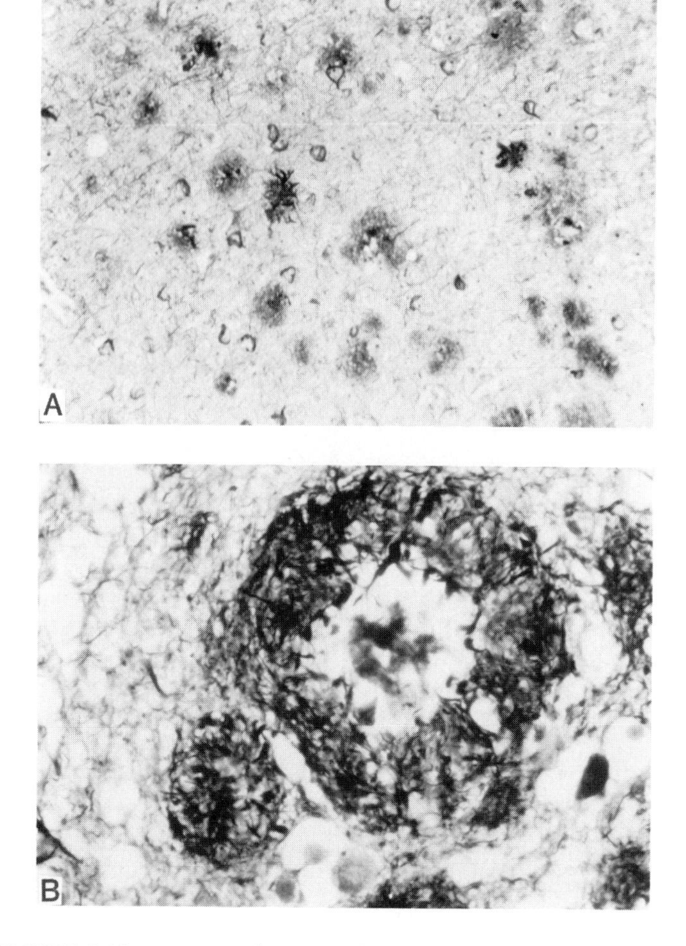

FIGURE 8-10 *Neuritic plaques in frontal cortex, silent area:* (A) presence of neuritic plaques, large, and neurofibrillary tangles, small (low power); (B) neuritic plaque (high power). (Courtesy of Dr. L. S. Forno.)

selected substances, and the passage of blood-borne substances may represent one of the causes of such dementias as those of the Alzheimer type. Studies of the structure of cerebral capillaries at progressive ages reveal an increase in the capillary wall thickness in old rats and a decrease in mitochondrial content of endothelial cells in old monkeys. Although direct evidence for increased permeability of blood-brain barrier as the primary pathogenic factor in AD is still lacking, defects of the blood-brain barrier may represent a common pathogenic mechanism linking many different risk factors.[24]

2 BIOCHEMICAL CHANGES

2.1 Neurotransmission and Cell Communication

Information processing in the nervous system involves neurons "talking" to each other or with target cells; therefore, research in neurotransmission, including neurotransmitter turnover, release, and binding, is central to our understanding of CNS aging. One of the most studied aspects of aging of the nervous system involves neurotransmitter changes at the synapse.[25-27]

Chemically, transmitters may be amines, amino acids, peptides, or gases (Table 8-3). Of these, some may be viewed as classic, our knowledge of them extending back 80 years; they are *acetylcholine,* the catecholamines — *norepinephrine, epinephrine and dopamine* — and *serotonin.* Others have been associated with neurotransmission for some time: these are certain of the amino acids such as *glycine, glutamate,* and *GABA.* Recent candidates are the peptides, among which a few of the more extensively studied are enkephalin, substance P, and the hypothalamic neurohormone thyrotropin-releasing hormone (TRH). Some peptides, such as cholecystokinin and somatostatin, are found both in the brain and the gastrointestinal tract.

Another quite new candidate is the gas *nitric oxide* (NO), viewed until recently as a toxic molecule (from cigarette smoke and smog) but now considered to act as a biologic messenger in mammals.[28] A soluble gas, NO is generated by the enzyme NO synthase, sensitive to calcium and calmodulin modulation. Unlike other neurotransmitters, NO is not stored nor released from vesicles in the neuron, but rather it simply diffuses out of the producing cell. NO is also found in genitalia (where it is essential for penile erection) and in the gut (where it is involved in peristalsis).[28] Another gas recently implicated in biological signaling is *carbon monoxide* (CO). Originally viewed as an exclusively toxic gas, CO would also possess physiologic actions in mediating odor perception

TABLE 8-3

Neurotransmitters and Modulators in the Nervous System

Amines	Amino acids	Peptides	Others
Acetylcholine	Glutamate	Enkephalin	Nitric oxide
Catecholamines	Aspartate	Cholecystokinin	Carbon monoxide
Norepinephrine	Glycine	Substance P	Zinc
Epinephrine	GABA[b]	VIP[c]	Synapsins
Dopamine	Taurine	Somatostatin	Miscellaneous
Serotonin[a]	Histamine	TRH[d]	

[a] Serotonin, 5-hydroxytryptamine, or 5-HT.
[b] GABA or gamma-amino butyric acid.
[c] VIP or vasoactive intestinal polypeptide.
[d] TRH or thyrotropin-stimulating hormone.

in olfactory neurons and learning in hippocampal neurons.[29,30] Both of these areas and functions are markedly affected with aging (Chapters 9 and 10). Metals, such as *zinc,* may also act as neuromodulators in some specific neurons (mossy fibers of hippocampus) at selected ages (during development, 14 postnatal days in the rat) and in association with the neurotransmitter GABA.[31]

A membrane-associated, synaptic vesicle protein, *synapsin,* would also influence neurotransmission by increasing the number of synapses, synaptic vesicles, and contacts.[32] There are four synapsins, generated by two gene splicing, and similar in amino acid sequences. All four are associated with the cell cytoskeleton and, like some of the other cytoskeletal proteins (e.g., tau proteins, Chapter 3), they may undergo phosphorylation (during synaptic transmission). By transfection, the production of synapsins has been increased in a cell line.[33] More studies must be conducted to establish these and possibly other molecules as true neurotransmitters and to investigate eventual changes with aging.

Different transmitters — usually a classic neurotransmitter and a peptide — may coexist within the same neuron. Such "coexistence", resulting in "co-utilization" of several transmitters, vastly expands the potential diversity of synaptic communications. Indeed, the peptide often "modulates" or "modifies" the action of the classic neurotransmitter. For example, serotonin coexists with substance P and TRH in neurons of the rat medulla and spinal cord where the three neurotransmitters collaborate to regulate some motor and behavioral responses.

A consequence of synaptic diversification and multiplicity of neurotransmitters is the increasing complexity of neuronal communication and the need for a fine tuning of all chemical messengers, transmitters, and their modulators. Neurotransmitters have excitatory or inhibitory actions, and, most important to the function of the nervous system, these actions must remain in balance. With aging, alterations in function are more likely to occur as a consequence of imbalance among neurotransmitters rather than as a consequence of a global alteration of a single neurotransmitter. Such imbalance could involve the classic neurotransmitters for which evidence is already available, or the sustaining peptides, or both.

Chemical Classification of Synapses

Many synapses are classified according to the neurotransmitter released at this site. Examples of currently recognized synaptic transmitters include acetylcholine (ACh), norepinephrine (NE), serotonin (5-hydroxytryptamine), and γ-aminobutyrate (GABA), and the synapses are named accordingly. Briefly summarized are their synthesis, storage, release, postsynaptic interactions, and inactivation (Figure 8-11).

In the human brain, age differences in neurotransmitter values, their synthesizing and degrading enzymes, and receptors are still fragmentary, tentative, and highly incomplete. The levels and activity of neurotransmitters and enzymes decline in many brain regions during normal aging; one classic example is the decreased levels of dopamine due to loss of neurons in the substantia nigra (see below). Neurotransmitter receptors may or may not respond to the neurotransmitter changes by correspondingly increasing number and/or affinity in response to reduction in neurotransmitter and *vice versa*.

Analysis of Synaptic Chemical Events: Difficulties of Measurement in Humans

Three tiers of chemical events should be considered to obtain a complete analysis of synaptic changes with aging (Table 8-4); they should include the study of the following:

- Presynaptic events comprising availability of precursors, enzymatic synthesis, and degradation of transmitters, their storage in the cell or their release in the synaptic space, transmitter re-uptake, and ionic regulation
- Transynaptic events comprising free neurotransmitters and enzymes for their degradation
- Postsynaptic events comprising receptor binding, enzymatic degradation, ionic regulation, and induction of second messenger

Unfortunately, in humans, information is limited to the measurement of the neurotransmitters and their metabolites in urine, blood, and cerebrospinal fluid and, more recently, to the use of labeled probes and imaging devices. The majority of these methods show how remote and indirect the sampling is from the area involved, in the central nervous system. An alternative approach is to use pharmacologic agonists and antagonists of the transmitter under study. Under all conditions, it is difficult to control the clinical setting and/or accurately measure subtle changes in physiological status. The intricacies of the synaptic complex make it currently impossible to develop a coherent description of the changes with aging in synaptic function, except in instances of pathologic events. Clearly, the current situation suggests that much is to be gained by the use of animal models from which to obtain information of the cellular and molecular mechanisms of synaptic function. Even in animals, the information on aging is still scarce and needs to be pursued vigorously.

2.2 Neurotransmitter Imbalance

Imbalance among neurotransmitters, rather than severe depletion or excess of an individual neurotransmitter, may underlie the functional changes with aging. The concentrations of serotonin, norepinephrine, and dopamine were measured in discrete brain regions of the rat brain from birth until 3 years of age, quite an old age for the rat with a median lifespan of 24 months.[34-36] We omit here, for brevity, the early age-related changes despite their significance for development, and compare the 6-month-old adult rat with rats at the old ages of 2 and 3 years. The data for the cerebral hemispheres, the hypothalamus, and the corpus striatum are presented in Figure 8-12. Overall, two major changes emerge: serotonin concentrations remain unchanged until very old age when they increase, and norepinephrine and dopamine concentrations progressively decrease starting at relatively less advanced ages. In the aging rat brain, then, the ratio of serotonin to the

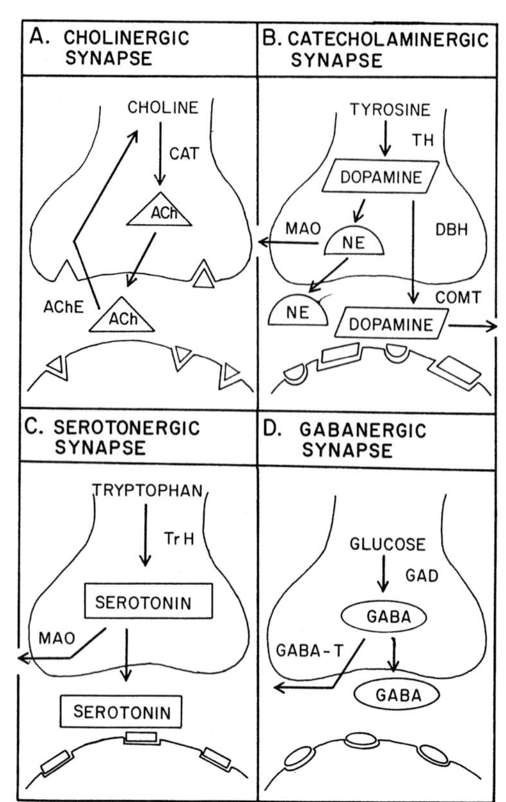

FIGURE 8-11 *Diagrams of established CNS synapses illustrating presynaptic synthesis, storage and metabolism, synaptic release, and postsynaptic binding.* (A) **Cholinergic synapse.** Choline is the precursor of the transmitter ACh through the actions of the synthesizing enzyme choline acetyltransferase (CAT). ACh is free in the cytoplasm of the prejunctional axonal ending as well as contained in vesicles. Release of cytoplasmic ACh initiates transmembrane events. Several distinct pre- and postsynaptic ACh receptors have been described. Receptor-bound ACh is inactivated by the enzyme acetylcholinesterase (AChE). (B) **Adrenergic synapse.** Norepinephrine's (NE) precursor is tyrosine, and the enzymes involved in synthesis are tyrosine hydroxylase (TH), dopa decarboxylase (DOD), and dopamine-beta-hydroxylase (DBH). The primary degradating enzyme is mitochondrial monamine oxidase (MAO). NE is stored in vesicles. Once released, transynaptic degradation is initiated by catechol-O-methyltransferase (COMT). Currently four types of adrenergic receptors have been identified, both pre- and postsynaptic. Released NE not degraded by COMT may be taken up by the presynaptic neuron. (C) **Serotonergic synapse.** The precursor of serotonin is tryptophan, and the important synthesizing enzyme is tryptophan hydroxylase (TrH); the degrading enzyme is MAO. Several different types of cell surface receptors are known. (D) **Gabanergic synapse.** GABA is formed in the GABA-shunt pathway of the Krebs cycle. It is formed by the action of glutamic acid decarboxylase (GAD) from glutamate and is metabolized by deamination. Two receptors for GABA have been identified in the CNS. (Drawing by S. Oklund.)

catecholamines progressively increases. For the resulting imbalance to become functionally manifest, it is sufficient that only one neurotransmitter be altered rather than an overall shift.

A final point to underline here is that each neurotransmitter has its own regional timetable of aging. For example, in rats, serotonin concentration remains unaltered in the cerebral hemispheres until 3 years, while dopamine levels in this same brain region are reduced at 1 year of age.[37] It is this differential rate of aging that may be responsible for creating an early neurotransmitter imbalance, well before significant decrements in each transmitter can be detected.

Examples of neurotransmitter systems markedly affected with aging, and leading to functional disorders, are the decline in dopamine content in the substantia nigra, discussed below, and in acetylcholine content in the nucleus basalis of Meynert postulated to have functional relevance in the cognitive deficits of AD (Chapter 9).

TABLE 8-4

Major Chemical Events at the Synapse

Presynaptic
Availability of precursors
Enzymatic synthesis and degradation
Storage or release
Transmitter re-uptake
Ionic regulation
Transynaptic
Free neurotransmitter
Enzymatic degradation
Postsynaptic
Receptor binding
Enzymatic degradation
Ionic regulation
Second messenger

2.3 Aging of Dopaminergic Systems and Parkinson's Disease

The dopaminergic system undergoes changes with normal aging and with aging-associated diseases. Dopamine, once considered an intermediary in the formation of norepinephrine, is now recognized as a distinct neurotransmitter. It is present throughout the brain, but its levels vary significantly from region to region. Currently four major dopaminergic pathways are recognized, the most prominent originates in the substantia nigra and extends to the corpus striatum (including the caudate nucleus and putamen), both structures of the so-called extrapyramidal system; the other dopaminergic pathways involve the limbic system, the hypothalamus, and the cortex (Table 8-5).

Major Dopaminergic Pathways

The nigral-striatal pathway participates in central control of motor functions. Dopamine released at the caudate nucleus inhibits the stimulatory (cholinergic? glutaminergic?) input from the motor cortex.

In the caudate, a balance is established between the inhibitory and excitatory control of motor activity. A deficit of dopamine and an increase in the cholinergic output results in rigor and tremor, whereas an increase of dopamine or a decrease of the excitatory input results in writhing movements, as in chorea.

Similar to the striatum and also involved in motor control is the nucleus accumbens in the septal region which controls locomotion (particularly forward movement).

Within the limbic system, other dopaminergic fibers synapse and mediate emotionality, sexual and aggressive behavior, and some neuroendocrine regulation.

In the hypothalamus, a rich dopaminergic network regulates various endocrine-releasing and inhibitory hormones. This dopaminergic activity will be discussed when the endocrine system is presented.

A less well-developed neocortical system is apparently activated by stress. Lesions of the system induce locomotor hyperactivity and inability to suppress behavioral "stereotypic responses". These responses can be

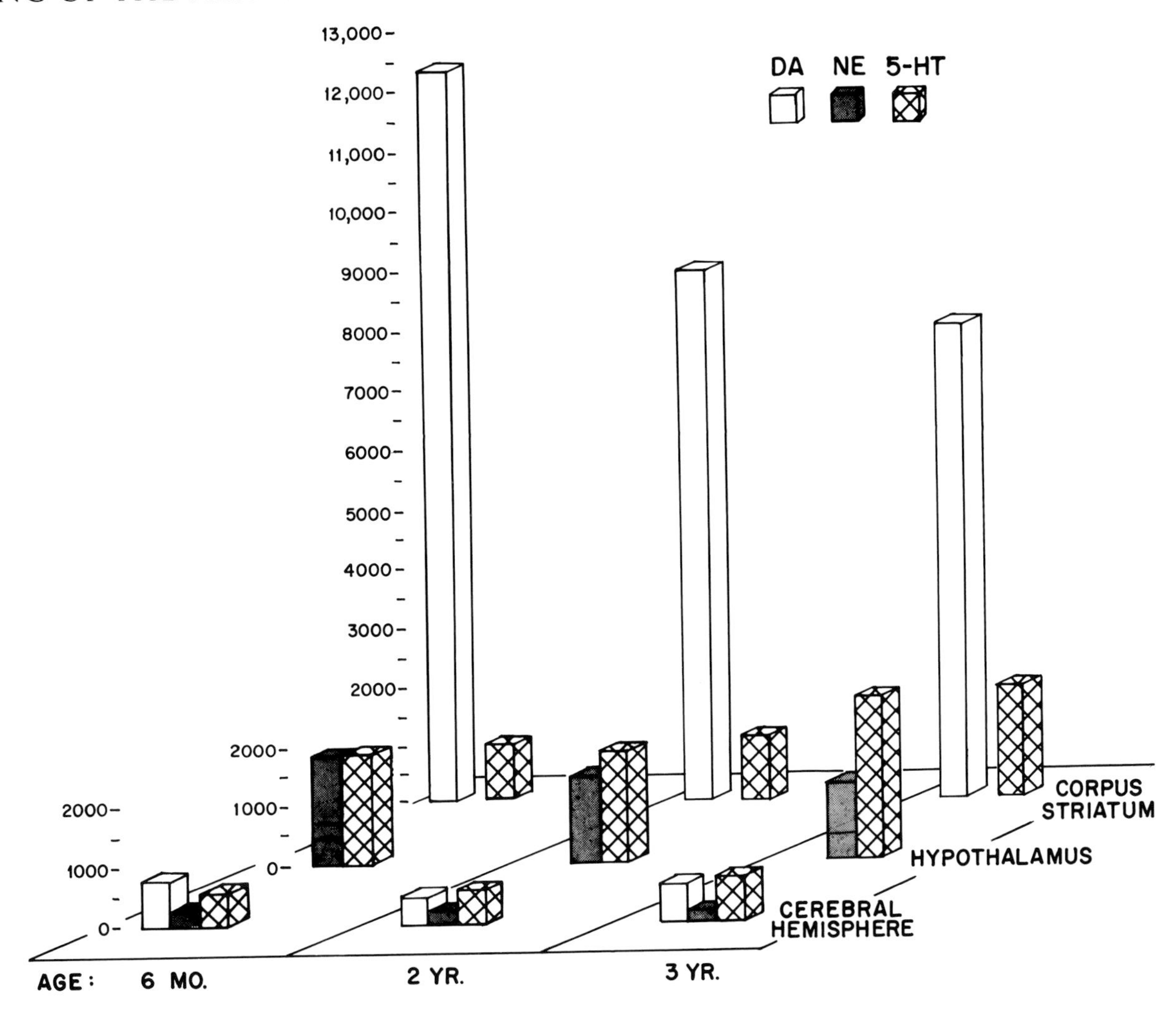

FIGURE 8-12 *Regional neurotransmitter changes with aging.* Dopamine (DA), NE, and serotonin (5-HT) have different patterns of aging in different brain areas.[34,35,37] While little change occurs with aging in any of the neurotransmitters in the cerebral hemispheres, NE in the hypothalamus and DA in the corpus striatum decrease while 5-HT increases in both areas.

blocked by drugs which act as dopamine inhibitors. Uncontrolled stimulation of dopamine receptors would have, on the one hand, a euphoric action but, on the other hand, would have a psychotic action, perhaps involved in manic and schizophrenic syndromes.

With normal aging, dopamine content decreases, particularly in the corpus striatum, probably as a consequence of the loss of dopaminergic neurons (perhaps due to free radical damage, Chapter 6). This decrease may or may not be associated with motor disturbances depending on the degree of neuronal loss or the severity of the action of excitotoxins (such as glutamate).[38] When neuronal loss is severe, the neurologic alterations become more pronounced and are associated with progressively more severe motor decrements and other neurologic and clinical manifestations; they are categorized as part of Parkinson's disease or parkinsonism. This disorder involves rigidity, tremor, and akinesia (loss of motor function).[39] It is largely a progressive disease of the elderly, about 60% of those affected are age 70 or older. In parkinsonism, the dopaminergic neurons projecting to the caudate nucleus and putamen degenerate due to reduced stimulation from the damaged

TABLE 8-5

Dopaminergic Pathways

Substantia nigra to corpus striatum
 Regulates control of motor function from cortex
Substantia nigra to nucleus accumbens
 Controls locomotion (forward movement)
Limbic System
 Regulates emotion, behavior (sexual and aggressive)
Hypothalamus
 Regulates endocrine-releasing and inhibiting hormones
Neocortex
 Regulates locomotor activity

substantia nigra (Figure 8-13). This damage is characterized by loss of neurons, gliosis, and the presence of specific inclusions, the so-called Lewy bodies (hyaline inclusions in neurons) (Figure 8-14).[40,41]

Lewy bodies may be found in normal aging, but their incidence is much greater not only in Parkinson's but also in AD patients.[42] There is an overlap in pathology between Parkinson's and Alzheimer's diseases.[41] The prevalence of Alzheimer-type histological changes and of dementia is higher

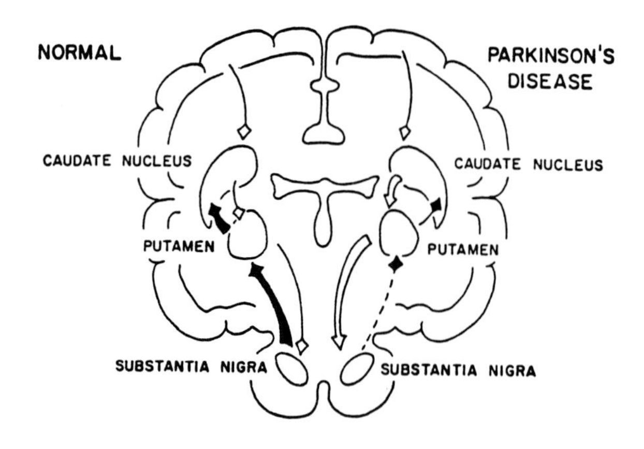

FIGURE 8-13 *Dopaminergic nigral–striatal pathway in the normal (left dark arrow) and in Parkinsonism (right dashed line).* In Parkinsonism, dopaminergic neurons of the substantia nigra are lost, reducing dopamine (right-dashed line) to putamen-caudate (striatum) and subsequent control of cortical stimulatory effects. Increased excitatory transmission to extrapyramidal system (broad open arrow) is associated with tremor and rigor. (Drawing by S. Oklund.)

in parkinsonians than in an age-matched unaffected population.[43]

Drug-Induced Parkinson's Disease

A contaminant of illegally synthesized heroin — 1-methyl-4-phenyl-1,2,3,6-tetrahydropyridine (MPTP) — acts specifically to induce a motor-deficit syndrome in experimental animals (mice, rabbits, monkeys) similar to Parkinson's disease in humans.[44] In young animals, the syndrome is only incompletely produced, whereas in old animals the neuropathology and behavior of Parkinson's disease are completely reproduced. Older neurons appear more vulnerable to MPTP, either because of lowered ability to recover from toxic damage or because the damage is cumulative with other damage which has occurred over time.[45] A variety of environmental agents and/or stress may be involved in inducing other CNS degenerative diseases, such as senile dementia of the Alzheimer's type (AD), in which age is a key factor (Chapter 9). This MPTP drug model is particularly useful in transplantation studies to assess the efficiency of the transplanted tissue in overcoming the MPTP-induced damage.

While motor abnormalities dominate the clinical picture of Parkinson's disease, current replacement therapy permits more patients to live long enough to develop more advanced features of the disease, that is, loss of memory, dementia, and postural instability, which are very difficult to treat.[46] Indeed, Parkinson's and Alzheimer's diseases may be viewed as two poles of a clinical spectrum of which the common feature is neuronal damage.

2.4 Treatment of Parkinson's Disease

A brief discussion of the treatment of Parkinson's disease is appropriate here, not only for its clinical importance but also because it provides a better understanding of the etiopathology of the disease (Table 8-6).

L-Dopa (dihydroxyphenylalanine, levodopa) is an intermediary product in catecholamine biosynthesis: it derives from the amino acid tyrosine which is converted to dopa by the enzyme tyrosine hydroxylase. L-Dopa is further converted to dopamine by the enzyme dopa decarboxylase. Treatment with L-dopa has emerged as a model for the effective therapy of aging-associated neurologic disorders. The loss of dopaminergic input to the striatum disrupts the balance with the mainly cholinergic excitatory activity. The original pharmacologic treatment was to reduce neural activity with cholinergic inhibitors. This is still used as an adjunct to replacement therapy with L-dopa. The relative success of the therapy for parkinsonism

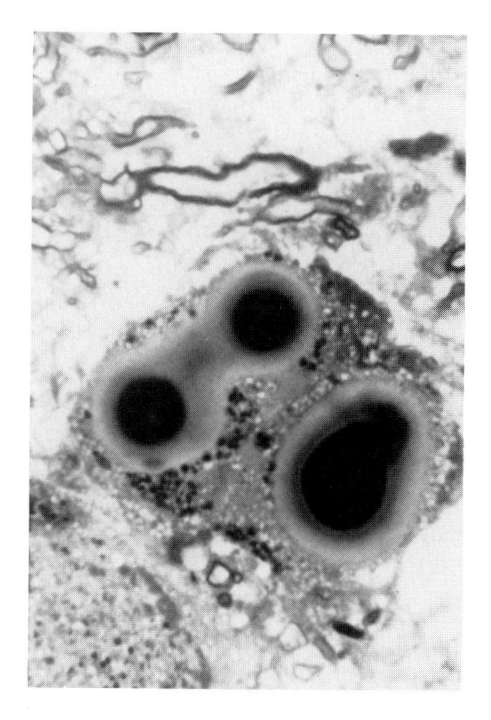

FIGURE 8-14 *Lewy bodies, intraneuronal cytoplasmic inclusions.* The dense core is composed of apparently random, tightly packed aggregations of filaments, vesicular profiles, and poorly resolved granular material containing neurofilament antigens. These bodies are found in substantia nigra as well as in other CNS regions. (Courtesy of Dr. L.S. Forno.)

TABLE 8-6
Treatment of Parkinson's Disease

1. **Administration of L-dopa** to correct dopamine deficiency consequent to loss of dopaminergic neurons. L-Dopa is the dopamine precursor. It is capable of crossing the blood–brain barrier and is converted to dopamine.
2. **Administration of monoamine oxidase (MAO) inhibitors** to block the action of the enzyme MAO to hydrolyse dopamine. MAO inhibitors (such as deprenyl) can be administered together with L-dopa, thereby permitting lower L-dopa dosage.
3. **Administration of antioxidants,** such as tocopherol, to prevent and reduce accumulation of free radicals (eventually capable of destroying dopaminergic cells).
4. **Administration of anticholinergic drugs** to reduce some of the symptoms related to parasympathetic stimulation. These agents are often administered in combination with L-dopa and MAO inhibitors.
5. **Administration of glial cell line–derived neurotrophic factor GDNF.** GDNF selectively promotes survival of dopaminergic neurons (but does not promote cells division). The effectiveness of this newly produced recombinant protein is promising *in vitro* and in animals *in vivo* but is still awaiting clinical confirmation.
6. **Transplantation of human fetal neurons, or adrenal medullary cells,** despite uncertain results and ethical controversy.

is encouraging pharmacologic interventions in other types of neurologic and psychiatric diseases of old age. Even though the treatment may not provide a "cure", it ameliorates some of the more disturbing symptoms. Thus, while most of the neurologic and psychiatric diseases of old age may not as yet be "curable", they certainly are "treatable" and must be treated.

In Parkinsonism, dopamine replacement is possible by the oral administration of L-dopa, natural precursor of dopamine (Table 8-6). Unlike dopamine, L-dopa passes through the blood-brain barrier. This is a rare instance when the administration of a neurotransmitter precursor is beneficial; however, this is not without its drawbacks. Side effects are specific and neurologic as well as general and systemic, and there is a gradual reduction of effectiveness. The latter is probably due to continuing loss of dopaminergic neurons and reduction of the decarboxylase enzyme which transforms L-dopa into dopamine. The treatment with L-dopa is often used in combination with other drugs such as dopamine agonists (e.g., bromocriptine), monoamine oxidase inhibitors (e.g., deprenyl), or a variety of anticholinergic drugs.[47]

Replacement of dopamine has also been accomplished in humans by transplantation of medullary tissue from the adrenal glands or of human fetal neurons from aborted fetuses to the substantia nigra in the brain. Although the patients with such transplants still require L-dopa to improve motor function, the dose of L-dopa is reduced, a sign that the transplanted cells do secrete dopamine.[47]

Transplant Techniques

These techniques have been derived from animal studies.[48,49] Basic techniques of neural transplantation have long been within the repertoire of the neuroscientist, but only since the 1970s have they been systematically applied to the mammalian CNS.[50] Transplants have shown promise in studying the development of various visual and olfactory centers by transplantation to novel locations in the brain, the reestablishment of endocrine function by fetal transplants into hypothalamic nuclei, and the repair of damaged functions by transplants of homologous fetal tissues in the hippocampus, nigrostriatal system, spinal cord and autonomic ganglia.[51] Another approach has taken advantage of genetically engineered cells selected to secrete growth or trophic factors (like GDNF, see below) and transplanted, either alone or with fetal neurons, into the brain of affected individuals.

A promising type of treatment involves the administration in the brain (by implanted cannula) of a neurotrophic factor of glial origin, the "glial cell line-derived neurotrophic factor" (GDNF) related to other members of the transforming growth factor superfamily.[52,53] GDNF would specifically act on dopaminergic cells in culture where it does not stimulate cell division but keeps dopaminergic cells alive. While the effectiveness of this treatment remains to be substantiated by clinical trials, the possibility that neutropic factors may benefit neuronal survival in CNS degenerative diseases represents a physiologic and hopeful therapeutic intervention.

3 METABOLIC CHANGES

In addition to neurotransmitters, other chemical constituents of the central nervous system are known to change with aging. A number of these are listed in Table 8-7.

From clinical and experimental data, some generalizations can be drawn: changes (1) are specific to discrete regions and structures; (2) follow differential timetables; and (3) differ according to the constituent considered. Because of the lack of consistent studies in humans during normal aging, only a few of these constituents will be discussed here. A good model of regional, age, and constituent specificity is presented by some amino acids which act as putative neurotransmitters. For example, of the excitatory amino acids, glutamic acid declines with aging in the cerebral cortex and brain stem of the rat, but not in the cerebellum and spinal cord.[36] Glycine and GABA, inhibitory amino acids, increase with age in the cerebral cortex, cerebellum, and brain stem, but remain unchanged in the spinal cord. These observations are supportive of an imbalance among neurotransmitters with aging.

TABLE 8-7

Possible Targets of Aging Changes in the Central Nervous System

Total water	Carbohydrates
Extra- and intracellular spaces	Circulation
Lipids	Energy metabolism
DNA, RNA, and protein	Oxygen uptake and glucose utilization
Amino acids	Blood-brain barrier

Other changes with normal aging include:

- Decreased water content of the brain (as of most other organs)
- Alterations in extra- and intracellular spaces while electrolyte distribution remains essentially unchanged
- Regional decrease in protein content and synthesis (perhaps related to learning and memory impairments), increase in complexity of RNA molecules; and either an increase (perhaps related to gliosis) or no change in DNA content
- Decreased lipid synthesis, primarily decreased synthesis of membrane phospholipids due to increased variation in the structure of lipid substrates rather than to reduction of synthesizing enzyme activity or concentration of substrates; changes in membrane lipids would alter membrane fluidity and, in turn, nerve conduction and receptor binding
- Circulatory changes characterized by a progressive reduction in cerebral blood flow and a corresponding decrease in oxygen uptake, glucose utilization, and, consequently, energy metabolism

Several parameters of cerebral energy metabolism such as cerebral blood flow, oxygen consumption, and glucose metabolism (cerebral metabolic rate) have been studied in humans and experimental animals utilizing tracers, autoradiography, computed tomography (CT) scans, positron emission tomography (PET), and nuclear magnetic resonance (NMR). In general, changes are not marked and vary with the animal species, the brain area considered, the ambient sensory state, and the health of the subject. In particular, the sensory state (visual, auditory, proprioceptive, etc.) in normal individuals influences cerebral metabolism.

Normal metabolism is dependent on normal glucose utilization, normal membrane function, and normal supplies of high-energy phosphates. With aging, overall cerebral metabolism is only slightly diminished or not at all. When present, impairment of brain energy metabolism (in the laboratory rat) has been related to loss of dendritic processes.[54]

Aged dogs, showing some neuropathologic changes with age, have significant reduction of glucose metabolism, especially in sensory-related structures and the frontal and temporal cortices.[2]

The relative stability of energy metabolism of the aging brain in the absence of neuropathology must be contrasted with the severe decrease that occurs in the aged brain affected by degenerative diseases such as AD.[55,56] In AD, the reduction in cerebral formation rate of adenosine triphosphate (ATP) from oxidized glucose and oxygen ranges from 7 to 20% in incipient AD and 35 to 50% in stable, advanced dementia[57] (Chapter 9).

4 PERSISTENCE OF CNS PLASTICITY IN AGING

Until recently most neuroscientists viewed much of the brain, with the exception of some specialized (mostly sensory) areas, as a complex of "hard-wired" centers and pathways. From birth on, "all the information-carrying pathways would be firmly and immutably formed."[58] This view was based on observations establishing that early experience (such as visual experience) during CNS ontogenesis would determine structure and function for life. Thus, factors, such as hormones (e.g., thyroid hormones, sex steroids), would determine CNS differentiation and maturation when present at "critical periods" during development.[59,60]

Today this view is being modified in the light of experiments supporting quite a different concept of the brain as "a network that is continually remodeling itself".[58] This continuing "plasticity" of the brain would originate during development and persist (possibly under the influence of continuing environmental, including hormonal, stimulation) throughout adult life and possibly old age. The "plastic" CNS would be capable of adapting to environmental changes or to changes within the organism itself and of controlling its responses accordingly.[12,13] Neural plasticity would be maximal during development and would continue throughout adulthood; thyroid hormones, for example, continue to influence brain function and other physiologic parameters well past the established critical periods and perhaps also in the aged animal.[61-63] However, to what extent plasticity is retained in the CNS of the elderly and how efficient it remains in old age are still to be definitely determined.

As indicated at the beginning of this chapter, adult neurons do not proliferate. This failure has been interpreted to mean that in the mature CNS, neurons lack the capacity of adaptive remodeling and, eventually, regenerating. Thus, in old age, any degenerative condition (such as occurs in AD) leading to cell loss would be irreparable. Current data seem to contradict this interpretation; rather, the evidence suggests that neurons would be capable of dividing and regenerating if their microenvironment resembled that present during early development. For example, inhibitory factors (e.g., gliahyalurinate-binding protein, GHAP), which prevent neuronal plasticity[64,65] and neurotropic factors that promote neuronal survival (e.g., GDNF) and division (e.g., nerve growth factor, NGF), have been identified and used for treatment of degenerative diseases of old age, such as Parkinson's discussed above, and Alzheimer's, which is discussed in the next chapter.

Neurotropic Agents

The best known among the neurotropic agents is *nerve growth factor (NGF)*. This is a basic protein, resembling in structure the hormone insulin (Chapter 15). NGF promotes growth and maintenance of sympathetic and sensory neurons as well as neurons in the brain during both developing and mature ages. In the brain, NGF is particularly effective in promoting growth and reducing or preventing damage of cholinergic neurons of the basal forebrain (nucleus basalis of Meynert) drastically affected by neuronal loss in AD.[66] In AD, NGF levels, receptors, and mRNA are reduced. Intraventricular administration of NGF prevents in young and adult rats these cholinergic neurons from dying after axonal transection or chemical damage and, in fact, may promote regeneration.[67-69]

NGF is abundant in salivary glands and is picked up by neuronal terminals and transported in retrograde fashion from the endings of the neurons to their cell bodies. NGF has been purified and sequenced; a number of similarly growth-promoting proteins have been identified, and a mouse fibroblast cell line capable of secreting recombinant NGF has been established.[66]

Other neurotrophic agents involved in development and/or maintenance or repair of neuronal tissue in adult and aged animals include, in addition to GDNF discussed above for its action on dopaminergic neurons, gangliosides[70,71] and several compounds of the fibroblast growth factor (FGF) family effective on neurons, glial cells, and specialized sensory cells (e.g., photoreceptors) in health and disease[72-75] and some components of the cytoskeleton.[76,77]

An attempt to generalize the current views suggests that neurons in old age are capable of displaying compensatory responses (reminiscent of those occurring in ontogenesis) when they are provided with an appropriate environment or when the conditions that curtail their growth are removed. With aging, surviving neurons are capable of remodeling their configuration (e.g., reactive dendritic sprouting,[17] axonal growth[73]) in response to functional challenges. However, neuronal loss, isolation of one structure from another, or neurotransmitter deficits are more difficult to repair in the aged brain unless growth promoting factors become available again and inhibitory factors are eliminated. As neuronal modeling is gene regulated during ontogenesis, likewise, differential gene expression in aged neurons may direct the morphological and biochemical compensatory events that follow aging or damage and lead to functional recovery.

REFERENCES

1. Terry, R. D. and Katzman, R., Eds., *The Neurology of Aging,* F. A. Davis, Philadelphia, 1983.
2. Duara, R., London, E. D., and Rapoport, S. I., Changes in structure and energy metabolism of the aging brain, in *Handbook of The Biology of Aging,* Finch, C. E. and Schneider, E. L., Eds., Van Nostrand Reinhold, New York, 1985, 595.
3. Haug, H., Are neurons of the human cerebral cortex really lost during aging? A morphometric examination, in *Senile Dementia of the Alzheimer Type: Early Diagnosis, Neuropathology and Animal Models,* Traber, Y. and Gispen, W. H., Eds., Springer-Verlag, New York, 1985, 150.
4. Coleman, P. E. and Flood, D. G., Neuron numbers and dendritic extent in normal aging and in Alzheimer's disease, *Neurobiol. Aging,* 8, 521, 1987.
5. Giacobini, E., Mussini, I., and Mattio, T., Aging of cholinergic synapses: fiction or reality?, in *Dynamics of Cholinergic Function,* Hanin, I., Ed., Plenum Press, New York, 1986, 177.

6. Bondareff, W., Neuropathology of Nucleus Basalis and Locus Ceruleus in Alzheimer's disease, in *The Biological Substrates of Alzheimer's Disease,* Scheibel, A. B., Wechsler, A. F., and Brazier, M. A. B., Eds., Academic Press, New York, 1986, 101.
7. Butcher, L. L. and Woolf, N. J., Central cholinergic systems: synopsis of anatomy and overview of physiology and pathology, in *The Biological Substrates of Alzheimer's Disease,* Scheibel, A. B., Wechsler, A. F., and Brazier, M. A. B., Eds., Academic Press, New York, 1986, 73.
8. Vogels, Q. J. M., Broere, C. A. J., Ter Laak, H. J., Ten Donkelaar, H. J., Nieuwenhuys, R., and Schulte, B. P. M., Cell loss and shrinkage in the Nucleus Basalis complex in Alzheimer's disease, *Neurobiol. Aging,* 11, 3, 1990.
9. Mervis, R., Cytomorphological alterations in the aging animal brain with emphasis on Golgi studies, in *Aging and Cell Structure,* Vol. 1, Johnson, J. E., Jr., Ed., Plenum Press, New York, 1981, 143.
10. Brizzee, K. R., Sherwood, N., and Timiras, P. S., A comparison of cell populations at various depth levels in cerebral cortex of young adult and aged Long-Evans rats, *J. Gerontol.,* 23, 289, 1968.
11. Brizzee, K. R., Neuron aging and neuron pathology, in *Relations Between Normal Aging and Disease,* Johnson, H. A., Ed., Raven Press, New York, 1985, 191.
12. Diamond, M. C., *Enriched Heredity, The Impact of the Environment on the Anatomy of the Brain,* The Free Press, New York, 1988.
13. Will, B. E., Schmitt, P., and Dalrymple-Alford, J. D., Eds., *Brain Plasticity, Learning and Memory,* Plenum Press, New York, 1985.
14. Koch, C., Zador, A., and Brown, T. H., Dendritic spines: convergence of theory and experiment, *Science,* 256, 973, 1992.
15. Scheibel, M. E., Lindsay, R. D., Tomiyasu, U., and Scheibel, A. B., Progressive dendritic changes in aging human cortex, *Exp. Neurol.,* 47, 392, 1975.
16. Scheibel, A. B. and Tomiyasu, U., Dendritic sprouting in Alzheimer presenile dementia, *Exp. Neurol.,* 60, 1, 1978.
17. Cotman, C. W. and Holets, V. R., Structural changes at synapses with age: plasticity and regeneration, in *Handbook of the Biology of Aging,* Finch, C. E. and Schneider, E. L., Eds., Van Nostrand Reinhold, New York, 1985, 617.
18. Brizzee, K. R., Cancilla, P. A., Sherwood, N., and Timiras, P. S., The amount and distribution of pigments in neurons and glia of the cerebral cortex. Autofluorescence and ultrastructural studies, *J. Gerontol.,* 24, 127, 1969.
19. Iqbal, K., Grundke-Iqbal, I., Merz, P. A., and Wisniewski, H. M., Age-associated neurofibrillary changes, in *The Aging Brain: Cellular and Molecular Mechanisms of Aging in the Nervous System,* Giacobini, E., Fiologamo, G., Giacobini, G., and Vernadakis, A., Eds., Aging, Vol. 20, Raven Press, New York, 1982, 247.
20. Pearson, R. C. A., Esiri, M. M., Hiorns, R. W., Wilcock, G. K., and Powell, T. S., Anatomical correlates of the distribution of the pathological change in the neocortex in Alzheimer's disease, *Proc. Natl. Acad. Sci. U.S.A.,* 82, 4531, 1985.
21. Jacobs, R. W. and Butcher, L. L., Pathology of the Basal Forebrain in Alzheimer's disease and other dementias, in *The Biological Substrates of Alzheimer's Disease,* Scheibel, A. B., Wechsler, A. F., and Brazier, M. A. B., Eds., Academic Press, New York, 1986, 87.
22. Terry, R. D., The fine structure of neurofibrillary tangles in Alzheimer's disease, *J. Neuropathol. Exp. Neurol.,* 22, 269, 1963.
23. Scheibel, A. B., Duong, T., and Tomiyasu, U., Microvascular changes in Alzheimer's disease, in *The Biological Substrates of Alzheimer's Disease,* Scheibel, A. B., Wechsler, A. F., and Brazier, M. A. B., Eds., Academic Press, New York, 1986, 177.
24. Mortimer, J. A. and Hutton, J. T., Epidemiology and etiology of Alzheimer's disease, in *Senile Dementia of the Alzheimer Type,* Hutton, J. T. and Kenny, A. D., Eds., Neurology and Neurobiology, Vol. 18, Alan R. Liss, New York, 1985, 177.
25. Giacobini, E., Cellular and molecular mechanisms of aging of the nervous system: toward a unified theory of neuronal aging, in *The Aging Brain: Cellular and Molecular Mechanisms of Aging in the Nervous System,* Giacobini, E., Filogamo, G., Giacobini, G., and Vernadakis, A., Eds., Aging, Vol. 20, Raven Press, New York, 1982, 271.
26. Rogers, J. and Bloom, F. E., Neurotransmitter metabolism and function in the aging central nervous system, in *Handbook of the Biology of Aging,* Finch, C. E. and Schneider, E. L., Eds., Van Nostrand Reinhold, New York, 1985, 645.
27. Strong, R., Wood, W. G., and Samorajski, T., Neurochemistry of ageing, in *Principles and Practice of Geriatric Medicine,* Pathy, M. S. J., Ed., John Wiley & Sons, New York, 1991, 69.
28. Culotta, E. and Koshland, D. E., Jr., Molecule of the year. NO news is good news, *Science,* 258, 1862, 1992.
29. Barinaga, M., Carbon monoxide: killer to brain messenger in one step, *Science,* 259, 309, 1993.
30. Verna, A., Hirsch, D. J., Glatt, C. E., Ronnet, G. V., and Snyder, S. H., Carbon monoxide: a putative neural messenger, *Science,* 259, 381, 1993.
31. Xie, X. and Smart, T. G., A physiological role for endogenous zinc in rat hippocampal synaptic neurotransmission, *Nature,* 349, 521, 1991.
32. Kelly, R. B., A system for synapse control, *Science,* 349, 650, 1991.
33. Han, H.-Q., Nichols, R. A., Rubin, M. R., Bähler, M., and Greengard, P., Induction of formation of presynaptic terminals in neuroblastoma cells by synapsin IIb, *Nature,* 349, 697, 1991.
34. Timiras, P. S., Cole, G., Croteau, M., Hudson, D. B., Miller, C., and Segall, P. E., Changes in brain serotonin with aging and modification through precursor availability, in *Aging Brain and Ergot Alkaloids,* Agnoli, A., Crepaldi, G., Spano, P. F., and Trabucci, M., Eds., Aging, Vol. 23, Raven Press, New York, 1983, 23.
35. Timiras, P. S., Hudson, D. B., and Miller, C., Developing and aging brain serotonergic systems, in *The Aging Brain: Cellular and Molecular Mechanisms of Aging in the Nervous System,* Giacobini, E., Filogamo, G., Giacobini, G., and Vernadakis, A., Eds., Aging, Vol. 20, Raven Press, New York, 1982, 173.
36. Timiras, P. S., Hudson, D. B., and Oklund, S., Changes in central nervous system free amino acids with development and aging, *Prog. Brain Res.,* 40, 267, 1973.
37. Timiras, P. S., Hudson, D. B., and Segall, P. E., Lifetime brain serotonin: regional effects of age and precursor availability, *Neurobiol. Aging,* 5, 235, 1984.
38. Olney, J. W., Excitotoxin-mediated neuron death in youth and old age, *Prog. Brain Res.,* 86, 37, 1990.
39. Hildick-Smith, M., Parkinson's disease, in *Principles and Practice of Geriatric Medicine,* Pathy, M. S. J., Ed., John Wiley & Sons, New York, 1991, 803.
40. Forno, L. S., Concentric hyalin intraneuronal inclusions of Lewy type in the brains of elderly persons (50 incidental cases): relationship to Parkinsonism, *J. Am. Geriatr. Soc.,* 17, 557, 1969.
41. Goldman, J. E., Yen, S. H.,Chiu, F. C., and Peress, N. S., Lewy bodies of Parkinson's disease contain neurofilament antigens, *Science,* 221, 1082, 1983.
42. Boller, F., Parkinson's disease and Alzheimer's disease: are they associated?, in *Senile Dementia of the Alzheimer Type,* Hutton, J. T. and Kenny, A. D., Eds., Neurology and Neurobiology, Vol. 18, Alan R. Liss, New York, 1985, 119.

43. Chen, K. and Yase, Y., Parkinsonism-dementia, neurofibrillary tangles and trace elements in the Western Pacific, in *Senile Dementia of the Alzheimer Type,* Hutton, J. T. and Kenny, A. D., Eds., Neurology and Neurobiology, Vol. 18, Alan R. Liss, New York, 1985, 153.
44. Langston, J. W., Ballard, P., Tetrud, J. W., and Irwin, I., Chronic Parkinsonism in humans due to a product of meperidine-analog synthesis, *Science,* 219, 979, 1983.
45. Lewin, R., Age factors loom in parkinsonian research, *Science,* 234, 1200, 1986.
46. Lees, A. J., Cognitive deficits in Parkinson's disease, in *Senile Dementia of the Alzheimer's type. Early Diagnosis, Neuropathology and Animal Models,* Traber, Y. and Gispen, W. H., Eds., Springer-Verlag, New York, 1985, 60.
47. James, D. G. and Hierons, D., Parkinson and his disease revisited, *Postgrad. Med. J.,* 67, 227, 1991.
48. Gage, F. H., Dunnett, S. B., Stenevi, U., and Bjorklund, A., Aged rats: recovery of motor impairments by intrastriatal nigral grafts, *Science,* 221, 996, 1983.
49. Gage, F. H., Bjorklund, A., Stenevi, U., Dunnett, S. B., and Kelly, P. A., Intrahippocampal septal grafts ameliorate learning impairments in aged rats, *Science,* 225, 533, 1984.
50. Harrison, D. E., Cell and tissue transplantation: a means of studying the aging process, in *Handbook of the Biology of Aging,* Finch, C. E. and Schneider, E. L., Eds., Van Nostrand Reinhold, New York, 1985, 322.
51. Sladek, J. R. and Gash, D. M., Eds., *Neural Transplants: Development and Function,* Plenum Press, New York, 1984.
52. Weiss, R., Promising protein for Parkinson's, *Science,* 260, 1072, 1993.
53. Lin, L.-F. H., Doherty, D. H., Lile, J. D., Bektesh, S., and Collins, F., GDNF: a glial cell line-derived neurotrophic factor for midbrain dopaminergic neurons, *Science,* 260, 1130, 1993.
54. Pettegrew, J. W., Panchalingam, K., Withers, G., McKeag, D., and Strychor, S., Changes in brain energy and phospholipid metabolism during development and aging in the Fisher 344 rat, *J. Neuropathol. Exper. Neurol.* 49, 237, 1990.
55. Blass, J. P. and Gibson, G. E., The role of oxidative abnormalities in the pathophysiology of Alzheimer's disease, *Rev. Neurol. (Paris),* 147, 513, 1991.
56. Hoyer, S., The biology of the aging brain. Oxidative and related metabolism, *Eur. J. Gerontol.,* 1, 25, 1992.
57. Hoyer, S., Oxidative energy metabolism in Alzheimer's brain. Studies in early-onset and late-onset cases, *Mol. Chem. Neuropathol.,* 16, 207, 1992.
58. Barinaga, M., The brain remaps its own contours, *Science,* 258, 216, 1992.
59. Fishman, R. B. and Breedlove, M. S., Sexual dimorphism in the developing nervous system, in *Handbook of Human Growth and Developmental Biology,* Vol. I, Part C, Meisami, E. and Timiras, P. S., Eds., CRC Press, Boca Raton, FL, 1988, 45.
60. Timiras, P. S., Thyroid hormones and the developing brain, in *Handbook of Human Growth and Developmental Biology,* Vol. I, Part C, Meisami, E. and Timiras, P. S., Eds., CRC Press, Boca Raton, FL, 1988, 59.
61. Tamasy, V., Meisami, E., Vallerga, A., and Timiras, P. S., Rehabilitation from neonatal hypothyroidism: spontaneous motor activity, exploratory behavior, avoidance learning and responses of pituitary-thyroid axis to stress in male rats, *Psychoneuroendocrinology,* 11, 91, 1986.
62. Ooka, H., Fujita, S., and Yoshimoto, E., Pituitary-thyroid activity and longevity in neonatally thyroxine-treated rats, *Mech. Ageing Dev.,* 22, 113, 1983.
63. Ooka, H. and Shinkai, T., Effects of chronic hyperthyroidism on the lifespan of the rat, *Mech. Ageing Dev.,* 33, 275, 1986.
64. Bignami, A., Mansour, H., and Dahl, D., Glial hialuronate-binding protein in Wallerian degeneration of dog spinal cord, *Glia,* 2, 391, 1989.
65. Bignami, A. and Perides, G., Brain extracellular matrix, *Adv. Struct. Biol.,* 1, 1, 1991.
66. Ebendal, T., Soderstrom, S., Hallbook, F., Ernfors, P., Ibanez, C. F., Persson, H., Wetmore, C., Stromberg, I., and Olson, L., Human nerve growth factor: biological and immunological activities, and clinical possibilities in neurodegenerative disease, in *Plasticity and Regeneration of the Nervous System,* Timiras, P. S., Privat, A., Giacobini, E., Lauder, J., and Vernadakis, A., Eds., Adv. Exp. Med. Biol. Vol. 296, Plenum Press, New York, 1991, 207.
67. Varon, S., Hagg, T., Fass, B., Vahlsing, L., and Manthorpe, M., Neurotropic factors in cellular functional and cognitive repair of adult brain, *Pharmacopsychiatry,* 22 (Suppl. 2), 120, 1989.
68. Werrbach-Perez, K., Jackson, G., Marchetti, D., Morgan, B., Thorpe, L., and Perez-Polo, J. R., Growth factor-mediated protection in the aging CNS, *Prog. Brain Res.,* Coleman, P., Higgins, G., and Phelps, C., Eds., Elsevier, Amsterdam, 86, 183, 1990.
69. Varon, S., Hagg, T., and Manthorpe, M., Nerve growth factor in CNS repair and regeneration, in *Plasticity and Regeneration of the Nervous System,* Timiras, P. S., Privat, A., Giacobini, E., Lauder, J., and Vernadakis, A., Eds., Adv. Exp. Med. Biol. Vol. 296, Plenum Press, New York, 1991, 267.
70. Geisler, F. H., Dorsey, F. C., and Coleman, W. P., Recovery of motor function after spinal-cord injury — a randomized, placebo-controlled trial with GM-1 ganglioside, *N. Engl. J. Med.,* 324, 1829, 1991.
71. Skaper, S. D., Katoh-Semba, R., and Varon, S., GM-1 ganglioside accelerates neurite outgrowth from primary peripheral and central neurons on the selected culture conditions, *Brain Res.,* 355, 19, 1985.
72. Faktorovich, E. G., Steinberg, R. H., Yasumura, D., Matthes, M. T., and LaVail, M. M., Photoreceptor degeneration in inherited retinal dystrophy delayed by basic fibroblast growth factor, *Nature,* 347, 83, 1990.
73. Bray, G. M., Vidal-Sanz, M., Villegas-Perez, M. P., Carter, D. A., Zwimpfer, T., and Aguayo, A. J., Growth and differentiation of regenerating CNS axons in adult mammals, in *The Nerve Growth Cone,* Letourneau, P. C., Kater, S. B., and Macagno, E. R., Eds., Raven Press, New York, 1991, 489.
74. Otto, D. and Unsicker, K., Basic FGF reverses chemical and morphological deficits in the nigrostriatal system of MTPT-treated mice, *J. Neurosci.,* 10, 1912, 1990.
75. Otto, D., Grothe, C., Westerman, R., and Unsicker, K., Basic FGF and its actions on neurons: a group account with special emphasis on the Parkinsonian brain, in *Plasticity and Regeneration of the Nervous System,* Timiras, P. S., Privat, A., Giacobini, E., Lauder, J., and Vernadakis, A., Eds., Adv. Exp. Med. Biol., Vol. 296, Plenum Press, New York, 1991, 239.
76. Kosik, K. S., D'Orecchio, L., Bruns, G. A., Benowitz, L. I., MacDonald, G. P., Cox, D. R., and Neve, R. L., Human GAP-43: its deduced amino acid sequence and chromosomal localization in mouse and human, *Neuron,* 1, 127, 1988.
77. Pasinetti, G. M., Cheng, H. W., Reinhard, J. F., Finch, C. E., and McNeil, T. H., Molecular and morphological correlates following neuronal deafferentiation: a cortico-striatal model, in *Plasticity and Regeneration of the Nervous System,* Timiras, P. S., Privat, A., Giacobini, E., Lauder, J., and Vernadakis, A., Eds., Adv. Exp. Med. Biol. Vol. 296, Plenum Press, New York, 1991, 249.

III
SYSTEMIC AND ORGANISMIC AGING

9 AGING OF THE NERVOUS SYSTEM: FUNCTIONAL CHANGES

Paola S. Timiras

The functional integrity of the nervous system is well maintained in most elderly persons despite the morphological and biochemical changes described in the previous chapter. This ability to conserve neurologic and intellectual competence into old age bears witness to the redundancy present in the system and possibly foretells of successful compensation.

We must recognize, however, that in a large proportion of the elderly, especially the very old, the boundaries between health and disease become less precise as incidence of neurologic and mental impairment increases. Thus, CNS disorders are the most common causes of disability in the aged, producing nearly half the disability in individuals beyond the age of 65 and causing more than 90% of the cases of total dependency. Impaired memory, intellect, strength, sensation, balance, and coordination represent some of the most frequent disabilities of old age.

One characteristic of the aging nervous system which emerged from the discussion in the previous chapter is the variable vulnerability or resistance of discrete structures or biochemical pathways to the aging process. Similarly, functions are differentially affected. In this chapter we have selected examples from three major functional areas of the nervous system:

- for motor function, gait and balance
- for electrical activity, sleep
- for cognitive function, memory

Disruption of these functions can lead to neurologic (e.g., locomotor) and mental (e.g., dementias) deficits. Dementias are characterized by global loss of cognitive function associated with neurologic deficits and vascular disturbances; they have catastrophic consequences for the demented individual, his family, and the society in which he/she lives. The prevalence of dementia is increasing rapidly. Currently the fourth leading cause of death in the 75 to 84 age group, dementia has been called "the disease of the century"; it has been predicted that it will be "the first killer of the 21st century". Some aspects of this cataclysmic disorder are presented in Chapters 3, 5, and 8 as well as briefly discussed here.

1 MOTOR CHANGES: GAIT AND BALANCE

In recent years, with an increase in the number of elderly and the lengthening of the average lifespan, altered motor function, involving pyramidal and extrapyramidal systems, has become more frequent. Alterations in motor function extend from borderline, barely perceptible and disabling, to overt manifestations of motor dysfunction with severe disability. Disorders of motility and gait in the aged are only second to impaired mental function as the most frequent accompaniments of the aging process.

1.1 Control of Posture and Movement

Skilled movements are regulated by nerve fibers that originate in the motor cortex and form the "pyramids" in the medulla, hence the term *pyramidal tracts.* Grosser movements and posture are regulated by CNS areas (e.g., basal ganglia) other than those connected with the pyramidal tracts, hence, by exclusion, the term *extrapyramidal system*. Coordination, adjustment, and smoothing of movements are regulated by the *cerebellum* which receives impulses from several sensory receptors. Impulses from all these brain structures ultimately determine the pattern and rate of discharge of the *spinal motor neurons* and neurons in motor nuclei of *cranial nerves,* thereby controlling somatic motor activity.

With aging, skilled motor movements are slowed and gross movements, particularly those related to maintenance of posture and gait, are altered. These alterations affect the speed of movement which may be accelerated (as in hyperkinesias, tremors, tics) or slowed (as in hypokynesias, akinesia), or the contraction of specific muscles resulting in abnormal movements (as in myoclonus, chorea, ballism, athetosis) or in abnormal posture (as in dystonias). Both alterations of movement and posture lead to imbalance and this to a higher incidence of falls, one of the most frequent and life-threatening accidents of old age.

With advancing age, the normal adult gait changes to a *hesitant, broad-based, small-stepped gait* with many of the characteristics of early Parkinsonism (Chapter 8), often including *stooped posture, diminished arm swinging, and turns performed en bloc* (i.e., in one single, rigid movement). Rarely do individuals over the age of 75 to 80 walk without the stigmata of age.[1] Indeed, it is the ability to walk without serious limitations or falling that distinguishes a normal aged gait from dysfunction. Alterations of locomotion are due to CNS impairment since the peripheral changes seen in normal aging — minor decreases in nerve conduction velocity, decrease in muscle mass, increased muscle tone (rigidity) — are insufficient to account for the disability.

Despite their frequency, relatively little is known of the mechanisms responsible for the changes with aging in neurologic and psychologic regulation of gait and balance. Yet, much can be learned about the overall physiologic competence of the aging individual by observing gait and balance; indeed, disturbances of gait and repeated falls, particularly in the very old, are signs of serious ill health.

1.2 Changes in Gait

Measurements of gait characteristics, such as speed and length of stride (stride is the distance between two successive contacts of the heel of the same foot with the ground), present no technical difficulty and can provide useful information on the degree of competence or damage of the central and peripheral nervous system.[2] Several pyramidal and extrapyramidal structures including visual, vestibular, and proprioceptive sensors impart finely graded instructions to the muscles of neck, trunk, and limbs for maintenance of normal posture, balance, and gait. Alteration in any of these functions

0-8493-8979-8/94/$0.00+$.50

will reflect impairment and failure of integration of one or more of the CNS structures involved. Further, disturbances of the psychologic and mental status, as well as the conditions of the skeleton and joints and orthopedic disorders (Chapter 21), may also be revealed by the degree of competence or impairment of gait and balance.[3]

Normal walking is with the the head erect (without spinal curvature), arms swinging reciprocally (without grabbing at furniture), regular stepping (without staggering or stumbling movements), and feet clearing the ground at each step. Several of these characteristics are altered with aging:

- Velocity and step length are decreased.
- Step-length variability and symmetry (e.g., between right and left foot) and double support ratio (i.e., a proportion of the total stride time) are increased.
- Step frequency is unaltered.

Such changes are indicative of impaired stability and may underlie other aspects of aging; for example, sex differences with females always performing less well than males on all gait parameters (Chapter 3), impaired vision or illumination (Chapter 10), cognitive loss — demented patients showing significant decrements in all gait parameters (see below). Gait changes may be useful in providing clinical insight; for example, gait asymmetry gives a clue to hemiplegia or arthritis, alterations of shoulder movements in walking may be a clue to Parkinsonism, increase in stride width relates to cerebellar disease and arthritis, and trunk flexion suggests unstable balance due to alterations in visual, vestibular, and propioceptive controls.[4]

1.3 Changes in Balance and Falls

As for gait, balance can be studied clinically by simple measurements such as observation of sway, sweep, and stagger when rising from a chair, standing, walking, or turning. In elderly subjects, obstacles to proper balance include the fear of falling and pain or limitation of joint movement.[5,6] Disturbances in balance are reflected in the pattern of the gait. The main adaptation to a balance disorder is the shortening of step length accompanied by slowing of gait and increase in double support ratio. This pattern is particularly noticeable in people who have fallen repeatedly, and, indeed, it is called "post-fall syndrome" or "3 Fs (fear of further falling) syndrome". The immediate consequence of alterations in balance is an increased frequency of falls.[7]

Falls are one of the commonest problems of old age, occurring most often indoors in the immediate environment of the house. Trips and accidents account for 45% of falls, but the prevalence of this type of fall decreases with aging, probably as a reflection of the reduced mobility of the very old. Other falls may occur without any external cause and may be due to alterations in peripheral (ocular, vestibular, and propioceptive) and central (cerebellar and cortical) coordination[8] or, especially in postmenopausal women, to bone fractures due to osteoporosis (Chapters 12 and 21).

There is considerable overall reserve in the locomotor system, and loss of one of the sources of control of postural information or maintenance may be of little consequence. However, in elderly individuals impairment at several functional levels may combine to deplete this reserve. It is not difficult to understand why a fall is the consequence of the simultaneous loss of a number of factors both neurologic (e.g., extrapyramidal damage) and extraneurologic (drugs, cardiovascular, skeletal, psychological). As our methods for the measurement of gait and balance become more sophisticated and quantitative, the usefulness of close examination of these two functions becomes increasingly more important in providing a comprehensive picture of the physiologic or pathologic condition of the old individual; at the same time, improvements of these two functions, although not vital for survival, will considerably increase the well-being and overall health of the elderly.

2 CHANGES OF SLEEP AND WAKEFULNESS

2.1 Biologic Clocks and Sleep Cycles

Alterations of several cyclic functions with aging may be ascribed to changes in so-called *biologic clocks*. The cessation of menstrual cycles and of ovarian function at menopause (Chapter 12) exemplifies such an alteration. Cyclic functions may depend on specific signals required for the progression of the clock through the synthesis of gene-regulated molecules (e.g., RNA, proteins, neurotransmitters).[9-11] A prime example of a neural cyclic function is the *sleep/wakefulness cycle,* which also undergoes characteristic changes with aging.[12]

A precise delineation of sleeping changes intrinsic to normal aging is still lacking. However, the incidence of sleep-related abnormalities, designated as pathologic in the young, becomes increasingly prevalent in old age.[13,14] Indeed, it may be suggested that a high percentage of the elderly may benefit from the diagnosis and treatment of sleeping/waking disorders and, *vice versa,* understanding of the aging changes may help to clarify the pathogenesis of sleep disorders in the young.

As we all know, sleep is a regular and, usually, necessary phenomenon of our daily lives. Yet, its mechanisms and purpose still elude us. Several interpretations of sleep function and characteristics have been suggested, but none has been proven. To quote only the most current:

- Sleep is a quiescent period during which the body recovers from the strains of the waking hours.
- Sleep represents a descent to a lower level of consciousness characterized by dreaming.
- Sleep depends on a specific stage — the rapid eye movement or REM sleep — and a selected sleep-wakefulness rhythmicity (as a type of biologic clock) to ensure homeostasis.

Other well-recognized characteristics of sleep include:

- Clear relationship of sleep to brain electrical activity, the so-called electroencephalogram (EEG), and association with whole-body visceral manifestations such as changes in heart rate, respiration, basal metabolic rate, endocrine function, etc.
- Sleep patterns change with age, i.e., during growth and development as well as aging.
- Sleep patterns may be modified by the administration of psychotropic drugs and by neuropsychiatric disorders.
- There are two kinds of sleep: (1) *the rapid eye movement (REM) sleep, with an EEG resembling the alert, awake state, and* (2) *a four-phase, non-REM, slow-wave (SW) sleep.*

2.2 Sleep EEG Changes with Aging

The EEG represents the background electrical activity of the brain with a particular rhythm represented by characteristic wave patterns with different frequencies: alpha, 8 to 12 waves per second; beta, 18 to 30 waves per second; theta, 4 to 7 waves per second; delta, less than 4 waves per second. The amplitude is highest in the alpha waves.

The alpha rhythm — eyes closed and mind at rest — slows with normal aging throughout the brain as well as focally (temporal region). The beta activity increases in aged persons.[15] With respect to evoked potential, the data suggest latency is prolonged after stimulation through a variety of sensory modalities, perhaps due to decreased conduction velocity[16] or a change in the sense organ itself (Chapter 10). These spontaneous and evoked potential alterations are aggravated in AD.[17-21]

When a person falls asleep, he/she goes from a state of quiet wakefulness through four consecutive stages of SWs and several episodes of REM sleep. These stages are characterized as follows:

- Stage 1 (the first of four stages of SW sleep), low-voltage, mixed-frequency EEG;
- Stage 2, which follows the appearance of the so-called "sleep spindles";
- Stages 3 and 4, a relative predominance of high-amplitude slow-frequency activity (Stage 3, 20 to 50%; Stage 4, >50%). These last two stages represent the period of deepest sleep and, possibly, of physiologic recuperation. Muscular electrical activity is low and ocular activity is generally quiescent.
- REM sleep, (1) an active brain wave pattern similar to the EEG of the waking brain; (2) rapid, binocularly synchronous eye movement (hence the name); and (3) absence of tonic muscular activity and failure to elicit spinal reflexes.

In the normal adult, the total amount of sleep per day is approximately 7 h, of which 23% is spent in REM sleep and 57% in stage 2 SW sleep. Of the other stages, stage 1 accounts for 7% and stages 3 and 4 for about 13% (Figure 9-1). Generally, the episodes of REM sleep occur at about 90-min intervals. They lengthen as the night progresses, and they are the stage wherein dreaming occurs.

With aging, sleep patterns undergo significant changes leading to deterioration of circadian organization (Figure 9-1). While the total amount of sleeping time changes little, sleep is interrupted and is distributed more widely over the 24-h period in short naps.[22-24] Likewise, while there is little change in REM sleep time, there are significant changes in the distribution of the other stages, i.e.,

- Lengthening of the period of quiet wakefulness; it takes longer to fall asleep and the number of wakenings per night increases
- Almost complete disappearance of stage 3 and disappearance of 4 so that there is little deep sleep time

The marked shortening of SW stages 3 and 4 may explain why older individuals complain subjectively of little sleep despite an overall near-normal sleeping time, and why they are much more easily aroused than the young. Additionally, although the statistics vary widely, there seems to be a progressive reduction of sleep during the night, and about 40% of the elderly suffer from insomnia so defined.

2.3 Respiratory, Cardiovascular, and Motor Changes during Sleep in the Elderly

Periods of *apnea* (cessation of respiration) or *hypoapnea* (slowing of respiration) during the sleeping period increase

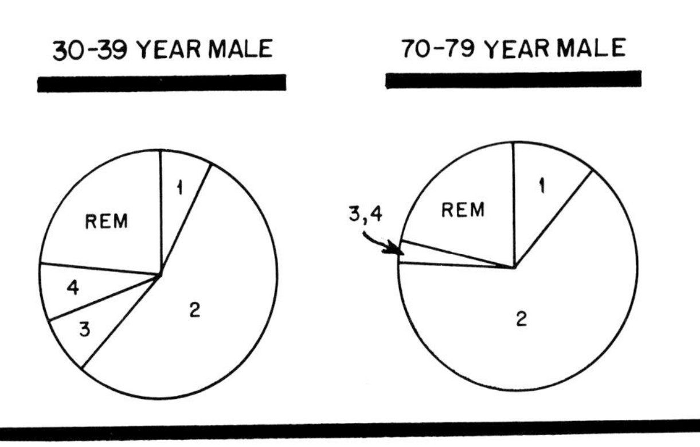

FIGURE 9-1 *Relative distribution of sleep stages in adult (30 to 39 years) and aged (70 to 79 years) male individuals.* Quiet wakefulness (preceding the onset of sleep) is followed by four (1 through 4) stages of slow-wave sleep with periods of rapid-eye-movement (REM) sleep. In the aged, the period of quiet wakefulness and stages 1 and 2 are lengthened, while stages 3 and 4 of deep sleep have almost completely disappeared.

with aging from an average of 5 respiratory disturbances per night at 24 years of age, to 50 disturbances per night at 74 years. These respiratory disturbances are terminated by an arousal and may account for a great deal of the fragmentation of sleep in the elderly. Since in young and adult individuals 5 to 8 apneas per hour of sleep are considered as the upper limit of normal, a number of elderly who are judged to be normal by other criteria would fall, according to their respiratory pattern during sleep, in the category of subjects with sleep apnea syndrome.[25,26]

Another aspect of impaired respiration, in this case of upper airway function, is the increased prevalence of *snoring* with aging. It seems that 60% of males and 45% of females in their 60s are habitual snorers.[27] The significance of respiratory dysfunction associated with sleep in otherwise normal elderly is not understood but has been related to an altered neural and chemical regulation of respiration (Chapter 18).

Cardiac arrhythmias and pulmonary hypertension are common during the apneic periods; apnea, arrhythmia, and hypertension appear to be most frequent during REM sleep and have been associated with an increased release of norepinephrine from sympathetic stimulation.[28] Periodic movements (sleep-related leg movements) occur in one third of elderly during large parts of the night and are associated with a brief arousal. The origin of these movements is little known but they seem (1) to be related to loss of coordination between motor excitation and inhibition, (2) to occur every 20 to 40 s during large parts of the night, and (3) to be associated with a brief arousal.[29]

2.4 The Reticular Activating and the Limbic Systems

The genesis of the sleep pattern seems to involve the reticular system. This system, formed of a network of interconnecting neurons distributed in the core midbrain, controls alertness. Alertness, in turn, modulates sensory, motor, and visceral inputs as well as consciousness. Changes in sleep with aging may, perhaps, be related to alterations in the reticular formation and the level of alertness as manifested in EEG and neurotransmitter changes. Serotonin, in particular, has been implicated in the generation of sleep. Insomnia, common in the elderly, might be related to several factors superimposed on the aging process itself. Anxiety, depression, and stress, which always affect sleep, are prevalent among the elderly and may account for some of the sleep disturbances.

Aging changes in the reticular activating system become manifest not only as changes in sleep, but also in alertness and behavior. Thus, a decrease in sensory input to the higher centers results from failure of the reticular formation to properly receive, integrate, and relay the signals to the sensory cortex as well as from decrements in peripheral sensory perception (Chapter 10). Impairment of sensory input would likewise impair motor responses and behavior as detected in EEG recordings and physiologic responses.[30] Such sensory-motor alterations indicate progressive declines in response time (i.e., speed of responding to a sensory stimulus) with normal aging.[31] The greatest slowing of performance is seen in demented individuals.[32]

Another brain system is also affected by aging; the limbic system, which regulates many types of behavior. One area so affected is the limbic lobe, primarily the amygdala and the hippocampus, both of which are involved in memory, mood, and motivation — all functions undergoing changes with aging. These changes involve not only the limbic system but also other modalities strongly operative in the elderly population, who may be profoundly affected by diminished socioeconomic status and are often exposed to isolation due to abandonment by the family and to bereavement upon loss of a spouse.

Because depression and insomnia in some people, and anxiety, restlessness, and aggressiveness in others, always complicate the clinical disorders of the elderly, a variety of drugs are administered either to stimulate or tranquilize. Drugs such as phenothiazines, barbiturates, dibenzoxazepines, monoamine oxidase inhibitors, etc. are a part of the medicinal armamentarium for old people. None of these drugs exists without side effects, and the metabolism of these drugs — as of most drugs — is impaired in the elderly (Chapter 23).

The Elderly and the Media

One of the consequences of lengthening of response time with aging is reduced ability of receiving and processing information when delivered at a relatively fast pace. This is exemplified by the difficulty of some elderly to follow and understand television and radio programs. Although in nursing homes particularly, and often also in private homes, television viewing is one of the most frequent pastimes, elderly individuals can only partially understand the programs. Such difficulty is not always due to decreased vision and hearing (Chapter 10); rather an overall decline in alertness, attention, and memory makes it difficult to keep up with the relatively fast rhythm of images and speech.

Research should be devoted to improve techniques for delivering information to the aged through optimal combinations of effective delivery and substantive content. Mass media have an important role not only in entertaining but also in educating and informing the elderly who have limited means of communication (e.g., due to isolation, confinement to bed, to chair, sensory deficts). Older people can be particularly susceptible to mass media influence; it should be the goal of the media services to use this susceptibility to provide informative and useful programming.[33]

3 MEMORY AND AGING

Adaptation to life events occurs during the lifetime. Environmental adjustments modify the nervous system and, as a result, animals can learn and remember. This ability, which can be viewed as an expression of neural plasticity, is altered with aging, *Memory* can be defined, in a large context, as a *processing-storage-retrieval function* of the brain/mind. Impairment of memory of varying severity, from benign forgetfulness to memory loss, seems to be a frequent, even regular, manifestation of old age. Nevertheless, the study of memory and aging is still a "fledgling field" despite the growing interest it generates, as demonstrated by the number of conferences, articles, reviews, and books since the mid-1980s focused on this subject.[34]

It is difficult to assimilate all the new data or even to discern whether progress is being made in understanding fundamental issues of memory changes during normal aging.[35] As memory is indispensable to normal cognitive function, in some degenerative diseases of aging (e.g., dementia), all cognitive functions, starting with memory, are lost. The discussion here addresses briefly the specific (memory) and the more general (cognitive function) forms under both normal and pathologic conditions. For more information, the reader is referred to some of the recent literature.[34-37]

3.1 Mechanisms of Memory Acquisition, Retention, and Recall

These mechanisms are little understood and still controversial. The original idea that information storage is widely and equally distributed throughout large brain regions has been displaced by the view that it is localized in specific areas of the brain.[38] Concepts of the process of memory — how we remember and how we recognize — should not be exclusive; rather, memory can be viewed as localized in discrete brain areas involved in specific aspects of short-term memory, and as widespread, with many areas articulating to form (long-term) memory. Among the prime areas involved are the cerebellum, the limbic system (amygdala and hippocampus), the thalamus, and the cerebral cortex.[39-41]

Indeed, memory is a complex process that involves the ability to sense (visually, audibly, tactily, by smell, etc.) an object or event, or to formulate a thought, retain this information, and recall it at will. Some current views[41] hold that memory processes may be classified as:

- Declarative, or explicit, memories of specific facts or event
- Nondeclarative, or implicit, memories as in learning, conditioning, habituation
- Adaptive, filtering memories (i.e., incoming sensory information is filtered by neurons according to how similar it is to information already acquired)
- Associative memories, (i.e., pairing of different sensory stimuli)

While neurotransmission is undoubtedly involved in memory and learning processes, the identification of which neurotransmitter or which combination of neurotransmitters is responsible for learning and memory continues to elude us. The view of localized discrete brain areas evokes the concept of cell groups producing specific neurotransmitters involved in the memory process.

The engram, that is, the collection of neural changes representing memory, has been identified differentially by various investigators. With respect to timing, different types of memory have been categorized distinguishing between short-term and long-term memory:

- In *sensory memory,* an image is recorded very briefly, less than a second
- The information may pass on to *short-term memory,* which endures for several minutes
- Then to an *intermediate memory* or labile storage, which lasts minutes to hours, and finally
- To *long-term memory,* which needs several hours to days to develop but which lasts a lifetime

With aging, impairment of memory is sporadic, not specifically related to how long information was retained or to the type of cognitive skills required. Aged normal subjects do not perform as well as young normals on many tasks with a significant memory component. Memory loss in the elderly appears to be restricted to memory for recent events leaving immediate and remote memory essentially intact. Age differences, with less proficiency for the elderly, are found for categorization of lists but not for semantic processing, association strategies, imagery, nor extracting main points from prose material,[42,43] or ability to suppress or inhibit relevant information.[44]

The demented patient does have progressive, severe, memory impairment. Memory loss for recent events is the first severe problem encountered. This is followed by language dysfunction, and, at this point, memory function is difficult to assess. It has been hypothesized that deficits in dementia occur in an order that is approximately the reverse of childhood developmental stages. As functional stages merge and overlap at certain points of human development, the broader lines between the various stages of dementia are subtle rather than abrupt. Clearly, severely affected patients have lost the ability to learn and to form new memories.

The relationship of memory events to structural and biochemical changes continues to be interpreted differently by investigators in this area. Of the many models purported to explain memory, one proposes three distinct temporal phases:

- A first phase, representing short-term memory, may result from ionic changes at the membrane such as increased potassium conductance and increased calcium flux.
- The second, intermediate, phase depends on the activity of membrane-bound enzymes involved in synaptic function such as Na+K+ATPase.
- The third phase, or long-term memory, is dependent on new protein synthesis.

In other models, using radioactive tracers (14C-labeled 2-deoxyglucose), increased metabolic activity occurred in response to the presentation of familiar (visual) cues; the distribution of the metabolic changes is compatible with the view of memory localization to specific brain areas, but the large number of neurons involved suggests that most "plastic" (i.e., metabolically responsive) cells participate in multiple memories.[45]

3.2 Neurotransmitters and Memory

Several neurotransmitters have also been implicated in memory processes where they would modulate and facilitate information acquisition and retention. One of the first to be considered was acetylcholine, implicated in the cholinergic neurons of the hippocampus. A "cholinergic hypothesis" states that acetylcholine, a major transmitter in the hippocampus, entorhinal area, and the nucleus of Meynert (all associated with memory), has a key role in the mechanism of memory. Indeed, in Alzheimer's dementia (AD) where memory loss is one of the signs of disease, several neurons are lost in these brain areas (Chapter 8). However, neither an increase in the precursor, choline, nor an increase in the activity of the synthesizing enzyme (choline acetlyltransferase) nor a decrease in the metabolizing enzyme (acetylcholinesterase) have normalized impaired memory due to old age, disease, drugs, or experimental lesions. These failures to resolve the impairment have dampened but not destroyed interest in acetylcholine. Early intervention in the disease with an acetylcholinesterase inhibitor and careful monitoring of blood levels thereof have shown some promising signs of slight improvement in Alzheimer's dementia.[46,47] Other agents continue to be studied, for example, the use of acetylcholine agonists, which bind to postsynaptic receptors.

As with the successful transplantation of fetal dopaminergic neurons into the neostriatum of aged rats with impaired motor coordination (Chapter 8), fetal septal tissue (rich in cholinergic neurons) has been transplanted to the hippocampus where it remains viable, secretes acetylcholine, and has improved learning in these aged rats.[48,49] Such success as has been attained with transplanted neural tissue is certainly encouraging and stimulates the extension of these studies to primates. If future studies provide favorable results, neural transplants to treat AD would appear to be a viable option.

Several neuropeptides have been implicated in normal memory processes and in AD.[50-52] Two often proposed candidates are vasopressin and cholecystokinin. Vasopressin, a hypothalamic hormone stored in the posterior pituitary, is claimed to facilitate both memory consolidation and retrieval. This action could be mediated through release of catecholamines — a presynaptic action — or through interference with the postsynaptic receptors. Cholecystokinin, an intestinal peptide found also in the brain, seems to possess memory-retaining properties in experimental animals and may act in synergy with catecholamines. While some memory improvement has been reported after administration of these neuropeptides in animals, including humans, the efficacy of neuropeptides in preventing or reducing memory loss in the elderly is controversial.

More recent studies have focused on still other neurotransmitters such as glutamate[53] and nitric oxide (Chapter 8).[54] The development of stem cell techniques for homologous recombination provides a useful tool for testing genetically whether a particular mechanism (e.g., long-term potentiation of synaptic activity) is important or only casually related to learning (and memory).[50] The possible involvement of excitotoxic amino acids such as glutamate in a number of neurodegenerative diseases including AD suggests both preventive and therapeutic interventions (administration of glutamate-receptor agonists, free radical scavengers, gangliosides, growth factors, proteases inhibitors, NO inhibitors, and others).

4 SENILE DEMENTIAS

4.1 Definitions and Prevalence

Dementia (from the Latin *de-mens,* without mind) is defined as a global deterioration of intellectual and cognitive functions characterized by a defect of all five major mental

TABLE 9-1

Reversible Causes of Dementia

D	Drugs
E	Emotional disorders
M	Metabolic or endocrine disorders
E	Eye and ear dysfunctions
N	Nutritional deficiencies
T	Tumor and trauma
I	Infections
A	Arteriosclerotic complications, i.e., myocardial infarction, stroke, or heart failure

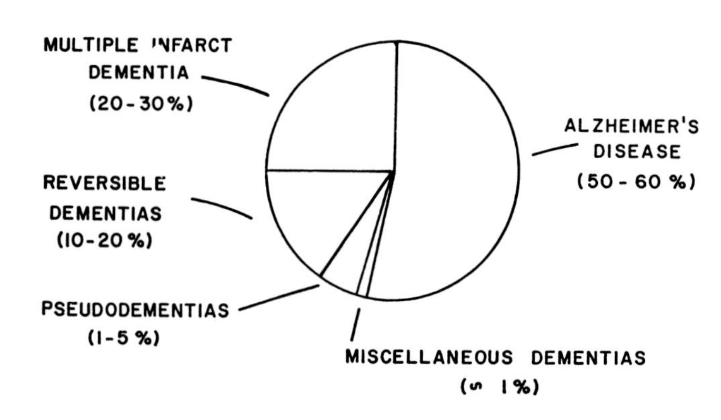

FIGURE 9-2 *Percentage of major forms of dementias in the elderly.*

functions, orientation, memory, intellect, judgment, and affect, but with persistence of a clear consciousness. Dementia (also referred to in the past as "organic brain syndrome") does occur at all ages and is caused by a variety of factors. It may be *reversible* or *irreversible and progressive.*

Reversible dementia is generally due to known causes and, once these are removed (e.g., drugs) or cured (e.g., infections), the dementia disappears. A handy mnemonic to list the main causes of reversible dementia is presented in Table 9-1.

In the elderly, "senile" dementia may be *secondary* to any of the factors listed in Table 9-1, or may be *primary,* due to unknown causes. According to etiopathogenesis and clinical manifestations, it has been categorized into several types, the distribution of which is illustrated in Figure 9-2:

- Senile dementia of the Alzheimer type (abbreviation: SDAT or shorter AD) accounts for 50 to 60% of all senile dementia cases
- Multiple infarct dementia accounts for 20 to 30%
- Reversible dementias for 10 to 20%
- Depression or pseudodementias for 1 to 5%
- The remainder is represented by miscellaneous disorders such as Parkinson's and Pick's diseases

The relative proportion of the dementias varies somewhat with the population, the age of the patient, and the particular period of the study, but, overall, AD remains the most frequent.

Differences between normal and pathologic aging of the brain may be essentially quantitative. In the normal brain, neurofibrillary tangles and neuritic plaques are few (Chapter 8), whereas they are numerous and widely distributed in AD, where their accumulation represents a definite diagnostic feature of the disease. Whether normal aging is simply an early stage of pathology without obvious clinical expression, and whether aging-associated diseases are merely *"accelerated aging"* or distinctive disease processes, remains to be evaluated.[55] Another, prevalent view is that while normal and pathologic aging are distinct entities, the aged cell may represent a favorable environment in which some pathological changes can be superimposed.

Undoubtedly, the larger proportion of the elderly remains lucid and mentally competent until death. That dementia is not an inevitable consequence of aging is evident from the current statistics: severe dementia can be found only in 4 to 5% of the population over 65 years and mild to moderate forms in 10%. However, the incidence of severe dementia increases with age from 0.01% per year at age 65, to 3.5% per year at age 85. Likewise, the prevalence also increases from less than 1 at age 65 to 70 to greater than 15% at age 85 and older. Indeed, *age is the single most important risk factor for dementia* in the older population[56] even though some studies seem to indicate a slight decline in AD after age 90.[57]

The relatively young age of the patient, first described by Alzheimer, indicates that Alzheimer's dementia does also occur at younger ages (40 to 50 years) than those so far discussed. In general, the *"adult"* or *"presenile" form of Alzheimer's dementia* presents the same symptoms and pathology as the "senile" form, but these are generally more severe and progress more rapidly.

It is predicted that, with the decrease in mortality in late age groups from other causes, the prevalence of dementia will double within 60 years. Currently, two thirds of the patients in nursing homes have dementia; therefore, 2/3 of the 21 billion dollars spent in nursing home care in the U.S. (1985) are expended on patients with dementia. Thus, the problem of caring for demented patients is a serious one which is likely to become worse as the segment of the affected population continues to grow.

4.2 A Capsule View of the Clinical Picture

Of all possible complaints from a patient or the family, none evokes (or should evoke) as much concern and anxiety than that of a change in mental function. "I'm lost, lost somewhere in the corridors of my mind..." an Alzheimer's victim explained to his devoted wife, in one of his more lucid moments. Part of the tragedy of this terrible dementia is that it takes a developed personality with family, friends, intelligence, and dignity, and slowly destroys it. It is a disease which kills the mind years before it takes the body.

Alzheimer's disease begins as simple forgetfulness that can be found in roughly 80% of the normal healthy elderly. For most, this is the extent of the problem, but in roughly 10% of those over 65 there is progression to a confusional phase of mild to moderate dementia. Many, perhaps most, of these cognitively impaired and confused elderly appear to decline further to the point where they can no longer care for themselves and are frankly demented. By the time dementia has developed, the life expectancy is about 2.5 additional years, and the whole clinical course of the disease averages about 8 years.

With rapidly increasing numbers of elderly with better medical care and significant inroads into deaths from the leading causes of cancer and heart disease, the num-

TABLE 9-2

Types of Cognitive Impairment in the Elderly

Delirium
Dementia
Depression (pseudodementia)
Benign senescent forgetfulness
Paranoid states and psychoses
Amnestic syndromes

bers of AD patients are growing steadily in the U.S. and Western Europe. Each year several hundred thousand more Americans develop the disease, and at least 100,000 die with it; we now have 1.5 to 2 million patients. AD is considered today the fourth leading cause of death in the 75 to 84 age group, an estimate considered quite conservative by many.[56]

4.3 Differential Diagnosis of Alzheimer's Disease

Confusional states in the elderly may indicate several conditions which must be differentiated from dementia. Some of the most common conditions to be differentiated from dementia are listed in Table 9-2 and include:

- *Delirium,* an acute or subacute alteration of mental status characterized by clouding of consciousness, fluctuation of symptoms, and improvement of mental function after the underlying medical condition has been treated (reversible dementia)
- *Depression* (or pseudodementia), a specific psychiatric entity which can precede or be associated with dementia, which can be differentially diagnosed and treated
- *Benign senescent forgetfulness,* which is not progressive and not of sufficient severity to interfere with everyday functions
- *Paranoid states and psychoses,* again with specifically diagnostic psychiatric correlates
- *Amnesic syndrome,* characterized by short-term memory losses without delirium or dementia

4.4 Multi-Infarct Dementia

Once dementia has been diagnosed, AD must be differentiated from several other forms, the most frequent, in the elderly, being *multi-infarct dementia.* This condition results from recurrent cortical or subcortical strokes. The infarcts can sometimes be detected with an X-ray CT scan (computerized tomography) or other imaging and metabolism-measuring devices (nuclear magnetic resonance, NMR, and positron emission tomography, PET) which have been suggested as diagnostic tools for AD.[58]

If the infarcts are too small to be visualized, a clinical diagnosis can be made by the presence of the following elements (Table 9-3):

1. A history of abrupt onset or stepwise deterioration related to the transient ischemic attacks and strokes
2. The presence of other symptoms of cardiovascular pathology, as hypertension and arrhythmias
3. The presence of focal neurologic signs or symptoms, related to the focal nature of the cardiovascular damage

Unfortunately, as in AD, once the diagnosis has been made, therapeutic measures to cure or improve the dementia

TABLE 9-3

Characteristics of Multi-Infarct Dementia

History of abrupt onset or stepwise deterioration
History of transient ischemic attack or stroke
Presence of hypertension or arrhythmia
Presence of any neurologic focal symptoms or signs

are not currently available. However, prevention of stroke through control of hypertension and cardiac arrhythmias continues to provide a means to reduce the incidence and progression of this disease (Chapters 15 and 23).

4.5 Pathogenesis of Alzheimer's Disease

In 1907, Alois Alzheimer, a German neurologist, described a case of a 51-year-old lady with a 5-year history of progressive dementia (i.e., the adult or presenile form) which led to her death. At autopsy, he found many neurofibrillary tangles and neuritic plaques in her cerebral cortex and hippocampus. Today, the diagnosis of AD remains essentially a clinicopathologic one, requiring dementia in life and the two pathologic lesions of tangles and plaques at autopsy (Chapter 8). Clinically, it remains a diagnosis of exclusion, that is, once all other possible causes of mental confusion and dementia have been examined and excluded.

Major signs of Alzheimers' disease have been summarized in Chapter 8. They include:

1. Decreased gross brain weight, as may occur in normal aging, but more severe and associated with flattening of gyri and widening of sulci and cerebral ventricles
2. Loss of neurons particularly in such areas as the nucleus basalis of Meynert (cholinergic neurons) and the locus ceruleus (adrenergic neurons) and the hippocampus (associated with memory)
3. In the remaining neurons, severe denudation (stripping) of dendrites and damage of axons
4. Accumulation of cell inclusions: lipofuscin, Hirano bodies, cytoskeletal proteins (e.g., tau proteins) and ubiquitin (viewed as a possible common denominator for many neurodegenerative diseases)
5. Accumulation of neurofibrillary tangles, neuritic plaques, and perivascular amyloid, the three most characteristic (diagnostic) pathologic signs of AD, generally distributed throughout the brain but highly concentrated in the same regions as in 2
6. Other associated alterations include:
   - Decreased oxidative and energy metabolism
   - Free radical accumulation
   - Lower enzyme (e.g., protein kinases, proteases) activity
   - Impaired iron homeostasis (causing oxidative injury)
   - Reduced levels/metabolism/activity of neurotransmitters
   - Increased levels of aluminum
   - Head trauma
   - Altered immune (to prions) and hormonal (hypothyroidism) responses
7. In a few cases of familial AD, amyloid gene expression

Despite active research, the origin and nature of the characteristic lesions, tangles, plaques, and perivascular amyloid remain controversial. One reason is the lack of animal or *in vitro* AD models; despite several promising models, none so far has proven entirely satisfactory (Chapter 3). Other reasons involve technical difficulties such as solubilization of the proteinaceous material of tangles. With better purification and the use of antibodies to neurofilament and microtubule proteins, it appears that the same or similar proteins are present in all three AD lesions.

Whether these proteins are unique to the lesion, such as the Alz-50 protein,[59] or are alterations of normal constituents, such as the microtubule-associated, low-molecular-weight "tau" protein, remains to be determined.[60,61] Equally uncertain at present is whether AD lesions (1) originate in the neurons and then spread to the extracellular and perivascular spaces, or (2) they are blood-born and carried to the brain through an altered blood-brain barrier,[62] or (3) they are consequent to altered brain oxidative and energy metabolism,[63,64] involving protein phosphorylation by protein kinases[65] and brain proteases,[66] or (4) they represent some altered immune response,[67,68] or (5) they are due to accumulation of free radicals[69] or toxic metals such as iron[70] and aluminum.[71,72] By far the most popular theory of AD pathogenesis today involves the overproduction of β-amyloid protein.[73-77]

The Amyloid Connection

Amyloid degeneration and amyloidoses have been discussed briefly in Chapter 5. The term "amyloid" given to these deposits over 100 years ago implies erroneously that they are formed of a starch-like substance (Latin *amylum* for starch). Actually, the amyloid molecules are normal or mutated proteins and protein fragments that differ among the various amyloidoses they generate (Chapter 5).

In the brain, β-amyloid, the major component of the neuritic plaque amyloid, does not inevitably result from some aberrent reaction, but is formed in healthy cells as well. What causes its accumulation in AD is not yet known. The β-amyloid protein, which contains just 42 amino acids, is made as part of a larger protein called the amyloid precursor protein (APP). Complete APP does not harm the cells; it is only when β-amyloid is clipped out of APP by protein splitting enzymes that the smaller molecule may lead to pathology. APP is imbedded in the cell membrane, and the β-amyloid molecule is astride of the membrane where it cannot be reached by the protein-splitting enzymes. Under normal circumstances, APP is split by the enzyme secretase, and this reaction does not produce β-amyloid. However, in some cases, APP yields fragments that contain β-amyloid, and such fragments become amyloidogenic (i.e., give rise to amyloid degeneration and accumulation, Chapter 5) in the lysosomes, the cells' organelles that are usually the site of protein breakdown.

In AD, presumably, the balance is shifted away from cleavage through the secretase pathway (which would not produce amyloid deposition) toward the lysosomal pathway (which would from β-amyloid deposition). The identity of the factor(s) capable of shifting the balance between the two degradation pathways remains unanswered. Tentative answers include: APP mutation, reduced secretase activity with aging, alterations in protein phosphorylation, abnormalities of lysosomes, and others.

Once β-amyloid has aggregated inside the lysosome, the cell has difficulty getting rid of it. β-amyloid accumulation would lead to cell damage and death followed by accumulation of the protein in the extracellular spaces and formation of the neuritic plaque (with the remnants of the neurofibrillary tangles from the dead cell). At this point, microglial cells from the immune system would surround the plaque; indeed, some investigators emphasize that brain damage starts with the formation of amyloid by perivascular macrophages and microglial cells.[67,68] Other investigators as well do not subscribe to the crucial role of β-amyloid in AD pathogenesis. Some believe that nerve cell injury is the primary cause and deposition of amyloid fibrils is secondary.

The amyloid connection was strengthened by the cloning of the gene-encoding β-amyloid protein and its mapping to chromosome 21. Chromosome 21 is altered in Down's syndrome and individuals with this trisomy are severely mentally impaired and develop AD at an early age (Chapters 3 and 5).[78-80] The association of mental disability with AD pathology suggested that overproduction of β-amyloid is the cause of the neural degeneration underlying dementia. Other genes (on chromosomes 14 and 19 in humans, 16 in mice) responsible for some forms of familial (inherited) AD have been identified, but the number of cases where a definitive hereditary linkage is present is still extremely small.

The "Amyloid Cascade Hypothesis" and Therapeutic Strategies

This hypothesis proposes β-amyloid or APP cleavage products are neurotoxic:

1. They disrupt calcium homeostasis by increasing neuronal intracellular calcium
2. Increased intracellular calcium would lead to tau phosphorylation and formation of paired-helical filaments, which, in turn would form neurofibrillary tangles
3. Accumulation of neurofibrillary tangles would induce neuronal death and formation of neuritic plaques with overproduction and deposition of amyloid.[75]

Based on this amyloid hypothesis, several therapeutic strategies are being proposed. Although clinical applications are still being investigated, a list of these strategies may help to acquire a better understanding of the role of amyloid in AD.[76] They intend: (1) to block delivery to the brain by blood stream of APP molecules responsible for β-amyloid deposits; (2) to inhibit proteases that cleave APP to produce the β-amyloid; (3) to delay the formation of β-amyloid deposits by interfering with the formation of amyloid filaments; (4) to interfere with the activity of macrophages, microglia, and other cells that contribute to the inflammatory reaction surrounding the plaques; and (5) to block the β-amyloid molecules on the surface of neurons to prevent their toxic action.

4.6 Etiology of Alzheimer's Disease

A few patients with Alzheimer dementia have a positive family history, with some instances of autosomal dominant inheritance.[76-80] For most patients the cause is still unknown. Epidemiological studies would exclude the influence of animal contacts, smoking, drinking, dietary habits, or prior viral infections as well as any correlation with the frequency of neoplasms. However, they have found an increased frequency of a history of head trauma, an association with Down's syndrome, and a less-convincing association with thyroid disease.

The hypothesis of thyroid insufficiency as a possible cause of AD may need to be reevaluated in view of current studies showing that thyroid hormone in the brain regulates NGF synthesis and actions. As discussed in Chapter 8, administration of NGF and other growth factors has proven effective in restoring neural plasticity in the aging brain.[81] Thus, NGF and thyroid hormone would provide complementary support for neuronal growth during early development and for eventual regeneration at later ages following normal or abnormal (e.g., in AD) neuronal death. Thyroid hormone would have a long, permissive role and NGF a local, short-term action.[82,83]

The etiology of Alzheimer's dementia remains unknown despite several theories proposed (Table 9-4). An autoimmune theory is supported by the increased incidence of head trauma in affected patients. According to this theory, an autoimmune process, possibly stimulated by a virus or damage to the blood-brain barrier, would induce deposition

TABLE 9-4

Proposed Etiologies of AD

Genetic
- Familial AD: gene on chromosome 21, 19?, 14?
- Mutations (base substitution) within APP gene

Infectious and Autoimmune
- Prions (spongiform encephalopathies)
- Autoimmune (consequent to head trauma?)
- Microglia and macrophages transport and secrete amyloid

Environmental
- Aluminum toxicity
- Iron toxicity

Metabolic
- β-amyloid overproduction
- Free radical accumulation
- Altered oxidative and energy metabolism
- Altered protein (cytoskeletal, amyloid) degradation

Neuroendocrine
- Adrenal steroids and neuronal death
- Hypothyroidism
- Cell stress
- Impaired cholinergic neurotransmission

Others?

of immunoglobulin-like material leading to amyloid production; or, alternatively, free radical damage of intracellular neurofilaments and microtubules would lead to the production of abnormal proteins (Chapter 6). Another theory involves the action of metals, such as aluminum, which accumulate in the brain of Alzheimer's patients; however, it is more likely that aluminum is just sequestered in the diseased cells rather than being a causative agent.[71,72]

Still another theory suggests that the dementia is caused by unconventional infectious agents such as a virus or virus particles called prions.[84,85] Spongiform encephalopathies in animals (e.g., scrapie in sheep) and associated with dementia in humans (e.g., Creutzfeld-Jakob disease) resemble, with important differences, some of the lesions in AD. Likewise, epidemiological evidence that the disease might be transmissible is lacking. The prions would either participate or induce the formation of the plaque amyloid and they would induce the formation of the paired helical filaments and be responsible for the loss of cholinergic neurons in certain brain areas.

AD and the Cellular Stress Response

The greater susceptibility of the aging brain to MPTP (discussed in the previous chapter) prompts consideration of SDAT as a consequence of nonspecific cellular stress. Indeed, the etiology of Alzheimer lesions has been related to a variety of superficially different types of damage or chronic stress such as persistent infection, metal poisoning, or genetic abnormalities. The most salient phenomenon of cellular stress (e.g., heat shock)[86-88] overlaps with those of AD as illustrated by reduced transcription and protein synthesis, nuclear alterations, and cytoskeletal disruption.[83] It is to be recalled that whole body stress also serves as a model of experimental aging (Chapters 3 and 11).

AIDS-Related Dementia

A pattern of neurologic symptoms occurs in individuals affected by acquired immune deficiency syndrome (AIDS). The exact incidence is unknown, but it is claimed 60% of AIDS patients will develop dementia. Early stages of the disease are characterized by forgetfulness, failure to concentrate, confusion, and slow mentation, rapidly followed by progressive loss of cognition. Simultaneously, motor problems appear such as leg weakness, unsteady gait, poor coordination, and trouble with handwriting. Neuropathology includes brain atrophy, ventricular enlargement, demyelination, and multinucleated giant cells. The virus has been cultured from CSF and brain tissue.[89]

4.7 AD Management

While there is presently no cure for AD, various treatments have been proposed to prevent or slow down its progression or to alleviate associated symptoms. Several treatments are promising, but none to date can provide a cure.

The basic medical workup for dementia consists of a complete history and physical examination and a formal mental status exam, in severe cases, with the help of family or friends. Laboratory tests should include, besides the usual blood tests, an electrocardiogram and a chest X-ray. The value of a head CT scan, NMR, and PET or cerebrospinal fluid analysis remains controversial and probably each case should be judged individually. The CT scan has been recommended in cases where reversible dementia is suspected, and the PET appears to provide some useful information, particularly with respect to eventual brain metabolic changes.

Once reversible causes of dementia have been excluded, the medical therapeutic modalities for intervening offer little immediate treatment. A wide range of drugs have been, or are being tried, but few appear to be clinically useful. Among those tried (in addition to those already mentioned in this and the previous chapter) are antioxidants, carnitine, antiviral agents, such as interferon, calcium channel blockers, and endorphin blockers, and new agents are continually being added to the list of potentially active drugs in AD.

The lack of a specific medical therapy for Alzheimer's dementia, however, should not discourage the physician from helping the patient and his/her family. The basic goals of management are

- To maintain the patient's safety while allowing as much independence and dignity as possible
- To optimize the patients' function by treating underlying medical conditions and avoiding the use of drugs with side effects on the nervous system
- To avoid catastrophic reactions or exacerbations by preventing stressful situations
- To identify and manage complications that arise such as agitation, depression, and incontinence
- To provide medical and social information to the patient's family in addition to any needed counseling

Many of these approaches to AD treatment will not be easy to follow, nor will they be successful, but the goal of finding a valid treatment is worth the effort. Dementia is a condition that requires the attention not only of the physician but also of the other branches of the health care system in order to be effectively managed.

REFERENCES

1. Klawans, H. L. and Tanner, C. M., Movement disorders in the elderly, in *Clinical Neurology of Aging*, Albert, M. L., Ed., Oxford University Press, New York, 1984, 387.
2. Klenerman, L., Dobbs, R. J., Weller, C., Leeman, A. L., and Nicholson, P. W., Bringing gait analysis out of the laboratory and into the clinic, *Age Ageing*, 17, 397, 1988.

3. Isaacs, B. (updated by M. S. J. Pathy), Gait and balance, in *Principles and Practice of Geriatric Medicine,* Pathy, M. S. J., Ed., John Wiley & Sons, Chichester, England, 1991, 1309.
4. Barbeau, A., Aging and the extrapyramidal system, *J. Am. Geriatr. Soc.,* 21, 145, 1973.
5. Lichtenstein, M. J., Shields, S. L., Schiavi, R. G., and Burger, M. C., Clinical determinants of biomechanics platform measures of balance in aged women, *J. Am. Geriatr. Soc.,* 36, 996, 1988.
6. Stelmach, G. E., Phillips, J., Di Fabio, R. P., and Teasdale, N., Age, functional postural reflexes and voluntary sway, *J. Gerontol. Biol. Sci.,* 44, B100, 1989.
7. Overstall, P. W., Falls, in *Principles and Practice of Geriatric Medicine,* Pathy, M. S. J., Ed., John Wiley & Sons, Chichester, England, 1991, 1331.
8. Luxon, L. M., Disorders of the vestibular system, in *Principles and Practice of Geriatric Medicine,* Pathy, M. S. J., Ed., John Wiley & Sons, Chichester, England, 1991, 1315.
9. Takahashi, J. S., Circadian clock genes are ticking, *Science,* 258, 238, 1992.
10. Takahashi, J. S., Kornhauser, J. M., Koumenis, C., and Eskin, A., Molecular approaches to understanding circadian oscillations, *Annu. Rev. Physiol.,* 55, 729, 1993.
11. Lloyd, D. and Rossi, E. L., Eds., *Ultradian Rhythms in Life Processes: An Inquiry into Fundamental Principles of Chronobiology and Psychobiology,* Springer-Verlag, New York, 1992.
12. Klein, L. J., Ulincy, L. D., and Monjan, A. A., *Sleep Disorders of Older People,* U.S. Dept. of Health and Human Services, Public Health Service, National Institutes of Health, National Library of Medicine, Reference Section, Washington, D.C., 1990.
13. Feinberg, I., Functional implications of changes in sleep physiology with age, in *Neurobiology of Aging,* Aging Series, Vol. 3, Gershon, S. and Terry, R. D., Eds., Raven Press, New York, 1976, 23.
14. Peter, J. H. et al., Eds., *Sleep and Health Risk,* Springer-Verlag, New York, 1991.
15. Muller, H. F. and Schwartz, G., Electroencephalograms and autopsy findings in geropsychiatry, *J. Gerontol.,* 33, 504, 1978.
16. Dorfman, L. J. and Bosley, T. M., Age-related changes in peripheral and central nerve conduction in man, *Neurology,* 29, 38, 1979.
17. Shaw, N. A. and Cant, B. R., Age-dependent changes in the latency of the pattern visual evoked potential, *Electroenceph. Clin. Neurophysiol.,* 48, 237, 1980.
18. Katz, R. I. and Horowitz, G. R., Electroencephalogram in the septagenerian: studies in a normal geriatric population, *J. Am. Geriatr. Soc.,* 30, 273, 1982.
19. Syndulko, K., Hansch, E. C., Cohen, S. N., Pearce, J. W., Goldberg, Z., Montan, B., Tourtellotte, W. W., and Potvin, A. R., Long-latency event-related potential in normal aging and dementia, *Adv. Neurol.,* 32, 279, 1982.
20. Coben, L. A., Danziger, W. L., and Berg, L., Frequency analysis of the resting awake EEG in mild senile dementia of Alzheimer type, *Electroenceph. Clin. Neurophysiol.,* 55, 372, 1983.
21. Coben, L. A., Danziger, W. L., and Hughes, C. P., Visual evoked potentials in mild senile dementia of the Alzheimer type, *Electroenceph. Clin. Neurophysiol.,* 55, 121, 1983.
22. Kales, A., Wilson, T., Kales, J. D., Jacobson, A., Paulson, M., Kollar, E., and Walter, R. D., Measurements of all-night sleep in normal elderly persons: effects of aging, *J. Am. Geriatr. Soc.,* 15, 405, 1967.
23. Miles, L. E., and Dement, W. C., Sleep and aging, *Sleep,* 3, 119, 1980.
24. Prinz, P. N., Peskind, E. R., Vitaliano, P. P., Raskind, M. A., Eisdorfer, C., Zemcuznikov, N., and Gerber, C. J., Changes in the sleep and waking EEG of nondemented and demented elderly subjects, *J. Am. Geriatr. Soc.,* 30, 86, 1982.
25. Guilleminalt, C. and Partinen, M., Eds., *Obstructive Sleep Apnea Syndrome: Clinical Research and Treatment,* Raven Press, New York, 1990.
26. International Symposium on Sleep and Respiration, *Sleep and Respiration in Aging Adults,* 2nd Int. Symposium, League City, TX, Elsevier, New York, 1991.
27. Lugaresi, E., Cirignotta, F., Coccagna, G., and Piana, C., Some epidemiological data on snoring and cardiocirculatory disturbances, *Sleep,* 3, 221, 1980.
28. Vitiello, M. V., Giblin, E. C., Schoene, R. B., Halter, J. B., and Prinz, P. N., Obstructive apnea, sympathetic activity, respiration and sleep, a case report, *Neurobiol. Aging,* 3, 263, 1982.
29. Coleman, R. M., Roffwarg, H. P., Kennedy, S. J., Guilleminault, C., et al., Sleep-wake disorders based on a polysomnographic diagnosis. A national cooperative study, *JAMA,* 247, 997, 1982.
30. Welford, A. T., Sensory, perceptual and motor processes in older adults, in *Handbook of Mental Health and Aging,* Birren J. E. and Sloane, R. B., Eds., Prentice-Hall, Englewood Cliffs, NJ, 1980, 192.
31. Carella, J., Aging, and information processing rate, in *Handbook of the Psychology of Aging,* Schaie, K. W. and Birren, J. E., Eds., 3rd ed., Academic Press, San Diego, 1990, 201.
32. Mahurin, R. K. and Pirozzolo, F. J., Chronometric analysis: clinical application in aging and dementia, *Dev. Neuropsychol.,* 2, 345, 1986.
33. Oyer, H. J. and Oyer, E. J., *Aging and Communication,* University Park Press, Baltimore, 1976.
34. West, R. L. and Sinnott, J. D., Eds., *Everyday Memory and Aging: Current Research and Methodology,* Springer-Verlag, New York, 1992.
35. Salthouse, T. A., *Theoretical Perspectives on Cognitive Aging,* Lawrence Erlbaum Assoc., Hillsdale, NJ, 1991.
36. Hess, T. M., Ed., *Aging and Cognition: Knowledge Organization and Utilization,* North-Holland, Amsterdam, 1990.
37. Lovelace, E. A., *Aging and Cognition: Mental Processes, Self-Awareness, and Interventions,* North-Holland, Amsterdam, 1990.
38. Polster, M. R., Nadel, L., and Schacter, D. L., Cognitive neuroscience analyses of memory — a historical perspective, *J. Cogn. Neurosci.,* 3, 95, 1991.
39. Thompson, R. F., The neurobiology of learning and memory, *Science,* 233, 941, 1986.
40. Squire, L. R., Mechanisms of memory, *Science,* 232, 1612, 1986.
41. Desimone, R., The physiology of memory: recordings of things past, *Science,* 258, 245, 1992.
42. Verhaeghen, P., Marcoen, A., and Goossens, L., Facts and fiction about memory aging: a quantitative integration of research findings, *J. Gerontol. Psychol. Sci.,* 48, P157, 1993.
43. Verhaeghen, P. and Marcoen, A., More or less the same? A memorability analysis on episodic memory tasks in young and older adults, *J. Gerontol. Psychol. Sci.,* 48, P172, 1993.
44. Stoltzfus, E. R., Hasher, L., Zacks, R. T., Ulivi, M. S., and Goldstein, D., Investigations of inhibition and interference in younger and older adults, *J. Gerontol. Psychol. Sci.,* 48, P179, 1993.
45. John, E. R., Tang, Y., Brill, A. B., Young, R., and Ono, K., Double-labeled metabolic maps of memory, *Science,* 233, 1167, 1986.
46. Farlow, M., Gracon, S. I., Hershey, L. A., Lewis, K. W., Sadowsky, C. H., and Dolan-Ureno, J., A control trial of tacrine in Alzheimer's disease (The Tacrine Study Group), *JAMA,* 268, 2523, 1992.

47. Davis, K. L., Thal, L. J., Gamzu, E. R., Davis, C. S., Woolson, R. F., et al., A double-blind, placebo-controlled multicenter study of tacrine for Alzheimer's disease (The Tacrine Collaborative Study Group), *N. Engl. J. Med.*, 327, 1253, 1992.
48. Arendash, G. W., Strong, P. N., and Mouton, P. R., Intracerebral transplantation of cholinergic neurons in a new animal model for Alzheimer's disease, in *Senile Dementia of the Alzheimer's Type*, Hutton, J. T. and Kenny, H. D., Eds., Alan R. Liss, New York, 1985, 351.
49. Geddes, J. W., Monaghan, D. T., Cotman, C. W., Lott, I. T., Kim, R. C., and Chui, H. C., Plasticity of hippocampal circuitry in Alzheimer's disease, *Science*, 230, 1179, 1985.
50. Husain, M. M. and Nemeroff, C. B., Neuropeptides in Alzheimer's disease, *J. Am. Geriatr. Soc.*, 38, 918, 1990.
51. Cotman, C. W., Geddes, J. W., Ulas, J., and Klein, M., Plasticity of excitatory amino acid receptors: implications for aging and Alzheimer's disease, *Prog. Brain Res.*, 86, 55, 1990.
52. Wikkelso, C., Ekman, R., Westergren, I., and Johansson, B., Neuropeptides in cerebrospinal fluid in normal-pressure hydrocephalus and dementia, *Eur. Neurol.*, 31, 88, 1991.
53. Choi, D. W., Bench to bedside: the glutamate connection, *Science*, 258, 241, 1992.
54. Kandel, E. R. and O'Dell, T. J., Are adult learning mechanisms also used for development?, *Science*, 258, 243, 1992.
55. Von Dras, D. D. and Blumenthal, H. T., Dementia of the aged: disease or atypical-accelerated aging? Biopathological and psychological perspectives, *J. Am. Geriatr. Soc.*, 40, 285, 1992.
56. Mortimer, J. A. and Hutton, J. T., Epidemiology and etiology of Alzheimer's disease, in *Senile Dementia of the Alzheimer's Type*, Hutton, J. T. and Kenny, H. O., Eds., Alan R. Liss, New York, 1985, 177.
57. Hagnell, O., Lanke, J., Rorsman, B., and Ojesjo, L., Does the incidence of age-psychosis decrease? A prospective, longitudinal study of a complete population investigated during the 25-year period 1947–1972: The Lundby Study, *Neuropsychobiology*, 7, 201, 1981.
58. Friedland, R. P., Budinger, T. F., Brant-Zawadski, M., and Jagust, W. J., The diagnosis of Alzheimer-type dementia. A preliminary comparison of positron emission tomography and proton magnetic resonance, *JAMA*, 252, 2750, 1984.
59. Wolozin, B. L., Pruchnicki, A., Dickson, D. W., and Davies, P., A neuronal antigen in the brains of Alzheimer patients, *Science*, 232, 648, 1986.
60. Kosik, K. S., Joachim, C. L., and Selkoe, D. J., Microtubule-associated protein, tau, is a major antigenic component of paired helical filaments in Alzheimer's disease, *Proc. Natl. Acad. Sci. U.S.A.*, 83, 4044, 1986.
61. Kosik, K. S., Alzheimer's disease, a cell biological perspective, *Science*, 256, 780, 1992.
62. Wisniewski, H. M. and Kozlowsi, P. B., Evidence for blood-brain barrier changes in senile dementia of the Alzheimer type (SDAT), *Ann. N.Y. Acad. Sci.*, 396, 119, 1982.
63. Blass, J. P. and Gibson, G. E., The role of oxidative abnormalities in the pathophysiology of Alzheimer's disease, *Rev. Neurol. (Paris)*, 147, 513, 1991.
64. Hoyer, S., Oxidative energy metabolism in Alzheimer's brain. Studies in early-onset and late-onset cases, *Mol. Chem. Neuropathol.*, 16, 207, 1992.
65. Saitoh, T., Masliah, E., Jin, L.-W., Cole, J. M., Wieloch, T., and Shapiro, I. P., Protein kinases and phosphorylation in neurologic disorders and cell death, *Lab. Invest.*, 64, 596, 1991.
66. Marx, J., A new link in the brain's defenses, *Science*, 256, 1278, 1992.
67. Wisniewski, H. M., Vorbrodt, A. W., Wegiel, J., Morys, J., and Lossinsky, A. S., Ultrastructure of the cells forming amyloid fibers in Alzheimer disease and scrapie, *Am. J. Med. Gen. Suppl.*, 7, 287, 1990.
68. Wisniewski, H. M. and Wigiel, J., The role of perivascular and microglial cells in fibrillogenesis of β-amyloid and PRP protein in Alzheimer disease and scrapie, *Res. Immunol.*, 143, 642, 1992.
69. Harman, D., Free radical theory of aging — a hypothesis on pathogenesis of senile dementia of the Alzheimer's type, *Age*, 16, 23, 1993.
70. Connor, J. R., Snyder, B. S., Beard, J. L., Dine, R. E., and Mufson, E. J., Regional distribution of iron and iron-regulatory proteins in the brain, in aging and Alzheimer's disease, *J. Neurosci. Res.*, 31, 327, 1992.
71. Martyn, C. N., Barker, D. J., Osmond, C., Harris, E. C., Edwardson, J. A., and Lacey, R. F., Geographical relation between Alzheimer's disease and aluminum in drinking water, *Lancet*, 1 (8629), 59, 1989.
72. Sherrard, D. J., Aluminum — much ado about something, *N. Engl. J. Med.*, 324, 558, 1991.
73. Yankner, B. A. and Mesulam, M.-M., Seminars in medicine of the Beth Israel Hospital, Boston. β-Amyloid and the pathogenesis of Alzheimer's disease, *N. Engl. J. Med.*, 325, 1849, 1991.
74. Marx, J., Boring in on β-Amyloid's role in Alzheimer's, *Science*, 255, 688, 1992.
75. Hardy, J. A. and Higgins, G. A., Alzheimer's disease: the amyloid cascade hypothesis, *Science*, 256, 184, 1992.
76. Selkoe, D. J., Amyloid protein and Alzheimer's disease, *Sci. Amer.*, Special Issue "Medicine", Vol. 4, Number 1, 54, 1993.
77. Wright, A. F., Goedert, M., and Hastie, N. D., Familial Atzheimer's disease. Beta amyloid resurrected, *Nature*, 349, 653, 1991.
78. Nalbantoglu, J., Lacoste-Royal, G., and Gauvreau, D., Genetic factors in Alzheimer's disease, *J. Am. Geriatr Soc.*, 38, 564, 1990.
79. Marx, J., Familial Alzheimer's linked to chromosome 14 gene, *Science*, 258, 550, 1992.
80. Goate, A., Chartier-Harlin, M. C., Mullan, M., Brown, J., Crawford, F., et al., Segregation of a missense mutation in amyloid precursor protein gene with familial Alzheimer's disease, *Nature*, 349, 704, 1991.
81. Clos, J. and Legrand, C., An interaction between thyroid hormone and nerve growth factor promotes the development of hippocampus, olfactory bulbs, and cerebellum: a comparative biochemical study of normal and hypothyroid rats, *Growth Factors*, 3, 205, 1990.
82. Charrasse, S., Jehan, F., Confort, C., Brachet, P., and Clos, J., Thyroid hormone promotes transient nerve growth factor synthesis in rat cerebellar neuroblasts, *Dev. Neurosci.*, 14, 282, 1992.
83. Isaeff, M., Goya, L., and Timiras, P. S., Alteration in the growth and protein content of human neuroblastoma cells in vitro induced by thyroid hormones, stress and ageing, *J. Reprod. Fert. Suppl.*, 46, 21, 1993.
84. Prusiner, S. B., Scott, M., Foster, D., Pan, K. M., Groth, D., Mirenda, C., Torchia, M., Yang, S. L., Serban, D., Carlson, G. A., et al., Transgenetic studies implicate interactions between homologous PrP isoforms in scrapie prion replication, *Cell*, 63, 673, 1990.
85. Weissmann, C., Spongiform encephalopathies. The prion's progress, *Nature*, 349, 569, 1991.

86. Ananthan, J., Goldberg, A. L., and Voellmy, R., Abnormal proteins serve as eukaryotic stress signals and trigger the activation of heat shock genes, *Science,* 232, 522, 1986.
87. Fleming, J. E., Guattrocki, E., Latter, G., Miquel, J., Marcuson, R., Zuckerkandl, E., and Bensch, K. G., Age-dependent changes in proteins in *Drosophila melanogaster, Science,* 231, 1157, 1986.
88. Cole, G. M. and Timiras, P. S., Aging-related pathology in human neuroblastoma and teratocarcinoma cell lines, in *Model Systems of Development and Aging of the Nervous System,* Vernadakis, A., Ed., Martinius Nijhoff, Boston, 1987, 453.
89. Barnes, D. M., AIDS-related brain damage unexplained, *Science,* 232, 1091, 1986.

10 AGING OF THE SENSORY SYSTEMS

Esmail Meisami

1 INTRODUCTION

Sensory impairments such as diminished vision and hearing occur so commonly with aging that they often tend to characterize the aged and the aging process. Some of these impairments are due to intrinsic aging processes occurring in the sense organs and their associated nervous systems. Others are caused by environmental effects and still others represent manifestations of aging diseases.

The study of the aging process in the sensory systems of humans and other mammals, in addition to its importance and applicability to geriatrics, also provides some of the most interesting and challenging cases of gerontological investigation as the elements comprising the various senses portray the entire spectrum of cellular, tissue, organ, and system aging. Thus, while the peripheral receptor cells of the ear's cochlea and the eye's retina are permanently established at birth with no turnover and regeneration in later life, those of the olfactory and cutaneous are endowed with these properties. Yet functional decrement with aging occurs in all of these systems. The aging changes in the eye's lens provide another interesting model system for the study of the aging process as they begin so early in life and lend themselves to a wide variety of investigations ranging from molecular biology to physiological optics.

This chapter describes normal aging changes in the various sensory structures and functions. The account is limited to the changes in humans and to the senses of vision, hearing, taste, and smell and certain aspects of somatic sensation such as touch, vibration, and complex somesthesia. Pain and temperature are not included mainly due to lack of space and also because less is known about the physiological and structural aspects of these senses in the elderly; however, appropriate references are provided for the interested reader.

In addition to normal aging changes, certain aspects of the sensory disorders and diseases associated with aging are also discussed. As shown in Figure 10-1, the incidence of sensory impairments increases markedly in people with aging. More than 25% of the population 85 years or older suffers from visual abnormalities; twice as many suffer from hearing impairments. Impaired vision and hearing reduce the capacity for social communication — one of humans' cardinal needs and functions — resulting in social isolation and deprivation. Reduced capacities in somatic sensation may interfere with professional activities requiring fine touch and manipulative sensations. Impairments in taste and smell may interfere with appetite, food selection, good nutrition, and the feelings of pleasure and well-being. The reduced sensory abilities may lead to depression; in those already suffering from depression, it may hinder the progress of recovery. Because sensory losses in old age are so common and their consequences so widespread, an understanding of these impairments is now essential in geriatrics and elderly care.

2 VISION

The eye, whose structure indicates exquisite adaptation for optical and nervous functions, is the sensory organ for vision (Figure 10-2). The cornea, lens, pupil, and the aqueous and vitreous humors participate in the optical functions of the eye while the retina carries out neural visual functions. Both the optical and neural compartments undergo aging changes, although those of the optical compartment are better known. Some of these changes in the eye's optical apparatus like those in the lens start very early in life. The changes in the optical compartment are probably the primary causes of the decline in the visual capacities of the elderly, while the degenerative changes in retina are one of the leading causes of old-age blindness (Table 10-1).[1-11]

2.1 The Eye's Optical Components

Cornea

The *cornea* is the anterior portion of the eye, and its curved surface together with the watery layer of tears is responsible for most of the refraction of the light rays. During aging, the cornea becomes thicker and less curved, mainly due to an increase in the horizontal diameter of the eye. These changes

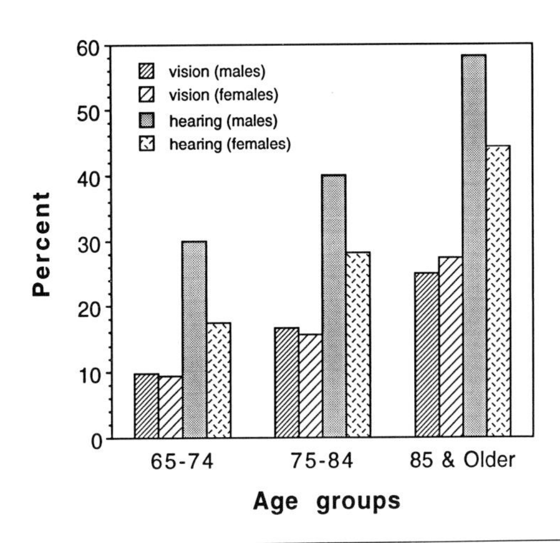

FIGURE 10-1 *Increase in the incidence of visual and hearing impairments in males and females with age in the general U.S. population.* Note that in general the males show a higher rate of sensory impairments, particularly for hearing. (Based on the data of National Center for Health Statistics, Publ. No. (PHS) 86-1250, 1986.)

0-8493-8979-8/94/$0.00+$.50

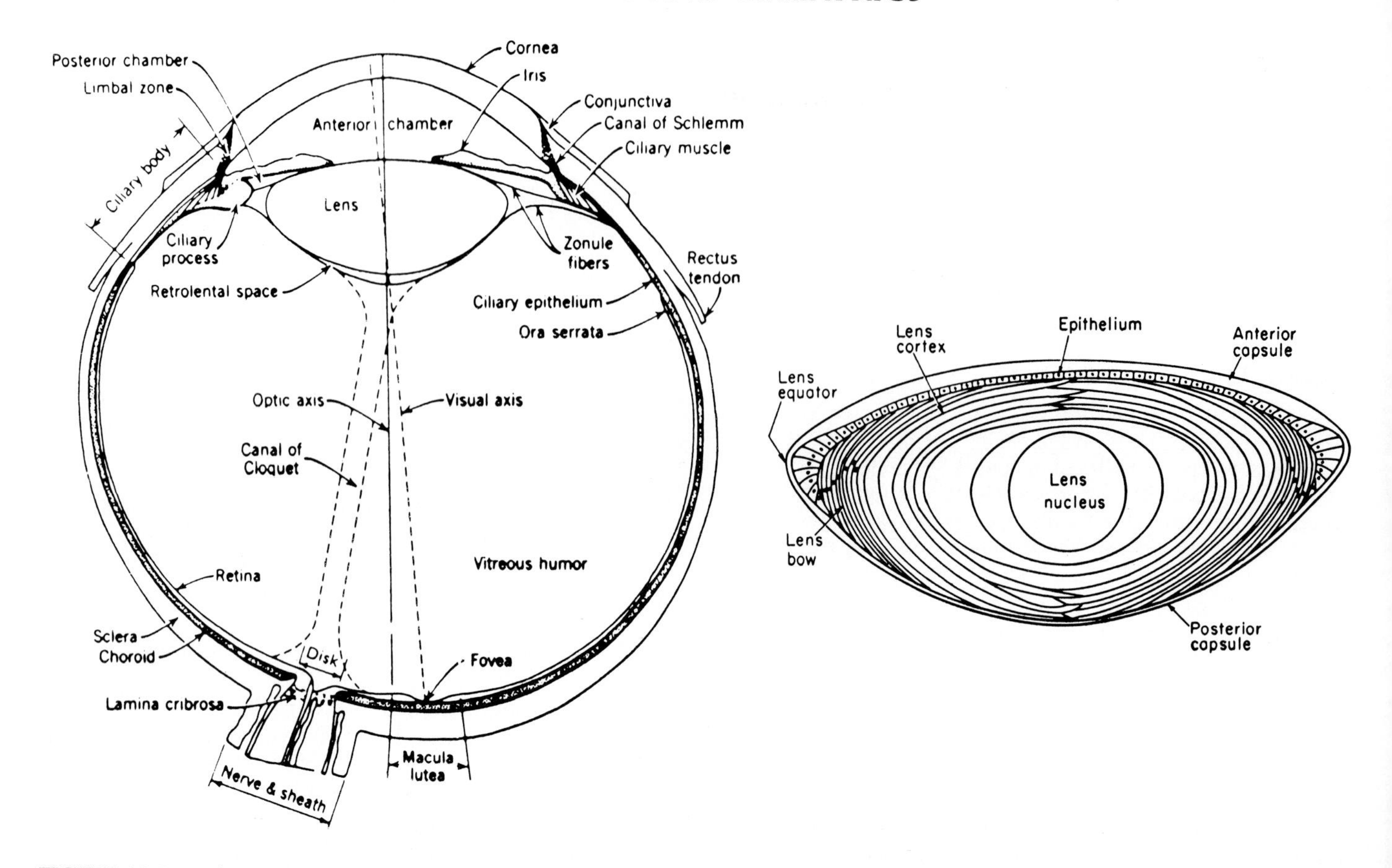

FIGURE 10-2 *Schematic drawing of the structure of the human eye* (from Ordy and Brizee[7]) *and the eye lens* (from Vaughan et al.[9])

alter the refractive properties of the cornea, leading to "against the rule" *astigmatism,* a condition characterized by defective corneal curvature and diffusion of light rays.[3] The cornea is also highly sensitive to irritable stimuli, a protective function for the eye. Corneal sensitivity declines by nearly one half between youth and old age.[2]

Other conditions in the cornea associated with aging are the *arcus senilis* (or lipid arc), the *Hudson–Stahli line,* and *spheroidal degeneration.* Arcus senilis increases in frequency and density with aging particularly after 60. It is a yellowish-white ring around the cornea's outer edge, formed by cholesterol ester deposits derived from plasma lipoproteins *(lipid arcus).* In the dilated pupil, this ring would interfere with passage of light rays; however, because of partial pupillary constriction in the elderly, arcus senilis is not detrimental to visual function.[1-4]

The Hudson–Stahli line is a horizontal brown line formed by iron deposits in corneal basal cells. Its frequency increases from 2% at 10 years to 14% at 30 years and 40% at 60 years but has no detrimental effect on vision. Spheroidal degeneration occurs in the Bowman's layer of cornea. It is observed frequently in aged populations exposed to high levels of ultraviolet radiation or ambient light reflected from snow or sand.[2]

The corneal endothelial cells number about one million per cornea at birth; this number declines to 70% by 20 years and to 50 and 30% by 60 and 80 years, respectively. Normally, pumping action of corneal cells removes water and helps keep the cornea transparent. Since these cells do not divide after birth, their loss due to aging and injury and after surgical treatments of the cornea or lens can lead to a decline in corneal transparency. Endothelial cells also secrete the cornea's basement membrane. With aging, warts *(cornea guttata)* appear in this membrane mainly in the cornea's periphery and cause marked increase in corneal permeability. Guttata are observed with increasing frequency with aging: 20% in youth, 60% in the 6th decade, and nearly 100% in very old age.[2]

The Lens

In the process of image formation, the crystalline *lens* of the eye performs two important functions, *refraction* and *accommodation.* For refraction, the lens requires an appropriate crystalline structure and transparency, while for accommodation it needs to be elastic, amenable to changes in its curvature. The increase in the opacity (optical density) and the hardness (loss of elasticity) of the lens with age are two of the best known change in the eye's optical properties which interferes with refraction and accommodation, respectively.[9]

A knowledge of the structure and development of the lens is essential for understanding its aging. The biconvex lens is basically a fibrous and relatively acellular structure, consisting of a core surrounded by a capsule (Figure 10-2). Anteriorly, the capsular epithelial cells form the fibers and other lens proteins. The collagen fibers of the lens capsule facilitate changes in lens shape during accommodation. The lens core packed with transparent protein fibers consists of an inner nuclear zone surrounded by a cortex (Figure 10-2).

The lens is formed during the embryonic period and is fairly spherical in the fetus and newborn. During postnatal development and throughout maturity, the lens continues to

TABLE 10-1

Summary of the Normal Aging Changes in the Human Eye

Structural Changes

Cornea: increased thickness; decreased curvature; some loss of transparency; pigment/lipid accumulation (arcus senilis); loss of epithelial cell; reduced epithelial regeneration
Anterior chamber: decreased volume and flow of aqueous humor
Iris: decreased dilator muscle cell number, pigment, and activity; mild increase in density of collagen fibers in stroma
Lens: increased size and anterior-posterior thickness; decreased curvature; increased pigment accumulation (yellowishness) and opacity (optical density); decreased epithelial cell number; decreased new fiber formation and antioxidant levels; increased cross-over in capsule collagens and len crystallins; increased hardness in capsule and body and lens nucleus
Vitreous body: increased inclusion bodies; decreased water content; lesser support to globe and retina
Ciliary body and muscles: decreased number of smooth muscles (radial and circular); increased hyaline substance and fiber in ciliary process; decreased ciliary pigment epithelial cells
Retina: decreased thickness in periphery; defects in rod outer segments and regeneration of discs and rhodopsin; loss of rods and associated nerve cells; some cone loss; reduced cone pigment density; expansion of Muller cells; increased cyst formation; formation of Drussen-filled lesions and degeneration of macular region in diseased condition
Pigment epithelium: loss of melanin; increased lipofuscin granules

Functional Changes

Corneal and lens functions: decreased accommodation power (presbyopia); increased accommodation reflex latency; increased near point of vision; increased lenticular light scattering; decreased refraction; decreased lens elasticity
Retinal function: decreased critical flicker frequency; decreased light sensitivity (increased light thresholds before and after dark adaptation); reduced color vision initially in yellow to blue and later in the green range
General optical functions: increased pupillary constriction (senile miosis); reduced visual acuity; presbyopia

Major Pathologies

Cornea: "against the rule" astigmatism
Lens: cataract
Retina: senile macular degeneration; glaucoma; diabetes retinopathy

grow by addition of new layers of protein fibers laid down by the capsular epithelial cells. As new fibers form, older fibers are pushed into the lens core. This mode of growth results in increased horizontal thickness of the lens together with increased compaction of the fibers in the nuclear zone.[2] The lens thickness increases from about 3.5 mm in infancy to 4.5 in middle age and 5.5 in the old age, growing at a steady linear rate of 25 μm per year.[2,6] Underlying this process of growth are the capsular epithelial cells which divide and differentiate, losing their nucleus and organelles, and eventually transform into an inert skeleton of fibrous proteins.

Increased Opacity of the Lens

Although many cytoskeletal proteins such as actin, tubulin, and vimentin are found in the lens core, the transparency of the lens is in principle due to a particular supra-molecular arrangement of the specific lens proteins, α-, β-, and γ-crystallins, within an ion- and water-free environment.[2] During aging, lens' opacity increases leading to decreased transparency and increased refraction. Since the crystallin fibers in

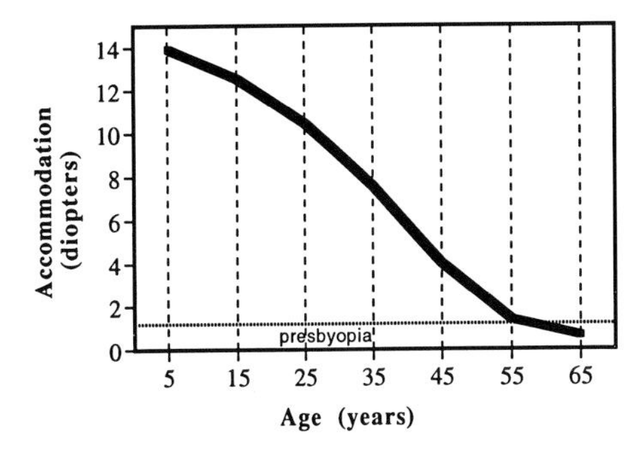

FIGURE 10-3 *Changes in visual accommodation in humans with age.* Note that the decline occurs throughout life resulting in presbyopia in the early fifties. (Adapted from various similar reports, e.g., Duane.[14])

the lens interior are not regenerated during growth and aging, they undergo many post-translational changes including glycation, carboamination, and deamidation. These changes increase cross-over and interdigitation among crystallins, making them less elastic, more dense, opaque, and yellowish.[2,8,12,13]

Some of these aging changes in the lens proteins occur as a consequence of oxidative damage to the proteins. Antioxidants like glutathione and ascorbate diminish in concentrations in the aged lens while yellow chromophores, particularly metabolites of tryptophan (β-OH-kynurenine, anthranilic acid, bityrosine), increase in frequency of occurrence and concentration. The net result is a 3-fold increase in lens optical density (at 460 nm, blue) between 20 to 60 years.[8] This results in decreased transmission and increased light scattering, particularly in the blue and yellow range but much less so in the red range. Percent transmission of light by the eye is about 75% at 10 years and 20% at 80 years. In addition to impairing transparency and refraction of light, these aging changes may also affect color perception.

Excess lens opacity as a consequence of extensive accumulation of pigments may result in a pathological condition known as *cataract* characterized by a cloudy lens.[1-4] If not operated on, this condition may result in reduced vision or blindness (see also below). In normal aging, the accumulation of yellow chromophores and the increased refraction of blue light is believed to protect the retina from the damaging effect of blue light, "blue-light-hazard".[2]

Decline in Accommodation and Presbyopia

Lens elasticity is critically important for the operation of the eye's *accommodation reflex.* During accommodation, the lens becomes more spherical in order to focus the image of the near objects on the retina. Contraction of the *ciliary muscles* relaxes the lens' *suspensory ligaments* and the lens itself, allowing it to have a higher curvature (Figure 10-2). This increases the refractive power of the lens and decreases the focal point, resulting in sharp focusing.

With aging the increased cross-over and interdigitation and compaction of collagen fibers in the capsule and crystallins in the lens nucleus result in increased hardness and reduced elasticity of the capsule and lens's interior.[2,9] These changes make the lens gradually less resilient to accommodate for near vision. Indeed the *point of near vision,* i.e., the minimum distance between the object and the eye for formation of a clear image, increases 10-fold during human life, from 9

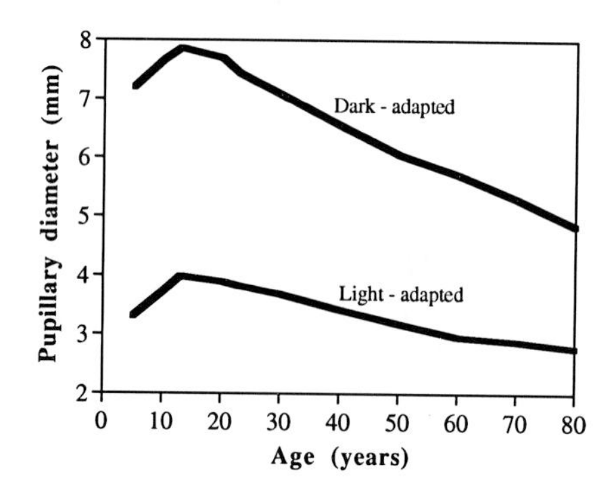

FIGURE 10-4 *Changes with age in pupillary diameter and area, measured in light-adapted and dark-adapted individuals.* Note more pronounced effects in dark-adapted eyes. (Adapted from Weale[8] based on the original data of Verriest.[11])

cm at the age of 10 years to 10 cm at 20 years, 20 cm at 45, and 84 cm at 60 years.

The loss of accommodation with aging can also be determined by studying the changes in the eye's *refractive power,* measured in diopters (reciprocal of the principle focal distance of the lens in meters). Thus, the newborn's lens, being more spherical, shows the highest refractive power (about 60 diopters). As shown in Figure 10-3, accommodation power of the human lens decreases to about 14 diopters at the age of 10, 5 at 40 years, reaching a minimum of 1 diopter by the sixth decade.[14] At this age, the lens becomes hard and nonresilient, essentially unable to accommodate for near vision tasks. This condition known as *presbyopia* (Figure 10-3) is of major clinical significance as practically everyone over 55 years needs corrective convex lenses or eyeglasses for reading and other near vision tasks.

Environmental effects such as heat and temperature can increase the rate of aging changes in the lens fibers, accelerating presbyopia. People from warmer climates show earlier presbyopia.[2,8] The aging changes in the suspensory ligaments, the ciliary muscles, and their parasympathetic nerve supply and the associated synapses may contribute to the decline in accommodation with aging. Latency of accommodation reflex decreases during aging.

Iris and Senile Miosis

The *iris* is a smooth muscular ring forming the *pupil* of the eye (Figure 10-2). Contraction and dilation of the pupil during the light reflex changes the amount of light entering the eye and is also important in the accommodation reflex. In the elderly, the iris appears paler in the middle mainly due to loss of pigmentation in the radial dilator muscles. With aging, there is a mild but constant increase in the density of collagen fibers in the iris stroma and noncellular perivascular zone.[2]

A characteristic ocular impairment in the elderly is a persistent reduction in the pupil size (diameter), the so-called *senile miosis.*[8,10] Senile miosis is particularly notable in the fully dark adapted eye; the reduction in diameter occurs gradually with aging, decreasing from a mean of 8 mm in the 3rd decade to 6 mm in the 7th and 5 mm in the 10th decade of life[9] (Figure 10-4). Senile miosis results from a relatively higher rate of aging atrophy in the *radial dilator* muscles which dilate the pupil compared to the *sphincter constrictor* muscles which constrict it. As a result, the sphincter is constantly dominant in the aged causing persistent constriction. Compared to youth, the reduced pupil aperture of the elderly results in one third reduction in amount of light entering the eye.[8,10]

Anterior Chamber and Vitreous Humor

The *anterior chamber* and its fluid the *aqueous humor* occupy a space between the cornea and the lens (Figure 10-2). The size and volume of the eye's anterior chamber decrease with age mainly due to the thickening of the lens. This growth occasionally exerts pressure on the *canal of Schlemm* (Figure 10-2), an outflow channel at the junction between the iris and cornea, causing decreased flow and increased pressure (intraocular pressure) of the aqueous humor. In normal aging, the increase in intraocular pressure is small and steady. Severe obstruction of the canal of Schlemm caused by degenerative changes in the endothelial cells of the trabecular sheets and meshwork leads to markedly increased intraocular pressure (>22 mm Hg) and the serious eye disease *glaucoma*[1,2,4] (see also below).

The *vitreous humor,* a mass of gel-like substance filling the eye's *posterior chamber,* gives the eye globe its shape and support (Figure 10-2). With age, the vitreous loses its gel-like structure and support, becoming more fluid and pigmented. The increasing inhomogeneity in its gel structure, a process called *syneresis,* can lead to vitreous collapse or its detachment from retina; often vitreous floaters *(inclusion bodies)* are released in the process which are responsible for occasional visual flashes. These physical changes in the vitreous may also be due to aging changes in its collagenous fibrous skeleton which has attachments to retina particularly in the vitreous base near the periphery. These attachments change with age, moving posteriorly and decreasing in number.[2,3,5]

2.2 Aging Changes in the Retina

The human *retina* shows considerable age-related structural changes, particularly in its *peripheral zones,* although the *macula* and its *fovea centralis* is not spared. The aged retinal periphery is thinner (10 to 30 μm), containing lesser number of *rods* and other nerve cell types. The aging loss of rods appears to be a very slow process beginning in the third and fourth decade and may be related to cumulated damage due to physiological exposure to light.[2,5] With aging rods, outer segments shorten and disengage from the microvilli of pigment epithelium, resulting in lesser amounts of *membranous discs* and their major constituent *rhodopsin,* the rods' photoreceptor molecule. These events may be related to changes in the turnover of rod discs with aging.[2] Normally the entire population of rod discs turn over every 2 weeks. This process and packing orderliness of discs slowly declines with aging, perhaps due to changes in the function of pigment epithelial cells which regulate the turnover of photoreceptor cells. The result is reduced efficiency of phototransduction. Pigment cells do not divide after maturity and are known to decrease in number. Those in the foveal region show increased accumulation of the aging pigment lipofuscin during aging.[2]

Recent studies indicate loss of cones in the retinas of humans and monkeys at a rate of 3% per decade.[2] The turnover rate of cones is believed to be about a year, making them more susceptible to accumulated effects of light damage

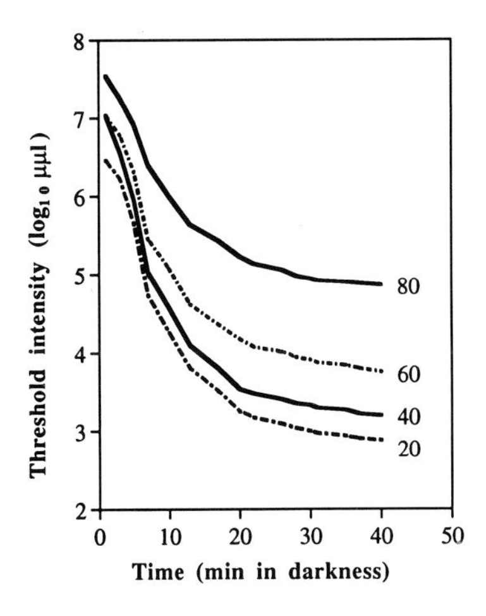

FIGURE 10-5 *Decline in light sensitivity with age.* Data show changes in visual threshold after dark adaptation in different age groups (20, 40, 60, 80 years). (Adapted and redrawn in part from Marsh,[12] based on original data of McFarland.[15])

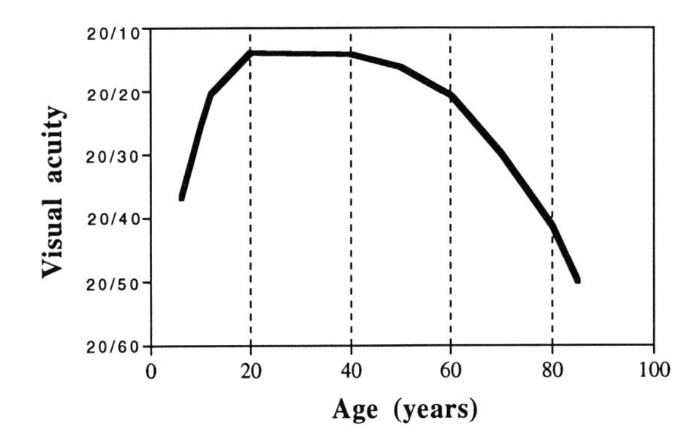

FIGURE 10-6 *Changes in visual acuity measured in Snellen index with age.* Note that the aging decline begins after 60 years resulting in serious deficits in acuity after 80 years. (From Pitts, D. G., *Aging and Visual Function,* Sekule, R., Kline, D., and Dismukes, D., Eds., Alan R. Liss, New York, 1982, 131. With permission.)

and post-translational modification of their photoreceptor proteins. Cone pigment density decreases with aging, presumably as a result of cone cell loss.[2] Remaining cones increase in average diameter. Since size and geometry of foveal cones are important in visual acuity, these changes may contribute to the observed losses in visual acuity with age (see below).

Nerve cell loss during aging in other neuronal types of retina, *bipolar, amacrine,* and *horizontal cells* is in the range of 30%. The loss for the *ganglion cells* is believed to be higher, about 50%, *Muller cells* (a type of glia cells in the retina) take over the space left by the lost neurons and form cysts which are common in the aged retina.[1,2,5] The degeneration of macula in advanced age is discussed below under diseases of eyes in the aged.

2.3 Aging of Visual Functions

Sensitivity and Visual Threshold

As discussed above, with age, the pupillary aperture decreases, resulting in lesser light input. The decline in the number of photoreceptor cells (rods) and other aging changes in the retina result in reduced availability and regeneration capacity of photoreceptor pigment (rhodopsin), leading to reduced light utilization in the aged eye. As a result, the *visual threshold,* i.e., the minimum amount of light necessary to see an object, increases with age. This is tested by measuring the change in visual threshold as a function of time spent in darkness (*dark adaptation*). It is known that the threshold for light detection decreases with increased duration of dark adaptation, because rhodopsin regeneration is enhanced in the dark. With advancing age, this regeneration is presumably deficient, resulting in higher light thresholds (i.e., lower sensitivity).

As illustrated in Figure 10-5, the enhanced light sensitivity after dark adaptation is markedly reduced with aging;[15] in fact, the visual thresholds in the completely dark-adapted eyes of the aged group (80 years) is 100 times higher than that of the young group (20 years). However, as evident from the data, the pattern of change in sensitivity during dark adaptation is basically similar in the different age groups. This indicates that the retinal function is quantitatively impaired not qualitatively. Some believe that the decline in threshold is due to reduced oxygenation of retina and the rods in the aged.[2,12]

Critical Flicker Frequency

The rate at which consecutive visual stimuli can be presented and still be perceived as separate is called the *critical flicker frequency* (CFF). Determination of CFF in the different age groups provides one way by which changes in visual function with age can be measured. These tests reveal a decline in CFF with aging from a value of 40 Hz (cycles/s) during the 5th decade to about 30 Hz in the 8th decade.[15,16] The persistent miosis in the elderly must contribute to this decline since the decline is less marked with fully dilated pupils.

Aging Changes in Visual Field and Spatial Vision

With aging, a loss in the size of *visual field* is observed, ranging from 3 to 3.5% in the middle age to 2 and 4 times as much at 60 and >65 years, respectively. Disturbances in visuo-motor performance are particularly observed when changes in "useful or effective visual field" are measured, in contrast to measures of visual field under standard clinical conditions.[17] *Visual acuity* reflects the ability to detect details and contours of objects; it is defined as the minimum separable distance between objects (usually fine lines) and is one of the measures of the visual system's *spatial* discrimination ability. Several studies reviewed by Pitts[18] (see Figure 10-6) have shown a decline in visual acuity, commencing at 50 years and particularly worsening during and after the 8th decade,[2,7,18] when it becomes detrimental to vision.

A more recently devised measure of spatial visual ability is *contrast sensitivity* where the test format and conditions more closely resemble the real world.[17] Aging studies in this category indicate that at low spatial frequencies, such as when grating of wide bars are presented, little aging changes are observed, while at high frequencies (fine bars), marked decline in contrast sensitivity is observed with aging, beginning at 30 years. This deficit may underlie certain reading disorders in the elderly such as reading of very small or very large repetitive characters, where the elderly show nearly 70% deficit.[17]

The decline in visual acuity and contrast sensitivity with age may be the result of alterations in the following optical and neural factors: (1) altered refraction of the light rays by the cornea and the lens; (2) decline in accommodation; (3) decreased light input due to constricted pupils (senile miosis); (4) decline in the density and number and function of visual receptor cells, particularly in the fovea; and (5) aging changes in central neural structures of the visual system. It is important to note that elimination of glare and improved illumination can help enhance visual acuity substantially in the aged population.

Color Perception

Beginning with the fourth decade of life, color perception shows a progressive decline with age. Women are relatively less affected than men. During the fourth and fifth decades, the deficiency in color perception is mostly in the short wavelength range, i.e., yellow to blue. This is explained by the changes in the lens, as it becomes more yellowish with age. In the later decades, i.e., after the 60s, a deficiency in the green range also becomes manifested, probably due to retinal or more central factors as it is evident even after removal of the lens.[12,17] Aging defects in color perception are exaggerated under reduced illumination.

Impaired Vision and Everyday Tasks

Recent studies indicate that impaired visual capacity is a major factor in performance of the elderly in such everyday visual tasks as driving.[17] Incidence of driving accidents was four times higher in elderly with reduced "useful field of vision", compared to the age-matched group with maximum useful field of vision. Indeed, in the category of accidents occurring at intersections, the incidence rate was 16 times higher in individuals with reduced useful field of vision.[17]

Changes in Central Visual Pathways

With advancing age, considerable changes in the *optic nerve*[19] and *visual cortex*[20] have been reported. Changes in the visual cortex have included thinning and cell loss. Earlier electrophysiological investigations had indicated marked flattening of the *evoked potential* responses to light flashes in the different parts of the cortex. A more recent study has found that the amplitude of some of the components of visual-evoked potentials are markedly reduced while the latency of response is increased.[21] The latter changes are significantly more marked in elderly men.

2.4 Aging and Eye Diseases

Various degrees of vision loss and blindness, commonly caused by senile cataract, glaucoma, macular degeneration, and diabetic retinopathy, represent the extreme consequence of age-related ocular pathologies (Table 10-1). Diabetic retinopathy will be discussed in Chapter 15. In the U.S., for patients in the age range of 75 to 85 years, the prevalence of cataracts is 46%, macular degeneration 28%, and glaucoma 7.2%.[2] Screening for these disorders includes testing visual acuity, ophthalmoscopic examination, and checking intraocular pressure.[1,4] As a result of increased occurrence of eye diseases with age, the incidence of blindness shows a 25-fold increase with age, from about 0.1% in the middle age group to 2.5% in the elderly (>75 years).[2]

Cataract

In some individuals, the accumulated normal aging changes in pigment and protein composition in the lens take pronounced and pathological dimensions, leading to a condition known as *cataract*.[1,2,4,22] Among the causes of cataract is the glycation of lens proteins, such as the Na-K-ATPase, resulting in abnormal ion-water balance, swelling, and breakdown of crystallin lens proteins. Also, water-insoluble crystallins and yellow chromophores accumulate excessively, particularly in the nuclear type.[2]

As a result of the above pathologies in senile cataract, the lens interior becomes cloudy and opaque, light refraction is greatly reduced, and light scattering markedly increased. These effects lead to loss of visual acuity, reduced patterned vision, and eventually to functional blindness with only a degree of light perception remaining. The occurrence of lens opacities markedly increases with age from about 4% in the middle age group to 12% at 50 years to nearly 60% in the 7th decade.[2] In the same age groups, the percent of individuals showing visual loss as a result of cataract increases from about 1 to 3% and 30%, respectively. Cataract occurs more frequently in diabetic individuals, especially females. Severe cataract is the third cause of blindness in the western world, after macular degeneration and glaucoma.

Cataracts may occur in the lens periphery *(cortical cataract)* or center *(nuclear cataract)*. The incidence of nuclear type is always higher than the cortical type (65% compared to 28% in the >75 years group). Cortical cataract is particularly detrimental to visual acuity, while nuclear cataract interferes more with color perception. A mild cataract can be managed with periodic examination and use of eyeglasses. However, when the reduction in visual acuity interferes with the patient's daily activities, cataract surgery may be necessary.[1-3]

In the U.S., nearly half a million cataract operations are performed per year resulting in improved vision in 97% of the cases. All cataract operations involve removal of opacified lens via a corneal incision followed by fine suturing. In most cases the operation is done under local anesthesia and sedation. The postoperative patient requires means for optical corrections due to loss of lens. Several options are available:

1. *Eyeglasses;* these are thick and heavy, increasing object size by 25%; they induce optical distortions and interfere with peripheral vision. Although they provide good central vision they cannot be used after surgery, if the other eye is normal.
2. *Contact lenses;* hard or soft extended-wear contact lenses have been used. They are more difficult to use and eyeglasses are required for reading. However, they do correct central and peripheral vision, increase image size by only 6%, and can be used after surgery on one or both eyes.
3. *Intraocular lens;* this is surgically placed inside the iris at the time of cataract surgery. It requires the use of bifocal eyeglasses and does have a higher incidence of surgical and postsurgical complications. However, it increases image size by only 1%, corrects central and peripheral vision, and can be used on one or both eyes.
4. *Refractive keratoplasty,* which is still an experimental procedure whereby the cornea is cut and reshaped.

Glaucoma

Another treatable eye disease is *glaucoma,* which is characterized by the following triad: the intraocular pressure increases (>21 mmHg); this leads to progressive excavation of the optic disc, the site where the optic nerve leaves the eye, and cupping of the disc. This leads to ischemic damage to the optic nerve fibers which may result in blindness. Two types of glaucoma are known, *chronic open-angle glaucoma* (COAG), which is the frequent type, and the *closed angle* type, which is rare.[1-4] The incidence of COAG in the population increases rapidly with age, from a low of 0.2% in the 5th decade to about 1% in the 7th decade, 3% in the 8th, and 10% in the 9th decade.

Open-angle glaucoma is characterized by an insidious and slow onset, initially asymptomatic and then gradually leading to blindness, if unchecked. Closed-angle glaucoma is rare and is characterized by an acute attack of severe eye pain and marked loss in vision due to a rapid increase in intraocular pressure compressing the entire retina.

Whereas cataracts are usually treated surgically, glaucoma is often treated medically with eyedrops. Miotics (substances which constrict the pupil) such as pilocarpine are most commonly used. Beta-blockers — which block specific sympathetic innervation — such as Timolol are also used, but caution is warranted since systemic side effects such as heart failure have been shown to occur. Sometimes systemic medications are used, such as carbonic anhydrase inhibitors, or surgery may be necessary in the case of an acute attack of closed-angle glaucoma.

Senile Macular Degeneration

An important cause of visual impairment in the elderly, often leading to legal blindness within 5 years after onset, is *senile macular degeneration* or ARMD (age-related macular degeneration).[1-3,23,24] The disease accounts for nearly half of the registered (legal) cases of blindness in the U.S. and England. The incidence of ARMD increases with increasing senescence, from about 4% in the 66 to 74 years group, to 17 and 22% in the 75 to 84, and >84 years groups, respectively.

The *macula* is an area of retina 6 mm in diameter located just posterior to the lens on the optic axis of the eye (Figure 10-2). Through its high density of cones and involvement in day and color vision, the macula and in particular its central zone, the fovea, provide the structural basis for high visual acuity. Hence, macular degeneration more than any other eye disease affects visual acuity and central vision. This disease occurs generally in both eyes and more often (50%) in women; it is believed to be a hereditary disorder, not caused by simple aging of the retinal nerve cells, but largely related to manifestation of inherited pathologies in the nonneural retinal elements such as the pigment epithelium. The patients also show an increased incidence of hyperopia (farsightedness). The disease may result from disturbances in the walls of subretinal capillaries or in the thickness of subretinal membrane or the retinal pigment epithelium itself.[1-6]

Senile macular degeneration may be accompanied with any of the following pathologic changes depending on the stage and extent of the disease;[1,23,24]

1. White excrescences in the subretinal membrane, called *Drusen* (nodules); these are hyaline deposits ranging in size from punctuate lesions to dome-shaped structures 0.5 mm in diameter and are often observed during the early stages of disease; the Drusen occur mostly around the macula but can be found dispersed throughout the fundus;
2. Atrophy of *retinal pigment epithelium;*
3. *Serous detachment* of retinal pigment epithelium;
4. Subretinal *neovascularization;* and
5. *Disciform scars* which result from the retinal pigment epithelium detachment.

Unfortunately, treatment of macular degeneration is not nearly as successful as cataract and glaucoma.[24] The only realistic goal of treatment is to prevent subretinal detachment and hemorrhage, disorders which lead to an acute loss of vision. To accomplish this, the sites of subretinal neovascular formation are located using fluorescin angiography followed by laser photocoagulation. This procedure is limited to certain cases and is still considered risky.[1]

3 HEARING

Changes in auditory functions with age provide some of the classic and important case studies in gerontology and the physiology of aging. Incidence of hearing disorders rapidly increases with aging afflicting nearly a third of people over 65 and one half of those over 85 years (Figure 10-1);[25] hearing impairments are 20% more frequent in elderly men than women. Hearing disorders interfere with the perception of one's own speech and that of others, creating behavioral and social disabilities. These conditions may lead to social withdrawal and isolation, particularly under the extreme condition of deafness.[26]

As in the visual system, the age-related problems of the auditory system may stem from structural and functional disorders of the peripheral auditory components, the central neural aspects, or both. Here, too, more is known about the aging of the ear than the central auditory system (Table 10-2).

Age-Related Hearing Loss (Presbycusis)

Age-related hearing loss is called *presbycusis.* Numerous studies have reported loss of hearing with age particularly for sounds in the high frequency range, i.e., above 1 kHz.[12,25,27,28] Presbycusis occurs in both ears but not necessarily at the same time. To assess auditory loss with age, pure tone audiograms of subjects in different age groups are determined. That is, the hearing threshold in decibels (unit of sound intensity, dB) for sounds of increasing frequency is determined and the results are presented as relative loss of decibels at different frequencies.

These studies indicate that in the low frequency range (0.125 to 1 kHz),* young subjects have essentially no hearing deficits while old subjects show deficits of about 10 to 15 dB. In the high frequency range, hearing loss for the young group is very mild while in the old group it becomes progressively worse with increasing sound frequency (Figure 10-7).[12,28,29] Typical magnitudes of hearing loss is 30 dB at 2 kHz, increasing by about 10 dB for each additional kHz;[25] in octogenarian men living in urban areas, the hearing loss may be as much as 80 dB. Another way of determining the age-related hearing loss is by measuring the maximum frequency of sound capable of being heard. This frequency is 20 kHz for children (10 years) decreasing to only 4 kHz for the elderly (80 years); the decline is steady and linear, occurring at the rate of about 2.3 kHz per decade.[25]

---

* Hz is the abbreviation for Hertz, a unit of frequency equal to one cycle per second.

TABLE 10-2

Summary of the Normal Aging Changes in the Human Ear

Structural Changes

Hair cells degeneration
Basal cochlea: frequent, especially in first quadrant; diffuse and patchy; main cause of sensory presbycusis. Apical cochlea: infrequent

Nerve cell degeneration
Observed in spiral ganglia often with basal cochlear hair cell loss but not with apical cases (involved in neural presbycusis). Is accompanied with loss of myelinated auditory nerve fibers

Atrophic changes
Generally occur in nonneural components (vascular and connective tissue) of cochlea and lead to strial or conductive types of presbycusis. In stria vascularis; frequent in the middle and apical turns of cochlea. In spiral ligaments; accompanied with devascularization. In inner and outer spiral vessels. In Reissner's membrane, due to vacuolization in basilar membrane leading to mechanical damage

Central neural changes
Little neuronal loss in lower auditory centers. Heavy loss in cortical auditory centers. Dendritic degeneration of cortical pyramidal neurons. Increased latency and decreased amplitude of auditory-evoked potentials; effects more marked in elderly males than in females

Functional Changes

Pure tone hearing:
Loss of hearing in the high-frequency range (presbycusis): loss progressively worsens with age; effects more pronounced in males; noise exposure enhances loss.
Decline in maximum sound frequency capable of being heard, from 20 kHz at 10 years to 4 kHz at 80 years

Speech perception:
Diminished ability to hear consonants; speech is heard but unintelligible. Diminished ability to perceive reverberated and interrupted speech

Sound localization:
Diminished ability to localize sound source, particularly high-frequency ones

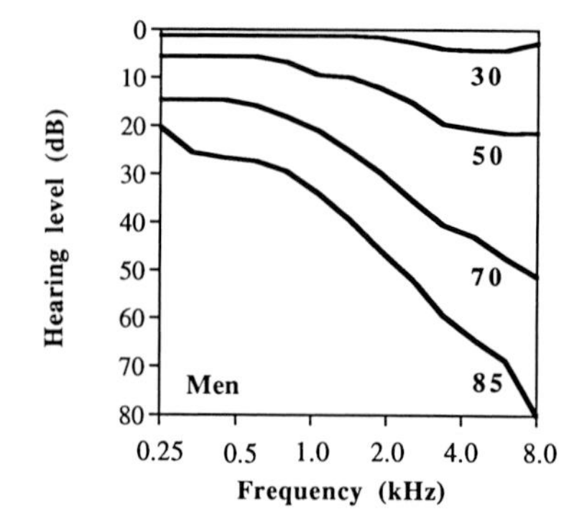

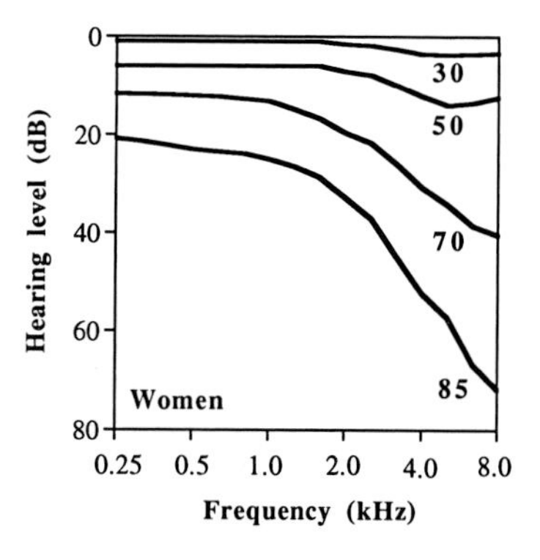

FIGURE 10-7 *Amount of hearing loss in decibels for different pure tones in various age groups of men and women.* Note higher loss in higher frequencies and higher ages and more pronounced hearing loss in men. (Adapted and redrawn from Marsh,[12] based on original data of Spoor.[29])

Typical pure tone audiograms for normal men and women at different age groups are shown in Figure 10-7 which is based on a study by Spoor.[29] In this study, the hearing loss was found to be higher in men than women; the sex difference, which begins after the age of 35 years, may possibly reflect the effects of higher exposure of men to work-related and environmental noise. The result of the latter study and another,[30] showing that the elderly from noise-free rural areas show lesser hearing loss compared to those in the urban environment, suggests that environmental noise is one of the determinants of presbycusis. Another cause of presbycusis may be vascular (atherosclerosis) and similar disorders related to elevated blood lipids (hyperlipoproteinemia). According to Spencer,[31] incidence of hearing loss and inner ear diseases is very high in patients with elevated cholesterol levels.

Presbycusis Related to Ear Structure

The major structures of the ear include: (1) the external ear (pinna, external auditory canal, and tympanic membrane); (2) the middle ear (the ossicular chain); and (3) the inner ear (cochlea and the organ of Corti). The *organ of Corti* contains the *hair cells,* which are the auditory receptors and the mechanoelectrical transducing organs (Figure 10-8a). The cell bodies of the primary auditory neurons are in the cochlea's *spiral ganglia* and the axons comprising the auditory nerve enter the medulla to synapse with central auditory neurons. Auditory pathways in the brain include the *medullary* and *midbrain* centers for signal transmission and auditory reflexes, the *inferior colliculi,* and the *auditory cortex* in the temporal lobe of the cerebral cortex. The selective nature of presbycusis indicates that it is probably not associated with aging changes in the outer or middle ear (tympanic membrane and ossicles) but more likely due to changes in the inner ear (cochlea) or the central auditory system.[27,28,32]

Types of Presbycusis

Presbycusis may occur due to damage to different parts of the auditory systems. Based on the source of damage, four types of presbycusis are recognized:[26,28,32] sensory, neural, metabolic (or strial), and cochlear conductive. The onset of presbycusis may be anytime from the third to sixth decade of life, depending on type. Individuals suffering from these disturbances show distinct and differing audiograms (Figure 10-8), which are clinically used to diagnose types of impairment. More complicated audiograms are produced when the pathology involves a combination of these disorders (Figure 10-8). The standard type of presbycusis with hearing loss at high Hz is often associated with *neural* or *sensory presbycusis* (Figure 10-8).

Sensory Presbycusis

Individuals with sensory presbycusis show a major and sudden loss of hearing in the high frequency range (>4 kHz), indicating a selective deficit in transduction mechanisms of high frequency sounds (Figure 10-8B). Speech discrimination is normal. Although the hearing deficit is observed from middle age, the histopathological problems believed to be mainly associated with the cochlear hair cells may start much earlier. Cochleas of humans with sensory presbycusis typically show loss of outer cells and less often of the inner hair cells of the organ of Corti.[28,32] The loss is diffuse or patchy and mainly limited to the first quadrant of the cochlea's lower basal turn. This part of the cochlea is specialized for detection of high frequency sounds. The affected sensory hair cells and other supporting cells (Hensen's and Claudius' cells) show accumulation of aging pigment lipofuscin, the amount of which corresponds with the degree of sensory deficits.

FIGURE 10-8 (A) *Schematic cross-section of human cochlea.* (B–E) *Typical pure-tone audiograms of aged individuals suffering from different types of presbycusis (sensory, neural, cochlear conductive, and strial).* Each type of presbycusis reflects damage to a different component of the peripheral auditory system: sensory to hair cells; neural to primary auditory neurons; cochlear to basilar membrane; and strial to cells of the stria vascularis. (Redrawn from Gulya.[28])

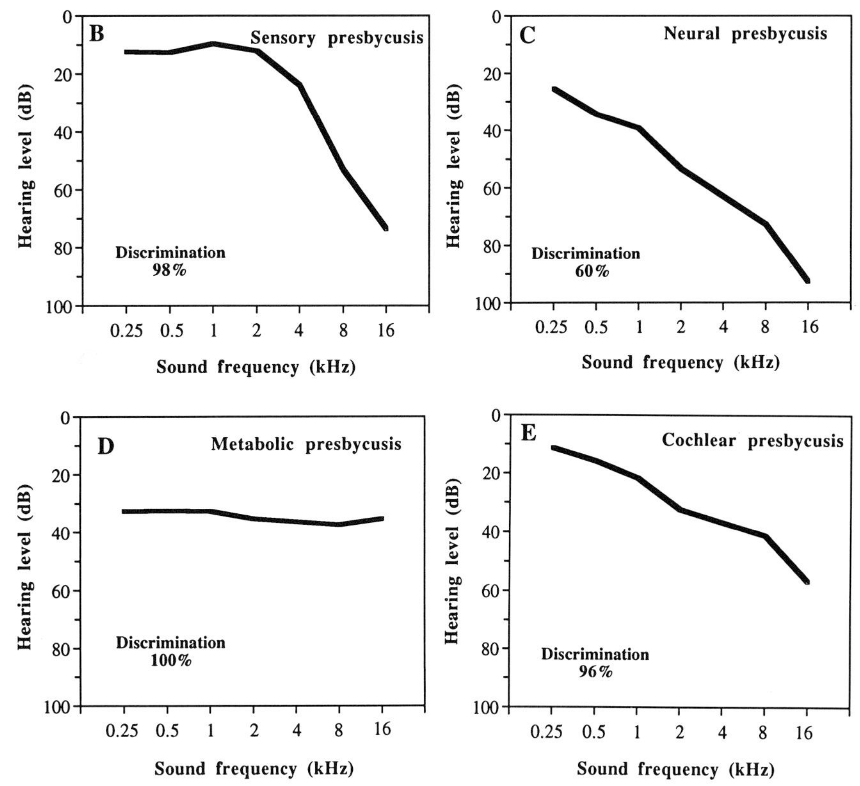

Neural Presbycusis

In this disorder, hearing of pure tones for all frequencies are affected, but the extent of hearing loss increases with increasing frequency of sound, the deficit being about 40 dB at 1 kHz and nearly 100 dB for high frequencies (>8 kHz) (Figure 10-8C). As a result, speech discrimination is reduced to 60% of the normal level. Structurally, abnormalities of the auditory system include many of those seen in the sensory system.[33] In addition, the first-order sensory neurons are also affected. This damage ranges from synaptic structures between the hair cells and the dendrites of the auditory nerve fibers, accumulation of lipofuscin, or signs of degeneration in the cell bodies of the spiral ganglion neurons. Disruption of myelin sheath of the auditory nerve fibers can by itself cause disordered transmission even if the nerve cells were present.[28,34]

Metabolic Presbycusis

In this case, also called the *strial* type, the audiogram is flat (Figure 10-8D), indicating a loss of about 30 to 40 dB at all frequencies. This type is believed to be associated with atrophic changes in the vascular supply, stria vascularis, to the cochlea.[27,28,32,34] The extent of hearing loss is correlated to the degree of degeneration in stria vascularis.[28]

Cochlear Conductive

In this type (also called mechanical presbycusis), the audiogram indicates some hearing loss at all frequencies with the loss magnitude increasing linearly with increasing tone frequency (Figure 10-8E). The magnitude of hearing loss is about half of that seen in the neural type, and speech discrimination is only slightly affected in most cases (96% of normal). This type of presbycusis is believed to be due to changes in the mechanical properties of cochlea's basilar membrane, hence its other name, "mechanical presbycusis". The basilar membrane is markedly thickened, especially at the basal cochlea, and shows calcified, hyaline, or fatty deposits. There is usually no change in the hair cells and sensory neurons.

Sound Localization

Tests of sound localization indicate a decline in this ability with aging, beginning in the fourth decade.[12] It is known that localization of low-frequency sounds depends on *temporal* discrimination (i.e., time of sound arrival) between the two ears, while in the high-frequency range, localization depends on discrimination of sound *intensity* between the two ears. Aging changes occur in both ears, though the rate of aging may be different in the two ears.[27,32] Thus, deficits in localization of sound will be apparent for all frequencies but may be more marked for sounds of higher frequency, as the perception of these are particularly impaired in the old age (see above).

Hearing Deficits and Speech Perception

The marked hearing deficits in the high-frequency range have important bearing on impaired speech perception in the elderly.[28,35] Thus, *vowels,* generated by low-frequency sounds, are heard better than the *consonants* which are produced by high-frequency sounds. Similarly, voices of men are heard better that those of women and children, which have a characteristic high pitch. Since consonants make speech intelligible while vowels make it more audible, a common complaint of the elderly is that they cannot understand spoken words although they can hear them.[12] Fortunately, lip reading, which is associated more with the expression of consonants, can greatly help this deficiency.

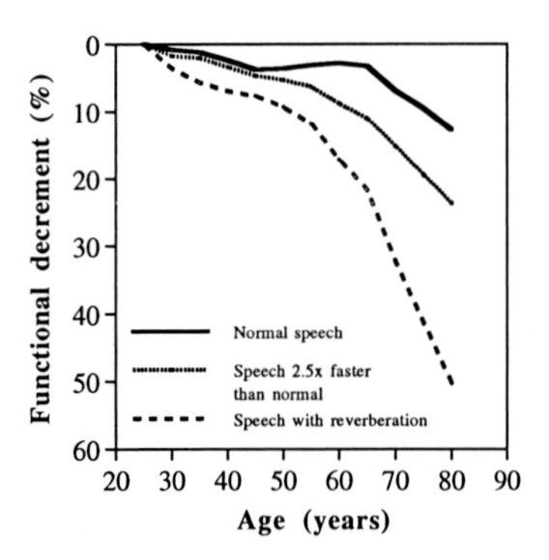

FIGURE 10-9 *Deficits in speech comprehension at different ages.* Note that normal aging deficits are exaggerated when speech is presented faster or with reverberations. (Adapted from Marsh,[12] based on original data of Bergman et al.[36])

As indicated in the data of Figure 10-9, if speech is presented to the aged subjects too fast or with reverberations (i.e., echoing or booming), its comprehension declines markedly.[36] The greatest decline in comprehension is observed if speech is presented with repeated interruptions, such as eight times per second, as is the case with many modern telephone systems.[12] This presents an unfortunate situation for the elderly, who rely so much on the telephone for their communication with the outside world.

The ability to mask sounds, important for speech comprehension in a crowd of talking people, is considerably diminished in the elderly. Indeed tests of hearing loss for pure tones provide usually conservative estimates of hearing deficits as the tests are usually performed under quiet laboratory conditions, eliminating the need for masking. One example of loss of masking ability with age is the increase, with advancing age, in reporting the incidence of ringing in the ear *(tinnitus)*, even though this problem is manifested at all ages. Perhaps the elderly cannot mask these unusual sounds, while the young can.[12]

Aging Changes in Central Auditory Pathways

In addition to the frequently encountered degenerative aging changes in the neurons of spiral ganglia discussed above,[32,34] varying degrees of aging changes in the central auditory structures have been reported[37] — although not by all investigators. These include loss of myelin and neuropil (dendrites and synapses) and degeneration and loss of neurons and lipofuscin accumulation in the cochlear nuclei and superior olives of medulla, medial geniculate of thalamus, and the inferior colliculi of the midbrain.[28]

In contrast to the lower centers, the human auditory cortex shows very marked and clear degenerative aging changes, consisting of a general thinning of the cortex and disrupted organization of the vertical columns; this is caused by heavy neuronal loss and degeneration. In his well-known study of the human cortex aging, Brody[38] found the highest degree of cell loss to occur in the superior temporal gyrus, the anatomical locus of the auditory association cortex; the loss, amounting to about 50% between the ages of 20 to 90, occurred across the cortical thickness. Golgi studies of pyramidal neurons of the auditory cortex by Scheibel et al.[39] revealed marked degenerative changes in the basal dendrites of layer III and V neurons in the old subjects (eighth to tenth decades).

Allison et al.[21] have recently described significant changes in the auditory-evoked potentials in the elderly, including increases in the latency and decreases in the amplitude of the response. The changes are more marked in the males than females and occur generally after 60 years. Undoubtedly these degenerative changes contribute to the observed disorders of hearing and speech comprehension in the elderly, although an exact cause and effect relationship is still not established.

4 SOMATIC SENSATIONS

Somatic sensory systems comprise a large group of sensations (touch, pressure, vibration, proprioception, heat and cold, pain) whose receptors are located in the skin, tendons, joints, and viscera. A detailed account of aging changes in all these systems is clearly beyond this chapter's scope; hence, only certain selected senses will be discussed.

4.1 Functional Changes

Touch

In humans, tactile sensitivity shows a wide variation in different parts, being very high in the palmar surface of fingertips, lips, and tongue tip while relatively low in the back, volar surfaces of hands and feet. Most studies have shown decreases in tactile sensitivity with aging as measured by increased thresholds to touch stimuli. The aging effects may be illustrated by a recent study by Bruce[40] who compared tactile thresholds on the different surfaces of the little finger in two groups of healthy elderly and young subjects and found the following significant elevations in thresholds with aging: 2.2 times on the palmar surface, 2.2 times on the lateral surface, and 2.6 times on the dorsal surface.

Although tactile studies are more readily associated with areas such as hands, the eye's cornea has been a favorite test case, as the contribution of aging changes in the connective tissue to threshold determinations is less critical in the cornea than in the skin; in fact, the skin shows great aging changes, including wide individual variation. Also, corneal tactile receptors are simple and uniform nerve nets. Boberg-Ans found that in subjects 10 to 90 years, corneal threshold to touch increases by threefold during the life span, the aging change being slow between 10 to 40 years and faster thereafter.[41] Similar patterns of aging loss were found by Millodot with different absolute threshold values (see Kenshalo[42] for details).

Vibrotactile Sensation

A number of studies have reported decline in sensitivity to vibrotactile stimulation. According to Cosh, in the great toe pad, the threshold to vibratory stimulation at 100 Hz frequency shows about a threefold increase between childhood (5 years) and extreme old age (90 years).[42] The decline is significantly less marked for the index finger pad. Indeed, most of the clinical investigations in this area have indicated that the decline in vibratory sensitivity is more conspicuous for the lower extremities than in the upper ones.[42] The reason for this differential loss may lie in the higher loss of sensory nerve fibers from the lower spinal nerves (see also below).

It appears that the loss of sensitivity to vibratory stimuli may be valid only for the high-frequency stimulation and not, as previously believed, for all frequencies. In a recent comprehensive study, Verrillo[43] also finds that vibrotactile sensitivity, measured at the thenar eminence of the right hand, progressively decreases with age. The decrease is negligible for low-frequency stimuli (25 to 40 Hz) and very pronounced for high-frequency ones (160 to 250 Hz). In Verrillo's study, the magnitude of the increase in vibrotactile threshold was found to be 3.0 dB per decade for 80 Hz stimulation; in the 160/250 range, the increase was 1, 1.5, 2.25, 3.5, and 6 dB in the 2nd to 6th decades of life, respectively.

Complex Somesthetic Abilities

Complex tactile abilities have also been tested in relation to aging. In the two-point limen test, which measures somatic spatial discrimination ability, the minimum separable distance between two-point stimuli, presented simultaneously on the skin and still perceived as two points, is determined for the various body parts. Lowest limens are found for fingertips and lips (1 to 2 mm) and largest for the back of trunk and neck (30 to 70 mm). As evident by the data in Table 10-3, during aging the two-point limen increases in all body parts, although not to the same extent. The largest change occurs in the great toe pad (15×) and the smallest in the palms (1.2×).

Another complex tactoperceptual ability markedly affected by aging is *stereognosis*. Thus, the elderly score consistently poorly in using blindfold manipulation to identify the objects.[12,42] Similarly, objects placed on the tongue are identified with less precision by the aged group. A third interesting ability involves the ability to recognize body parts. When blindfolded subjects are touched simultaneously on the cheek and hand and asked to identify the parts contacted, they often reply the face. Upon successive testing, this "face-dominant" response diminishes in the young adults but not in the elderly and young children.[44] Also brain-damaged patients retain face-dominant responses. The decline in stereognosis and perception of body parts are clearly connected to central neural changes related to the aging process.

TABLE 10-3

Comparison of Some Parameters of Somatic Sensation Between the Young and Elderly Humans

Parameter	Young	Elderly	E/Y ratio
Two-point limen (mm)[a]			
Hand palm	6	8	1.24
Thumb pad	2	4	1.7
Volar ring finger	2	4	2
Dorsal middle finger	3	7	2.5
Little finger pad	2	6	3
Thumb-index web space	6	18	3
Great toe pad	2	30	15
Vibrotactile thresholds[b]			
Index finger pad	13	18	1.2
Great toe pad	16	34	2.1
Tactile thresholds			
Cornea (mg/sq mm)[c]	12	30	2.5
Little finger (g)[d]			
Lateral surface	0.16	0.36	1.9
Dorsal surface	0.19	0.50	2.6
Palmar surface	0.07	0.19	2.7

Note: Young group, 20 to 30 years; elderly group, 60 to 80 years; the change is expressed as the elderly/young (E/Y)ratio of means.

[a] Two-point limen data are from Bolton et al. (1966) and Gellis and Pool (1977) as quoted in Kensahalo (1977); means are rounded off to nearest mm.

[b] Means in decibels (re: 1 micron displacement) from Cosh (1953) measured at 100 Hz vibration frequency as quoted in Kensahalo (1979);[42] according to Verrillo (1980),[43] sensitivity loss is much greater at higher frequency, 160/250 Hz (19 decibels in the thenar eminence between 2nd and 7th decade).

[c] From Boberg-Ans (1956) as quoted in Kensahalo, 1977.

[d] From Bruce (1980).

Aging Loss of Cutaneous Mechanoreceptors

The Meissner end organs and the Pacinian corpuscles are believed to mediate the sensations of fine touch and pressure/vibration, respectively. The Meissner end organs are found mainly in the glabrous skin of the hand and feet; they show differential distribution even within the hand's palmar surface; their density is high in the fingertips (particularly in the index finger) and low in the thenar eminence. The high density of Meissner end organs in fingertips is believed to be responsible for the great tactile sensitivity of these areas. Pacinian corpuscles are found not only in the deeper layers of the palmar and solar skin but also around the joints and tendons.

Many studies, dating back to Meissner's own works, have indicated a continuous decrease in the density of Meissner corpuscle with aging. According to Bolton et al.,[45] the density in the little finger declines nearly 3 times, from about 25 per mm^2 during the 2nd to 3rd decades to about 8 in the 8th decade; this is confirmed in a recent study by Bruce[40] who notes similar declines in the lateral and dorsal surfaces of the finger in both men and women. In the great toe pad where the density is 10 per mm^2 in early adulthood, a comparable decline with aging occurs. Loss of similar magnitudes in the index finger has been reported by others.[46] Such declines are not limited to humans and have been observed in other primates as well. In the pigtail macaque monkey, between the ages of 5 to 20 years, a reduction of more than 50% in Meissner corpuscle density is observed in the thumb, index finger, and great toe.[47]

Aging changes in the Meissner corpuscles are not limited to a decline in number. They also show qualitative changes with aging.[42,47] In the newborn infant they are nearly spherical, growing in size and becoming elongated with age. In the young adult, each receptor organ being cylindrical in shape is attached to the epidermis superficially and to the elastic elements of the dermis basally. In old age, the receptors appear more loosely anchored and become dissociated from the epidermis.

Loss of Pacinian corpuscles with age has also been demonstrated. During infancy, the Pacinian corpuscle is small and oval containing few concentric lamellae; with aging, addition of new lamellae increases the size and rigidity of the corpuscle and distorts its shape. There is also a gradual decrease in the number of the Pacinian corpuscles with aging in the glabrous skin.[47]

The causes of aging changes in the tactile receptors is not well understood. Clearly these receptors are able to regenerate throughout life as evident by the return of tactile sensibility following repair of skin damage. On the other hand, the regenerative capacity of many of these receptors depends on the integrity of their innervation. Therefore, the age-related loss in number may reflect changes in fiber branching or fiber number. Physiological declines in fiber activity or axoplasmic transport of trophic substances may also be involved. Declines in the number of sensory fibers

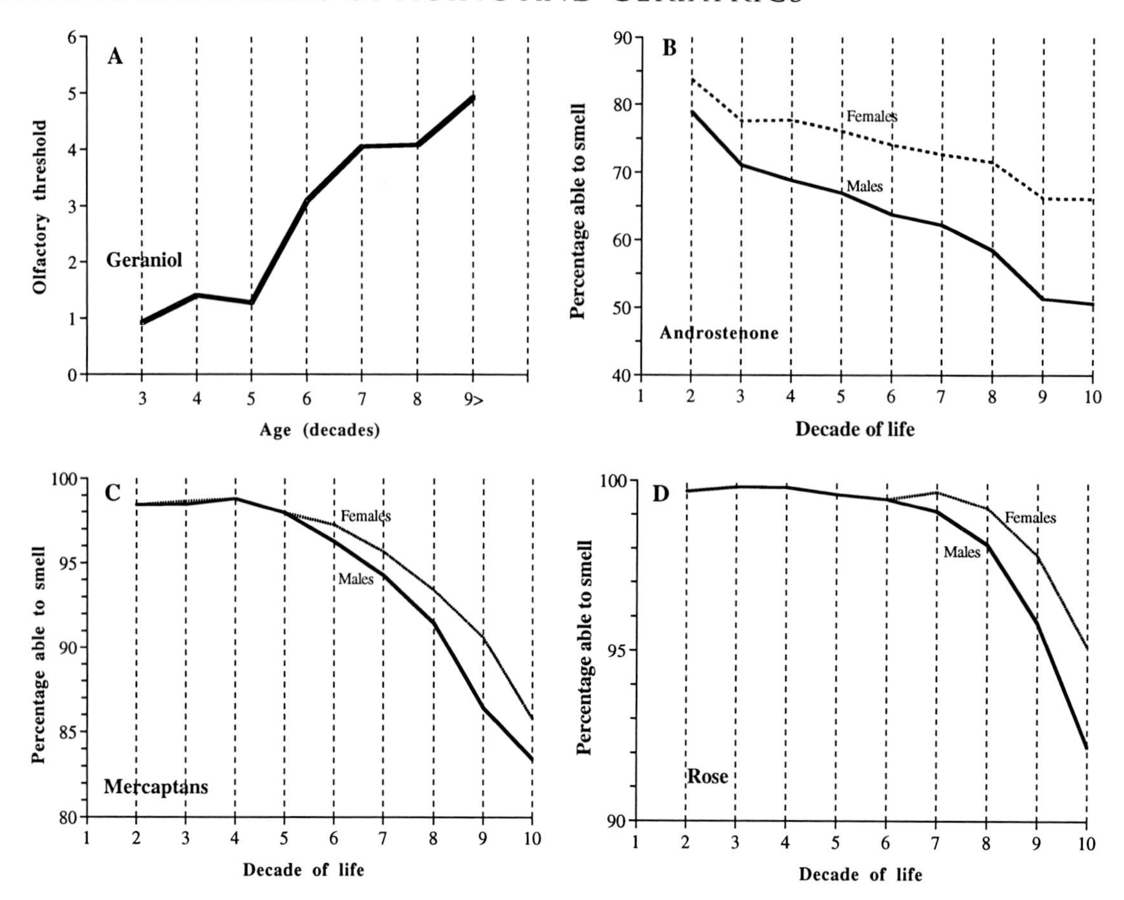

FIGURE 10-10 (A) *Changes in olfactory threshold for the odor of geraniol with age.* (Based on the data of Serby et al.[67]); (B–D) Changes in sensitivity (ability to detect very low concentrations of substances) for mercaptans (natural gas odor), rose, androsterone (musky, sweaty), measured as changes with age in percent of individuals able to detect the odor. In general, females perform significantly better than males. (Redrawn from data of Wysocki and Pelchat.[57])

innervating the periphery may also occur with old age. A decrease of up to 30% in the number of myelinated fibers in the sensory nerve and in the spinal sensory roots has been observed in the old age.[48] In extreme cases, these declines may underlie the observed senile peripheral neuropathies.

Although morphological aging changes in the Meissner and Pacinian corpuscles clearly influence their physiologic properties, it is generally believed that, at least with regard to simple tactile abilities such as sensitivity, the loss in receptor number can best explain the observed functional deficits in the aged, particularly at the peripheral level. Specifically, a reduction in tactile receptor number, by decreasing receptor recruitment, can cause increased tactile thresholds; also, decline in receptive field overlap may explain the decreased capacity for spatial discrimination as revealed by the two-point discrimination test. Qualitative aging changes in the morphology of the tactile receptors may be involved in causation of the more complex functional changes.

Central Neural Changes

As mentioned above, loss of sensory fibers and neurons in peripheral nerves and spinal sensory roots has been reported particularly for the lower extremities.[48] Thinning of the cortex and cell loss in the somatic sensory area are also observed, although not to the same extent as in the auditory cortex.[20] Nevertheless, electrophysiological studies indicate that somatic sensory-evoked potentials show diminished amplitude and increased latency in the elderly (after 60 years), suggesting significant structural and functional changes, such as fiber loss, reduced myelination, and decreased conduction velocity, as well as synaptic changes, in the peripheral and central sensory pathways and sensory cortex.[21] Here, as in visual and auditory systems, the changes are more marked in males than in females.

5 ▬ OLFACTION

Olfaction is important in food selection and nutrition, social interaction, and enjoyment of good life.[49] Many of these olfactory functions begin early in life.[50] Olfaction aids in promotion of health by avoiding the putrid smell of rotten foods and infected matter. In addition to these values, olfaction warns against certain dangers posed by modern living and industrial odorous pollutants. The value of olfaction for the elderly is often shown by the consequences of its deficiency or loss. Anand[51] found that a high proportion of the elderly die of accidents involving household gas poisoning, indicating that they are unable to smell gas leakage. Stevens and Cain[52] have confirmed this and have shown that nearly half of the elderly fail to detect the odor of ethyl mercaptan at concentrations normally found in the bottled propane gas and have thresholds 10 times higher than the young for mercaptan substances added to natural gas for warning. Even when they smell gas, their perception of its intensity and unpleasantness is markedly reduced.

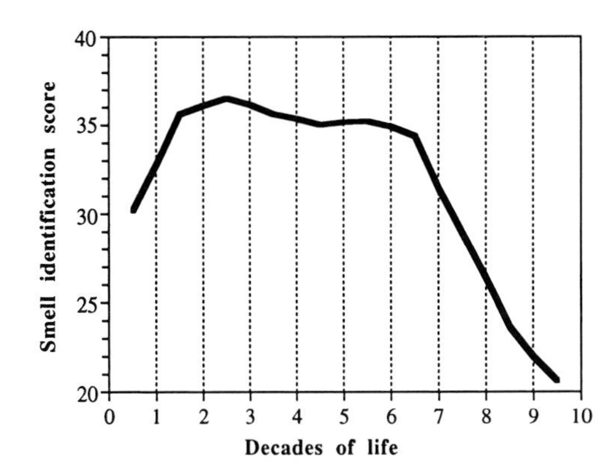

FIGURE 10-11 *The effect of aging on smell identification performance of humans at different age groups.* Testing is based on the UPSIT test whereby subjects are asked to identify 40 odorants using name cues provided in a forced-choice system. Maximum score is 40. Note marked decline in smell identification ability after the seventh decade. (Redrawn from data of Doty et al.[58])

Declining olfactory ability with age *(hyposmia)* leading to complete loss *(anosmia)* in some of the highly senescent elderly is now generally accepted. Such deficiencies may be one of the causes of poor appetite and irregular eating habits, which result in malnutrition and weight loss, as well as a lack of interest in the good life and in seeking pleasure.

5.1 Functional Changes

Two kinds of evaluations have been employed in studies of human olfactory function in aging.[53] One type measures changes in olfactory sensitivity by usually determining the threshold for various odors in subjects at different ages. Another type evaluates olfactory recognition and discrimination and the changes with age in the ability to identify different odors in a mixture.

Olfactory Sensitivity and Odor Thresholds

Several early studies at the turn of the century indicated a decline in odor sensitivity with age.[49] Later, better controlled studies have generally confirmed this conclusion.[54,55] For example, Venstrom and Amoore's[54] compared odor thresholds of people at different ages to that of a middle age group (early 40s), finding better sensitivity in the younger group and lower sensitivity in the older group; mean sensitivity declined throughout adult life from the third decade to the eighth. More recently, Stevens and Cain[56] found higher absolute thresholds for odorants in the elderly compared to the young group; the differences varied depending on the odorant (ninefold for limonene —lemon odor, threefold for benzaldehyde — almond odor, and twofold for isoamylbutyrate — fruity odor). Aging changes in odor sensitivity during the lifespan is depicted in Figure 10-10A, showing increase in olfactory threshold for geraniol (geranium odor) in a population of normal individuals of different ages.

In a very interesting study involving thousands of readers of *National Geographic* magazine as subjects, Wysocki,[57] using a scratch and sniff test, measured the changes in odor detection ability (sensitivity) for several odorants by individuals of different ages. The results shown in Figure 10-10B–D indicate that for natural and common odors like amyl acetate (banana, pears), Eugenol (clove oil), and rose (Figure 10-D), a decline in sensitivity becomes evident in the seventh decade, with females scoring better than males; 10 to 15% of subjects over 80 years failed to detect any of these odors. For odor of mercaptan (the odor added to natural gas), the deficit was more marked and the decline began earlier, in the sixth decade (Figure 10-C). Of men over 80 years, 50% failed to detect the odor of androsterone (sweaty, musky) (Figure 10-B) and galaxolide (a synthetic musk). When readers were asked to rate their own olfactory ability from extremely good to extremely poor, the results indicated significant reduction with age. When asked to rate intensity of suprathreshold odors, mean scores for intensity rating showed a gradual but definite reduction with age.[57]

Odor Identification Ability

Olfactory discrimination may be measured by the ability to identify the constituent odors of a mixture or identify by name or by visual association a host of odors presented to a subject. In one of the earliest studies, Anand[51] presented to subjects in different age groups a mixture containing odors of coffee, peppermint, coal-tar, and almond oil. He found that with increasing age the ability to identify the constituent odors of the mixture diminished. Thus, whereas nearly 70% of the subjects in the 20 to 40 years age group were able to detect all four odors in the mixture, in the 40 to 60 age group, only 50% could do the same, in people over 60 years, 10%, and in those over 70, a mere 3% identified all four odors.

In a comprehensive study of changes in odor identification throughout the life span by Doty et al.,[58] 2000 subjects in different age groups were tested for their smell identification ability according to the University of Pennsylvania Smell Identification Test (UPSIT). The results shown in Figure 10-11 indicated that the ability to identify the odors of different substances improves from childhood to adolescence, reaches a peak by the late teens, and begins to decline in the sixth decade. The decline during the 7th and 8th decades is quite drastic; in the latter group nearly 60% show impairment in olfactory function and 25% are actually anosmic, in terms of identifying smells. In those over 80, 80% show impairments and 50% are anosmic. A major finding of this and previous similar studies is that women outperform men in olfactory abilities across all age groups, although they too show a similar general pattern of aging decline. Similar conclusions have been found by other investigators using the same or different methods.[59]

5.2 Neural Changes

The physiological and anatomical bases of the decline in olfactory abilities are not well understood, although cell loss seems to play a major part. Bipolar sensory neurons located in the *olfactory mucosa,* a sheet of neuroepithelium covering the walls of superior nasal cavities, mediate olfactory transduction. The cilia of dendrites of the *olfactory neurons* contain the receptor molecules and the chemoelectrical transducing elements protrude into the nasal cavity. The axons of receptor cells, after passage through the cribriform plate pores, enter the *olfactory bulb* to synapse, in the *olfactory glomeruli,* with the dendrites of the mitral cells, the bulb's principal relay neurons. The axons of the mitral cells move through the *olfactory tract* into the brain to end in the *cortical olfactory areas.* The *internal granule cells* and the *periglomerular cells* of the olfactory bulb are small inhibitory neurons involved in integrative activity in the bulb.[60]

As evident from this account, the *cilia* of the olfactory neurons are directly exposed to the environment of the nasal cavity and the hazards of smoking and other airborne pollutants. Perhaps for this reason, olfactory neurons are constantly renewed with a turnover cycle of about a month; also, after transection of the olfactory nerve, which leads to the death of olfactory neuron, new neurons replace the degenerated ones.[61] This unique regeneration ability is well proven for rodents and also occurs in monkeys but is not clearly proven in humans. Some recent studies in rats have indicated that the regenerative capacity of the olfactory mucosa may diminish in old age.

The human olfactory mucosa begins to show signs of aging rather early in life. Thus, while the mucosa in the fetus has a very healthy appearance, in the adult it contains patches of respiratory epithelium; even the mature adult human olfactory epithelium shows diminished number of receptor neurons and of other cellular elements like the basal and supporting cells.[62] With aging, this degeneration becomes so marked that in some aged humans there is complete loss of the olfactory sensory neurons.[63] Presumably, the neuronal replacement process occurring in lower mammals does not occur in humans, or, if it does, it is not successful in maintaining the olfactory mucosa of the adult and elderly in a healthy state.

Loss of neurons has been also demonstrated in the higher nerve centers associated with smell. Thus, Smith[64] showed that in the human olfactory bulbs a gradual atrophy of olfactory *glomeruli* occurs with age, at a rate of about 10% per decade so that in very old people only 20% of the glomeruli remain. Bhatnagar et al.[65] found a substantial loss of *mitral cells*, the principal neurons of the human olfactory bulb, from 50,000 at age 25 to 30,000 at 60 years and 15,000 at 90 years. Although the loss in mitral cells amounts to about 5000 per decade, most of the decline occurs after the 6th decade. These atrophic changes in the olfactory bulb are likely to be secondary to transneuronal effects caused by gradual loss of olfactory receptor neurons, as suggested by Liss and Gomez[66] and Naessen,[63] although degenerative factors intrinsic to the aging bulb and brain cannot be ruled out.

As the extremely high sensitivity of the olfactory system in odor detection is in part due to the great number of the olfactory receptor neurons and the high convergence ratio of these on the mitral cells, then a loss in these neuronal components may underlie the gradual decline in olfactory sensitivity described above. The causes of decline in odor identification ability may be more complex and relate to degenerative aging changes in central neural structures.

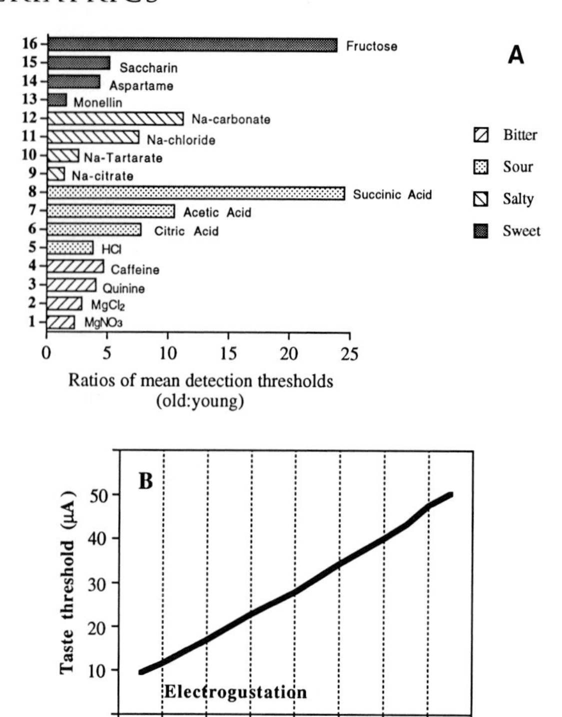

FIGURE 10-12 (A) *Increase in taste threshold for different sweet, salty, sour, and bitter compounds with age.* (Bars indicate the ratios of thresholds of the elderly to young.) Note marked differences in individual compound within each category; also note the aging decline is more for the salty and bitter compared to the sweet compounds, which are least affected. (Based on the data of Schiffman.[76]) (B) *Changes in taste threshold as determined by minimum current in microamperes necessary to perceive a "taste" sensation.* Note the gradual increase in threshold, indicating a decrease in taste sensitivity with age. (Redrawn from data of Hughs.[79])

5.3 Olfaction and Aging-Related Brain Diseases

Olfactory dysfunction has been found in numerous diseases and medical conditions. Some of the aging-related brain diseases that are associated with varying degrees of olfactory loss are *Alzheimer's disease*,[67,68] *Parkinson's disease*,[69,70] *Korsakoff's syndrome*,[71] and *senile depression*.[67]

Olfaction and Alzheimer's Disease

Many studies have linked olfactory dysfunction to Alzheimer's disease, the major dementing disease of the aging brain.[67,68] Olfactory dysfunction in these patients involves both altered sensitivity and identification of odors. According to Serby's studies, patients in the early stages of the disease are able to identify only 60% of the odors identified by age-matched normal controls (60 to 69 years); in the more advanced stages of the disease, this percentage declined to 40% of the age-matched controls (70 to 79 years). Thresholds for odors were only slightly raised in the early stages of the disease but were increased nearly twofold in the more advanced stages.[67] Other investigators have reported significant loss of sensitivity even in the early stages. The deficiency in olfactory functions has suggested a potential use for olfactory tests in early antemortem diagnosis of Alzheimer's disease.[67,68]

The presence of olfactory sensitivity and identification deficits in Alzheimer's patients has implicated both peripheral and central lesions of olfactory structures. Evidence for Alzheimer's type of molecular and structural lesions has been found in the olfactory epithelium.[72] Aging-related degenerative brain lesions (neuritic plaques and neurofibrillary tangles) have been found abundantly in the primary olfactory cortical areas (olfactory bulb, anterior olfactory nucleus, prepyriform cortex)[68,73] and secondary olfactory cortical regions of the brain (amygdala, hippocampus, entorhinal cortex);[68,74] these lesions are usually absent or far less extensive in the primary visual or auditory areas. Also, the olfactory structures show the same marked deficiency in cholinergic transmission found for cortical association areas and basal forebrain structure.[67] One theory states that potential toxic agents such as aluminosilicates or viruses may enter the brain through the olfactory pathways, causing the Alzheimer's disease; the aging atrophy of the olfactory system may facilitate this transmission.[75]

6 TASTE

Taste is served by special *taste buds* located in the *tongue's papillae* and other regions of the oral cavity. Taste buds contain the *taste cells,* which have no axons and show constant renewal with a turnover cycle of about 10 days. Taste cells are in synaptic contact with the fibers of the *gustatory sensory neurons* which transmit taste signals to the medullary taste centers via the fibers of the *taste nerves*. From the lower gustatory centers, signals are relayed to the brain's higher gustatory centers in the pons, thalamus, and cerebral cortex.[60] Classically, four primary taste qualities — *sour, sweet, salty,* and *bitter* — are recognized for which specialized receptor sites within the taste cells are believed to exist. Also, the tongue is believed to be regionally specialized so that sweet sensation is best detected in the tip, bitter in the back, and salty and sour on the sides. More recent thinking suggests a more complex spectrum of taste primaries and less emphasis on regional variation.[76,77]

Recent studies indicate a complex spectrum of changes in taste with aging, although it is generally believed that compared to olfaction, taste is relatively less affected.[59,76,78]

Taste Sensitivity

Hughs,[79] Murphy,[80] and Schiffman,[76] have summarized the studies on the effects of age on taste sensitivity. Although the findings of the different investigators are not as consistent as in olfaction studies, many report declines in sensitivity. Murphy measured absolute thresholds for various taste modalities in subjects ranging in age from 6 to 73, and found the taste sensitivity to vary as a function of age and tastant, with no sex difference. Loss in sensitivity was steady and gradual, being in the order of 1 log unit, from a mean of about 1 part per 10,000 in the young adults to 1 per 1000 in the elderly group. The decline in sensitivity, which is reflected in increasing thresholds, varies in magnitude depending on the substance being measured.

Figure 10-12A, based on the data of Schiffman,[76] shows the elderly/young ratios of taste thresholds for a number of compounds with different taste qualities. The average increase in threshold with aging was 2.7 times for sweet compounds, 4.3 times for sour acids, 6.9 times for bitter compounds, and 11.6 times for salty compounds. Data show marked variation for compounds within the same taste categories (compare 4 times for sodium carbonate to 24 times for sodium citrate or 1.5 times for caffeine with 24 times for magnesium nitrate). The range of aging changes in thresholds for amino acids, which produce a complex gustatory sensation, is from no change (arginine) to a 6-fold increase (cysteine) with an average of 2.5 times.

Quantitative measurement of taste threshold may be carried out in a different way by determining the amount of bitter–sour sensation produced by the passage of a mild galvanic current over the tongue surface;[79] the minimum current causing sensation being proportional to the taste threshold. This threshold current is found to increase linearly with age from a value of about 10 μA during the third decade to 60 μA in the elderly (Figure 10-12B).

In addition to detection, the elderly also show difficulty in discriminating between different intensities of substances presented at *suprathreshold* concentrations. For example, a 6 to 12% change in the concentration of salt was sufficient for the young subjects to detect a difference; the elderly needed an increment of 25% to perceive a change in concentration.[81]

Taste of Mixtures and Flavor Identification

In a study related to taste sensitivity for salts in mixtures, Stevens and Cain[81] studied aging changes in the perception of salt in presence of other tastants. Thus, when the presence or absence of suprathreshold amounts of salt flavoring in the tomato soup was taste tested, the elderly performed significantly poorly compared to the middle age, which in turn performed poorly compared to the young. In fact, the elderly's taste thresholds for sodium chloride was found to be several times higher in the tomato soup than in water. These results imply that the elderly may not be able to regulate their dietary salt needs when using salts in mixtures.

Although the aging decline in taste sensitivity is less marked when suprathreshold amounts of tastants are used, a study by Schiffman[82] on blindfolded elderly concluded that the elderly's performance in detection and recognition of various foods and complex flavors, presented in real life situations, is considerably diminished compared to the young. Furthermore, while the elderly seem to rate the foods' qualities mainly on hedonic dimensions, the young group uses both the hedonic and tactile dimensions to describe their ratings. The elderly are also less able to identify the flavors of foods in a mixture. This defect has been ascribed to the decline in olfactory abilities which have an extremely important role in flavor identification. A discussion of the complex relationship between taste and olfaction and nutrition in the elderly is beyond the scope of this chapter and has been reviewed by experts elsewhere.[59,76,81,82]

Neural Basis of Gustatory Decline

Causes of the decline in taste sensitivity and discrimination with aging have been variously ascribed to changes in the peripheral and/or central components of the gustatory system. Classical studies have indicated a decline in the number of taste buds and papillae (see Bradley[83] for review). Thus, the number of *fungiform papillae* on the tongue tip shows a steady decline (50%) throughout the adult life, from about 60 in the young to about 30 in the 6th decade. Also, the number of taste buds on the *circumvallate papillae* in the back of the tongue was found to decrease by more than 50% but only after 50 years. Taste buds in the *foliate papillae* (on the lateral regions) show little aging decline. According to reports cited by Marsh[12] the major loss of taste buds occurs earlier in women (40 to 45 years) compared to men (50 to 60), suggesting involvement of gonadal hormones. It is often assumed that the changes in number of taste buds are one of the underlying causes of decline in sensitivity. These reported losses are contradicted by more recent studies (see Mistretta[84] for review), which do not indicate any significant loss of taste cells in human fungiform papilla and in the Rhesus monkey.

The reduction in taste buds and cells, even if present, may not affect the behavioral and complex perceptual capacities related to taste which are normally evoked with suprathreshold amounts of tastants.[77] In addition, aging changes in the tongue tissue (surface texture) and salivary secretion may play a part. Indeed, the mouth in the elderly is drier due to decreased volume of salivary secretion and increased mucin content.[80] Activity of the salivary enzyme amylase, needed for sweet sensation, is reduced in old age and may contribute to the functional results.[12] Less is known about the aging changes in the cerebral centers for the gustatory system. Nevertheless, these changes must be occurring since perception of complex flavors relating to foods and recognition of food qualities is considerably diminished in the old age[76,82] and it is reasonable to ascribe these effects to aging changes in the higher nervous centers.

REFERENCES

1. Graham, P., The eye, in *Principles and Practice of Geriatric Medicine,* 2nd ed., Pathy, M. S. J., Ed., John Wiley & Sons, London, 1991, 985.
2. Bron, A. J., The aging eye, in *Oxford Textbook of Geriatric Medicine,* Evans, J. G. and Williams, T. F., Eds., Oxford University Press, Oxford, 1992, 557.
3. Leighton, D. A., Special senses — aging of the eye, in *Textbook of Geriatric Medicine and Gerontology,* Brocklehurst, J. C., Ed., Churchill-Livingstone, Edinbrough, 1985, chap. 21.
4. Stefansson, E., The eye, in *Principles of Geriatric Medicine and Gerontology,* Hazzard, W. R., et al., Eds., McGraw-Hill, New York, 1991, chap. 42.
5. Kuwabara, T., Age related changes of the eye, in *Special Senses in Aging,* Han, S. S. and Coons, D. H., Eds., Institute of Gerontology, University of Michigan, Ann Arbor, 1979, 46.
6. Sekuler, R., Kline, D., and Dismuskes, K., Eds., *Aging and Human Visual Functions,* Alan R. Liss, New York, 1982.
7. Ordy, J. M. and Brizee, K. R., Functional and structural age differences in the visual system of man and non-human primate monkeys, in *Sensory Systems and Communication in the Elderly,* Ordy, J. M. and Brizzee, K., Eds., Raven Press, New York, 1979, 13.
8. Weale, R. A., *Focus on Vision,* Harvard University Press, Cambridge, MA, 1982, chap. 3.
9. Vaughan, W. J., Schmitz, P., and Fatt, I., The human lens: a model system for the study of aging, in *Sensory Systems and Communication in the Elderly,* Ordy, J. M. and Brizzee, K., Eds., Raven Press, New York, 1979, 51.
10. Weale, R. A., *The Aging Eye,* Lewis, London, 1963.
11. Verriest, G., L'influence de l'age sur les fonctions visuelles de l'homme, *Bull Acad. R. Med. Belg.,* 11, 527, 1971.
12. Marsh, G., Perceptual changes with aging, in *Handbook of Geriatric Psychiatry,* Busse, E. W. and Blazer, D. G., Eds., Van Nostrand, New York, 1980, 147.
13. Dark, A. J., Streeten, B. W., and Jones, D., Accumulation of fibrillar protein in the aging human lens capsule, *Arch. Ophthal.,* 82, 815, 1969.
14. Duane, A., Accommodation, *Arch. Ophthalmol.,* 5, 1, 1931.
15. McFarland, R. A., Domey, R. G., Warren, A. B., and Ward, D. C., Dark adaptation as a function of age. I. A statistical analysis, *J. Gerontol.,* 15, 149, 1960.
16. McFarland, R. A., Warren, A. B., and Karris, C., Alterations in critical flicker frequency as a function of age and light-dark ratio, *J. Exp. Psychol.,* 56, 529, 1958.
17. Sekuler, R. and Sekuler, A. B., Visual perception and cognition, in *Oxford Textbook of Geriatric Medicine,* Evans, J. G. and Wiliams, T. F., Eds., Oxford University Press, Oxford, 1992, 575.
18. Pitts, D. G., The effects of aging on selected visual function: dark adaption, visual acuity, stereopsis and brightness contrast, in *Aging and Visual Function,* Sekule, R., Kline, D., and Dismukes, D., Eds., Alan R. Liss, New York, 1982, 131.
19. Dolman, C. L., Mccormick, A. O., and Drance, S. M., Aging of the optic nerve, *Arch. Ophthalmol.,* 98, 2052, 1980.
20. Brody, H. and Vijayashankar, N., Anatomical changes in the nervous system, in *Handbook of the Biology of Aging,* Finch, C. E. and Hayflick, L., Eds., Van Nostrand, New York, 1977, 241.
21. Allison, T., Hume, A. L., Wood, C. C., and Goff, W. R., Developmental and aging changes in somatosensory, auditory and visual evoked potentials, *Electroenceph. Clin. Neurophysiol.,* 58, 14, 1984.
22. Friedenwald, J. S., Permeability of the lens capsule, with special reference to etiology of senile cataract, *Arch. Ophthalmol.,* 3, 182, 1930.
23. Lewis, R. A., Macular degeneration in the aged, in *Special Senses in Aging,* Han, S. S. and Coons, D. H., Eds., Institute of Gerontology, University of Michigan, Ann Arbor, 1979, 93.
24. Sarks, S. H., Aging and degeneration in the macular region. A clinicopathological study, *Br. J. Ophthalmol.,* 60, 324, 1976.
25. Weiss, A. D., Auditory perception in relation to age, in *Human Aging,* Birren, J., et al., Eds., U.S. Government Printing Office, Washington, D.C., 1963, 111.
26. Mauer, J. F. and Rupp, R. R., *Hearing and Aging: Tactics for Intervension,* Grune & Stratton, New York, 1979.
27. Schuknett, H. F., Further observations on the pathology of presbycusis, *Arch. Otolaryngol.,* 80, 369, 1964.
28. Gulya, A. J., Disorders of hearing, in *Oxford Textbook of Geriatric Medicine,* Evans, J. G. and Williams, T. F., Eds., Oxford University Press, Oxford, 1992, 580.
29. Spoor, A., Presbycusis values in relation to noise induced hearing loss, *Inter. Audiol.,* 6, 48, 1967.
30. Rosen, S., et al., Presbycusis study of a relatively noise-free population in the Sudan, *Ann. Otol.,* 71, 727, 1962.
31. Spencer, J. T., Jr., Hyperlipoproteinemia in the etiology of inner ear disease, *Laryngoscope,* 83, 639, 1973.
32. Schuknett, H. F., *Pathology of the Inner Ear,* Harvard University Press, Cambridge, MA, 1974, 388.
33. Johnsson, L. and Hawkins, J. E., Jr., Sensory and neural degeneration with aging, as seen in microdissections of the human inner ear, *Ann. Otol.,* 81, 1, 1972.
34. Johnsson, L. and Hawkins, J. E., Jr., Age related degeneration of the inner ear, in *Special Senses in Aging,* Han, S. S. and Coons, D. H., Eds., Institute of Gerontology, University of Michigan, Ann Arbor, 1979, 119.
35. Jerger, J. and Hays, D., Diagnostic speech audiometry, *Arch. Otolaryngol.,* 103, 216, 1977.
36. Bergman, M., Blumensfeld, V. G., Cascardo, D., Dash, B., Levitt, H., and Margulies, M. K., Age related decrement in hearing for speech: sampling and longitudinal studies, *J. Gerontol.,* 31, 533, 1976.
37. Feldman, M. L. and Vaughan, D. W., Changes in the auditory pathway with age, in *Special Senses in Aging,* Han, S. S. and Coons, D. H., Eds., Institute of Gerontology, University of Michigan, Ann Arbor, 1979, 143.
38. Brody, H., Organization of the cerebral cortex. III. A study of aging in the human cerebral cortex, *J. Comp. Neurol.,* 102, 511, 1955.
39. Scheibel, M., Lindsay, R. D., Tomiyasu, V., and Scheibel, A. B., Progressive dendritic changes in aging human cortex, *Exp. Neurol.,* 47, 392, 1975.
40. Bruce, M. F., The relation of tactile thresholds to histology in the fingers of the elderly, *J. Neurol. Neurosurg. Psychiatry,* 43, 730, 1980.
41. Boberg-Ans, J., On the corneal sensitivity, *Acta Ophthalmol.,* 34, 149, 1956.

42. Kenshalo, D. R., Sr., Aging effects on cutaneous and kinesthetic sensibilities, in *Special Senses in Aging,* Han, S. S. and Coons, D. H., Eds., Institute of Gerontology, University of Michigan, Ann Arbor, 1979, 189.
43. Verrillo, R. T., Age related changes in the sensitivity to vibration, *J. Gerontol.,* 35, 185, 1980.
44. Bender, M. B., Fink, M., and Green, M., Patterns in perception on simultaneous tests of the face and hand, *Arch. Neurol. Psychiatry,* 66, 355, 1952.
45. Bolton, C. F., Winkelman, R. K., and Dyke, P. J., A quantitative study of Meissner's corpuscles in man, *Neurology,* 16, 1, 1966.
46. Cauna, N., The effects of aging on the receptor organs of the human dermis, in *Advances in Biology of Skin Aging,* Montagna, W., Ed., Vol. 6, Pergamon Press, New York, 1965.
47. Witkin, J., Peripheral tactile innervation, in *Aging in Non-Human Primates,* Bowden, D. M., Ed., Van Nostrand, New York, 1979, 158.
48. Gardner, E. D., Decrease in human neurons with age, *Anat. Rec.,* 77, 529, 1940.
49. McCartney, W., *Olfaction and Odours,* Springer-Verlag, Berlin, 1968, 176.
50. Meisami, E., Olfactory development in the human, in *Handbook of Human Growth and Developmental Biology,* Meisami, E. and Timiras, P. S., Eds., Vol. 1, Part B, CRC Press, Boca Raton, FL, 1988, 33.
51. Anand, M. P., Accidents at home, in *Conference on Medical and Surgical Aspects of Aging. Current Achievements in Geriatrics,* Anderson, W. F. and Isaacs, B., Eds., Cassell, London, 1964, 239.
52. Stevens, J. C., Cain, W. S., and Weinstein, D. E., Aging impairs the ability to detect gas odor, *Fire Technology,* 23, 198, 1987.
53. Murphy, C., Olfactory psychophysics, in *Neurobiology of Taste and Smell,* Finger, T. E. and Silver, W. L., Eds., John Wiley & Sons, New York, 1987, 251.
54. Venstrom, D. and Amoore, J. E., Olfactory thresholds in relation to age, sex or smoking, *J. Food Sci.,* 33, 264, 1968.
55. Strauss, E. L., A study on olfactory acuity, *Ann. Otol.,* 79, 1, 1970.
56. Stevens, J. C. and Cain, W. S., Old age deficits in the sense of smell as gauged by thresholds, magnitude matching and odor identification, *Psychol. Aging,* 2, 36, 1987.
57. Wysocki, C. J. and Pelchat, M. L., The effects of aging on the human sense of smell and its relationship to food choice, *Crit. Rev. Food Sci. Nutr.,* 33, 63, 1993.
58. Doty, R. L., Shaman, P., Applebaum, S. L., Giberson, R., Sikorski, L., and Rosenberg, L., Smell identification ability changes with age, *Science,* 226, 1441, 1984.
59. Murphy, C., Nutrition and chemosensory perception in the elderly, *Crit. Rev. Food Sci. Nutr.,* 33, 3, 1993.
60. Meisami, E., Chemoreception, in *Neural and Integrative Animal Physiology,* 3rd ed., Prosser, C. L., Ed., John Wiley & Sons, New York, 1991, chap. 7.
61. Graziadei, P. P. C. and Monti-Graziadei, G. A., Continuous nerve cell renewal in the olfactory system, in *Handbook of Sensory Physiology,* Vol. 9, Development of Sensory Systems, Jacobson, M., Ed., Springer-Verlag, Berlin, 1978, 55.
62. Nakashima, T., Kimmelman, P., and Snow, J. B., Structure of human fetal and adult olfactory neuroepithelium, *Arch. Otolaryngol.,* 10, 641, 1984.
63. Naessen, R., An enquiry on the morphological characteristics and possible changes with age in the olfactory region of man, *Acta Otolaryngol.,* 71, 49, 1971.
64. Smith, C. G., Age incidence of atrophy of olfactory nerves in man. A contribution to the study of the process of aging, *J. Comp. Neurol.,* 77, 589, 1942.
65. Bhatnagar, K. P., Kennedy, R. C., Baron, G., and Greenberg, R. A., Number of mitral cells and the bulb volume in the aging human olfactory bulb: a quantitative morphological study, *Anat. Rec.,* 218, 73, 1987.
66. Liss, L. and Gomez, F., The nature of senile changes of the human olfactory bulb and tract, *Arch. Otolaryngol.,* 67, 157, 1958.
67. Serby, M. J., Larson, P. M., and Kalkstein, D., Olfaction and neuropsychiatry, in *Science of Olfaction,* Serby, M. J. and Chobor, K. L., Eds., Springer-Verlag, New York, 1992, 559.
68. Schiffman, S. S., Olfaction in aging and medical disorders, in *Science of Olfaction,* Serby, M. J. and Chobor, K. L., Eds., Springer-Verlag, New York, 1992, 500.
69. Doty, R. L., Deems, D. A., and Stellar, S., Olfactory dysfunction in Parkinsonism: a general deficit unrelated to neurologic signs, disease stage or disease duration, *Neurology,* 38, 1237, 1988.
70. Doty, R. L., Iklan, M., Deems, D. A., Reynolds, C., and Stellar, S., The olfactory and cognitive deficits of Parkinson's disease: evidence for independence, *Ann, Neurol.,* 25, 166, 1989.
71. Mair, R. G. and Flint, D. L., Olfactory impairment in Korsakoff's syndrome, *Science of Olfaction,* Serby, M. J. and Chobor, K. L., Eds., Springer-Verlag, New York, 1992, 526.
72. Talamao, B. R., Rudel, J. S. R., Kosik, K. S., Lee, V. M., Neff, S., Adelman, L., and Kauer, J. S., Pathological changes in olfactory neurons in patients with Alzheimer's disease, *Nature,* 337, 736, 1989.
73. Ohm, T. G. and Braak, H., Olfactory bulb changes in Alzheimer's disease, *Acta Neuropathol. (Berlin),* 73, 365, 1987.
74. Feyes, P. F., Deems, D. A., and Suarez, M. G., Olfactory related changes in Alzheimer's disease: a quantitative neuropathologic study, *Brain Res. Bull.,* 32, 1, 1993.
75. Roberts, E., Alzheimer's disease may begin in the nose and may be caused by aluminosilicates, *Neurobiol. Aging,* 7, 561, 1986.
76. Schiffman, S. S., Perception of taste and smell in elderly persons, *Crit. Rev. Food Sci. Nutr.,* 33, 17, 1993.
77. Bartoshuk, L. M. and Weifenbach, J. M., Chemical senses in aging, in *Handbook of the Biology of Aging,* 3rd ed., Schneider, E. D. and Rowe, J. W., Eds., Academic Press, San Diego, 1990, 429.
78. Stevens, J. C., Bartoshuk, L. M., and Cain, W. S., Chemical senses and aging: taste versus smell, *Chem. Senses,* 9, 167, 1984.
79. Hughs, G., Changes in taste sensitivity with advancing age, *Gerontol. Clin.,* 11, 224, 1969.
80. Murphy, C., The effect of age on taste sensitivity, in *Special Senses in Aging,* Han, S. S. and Coons, D. H., Eds., Institute of Gerontology, University of Michigan, Ann Arbor, 1979, 21.
81. Stevens, J. C. and Cain, W. S., Changes in taste and flavor in aging, *Crit. Rev. Food Sci. Nutr.,* 33, 27, 1993.
82. Schiffman, S. S., Food recognition by the elderly, *J. Gerontol.,* 32, 586, 1977.
83. Bradley, R. M., Effects of aging on the sense of taste: anatomical considerations, in *Special Senses in Aging,* Han, S. S. and Coons, D. H., Eds., Institute of Gerontology, University of Michigan, Ann Arbor, 1979, 3.
84. Mistretta, C. M., Aging effects on anatomy and neurophysiology of taste and smell, *Gerontology,* 3, 131, 1984.

III
SYSTEMIC AND ORGANISMIC AGING

11 AGING OF THE ADRENALS AND PITUITARY

Paola S. Timiras

Together with the nervous system, the endocrine system coordinates homeostatic responses to environmental signals for the protection of the individual organism and regulates reproductive functions for the perpetuation of the species. With aging, the main expressions of declining functional competence are

- Cessation of reproduction
- Diminishing capacity to adapt to external demands, especially those involving stress conditions (Chapters 1 to 3).

The decline in physiologic competence with aging has been related to endocrine deficiency because of the broad range of hormonal actions, many of them necessary for reproduction and survival, and because of the possibility of restoration of "vitality" and "well-being" by hormonal replacement therapy.

1 ENDOCRINE GLANDS, HORMONES, AND CHEMICAL MESSENGERS

Endocrine glands are organs which produce and secrete hormones. Reciprocally, hormones are molecules synthesized and secreted by specialized (endocrine) cells and released into the blood. They act on target cells at a distance from their site of origin. This view has been expanded: we are now aware of cells or groups of cells which do not form discrete organs but are interspersed with other cells and which also produce and secrete hormones. Examples of classic endocrines include the pituitary, adrenals, thyroid and parathyroids, testes, and ovaries. Examples of the more diffuse hormone-producing cells are those of the pancreas and intestinal mucosa.

Some of the *neurotransmitters* which we have described previously (Chapter 8) are also considered hormones, *Locally acting substances* released in the extracellular fluid (paracrine communication) either directly stimulate (or inhibit) neighboring cells (e.g., prostaglandins) or promote (or block) growth (growth factors); these are also considered hormones and, therefore, part of the endocrine system. The wide distribution of hormone-secreting cells, peripheral nerves and neural ganglia, and the multiciplicity and diversity of secretory products underline the critical role of endocrine and neural coordination of body functions.

Among the endocrine glands, the *adrenals* regulate certain aspects of metabolism, behavior, and nervous function and, thus, play a key role in homeostasis. The *pituitary* secretes several hormones and peptides that stimulate peripheral targets, either other endocrine glands or specific tissues and organs; its location in close vicinity of the *hypothalamus* with which it articulates through a vascular net, *the portal system,* and through *direct neuronal connections,* gives the pituitary the unique function of intermediary between nervous and endocrine systems. As all endocrine glands, the adrenals and pituitary do not act in isolation but are dependent both on *neuroendocrine signals,* usually relayed through or initiated in the hypothalamus, and on the functional status of the *target cells*. Thus, aging of all endocrines depends on changes in

- The gland itself
- Other endocrines
- The nervous system
- Metabolism and body composition
- Cellular and molecular responses
- The complicating occurrence of disease, medications, drugs
- The contributing influence of diet and exercise (Table 11-1)

Therefore, in this, as in the subsequent chapters on endocrine aging, we will consider (1) changes in the endocrine glands themselves, here, adrenals and pituitary, and (2) their neuroendocrine and metabolic interrelations, here, their role in homeostatic adjustments and adaptation to stress.

Endocrine and neural involvement in aging has led to the formulation of *neuroendocrine theories of aging* (Chapter 4). These theories propose that aging is not due to intrinsic deteriorative processes in cells or molecules but rather to a *programmed regulation by "pacemaker cells",* perhaps situated in the brain and acting through *neural and endocrine signals*.[1] These signals would orchestrate the passage from one stage of the lifespan to the other, thereby timing the entire life cycle, including development, growth, maturation, and aging. While these theories remain for the most part speculative and their general applicability remains difficult in view of the large range and diversity of senescent phenomena among multicellular organisms, they are amenable to experimental manipulation. Therefore, they provide a useful tool to investigate the progression and reversibility of aging processes. Some of the clinical observations and experimental evidence in support of the involvement of neural

TABLE 11-1
Factors Contributing to Endocrine Aging

Endogenous factors
Endocrine
Neural
Body composition
Target cells and molecules
Disease
Exogenous factors
Disease
Medications
Drugs, including smoking and alcohol
Diet
Exercise
Social factors

0-8493-8979-8/94/$0.00+$.50

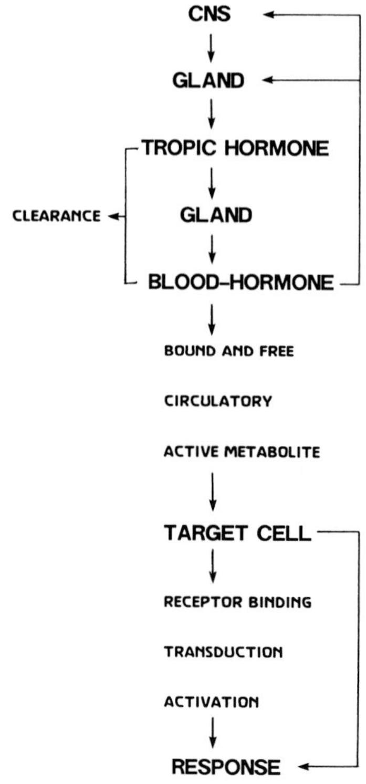

FIGURE 11-1 *Diagrammatic representation of a typical sequence of hormone action and regulation.*

CNS
↓
GLAND
↓
TROPIC HORMONE
↓
CLEARANCE ← GLAND
↓
BLOOD-HORMONE
↓
BOUND AND FREE
CIRCULATORY
ACTIVE METABOLITE
↓
TARGET CELL
↓
RECEPTOR BINDING
TRANSDUCTION
ACTIVATION
↓
RESPONSE

and endocrine factors in aging and some of the clinical interventions they inspire are presented in this and in the subsequent chapters on aging of endocrines.

2 ASSESSMENT OF ENDOCRINE FUNCTION

Evaluation of endocrine function in humans often relies on measurements of blood levels of appropriate hormones under basal (resting or steady state) conditions and stress (increased intrinsic and extrinsic demands), often leading to incomplete and erroneous conclusions.

An adequate endocrine evaluation must include several levels of endocrine action as well as consider interendocrine and neuroendocrine relations, hormone receptor interaction at the target cell, and molecular events inside the cell. A prototypic endocrine system, illustrated in Figure 11-1, shows hierarchical levels of endocrine regulation and provides a framework on which to investigate endocrine changes with aging. Aging changes may occur at several levels of the CNS–endocrine axis. While none of these changes may, by itself, be sufficient to alter homeostatic competence, a cascade of minor alterations may desynchronize the appropriate signal at the target cell/molecule and alter hormonal actions.

This more "global" approach to the study of endocrine aging cannot be currently undertaken in humans. It can be explored more easily in experimental animals and in cultured tissues or cells; such *in vivo* and *in vitro* models represent an important corollary to human studies.

2.1 Some Generic Endocrine Changes with Aging

As illustrated diagrammatically in Figure 11-1, changes with aging may occur at all levels of the endocrine system.

- *At the endocrine gland level,* weight loss with atrophy and vascular changes and fibrosis occur in most cases with or without the accompaniment of adenomas.
- *Basal hormone levels* (free, biologically active hormones or hormones bound to plasma proteins) in man and animals are generally not influenced by age, although some hormones decrease significantly.
- Hormone release depends on nervous and environmental stimuli as well as positive and negative feedback from circulating hormones.
- Some hormones act exclusively on *one target;* other hormones act on *many cell types (targets) and by several mechanisms.* Thus, the same hormone may have different actions on different tissues.
- With aging, one of the many actions or *one of the many targets may be selectively affected* while other actions and targets are preserved.
- *Secretory and clearance rates* often decrease, although it is not clear whether the primary defect involves hormone secretion or hormone clearance, but the question is to what extent the capacity to maintain stable levels of plasma hormones is preserved. The study of the generation of active metabolites may reveal how their rate of production is altered with advancing age.
- *Receptors* located on target cells mediate the specificity of hormones on the particular cells, and their number may increase (up-regulation) or decrease (down-regulation) depending on the stimulus. Hormone-receptor complexes are *internalized by endocytosis* and stimulate or inhibit transcription of selected RNAs or activity of specific enzymes. Cellular responses are determined by the *genetic programming* of the particular cell. With aging, receptor binding and intracellular responses vary greatly depending on the hormone and target cell.

Early reviews of the aging of endocrines have been presented in a previous edition.[2] In this and in the following four chapters, the salient actions of each endocrine gland will be summarized in the adult, and the major changes during aging will be outlined with emphasis on neuroendocrine regulation and metabolic interactions.

3 THE ADRENAL CORTEX

The adrenals are paired glands that lie above the kidneys (Figure 11-2). They have an inner medulla and an outer cortex (Figure 11-3). The medulla can be considered to be a sympathetic ganglion; it secretes catecholamines, epinephrine, and norepinephrine. The cortex secretes several steroid hormones distinguished in three categories:

- Glucocorticoids, e.g., cortisol in humans, corticosterone in rats
- Sex hormones, e.g., dehydroepiandrosterone (DHEA). Cortisol and DHEA are secreted by the cells of the zona fasciculata and zona reticularis, and corticosterone is secreted by these and also the zona granulosa
- Minerolocorticoids, e.g., aldosterone, are secreted by the cells of the zona glomerulosa

3.1 Glucocorticoids and Sex Hormones

Glucocorticoids have several actions. *The major metabolic actions* of glucocorticoids include

- Increased amino acid uptake and gluconeogenesis in liver
- Decreased amino acid uptake and protein synthesis in muscle
- Inhibition of somatic growth due to a general catabolic (decreased protein synthesis) action
- Suppression of pituitary growth hormone secretion
- Anti-insulin effect with consequent exacerbation of diabetes
- Mobilization of serum lipids and cholesterol

Glucocorticoids have anti-inflammatory activity. When administered in supraphysiologic doses, they inhibit inflammatory and allergic reactions. These immunosuppressive actions,

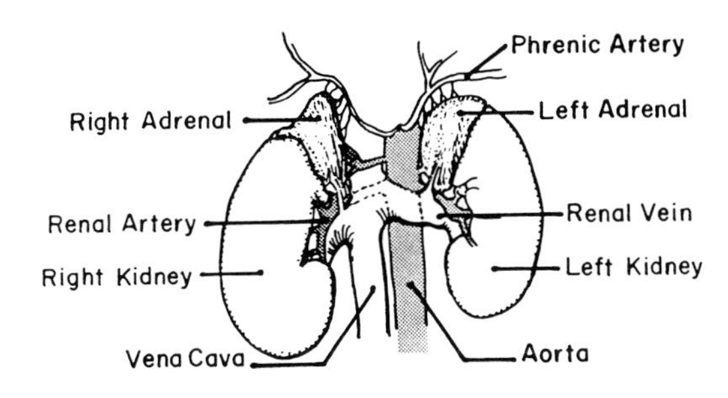

FIGURE 11-2 *Diagram of the kidney and adrenals.* (Drawings by S. Oklund.)

utilized clinically for the prevention of transplant rejections and the symptomatic treatment of allergies, have been explained by several mechanisms (e.g., inhibition of leucocyte migration and recruitment from blood to tissues, and changes in rates of lymphocyte formation and destruction.

Glucocorticoids have also effects on the nervous system, primarily of an excitatory nature, evidenced by the presence of hormone receptors on neurons and glia, induction of neurotransmitter enzymes, EEG changes and increased convulsibility, and toxicity for specific neurons.

Another important action of glucocorticoids is *their negative feedback on hypothalamus and pituitary.* They inhibit the secretion of adrenocorticotropic hormone (ACTH) from the pituitary and of corticotropin-releasing hormone (CRH) from the hypothalamus: the higher the levels of the circulating hormones, the lower the secretion of ACTH and CRH; the lower their levels, the greater the ACTH and CRH release.

Other actions of glucocorticoids include: retention of sodium at renal tubules (but much less efficacious than aldosterone), the reabsorption of bone and altered calcium metabolism, promotion of appetite and increased stomach acid and pepsin secretion, and maintenance of work capacity.

Sex hormones include both androgens (with masculinizing and anabolic actions) and estrogens (with feminizing actions). Their secretion decreases markedly in old age (see below).

3.2 Mineralocorticoids

Aldosterone, the major mineralocorticoid, is indispensable for survival. It increases *sodium reabsorption from the renal tubular fluid,* saliva, and gastric juice. In the kidney, aldosterone acts on the epithelium of the distal tubule and collecting duct where it facilitates the exchange of sodium (reabsorbed) for potassium and hydrogen ions (excreted). It may also increase potassium and decrease sodium in muscle and brain cells.

Secondary actions include: maintenance of blood pressure (as a consequence of sodium-water retention and increased blood volume); moderate potassium diuresis and increased urine acidity (sodium taken up is exchanged for potassium and hydrogen ions which are excreted).

3.3 Changes Under Basal Conditions

With aging, the adrenal cortex decreases in weight in men and women and in the several animals that have been examined. It shows the frequent occurrence of nodules (i.e., localized hyperplastic changes perhaps reactive to a reduced blood supply or a consequence of multifocal adenomas). The adrenocortical cells, typical secretory cells, rich in mitochondria and endoplasmic reticulum with numerous lipid droplets where the steroid hormones are stored undergo several changes. Of these, the most important are accumulation of lipofuscin granules (Chapter 5) and the thickening of the connective support tissue as shown by the thick capsule and the perivascular fibrous infiltrations.

Despite these anatomical and histologic changes, basal plasma glucocorticoid levels do not show any significant difference between maturity and old age in man and some animal species (cows, goats, dogs, monkeys).[2] However, in some species (rats,[3] vervet monkeys,[4] tree shrews,[5] baboons,[6]) glucocorticoids levels are slightly increased in some (but not all) animals.

In contrast, levels of the other adrenocortical steroids appear to decline with aging: this is the case for aldosterone, with values almost undetectable past age 65,[7] and DHEA, with values at age 60 and older approximately one third of those at age 30.[8]

The secretion rate appears to be reduced for all hormones.[9-11] In the case of glucocorticoids for which normal plasma levels are maintained with aging, the lower secretory rate is compensated by decreased clearance, that is, reduced excretion and metabolism. The reduced urinary excretion rate, however, is explainable when adjusted to the reduced creatinine excretion which also occurs with aging (Chapter 19). Compensation, therefore, relies primarily on a *decreased rate of removal of cortisol from the circulation,* well demonstrated in individuals aged 60 years and older and ascribed to a number of factors, singly or in combination:

- Decreased blood flow to the liver (a metabolizing organ)
- Decreased liver size associated with aging (Chapter 20)
- Decreased activity of liver enzymes involved in cortisol metabolism (Chapter 20)
- Decreased cortisol binding to liver cytosol

It is apparent that, at least for cortisol, metabolic compensatory mechanisms remain intact during aging, with the

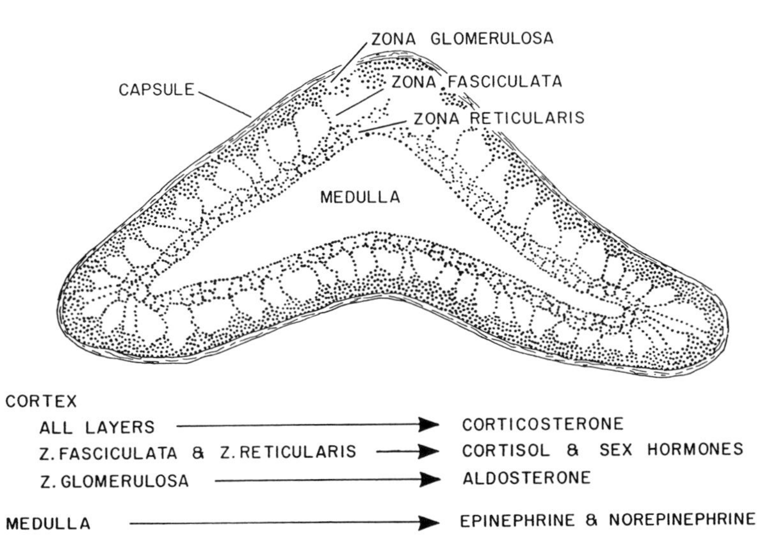

FIGURE 11-3 *Diagram of a section of the adrenals illustrating the various zones and hormones.* (Drawings by S. Oklund.)

body adapting either to the decline in rate of removal of the hormone by decreasing production rates of this compound or, vice versa, reducing the rate of removal to maintain normal circulating hormonal levels. The latter explanation may be the most appropriate in view of the declining secretory rate and levels of aldosterone and DHEA; with these hormones, however, metabolic compensatory adjustments fail to occur.

DHEA Replacement Therapy

The marked reduction in DHEA secretion and levels with aging has led to speculation that reduction of this major androgen may be responsible for some of the aging decrements. Historically, it may be recalled that one of the first associations between endocrines and aging, as well as one of the first secretory actions attributed to an organ (the testis), was drawn by the French physiologist Brown-Sequard who extolled the antiaging properties of testicular secretions (androgens).[12] Testicular transplants and administration of androgens as rejuvenating measures have been attempted repeatedly to delay or reverse aging but with little success. On this same basis, replacement therapy with DHEA has been suggested and its effects explored in animals. Indeed, long-term DHEA administration in mice has reduced the incidence of mammary cancer, increased survival, and delayed onset of immune dysfunction. The simultaneous reduction in body weight and food intake suggests that the hormone may act in a manner similar to caloric restriction which also retards tumor development and immunologic senescence (Chapter 24). Use of this hormone in humans as an antiaging agent is, until now, not warranted without adequate proof of its direct efficacy.

Regulation of Adrenocortical Secretion

As illustrated diagrammatically in Figure 11-4, the adrenal cortex has a hierarchy of regulation, from the hypothalamus and the pituitary, to the adrenal and then the target cell or molecule. With aging and under conditions of stress, a disruption of this complex regulatory system at one or more levels may result in failure of homeostasis and adaptation.

CRH, a polypeptide, is released from the median eminence of the hypothalamus; it is transported via the portal system, a special capillary system carrying directly secretory products from the hypothalamus to the pituitary, to the anterior pituitary where it stimulates synthesis and release of the *ACTH.*

ACTH, a protein, stimulates the adrenal cortex to synthesize and release the glucocorticoids and sex hormones from the two inner zones of the cortex. Thus, after ablation of the pituitary, these two zones atrophy and the levels of the corresponding hormones decrease. Conversely, in tumors of the pituitary in which ACTH levels are increased (as may occur in Cushing's disease), the two adrenocortical zones hypertrophy and the hormonal levels are increased.

CRH, ACTH, and glucocorticoid secretions are interregulated by feedbacks operating at each level, with low levels of adrenocortical hormones acting to increase CRH secretion and CRH stimulating ACTH release, which in turn stimulates adrenal cortex secretion.

Aldosterone secretion is stimulated only to a minor degree by ACTH, and, therefore, removal of the pituitary does not produce atrophy of the outer adrenal layer. Aldosterone secretion is regulated in part by the circulating levels of sodium

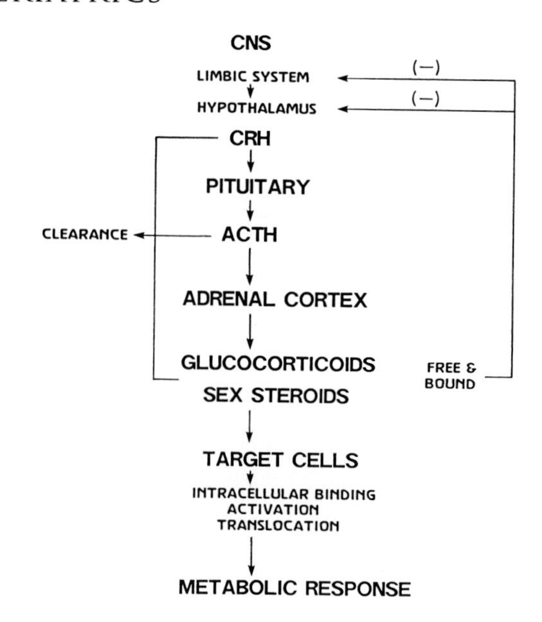

FIGURE 11-4 *Diagrammatic representation of the hypothalamo-pituitary–adrenocortical axis.*

and potassium; that is, low sodium levels and high potassium levels stimulate secretion. Most important in the control of aldosterone secretion is the *renin–angiotensin system,* that is, the secretion of the enzyme renin from the juxtaglomerular cells of the kidney and its action to transform the circulating hepatic protein, angiotensinogen, into the decapeptide angiotensin I. This, in turn, is transformed into the octapeptide angiotensin II, which then stimulates the release of aldosterone.

Excess glucocorticoids, as in Cushing's disease, leads to abnormalities of intermediary metabolism; *excess aldosterone,* as in Conn's disease, leads to hypertension and electrolyte disturbances; *adrenocortical insufficiency,* involving both gluco- and mineralocorticoids, as in Addison's disease, leads to hypotension, shock (from sodium loss), and death.

3.4 Changes Under Dynamic Conditions

The response of adrenocortical hormones to stimulation has been studied extensively because of its relation to adaptative capability (which is decreased) and survival (which is endangered) in the elderly.

A "cornerstone of stress physiology is that widely different stressors converge" to induce a series of responses depending on adrenocortical stimulation.[3] Thus, a number of animal species show increased glucocorticoid levels in response to administered CRH or ACTH or to endogenously increased levels of these hormones as a consequence of stress. Increased glucocorticoid levels in response to stress occur in humans and several animal species. Some old animals respond with less elevated glucocorticoid levels than young, but, in many, differences with aging are absent or small.[13]

In humans, all the surveys of old individuals seem to agree that the increase in plasma cortisol in response to ACTH is preserved.[14] In contrast, the response of DHEA to ACTH appears significantly reduced with aging.[15] Similarly, stimulation of aldosterone by sodium restriction is less efficient in old as compared to young and adult individuals.[7]

ACTH is secreted in irregular bursts throughout the 24-hour day, with the bursts being more frequent in early morning and least frequent in the evening. This circadian (diurnal) rhythm in ACTH secretion is associated with a corresponding circadian rhythm in cortisol secretion, and the rhythm is preserved during aging in humans.[16] However, sustained nighttime cortisol levels (i.e., blunting of the nocturnal drop in cortisol levels compared with daytime levels) have also been reported[17] and have been ascribed to

- Reduced renal clearance of the hormone
- Reduced muscle mass and generally reduced basal metabolism (thereby reducing need of the organism for cortisol)
- Alterations in sleep patterns and insomnia (Chapter 9)

Another important observation is the longer persistence of circulating high glucocorticoid levels after ACTH administration or stress, as illustrated in the rat (Figure 11-5).[18] The slower metabolism of glucocorticoids in older animals may be responsible for the prolonged high hormonal levels (see below). In addition, glucocorticoids would habituate to mild stress more slowly in aged than in young rats with consequent slower habituation of metabolic responses associated with glucocorticoid hypersecretion. Aged rats tend to be less capable of exerting an efficient glucocorticoid negative feedback on the pituitary and hypothalamus: after stress, the adrenal cortex continue to secrete glucocorticoids at a time when this secretion has ceased in young animals.[19]

In vitro experiments show that isolated adrenal cortical cells from old (24 months of age) rats are less responsive than those of young (3 months of age) rats to a synthetic subunit of ACTH and to cAMP; the marked reduction in corticosterone production in adrenal tissue from old rats suggests intracellular changes involving either impaired steroidogenesis or receptor function, perhaps secondary to increased free radical production and membrane alterations (Chapter 6).

Other tests (e.g., metyrapone-induced decreases in plasma cortisol) indicate that the feedback mechanism for ACTH control as well as the stimulatory action of ACTH on adrenocortical secretion is maintained in old men and women. Cortisol production adjusts to challenging stimuli. Given the extreme heterogeneity of aging processes, in the adrenocortical function as in many others, some individuals, but not all, retain a functional competence under basal and stress conditions. The challenge here is to identify those factors that separate the successful elderly from the average (Chapter 2).

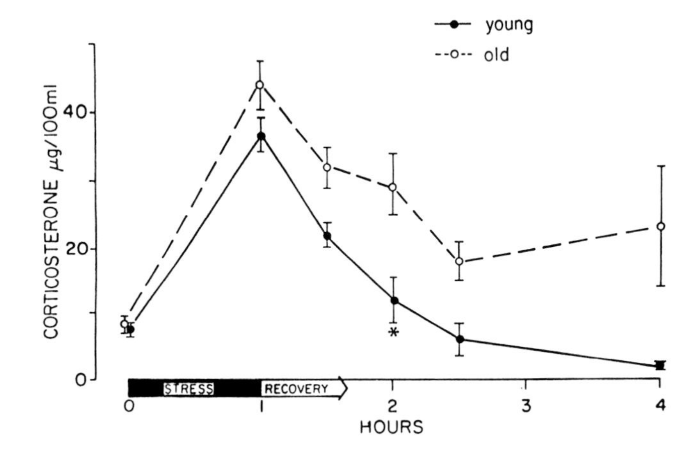

FIGURE 11-5 *Corticosterone titers in young (3 to 5 months) and aged (24 to 28 months) Fischer 344 rats during 1 h of immobilization stress, followed by 4 h of post-stress recover.* *Indicates time when titers are no longer significantly elevated above baseline (determined by two-tailed paired to test). In the case of young subjects, this was after 1 h of the recovery period; for aged subjects, such recovery did not occur within the monitored time period. (Reproduced with permission from Sapolsky et al., *Endocr. Rev.*, 7, 284, 1896.)

All molecular events subsequent to receptor binding are subject to alteration with age, although the nature and magnitude of these age-related changes are variable depending on the hormone, the target cell, and the animal species. In general, the concentration of corticosteroid receptors decreases either in early adulthood or during senescence.[20-22] For example, in the rat brain, glucocorticoid receptors are detectable on day 17 of gestation. They increase gradually after birth to adult levels by 15 days of age and are significantly reduced by comparison with adults in aged (24-month-old) animals.[23] In the rat hippocampus, the primary glucocorticoid-concentrating region of the brain, receptors decrease with aging due to decreased concentration of cytosolic corticosterone receptors with no change in receptor affinity or capacity for nuclear translocation.[24,25] Some of the physicochemical properties (e.g., activation, transformation) of the receptors seem to be more susceptible to aging than receptor number. Such age-related changes have been reported in

3.5 Changes in Adrenal Steroid Receptors

Adrenal steroids exert their cellular and molecular actions through their binding to cytoplasmic (cytosolic) and nuclear receptors, the degree of cellular responsiveness being directly proportional to the number of specific receptor molecules. Hormone-mediated responses are controlled partly by the binding of the hormone to specific intracellular receptors and, then, by translocation of the hormone–receptor complex to nuclear receptor sites (Figure 11-6).[20,21]

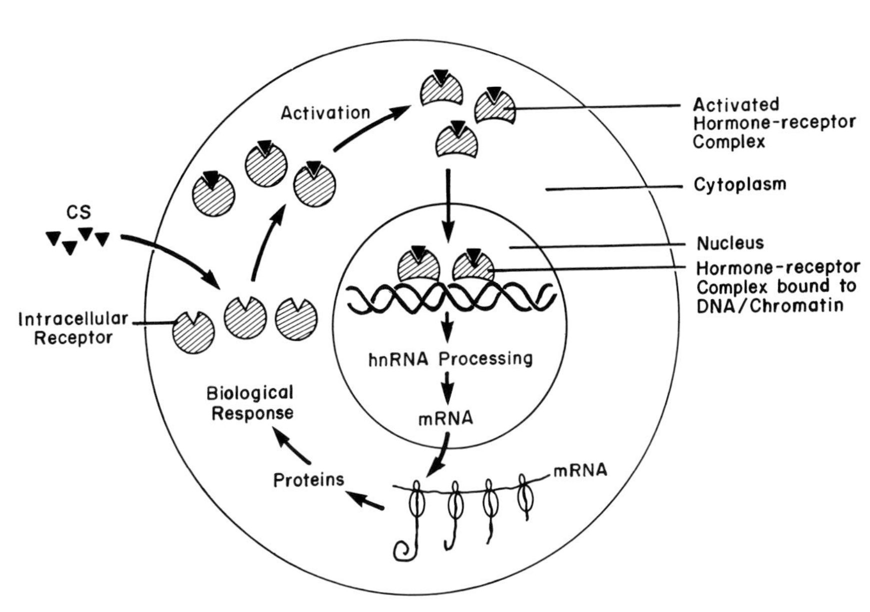

FIGURE 11-6 *Schematic diagram of corticosteroid (CS) action in a target cell.*

glucocorticoid receptors in liver,[21] skeletal muscle,[26] and cerebral hemisphere.[22] Aging changes in corticosteroid receptors, by altering the responsiveness of target cells and molecules to hormones, may contribute to the decline in the effectiveness of adrenocortical responses to stress.

4 THE ADRENAL MEDULLA

The adrenal medulla is part of the sympathetic division of the autonomic nervous system, and, as such, functions in unison with the other sympathetic structures. It is also interrelated anatomically by a rich vascular network to the cortex and functionally by interactions of glucocorticoids with some of the metabolic actions of medullary hormones (e.g., mobilization of free fatty acids in emergency situations) and by the induction of some of the enzymes for catecholamine synthesis (e.g., the enzyme phenylethanolamine-*N*-methyltransferase, PNMT, that catalyzes the formation of epinephrine from norepinephrine).

The sympathetic system, including the adrenal medulla, is not essential for life. However, one of the functional characteristics of the system is that, under emergency conditions of stress, *it can discharge as a unit,* as in rage and fright, when sympathetically innervated structures are affected simultaneously over the entire body. The heart rate is accelerated, the blood pressure rises, red blood cells are poured into the circulation from the spleen (in certain species), the concentration of blood glucose rises, the bronchioles and pupils dilate, and, on the whole, the organism is prepared for "fight or flight". The contribution of the adrenal medulla to "the emergency function" of the sympathoadrenal system varies with the animal species and the type of stress. It also seems to vary with the age of the animal, reaching optimal efficiency in adulthood and showing some selective decline in old age.

Structure and Function

The medullary cells (also known as chromaffin cells) are considered to be postganglionic neurons (innervated by preganglionic cholinergic fibers) which have lost their axons and become secretory cells. Major secretory products are the *catecholamines, norepinephrine,* which also is produced and released by neurons in the CNS and by sympathetic neurons, *epinephrine,* formed primarily in the medulla by the methylation of norepinephrine, and some *dopamine,* also a CNS neurotransmitter (Chapter 8). Opioid peptides are also secreted (most of the circulating encephalin comes from the medulla), although they do not pass the blood–brain barrier.

Norepinephrine and epinephrine mimic the effects of sympathetic discharge, stimulate the nervous system, and exert metabolic effects that include glycogenolysis in liver and skeletal muscle, mobilization of free fatty acids, and stimulation of metabolic rate. They act through the binding to two classes of receptors, α_1- and α_2-receptors, which induce vasoconstriction in most organs, and β-receptors, which mediate the metabolic effects and stimulate rate and force of cardiac contraction (β_1) and dilate blood vessels in muscle and liver (β_2). Their secretion is under neural control and is also influenced by other hormones, primarily glucocorticoids and thyroid hormones.

4.1 Changes With Aging

Under basal conditions, in humans, plasma levels and urinary excretion of catecholamines vary with aging depending on the individual. They may:

- Remain unchanged
- Show a reduction in absolute and averaged circadian amplitude
- Show an increase, the increase being greater after standing and isometric exercise

Plasma and urinary catecholamine elevation, reported after a variety of stimuli, has been interpreted as a compensatory reaction to the apparently increasing refractoriness (perhaps due to receptor down-regulation) of target tissues to catecholamines with aging. However, the apparent increase in norepinephrine plasma levels does not occur with all stimuli, and, additionally, the return to baseline levels is prolonged in the elderly. Also, the two catecholamines often show differential responses: plasma norepinephrine may be elevated in elderly subjects during a mental stress test, plasma epinephrine is not; this suggests a specific hyperactivity of the sympathetic system in general rather than of the adrenal medulla.[27]

4.2 Target Differential Responsiveness

An unresolved issue is whether the elderly release more norepinephrine than the young or whether the peripheral clearance of the hormone is reduced. Here, again, the observations vary, with some detecting no changes in clearance[28] whereas others report a decrease.[29]

One proposed explanation for the overall increase of sympathoadrenal activity with aging is an increased refractoriness (or decreased sensitivity) of target tissues to catecholamines due to alterations in transport and/or binding. Clinical and experimental observations show that the responsiveness of several tissues is diminished in some cases, although in many instances there is no decrease and in others there is increase. Decreased responsiveness may be found in the diminished efficiency of hemodynamic and cardiovascular responses to changes in posture (tilting, standing)(Chapters 21, 23) and to exercise and in the case of slower dark adaptation of pupillary size (Chapter 10). Conversely, increased responsiveness is manifested in those organs and tissues regulating blood pressure, which progressively increases with aging leading to hypertension (Chapter 16). Other examples show different responses. The ability of dopamine to stimulate adenylate cyclase in the corpus striatum decreases progressively with age as does dopamine binding in rats, mice, and rabbits but remains unchanged in adipocytes or is increased in hepatocytes Even the suggested decreased receptor density as an explanation of target refractoriness is not clear-cut. A decrease in adrenergic receptor number with aging has been reported in some organs and cells (e.g., cerebellum, adipocytes) and a decrease in receptor affinity in others (e.g., lung), but, in many cases, changes are not observed.

This lack of uniformity in pattern has been ascribed to the differential responsiveness of tissues to catecholamines. Sensitivity to catecholamines may also be influenced by other hormones, such as glucocorticoids and thyroid hormones, known to affect catecholamine metabolism and receptors and undergoing themselves age-related changes. Other factors such as alterations in structure and function of membranes and intracellular molecules may also be operative in changing cell responsiveness to adrenal medullary hormones.

5 THE PITUITARY GLAND

The pituitary gland (also called hypophysis) regulates several peripheral endocrines (adrenal cortex, thyroid, gonads) and target tissues (through the secretion of growth hormone and prolactin), has close functional ties with the hypothalamus and other CNS centers, and produces a number of peptides with hormone-precursor and behavioral activities. Relatively few studies have explored systematically the changes with aging in these important functions; the current observations reveal that the major pituitary functions show with aging:

1. Great variability among individuals of the same species and among different species
2. Differential alterations with aging, with some functions being greatly affected whereas others remain practically unaltered
3. The possibility of modifying aging-related changes and influencing the lifespan by ablation of the gland in rats
4. In view of the selective rather than global nature of these changes and their importance for survival, the formulation of a number of neuroendocrine theories of aging based on the hypothesis that disorganization or desynchronization of hypothalamo–pituitary signals may be a cause, or at least be closely involved, in the aging processes

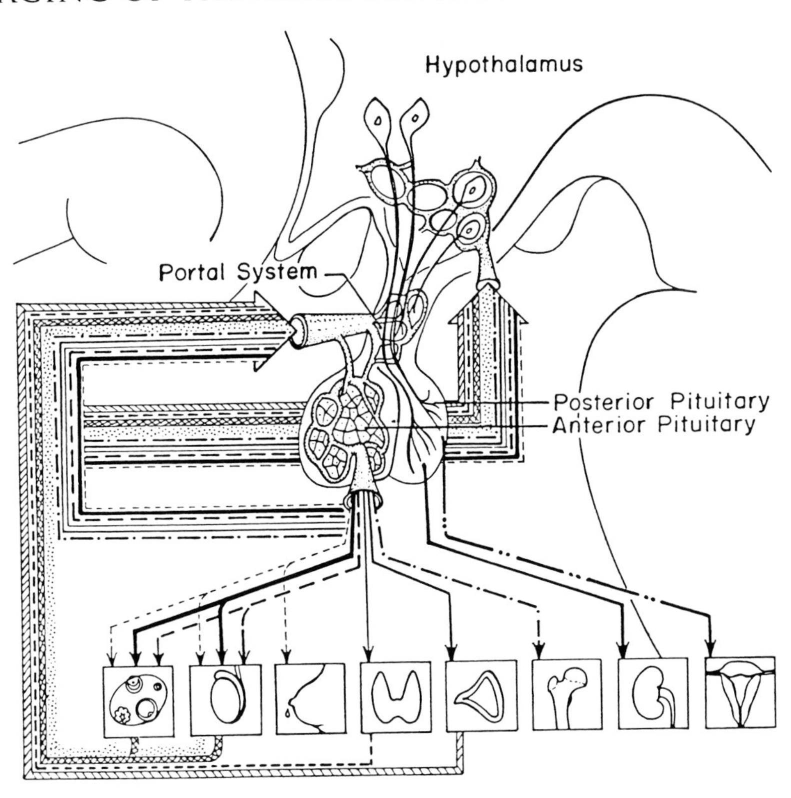

FIGURE 11-7 *Schematic representation of the hypothalamus, pituitary, and peripheral target sites, drawn to show the multiplicity of tropic hormones, targets, and feedbacks.* Hypothalamic hormones are carried to the anterior pituitary through the portal system and to the posterior via axoplasmic flow in nerve fibers. From the anterior pituitary, a number of tropic hormones are secreted. The tropic organs and their hormones are, from left to right:

ovary: prolactin, PL (- - - - -), luteinizing hormone, LH (▬▬▬), follicle-stimulating hormones, FSH (– – – –)
testes: primarily interstitial-cell–stimulating hormone, ISCH (similar to LH), (▬▬▬), and to a lesser extent PL (- - - - -) and FSH (– – – – –)
mammary gland: PL (- - - - -);
thyroid gland: thyroid-stimulating hormone, TSH (————);
adrenal cortex: adrenocorticotropic hormone, ACTH (————);
long bones; growth hormone, GH (– · – · –).

From the posterior pituitary, continuing from left to right, kidney, antidiuretic hormone or vasopressin, ADH (▬▬▬), smooth muscle, uterus, oxytocin (— · · —). The hormones secreted by their peripheral endocrines [sex (░░░) and adrenocortical (▧▧▧) steroids and thyroid hormones (- - - - -)] feedback to stimulate or inhibit secretion from the anterior pituitary or hypothalamus. Likewise, tropic hormones from the pituitary can feedback and regulate secretion by the hypothalamus. In this manner, short and long feedback loops are established.

Structure and Major Hormones

The pituitary gland is comprised of three more or less separate endocrine glands that produce a relatively large number of hormonally active substances (Figure 11-7). These include six well-characterized hormones secreted from the *anterior lobe* of the gland: ACTH, discussed above, the two gonadotropins, thyroid-stimulating hormone, growth hormone, and prolactin. In addition, the anterior lobe secretes lipotropin and pro-opio-melanocortin, precursors of ACTH, opioid peptides, and melanocyte-stimulating hormone (MSH).

The *intermediate lobe,* practically indistinguishable from the anterior lobe in humans, and formed of scattered cells in the adult secretes MSH, lipotropin, opioid peptides, and the large precursor, pro-opio-melanocortin.

The *posterior lobe* secretes two peptides, vasopressin (sometimes called antidiuretic hormone, ADH) and oxytocin.

The hormones tropic to a peripheral endocrine are discussed in the chapter on the gland, for example, ACTH with the adrenal cortex in this chapter. Of the other hormones, growth hormone, prolactin, ADH, and oxytocin will be discussed below, briefly.

5.1 Structural Changes

In the anterior lobe, changes with aging are relatively few, and include those cellular changes (accumulation of lipofuscin) characteristic of aging cells, in general (Chapter 5), and of neural and endocrine cells (changes in neurotransmitters, secretory granules, and ventricles), in particular (Chapter 8). The number of the specific cells that secrete the different hormones is not markedly altered except for the gonadotropes (i.e., gonadotropin-secreting cells), which show changes similar to those observed after castration in experimental animals (Chapters 12 and 13). Tumor incidence, illustrated primarily by prolactinomas, increases with aging in rats and mice, more so in females than in males.

In the posterior pituitary, histochemical studies reveal a decrease in neurosecretory material, according to some, more marked in oxytocin- than ADH-secreting cells, but the reverse according to others in old rats: this decrease was not seen in humans and other animals (e.g., cattle). Despite these contradictions, with increasing age, decreased amounts of neurohypophyseal hormones are stored in the pituitary, testifying to the depressed functional state of the gland. This conclusion is justified on the basis of quantitative ultrastructural analysis of old rodent pituitary that show increased autophagic activity, increased perivascular space, decline in cell volume, decreased size and number of neurosecretory granules, and reduction in endoplasmic reticulum activity.[30]

5.2 Growth Hormone

Aging may be associated, in very general terms, with decreased protein synthesis, lean body mass, and bone formation as well as increased adiposity, suggesting the involvement of GH because of its anabolic (i.e., promoting protein synthesis) and metabolic actions. Nevertheless, in man and other species, the number of somatotropes (i.e., pituitary cells secreting GH), pituitary content of GH, basal plasma levels of the hormone, and its clearance remains essentially unchanged into old age. Some studies, however, have reported a decrease in GH basal levels in rats[31,32] and in humans.[33] Obesity is known to suppress circulating GH levels in young individuals, and the frequent obesity of the elderly may contribute to the decreased GH levels[33] although not all reports agree.[34]

Similarly, contradictory results have been reported for GH levels after a variety of stimuli that are known to cause GH release in rats. GH elevation after stimulation is less marked in old than in young animals, whereas in man, it may be decreased or remain unaltered.

GH secretion undergoes in humans a nocturnal peak during the first 4 h of sleep coinciding with the stages 3 and 4 of slow-wave sleep; these are the stages most affected in aging (Chapter 9). Studies in older humans have shown a decrease in sleep-related GH secretion,[35] although the relation of GH changes with sleep remains as controversial as that with obesity.

Some of the effects of GH are not due to a direct action of the hormone on target tissues and cells. Rather, GH would stimulate the liver and other tissues to produce growth-promoting polypeptides, the so-called somatomedins, such as insulin-like growth factors I and II, nerve growth factor, epidermal growth factor, fibroblast growth factor, thymosin, and other less-known factors. A decreased level of somatomedins has been reported in old rats[36] and in humans[32] as well as a decreased responsiveness of somatomedins to GH administration.[32]

The beneficial effects of GH administration on muscle size and strength and on lean body mass in elderly men with low GH levels are discussed in Chapter 21 with aging of skeletal muscle.

5.3 Growth Hormone–Releasing Hormone (GRH) and Growth Hormone–Inhibiting Hormone (Somatostatin or GIH)

From the foregoing discussion it emerges that, during aging, GH release declines, particularly during the night. Whether this decrease is due to a decline in the releasing hormone GRH or to an increase in the inhibiting hormone somatostatin is not known. Although the major function of GH is to promote growth of the whole body and several organs during childhood and adolescence, GH, GRH, and somatostatin continue to be secreted throughout life, GH exerting metabolic effects and GRH and somatostatin regulating GH secretion.

GH secretion depends not only on hypothalamic control (through GRH and somatostatin) but also on several metabolic factors (sensed in the hypothalamus), such as hypoglycemia and fasting, increase in some amino acids in plasma (e.g., arginine), and on stressful stimuli. It is also influenced by neurotransmitter release, e.g., stimulated by increased norepinephrine and dopamine discharge from brain catecholaminergic neurons. It is not clear whether the decline in these neurotransmitters which occurs during aging (Chapter 8) accounts for the decrease in GH by way of a reduction in GRH release, and more research is needed to identify GRH changes with aging.

A number of studies have been undertaken to discern whether and how somatostatin synthesis and release may be influenced by aging. Somatostatin is secreted by hypothalamic cells, but is present also in several brain areas, the gastrointestinal tract, and the pancreas. Its levels in plasma are negligible and its actions are exerted locally in vicinity of the secretory cells; as such, it is representative of that large group of substances produced by diffuse (paracrine) cells in several locations. In the hypothalamus, its main function is to inhibit GH release from the anterior pituitary. It also inhibits TSH release.

Hypothalamic content of somatostatin decreases with age in rats, and inhibition of its release (by administration of antibodies to somatostatin) causes a greater rise in GH in old than young rats.[31,32 37,38] This latter observation suggests a heightened sensitivity of the pituitary GH-secreting cells to the inhibitory action of somatostatin. The hormone content in several brain areas is not altered with aging in normal elderly subjects. However, in individuals affected by senile dementia of the Alzheimer type, brain somatostatin levels are severely depressed in association (but not necessarily in parallel) with the cholinergic deficits (as measured by choline acetyltransferase activity). The functional significance of this decrease is not understood. Behaviorally, somatostatin administration to aged monkeys had little ability to reverse their deficits on a memory task.[39]

In contrast to the hypothalamic decrease, pancreatic somatostatin levels are significantly higher in old than in young rats. This increase has been held responsible for, or at least considered as contributing to, the impairment of glucose-stimulated release of insulin and is reversible with antisomatostatin antibodies (Chapter 15).

5.4 Vasopressin (ADH) and Oxytocin

These nonapeptides are secreted by neurons of the supraoptic and paraventricular nuclei of the hypothalamus and transported within the axons to the posterior lobe of the pituitary where they are stored before being released in the circulation. Within each hypothalamic nucleus, some neurons produce oxytocin and others ADH. The nonapeptides are synthesized as part of larger precursor molecules and are stored in association with neurophysins which are also part of the precursor molecule. The principal action of ADH is retention of water by the kidney (Chapter 19) and that of oxytocin is to stimulate contraction of smooth muscle of the lactating mammary gland and pregnant uterus.

In association with the morphologic alterations described above, normal, aged rats often suffer from functional impairment such as mild diabetes insipidus (resulting from ADH deficiency and failure of water retention by the renal collecting ducts). In addition, the redistribution of water between intracellular and extracellular sites, characteristic of old age, has been ascribed to a decline in ADH levels in humans and rats (Chapter 19).

6 CHANGING ADAPTIVE RESPONSES

What are the causes of the decreasing ability to respond to stress in old age? Among many hypotheses, one focuses on the hypothalamo–pituitary–adrenal axis as a key regulator of homeostasis and adaptation (Chapter 4). However, other organs, such as lungs, kidneys, liver, and gut as well as organs/cells of the immune system, are also involved.

Maintenance of homeostasis involves complex functions in which organs and systems of the body participate, albeit to varying degrees. Homeostatic competence provides a "panoramic view" of overall physiologic performance. This explains the current interest among physiologists to identify control mechanisms responsible for this integrated activity.

Under resting (basal) conditions few changes occur in the hypothalamo–pituitary–adrenal function in old age. However, under stress, evidence of decreased physiologic competence is plentiful. The increased risk of death following stress in the elderly is well acknowledged and needs no documentation here. Older people are less resistant than younger individuals to excessively cold or warm temperatures because of the progressive deterioration of thermoregulatory mechanisms that occurs with age (Chapter 14). Equally diminished is the capacity of old people to adapt to hypoxia, traumatic injury, exercise, and physical work, all representing types of stresses that require complex physiologic adjustments. Emotional stress in the old is also capable of triggering or aggravating a series of physical ailments that, superimposed on an already debilitated state, may be conducive to disease and death.

Some of the changes with aging in adrenal cortex and medulla described here, and those in the nervous system described in Chapters 8 to 10, have been implicated in the decline of adaptive capability. Stressful stimuli activate both the hypothalamo–pituitary–adrenocortical and sympatho–adrenal systems which coordinate physiologic reactions to maintain homeostasis. This coordination is maintained through a series of neuroendocrine signals which involve several hormones and brain centers. With aging, the synchrony of these signals is disrupted; the capacity to restore the original homeostasis is altered and the hypothalamus is unable to undergo the *major remodeling of its circuitry necessary for adaptation*.

6.1 The Neuroendocrinology of Stress and Aging

The role of the hypothalamo–pituitary–adrenal system in adaptive responses was first emphasized by Cannon (1929)[40] for the adrenal medulla and by Selye (1950)[41] for the adrenal cortex. This role has been both supported and criticized by many investigators and continues to be controversial. We shall briefly summarize here some aspects of the role of the adrenal cortex in homeostasis and aging, as originally formulated by Selye in his "General Adaptation Syndrome" and as currently modified by Sapolsky.[18] In this, as in other important functions and pathologies (see atherosclerosis, Chapter 16), the early concept of inevitability of progressive functional failure has been replaced by the perspective of successful preventive and therapeutic interventions.

6.2 The General Adaptation Syndrome and Aging

Four fundamental observations comprise the basis of Selye's theories:

1. The now classic observation that the adrenal cortex is indispensable for life, the mineralocorticoids, of which aldosterone is the most physiologically active, being of primary importance.
2. The adrenalectomized (or hypophysectomized) animal can survive indefinitely *if* it receives additional dietary salt (to compensate for the excessive loss of sodium due to failure of renal reabsorption of sodium in the absence of aldosterone) and *if* it is maintained under optimal environmental conditions.
3. The capacity to survive various types of stress is severely impaired in an adrenalectomized (or hypophysectomized) animal compared to its intact counterpart.
4. Maintenance doses of adrenocortical hormones to adrenalectomized (or hypophysectomized) animals, although sufficient to maintain life under no-stress conditions, are incapable of supporting adaptive reactions when stress is present — a finding that underlines the necessity *for activation of the pituitary–adreno–cortical system above normal levels* in order for adaptation to occur.

Selye further demonstrated that all harmful agents that act on the body, generally or locally, exert dual actions, i.e., specific and nonspecific. *Specific actions* are different for each agent, e.g., ether induces anesthesia, and *Salmonella typhosa* causes typhoid fever. *Nonspecific actions* are represented by stereotypical reactions, i.e., reactions that are always the same, regardless of the causative agent; both cold and trauma, for example, will cause, in rats, increased adrenocorticoid secretion, involution of thymus and lymph nodes, and eosinopenia, atherosclerotic-like vascular lesions (Chapter 16) (Figure 11-8).

When the organism is challenged by an adverse event — chemical, physical, biologic, or psychologic in nature — it responds both by mobilizing specific physiologic defenses to counterbalance the damaging effects of the stress and by stimulating the adrenal cortex to secrete larger amounts of its hormones; it is on the efficiency of these two sets of responses that the success or failure of the organism to adapt to the challenging stress will depend. However, it has been recognized since Selye's early experiments that not all responses to stress can be traced directly to the increased circulating levels of adrenocortical hormones, for some responses of still unknown etiopathology are found to occur also in adrenalectomized animals.

The general adaptation syndrome is characterized by three phases:[41]

1. An initial phase in which defense mechanisms are acutely challenged (alarm reaction)
2. A period of enhanced adaptive capacity (stage of resistance)
3. The loss of the capacity to adapt (stage of exhaustion)

This general response pattern assumes that some form of energy, finite in amount, is necessary for the performance of adaptive work and can be exhausted under the influence of stress (Figure 11-9).

These considerations have given rise to an analogy between the loss of adaptive capacity in the adrenalectomized animal and its decline in the old individual.[42] Old age is regarded as the accumulated result of the stresses and strains that the organism has supported during its lifespan; it would correspond to the stage in which the adaptive energy available to the organism from early development has been exhausted. Thus, not only is the resistance of the elderly generally low, but when adaptive energy is employed for resistance against one stressor, resistance to other types of stressors is

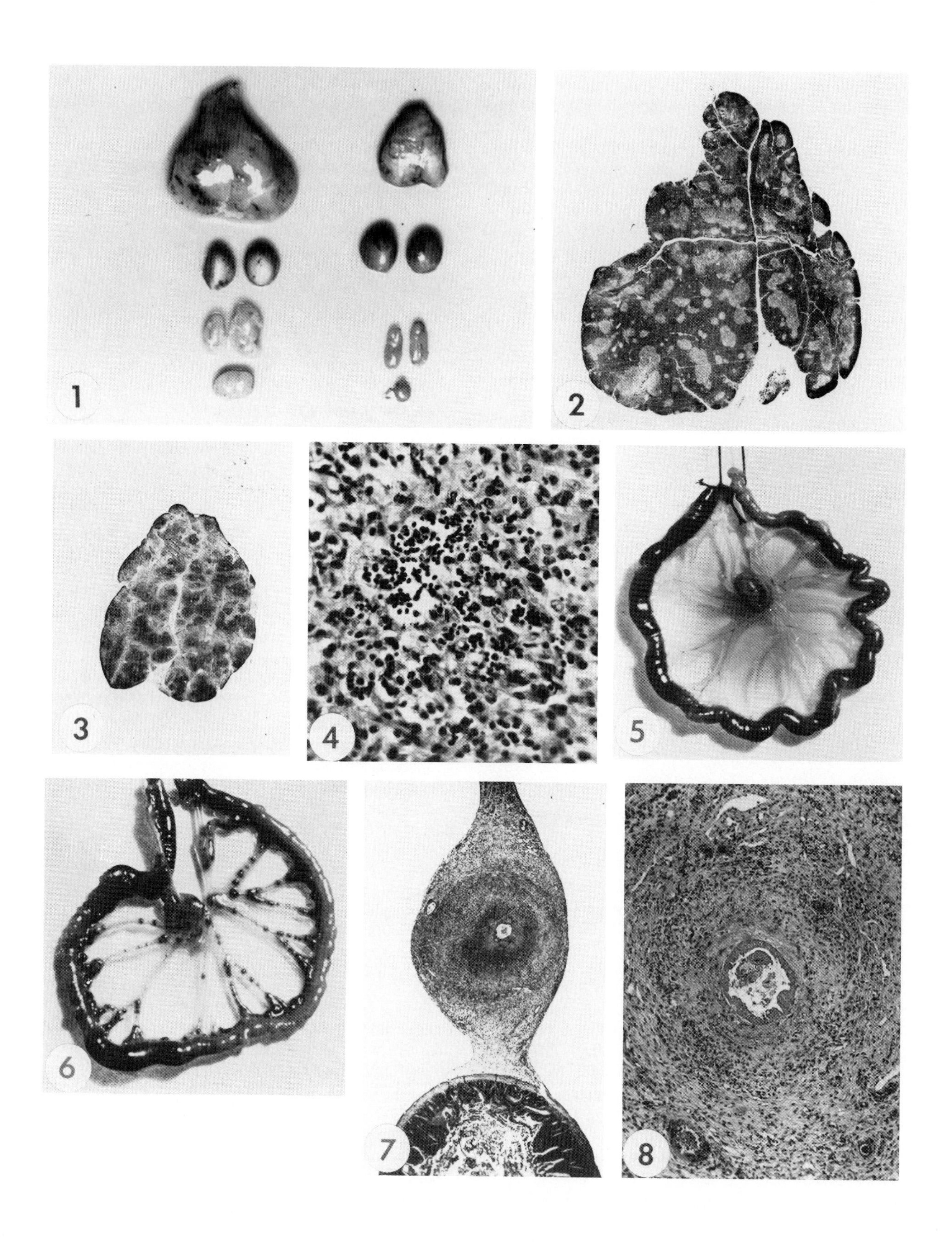
1
2
3
4
5
6
7
8

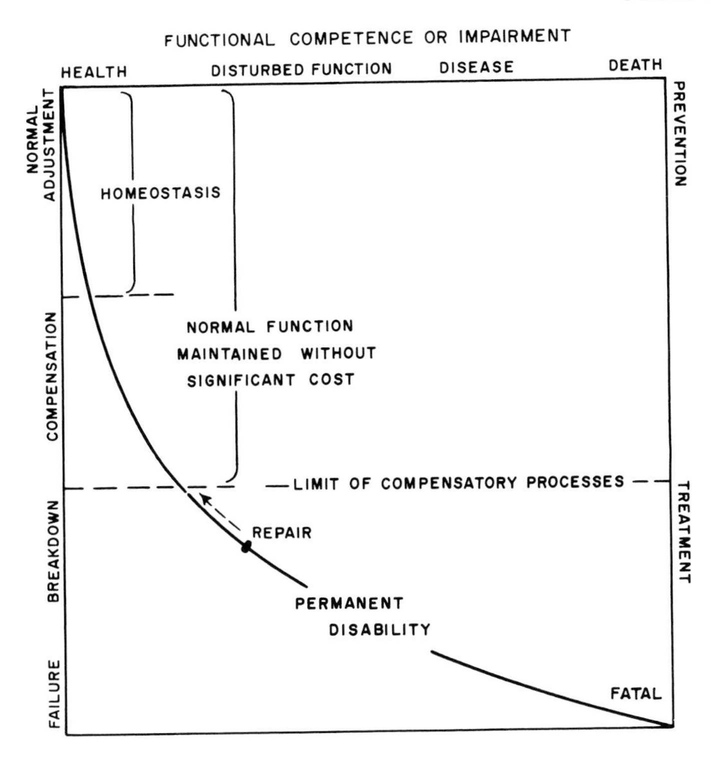

FIGURE 11-9 *Progressive stages of homeostasis from adjustment (health) to failure (death).* In the healthy adult, homeostatic processes ensure adequate adjustments in response to stress, and, even for a period beyond this stage, compensatory processes are capable of maintaining overall function without serious disability. When stress is exerted beyond the compensatory capacities of the organism, disability ensues in rapidly increasing increments to severe illness, permanent disability, and death. When this model is viewed in terms of homeostatic responses to stress imposed on the aged, and to aging itself, a period when the body can be regarded as at the point of "limit of compensatory processes", it is evident that even minor stresses are not tolerable and the individual moves rapidly into stages of breakdown and failure.

diminished. This concept could explain, for example, how a certain trauma — physical or psychologic, especially in the elderly — might significantly decrease the resistance of the organism to further stress and lead to disease and death. Indeed, the progressive loss and ultimate exhaustion of adaptive energy may explain the exponential increase in death risk.

It is a common observation that people living in impoverished areas — with poor hygiene, a low level of education, and insufficient food and medical care — have a shorter lifespan and that they not only look older but are physiologically older than people of the same age whose standard of living is high. Accounts of individuals held for relatively long periods of time in concentration camps under extremely unfavorable conditions have also shown that even though they have survived, they have suffered marked functional impairment. According to Selye's stress theory of aging outlined here, the continuing problem in aging studies of distinguishing between the alterations produced by aging and those produced by previous diseases is not an issue. If every stressful episode represents a fundamental component of the aging process and leaves an indelible scar, no distinction needs to be made between these two groups of phenomena; the organism pays for surviving the stress by becoming a little older. Within this theoretic framework, it is assumed that the newborn (indeed, the fertilized egg) is endowed with a finite quantity of adaptive energy, and, thus, his lifespan is governed by the amount he has inherited and the rate at which he spends it. In this sense, the stresses that draw on his/her adaptive energy throughout life are of paramount concern to the individual's health and longevity.

6.3 Glucocorticoids, Stress, and Neuronal Damage

Stress-elevated glucocorticoids induce predominantly catabolic reactions (e.g., increased gluconeogenesis from protein) to provide readily available energy in response to acute emergency conditions. These responses impose a cost on the organism as already suggested by Selye in the "alarm reaction". They may result in diabetes, hypertension, myopathy, immunosuppression, infertility, loss of weight, or inhibition of growth in children. Thus, while glucocorticoids (like thyroid hormones, Chapter 14) may have temporary value for survival, it is important that these actions be terminated rapidly lest they cause damage. Repeated exposure to stress would continuously stimulate adrenocortical secretion, and the ensuing elevated hormone levels would induce, contribute to, or aggravate existing physiologic impairment or disease, the so-called *diseases of adaptation* (atherosclerosis, gastro-duadenal ulcers, immunosuppression, etc.) outlined by Selye.

As discussed above, upon exposure to stress, aged rats are as capable as young ones of initiating an appropriate increase in secretion and levels of corticosterone. However, after stress, it is more difficult for the old than the young animals to terminate the stress-induced rise in corticosterone (Figure 11-5).[43] Also altered in the aged rats is the ability to adapt to mild sustained stress. The

FIGURE 11-8 (preceding page)

Some physiologic and pathologic responses to stress in selected organs. The experimental animal chosen is the rat. (1) (left) Naked view of the adrenal and lymphatic organs of normal rat; (right) same organs during alarm reaction. Three iliac lymph nodes (bottom) and thymus (top) are significantly decreased in size in the stressed animal, whereas adrenals are enlarged and hyperhemic. (2) Low magnification of cross-section of the thymus of a normal rat and (3) of a stressed rat. Note the inversion of the thymus pattern (light-dark areas) due to depletion of the cortex of thymocytes and migration of thymocyte debris into the medulla in the stressed animal. (4) Higher magnification of an area of the thymus during the alarm reaction, dark patches representing nuclear debris from the degenerate thymocyte. Before direct chemical measurement of adrenocortical hormones in the blood was possible, the involution of the thymus and lymph nodes was taken as an index of adrenocortical activity. (5–8) Hormonally induced vascular lesions approximate those occurring in diseases of adaptation. (5) Normal mesenteric vessels. Macroscopic view of a normal intestinal loop of a rat. Note thin and regular mesenteric vessels. (6) Abnormal mesenteric vessels. Macroscopic view from an animal chronically treated with desoxycorticosterone acetate. Note numerous beadlike periarteritis nodosa (resembling atherosclerosis) nodules along the mesenteric vessel. (7 and 8) Low and high magnification of periarteritis nodosa nodules in the mesenteric vessels. Note the thick layer of hyalinized fibrin lining vascular lumen and the partial necrosis of the arterial wall.[41]

prolongation of poststress corticosterone elevation has been ascribed to

1. Structural (e.g., loss of neurons) and functional (e.g., increased excitability) changes in the brain
2. Consequent increase in CRH, ACTH, and some of the opioid peptides (e.g., endorphin)
3. Decreased sensitivity of the adrenal cortex to ACTH (perhaps as a compensatory response to limit the effects of increased ACTH)
4. Dampened sensitivity of the pituitary to CRH, probably associated with (but not proven) higher CRH levels

A decreased, age-related brain sensitivity to corticosterone is not unique to the hypothalamo–pituitary–adrenal axis. Similar sensitivity changes to hormonal negative feedbacks in hypothalamic and other CNS areas have been described for other pituitary hormones (Chapter 12). The causes for the decreased glucocorticoid sensitivity are in part the loss of neuronal hormone receptors and in part the shift in the brain cell populations: decreased neuronal number and increased glial cells (Chapter 8).[44] Glucocorticoid receptors in the hippocampus, an area rich in these receptors, fall with aging to 50% of the adult values, whereas affinity of binding and translocation are unaltered. With aging, the glial population increases (Chapter 8). Glial cells do not have or have few glucocorticoid receptors; therefore, the progressive loss of neurons with aging and the increase in glial cells may explain the loss of hippocampal glucocorticoid receptors limited to the neuronal population.

Glucocorticoid receptors are lost with aging not only in the hippocampus but also in the amygdala, another limbic structure regulating hypothalamic functions including CRH secretion. If these limbic structures have an inhibitory action on the hypothalamus and CRH secretion, and indirectly also on the adrenal cortex,[45] then their loss with aging may reduce inhibition and cause hypersecretion. Increased ACTH and glucocorticoid secretion after lesioning of the hippocampus supports this concept of decreased inhibition.[46]

Glucocorticoid toxicity on hippocampal and amyloid neurons may be mediated by

1. Accumulation of excitatory neurotransmitters such as glutamate (Chapter 8)[47]
2. Enhanced mobilization and slower removal of calcium in postsynaptic neurons[48]
3. A series of degenerative events (e.g., cytoskeleton breakdown, microtubule degeneration, increased tau protein, decreased glucose and energy utilization) consequent to excessive calcium mobilization and leading to neuronal injury and death[49]

Glucocorticoids could damage neurons directly or may act through general metabolic actions, e.g., by decreasing glucose utilization (by inhibition of transcription of the gene for glucose transporter, Chapter 15),[50] thereby reducing energy availability and depleting energy stores. In addition to a direct neurotoxic action, it is possible that glucocorticoids may endanger the aging neurons by exacerbating everyday insults, themselves leading to neuronal damage.[51] These complex, interrelated actions may be viewed as creating a cascade of cumulatively altered neuroendocrine signals and responses (Figure 11-10).

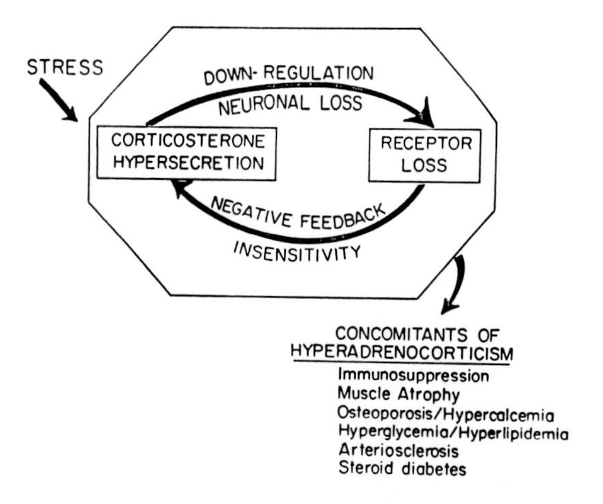

FIGURE 11-10 *Schematic representation of glucocorticoid cascade hypothesis.* (Reproduced with permission from Sapolsky et al., *Endocr. Rev.*, 7, 284, 1986.)

Is the central role of the adrenal cortex in stress-induced responses seen in rodents also observed in other animal species? Which and how many of these adrenocortical actions occur in humans? Of the major observations reported in rats after severe and prolonged stress and illustrated in Figure 11-5, at least the first two are not relevant to humans, in whom basal glucocorticoid level and responses to ACTH and stress remain normal well into old age. In primates, glucocorticoid excess damages hippocampal neurons leading to adrenocortical hypersecretion.[52] In humans, some of the features of the adrenocortical "cascade", while absent in normal aging, may be coupled with pathologic states, such as Alzheimer disease (where hippocampal damage may be associated with adrenocortical hypersecretion) and depression (an affective disorder frequent in the elderly and associated with several hormonal dysfunctions).

6.4 Strategies for Prevention and Delay of Glucocorticoid Excess and Stress

With better understanding of the role of the adrenals in adaptation and survival, and of the nature of the physiologic toll the organism pays for adaptation, it is possible to envision strategies for prevention and delay of the pathologic consequences of stress. As indicated for many other functions, many individuals age successfully and maintain an adequate capability for adaptation well into old age (Chapter 2). A study of these "successful agers" shows that the quality of the environment during the early stages of development may be particularly important.[53] For example, in rats, "handling" from birth until weaning induces long-term changes in adrenocortical secretion (lower basal levels, faster recovery after stress, enhanced sensitivity of glucocorticoid feedback) which become manifest in adulthood.[54,55] The possibility is raised that "stress management" may lead to "salutary states of adaptation" by several approaches: physical (e.g., early handling experiences), chemical (e.g., administration of neurotransmitter agonists/antagonists, calcium blockers), and/or psychologic (e.g., improvement of social environment). Although this type of intervention is just beginning to be considered on a scientific and rational basis and remains quite controversial, it is likely that it will yield considerable practical results in the near future for delaying pathology and improving life in old age.

7 NEUROENDOCRINE–IMMUNE RESPONSES TO STRESS AND AGING

As already indicated in Chapters 4 and 7 and in this chapter, the neuroendocrine and immune systems have many crossover responses to internal and external stimuli. The primary goal of such responses is a protective one, that is, the defense of the organism by maintenance of homeostasis. However, with time and repeated stresses, such protective mechanisms may fail or be vitiated, thereby leading to impaired function and increased pathology. Evidence is now pointing to a close interrelation between the two systems, which would advocate for a merging of the two into a single neuroendocrine–immune theory of aging.[53]

The actions of hormones (e.g., glucocorticoids) on the immune system, already underlined by Selye (Figure 11-8), are now extended to include a number of hormones known to affect lymphoid organs and cells, especially the thymus. While some hormones, glucocorticoids, and sex steroids induce thymic involution, others, such as growth hormone and prolactin, promote thymic development and maturation. Reciprocally, thymic hormones and cytokines from lymphocytes and other cells of the immune system stimulate or inhibit hormonal secretions by direct action on the peripheral endocrine gland or indirect action through the pituitary and hypothalamus.[54]

The thymus has both endocrine (i.e., secretion of thymic hormones in the circulation) and paracrine (i.e., secretion of cytokines in the extracellular space) functions. It attains its maximal size at puberty, and once puberty has been reached it starts to involute. It is very small in the adult and is almost entirely replaced by adipose tissue in the aged human. While it was originally thought that this involution was irreversible, there is now considerable evidence to show that under the influence of hormones (e.g., growth hormone and prolactin), promoters of development at an early age, involution may be reversed and endocrine and immunologic actions restored. As suggested for neuronal death with aging (Chapter 8), thymic involution may be viewed as a phenomenon secondary to alterations in neuroendocrine–thymus interrelations (the term "thymic menopause" has been introduced).[55] Such alterations would be responsible for several of the aging-related pathology. However, continuing research shows that it may be possible to revert such pathology by restoring immune function through appropriate neuroendocrine signals.

REFERENCES

1. Finch, C. E., Neural and endocrine determinants of senescence: investigation of causality and reversibility by laboratory and clinical interventions, in *Modern Biological Theories of Aging*, Warner, H. R., et al., Ed., Raven Press, New York, 1987, 261.
2. Timiras, P. S., *Physiological Basis of Aging and Geriatrics*, Macmillan, New York, 1988.
3. Sapolsky, R., Neuroendocrinology of the stress response, in *Behavioral Endocrinology*, Becker, J., Breedlove, S. M., and Crews, D., Eds., MIT Press, Cambridge, MA, 1992, 287.
4. Uno, H., Tarara, R., Else, J. G., Suleman, M. A., and Sapolsky, R. M., Hippocampal damage associated with prolonged and fatal stress in primates, *J. Neurosci.*, 9, 1705, 1989.
5. Uno, H., Flugge, G., Thieme, C., Johren, O., and Fuchs, E., Degeneration of the hippocampal pyramidal neurons in the socially stressed tree shrew, presented at the 21th Annual Meeting of the Soc. Neurosci., New Orleans, LA, Nov. 10–15, 1991 (session on Limbic System, I, 52.20), Abstr. 17, Part 1, 129, 1991.
6. Sapolsky, R. and Altmann, J., Incidences of hypercortisolism and dexamethasone resistance increase with age among wild baboons, *Biol. Psychiatry*, 30, 1008, 1991.
7. Flood, C., Gherondache, C., Pincus, G., Tait, J. F., Tait, S. A., and Willoughby, S., The metabolism and secretion of aldosterone in elderly subjects, *J. Clin. Invest.*, 46, 961, 1967.
8. Migeon, C., Keller, A., Lawrence, B., and Shepard, T., Dehydroepiandrosterone and androsterone levels in human plasma. Effect of age and sex, day to day and diurnal variation, *J. Clin. Endocrinol. Metab.*, 17, 1051, 1957.
9. Samuels, L. T., Factors affecting the metabolism and distribution of cortisol as measured by levels of 17-hydroxycorticosteroids in blood, *Cancer*, 10, 746, 1957.
10. Romanoff, L. P., Morris, C. W., Welch, P., Rodriguez, R. M., and Pincus, G., The metabolism of cortisol 4-C^{14} in young and elderly men. I. Secretion rate of cortisol and daily excretion of tetrahydrocortisone, allotetrahydrocortisol, tetrahydrocortisone, and cortalone (20 alpha and 20 beta), *J. Clin. Endocrinol. Metab.*, 21, 1413, 1961.
11. Serio, M., Piolanti, P., Cappelli, G., De Magistris, L., Ricci, F., Anzalone, M., and Giusti, F., The miscible pool and turnover rate of cortisol in the aged and variations in relation to time of day, *Exp. Gerontol.*, 4, 95, 1969.
12. Timiras, P. S., Neuroendocrinology of aging, retrospective, current, and prospective views, in *Neuroendocrinology of Aging*, Meites, J., Ed., Plenum Press, New York, 1983, 5.
13. Severson, J. A., The adrenal gland, in *CRC Handbook of Physiology in Aging*, Masoro, E. J., Ed., CRC Press, Boca Raton, FL, 1981.
14. Blichert-Toft, M., Secretion of corticotrophin and somatotrophin by the senescent adenohypophysis in man, *Acta Endocrinol.*, 78 (Suppl. 195), 15, 1975.
15. Vermeulen, J. P., Deslypere, J. P., Shelfhout W., Verdonck, L., and Rubens, R., Adrenocortical function in old age: response to acute adrenocorticotropin stimulation, *J. Clin. Endocrinol. Metab.*, 54, 187, 1982.
16. Colucci, C. F., D'Allesandro, B., Bellastella, A., and Montalbetti, N., Circadian rhythm of plasma cortisol in the aged (Cosinor method), *Gerontol. Clin.*, 17, 89, 1975.
17. Murray, D., Wood, P. J., Moriarty, J., and Clayton, B. E., Adrenocortical function in old age, *J. Clin. Exp. Gerontol.*, 3, 255, 1981.
18. Sapolsky, R. M., Krey, L. C., and McEwen, B. S., The neuroendocrinology of stress and aging: the glucocorticoid cascade hypothesis, *Endocr. Rev.*, 7, 284, 1986.
19. Sapolsky, R. M., Krey, L. C., and McEwen, B. S., The adrenocortical axis in the aged rat: impaired sensitivity to both fast and delayed feedback inhibition, *Neurobiol. Aging*, 7, 331, 1986.
20. Roth, G. S., Changes in hormone receptors during adulthood and senescence, in *CRC Handbook of Biochemistry in Aging*, Florini, J. R., Ed., CRC Press, Boca Raton, FL, 1981, 257.
21. Kalimi, M., Glucocorticoid receptors: from development to aging. A review, *Mech. Ageing Dev.*, 24, 129, 1984.
22. Sharma, R. and Timiras, P. S., Changes in glucocorticoid receptors in different regions of brain of immature and mature male rats, *Biochem. Int.*, 13, 609, 1986.
23. Kitraki, E., Alexis, M. N., and Stylianopoulou, F., Glucocorticoid receptors in developing rat brain and liver, *J. Steroid Biochem.*, 20, 263, 1984.

24. Sapolsky, R. M., Krey, L. C., and McEwen, B. S., Prolonged glucocorticoid exposure reduces hippocampal neuron number: implications for aging, *J. Neurosci.*, 5, 1222, 1985.
25. Pfeiffer, A., Barden, N., and Meaney, M. J., Age-related changes in glucocorticoid receptors binding and mRNA levels in the rat brain and pituitary, *Neurobiol. Aging*, 12, 475, 1991.
26. Sharma, R. and Timiras, P. S., Regulatory changes in glucocorticoid receptors in the skeletal muscle of immature and mature male rats, *Mech. Ageing Dev.*, 37, 249, 1987.
27. Barnes, R. F., Raskind, M., Gumbrecht, G., and Halter, J. B., The effects of age on the plasma catecholamine response to mental stress in man, *J. Clin. Endocrinol. Metab.*, 54, 64, 1982.
28. Rubin, P. C., Scott, P., McLean, K., and Reid, J. L., Noradrenaline release and clearance in relation to age and blood pressure in man, *Eur. J. Clin. Invest.*, 12, 121, 1982.
29. Esler, M., Skews, H., Leonard, P., Jackman, G., Bobik, A., and Korner, P., Age-dependence of noradrenaline kinetics in normal subjects, *Clin. Sci.*, 60, 217, 1981.
30. Choy, V. J., Structure and function of the aged hypothalamo-neurohypophysial system, in *Experimental and Clinical Interventions in Aging*, Walker, R. F. and Cooper, R. L., Eds., Marcel Dekker, New York, 1983, 215.
31. Sonntag, W. E., Forman, L. J., Miki, N., Steger, R. W., Ramos, T., Arimura, A., and Meites, J., Effect of CNS active drugs and somatostatin antiserum on growth hormone release in young and old male rats, *Neuroendocrinology*, 33, 73, 1981.
32. Sonntag, W. E., Forman, L. J., Miki, N., Trapp, J. M., Gottschall, P. E., and Meites, J., L-Dopa restores amplitude of growth hormone pulses in old male rats to that observed in young male rats, *Neuroendocrinology*, 34, 163, 1982.
33. Rudman, D., Kutner, M. H., Rogers, C. M., Lubin, M. F., Fleming, G. A., and Bain, R. P., Impaired growth hormone secretion in the adult population: relation to age and adiposity, *J. Clin. Invest.*, 67, 1361, 1981.
34. Elahi, D., Muller, D. C., Tzankoff, S. P., Andres, R., and Tobin, J. D., Effect of age and obesity on fasting levels of glucose, insulin, glucagon, and growth hormone in man, *J. Gerontol.*, 37, 385, 1982.
35. Prinz, P. N., Weitzman, E. D., Cunningham, G. R., and Karacan, I., Plasma growth hormone during sleep in young and aged men, *J. Gerontol.*, 38, 519, 1983.
36. Florini, J. R., Harned, J. A., Richman, R. A., and Weiss, J. P., Effect of rat age on serum levels of growth hormone and somatomedins, *Mech. Ageing Dev.*, 15, 165, 1981.
37. Sonntag, W. E., Hylka, W., and Meites, J., Impaired ability of old male rats to secrete GH *in vivo* but not *in vitro* in response to hpGRF 1-44, *Endocrinology*, 113, 2305, 1983.
38. Sonntag, W. E., Steger, R. W., Forman, L. J., and Meites, J., Decreased pulsatile release of growth hormone in old rats, *Endocrinology*, 107, 1875, 1980.
39. Bartus, R. T., Dean, R. L., and Beer, B., Neuropeptide effects on memory in aged monkeys, *Neurobiol. Aging*, 3, 61, 1982.
40. Cannon, W. G., Organization for physiological homeostasis, *Physiol. Rev.*, 9, 399, 1929.
41. Selye, H., *The Physiology and Pathology of Stress; a Treatise Based on the Concepts of the General-Adaption-Syndrome and the Diseases of Adaption*, Acta, Montreal, 1950.
42. Selye, H. and Prioreschi, P., Stress theory of aging, in *Aging — Some Social and Biological Aspects*, Shock, N. W., Ed., American Association for the Advancement of Science, Washington, D.C., 1960, 261.
43. Sapolsky, R. M., Krey, L. C., and McEwen, B. S., The adrenocortical stress response in the aged male rat: impairment of recovery from stress, *Exp. Gerontol.*, 18, 55, 1983.
44. Wilson, M., Greer, M., and Roberts, L., Hippocampal inhibition of pituitary adrenocortical function in female rats, *Brain Res.*, 197, 433, 1980.
45. Stein-Behrens, B., Elliot, E., Miller, C., Schilling, J., Newcombe, R., and Sapolsky, R., Glucocorticoids exacerbate kainic-acid-induced extracellular accumulations of excitatory amino acids in the rat hippocampus, *J. Neurochem.*, 58, 1730, 1992.
46. Elliott, E. and Sapolsky, R., Corticosterone impairs hippocampal neuronal calcium regulation: possible mediating mechanisms, *Brain Res.*, in press.
47. Elliott, E., Mattson, M., Vanderklish, P., Lynch, G., Chang, I., and Sapolsky, R., Corticosterone exacerbates kainate-induced alterations on hippocampal tau immunoreactivity and spectrin proteolysis *in vivo*, *J. Neurochem.*, in press.
48. Garvey, W. T., Hueckstеadt, T. P., Lima, F. B., and Birnbaum, M. J., Expression of a glucose transporter gene cloned from brain in cellular models of insulin resistance: dexamethasone decreases transporter mRNA in primary cultured adipocytes, *Mol. Endocrinol.*, 3, 1132, 1989.
49. Krieger, D. T., Ed., *Cushings Syndrome*, Springer-Verlag, Berlin, 1982.
50. Meaney, M., Aitken, D., Bhatanager, S., van Berkel, C., and Sapolsky, R., Effects of neonatal handling on age-related impairments associated with the hippocampus, *Science*, 239, 766, 1988.
51. Meaney, M., Aitken, D., Bhatnagar, S., and Sapolsky, R., Postnatal handling attenuates neuroendocrine, anatomical and cognitive dysfunctions associated with aging in female rats, *Neurobiol. Aging*, 12, 31, 1991.
52. Meaney, M., Viau, V., Bhatnager, S., Betitio, K., Iny, L., O'Donnell, D., and Mitchell, J. B., Cellular mechanisms underlying the development and expression of individual differences in the hypothalamic-pituitary-adrenal stress response, *J. Steroid Biochem. Mol. Biol.*, 39, 265, 1991.
53. Fabris, N., Biomarkers of aging in the neuroendocrine-immune domain. Time for a new theory of aging?, *Ann. N.Y. Acad. Sci.*, 663, 335, 1992.
54. Morley, J. E., Kay, N. E., Solomon, G. F., and Plotnikoff, N. P., Neuropeptides: conductors of the immune orchestra, *Life Sci.*, 41, 527, 1987
55. Hadden, J. W., Malec, P. H., Coto, J., and Hadden, E. M., Thymic involution in aging: prospects for correction, *Ann. N.Y. Acad. Sci.*, 673, 231, 1992.

III
SYSTEMIC AND ORGANISMIC AGING

12 AGING OF THE FEMALE REPRODUCTIVE SYSTEM: THE MENOPAUSE

Brian J. Merry and Anne M. Holehan

1 DEFINITIONS

The "menopause" is the permanent cessation of menstruation which is preceded, in the majority of women, by a period of approximately 12 months of perimenopause when cycles are irregular. Spontaneous menopause follows the loss of ovarian follicular function, but the term menopause also refers to the cessation of menstruation due to surgical procedures. As normal ovarian function may persist for some time after simple hysterectomy, the World Health Organization[1] suggested that the term "surgical menopause" be applied only to the procedure of bilateral ovariectomy with or without hysterectomy. The WHO recommended the following definitions (Figure 12-1):

Menopause — the permanent cessation of menstruation resulting from a loss of ovarian follicular activity.

Perimenopause of Climacteric — the period immediately prior to and at least 1 year after the menopause, characterized by physiologic and clinical features of ovarian involution.

Postmenopause — the period of life remaining after the menopause.

Premenopause — an ambiguous term used variously to describe the 1 to 2 years previous to the menopause *or* the whole reproductive period prior to the menopause. The WHO recommended that users define the term precisely and it is used in this chapter in the latter sense.

Evolutionary Considerations

The menopause appears restricted to the human female although, if they survive long enough in captivity, *Macaca mulatta, M. nemestrina,* and possibly *Papio* spp. undergo hormonal and other physiologic changes similar to those seen in menopausal *Homo sapiens*. However, the *Pongidae* (evolutionarily our closest living relatives) do not exhibit a menopause.[2,3]

The origin of the menopause is a source of controversy. In the wild state the reduction or cessation of reproduction precedes senescence, and the postreproductive period is theoretically irrelevant to natural selection. Consequently, selection against genes that have a potentially deleterious effect later in life will be very weak. The lifespan of a species depends on a balance between positive selection pressures, which favor long life, and negative factors, which cause a decline in reproductive function in later life. The evolutionary value of an individual will rest on its ability to produce the maximum number of surviving offspring; therefore, selective pressures favor survival of young reproductively active individuals. The menopause evolved because women survived to an age where natural selection no longer favored the maintenance of reproductive capacity.[4] The menopause may be a pleiotropic effect of a gene or gene cluster that has selection value earlier in life.

The contrasting argument suggests that the menopause is advantageous to the individual female; as early hominids increased population growth, the greater number of children required more parental care and protection.[5] In the hunter-gatherer hominids, this would have required the presence of nonreproductive group members. Selection in older females favored cessation of reproduction, and these postreproductive women became the "surrogate" mothers. Alternatively, the costs of the menopause, i.e., increased mortality and morbidity of women and their offspring in the period of declining reproduction, would far outweigh any benefits that might accrue from "grandmotherly" behavior: "the menopause is an artefact of human civilization which emerged when our growing mastery of the environment increased our survivorship".[6]

2 FUNCTIONAL CHARACTERISTICS OF MENOPAUSE

2.1 Age at Natural Menopause

Determining the age at menopause is subject to analytic inaccuracies. Serious sampling errors may occur in a population study, and two populations may not be comparable on the basis of age and birth cohort. Random sampling is essential. Hospital patients are already self-selected on the basis of illness and may introduce bias into the data. Retrospective sampling (asking women to recall the date of their last menstrual period) can be biased by memory error, understating the age at menopause, recalling age at last birthday in menopause year rather than actual date of last menstrual flow (which can lead to an underestimation of up to six months), and the tendency to round up age to the nearest 5 years resulting in an artificially high incidence of menopause at ages 40, 45, and 50.[7]

Methods to Detect Menopausal Age

Two recent methods, although still associated with the risk of bias by including more women who have an early menopause, have avoided the problems associated with retrospective studies. The first is a cross-sectional design where age and menopausal status of women are determined at the time of interview and the second, a longitudinal cohort design in which groups of women are followed over time until they reach natural menopause.

The frequency distribution of age at menopause is asymmetric as there is a wider scatter of women at a lower menopausal age (i.e., negatively skewed data); therefore, use of the mean age at menopause leads to an underestimate of the age. It is statistically more accurate to use the median age, i.e., the age at which 50% of the population has reached menopause. A probit or logit transformation of the percentage of postmenopausal women at each age makes the distribution more symmetric (Table 12-1). In Caucasian women, the median age of menopause is between 49 and 51, a value which has remained remarkably constant since medieval times.[8]

0-8493-8979-8/94/$0.00+$.50

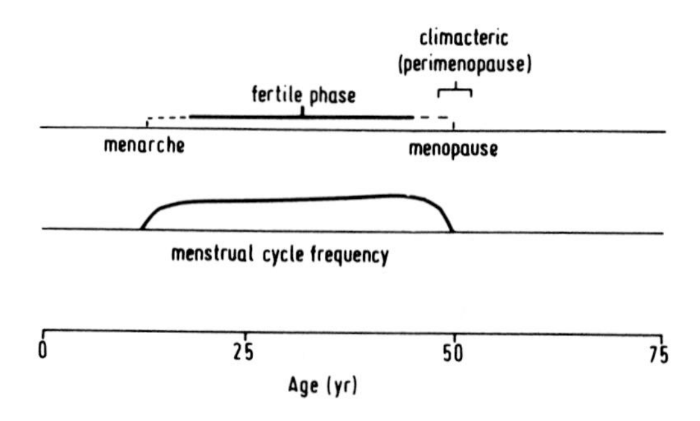

FIGURE 12-1 *Schematic diagram illustrating definitions and timing of events in the reproductive lifespan of women.* (From Gosden, R. G., *Biology of the Menopause: The Causes and Consequences of Ovarian Ageing,* Academic Press, London, 1985. With permission.)

2.2 Factors Influencing the Age at Menopause

Race

The median age at menopause in Caucasian women is remarkably similar, but less is known about other racial groups. The median age at menopause for Caucasian and Negro women in the U.S. and South Africa differs by less than 1 year (Table 12-1). In the U.S. today, one third of all women are over 50 years of age, which is the approximate age of the menopause. In a 1987 survey of 13,996 Japanese women (age range 22 to 86 years), the mean age of menopause was 49.3 years, very similar to the value obtained for the U.S., Scotland, and England (Table 12-1). No significant difference was observed between urban and rural populations.[9] The mean age at menopause for educated central Javanese women is 50.2 years, again very similar to the values reported above.[10] However, the Punjabis in India, the Melanesian women of New Guinea, the Indians in Mexico, and the migrant educated or noneducated women and rural women in Central Java all have lower median ages at the menopause.[10,11] This difference may occur because of poor nutrition, health, or socioeconomic situation.

Nutrition

In the Melanesian population of New Guinea, median age at the menopause (43.6 years) is lower in women suffering protracted malnutrition with mean height of only 144.5 cm and mean weight 40.22 kg. The nonmalnourished group had a median age at menopause of 47.3 years, a mean height of 153.8 cm, and a mean weight of 51.14 kg.[12] Identifiable malnutrition appears the important factor for determining early onset menopause, for a comparison of three socioeconomic urban groups in Karachi, Pakistan (poor slum dwellers, middle-class clinic attendees, and wives of retired military officers) revealed no significant difference in the mean age at menopause, i.e., 47 years.[13] Thinner women, particularly with small stature, tend to have an earlier menopause,[14] and, in well-nourished women, age at menopause is associated with percentage of body fat in early adulthood.[15]

Marital Status and Age at Parity

Single and/or employed women may reach the menopause at a slightly earlier age, an effect that cannot be accounted for by parity (i.e., bearing of children) or age at first pregnancy.[16] The age at menopause may also be slightly lower in nulliparous (not having borne children) women.[17,18] The relationship between parity and age at menopause is unclear; some studies report no relationship,[19,20] while others demonstrate a positive correlation between higher parity and later menopause in women of upper but not lower socioeconomic status.

Use of Oral Contraceptives

Little is known on this subject. Oral contraception could mask the transition from menstruation to menopause. A slightly higher median age at menopause follows use of oral contraceptives.[14] Factors which affect the menopause do so to advance its appearance rather than to delay it. Within this context the widespread use of those oral contraceptives that inhibit ovulation may be expected to delay the onset of menopause if its timing is related to absolute oocyte number. No consistent association between oral contraceptive use and a delay in menopause onset has been demonstrated.[21] It has been proposed that ovarian failure results primarily from the rate of oocyte atresia rather than being dependent on absolute oocyte number.[22]

Smoking

Smoking tobacco is associated with an earlier menopause.[14,23,24] Naturally postmenopausal women aged 60 to 69 from the

TABLE 12-1

Estimates of the Age at Menopause from Selected Studies

Country and year of study	Race	Mean or median age at menopause (years)		Study design
Scotland 1970	Caucasian	50.1	Median	Cross-sectional
England 1965	Caucasian	50.78	Median	Cross-sectional
		47.49	Mean	
England 1951–1961	Caucasian	49.82	Median	Cross-sectional
U.S. 1934–1974	Caucasian	49.8	Median	Cohort
		49.5	Mean	
U.S. 1966	Caucasian	50.02	Median	Cross-sectional
	Negro	49.31	Median	
	Both races	49.8	Median	
Germany 1972	Caucasian	49.06	Mean	Retrospective
Finland 1961	Caucasian	49.8	Mean	Retrospective
Switzerland 1961	Caucasian	49.8	Mean	Retrospective
Israel 1963	Caucasian	49.5	Mean	Retrospective
Netherlands 1969	Caucasian	51.4	Median	Cross-sectional
New Zealand 1967	Caucasian	50.7	Median	Cross-sectional
South Africa 1971	Caucasian	50.4	Median	Cross-sectional
	Negro	49.7	Median	
South Africa 1960	Negro	48.1	Median	Retrospective
		47.7	Mean	
South Africa 1960	Caucasian	48.7	Mean	Retrospective
Punjab 1966	Asian	44.0	Median	Cohort and cross-sectional
New Guinea 1973	Melanesian	47.3 (nonmalnourished)	Median	Cross-sectional
		43.6 (malnourished)	Median	

From Gray, R. H., in *The Menopause,* Beard, R. J., Ed., MTP Press Ltd., Lancaster, England, 1976. With permission.

U.S., Canada, and Israel who had never smoked had a mean age at menopause of 49.4 years, whereas in those who smoked 1 to 14 cigarettes per day from before the age of 35, the age was 48.0 years and fell to 47.6 when cigarette consumption was 15 or more per day. The tar constituents and tobacco smoke reduce the oxygen carrying capacity of hemoglobin, have adverse effects on ovarian blood vessel walls, may affect neuroendocrine status and steroid metabolism, and one component, benzo(a)pyrene, destroys mouse oocytes and affects follicular reserve.[25] One alternative explanation as to why smokers experience the menopause on average 1 to 2 years earlier than nonsmokers is a possible anti-estrogenic effect of cigarette smoking. It has been reported that women smokers metabolize exogenous estrogen more rapidly than nonsmokers, and lower circulating levels of estrogen have been reported in smokers compared to nonsmokers.[26,27]

Handedness

Associations have been reported between handedness, autoimmune disorders, and reproductive hormonal changes. Two national survey data sets were used to explore whether an association existed between age at menopause and handedness. Mexican–American women, ages 35 to 74, and white and black women were selected from the Hispanic Health and Nutrition Examination Survey (HHANES) and the Nation Health and Nutrition Examination Survey (NHANES-I), respectively. The mean age at menopause was found to be earlier among left-handed women than right-handed women from the NHANES-I data set, and in the HHANES data set the mean age at menopause was significantly earlier among left-handed women.[28] The reasons for this association of age of menopause with handedness are unclear.

3 THE ENDOCRINOLOGY OF MENOPAUSE AND POSTMENOPAUSE PERIODS

The Female Reproductive Tract

The major components of the female reproductive tract are the two ovaries, the two oviducts (or Fallopian tubes), the uterus, the cervix, and the vagina. The ovaries, located in the pelvic cavity, are small (in humans, walnut-sized), oval structures containing the germinal cells, the ova, and the endocrine cells which secrete the two major hormones, estrogen(s) and progesterone. Estrogens are secreted from granulosa and thecal cells lining the follicles and by cells of the corpus luteum; these latter cells are formed at the site of the ruptured follicle at ovulation and produce both estrogens and progesterone. Thecal cells also secrete weak androgens. The ovaries lie on either side of the uterus and are covered by the fimbriae, the fringed ends of the oviducts, the tubes that lead to the uterus and in which fertilization occurs. The uterus serves as a gestation sac for the developing embryo and fetus and the vagina as the receptive organ during intercourse and the birth canal at parturition. The ovary is the primary sex organ or gonad, and the other structures are secondary sex organs, which also include the breast or mammary gland and the external genitalia (vulva). Dependent on the ovarian hormones are the secondary sex characteristics, which include hair distribution, voice pitch, adipose tissue distribution, stature, muscle development, and so forth.

Major Hormones Regulating Sexual and Reproductive Function

These hormones are operative at four different levels, which are all affected by aging:

- In the brain and, particularly, the hypothalamus, the gonadotropin-releasing hormone, GnRH, is a polypeptide secreted into the portal blood vessels, which carry it to the anterior pituitary where it stimulates the synthesis and release of the two gonadotropins: follicle-stimulating hormone, FSH, and luteinizing hormone, LH. The midbrain and the limbic system (specifically outputs from the amygdala) sustain ovulation and an increase in LH secretion.
- In the anterior pituitary, the two glycoproteins FSH and LH are synthesized and released in response to stimulation by the GnRH and are also stimulated or inhibited by the positive and negative feedback from the plasma ovarian hormones.
- In the ovary, the major hormones are steroids, estrogens (in humans, the most potent estrogen is 17-β estradiol followed in decreasing order of potency by estrone and estriol), progesterone, and androgens (e.g., testosterone, in very small amounts). Androgens, especially weak androgens, i.e., dehydroepiandrosterone (DHEA), are synthesized by the ovary; this is especially true in the postmenopausal ovary. During pregnancy the ovary also secretes the polypeptides relaxin and inhibin.
- In the periphery, ovarian steroid hormones are bound to plasma proteins, metabolized in the liver, and excreted in the urine. At the target cells, estrogens and progesterone follow the same mechanism of action as the other steroid hormones (from the adrenal cortex) and bind to the nuclear receptors and stimulate the RNA and protein synthesis responsible for the many actions of these hormones (Chapter 14).

Reproductive Cyclicity

In the female, but not in the male, the reproductive function shows cyclic changes viewed as periodic preparations for fertilization and pregnancy and is associated with underlying cyclic endocrine and behavioral changes. The reproductive cycle of varying length in lower animals is called the estrous cycle; in primates, the reproductive cycle is called the menstrual cycle with ovulation at midcycle. The major function of the menstrual cycle is ovulation, but another feature is menstruation.

Endocrine cyclicity involves all hormones active in reproductive function. In brief: (1) at the beginning of the menstrual cycle (the first day of menstruation) several follicles begin to grow, but only one is dominant and continues to grow and will ovulate. The others degenerate, forming atretic follicles. Levels of ovarian hormones FSH and LH are low, but FSH levels are rising due to the low levels of sex hormones and therefore the reduced negative feedback of these hormones on the hypothalamo-pituitary axis; (2) with the maturation of the follicle in the ovary under the influence of FSH stimulation, estrogen levels increase and, through a positive feedback, stimulate the release of LH, which triggers ovulation, at about the 14th postmenstrual day; (3) at ovulation, the ovum is released into the oviduct to be fertilized. The site of the ruptured follicle is transformed into the corpus luteum, which secretes abundant amounts of both estrogens and progesterone, and finally (4) in case of pregnancy, the corpus luteum continues to secrete estrogens and progesterone necessary for survival and implantation of the new organism; in the absence of fertilization, the corpus luteum recedes and is replaced by scar tissue (corpus albicans). Estrogens and progesterone levels fall and the hypertrophic and hyperemic uterine mucosa is shed and menstruation occurs.

The transfer from regular menstrual cycles to amenorrhea (i.e., cessation of menstruation) is preceded by a period of menstrual cycle irregularity characterized by a decline in length of the follicular phase in the two decades prior to the menopause[29] and an increase in anovulatory cycles (3 to 7% of cycles at 26 to 40 years and 12 to 15% at 41 to 50 years).[30] In normal regularly cycling menstruating women over 45 years of age, cycle length is significantly shorter than in younger women and estradiol levels are lower (50 to 120 ng/l as opposed to 150 ng/l). In the 2- to 8-year period preceding the menopause, cycles become highly variable, ranging from very short to very long, and this is characteristic of the perimenopause.

FSH levels are increased particularly in the follicular phase of the cycle,[31] but LH remains unchanged when compared to younger women. When menstrual irregularities occur, a variety of hormonal patterns are recorded, including phases of rising and falling estradiol concentrations with or without elevations in progesterone, elevated levels of FSH alone, or FSH and LH.[32,33] These data suggest that ovarian follicles are maturing in an irregular fashion and ovulation is no longer occurring in each cycle.

The menopause has been described as resulting from a deficiency of estrogen because of a reduction in the number of primary follicles responding to the rising levels of gonadotropins.[34] Prior to menopause, estrogen is mainly produced by aromatization of weak androgens to estradiol in the granulosa cells of the ovarian follicle with smaller amounts from extraglandular aromatization of plasma androstenedione. In postmenopausal women, estrogen production is almost exclusively due to peripheral aromatization of circulating adrenal androstenedione.[35,36] The ratio of androstenedione to estrone is inversely related to the androstenedione level, which falls after the menopause. Adipose tissue is an important site of peripheral aromatization, but the increase seen in women associated with the menopause is not correlated with weight gain.[37,38] The androstenedione is aromatized to estrone in stromal tissue and bone, muscle, hair, and brain. Little, if any, ovarian or adrenal secretion of estrogen occurs in postmenopausal women. The metabolic clearance rates and the interconversion rates between estrone and 17-β estradiol remain similar in pre- and postmenopausal women.[36]

During the first 6 months after the menopause, significant amounts of estrogen may be secreted intermittently from the ovary, probably from residual follicles,[15] but, subsequently, plasma estradiol levels fall to low stable values for the remainder of life. The production rates of estradiol and estrone fall at different rates so that the mean blood production rate (BPR) of estradiol in postmenopausal women is 0.006 mg/24 h, which is less than 1% of the level of menstruating midcycle females. The BPR of estrone only falls by approximately 50% to 0.04 mg/24 h and estrone is, therefore, the most abundant estrogen of the postmenopausal period.[39] Estrone can be produced by peripheral conversion of androstenedione via estradiol and testosterone. This pathway is, however, much less active than the direct conversion of androstenedione to estrone, as levels of testosterone are lower than those of androstenedione. Despite low levels, testosterone becomes the principal steroid hormone secreted by the ovarian stroma. As the estrogen levels fall, testosterone action becomes unopposed and may be responsible for hair growth on the chin and upper lip of many elderly women.

With the onset of menopause, serum levels of FSH and LH rise so that, 1 year after the menopause, FSH is 10 to 15 times higher than in young women, and LH is about three times higher.[33,40-42] The maximum levels of LH are found 2 to 3 years after the menopause and remain constant thereafter[43] or show a steady downward trend after the age of 60.[40]

Associated with this increase in serum levels of FSH is a change in molecular heterogeneity with two FSH peaks of immunoreactivity.[44] The high levels of FSH are presumably due to the loss of negative feedback of estradiol at the hypothalamic–pituitary level. The levels of both FSH and LH show a marked pulsatility, pulsing every 1 to 2 h in a frequency similar to that seen in young women, although the amplitude of the pulses is much greater. This increased amplitude is believed to be due to increased GnRH. The amplitude of the pulses of GnRH is coincident with LH secretion, and the increased amplitude is consistent with increased hypothalamic activity after ovarian failure.[45]

β-Endorphin is considered to contribute to the tonic inhibition of LH secretion through its influence on norepinephrine and dopaminergic pathways controlling the pulsatile release of GnRH. Cessation of ovarian function has been associated with depressed levels of β-endorphin in serum, peritoneal, and ovarian fluids.[46]

Follicular phase plasma levels of inhibin from women in the perimenopause were lower than those recorded from young women with normal menstrual cycles. The relationship between FSH, estradiol, and inhibin at the time of the perimenopause has revealed a closer association between FSH and estradiol than between FSH and inhibin.[47] MacNaughton et al.[48] confirmed these findings and in addition demonstrated a significant negative correlation between serum immunoreactive inhibin and FSH and between estradiol and FSH. They observed a significant negative relationship between immunoreactive inhibin and age, with serum inhibin decreasing 49.3 U/l for every 10-year period.

4 SYMPTOMS OF THE MENOPAUSE

The most common symptoms of the menopause can be subdivided into three groups (Table 12-2). Not all these symptoms are present in women undergoing the menopause and, when present, they are of differing intensities.

4.1 Hot Flushes/Flashes

The most common and well-known symptom is the hot flush or flash, usually associated with inappropriate sweating.[49] Some authors make a distinction between these terms, the "hot flush" being associated with the blushing and sensation of heat while the "hot flash" describes the sweating. There is no sound physiological basis to separate these phenomena, and the terms will be used interchangeably in this chapter. They can vary in intensity from the occasional, transient sensation of warmth to periodic episodes of heat, drenching sweats, and tachycardia culminating in disturbed sleep, fatigue, and irritability. Women in Aberdeen, Scotland who were approaching the menopause were examined, and 48% experienced flushes close to the time of the menopause, the peak incidence being the first 2 postmenopausal years (Table 12-3). In a Netherlands study of women ages 42 to 62 years, 17% of women with regular cycles were already experiencing flushes, in women with irregular cycles this figure rose to 40%.[19] The incidence was 65% 1 to 2 years after menopause and remained 35% even 5 to 10 years after menopause. The epidemiology of hot flushes and its associated methodological problems have been recently reviewed.[50,51] The frequency of flushes varies from several each day to one to two per week. Little understanding exists as to what factors determine the frequency of hot flushes, although a correlation has been established between frequency and lower body weight.[52]

After a sensation of pressure in the head, a sudden feeling of heat is felt in the face, neck, and chest associated

TABLE 12-2

The Symptoms of the Menopause

Autonomic nervous system imbalance	Psychogenic	Features related to metabolic changes
Hot flushes	Headaches	Senile vaginitis
Perspiration	Insomnia	Atheromatosis and thrombosis
Palpitations	Mood changes	Skin atrophy
Angina pectoris	Irritability	Breast atrophy
	Depression	Osteoporosis
	Frigidity	Degenerative arthropathy
	Apprehension	Hirsutism

From Utian, W. H., in *The Menopause*, Beard, R. J., Ed., MTP Press Ltd., Lancaster, England, 1976. With permission.

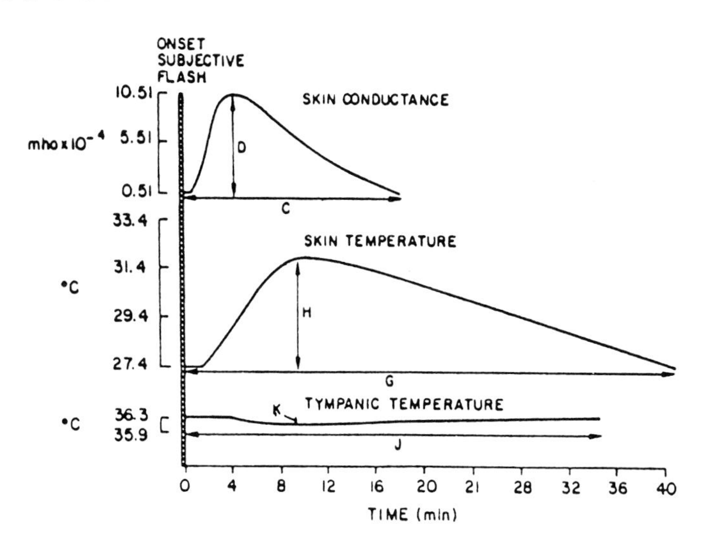

FIGURE 12-2 *Characteristics of the changes in skin conductance, skin temperature, and tympanic membrane temperature during a hot flush (n = 25).* Measurements commenced when the onset of the subjective flush was signaled by the patient. (From Tataryn, I. V., Lomax, P., Bajorek, J. G., Chesarek, W., Meldrum, D. R., and Judd, H. L., *Maturitas,* 2, 101, 1980. With permission.)

TABLE 12-3

Flushing Experience of Women in Aberdeen by Menopausal Group

Menopausal status	Flushing: Current	Flushing: Stopped	Never	Number of women
Premenopausal	11	—	87	98
Menopausal	6	1	9	16
Postmenopausal	37	41	29	107
Years since menopause				
<2 years	11	2	1	14
2–3 years	7	2	1	10
3–4 years	4	3	1	8
4–6 years	8	7	11	26
7–9 years	5	11	8	24
10+ years	1	15	7	23
Not stated	1	1		2
Total number of women	54	42	125	221

From Thompson, B., Hart, S. A., and Durno, D., *J. Biosocial Sci.*, 5, 71, 1973. With permission.

with cutaneous vasodilation and sweating in most regions of the body. Skin conductance rises beginning about 45 sec after the pressure sensation, is maximal 4 min later, and by 18 min has declined to baseline. The concomitant rise in skin temperature (Figure 12-2) has a longer duration and is greatest in the fingers and toes.[53] Because of the sweating and peripheral vasodilation, heat loss is increased with a fall in core body temperature (0.1 to 0.9°C). Body temperature reaches a nadir between 5 to 9 min after the onset of the hot flash, and the fall may be sufficient to induce shivering. The flush is accompanied by an increase of 20% of heart rate over resting values and a 4- to 30-fold increase in blood flow to the hand in the first 2 to 3 min.[54,55]

The flushing is not a direct or specific effect of estrogen withdrawal. Flushes have been experienced by boys and girls with hypothalamic damage in the early stages of puberty before exposure to adult estrogen levels[56] and by hypogonadal men following luteinizing hormone–releasing hormone (LHRH) agonist treatment.[57] Although estrogen treatment can alleviate flushing, other compounds (progestogens, placebos, Clonidine, and Nalazone) are also effective.[6,58]

Etiology of Hot Flushes

Sweating and hot flushes occur after hypophysectomy, and neither LHRH nor exogenous gonadotropins trigger flushes in young women. Rather, the rise in pituitary gonadotropins during the menopause would be responsible for the hot flushes.[59] The rise in finger temperature (indicative of flush) is related to a pulsatile release of LH[60] but not estrogen (Figure 12-3). Using LHRH analogs to inhibit pulsatile release of LH by down-regulation of pituitary LHRH receptors, flushing is induced in premenopausal women by lowering gonadotropins,[61] but flushing is not relieved in menopausal women.[62] Therefore, LH pulses are not responsible for the flushing. The flushing may have a central origin as each pulse of LH corresponds to a pulse of LHRH neurosecretion. This would explain the relationship between vasodilation (under adrenergic control) and sweating (sympathetic cholinergic fibers), which is consistent with increased sympathetic reaction when core body temperature must be lowered. The decrease in core body temperature experienced

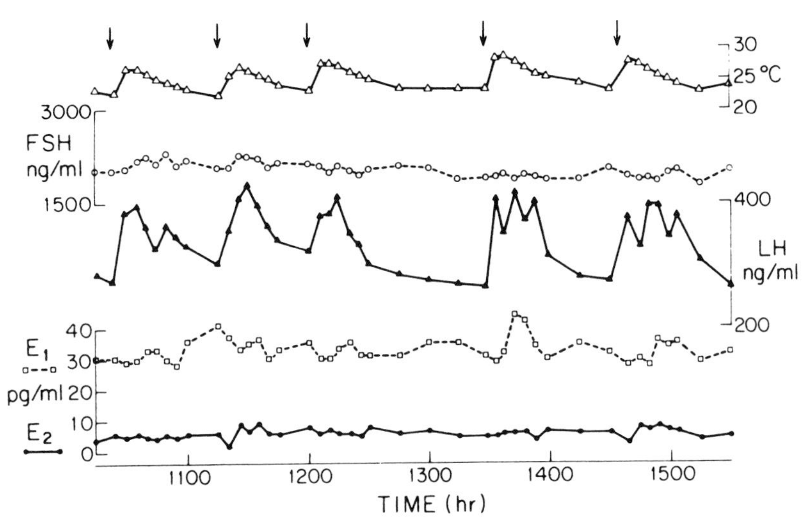

FIGURE 12-3 *Serial measurements of finger temperature and serum LH, FSH, estrone (E_1), and estradiol (E_2) in an individual subject.* The onset of the temperature rises is marked by the arrows. (From Meldrum, D. R., Tataryn, I. V., Frumar, A. M., Erlik, Y., Lu, K. H., and Judd, H. L., *J. Clin. Endocrinol. Metab.*, 50(4), 685, 1980. © The Endocrine Society. With permission.)

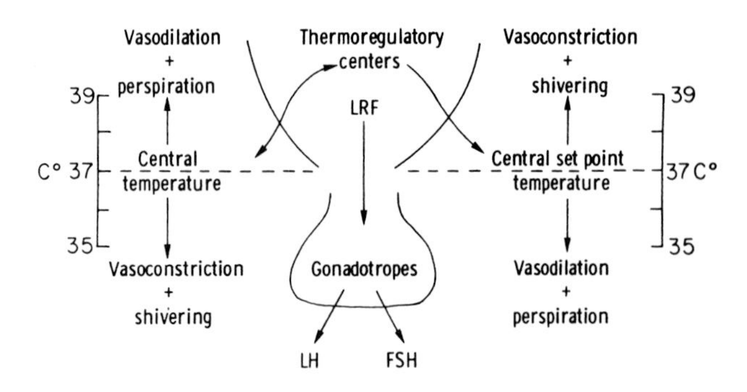

FIGURE 12-4 *The proposed mechanism of hot flushes is a sudden downward setting of the central set point temperature in the hypothalamic regulatory centers.* Since central temperature would be higher this would trigger vasodilation and perspiration (hot flush) to dissipate heat. (From Judd, H. L., in *Neuroendocrinology of Aging*, Meites, J., Ed., Plenum Press, New York, 1983. With permission.)

during the flush is itself consistent with central disorder as, under normal conditions, the peripheral cooling would be compensated by thermoregulatory adjustments to prevent a decrease in core temperature.[63]

An animal model of the hot flash using the tail-skin temperature has been developed in normal and morphine-addicted rats in an attempt to elucidate the biochemical mechanism underlying this phenomenon.[64] Induction of hypoglycemia by intravenous injection of insulin or preventing the cellular use of glucose by inhibiting glucose oxidation with 2-deoxyglucose (2DG) has resulted in a consistent flushing response. Tail temperature in normal rats increased by 4.7 ± 0.4°C and by 5.9 ± 0.6°C in morphine-addicted rats in response to 2.5 IU of Na insulin/kg. When glucose oxidation was inhibited by a dose of 750 mg 2DG/kg, tail temperature in normal rats was raised 4.76 ± 0.55°C and in morphine-addicted rats by 6.23 ± 0.46°C.[65] These observations taken together with the reported inverse relationship between blood glucose levels and the frequency of flushing episodes in women,[66] suggest that hypoglycemia is a factor underlying hot flushes. The hypothesis has been proposed that the vasomotor flushing of postmenopausal women is a result of enhanced sympathetic nervous system activity aimed at elevating blood glucose levels.[65]

Hypothalamus and Thermoregulation

The hypothalamic LHRH-containing neurons are located adjacent to the central thermoreceptors in the posterior preoptic–anterior hypothalamic areas, and inappropriate activity of these central thermoregulatory mechanisms would be responsible for the vasomotor symptoms (Figure 12-4). When the hot flush is initiated, the set point of the thermoregulatory systems falls transiently; this indicates the need for heat dissipation and triggers the heat dissipation responses including changes in behavior. This dissipation of heat leads to lower core temperature.[51,53] When the central thermoreceptors reset to normal, a sensation of chill is induced and heat conservation (vasoconstriction), heat generation (shivering), and behavioral responses are induced to raise the core temperature.

Although a significant correlation between low estrogen levels and the occurrence of hot flashes has been demonstrated, it is not understood how estrogen withdrawal can induce a transient resetting of the thermoregulatory center.[52] Indeed, not all postmenopausal women with low serum estrogen levels experience hot flashes.

It has been proposed that catechol estrogens may modify the autonomic nervous system. These metabolic products of estradiol and estrone are competitive inhibitors of the metabolism of catecholamines by catechol-*o*-methyltransferase.[67,68] They displace catecholamines from cellular binding sites in nonneural tissue and inhibit the rate-limiting enzyme of catecholamine synthesis, tyrosine hydroxylase.[68,69] The catechol estrogens formed in the hypothalamus might affect menopausal hot flashes through their actions on synaptic activity. Estrogen is known to alter the firing rate of neurons in the preoptic area of the hypothalamus of the rat,[70] and both estrogen and progesterone are recognized to modify the core temperature during the normal menstrual cycle. Neuronal hypertrophy occurs in a subpopulation of neurons in the infundibular nucleus of postmenopausal women which is considered not to be a sign of central nervous system degeneration. These neurons contain receptor gene transcripts for estrogen and a marked increase in the gene expression of tachykinin, a potent local vasodilator. It has been proposed that the enhanced tachykinin expression in these neurons may be involved in the initiation of hot flashes.[71]

A number of compounds have been found to increase in the peripheral circulation during a hot flash. These include β-endorphin, adrenocorticotropic hormone (ACTH), cortisol, β-lipotropin, dehydroepiandrosterone, and androstendione. This association is not unexpected, for the peptides are derived by post-translational cleavage of the POMC gene transcript, and synthesis of the steroids within the adrenal cortex will be enhanced by ACTH.[72,73]

4.2 Psychological Symptoms

Many menopausal women experience anxiety (apprehension, loss of self-image, fear), nervousness (becoming easily excitable, having mental and physical unrest), irritability (frequent rage or anger, uncontrollable crying fits), and depression (apathy, inability to make decisions, loss of emotional reaction and libido, psychomotor retardation). The depression appears not to be hormone dependent.[74] Several studies report little difference in overall psychological and psychosomatic symptoms at any age (13 to 64), but certain symptoms (irritability, nervousness, depression, headaches, and palpitations) were more common at the menopause than at other ages. The most frequent symptoms reported by menopausal women after flushing were headache, giddiness, obesity, and nervous instability, the latter being experienced by just over 30% of the women.[75] In a study in the Netherlands, irritability was reported by 28% of premenopausal women, 37% of menopausal women, and 24% 2 years later.[19] The corresponding figures for pre- and postmenopausal women for depression were 20 and 24% and for mental imbalance, 17 and 23%. Fatigue and headache also peaked at menopause, but the frequency of shortness of breath, palpitations, insomnia, and dizziness did not alter.

Sexual behavior has been extensively studied.[76] Sexual interest and activity in men decline continuously from puberty, but the sexual activities of (unmarried) women do not decline until 55 to 60 years. Therefore, the decline in marital coital rates (50% of age 20 rates at age 40) is due to declining libido in the male.[77] For every 5 additional years of age of the wife, the decline in coitions per month when other variables were controlled was 0.1 but that of the husband was 0.8. Decline in sexual interest in postmenopausal women cannot be explained simply by the decline in their husbands' interest.[78] Rather, postmenopausal women would retain the potential for effective sexual response, particularly when regularly stimulated.[20] In most of these women, however, the intensity and duration of each phase of the sexual response is reduced. In 3 to 5% of women, vaginal dryness and associated dyspareunia (painful intercourse) can lead to a secondary decline in sexual enjoyment and interest.[78,79]

5 EFFECTS OF ESTROGEN DEPRIVATION ON TARGET ORGAN RESPONSE

5.1 Vulva

At menopause, pubic hair decreases and the labia majora become very small. The vulva skin shows atrophy and reduction in the amount of glycogen in the epithelial cells.[80] Skin conditions (e.g., lichen sclerosis and hyperplastic dystrophy) are prevalent.

5.2 Vagina

The vaginal epithelium is highly responsive to estrogen and at the menopause when levels decrease, atrophy occurs, leading to shortening of the vagina and loss of elasticity.[81] Histologically, the vaginal epithelium becomes flattened with a loss of glycogen. The glycogen from vaginal epithelial cells provides energy for the lactobacilli present in the vagina. After the menopause, the population of lactobacilli decreases, reducing the release of lactic acid, with a rise in vaginal pH from 3.5 to 4.5 to above 5.0.[82,83] The microbial population often includes staphylococci and streptococci, and the postmenopausal vagina is at a greater risk of bacterial infections, which, if treated with antibiotics, may encourage the growths of yeasts responsible for postmenopausal vaginitis.[84] The reduction in thickness of the epithelium (from eight to ten cell layers to three to four layers) associated with the withdrawal of estrogen results in a vagina that is vulnerable to mechanical injury and bleeds readily (atrophic vaginitis). This condition is aggravated by vaginal dryness and may lead to painful intercourse but is responsive to exogenous estrogen treatment. Sexually inactive women may also complain of vaginal irritation, burning, or discomfort resulting from a lack of vaginal secretion. The vaginal health of sexually active postmenopausal women not on hormone replacement therapy has been observed to exceed that of coitally inactive women when assessed on six parameters.[85]

5.3 Cervix

The uterine cervix becomes atrophic with epithelial atrophy and loss of fibro-muscular stroma during the menopause.[86] The cervix often becomes flush with the apex of the vagina as it shortens and the cervical os (opening) is visible. The epithelium becomes thinner with a flattened superficial layer, and the glandular tissue is much less active, producing little mucus. Cytological examination of the cervix becomes difficult at the menopause due to the predominance of parabasal cells as estrogen is withdrawn,[87] and the atrophic vaginitis and accompanying inflammation often exacerbate this problem. A more reliable interpretation of the smears can be obtained after a short course of vaginal or systemic estrogen.

5.4 Uterus

As estrogen levels fall, the uterus gradually reduces in size. The total collagen and elastin content as well as the wet weight of uterine tissue reaches a maximum around age 30 (approximately 120 g wet weight) and remains constant until about age 50. Over the next 15 years, the fibrous tissue content and wet weight decline by 30 to 50%[88] and the uterus atrophies to as little as 1 cm in very old women.[81] The uterine endometrium become atrophic with a residual cystic dilation of the glands; these show no cyclic changes of activity and very little sign of secretion as the columnar epithelium becomes flattened to a cuboidal shape. In the premenopausal period, the estrogen stimulation of the uterus, unopposed by progesterone, can stimulate proliferation of the atrophic endometrial glands that can lead to cystic changes. At this time the stromal and epithelial cells hypertrophy, and glandular hyperplasia is seen in some women. This can lead to confusion when endometrial curettings are taken, as the disordered proliferative pattern can be confused with adenocarcinoma.[89] Active hyperplasia after the menopause is confined almost exclusively to the first 5-year period; in curettage of postmenopausal women with no irregular bleeding, 31.5% had simple atrophy, 40.7% had active cystic glands, and only 1% showed actively secreting glands indicative of progesterone stimulation.[90]

5.5 Oviducts

Postmenopausally the oviducts shorten and their diameter decreases, conditions ameliorated by estrogen. As the epithelium becomes atrophic over a period of years, the stromal layer underneath forms blunted villi and the smooth muscle layer becomes replaced by fibrous tissue.[91] After the age of 60, cilia are gradually lost and tubal secretion and peristalsis are decreased or absent.[92]

5.6 Ovary

As the ovary ages, it becomes smaller and more fibrotic. The number of primordial and Graafian follicles diminishes markedly, but the follicles persist in small numbers although most of those present are atretic.[93,94] The stromal tissue becomes fibrotic with a high incidence of hyperplasia, usually attributed to stimulation by the high level of gonadotropins.[95] Postmenopausally, the ovarian surface becomes convoluted with deep clefts where the germinal epithelium can be preserved; inclusion cysts, which may become cystoadenomas, form from the surface epithelium when the clefts become covered (Figure 12-5).[80] The postmenopausal medulla has a characteristic accumulation of corpora albicanta (scar tissue), which are composed of fine collagen fibers with very few cells. These may be formed from degenerating corpora lutea or from relatively recent atretic follicles.[96] The blood vessel walls of the medulla and hilus become sclerotic; there is hyalinization of collagen elements[97] and stromal calcification and/or hemorrhage.[98]

5.7 Skin

The skin contains estrogen receptors and can actively metabolize estrogens which affect the hyaluronic acid and water content of the sex skin and increase its thickness.[99] With aging, the dermis and epidermis become thinner, lose water and elasticity, and accumulate increasing amounts of pigment giving rise to "aging spots" and lose pigmentation (vitiligo), producing pale areas (Chapter 22). The reduced activity of the sebaceous and sweat glands, together with the thinning of the epidermis, makes the skin more sensitive to temperature, humidity, and trauma.

5.8 Bladder and Urethra

The atrophy of the urethra parallels that of the vagina as both tissues have a common embryonic origin. Both the urethra and the trigone region of the bladder are rich in estrogen receptors and are responsive to estrogen withdrawal. As the bladder and urethra atrophy, the incidence of cystitis, urinary atrophy, dysuria, and noninflammatory urethritis increases (Chapter 19). Problems with incontinence can also be caused by relaxation of the pelvic floor and atrophy of pelvic structures, but this correlates more closely with other factors (parity, obesity, constipation, race) than with withdrawal of estrogen (Chapter 19).[100,101]

5.9 Cardiovascular System

The cessation of menstruation is associated with a rise in hemoglobin and serum cholesterol, an increase in low-density

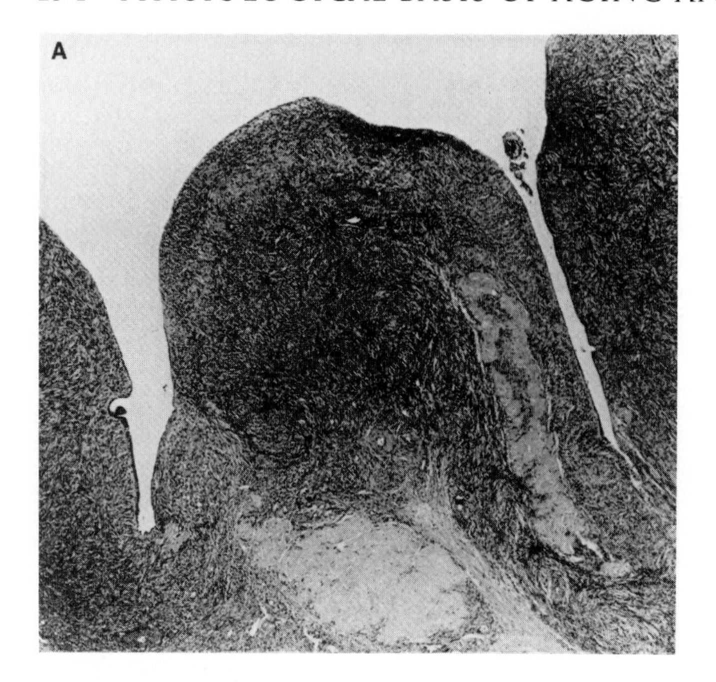

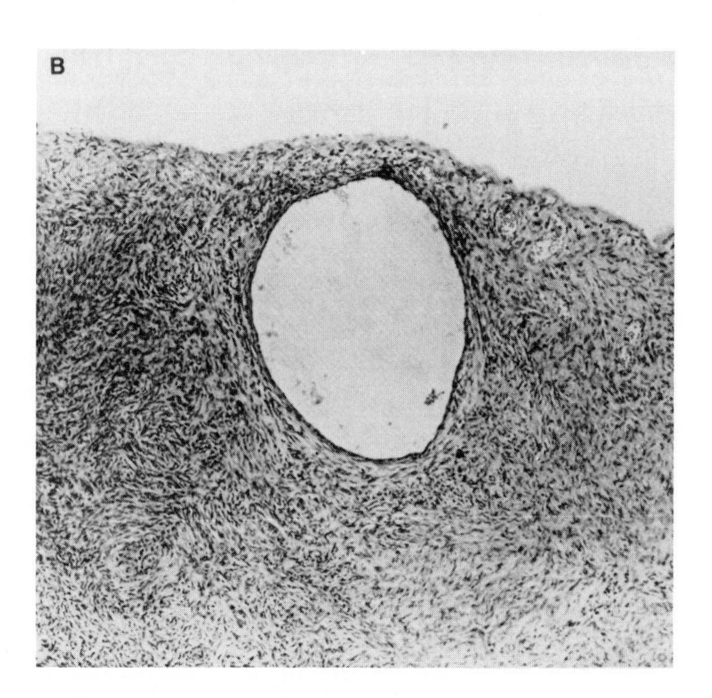

FIGURE 12-5 (A) *Section of postmenopausal ovary showing the deep convolutions that give rise to the cerebriform convolutions characteristic of the postmenopausal ovaries.* These are scattered corpora albicanta with preservation of the germinal epithelium in some of the clefts. HE stain × 45. (B) *Section of a postmenopausal ovary demonstrating a germinal inclusion cyst.* The convolutions shown in (A) that retain the germinal epithelium can be covered and produce the inclusion cysts from the surface epithelium. He stain × 108. (From Voet, R. L., in *The Menopause,* Buchsbaum, H.J., Ed., Springer-Verlag, New York, 1983. With permission.)

lipoproteins (LDL), and a gradual rise in high-density lipoproteins (HDL) possibly associated with the high gonadotropin levels of postmenopausal women[102] (Chapter 17). Postmenopausal women aged 45 to 54 years have a coronary heart disease (CHD) incidence 2.4 to 3.0 times greater than that of premenopausal women of the same age. However, women with a history of ischemic heart disease had a tendency to early menopause with myocardial infarct before age 48 and were still menstruating at the time of the cardiovascular incident.[103] On average, women develop CHD 7 to 8 years later than do men, and the lower rate of CHD observed in premenopausal women has been attributed to a protective effect conferred by the high levels of circulating estrogen. It has been argued that the increased rates of CHD observed in postmenopausal women are the result of cessation of steroid secretion at the menopause. Early studies using small sample numbers have suggested that early natural menopause may increase the risk of CHD, but the existence of cigarette smoking has often confounded the interpretation of such studies.[104] Smoking is known to be associated with CHD and early natural menopause (see above). Reexamination of the morbidity data for CHD has not revealed any evidence of an increase in the risk of CHD resulting from the natural menopause alone, if this factor is separated from that of age.[105] The CHD mortality curve for premenopausal women (30 to 49 years) is parallel with that from postmenopausal women (50 to 69 years); the major risk factor is age rather than reproductive status (Figure 12-6). In contrast, early surgical menopause induced by bilateral oophorectomy is associated with an increased risk of CHD, but it is not known if this enhanced risk results from estrogen withdrawal. Several studies of surgical menopause have shown that incidence of CHD and angina was more prevalent in ovariectomized women due to loss of estrogen, although the subjects studied were few, other risk variables not accounted for, and thus the interpretation of the data was uncertain.

In both men and women, an age-related rise in LDL is observed with age, with the rise in premenopausal women being less rapid than for males[106] (Figure 12-7). This age-associated rise in LDL is thought to result from the loss of LDL receptors on cells which remove LDL particles from the circulation.[107] Studies using rat hepatocytes and porcine granulosa cells have shown an increase in LDL receptors in response to estrogen, suggesting that estrogen may exert a

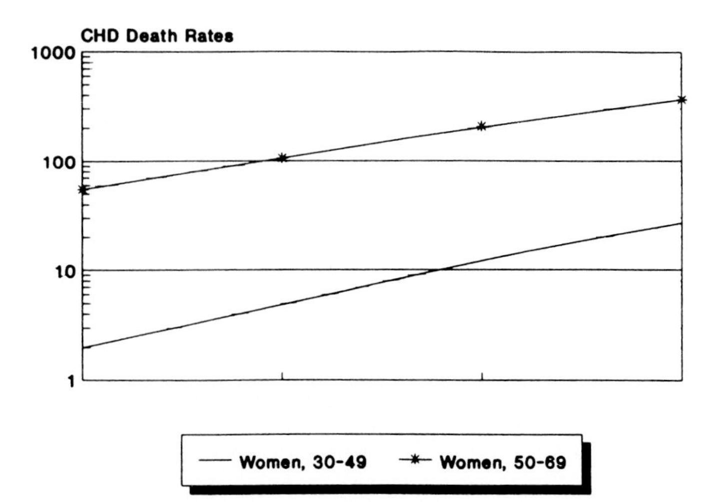

FIGURE 12-6 *Death rates from coronary heart disease in women 30 to 59 years and 50 to 69 years of age (United States, 1986).* (From Bush, T.L., in *Multidisciplinary Perspectives on Menopause,* Flint, M., Kronenberg, F., and Utian, W., Eds., New York Academy of Sciences, 1990, 263. With permission.)

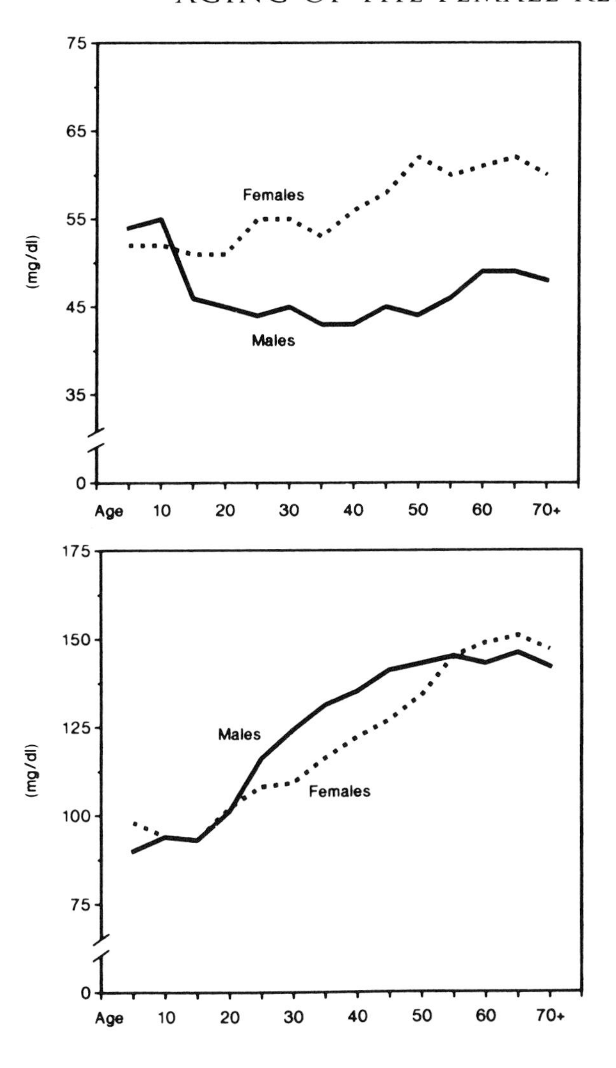

FIGURE 12-7 Top panel: *Relationship between age and plasma HDL in men and women.* Bottom panel: *Relationship between age and plasma LDL in men and women.* (From Sacks, F. M. and Walsh, B. W., in *Multidisciplinary Perspectives on Menopause,* Flint, M., Kronenberg, F., and Utian, W., Eds., New York Academy of Sciences, 1990, 272. With permission.)

protective effect against the age-related loss of LDL receptors (Chapter 17).[108,109]

Factors other than estrogen withdrawal may account for the progressive increase in CHD deaths in postmenopausal women. In ovariectomized women, serum cholesterol levels did not increase significantly for at least 2 years after surgery, no difference could be demonstrated in the level of coronary atherosclerosis when control and castrated women were compared, and the age at onset of myocardial infarction did not correlate with age at menopause.[79]

The effects of estrogen therapy on LDL and HDL levels have been inconsistent between different studies. A comparison of postmenopausal women in the U.S. from a number of geographical locations revealed lower LDL levels, higher HDL levels and higher triglycerides in response to the use of oral estrogen.[110-113] Interpretation of the effects of oral estrogen on HDL and LDL levels is complicated by the response of the liver to the ingested estrogen absorbed into the portal circulation. Ovarian estrogen, in contrast, is secreted directly into the systemic circulation. Because the liver is central to the synthesis and catabolism of LDL and HDL, exposure to high levels of estrogen may induce changes in LDL and HDL metabolism, which do not reflect normal physiological effects.[114] Where estrogen has been delivered by a systemic route through the use of subdermal implants or cutaneous application, a rise in HDL levels has been reported, but the effect on LDL is often inconsistent and the effect observed is not statistically significant.[115,116]

There is a general consensus from the published studies that the risk of CHD in postmenopausal women is significantly reduced by estrogen use.[105,117] The protective mechanism may be through the induction of higher HDL levels since high HDL levels are recognized to reduce the risk of CHD[118] (Figure 12-8). It is not known if adding a progestogen to protect against endometrial hyperplasia will oppose the protective effect of estrogen against CHD. Adding a progestogen to the estrogen can result in a blood profile characteristic of atherosclerosis, i.e., lowered HDL, increased LDL, and increased cholesterol esters, and certain combinations of estrogen/progestogen may greatly increase the risk of cardiovascular disease in the postmenopausal woman.[100] Present evidence does not suggest that estrogen therapy produces the atherosclerotic pattern, and it is insufficient to ascribe CHD development to reduction in estrogen. Oral estrogen does increase serum triglyceride and very-low-density lipoprotein (VLDL), the precursor of LDL. However, estrogen induces an increase clearance of VLDL and of LDL from the blood, and the higher VLDL levels do not result in increase levels of LDL.[114]

The progressive decrease in male:female deaths from CHD from 8:1 at age 45 to 1:1 at age 85[100] may be due not to a rapid acceleration in the rate of mortality in women after the menopause, but to a slowing of the rate in men at this age (Figure 12-9).[119] Around age 50, men appear to begin to lose some factor that had increased their risk of developing CHD, possibly the male sex hormones or an alteration in the estradiol-to-testosterone ratio. Puberty in the male is associated with a sharp fall in HDL concentration associated with an increase in circulating testosterone.[120] During adult

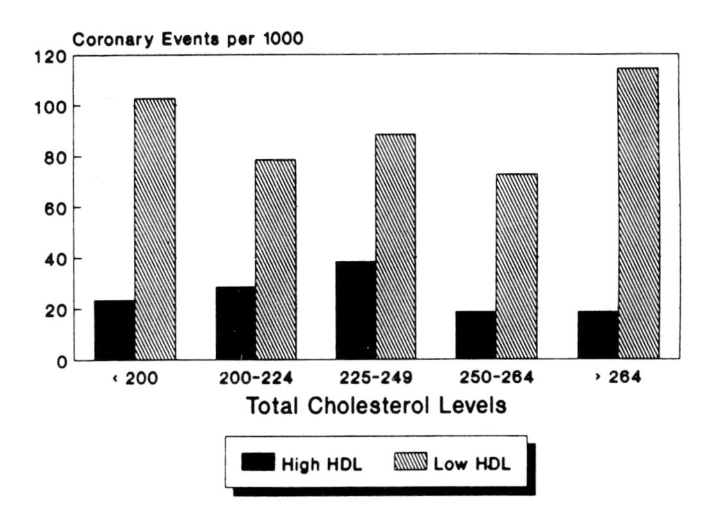

FIGURE 12-8 *Age-adjusted coronary rates (per 1000) by total cholesterol level in women with high and low HDL concentrations: The Donolo–Tel Aviv Study.* (From Bush, T. L., in *Multidisciplinary Perspectives on Menopause,* Flint, M., Kronenberg, F., and Utian, W., Eds., New York Academy of Sciences, 1990, 263. With permission.)

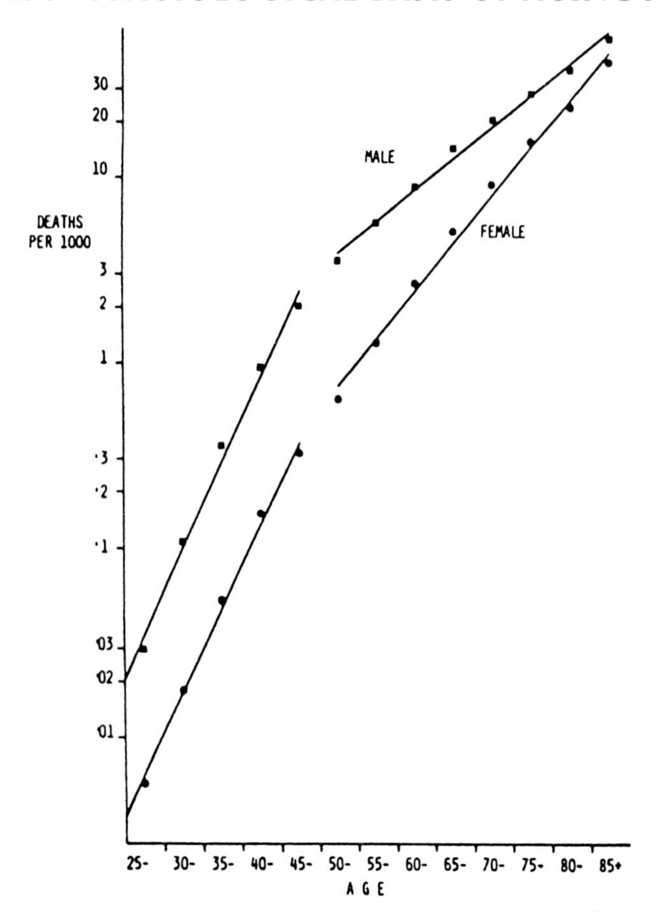

FIGURE 12-9 *Death rates from ischemic heart disease in men and women according to age (England and Wales, 1970 to 1974).* The change occurring at the menopausal age is in men rather than women. (From Heller, R. F. and Jacobs, H. S., *Br. Med. J.*, 1, 472, 1978. With permission.)

life the mean HDL levels for men are about 20% lower than in women and show little change. It is considered that the sex difference in HDL concentrations is determined primarily by the circulating levels of testosterone. Endogenous estrogens appear to have little effect on HDL levels.[114]

5.10 The Skeletal System

Postmenopausal osteoporosis is manifest approximately 4 to 5 years after the menopause and the frequency increases thereafter. It may produce a variety of symptoms associated with the musculoskeletal system, one of the most common being low back pain. In osteoporosis, the total bone mass is reduced while retaining the normal mineral-to-matrix ratio, resulting from a greater amount of bone resorption than formation. This leads to an increased incidence of fractures, particularly in bones which have a high proportion of trabecular bone and/or are weight bearing (spinal column, particularly the lumbar area; proximal femur; pelvis and distal radius). The rate of bone resorption is greater in women than in men. Black women have a higher bone density and lower incidence of osteoporosis than Caucasian women, but bone density is lowest and osteoporosis most prevalent in Asian women. By age 60, the incidence of compression fractures of the lumbar area of the spinal column in white women is 25% and distal radius fractures are 10 times more frequent in women than men of the same age. In women, for every 10 years of life after 60, the incidence of proximal femur fractures doubles, and by age 75, 50% of all women have compression fractures of the spinal column.[100] The bone may become so fragile that the stresses generated from normal everyday activity may be sufficient to induce fractures.

While genetic factors may have some influence on bone density and increased risk of osteoporosis, nutritional and hormonal factors may be more important. Two distinct types of osteoporosis can be identified on the basis of the pattern of loss of cortical and trabecular bone. Type II (senile osteoporosis) is associated with accelerated bone loss and poor absorption of calcium and is seen in women older than 75 years. Type I osteoporosis is associated with estrogen deficiency after menopause.[121] The rate of bone formation and loss involves a complex relationship between dietary intake of calcium and its absorption, the serum levels of calcium and phosphorus, and the secretion of calcitonin, parathyroid hormones, and 1,25-dihydroxy-vitamin D (Chapters 14, 20, and 21).

The menopause is associated with increased skeletal remodeling and calcium homeostasis in which bone resorption exceeds bone formation.[122,123] Total body calcium levels are lower in women than in men and decline from middle age at a much faster rate in women. Bone loss is aggravated by inadequate calcium intake due to a poor diet or an impaired ability of the gut to absorb calcium due to vitamin D deficiency[6] (Chapter 19). This leads to an increased calcium requirement with increasing age. The levels of plasma estrogen are inversely related to fasting urine levels of calcium hydroxyproline and the degree of metacarpal bone loss. The transferral of calcium from bone to blood may be inhibited by estrogen. This increases parathyroid hormone, which acts on the kidney to stimulate 1α-hydroxylase activity producing 1,25-dihydroxy-cholecalciferol, the active metabolite of vitamin D. Lack of estrogen would cause a reduction in the synthesis of this active metabolite. Giving estrogen to patients with postmenopausal osteoporosis results in retention of calcium and phosphorus, decreased serum and urinary calcium and phosphorus, urinary hydroxyproline, fasting urinary calcium and bone resorption, and increased serum immunoreactive parathyroid hormone.[124] These effects support the concept that estrogen decreases the responsiveness of bone to endogenous parathyroid hormone (which promotes bone resorption) and that the development of bone disease in postmenopausal females may be due in part to a loss of the protective effect of estrogen. Further evidence is seen in the finding that postmenopausal osteoporosis increases in nonobese women, while obese women have a greater bone mass and are less likely to develop symptomatic osteoporosis presumably because of the higher circulating estrogen levels.[124]

Not all investigators, however, accept the concept of estrogen withdrawal as a cause of osteoporosis. While circulating estrone levels are decreased in osteoporotic women compared to age-matched controls, plasma estrogen levels do not differ when women with and without osteoporotic fractures are compared.[100] Until recently it had not been possible to show the existence of estrogen receptors in bone. The demonstration that osteoblast-like osteosarcoma cells exhibit estrogen binding and express receptor mRNA,[125] and that normal human osteoblast-like cells possess estrogen receptors,[126] suggests that osteoporosis is a primary response of the skeletal system to estrogen withdrawal.[127] Estrogen may modify bone resorption rates by inhibiting the synthesis of prostaglandins such as PGE_2 and interleukin-1, which are recognized to increase bone resorption.[128,129] Bone formation may be stimulated by estrogen through its action on growth factors such as TGF-β and IGF-1.[128]

6 THE RISKS AND BENEFITS OF HORMONE REPLACEMENT THERAPY

Hormone replacement therapy, particularly with estrogen, is used to alleviate menopausal symptoms. Not all these symptoms are eliminated or relieved by the therapy, and concern is expressed over the dangers of uninterrupted estrogen therapy in postmenopausal women. Certain specific symptoms of the menopause are related to estrogen withdrawal, i.e., hot flushes, atrophic vaginitis; short-term estrogen therapy has been used to treat these successfully in some women. Both the frequency and duration of the flushes can be reduced and associated insomnia relieved by estrogen, and it is possible to titrate the dose so that the minimum possible is given to ensure relief of symptoms.

The beneficial effect of estrogen treatment on hot flushes and night sweats, by improving the sleep pattern, may be responsible in part for the reported action of estrogen in relieving some of the minor psychiatric symptoms of the menopause (i.e., irritability, minor depression, insomnia). Estrogens may directly reduce the latency period required to fall asleep and increase the amount of rapid eye movement sleep,[130] and the resulting enhanced ability to sleep will improve mood and enhance concentration.

The protective effect of estrogen on bone resorption is dose dependent. If therapy is started shortly after the menopause, bone resorption is retarded for up to 10 years, the incidence of fractures is reduced, and the development of bone fragility is prevented. The use of photon absortiometry now allows the effect of estrogen therapy to be determined on the axial skeleton. It has been demonstrated that estrogen therapy is effective in preventing the loss of cortical bone irrespective of the time elapsed from endogenous estrogen withdrawal[131] (Figure 12-10). The effects on bone density, however, are reversed, if the therapy is discontinued. Because of the risks attendant on long-term estrogen therapy, it is recommended only for those women most at risk from osteoporosis, i.e., small-framed Caucasian women who smoke or who had a surgical menopause before the age of 40. Increasing the levels of calcium is advisable in postmenopausal women with negative calcium balance. As the requirement for calcium rises with age, a supplement is likely to be beneficial to many postmenopausal women. The rate of vertebral fracture was reduced by 36% when 1500 mg/day calcium carbonate was given,[132] and this figure increased to 50% with the addition of vitamin D and 65% with sodium fluoride. The best results were obtained with combinations of calcium and estrogen (78% reduction in fracture rate) or calcium, estrogen, and sodium fluoride (94% reduction). Vitamin D should not be used alone as it induces hypercalcemia and hypercalciuria,[133] thus accelerating mineral bone loss. A daily regime of light exercise can promote deposition of calcium in bone and increase muscle tone. It also has beneficial effects on atherosclerosis by increasing the HDL cholesterol levels.[6]

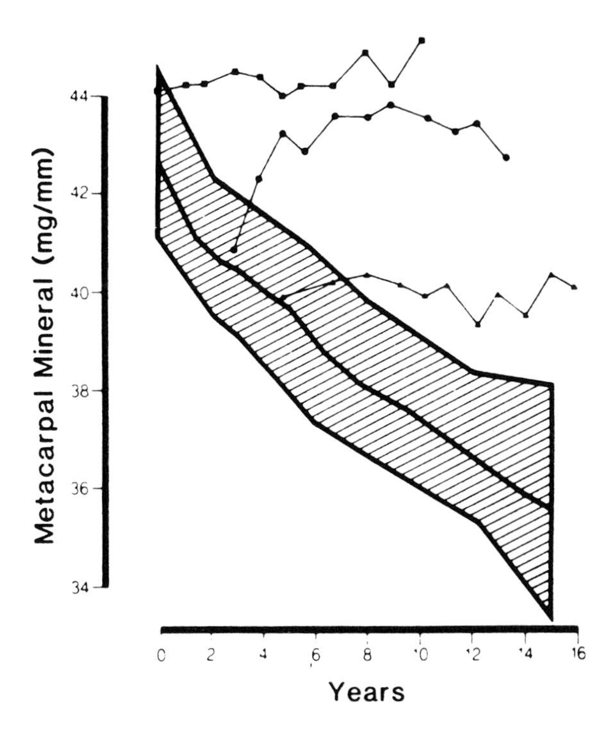

FIGURE 12-10 *Long-term prevention of bone loss by estrogen.* The hatched area represents the placebo used (mean ± SD) in a prospective controlled study in oophorectomized women. The three lines show the mean values only for three estrogen-treated groups: treatment initiated at the time of oophorectomy (squares), 3 years (circles), or 6 years (triangles) after oophorectomy. Bone loss is prevented in all three situations. However, the earlier treatment is begun, the better the outcome, in terms of bone mass after 10 years of therapy. (From Linsay, R., in *Osteoporosis: Etiology, Diagnosis, and Management,* Riggs, B. L. and Melton, L. J., III, Eds., Raven Press, New York, 1988. With permission.)

Dangers of Estrogen Replacement Therapy

The topic of ERT, in particular the weighing of the benefits and risks of unopposed estrogen therapy in women who have not had a hysterectomy, is recognized to be one of the most complex issues arising in present-day medical practice. The issues have been addressed in some detail by Meade and Berra.[134] These authors draw attention to the lack of randomized controlled trials to accurately assess the balance between the benefits conferred, i.e., protection against CHD and osteoporosis, against the increased risk of morbidity from uterine cancer.

Clearly estrogen replacement must not be given to any woman who has breast or genital tract carcinoma, chronic liver disease, neurophthalmologic vascular disease, undiagnosed vaginal bleeding, or who has a history of thromboembolism. Caution should be observed in prescribing this treatment when endometriosis, familial hyperlipidemia, gall bladder disease, fibrocystic breasts, or porphyria is diagnosed and when the woman is a heavy cigarette smoker, severely overweight, or has severe varicose veins or preexisting hypertension.

One major problem associated with estrogen therapy is endometrial hyperplasia and the risk of developing endometrial cancer. A dose-dependent effect results in a 16 to 32% incidence of cystic glandular and atypical hyperplasia. The relative risk of developing endometrial cancer has been estimated to increase between 3 and 25 times with unopposed estrogen therapy,[135,136] and hyperplastic changes develop rapidly, sometimes within 6 to 7 months of the start of therapy. The increased risk increases with both the dose and duration of unopposed estrogen use and persists for many years after treatment ceases.[137] However, the development of estrogen-dependent endometrial cancer is a localized process and early diagnosis and treatment result in a 5-year survival rate of 95%. Indeed, estrogen therapy has been reported to enhance survival in women with endometrial cancer.[138,139] However, invasive and extrauterine cancer is also increased in response to long-term use of high dose estrogen, and with these conditions the prognosis is poor.[137] The addition of cyclic progestational therapy to the estrogen reduces the risk and incidence of endometrial cancer. Administration of progestins for 10 or more days per month to mimic the characteristic pattern of the normal cycle results in withdrawal bleeding for 3 to 4 days per month in the majority of women. This treatment has a protective effect on the endometrium.

It is unclear whether long-term estrogen or estrogen–progestogen therapy has an adverse effect on the incidence of carcinoma of the breast. An association is recognized between the risk of breast cancer and early menarche and late menopause, and it is reduced by early oophorectomy. Such observations are indicative of a promotional role for chronic exposure to estrogen. The presence of fibrocystic breast disease is a contraindication for estrogen therapy as there is some evidence of breast tumor formation in these women when doses of estrogen (0.62 mg conjugated estrogen) are given that have no effect on healthy breasts.[140] The evidence that breast tumor formation in women with no fibrocystic breast disease is linked to estrogen therapy is inconclusive. Two recent prospective studies have found no increased risk of breast cancer in women using oral contraceptives who experienced long-term exposure to elevated estrogen.[141] An analysis of 28 studies failed to reveal an increase risk of breast cancer in estrogen-treated women with the exception of those women exposed to high dose preparations.[142] However, an increased risk of breast cancer (1.3- to 1.4-fold) has been found after long-term use of estrogen subsequent to the menopause.[143] This finding relates to the use of unopposed conjugated estrogen. Other studies report a greater increase in the risk of breast cancer after treatment with both an estrogen and a progestin and the interpretation of the data is unclear (for review see Barrett-Connor[144]).

An association between estrogen therapy in postmenopausal women and an increased risk of ovarian cancer has not been clearly established. Contradictory conclusions have been drawn from a number of studies which have addressed this issue.[144]

Short-term, low-dose estrogen therapy does not apparently increase hypertension and thrombosis or alter glucose metabolism,[145] but postmenopausal women using this treatment are more at risk of developing gallstones (Chapter 20), particularly when overweight.[146]

Estrogen replacement therapy has a role to play in the management of menopausal symptoms, particularly hot flushes, atrophic vaginitis, and osteoporosis. It is important that the risks associated with this treatment are understood and proper examination of the patient undertaken while therapy is administered. It is recommended that regular checkups should assure no changes in blood pressure, regular breast and pelvic examinations should be undertaken, and histological evaluation of the endometrium is necessary, particularly if any unexplained vaginal bleeding has occurred. Doses of estrogen should be kept as low as possible to ensure relief of symptoms, and the therapy should be given in a cyclic fashion with the addition of progestin in order to protect the uterus if present.[133,135] Developments in the treatment and management of the menopause have been the subject of recent reviews.[147,148]

TABLE 12-4

Age-Specific Fertility Rates for the Ethnic Hutterites: 1926–1930, 1936–1940, 1946–1950

	Annual number of live births per 1000 women		
Age of mother (years)	**1926–1930**	**1936–1940**	**1946–1950**
15–19	16.9	13.1	12.0
20–24	268.0	259.1	231.0
25–29	417.4	465.6	382.7
30–34	397.1	461.9	391.1
35–39	355.0	430.6	344.6
40–44	238.9	202.9	208.3
45–49	23.5	47.6	42.1
Total	236.5	258.0	226.6

From Eaton, J. W. and Mayer, A. J., *Hum. Biol.*, 25, 206, 1953.

7 AGE-RELATED LOSS OF FERTILITY

7.1 Reproductive Lifespan in Humans

Fertility patterns in modern human populations are subject to the outside influences of contraception and abortion, but one group has been studied in which these confounding parameters are eliminated. The Hutterite communities of North America are an Anabaptist sect whose beliefs prohibit extramarital coitus, contraception, and abortion, and whose diet and standards of medical care produce high fertility rates.[149,150] Early marriage is not usual among the Hutterites and the degree of fertility is high among the older age groups (Table 12-4). The average age at birth of the last child is 40.9 years, and the number of births per 1000 women falls rapidly once the fourth decade is reached with a rapid increase in the percentage of sterile couples from 33% at age 35 to 39 to 100% a decade later.[6] These figures are in agreement with other data from earlier populations in which contraception was less common[151] and would indicate that births after the age of 50 are extremely rare, although the oldest documented age at birth is 57 years and 129 days.[152] In the Hutterite society, fertility is highest between ages 25 and 34, whereas highest fertility in a modern society where contraceptive methods are practiced lies between the ages of 20 and 29, with a rapid decline in the older age groups. Comparison of the birth rate for 1955 and 1985 in Scotland reveals a decline in all age groups except the 15- to 19-year-olds, with the greatest rate of decline seen from age 30 onward. This apparent drop in fertility was confirmed by data from an AID (Artificial Insemination by Donor) program in which the average time to achieve a pregnancy by donor sperm rose from 6 months below age 30 to over a year when the woman was older than 35.[153] In this study any possibility of male infertility or alteration in coital rates was eliminated and only the fertility of the female was studied. The results seem to indicate a drop in human fertility at a relatively early age.

7.2 Reproductive Lifespan in Laboratory Animals

In the typical laboratory rat or mouse there is an initial juvenile period of 3 months, followed by a fertile period of approximately 12 months. In these polytocous species (i.e., producing many eggs or young at the same time), the number of ovulations increases at the second or third litter, resulting in a greater litter size.[154] Subsequently, after a plateau period, the size of the litters declines progressively, the interval between litters increases, delivery of young becomes difficult, and lactation becomes poor in the older mothers.[155,156]

In mice, the genetic constitution of the strain may influence the decline in litter size and the eventual termination of fertility.[157] Inbreeding represents a definite disadvantage with respect to reproductive capacity, leading to a significant postreproductive portion of the lifespan in hamsters, mice, rats, guinea pigs, rabbits, and gerbils.[158]

Although the ability of Long-Evans rats to reproduce successfully falls sharply and continuously with age, the age at which repeated pregnancies were initiated (2 or 9 months) did not influence the litter size or average pup weight at birth.[159] This is in marked contrast to the results of other studies, where decreases in average litter size were found in the aged rat.[155,160] The parental age of mice may influence the length of the fertilization plus gestation period, litter size, and the stillborn mortality rate.[156] In C57BL/6J mice, age did not alter the fraction of matings in previously parous mice (>80% at 3 to 7 and 11 to 12 months), rather the effects of age were smaller than the effects of virgin status.[161] In another study, the reproductive performance of mothers kept virgin or continuously pseudopregnant for 9 months before mating was not significantly different to that of mothers bred continuously from puberty. It was concluded that in these mice, the decline in litter size with age, which was associated with a high level of embryonic loss after implantation, was an effect of chronological aging and not of parity,[162] a finding since confirmed in rats.[163,164]

7.3 Reproductive Lifespan in Domesticated Animals

Few data are available on total reproductive capacity in domestic animals, as it is unusual to keep breeding stock once reproductive capacity begins to decline. The most accurate records are those kept for thoroughbred horses, which show a decline in fertility from about 19 years, although some horses have been bred successfully at 27 to 33 years, almost their normal lifespan. Reproduction also declines in cattle and sheep as well as in dogs, with the peak reproductive capacity followed by a plateau and then decline.[158,165] All these animals are in a protected environment, and it is impossible to draw conclusions from these data concerning wild animals, as reproductive senescence is rare and disease and predators remove animals from the population before reproductive potential is exhausted.

7.4 Reproductive Lifespan in Primates

As with domestic animals, data concerning the end of the reproductive lifespan in primates are scarce. Only a few centers maintain primate colonies, and the length of life is such that these data are only just becoming available. At the Wisconsin Primate Research Center, pregnancy in Rhesus monkeys has been reported to occur as late as 26 years,[166] although in other studies, menopause in this species occurred between the ages of 25 and 30 and the latest age for carrying a pregnancy was 19 years.[167] Chimpanzees continue to menstruate at the age of 44 years,[158] but fertility at this age is poor.[168]

7.5 The Basis for Declining Fertility

Various explanations may account for the decline in fertility with advancing age: a reduction in the number of ova shed, an increase in abnormal ova or a wearing out of the accessory organs so that the ova cannot be fertilized or embryos carried to term, or disruption of the hormonal control of the ovary or the uterine environment.[155,157,169] Here, again, comparative animal physiology provides useful models for human aging.

Oocyte Depletion

Most mammals begin extrauterine life with a finite stock of oocytes that is not added to in adult life.[170] Aging is accompanied by a progressive loss of oocytes at a rate apparently determined by the genetic makeup of the species or strain.[157] Normal disappearance of oocytes by ovulation after puberty accounts for a small proportion of the total loss, but this is insignificant when compared to the effects of atresia (involution of follicles that failed to mature).

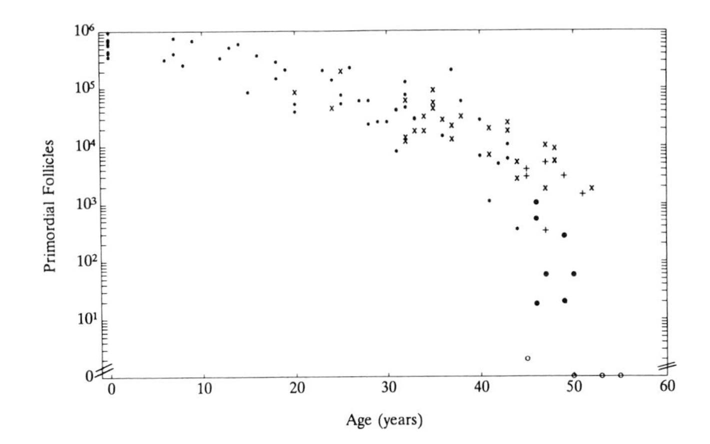

FIGURE 12-11 *The relationship between age and primordial follicle number is compared using data from four studies.* Follicle depletion appears to accelerate in the decade preceding menopause. *, Block's study of stillborn[171] and of girls and women with regular menses aged 6 to 44 years;[172] ×, extrapolation from Figure 8 in Gougeon's study of women aged 20 to 52, all with regular menses;[176] and in the study of Richardson et al.[175] of women aged 45 to 55 years, +, women with regular menses, ●, perimenopausal women, and ○, postmenopausal women. (From Richardson, S. J. and Nelson, J. F., in *Multidisciplinary Perspectives on Menopause,* Flint, M., Kronenberg, F., and Utian, W., Eds., New York Academy of Sciences, 1990, 13. With permission.)

In the human female there is a steady decline in oocyte number from birth to 25 or 30 years.[171,172] Very little data are available on women over 45 years old. A qualitative assessment in a group of women past the age of 50 observed only an occasional follicle or corpus luteum, indicating that few ova may persist to the sixth decade.[173] However, primordial follicles and remnants of old tertiary follicles and corpora lutea were found in ovarian samples from women up to 80 years of age.[174] A recent study has attempted to clarify the relationship between follicle number and the perimenopausal transition from regular to irregular menses.[175] Ovaries removed at surgery from 17 women aged 45 to 55 years were serially sectioned to obtain estimates of follicle number. The women were divided into three age-matched groups: (1) those experiencing regular menstrual cycles with intervals between 3 and 5 weeks with no hot flushes; (2) those in the perimenopausal transition experiencing irregular menses with intervals of less than 3 weeks and/or more than 5 weeks for more than 12 months, with or without hot flushes; and (3) a postmenopausal group with no menses for at least a year. Little overlap between groups was observed for the mean number of follicles per ovary. In ovaries from women experiencing normal menstrual cycles, the mean number of primordial follicles was tenfold greater than for ovaries from women in the perimenopause of the same age. In the ovaries of the four menopausal women, very few primordial follicles were observed.[175] It was concluded that the size of the follicular reserve was the important factor in controlling the transition from regular menses through to the menopause.

Richardson[29] has combined the data from three studies which have made estimates of primordial follicle number in human ovaries as a function of age (Figure 12-11). These estimates of primordial follicle number were very similar in the three studies and all showed an acceleration in follicle loss in the last decade before the menopause.[172,175,176] The mechanism for this accelerated loss is not known, but Richardson has speculated that the increased circulating levels of FSH may stimulate a greater proportion of the primordial follicles to enter the growing pool and subsequently become atretic.[175] This effect of FSH to enhance recruitment of primordial follicles has been demonstrated *in vitro* but not *in vivo*.[177] This author draws attention to the effects of hypophysectomy, calorie restriction (see Chapter 24), and the chronic administration of an analog of LHRH, all strategies that lower gonadotropin levels in laboratory rodents, as effective means of retarding the rate of follicular depletion.

Despite this progressive loss of oocytes, reproduction in most rodents, unlike humans, comes to an end long before the ovaries are depleted of oocytes. The ovaries of rats that ceased breeding around 550 days still had 1200 to 1500 oocytes at this age,[178] and in mice of different strains between 500 and 1000 oocytes were still present when the last litter was born. The only known exception among the rodents are CBA mice in which total depletion of oocytes occurs soon after reproduction ends. However, experimental radical resection of 75 or 90% of the follicular reserves from the ovaries of 5-month-old C57BL/6J mice advances the age of onset of acyclicity by 4 and 7 months, respectively. The advance in the timing of onset of acyclicity was in agreement with the known rate of oocyte depletion with age in this mouse strain. A minimum follicular reserve appears to be required to sustain normal estrous cycles.[179]

Unilateral ovariectomy in rodents leads to compensatory hypertrophy of the remaining ovary, which becomes more responsive to gonadotropin and produces approximately double the normal number of ovulations.[180,181] The overall number of litters born and the mean litter size were reduced and reproduction ceased earlier than in the intact animals, but the early cessation of breeding was not due to premature exhaustion of the oocytes. Intensive breeding had no effect on the rate at which oocyte number declined, and prevention of ovulation after prolonged steroid-induced infertility in mice did not counteract the decline in fertility.

Ovulation does not significantly affect the rate of loss of oocytes since only 1 ovum is lost by ovulation for every 1000 lost by atresia in the human. In virgin rats and mice, which ovulate regularly every 4 to 5 days throughout their reproductive lifespan, the rate of loss of oocytes is no greater than in repeatedly bred females which ovulate less frequently. The greatest influences on the rate of loss of oocytes are the factors that control the rate of atresia.

Developmental Abnormalities

With advanced age, increased losses prior to implantation and increased intrauterine mortality lead to resorption of fetuses in polytocous species or to spontaneous abortion in humans. Although one in every 200 live births in older women is associated with a chromosomal disorder, about 50% of all spontaneous abortions are genetically abnormal. As the spontaneous termination rate of clinically recognizable pregnancies in humans is 10 to 15%, this would indicate that at least 5 to 10% have abnormal chromosome complements mainly reflecting an increasing maternal age.[182]

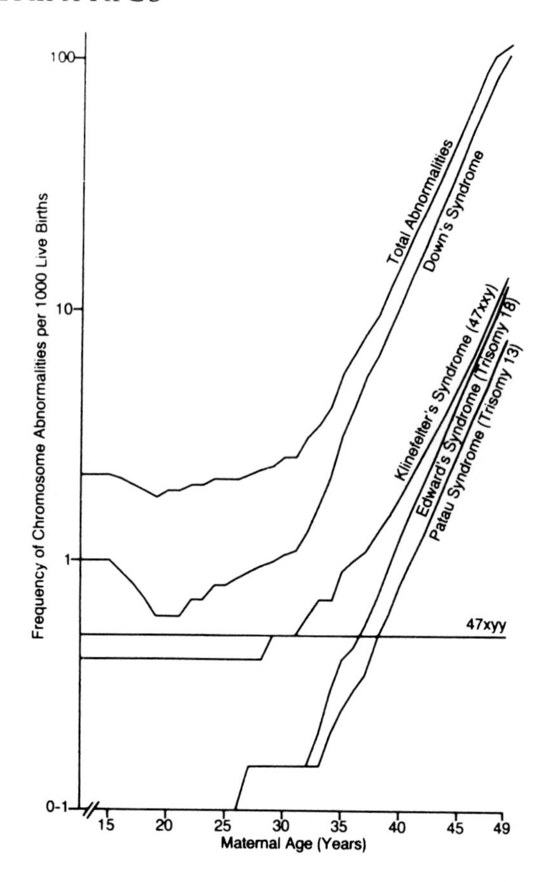

FIGURE 12-12 *Incidence of autosomal and sex chromosome trisomies in American live births according to maternal age.* Data presented for autosomal trisomies represents the midpoint of the range at each age. (From Hook, E. B., *Obstet. Gynecol.*, 58, 282, 1981. With permission.)

Chromosomal Abnormalities and Down's Syndrome

Although the incidence of clinically recognizable chromosomal disorders rises with increasing maternal age in humans, the distribution of some of these disorders is J-shaped because of a small, but significant, increase among the offspring of teenage mothers. This increase, in the case of Down's syndrome, is thought to arise from environmental effects on the ova or from hereditary causes. Down's syndrome (trisomy of chromosome 21) is the most common age-related disorder in humans, rising from an incidence of 1:2000 to 2500 in women under 30 years to 1:300 between the ages of 35 and 39 and as high as 1:45 after 45.[183] Edwards' and Patau's syndromes involving chromosomes 18 and 13, respectively, are the two other most common autosomal trisomies showing a dramatic increase with maternal aging (Figure 12-12). Also associated with advanced maternal age are two sex chromosome trisomies, XXY, the Klinefelter syndrome, and XXX, the triple X syndrome. Trisomies arise in the majority of cases at the first meiotic division of oocytes because of maternal chromosome nondisjunction. It has been suggested that trisomy risk is a function of the size of the oocyte pool and that risk is increased in women with accelerated rates of oocyte atresia. The known association between the rate of atresia and the age of onset at menopause would suggest that women who had trisomic pregnancies should reach the menopause earlier than women who had not.[184]

The association of paternal nondisjunction is thought to be responsible for the most common chromosome monosomy XO (Turner's syndrome), which is not related to maternal age.

The mechanism(s) involved in the production of genetically abnormal offspring with increased maternal age are little known; they include increased exposure to radiation or chemicals, autoantibodies, infectious agents, genetic predisposition, and pre- and postovulatory aging of oocytes. Mothers of Down's syndrome offspring may have been exposed to a greater cumulative dose of radiation than controls, and X-irradiation has been shown to affect sex-chromosome nondisjunction in *Drosophila* and to produce decreased chiasmata frequency (visible cross-formation seen in homologous chromosomes during prophase, thought to be due to crossing over) in mammalian oocytes. Chronic or acute exposure to specific chemicals (e.g., methadone, mercury compounds, methotrexate, diethylstilbestrol diphosphate, 6-aminonicotinamide) may lead to the production of abnormal offspring. There are strong correlations between Down's syndrome and thyroid autoantibodies, although these may be involved in postzygotic chromosomal abnormalities rather than maternal age-dependent meiotic nondisjunction. It seems likely that infectious agents are involved in the etiology of Down's syndrome or that there is a strong genetic influence.[182]

TABLE 12-5

Effects of Neutralization of Circulating Estrogen Early in the Estrous Cycle on Early Embryonic Development

	% Recovered Embryos			
Groups[a]	Normal blastula	Normal morula	Abnormal	Degenerating
Control	89.7 ± 2.5	5.6 ± 2.2	3.1 ± 1.4	1.5 ± 0.8
Nemb[b]	64.5 ± 8.5	13.4 ± 4.2	11.1 ± 3.3	11.1 ± 7.7
ASE[c] + Nemb	83.6 ± 4.2	7.7 ± 2.9	5.9 ± 1.5	2.9 ± 1.4
DES[d] + ASE + Nemb	50.0 ± 6.8	22.5 ± 7.4	20.2 ± 3.5	7.6 ± 2.2

[a] n = 15 pregnant rats per group on day 4 of gestation.
[b] Sodium pentobarbital (Nembutal) delay of ovulation for 48 h.
[c] ASE = antisera to estradiol on day 1 and 2 of the cycle.
[d] DES = diethylstilbestrol (0.25 μg day 1 and 0.5 μg day 2).

From Butcher, R. L. and Pope, R. S., *Biol. Reprod.*, 21, 491, 1979. With permission.

The study of the relationship between age and frequency of chromosomal abnormalities in mouse oocytes revealed a higher number of alterations at the first meiotic division in aged mice than young adults.[185] A gradient in oocyte development was proposed, the oocytes formed earliest in the ovary being the one to ovulate first. When the total number of oocytes available for ovulation is reduced later in life, the frequency of ovulation of chromosomally abnormal oocytes would be increased. This, however, has not been confirmed in the fetal mouse ovary.[186] It would be expected, if a gradient existed, that the frequency of abnormal offspring should increase as the age of the mother increases. It has been pointed out, however, that many spontaneously aborted human embryos or fetuses are physically malformed or chromosomally aberrant, or both.[187] Loss of such conceptuses would reduce the frequency of liveborn offspring with congenital malformations and chromosomal abnormalities to a small fraction of what it would be otherwise, and the existence of a gradient of oocyte development is uncertain. The decreased incidence of liveborn malformed offspring with increasing age of mother may be due to a decreased ability of the uterus to support embryos, the abnormal fetuses being resorbed.[188]

Postovulatory Aging

In humans and rats, preovulatory aging of oocytes is detrimental to development. Since coital activity in the human is not restricted to the period of ovulation, the human ovum is vulnerable to fertilization late in the menstrual cycle, leading to malformation and abortion. Aging of human ova is associated with a high risk of abortion.[189,190] The oocyte undergoes preovulatory aging as a consequence of prolonged menstrual cycles during late reproductive life. An examination of embryos from patients with dated coitus revealed that of those women who ovulated after day 14 of the cycle, over 50% exhibited developmental abnormalities. In a study of over 1000 spontaneously aborted fetuses from women with known reproductive histories, 62% of these abortions were classified as relating to factors in the ovum. The similarity of the anatomic and chromosomal anomalies found in human spontaneous abortions to those observed in laboratory animals has led to the suggestion that the fertilization of overripe ova (probably intrafollicular overripeness) is a cause of abortion.[191] The incidence of human trisomies precedes any extension in cycle length; thus, extension is unlikely to be the primary causative factor. However, the endocrine changes prior to and consequent on any alterations in cycle length may have a considerable effect, for ovarian steroids are known to impair segregation of meiotic chromosomes when ova are matured *in vitro*.[192] Teenage women with irregular cycle length have an increased incidence of Down's syndrome among their offspring, and hormonal imbalance due to oral contraception has also been implicated.[193]

Role of Estrogen in Preovulatory Aging

Aged rats with 4 or 5 days of persistent estrus appear to have an increase in developmental abnormalities when compared to younger controls. Extension of the estrous cycle to 6 days

TABLE 12-6

Effects of Neutralization of Estrogen Early in the Estrous Cycle on Embryonic Development by Midgestation

			%Surviving Embryos			
Groups[a]	Implantation Rate (%)	Postimplantation Death (%)	Normal	Abnormal	<3/4 Size	< 1 to 3/4 Size
Control	84.7 ± 1.9	4.7 ± 1.4	91.9 ± 1.5	0.8 ± 0.5	2.7 ± 0.9	4.6 ± 1.5
Nemb[b]	43.4 ± 5.3	21.6 ± 6.2	57.0 ± 5.8	8.9 ± 5.5	22.7 ± 4.7	11.4 ± 3.7
ASE[c] + Nemb	89.1 ± 3.5	7.5 ± 2.4	80.6 ± 4.6	1.3 ± 0.7	6.4 ± 1.7	11.7 ± 3.8
DES[d] + ASE + Nemb	75.0 ± 5.4	18.2 ± 5.0	70.4 ± 5.6	7.4 ± 3.8	11.9 ± 2.7	10.3 ± 2.8

[a] n = 27 to 29 litters per group at day 11 of gestation.
[b,c,d] See Table 12-5 for definitions.

From Butcher, R. L. and Pope, R. S., *Biol. Reprod.*, 21, 491, 1979. With permission.

markedly increases the number of abnormal and unfertilized one-cell ova, degenerating embryos, and conceptuses.[194] Artificial extension of the estrous cycle in the mature rat by injections of pentobarbital is also associated with developmental defects, decreased implantation and fertilization rates, increased polyspermy, and a threefold increase in the number of aneuploid fetuses, including trisomies, polyploidy, and mosaicism. During the naturally and artificially prolonged period of ovulatory delay, the oocytes remain in meiotic arrest. In these rats there is an early rise in the plasma level of 17-β estradiol in relation to ovulation. Therefore, a role for estrogen in the alteration of follicularly aged oocytes seems probable. Detrimental effects of delayed ovulation were prevented by binding of the endogenous estrogen with an antiserum[195] but were reinstituted by administration of diethylstilbestrol, an estrogenic compound not bound by the antiserum (Tables 12-5 and 12-6). The early rise in estrogen has the consequence that the preovulatory oocyte is contained for a prolonged time in an environment of elevated estrogen, which appears to result in an increased incidence of subsequent abnormal development or death of the embryo; such an endocrine change also has the effect of producing an altered uterine environment and a decreased implantation rate.[194]

7.6 Aging of the Uterus

Alterations of the uterus with aging in humans were described earlier in this chapter. In old C56BL/6J mice, the reduced number of offspring at term may be ascribed, at least in part, to aging-related changes in uterine structure and function. Goodrick and Nelson[196] consider uterine aging rather than aging of the oocyte to be the major cause of loss of fertility in mice, but other workers disagree with this conclusion.[197] In rats, the age-associated increase in pregnancy wastage has been ascribed to an age-related reduction in the viability of the ovulated eggs.[198,199] In both rats and some strains of mice the ovulation rate falls with aging. After 240 days of age, a steady decline in the number of ova was observed (11.4 ova at 240 days to 4.8 ova at 360 days) for IVCS mice.[200] After 12 months of age a decline in ovulation rate has been recorded for the rat (12.8 ova compared to 9.0). Conversely there was no age-related decline in the fertilization rate, but embryos from rats 10 months or older displayed a delayed pattern of development and increased morphological abnormalities. Progesterone levels are low in early (and then again in late) pregnancy, and the rise in estradiol that precedes parturition is delayed and reduced.[201] Most of the hormonal alterations occurring in middle and late pregnancy are associated with increased resorption of fetuses at midgestation. If blastocysts from old mothers are transplanted to the uteri of young, hormonally prepared females, the survival rate ranges from 48 to 50% in rabbits, hamsters, and mice. In the reverse procedure, blastocysts from young to old mothers, survival rate is reduced to 1.5 to 14% in the same species.[202-204] The old uterus in these species is clearly a hostile environment for the fertilized egg. Similarly reduced is the capacity of old hamsters, rats, and mice to respond to hormonal and mechanical stimuli to simulate pregnancy, although in a recent study no age-related reduction in the decidual cell response of aging C57BL/6J mice could be demonstrated.[196] This unfavorable environment offered by the uterus may be ascribed to characteristic structural changes,[205] decreased responsiveness to hormones,[206] and alteration in blood flow.[207,208]

8 REPRODUCTIVE DECLINE IN FEMALE RODENTS

The foregoing discussion underlines the usefulness of laboratory animals, especially rodents, to serve as models for understanding the aging of reproductive function. Keeping in mind the notable difference between animals and humans as well as among animals, the following section focuses on reproductive aging in the female rodent, perhaps the most extensively studied animal in this area.

Reproductive function in rodents follows the estrous cycle with respect to morphology and function of ovary and secondary sex organs, role of hormones in the regulation of estrous cyclicity, fertilization, and pregnancy. Rats, mice, hamsters, and guinea pigs are polyestrous species that, unlike the "reflex ovulators" such as the rabbit, repeat the cycles (lasting 4 to 5 days in the rat) throughout the year without much variation unless interrupted by pregnancy or pseudopregnancy (the latter, as the name indicates, with hormonal signs of pregnancy but without the occurrence of fertilization). In these species, eggs are normally released from the ovary during "heat" or estrus when the sexual interest of the female is aroused. The days of the cycle are numbered from the day(s) of estrus (in the rat: estrus, diestrus, proestrus, and again estrus). A number of interventions — hormonal, nutritional, psychological — can alter the rhythmicity and duration of reproductive function and interact with the effects of age in directing both onset and cessation of reproduction.

8.1 Aging and the Estrous Cycle in Female Rodents

Beginning at approximately 10 to 12 months of age, female rats show gradual changes in the estrous cycle. The cycles become irregular and lengthened, usually characterized by an increase in the number of days of diestrus or estrus. At this age, circulating estradiol levels are high on day 2 of diestrus and on proestrus (Figure 12-13), which is of interest in view of the work on preovulatory aging of oocytes already discussed.

In 8- to 10-month-old rats, the LH surge is delayed[209,210] and its magnitude significantly decreased.[211] In rats of 7 to 9 months of age, median eminence GnRH concentrations do not exhibit the rise prior to the onset of LH release observed in young rats.[212] The release of LH by GnRH appears to be controlled by a diurnal pattern of turnover of norepinephrine in specific hypothalamic nuclei. Such diurnal activity is seen only on the days of LH release and does not occur when LH is suppressed The density of norepinephrine α_1-adrenergic receptors exhibits a diurnal rhythm in the suprachiasmatic and the medial preoptic nuclei. These are nuclei which function either as an endogenous circadian oscillator or control the cyclic release of LH, respectively. The density of receptors to norepinephrine in these hypothalamic nuclei is sensitive to fluctuations in steroid hormones[213] and is modified by aging.[214] Wise et al.[211] consider that a decline in α_1-receptor concentrations in the medial preoptic nucleus and median eminence may contribute to the changes in LH secretion and loss of reproductive function observed in middle-aged female rats. These workers have identified a shift toward longer interpeak intervals of LH pulses in middle-aged animals exhibiting regular estrous cycles. This process is progressive, and a lesser mean pulse frequency is observed in middle-aged rats exhibiting estrous cycle irregularities. The inference from these studies is that changes induced in the

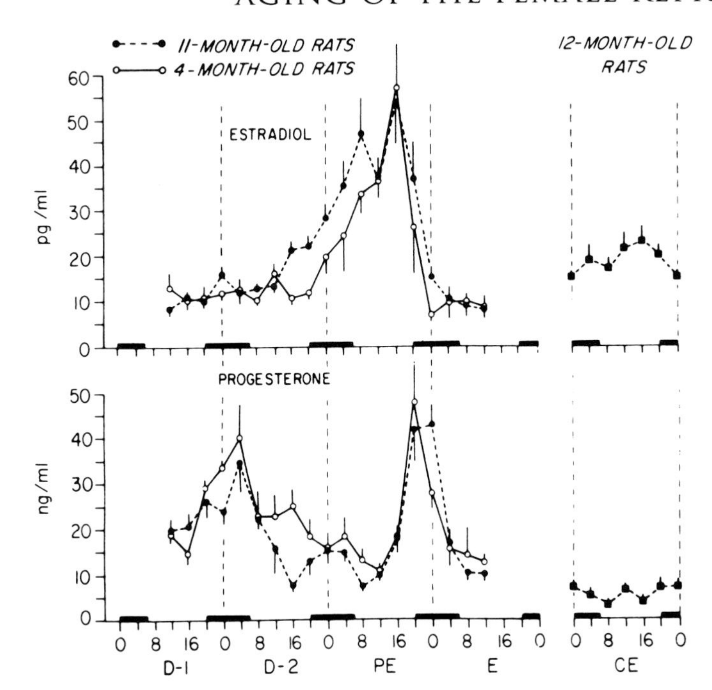

FIGURE 12-13 *Circulating concentrations of estradiol and progesterone in 4- and 11-month-old regularly cycling female rats and in 12-month-old constant-estrus (CE) females.* D-1, diestrus day one; D-2, diestrus day two; PE, proestrus; E, estrus. The solid bars on the abscissa indicate the dark phases of the dark-light photoperiod, and the numbers are hours of the day. From D-2 afternoon until PE noon, estradiol values appear to be higher in old than in young cyclic animals, and estradiol levels in CE are similar to those of old cyclic animals on D-2 evening. By contrast, progesterone values in CE rats are persistently low. (From Lu, J. K. H., in *Neuroendocrinology of Aging,* Meites, J., Ed., Plenum Press, New York, 1983, 103. With permission.)

hypothalamic centers controlling LH pulse frequency occur prior to identified irregularity in estrous cycles.[215] Isolated hypothalami from middle-aged rats perfused *in vitro* show a capacity to respond to depolarizing stimuli and release LHRH at a time when regular estrous cycles cease. This finding again suggests that aging moderates the neurosecretion of LHRH and that hypothalamic neurons retain the capacity to synthesize and release LHRH.[216]

Serum progesterone is decreased at proestrus in the middle-aged rat,[217] and FSH levels are elevated and sustained on the day of estrus, probably in association with increased estradiol.[218] The progressive cessation of regular estrous cycles may be preceded in middle-aged rats undergoing regular cycles by an enhanced FSH secretion, resulting in advanced follicular growth and increased ovarian estrogen production. Repeated exposure to high estrogen levels may eventually reduce the responsiveness of the hypothalamus to the feedback effect of ovarian steroids; LH surge is delayed and its magnitude is decreased; the timing of the follicular maturation and ovulation in the estrous cycle is altered and the regular cycles gradually become irregular.

This period of irregular cycling is followed by a constant estrus (or persistent vaginal cornification, (that is, continuous estrus with only an occasional diestrus) or persistent pseudopregnancy (characterized by long periods of diestrus with occasional estrus and ovulation) and finally an anestrus state (characterized by continuous diestrus). Not all rats follow this pattern; some may continue to cycle regularly until approximately 2 years old, and some may proceed directly from irregular cycles to persistent pseudopregnancy.[219,220]

While rats in constant estrus or persistent pseudopregnancy may revert to irregular cycling, reversion of the anestrus state to any other form has not been reported. Anestrus is seen in the oldest rats, represents the final form of reproductive decline, and could be a pathological state.[221] Its manifestation in mice has been likened to the postmenopausal period in humans.[6]

8.2 Steroid Feedback Regulation of the Hypothalamus

In ovariectomized old rats, injection of estradiol or estradiol–progesterone increases LH secretion only in rats that were formerly in persistent pseudopregnancy and not in those showing constant estrus. However, 5 weeks after ovariectomy, the positive feedback effect of estrogen and progesterone on LH secretion was present in constant-estrus females. Chronic estrogen implants abolished the ability of both groups to produce a surge in LH in response to administered ovarian steroids. The constant-estrus syndrome is associated with sustained, but moderate, amounts of estrogen in the presence of low progesterone, and, in persistent-pseudopregnancy animals, estrogen levels are low and progesterone values high. In constant estrus rats, the positive feedback mechanism of steroids on LH is present, but the high estrogen levels may act through a neurotoxic action on the hypothalamus to prevent pituitary secretion of LH. This action is perhaps alleviated by the high progesterone levels of persistent pseudopregnancy animals in which the central nervous system mechanisms that regulate the positive feedback of steroids on pituitary secretion of LH are still functioning.[222]

In young, cycling female rats, chronic elevation of circulating estrogen alters dopamine neurotransmission in the tuberinfundibular region associated with high prolactin secretion and causes neuronal degeneration accompanied by astrocytic and microglial hyperactivity (histologic indices of hypothalamic aging) in the hypothalamic arcuate nucleus.[223,224] This is accompanied by impairment of the LH surge. In Wistar rats and C57BL/6J mice, astrocytic and microglial activity increased significantly from 6 to 14 and 4 to 13 months, respectively. Ovariectomy at 2 months of age markedly retarded the development of gliosis in 14-month-old rats and 13-month-old mice.[225] These authors suggested that the changes seen in these old animals were similar to the arcuate lesion found in young estrogen-treated rats.[223] This was thought to represent a chemical deafferentation (i.e., separation) of the arcuate nucleus from the medial preoptic area. Arcuate neurons of the rat hypothalamus have a sexual dimorphic phenotype with females exhibiting an enriched population of intramembrane particles (IMPs) (<10 nm), whereas in neurons from the male, the predominant IMPs are greater than 10 nm. In senescent females, the number of small IMPs decreases in the perikarya and dendritic shafts compared to young females, while the distribution and number of large IMPs in the membrane of dendritic shafts increased and resembled the male phenotype.[226] Therefore, reproductive aging in the female rat is associated with significant changes in the plasma membrane of arcuate neurons.

In senescent female rodents, the age-related functional impairment of reproduction may be due to deterioration of the preoptic arcuate pathway although the disruption may be incomplete as it is possible to reinitiate ovarian cyclicity by stimulation of the hypothalamic–pituitary axis, as discussed below.

In rodents, sex steroids, like glucocorticoids (Chapter 11), may initiate a series of functionally damaging events whereby the hormones induce neuronal and endocrine cell loss in the hypothalamus and pituitary throughout the period of reproductive activity, culminating with cessation of reproduction. These losses, in turn, would lead to alterations of hypothalamopituitary responsiveness to hormonal feedbacks with consequent impairment or cessation of cyclicity and reproduction. Thus, while in humans menopause is essentially due to termination of ovarian function, in rodents failure of the hypothalamopituitary axis may be prevented or modified by a number of interventions that may delay or accelerate reproductive aging or may reactivate damaged reproductive function.

8.3 Reactivation of Regular Estrous Cycles

A variety of agents have been used to reactivate regular cycles in irregularly cycling rodents. They include progesterone, ACTH, vitamin E, ether or cold stress, L-dopa, ergotrile mesylate, iproniazid, and epinephrine. In constant-estrus rats, regular or irregular cycling is reinitiated by L-dopa, progesterone, ACTH, ether, stress, and iproniazid, which increase hypothalamic catecholamines.[227] Old persistent-pseudopregnancy rats respond to ergotrile mesylate (a dopamine agonist) and prostaglandin, which reduce progesterone secretion.[228] LH release, and ovulation in constant-estrus rats can be induced by electrical stimulation of the anterior hypothalamus or preoptic area or administration of synthetic LHRH or by injection of LH.[229]

8.4 Experimental Delay of the Age-Related Changes in Reproductive Function

When rats were ovariectomized at 6 to 12 months and given kidney capsule grafts of young ovaries at 2 years, reproductive cycles were reinstated.[230] This work has been repeated and extended in C57Bl/6j mice.[231,232] In these long-term ovariectomized mice with grafts, normal numbers of corpora lutea and growing follicles were produced, and apparently normal ova were shed. The capacity for cyclic LH release was retained 6 to 12 months beyond the usual age, and the magnitude of the responses was equal to that found in young controls. Histological changes in the arcuate nucleus thought to be caused by estrogen can be retarded by long-term ovariectomy and ovarian grafting.

A mechanism has been proposed whereby cumulative steroid impact is envisaged to have a detrimental effect on unspecified hypothalamic and pituitary loci during some or all of the estrous cycles, resulting in a loss of neuroendocrine sensitivity to estrogen, a smaller LH surge, and cycle lengthening. After cessation of estrous cyclicity, further exposure to estrogen in constant-estrous rodents causes additional damage to the neuroendocrine system. The lengthening of the estrous cycle can be prevented if animals are ovariectomized at a young age, and some of the additional neuroendocrine impairments in constant estrus animals can be reversed by prolonged ovariectomy. Indeed, aging of the neuroendocrine regulating system may be viewed as suspended, at least temporarily, during the postovariectomy period by lack of steroidal action. Successive treatments of young female rats with progesterone implants delay reproductive senescence. Progesterone implants induce low circulating estradiol concentrations. This effect of progesterone in delaying the onset of reproductive aging can be opposed by exogenous estradiol.[233]

The period of ovarian cyclicity has been extended in rats by addition of L-tyrosine to the diet from 7.5 months of age.[234] This probably resulted in the achievement of an appropriate catecholamine to serotonin balance within the central nervous system. Dietary restriction of rats from weaning also slows down the rate of reproductive aging and is associated with reduced circulating levels of estradiol (Chapter 24).

REFERENCES

1. Research on the menopause, in *WHO Technical Report Series,* No. 670, World Health Organization, Geneva, 1981.
2. Graham, C. E., Menstrual cycle of the great apes, in *Reproductive Biology of the Great Apes,* Gram, E. C., Ed., Academic Press, New York, 1981.
3. Hogden, G. D., Goodman, A. L., O'Connor, A., and Johnson, D. K., Menopause in rhesus monkeys: model for study of disorders in the human climacteric, *Am. J. Obstet. Gynecol.,* 127, 581, 1977.
4. Charlesworth, B., *Evolution in the Age-Structured Populations,* Cambridge University Press, Cambridge, 1980.
5. Mayer, P. J., Evolutionary advantage of the menopause, *Hum. Ecol.,* 10, 477, 1982.
6. Gosden, R. G., *Biology of the Menopause: The Causes and Consequences of Ovarian Aging,* Academic Press, London, 1985.
7. McKinley, S., Jefferys, M., and Thompson, B., An investigation of the age at menopause, *J. Biosocial Sci.,* 4, 161, 1972.
8. Amundsen, D. W. and Diers, C. J., The age of menopause in medieval Europe, *Hum. Biol.,* 45, 605, 1973.
9. Kono, S., Sunagawa, Y., Higa, H., and Sunagawa, H., Age of menopause in Japanese women: trends and recent changes, *Maturitas,* 12(1), 43, 1990.
10. Flint, M. and Samil, R. S., Cultural and subcultural meanings of the menopause, *Ann. N.Y. Acad. Sci.,* 592, 134, 1990.
11. Chavez, A. and Martinez, C., *Growing up in a Developing Community,* Institute of Nutrition of Central America and Panama, Mexico City, 1982.
12. Scragg, R. F. R., Menopause and reproductive span in rural Niugini, in Proceedings of the Annual Symposium of the Papua New Guinea Medical Society, Port Moresby, 1973.
13. Wasti, S., Robinson, S. C., Akhtar, Y., Khan, S., and Badaruddin, N., Characteristics of menopause in three socioeconomic urban groups in Karachi, Pakistan, *Maturitas,* 16(1), 61, 1993.
14. Van Keep, P. A., Brand, P. C., and Lehert, P., Factors affecting the age at menopause, *J. Biosocial Sci. Suppl.,* 6, 37, 1979.
15. Sherman, B. J., Wallace, R. B., and Treloar, A. E., The menopausal transition: endocrinological and epidemiological considerations, *J. Biosocial Sci. (Suppl.),* 6, 19, 1979.
16. Brand, P. C. and Lehert, P., A new way of looking at environmental variables that may affect the age at menopause, *Maturitas,* 1, 121, 1978.
17. Benjamin, F., The age of menarche and of the menopause in White South African Women and certain factors influencing these times, *S. Afr. Med. J.,* 34, 316, 1960.
18. Hauser, G. A., Remen, U., Valaer, M., Erb, H., Muller, T., and Oribi, J., Menarche and menopause in Israel, *Gynaecologia,* 155, 39, 1963.
19. Jaszmann, L., van Lith, N. D., and Zaat, J. C. A., The perimenopausal symptoms: the statistical analysis of a survey, *Med. Gynaecol. Sociol.,* 4, 268, 1969.

20. Masters, W. H. and Johnson, V. E., *Human Sexual Response,* Churchill, Livingstone, London, 1966.
21. Stanford, J. L., Hartge, P., Brinton, L. A., Hoover, R. N., and Brookmeyer, R., Factors influencing the age at natural menopause, *J. Chron. Dis.,* 40, 995, 1987.
22. Thomford, P. J., Jelovsek, F. R., and Mattison, D. R., Effect of oocyte number and rate of atresia on the age of menopause, *Reprod. Toxicol.,* 1, 41, 1987.
23. Lindquist, O. and Bengtsson, C., The effect of smoking on menopausal age, *Maturitas,* 1, 191, 1979.
24. Kaufman, D. W., Slone, D., Rosenberg, L., Miettinen, O. S., and Shapiro, S., Cigarette smoking and age at natural menopause, *Am. J. Public Health,* 70, 420, 1980.
25. Mattison, D. R. and Thorgeirsson, S. S., Smoking and industrial pollution, and their effects on menopause and ovarian cancer, *Lancet,* 1, 187, 1978.
26. Michnovicz, J. J., Herschcopf, R. J., Naganuma, H. L., and Fishman, J., Increased 2-hydroxylation of estradiol as a possible mechanism for the anti-estrogenic effect of cigarette smoking, *N. Engl. J. Med.,* 315, 1305, 1986.
27. Baron, J. A., La Vecchia, C., and Levi, F., The antiestrogenic effect of cigarette smoking in women, *Am. J. Obstet. Gynecol.,* 162, 502, 1990.
28. Leidy, L. E., Early age at menopause among left-handed women, *Obstet. Gynecol.,* 76(6), 1111, 1990.
29. Richardson, S. J. and Nelson, J. F., Follicular depletion during the menopausal transition, *Ann. N.Y. Acad. Sci.,* 592(13), 13, 1990.
30. Doring, G. K., The incidence of anovular cycles in women, *J. Reprod. Fertil. (Suppl.),* 6, 77, 1969.
31. Sherman, B. M., West, J. H., and Korenman, S. C., The menopausal transition: analysis of LH, FSH, estradiol and progesterone concentrations during menstrual cycles of older women, *J. Clin. Endocrinol. Metab.,* 42, 629, 1976.
32. Steger, R. W. and Peluso, J. J., Sex hormones in the aging female, *Endocrinol. Metab. Clin. North Am.,* 16(4), 1027, 1987.
33. Lee, S. J., Lenton, E. A., Sexton, L., and Cooke, I. D., The effect of age on the cyclical patterns of plasma LH, FSH, oestradiol and progesterone in women with regular menstrual cycles, *Hum. Reprod.,* 3(7), 851, 1988.
34. Al-Azzawai, F., Endocrinological aspects of the menopause, *Br. Med. Bull.,* 48 (2), 262, 1992.
35. Judd, H. L., Shamonki, I. M., Frumar, A. M., and Lagasse, L. D., Origin of serum estradiol in postmenopausal women, *Obstet. Gynecol. (N.Y.),* 59, 680, 1982.
36. Longcope, C., Hormone dynamics at the menopause, in *Multidisciplinary Perspectives on Menopause,* Flint, M., Kronenberg, F., and Utian, W., Eds., The New York Academy of Sciences, New York, 1990, 21.
37. Siiteri, P. K. and MacDonald, P. C., Role of extraglandular estrogen in human endocrinology, in *Handbook of Physiology: Endocrinology,* Greep, R. O. and Astwood, E., Eds., American Physiology Society, Washington, D.C., 1973, 615.
38. Simpson, E. R., Merrill, J. C., Hollub, A. J., Graham-Lorence, S., and Mendelson, C. R., Regulation of estrogen biosynthesis by human adipose cells, *Endocr. Rev.,* 10, 136, 1989.
39. Roberts, K. D., Rochefort, J. G., Blean, G., and Chapdelaine, A., Plasma estrone sulfate levels in postmenopausal women, *Steroids,* 35, 179, 1980.
40. Chakravarti, S., Collins, W. P., Forecast, J. S., Newton, J. R., Oram, D. H., and Studd, J. W. W., Hormonal profiles after the menopause, *Br. Med. J.,* 2, 784, 1976.
41. Wide, L., Nillius, S. J., Gensell, C., and Roos, P., Radio immunosorbent assay of follicle-stimulating hormone and luteinizing hormone in serum and urine from men and women, *Acta Endocrinol. (Copenhagen) Suppl.,* 174, 1, 1973.
42. Musey, V. C., Collins, D. C., Musey, P. I., Martino-Saltzman, D., and Preedy, J. R., Age-related changes in the female hormonal environment during reproductive life, *Am. J. Obstet. Gynecol.,* 157(2), 312, 1987.
43. Scaglio, H. M., Medina, M., Pinto-Ferreira, A. L., Vazques, C. G., and Perez-Palacios, G., Pituitary LH and FSH secretion and responsiveness in women of old age, *Acta Endocrinol. (Copenhagen),* 81, 673, 1976.
44. Mason, M., Fonseca, E., Ruiz, J. E., Moran, C., and Zarate, A., Distribution of follicle-stimulating hormone and luteinizing hormone isoforms in sera from women with primary ovarian failure compared with that of normal reproductive and postmenopausal women, *Fertil. Steril.,* 58(1), 60, 1992.
45. Yen, S. S. C., Tsai, C. C., Naftolin, F., Vandenberg, G., and Ajabor, L., Pulsatile patterns of gonadotropin release in subjects with and without ovarian function, *J. Clin. Endocrinol. Metab.,* 34, 671, 1972.
46. Seifer, D. B. and Collins, R. L., Current concepts of beta-endorphin physiology in female reproductive dysfunction, *Fertil. Steril.,* 54(5), 757, 1990.
47. Lenton, E. A., de-Kretser, D. M., Woodward, A. J., and Robertson, D. M., Inhibin concentrations throughout the menstrual cycles of normal, infertile, and older women compared with those during spontaneous conception cycles, *J. Clin. Endocrinol. Metab.,* 73(6), 1180, 1991.
48. MacNaughton, J., Banah, M., McCloud, P., Hee, J., and Burger, H., Age related changes in follicle stimulating hormone, luteinizing hormone, oestradiol and immunoreactive inhibin in women of reproductive age, *Clin. Endocrinol. (Oxford),* 36(4), 339, 1992.
49. Thompson, B., Hart, S. A., and Durno, D., Menopausal age and symptomatology in general practice, *J. Biosocial Sci.,* 5, 71, 1973.
50. Kaufert, P., Lock, M., McKinlay, S., Beyenne, Y., Coope, J., Davis, D., Eliasson, M., Gognalons-Nicolet, M., Goodman, M., and Holte, A., Menopause research: the Korpilamip workshop, *Soc. Sci. Med.,* 22, 1285, 1986.
51. Kronenberg, F., Hot flashes: epidemiology and physiology, *Ann. N.Y. Acad. Sci.,* 592, 52, 1990.
52. Erlik, Y., Meldrum, D. R., and Judd, H. L., Estrogen levels in postmenopausal women with hot flashes, *Obstet. Gynecol.,* 59, 403, 1982.
53. Tataryn, I. V., Lomax, P., Bajarek, J. G., Chesarek, W., Meldrum, D. R., and Judd, H. L., Postmenopausal hot flushes: a disorder of thermoregulation, *Maturitas,* 2, 101, 1980.
54. Ginsburg, J., Sinhoe, J., and O'Reilly, B., Cardiovascular responses during the menopausal hot flush, *Br. J. Obstet. Gynaecol.,* 88, 925, 1981.
55. Sturdee, D. W., Wilson, K. A., Pipili, E., and Crocker, A. D., Physiological aspects of menopausal hot flush, *Br. Med. J.,* 2, 79, 1978.
56. Witt, M. F. and Blethen, S. L., The endocrine evaluation of three children with vasomotor flushes following hypothalamic surgery, *Clin. Endocrinol. (Oxford),* 18, 551, 1983.
57. Ginsburg, J. and O'Reilly, B., Climacteric flushing in a man, *Br. Med. J.,* 287, 262, 1983.
58. Judd, H. L., Pathophysiology of menopausal hot flushes, in *Neuroendocrinology of Aging,* Meites, J., Ed., Plenum Press, New York, 1983, 173.

59. Meldrum, D. R., Erlik, Y., Lu, J. K. H., and Judd, H. L., Objectively recorded hot flushes in patients with pituitary insufficiency, *J. Clin. Endocrinol. Metab.*, 52, 684, 1981.
60. Meldrum, D. R., Tataryn, I. V., Frumar, A. M., Erlik, Y., Lu, J. K. H., and Judd, H. L., Gonadotropins, estrogens and adrenal steroids during the menopausal hot flush, *J. Clin. Endocrinol. Metab.*, 50, 685, 1980.
61. DeFazio, J., Meldrum, D. R., Laufer, L., Vale, W., Rivier, J., Lu, J. K. H., and Judd, H. L., Induction of hot flashes in premenopausal women treated with long-acting GnRH agonist, *J. Clin. Endocrinol. Metab.*, 56, 445, 1983.
62. Casper, R. F. and Yen, S. S. C., Menopausal flushes: effect of pituitary gonadotropin desensitization by a potent luteinizing hormone-releasing factor agonist, *J. Clin. Endocrinol. Metab.*, 53, 1056, 1981.
63. Judd, H. L. and Korenman, S. G., Effects of aging on reproductive function in women, in *Endocrine Aspects of Aging*, Korenman, S. G., Ed., Elsevier, New York, 1982, 163.
64. Rahimy, M. H., Bodor, N., and Simpkins, J. W., Effects of a brain-enhanced estrogen delivery system on tail-skin temperature of the rat: implications for menopausal hot flush, *Maturitas*, 13(1), 51, 1991.
65. Simkins, J. W. and Katovich, M. J., Hypoglycemia causes hot flashes in animal models, *Ann. N.Y. Acad. Sci.*, 592, 436, 1990.
66. Simpkins, J. W. and Katovich, M. J., Relationship between blood glucose and hot flushes in women and an animal model, in *Thermoregulation: Research and Clinical Applications*, Lomax, P. and Schonbaum, E., Eds., S. Karger, Basel, 1989, 95.
67. Ball, P., Knuppen, R., Haupt, M., and Brueur, H., Interactions between estrogens and catechol amines. III. Studies on the methylation of catechol estrogens, catechol amines and other catechols by the catechol-o-methyl-transferase of human liver, *J. Clin. Endocrinol. Metab.*, 34, 736, 1972.
68. Lloyd, T. and Weisz, J., Direct inhibition of tyrosine hydroxylase activity by catechol estrogens, *J. Biol. Chem.*, 253, 4841, 1978.
69. Schaeffer, J. M. and Hsueh, A. J. W., 2-Hydroxyestradiol interaction with dopamine receptor binding in rat anterior pituitary, *J. Biol. Chem.*, 254, 5606, 1979.
70. Silva, N. L. and Boulant, J. A., Effects of testosterone, estradiol, and temperature on neurons in preoptic tissue slices, *Am. J. Physiol: Regulatory Integrative Comp. Physiol.*, 250, R625, 1986.
71. Rance, N. E., Hormonal influences on morphology and neuropeptide gene expression in the infundibular nucleus of postmenopausal women, *Prog. Brain Res.*, 93, 221, 1992.
72. Genazzani, A. R., Petraglia, F., Facchinetti, F., Facchini, A., Volpe, A., and Alessandrini, G., Increase of proopiomelanocortin-related peptides during subjective menopausal flushes, *Am. J. Obstet. Gynecol.*, 149, 775, 1984.
73. Meldrum, D. R., DeFazio, J. D., Erlik, Y., Lu, J. K. H., Wolfsen, A. F., Carlson, H. E., Hershman, J. M., and Judd, H. L., Pituitary hormones during menopausal hot flash, *Obstet. Gynecol.*, 64, 752, 1984.
74. Winokur, G., Depression in the menopause, *Am. J. Psychiatry*, 130, 1, 1973.
75. Federation, M. W., An investigation of the menopause in 1000 women, *Lancet*, 1, 106, 1933.
76. Kinsey, A. C., Pomeroy, W. B., Martin, C. E., and Gebhard, P. H., *Sexual Behavior in the Human Female*, W.B. Saunders, Philadelphia, 1953.
77. James, W. H., Marital coital rates, spouses' ages, family size and social class, *J. Sex. Res.*, 10, 205, 1974.
78. Hallstrom, T., Sexuality of women in middle age: the Goteborn study, *J. Biosocial Sci. Suppl.*, 6, 165, 1979.
79. Utian, W. H., The true clinical features of postmenopause and oophorectomy, and their response to estrogen therapy, *S. Afr. Med. J.*, 46, 732, 1972.
80. Voet, R. L., End organ response to estrogen deprivation, in *The Menopause*, Buchsbaum, H. J., Ed., Springer-Verlag, New York, 1983.
81. Lang, W. R. and Aponte, G. E., Gross and microscopic anatomy of the aged female reproductive organs, *Clin. Obstet. Gynecol.*, 10, 454, 1967.
82. Semmens, J. F. and Wagner, G., Estrogen deprivation and vaginal function in postmenopausal women, *JAMA*, 248, 445, 1982.
83. Tsai, C. C., Semmens, J. P., Semmens, E. C., Lam, C. F., and Lee, F. S., Vaginal physiology in postmenopausal women: pH value, transvaginal electropotential difference and estimated blood flow, *South. Med. J.*, 80, 987, 1987.
84. Ross, C. A. C., Post-menopausal vaginitis, *J. Med. Microbiol.*, 11, 209, 1978.
85. Leiblum, S., Bachmann, G., Kemmann, E., Colburn, D., and Swartzman, L., Vaginal atrophy in the postmenopausal woman, *JAMA*, 249, 2195, 1983.
86. Singer, A., The uterine cervix from adolescence to the menopause, *Br. J. Obstet. Gynaecol.*, 82, 81, 1975.
87. Tweeddale, D. N., Cytopathology of cervical squamous carcinoma *in situ* in postmenopausal women, *Acta Cytol.*, 14, 363, 1970.
88. Woessner, J. F., Age-related changes of the human uterus and its connective tissue framework, *J. Gerontol.*, 18, 220, 1963.
89. Hendrickson, M. R. and Kempson, R. L., Endometrial epithelial metaplasia: Proliferations frequently misdiagnosed as adenocarcinoma, *Am. J. Surg. Pathol.*, 4, 525, 1980.
90. McBride, J. M., The normal postmenopausal endometrium, *J. Obstet. Gynaecol. Brit. Empire*, 61, 691, 1954.
91. Gaddum-Rosse, P., Rumery, R. E., Blandau, R. J., and Thiersch, J. B., Studies on the mucosa of postmenopausal oviducts: Surface appearance, ciliary activity, and the effect of estrogen treatment, *Fertil. Steril.*, 26, 951, 1975.
92. Hafez, E. S. E., Scanning electron microscopy of female reproductive organs during menopause and related pathologies, in *The Menopause: Clinical Endocrinological and Pathophysiological Aspects*, Fiorette, P., Martini, L., Melis, G. B., and Yen, S. S. C., Eds., Academic Press, New York, 1982, 201.
93. Chang, R. J. and Judd, H. L., The ovary after menopause, *Clin. Obstet. Gynaecol.*, 24, 181, 1981.
94. Talbert, G. B., Effect of aging of the ovaries and female gametes on reproductive capacity, in *The Aging Reproductive System*, Schneider, E. L., Ed., Raven Press, New York, 1978, 59.
95. Rakoff, A. E. and Nowrooz, K., The female climacteric, in *Geriatric Endocrinology*, Greenblatt, R. B., Ed., Raven Press, New York, 1978, 165.
96. Mossman, H. W. and Duke, K. L., *Comparative Morphology of the Mammalian Ovary*, University of Wisconsin Press, Madison, 1973.
97. Thung, P. J., Aging changes in the ovary, in *Structural Aspects of Aging*, Bourne, G. H., Ed., Pitman, London, 1961, 110.
98. Kuppe, G., Metzger, H., and Ludwig, H., Aging and structural changes in the female reproductive tract, in *Aging and Reproductive Physiology*, Hafez, E. S. E., Ed., Ann Arbor Science, MI, 1976, 21.
99. Marks, R. and Shahrad, P., Aging and the effects of oestrogens on the skin, in *The Menopause: A Guide to Current Research and Practice*, Beard, R. J., Ed., MTP Press, Lancaster, 1976, 143.

100. Edman, C. D., The climacteric, in *The Menopause,* Buchsbaum, H. J., Ed., Springer-Verlag, New York, 1983.
101. Smith, P., Age changes in the female urethra, *Br. J. Urol.,* 44, 667, 1972.
102. Krauss, R. M., Regulation of high density lipoprotein levels, *Med. Clin. N. Am.,* 66, 403, 1982.
103. Bengtsson, C., Ischaemic heart disease in women, *Acta Med. Scand.,* 49 (Suppl. 549), 1, 1973.
104. Gordon, T., Kannel, W. B., Hjortland, M. C., and McNamara, P. M., Menopause and coronary heart disease: the Framingham Study, *Ann. Intern. Med.,* 89, 157, 1978.
105. Bush, T. L., The epidemiology of cardiovascular disease in postmenopausal women, *Ann. N.Y. Acad. Sci.,* 592, 263, 1990.
106. U.S. Dept. Health Services, Public Health Service, *The Lipid Research Clinics' Populations Data Book,* Vol. 1. *The Prevalence Study,* National Institutes of Health, Bethesda, MD, 1980.
107. Miller, N. E., Why does plasma LDL concentration in adults increase with age?, *Lancet,* 1, 263, 1984.
108. Veldhuis, J. D., Gwynne, J. T., Azimi, P., Garmey, D., and Juchter, D., Estrogen regulates LDL metabolism by cultured swine granulosa cells, *Endocrinology,* 117, 1321, 1985.
109. Windler, E., Kovanen, P. T., Chao, Y. S., Brown, M. S., Havel, R. J., and Goldstein, J. L., The estradiol-stimulated lipoprotein receptor of rat liver, *J. Biol. Chem.,* 255, 10464, 1980.
110. Bush, T. L., Barrett-Connor, E., Cowan, L. D., Criqui, M. H., Wallace, R. B., Suchindran, C. M., Tyroler, H. A., and Rifkind, B. M., Cardiovascular mortality and noncontraceptive use of estrogen in women: results from the Lipid Research Clinics Program Follow-up Study, *Circulation,* 75, 1102, 1987.
111. Barrett-Conor, E., Wingard, D. L., and Criqui, M. H., Postmenopausal estrogen use and heart disease risk factors in the 1980s, *J. Am. Med. Assoc.,* 261, 2095, 1989.
112. Krauss, R. M., Perlman, J. A., Ray, R., and Petitti, D., Effects of estrogen dose and smoking on lipid and lipoprotein levels in postmenopausal women, *Am. J. Obstet. Gynecol.,* 158, 1606, 1988.
113. Cauley, J. A., LaPorte, R. E., Kuller, L. H., Bates, M., and Sandler, R. B., Menopausal estrogen use, HDL cholesterol subfractions and liver function, *Atherosclerosis,* 49, 31, 1983.
114. Sacks, F. M. and Walsh, B. W., The effects of reproductive hormones on serum lipoproteins: unresolved issues in biology and clinical practice, *Ann. N.Y. Acad. Sci.,* 592, 272, 1990.
115. Stanczyk, F. Z., Shoupe, D., Nunez, V., Macias-Gonzalez, P., Vijod, M. A., and Lobo, R. A., A randomized comparison of nonoral estradiol delivery in post-menopausal women, *Am. J. Obstet. Gynecol.,* 159, 1540, 1988.
116. Jensen, J., Riss, B. J., Strom, V., Nilas, L., and Christiansen, C., Long-term effects of percutaneous estrogens and oral progesterone on serum lipoproteins in postmenopausal women, *Am. J. Obstet. Gynecol.,* 156, 66, 1987.
117. Wren, B. G., The effect of oestrogen on the female cardiovascular system, *Med. J. Aust.,* 156(3), 204, 1992.
118. Brunner, D., Weisbort, J., Meshulam, N., Schwartz, S., Gross, J., Salt-Rennert, H., Altman, S., and Loebl, K., Relationship of serum total cholesterol and high-density lipoprotein cholesterol percentage to the incidence of definite coronary events: twenty-year follow-up of the Donolo-Tel Aviv Prospective Coronary Artery Disease Study, *Am. J. Cardiol.,* 59, 1271, 1987.
119. Heller, R. J. and Jacobs, H. S., Coronary heart disease in relation to age, sex and menopause, *Br. Med. J.,* 1, 472, 1978.
120. Kirkland, R. T., Keenan, B. S., Probstfield, J. L., Patsch, W., Lin, T.-L., Clayton, G. W., and Insull, W., Decrease in plasma HDL cholesterol levels at puberty in boys with delayed adolescence: correlation with plasma testosterone levels, *J. Am. Med. Assoc.,* 257, 502, 1987.
121. Riggs, B. L. and Melton, L. J. I., Evidence for two distinct syndromes of involutional osteoporosis, *Am. J. Med.,* 75, 899, 1983.
122. Heaney, R. P., Recker, R. R., and Saville, P. D., Calcium balance and calcium requirements in middle-aged women, *Am. J. Clin. Nutr.,* 30, 1603, 1978.
123. Heaney, R. P., Recker, R. R., and Saville, P. D., Menopausal changes in bone remodeling, *J. Lab. Clin. Med.,* 92, 964, 1978.
124. Pak, C. Y. C., Post menopausal osteoporosis, in *The Menopause,* Buchsbaum, H. J., Ed., Springer-Verlag, New York, 1983.
125. Komm, B. S., Terpening, C. M., Benz, D. J., Graeme, K. A., Gallegos, A., Korc, M., Greene, G. L., O'Malley, B., and Haussler, M. R., Estrogen binding, receptor mRNA, and biologic response in osteoblast-like osteosarcoma cells, *Science,* 241, 81, 1988.
126. Eriksen, E. G., Colvard, D. S., Berg, N. J., Graham, M. L., Mann, K. G., Spelsberg, T. C., and Riggs, B. L., Evidence of estrogen receptors in normal human osteoblast-like cells, *Science,* 241, 84, 1988.
127. Linsay, R. and Cosman, F., Estrogen in prevention and treatment of osteoporosis, *Ann. N.Y. Acad. Sci.,* 592, 326, 1990.
128. Ernst, M., Schmid, C., Frankenfoldt, C., and Froesch, E. R., Estradiol stimulation of osteoblast proliferation in vitro: mediator roles for TGFβ, PGE_2, IGF1, *Calcif. Tissue Res. (Suppl. 1),* 42, 117, 1988.
129. Pacifici, R., Rifas, L., McCracken, R., and Avioli, L. V., The role of interleukin-1 in postmenopausal bone loss, *Exp. Gerontol.,* 25(3–4), 309, 1990.
130. Thomson, J. and Oswald, I., Effect of oestrogen on the sleep, mood and anxiety of menopausal women, *Br. Med. J.,* 2, 1317, 1977.
131. Lindsay, R., Sex steroids in the pathogenesis and prevention of osteoporosis, in *Osteoporosis: Etiology, Diagnosis, and Management,* Riggs, B. L. and Melton, L. J., Eds., Raven Press, New York, 1988, 353.
132. Riggs, B. L., Seeman, E., and Hodgson, S. F., Effect of the fluoride/calcium regimen on vertebral fracture occurrence in postmenopausal osteoporosis, *New Engl. J. Med.,* 306, 446, 1982.
133. Edman, C. D., Estrogen replacement therapy, in *The menopause,* Buchsbaum, H. J., Ed., Springer-Verlag, New York, 1983, 77.
134. Meade, T. W. and Berra, A., Hormone replacement therapy and cardiovascular disease, *Brit. Med. Bull.,* 48(2), 276, 1992.
135. Utian, W. H., *Menopause in Modern Perspective: A Guide to Clinical Practice,* Appleton-Century-Crofts, New York, 1980.
136. Henderson, B. E., The cancer question: an overview of recent epidemiologic and retrospective data, *Am. J. Obstet. Gynecol.,* 161, 1859, 1989.
137. Rubin, G. L., Peterson, H. B., Lee, N. C., Maes, E. F., Wingo, P. A., and Becker, S., Estrogen replacement therapy and the risk of endometrial cancer: remaining controversies, *Am. J. Obstet. Gynecol.,* 162, 148, 1990.
138. Collins, J., Allen, L. H., Donner, A., and Adams, O., Oestrogen use and survival in endometrial cancer, *Lancet,* 2, 961, 1980.
139. Schwartzbaum, J. A., Hulka, B. S., Fowler, W. C., Kaufman, D. G., and Hoberman, D., The influence of exogenous estrogen use on survival after diagnosis of endometrial cancer, *Am. J. Epidemiol.,* 126(5), 851, 1987.

140. Ross, P. K., Paganini-Hill, A., and Gerkins, V. R., A case-control study of menopausal estrogen therapy and breast cancer, *JAMA*, 243, 1635, 1980.
141. Thomas, D. B. and Chu, J., Nutritional and endocrine factors in reproductive organ cancers: opportunities for primary prevention, *J. Chron. Dis.*, 39(12), 1031, 1986.
142. Dupont, W. D. and Page, D. L., Menopausal estrogen replacement therapy and breast cancer, *Arch. Intern. Med.*, 151, 67, 1991.
143. Colditz, G. A., Stampfer, M. J., Willett, W. C., Hennekens, C. H., Rosner, B., and Speizer, F. E., Prospective study of estrogen replacement therapy and risk of breast cancer in postmenopausal women, *J. Am. Med. Assoc.*, 264(20), 2648, 1991.
144. Barrett-Connor, E., Hormone replacement and cancer, *Br. Med. Bull.*, 48(2), 345, 1992.
145. Studd, J. W. W. and Thom, M. H., Ovarian failure and aging, in *Endocrinology and Aging*, Green, M., Ed., W.B. Saunders, Eastbourne, 1981, 89.
146. Boston Collaborative Drug Surveillance Program, Surgically confirmed gallbladder disease, venous thromboembolism, and breast tumours in relation to postmenopausal estrogen therapy, *New Engl. J. Med.*, 290, 15, 1974.
147. Ellerington, M. C., Whitecroft, S. I. J., and Whitehead, M. I., HRT, Developments in therapy, *Br. Med. Bull.*, 48(2), 401, 1992.
148. Marsh, M. S. and Whitehead, M. I., Management of the menopause, *Br. Med. Bull.*, 48(2), 426, 1992.
149. Eaton, J. W. and Mayer, A. J., The social biology of very high fertility among the Hutterites. The demography of a unique population, *Hum. Biol.*, 25, 206, 1953.
150. Laing, L. M., Declining fertility in a religious isolate: the Hutterite population of Alberta, Canada 1950–1971, *Hum. Biol.*, 52, 288, 1980.
151. Henry, L., Some data on natural fertility, *Eugen. Q.*, 8, 81, 1961.
152. Fergusson, I. L. C., Taylor, R. W., and Watson, J. M., *Records and Curiosities in Obstetrics and Gynaecology*, Bailliere, London, 1982.
153. Schwardtz, D. and Mayaux, M. J., Female fecundity as a function of age. Results of artificial insemination in 2193 nulliparous women with azoospermic husbands, Federation CECOS, *New Engl. J. Med.*, 306, 404, 1982.
154. Kennedy, T. G. and Kennedy, J. P., Effects of age and parity on reproduction in young female mice, *Biol. Reprod.*, 16, 286, 1972.
155. Asdell, S. A., Bogart, R., and Sperling, G., The influence of age and rate of breeding upon the ability of the female rat to produce and raise young, *Mem. Cornell Univ. Agric. Exp. Stn.*, 238, 3, 1941.
156. Roman, L. and Strong, L. C., Age, gestation mortality and litter size in mice, *J. Gerontol.*, 17, 37, 1962.
157. Jones, E. C. and Krohn, P. L., The relationships between age, numbers of oocytes and fertility in virgin and multiparous mice, *J. Endocrinol.*, 21, 469, 1961.
158. Talbert, G. B., Aging of the reproductive system, in *Handbook of the Biology of Aging*, Finch, C. E. and Hayflick, L., Eds., Van Nostrand Reinhold, New York, 1977, 318.
159. Miller, A. E., Wood, S. M., and Riegle, G. D., The effect of age on reproduction in repeatedly mated female rats, *J. Gerontol.*, 34, 15, 1979.
160. Ingram, D. L., Mandl, A. M., and Zuckerman, S., The influence of age on litter size, *J. Endocrinol.*, 17, 280, 1958.
161. Holinka, C. F. and Finch, C. E., Efficiency of mating in C57BL/6J female mice as a function of age and previous parity, *Exp. Gerontol.*, 16, 393, 1981.
162. Finn, C. A., Reproductive capacity and litter size in mice: effect of age and environment, *J. Reprod. Fertil.*, 6, 205, 1963.
163. Holehan, A. M., *The Effect of Ageing and Dietary Restriction upon Reproduction in the Female CFY Sprague-Dawley Rat*, University of Hull, Hull, U.K., 1984.
164. Holehan, A. M. and Merry, B. J., Modification of the oestrous cycle hormonal profile by dietary restriction, *Mech. Ageing Dev.*, 32, 63, 1985.
165. Anderson, A. C., Reproductive ability of female beagles in relation to advancing age, *Exp. Gerontol.*, 1, 189, 1965.
166. Dierschke, D. J., Koening, J., Krueger, G., and Robinson, J. A., Reproductive and hormonal patterns in perimenopausal rhesus monkeys, *Program Endocr. Soc.* Abstr. No. 678, 1983.
167. Van Wagenen, G., Vital statistics from a breeding colony: reproductive and pregnancy outcome in *Macaca mulatta*, *J. Med. Primatol.*, 1, 3, 1972.
168. Graham, C. E., Reproductive function in aged female chimpanzees, *Am. J. Phys. Anthropol.*, 50, 291, 1979.
169. King, M. D., The relation of age to fertility in the rat, *Anat. Rec.*, 11, 286, 1916.
170. Franchi, L. L., Mandl, A. M., and Zuckerman, S., The development of the ovary and the process of oogenesis, in *The Ovary*, Zuckerman, S., Ed., Academic Press, New York, 1962, 1.
171. Block, E., Quantitative morphological investigations of the follicular system in women. Variations at different ages, *Acta Anat.*, 14, 108, 1952.
172. Block, E., A quantitative morphological investigation of the follicular system in newborn female infants, *Acta Anat.*, 17, 201, 1953.
173. Novak, E. R., Ovulation after fifty, *Obstet. Gynecol.*, 36, 903, 1970.
174. Costoff, A. and Mahesh, V. B., Primordial follicles with normal oocytes in the ovaries of postmenopausal women, *J. Am. Geriatr. Soc.*, 23, 193, 1975.
175. Richardson, S. J., Senikas, V., and Nelson, J. F., Follicular depletion during the menopausal transition: evidence for accelerated loss and ultimate exhaustion, *J. Clin. Endocrinol. Metab.*, 65(6), 1231, 1987.
176. Gougeon, A., Caractère qualitative et quantatif de la population foliculaire dans l'ovaire humain adulte, *Contracep. Fertil. Sexual.*, 12, 527, 1984.
177. Ryle, M., The growth in vitro of mouse ovarian follicles of different sizes in response to purified gonadotropins, *J. Reprod. Fertil.*, 30, 395, 1972.
178. Mandl, A. M. and Shelton, M., A quantitative study of oocytes in young and old nulliparous laboratory rats, *J. Endocrinol.*, 18, 444, 1959.
179. Nelson, J. F. and Felicio, L. S., Radical ovarian resection advances the onset of persistent vaginal cornification but only transiently disrupts hypothalamic-pituitary regulation of cyclicity in C57BL/6J mice, *Biol. Reprod.*, 35(4), 957, 1986.
180. Biggers, J. S., Finn, C. A., and McLaren, A., Long term reproductive performance of female mice. I. Effect of removing one ovary, *J. Reprod. Fertil.*, 3, 303, 1962.
181. Jones, E. C. and Krohn, P. L., Effect of unilateral ovariectomy on reproductive lifespan of mice, *J. Endocrinol.*, 20, 129, 1960.
182. Schneider, E. L., Aging and reproductive performance: models for the study of age-related chromosomal and point mutations. Maternal age and aneuploidy, in *Biological Mechanisms in Aging, Conference Proceedings*, Schimke, R. T., Ed., NIH Publication No. 81-2194, U.S. Dept. Health and Human Services, Washington, D.C., 1981.

183. Roberts, J. A. F., *An Introduction to Medical Genetics*, Oxford University Press, London, 1967.
184. Kline, J. and Levin, B., Trisomy and age at menopause: predicted associations given a link with rate of oocyte atresia, *Paediatr. Perinat. Epidemiol.*, 6(2), 225, 1992.
185. Henderson, S. A. and Edwards, R. G., Chiasma frequency and maternal age in mammals, *Nature (London)*, 218, 22, 1968.
186. Speed, R. M. and Chandley, A. C., Meiosis in the foetal mouse ovary. II. Oocyte development and age-related aneuploidy. Does a production line exist?, *Chromosoma*, 88, 184, 1983.
187. Kalter, H., Elimination of fetal mice with sporadic malformations by spontaneous resorption in pregnancies of older females, *J. Reprod. Fertil.*, 53, 407, 1978.
188. Parsons, P. A., Congenital abnormalities and competition in man and other mammals at different maternal ages, *Nature (London)*, 198, 316, 1963.
189. Guerrero, R. and Rojas, O. I., Spontaneous abortion and aging of human ova and spermatozoa, *New Engl. J. Med.*, 293, 573, 1975.
190. Hertig, A. T., Rock, J., and Adams, E. C., A description of 34 human ova within the first 17 days of development, *Am. J. Anat.*, 989, 435, 1956.
191. Mikamo, K., Anatomic and chromosomal anomalies in spontaneous abortion, *Am. J. Obstet. Gynecol.*, 106, 243, 1970.
192. McGaughey, R. W., The culture of pig oocytes in minimal medium, and the influence of progesterone and estradiol-17β on meiotic maturation, *Endocrinology*, 100, 38, 1977.
193. Read, S. G., The distribution of Down's syndrome, *J. Ment. Defic. Res.*, 26, 215, 1982.
194. Butcher, R. L. and Page, R. D., Role of the aging ovary in cessation of reproduction, in *Dynamics of Ovarian Function*, Schwartz, N. R. and Hunzicker-Dunn, M., Eds., Raven Press, New York, 1981, 253.
195. Butcher, R. L. and Pope, R. S., Role of estrogen during prolonged oestrous cycles of the rat on subsequent embryonic death or development, *Biol. Reprod.*, 21, 491, 1979.
196. Goodrick, G. J. and Nelson, J. F., The decidual cell response in aging C57BL/6J mice is potentiated by long-term ovariectomy and chronic food restriction, *J. Gerontol.*, 44(3), B67, 1989.
197. Xu, Z. and Clark, J., Characterization of uterine cytosol and nuclear sex steroid receptors in aging female ICR mice, *Proc. Chin. Acad. Med. Sci. Peking Union Med. Coll.*, 5(4), 207, 1990.
198. LaPolt, P. S., Day, J. R., and Lu, J., Effects of estradiol and progesterone on early embryonic development in aging rats, *Biol. Reprod.*, 43(5), 843, 1990.
199. Mattheij, J. A. M. and Swarts, J. J. M., Quantification and classification of pregnancy wastage in 5-day cyclic young through middle-aged rats, *Lab. Anim.*, 25, 30, 1991.
200. Tappa, B., Amao, H., Ogasa, A., and Takahashi, K. W., Changes in the estrous cycle and number of ovulated and fertilized ova in aging female IVCS mice, *Jikken Dobutsu*, 38(2), 115, 1989.
201. Holinka, C. F., Tseng, Y.-C., and Finch, C. E., Impaired preparturitional rise of plasma estradiol in aging C57BL/6J mice, *Biol. Reprod.*, 21, 1009, 1979.
202. Adams, C. E., Aging and reproduction in the female mammals with particular reference to the rabbit, *J. Reprod. Fertil. (Suppl.)*, 12, 1, 1970.
203. Blaha, G. C., Effect of age of the donor and recipient on the development of transferred golden hamster ova, *Anat. Rec.*, 150, 413, 1964.
204. Talbert, G. B. and Krohn, P. L., Effect of maternal age on viability of ova and uterine support of pregnancy in mice, *J. Reprod. Fertil.*, 11, 399, 1966.
205. Finn, C. A. and Martin, L., The cellular response of the uterus of the aged mouse to oestrogen and progesterone, *J. Reprod. Fert.*, 20, 545, 1969.
206. Holinka, C. F. and Finch, C. E., Age-related changes in the decidual response of the C57BL/6J mouse uterus, *Biol. Reprod.*, 16, 385, 1977.
207. Sorger, T. and Soderwall, A., The aging uterus and the role of edema in endometrial function, *Biol. Reprod.*, 24, 1134, 1981.
208. Finch, C. E. and Holinka, C. F., Aging and uterine growth during implantation in C57BL/6J mice, *Exp. Gerontol.*, 17, 235, 1982.
209. Cooper, R. L., Conn, M., and Walker, R. F., Characterization of the LH-surge in middle-aged female rats, *Biol. Reprod.*, 23, 611, 1980.
210. Van der Schoot, P., Changing pro-oestrous surges of luteinizing hormone in aging 5-day cyclic rats, *J. Endocrinol.*, 69, 287, 1976.
211. Wise, P. M., Weiland, N. G., Scarbrough, K., Larson, G. H., and Lloyd, J. M., Contribution of changing rhythmicity of hypothalamic neurotransmitter function to female reproductive aging, *Ann. N.Y. Acad. Sci.*, 592, 31, 1990.
212. Wise, P. M., Alterations in the proestrous pattern of median eminence LHRH, serum LH, FSH, estradiol and progesterone in middle-aged rats, *Life Sci.*, 31, 165, 1982.
213. Weiland, N. G. and Wise, P. M., Estrogen alters the diurnal rhythm of alpha-1 adrenergic receptor densities in selected brain areas, *Endocrinology*, 121, 1751, 1987.
214. Weiland, N. G. and Wise, P. M., Age-associated alterations in catecholaminergic concentrations, neuronal activity, and alpha-1 receptor densities in female rats, *Neurobiol. Aging*, 10, 323, 1989.
215. Scarbrough, K. and Wise, P. M., Age-related changes in pulsatile luteinizing hormone release precede the transition to estrous acyclicity and depend upon estrous cycle history, *Endocrinology*, 126(2), 884, 1990.
216. Rubin, B. S., Isolated hypothalami from aging female rats do not exhibit reduced basal or potassium-stimulated secretion of luteinizing hormone-releasing hormone, *Biol. Reprod.*, 47(2), 254, 1992.
217. Miller, A. E. and Riegle, G. D., Endocrine factors associated with the initiation of constant estrus in aging female rats, *Fed. Proc.*, 38, 1248, 1979.
218. Lu, J. K. H., Changes in ovarian function and gonadotropin and prolactin secretion in aging female rats, in *Neuroendocrinology of Aging*, Meites, J., Ed., Plenum Press, New York, 1983, 103.
219. Clemens, J. A. and Meites, J., Neuroendocrine status of old constant estrous rats, *Neuroendocrinology*, 7, 249, 1971.
220. Huang, H. H. and Meites, J., Reproductive capacity of aging female rats, *Neuroendocrinology*, 17, 289, 1975.
221. Aschheim, P., Relation of neuroendocrine system to reproductive decline in female rats, in *Neuroendocrinology of Aging*, Meites, J., Ed., Plenum Press, New York, 1983, 73.
222. Lu, J. K. H., Damassa, D. A., Gilman, D. P., Judd, H. L., and Sawyer, C. H., Relationship between circulating estrogens and the central mechanism by which ovarian steroids stimulate luteinizing hormone secretion in aged and young female rats, *Endocrinology*, 108, 836, 1981.
223. Brawer, J. R., Schipper, H., and Naftolin, F., Ovary dependent degeneration in the hypothalamic arcuate nucleus, *Endocrinology*, 107, 274, 1980.
224. Casanueva, F., Cocchi, D., Locatelli, V., Flavo, C., Zambotti, F., Bestetti, G., Rossei, G. L., and Muller, E., Defective central nervous system dopaminergic function in rats with estrogen-induced pituitary tumours, as assessed by plasma prolactin concentrations, *Endocrinology*, 110, 590, 1982.

225. Schipper, H., Brawer, J. R., Nelson, J. F., Felicio, L. S., and Finch, C. E., Role of the gonads in the histologic aging of the hypothalamic arcuate nucleus, *Biol. Reprod.,* 25, 413, 1981.
226. Garcia-Segura, L. M., Perez, J., Jones, E., and Naftolin, F., Loss of sexual dimorphism in rat arcuate nucleus neuronal membranes with reproductive aging, *Exp. Neurol.,* 112(1), 125, 1991.
227. Huang, H. H., Marshall, S., and Meites, J., Induction of estrous cycles in old noncyclic rats by progesterone, ACTH, ether stress or L-dopa, *Neuroendocrinology,* 20, 21, 1976.
228. Clemens, J. A. and Bennett, D. R., Do aging changes in the preoptic area contribute to loss of cyclic endocrine function?, *J. Gerontol.,* 32, 19, 1977.
229. Meites, J., Huang, H. H., and Simpkins, J. W., Recent studies on neuroendocrine control of reproductive senescence, in *The Aging Reproductive System,* Schneider, E. L., Ed., Raven Press, New York, 1978, 213.
230. Aschheim, P., La reactivation de l'ovaire des rattes seniles en oestrus permanent au moyen de'hormones gonadotropes ou de la mise a l'obscurite, *C.R. Acad. Sci.,* 260, 5627, 1965.
231. Finch, C. E., Felicio, L. S., Flurkey, K., Gee, D. M., Mobbs, C., Nelson, J. F., and Osterburg, H. H., Studies on ovarian-hypothalamic-pituitary interactions during reproductive aging in C57BL/6J mice, *Peptides,* 1 (Suppl. 1), 163, 1980.
232. Finch, C. E., Felicio, L. S., Mobbs, C. V., and Nelson, J. F., Ovarian and steroidal influences on neuroendocrine aging processes in female rodents, *Endocr. Rev.,* 5(4), 467, 1984.
233. Lapolt, P. S., Yu, S. M., and Lu, J. K., Early treatment of young female rats with progesterone delays the aging-associated reproductive decline: a counteraction by estradiol, *Biol. Reprod.,* 38(5), 987, 1988.
234. Cooper, R. L. and Linnoila, M., Effects of centrally and systematically administered L-tyrosine and L-leucine on ovarian function in the old rat, *Gerontology,* 26, 270, 1980.

III
SYSTEMIC AND
ORGANISMIC
AGING

13 AGING OF THE MALE REPRODUCTIVE SYSTEM

Brian J. Merry and Anne M. Holehan

The aging of the testis does not culminate in the cessation of function as in the ovary, but is rather a long, gradual process. It has been argued that testicular function continues indefinitely, or, alternatively, that an "andropause" begins at about age 60. The andropause would be associated with some clinical and emotional symptoms and with minimal endocrine alterations, involving a moderate decline in testosterone and a minimal increase in gonadotropins. Thus, the andropause, if it exists, differs endocrinologically from the menopause, but shares with the menopause some of the changes involving nonreproductive organs and functions.

1 THE REPRODUCTIVE SYSTEM

The reproductive system in human males includes the primary organ, the two testes, which produce the germ cells, or sperm, as well as the principal male hormones and a number of secondary sex organs — epididymi, vas deferens — which serve for transport of the sperm, and others — seminal vesicles, prostate — for the production of the seminal fluid.

As for the ovary, the hormones involved in testicular function are operative at four different levels but, unlike the ovary, these are generally little affected by aging:

- In the hypothalamus, GnRH (or luteinizing hormone–releasing hormone, LHRH), equivalent to the same polypeptide in females, is secreted and carried through the portal system to the anterior pituitary where it stimulates the synthesis and release of the two gonadotropins, FSH and LH, also similar to those in females.
- In the anterior pituitary, the synthesis and release of the two gonadotropins FSH and LH (also called ICSH, interstitial cell–stimulating hormone) are regulated by hypothalamic LHRH and, through a negative feedback, by circulating levels of testosterone, for LH, and inhibin for FSH.
- In the testis, the major hormones are steroids: testosterone, secreted by the Leydig cells (the major type of interstitial cell) and extremely small amounts of estradiol, probably secreted by both the Leydig and Sertoli cells. Another hormone, inhibin, is a protein secreted by the Sertoli cells; its secretion is stimulated by FSH and its levels inhibit FSH release from the pituitary and may also decrease LHRH.
- At the periphery, testosterone, as with the ovarian steroids, is transported in the plasma partly (99%) bound to the steroid-binding protein (steroid hormone–binding globulin, SHBG) and partly (1%) transported in the free, biologically active form; it is metabolized in the liver and excreted in the urine.

At the target cells, testosterone has the same mechanism of action as the other steroid hormones such as corticosteroids and ovarian hormones, already described. It freely passes the plasma membrane, forms a cytoplasmic activated steroid–receptor complex, which is carried to the nucleus where it stimulates protein synthesis to induce the characteristic masculinizing effects and growth-promoting actions of the hormone.

In the prostate and several other tissues that are primary targets for testosterone, testosterone is converted to dehydrotestosterone (DHT) and it is DHT rather than testosterone that is physiologically active. A small amount of testosterone is also converted (by aromatization) to estrogens.

It should be recalled that androgens are also secreted by the adrenal cortex and, in fact, these androgens represent the major portion of the adrenal sex hormones. In general, their androgen activity is 20% less than that of testosterone. Changes with aging have been discussed in Chapter 11.

2 AGE-RELATED CHANGES IN THE TESTIS

2.1 Histology, Sperm Count, and Fertility

Testis weight does not change significantly with age in men over 40. Data from several strains of rat have opposite results; some studies indicate no change in testis weight although an upward trend with age was seen, while others show a decline in absolute testis weight and volume. In C57BL/6J mice, testis weight did not change between 12 and 28 months, unless disease was present.[1-3] Spermatozoa from C57BL/6NNia mice from 7- and 25-month-old males were assessed in their ability to fertilize oocytes *in vitro*. The spermatozoa from 7-month-old males fertilized the largest number of oocytes (80 to 86%) *in vitro,* and 79% of these developed into blastocytes in culture. In contrast, spermatozoa from aged males which had failed to mate, fertilized only 11 to 19% of the oocytes with 48% developing to blastocytes in culture. These old mice which failed to mate had the lowest number of spermatozoa in the cauda epididymis, with fewer mobile spermatozoa than the younger males. Superovulated mice, which were artificially inseminated with spermatozoa from 25-month-old mice that had not mated over a 1-month period with a proven fertile female, did not become pregnant.[4]

Spermatogenesis in men continues until old age; sperm is found in the ejaculate of 48% of men between the ages of 80 and 90, although the percentage of seminiferous tubules containing sperm drops from 90% at 20 to 30 years to 10% in men over 80. The number of spermatocytes is reduced in older rats, but the spermatogenic cycle is unchanged at 22 months. Epididymi from rats aged 5 to 24 months have been subjected to density gradient centrifugation to separate gametes

0-8493-8979-8/94/$0.00+$.50

according to different stages of maturity.[5] The lifetime study revealed a pattern of decreasing reproductive competence. There was an age-associated decline in absolute numbers of sperm, circulating levels of testosterone, and a shift in the sperm profile toward more mature gametes at around middle age. Testosterone supplementation (400 mg/kg/bw per day for 30 days) did not restore absolute sperm number but returned the sperm profile to that more characteristic of a young adult male.

Atrophy of the seminiferous tubules which occurs with aging in rodents and humans begins at discrete foci with atrophic tubules observed adjacent to normal tubules.[6,7] In rodents, the atrophy of the seminiferous epithelium is thought to be primarily from a loss of spermatogenic cells for the tubules in aged rats exhibit high number of Sertoli cells,[8] whereas in the human a loss of Sertoli cells and spermatids per Sertoli cell has been reported.[9] Detailed histological examination of the seminiferous epithelium with age in the Brown Norway rat has shown that no significant death of Sertoli cells occurs.[10] As an indication of Sertoli cell function, total testis content was determined for the transcripts for SGP-2, transferrin, and cyclic protein-2/cathepsin L. No change in total testis content of SGP-2 mRNA occurred between 6 and 24 months, whereas transferrin mRNA increased between 18 and 24 months to over 3 times the tissue content of a 6-month-old testis. Conversely, the tissue content of CP-2/cathepsin L decreased between 12 and 24 months to 58% of the content found in the 6-month-old testis.[10] Because these authors found no significant cell loss in Sertoli cells, this change in transcription profile and gene expression must indicate a change in Sertoli cell function with age.

In the aging human testis, the tunica propria and basement membrane of the seminiferous tubules thicken with a progressive intertubular fibrosis and a thinning of the spermatogenic epithelium, as well as a decrease in capillary numbers. In 22-month-old Wistar rats, limited thickening of the basement membrane and tubular fibrosis occurs.

Alterations in number and volume of Leydig cells (the major producers of testosterone) vary with age and animal species. In humans, the number of these cells declines or is unchanged in old men, but it increases in Sprague–Dawley rats and stallions.[11] In the rat, the total Leydig cell volume per testis and volume of interstitial fluid are constant with age, but in the stallion, total volume of Leydig cells increases with age and is seasonally dependent. Total Leydig cell volume and number increase by approximately 40% in the midbreeding season.

Barring dysfunction in the reproductive system, fertility in males is retained into old age. The oldest reported case of successful paternity was in a 94-year-old man,[12] and in male rats as old as 24 months, 50% were still capable of siring litters, although the capacity for repeated matings was reduced.[13]

2.2 Testicular Steroids

In the Baltimore Longitudinal Study, using healthy, well-educated volunteers, no decrease was noted in serum testosterone and serum estrogens between the ages of 25 and 90 years (Figure 13-1). The free testosterone index (a measure of biologically active serum testosterone) shows no alteration with age, although the binding of testosterone to sex hormone

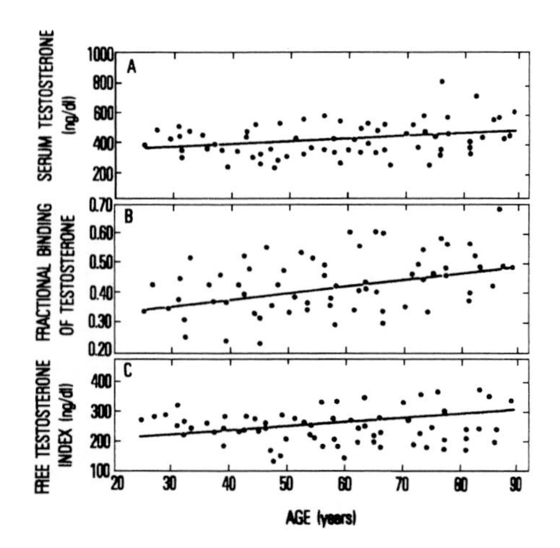

FIGURE 13-1 (A) *Total serum testosterone concentrations in healthy male participants in the Baltimore Longitudinal Study (n = 69, ages = 25 to 80).* (B) *Fractional binding of testosterone to sex-hormone binding globulin in these men.* (C) *Free testosterone indices (percent binding × total testosterone concentration).* (From Harman, S. M. and Tsitouras, P. D., *J. Clin. Endocrinol. Metab.*, 51, 35–40, 1980. © The Endocrine Society. With permission.)

binding globulin (SHBG) is increased. Levels of 5α-dihydrotestosterone (DHT) remain constant.[14] These findings contrast with other studies in which plasma testosterone declined steadily from around age 50.[15] SHBG capacity increased with age, thus producing a decrease in free plasma testosterone.[16] Basal serum testosterone and inhibin levels were similar in young and old men of proven fertility, but after stimulation with human chorionic gonadotropin (hCG), there was a decreased response of serum testosterone and an insignificant increase in inhibin in the older men. These changes to hCG stimulation indicate a decreased secretory capacity of Leydig as well as of Sertoli cells in older men.[17] Plasma levels of unconjugated estradiol and estrone are increased in older men because of increased peripheral androgen conversion. DHT levels decrease in testicular vein and peripheral vein blood with age and increase in serum in men with benign prostatic hyperplasia.

Discrepancies between reports of testosterone production with age are attributed to the inclusion of obese, alcoholic, and/or chronically ill subjects without allowing for medications taken by the subjects. Obesity is associated with low serum testosterone and reduced SHBG binding capacity. Hepatic metabolism of testosterone is enhanced by barbiturates, benzodiazepines, and alcohol, and hypothyroidism is known to alter A-ring reduction of testosterone, producing more inactive 5β- rather than active 5α-DHT. In old men, the target tissue response to androgens is reduced. Benign prostatic hyperplasia is perhaps due to a disorder of DHT metabolism resulting in accumulation of DHT in the prostate, and possible overstimulation of the gland[18] (Chapter 19).

Comparative Androgen Levels with Aging

Unless the animals were diseased, serum testosterone levels did not decline in old C57BL/6J mice. In the long-lived White-Footed mouse *Peromyscus leucopus,* serum testosterone levels declined steadily after 20 months, although differences between young and old were not statistically significant.[19] The aging male golden hamster showed a comparable endocrine profile, in that serum testosterone levels had no age-related decrement, although there was a greater variance in values at 31 months than at 4 to

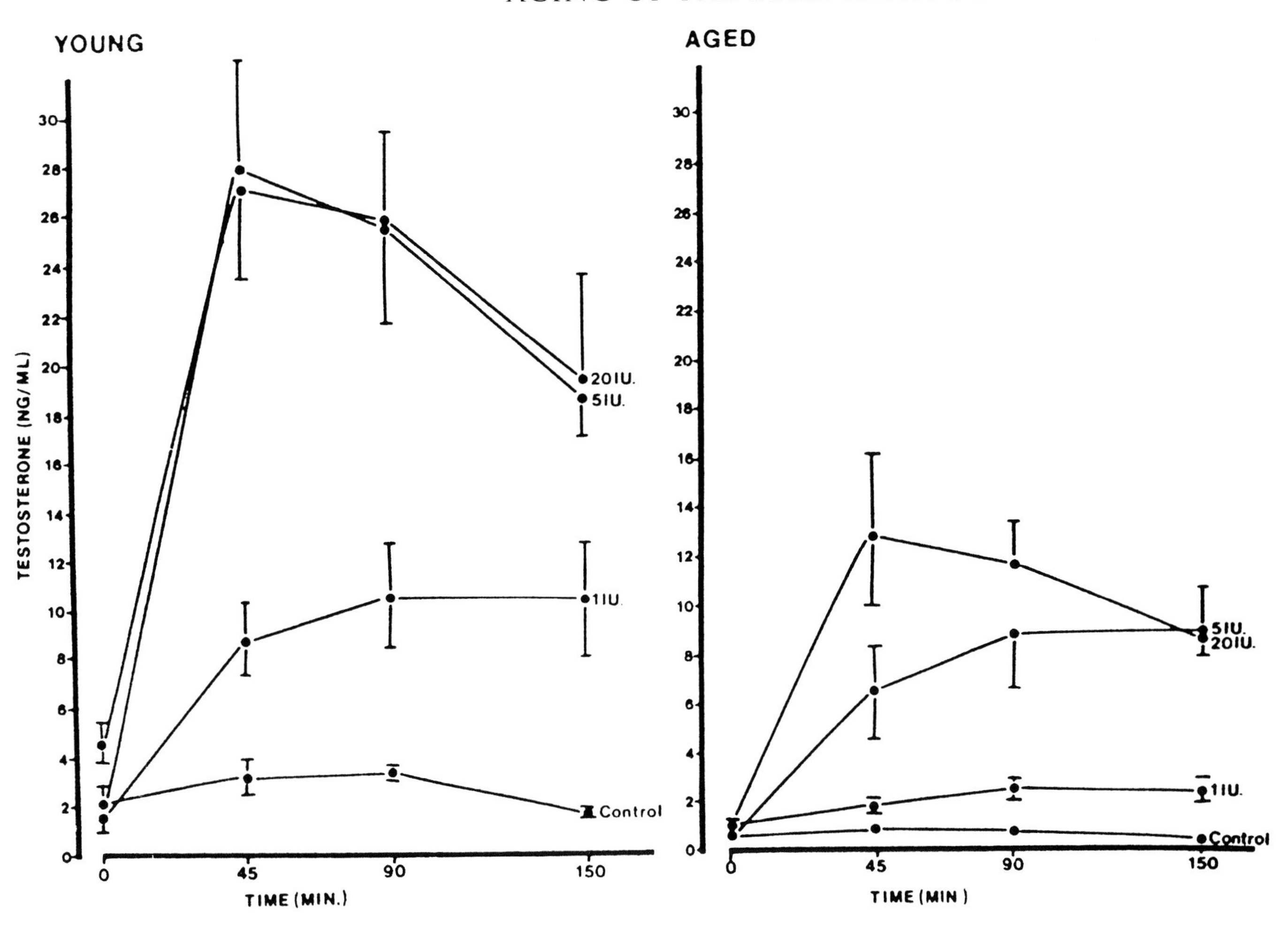

FIGURE 13-2 *Effect of intravenous injections of 1, 5, and 20 IU hCG on serum testosterone in young (4-month) and aged (22- to 30-month) male Long–Evans rats.* Testosterone levels are shown as the group mean and SEM from blood samples taken under light ether anesthesia before and at 45, 90, and 150 min after hCG injection. (From Riegle, G. D. and Miller, A. E., in *The Aging Reproductive System*, Schneider, E.L., Ed., Raven Press, New York, 1978. With permission.)

24 months.[20] In aging male rats, however, testosterone levels declined with age and the diurnal variations found in plasma testosterone levels at 3 to 4 months were not present at 19 to 20 months.[2,13,21] A slight decline in androstenedione and a significant decline in dehydroepiandrosterone with age were noted. 5α-DHT was unchanged, as were levels of estrone and estradiol, unless Leydig cell tumors were present and estradiol values rose. Increased serum pregnenolone and progesterone were present in the old male rat.[22] Reexamination of *in vivo* sex steroid levels in aging Fischer 344 rats has confirmed a significant increase in circulating progesterone and to a lesser extent estradiol levels while testosterone and gonadotropins were decreased.[23] Elevated circulating progesterone levels were inversely correlated with those of gonadotropins and testosterone. That the progesterone was of testicular origin was demonstrated by orchidectomy following which a significant decrease was observed. All the 24-month-old rats exhibited Leydig cell hyperplasia or tumors, which is frequently associated with increased synthesis of estrogens. The age-associated increase in progesterone synthesis and to a lesser extent estradiol synthesis may act to suppress gonadotropin release and testosterone function in aging male Fischer 344 rats.

Leydig cell steroidogenic activity in Brown Norway rats, assessed by determining the capacity of the testes to produce testosterone when perfused *in vitro* with a maximally stimulating dose of LH, diminishes with age.[24] Unlike other strains of rat (Wistar, Long–Evans, and Fischer), which show a decrease in plasma gonadotropin concentrations with age, the serum levels of follicle-stimulating hormone rose significantly in the Brown Norway rat, consistent with the defect in testosterone synthesis intrinsic to the Leydig cell, a situation similar to that of the human.

3 RESPONSE OF THE TESTIS TO GONADOTROPIN STIMULATION

The ability of the Leydig cells to produce testosterone in response to gonadotropin (LH) declines in aging men. Stimulation of these cells by hCG (a substitute for LH, easily obtainable from the urine of pregnant women), at a dose of 1500 IU for 3 days, elevates testosterone levels significantly less in elderly males than in young men.[25] Although the percentage response is unchanged with age, the levels of testosterone, both before and after hCG stimulation, are lower in elderly men. This indicates that the number of responsive Leydig cells in older men is reduced. In recent work, a reduction in levels of testosterone and percentage response to hCG[15] followed a characteristic time course with age, showing a reduced response to testosterone 15 h after hCG stimulation in both middle-aged (50 to 69 years) and elderly (70 to 89 years) men; 24 h later, only in the elderly group, testosterone values remained reduced when compared to young men (25 to 49 years).[14] Thus, there may be a reduction in the reserve capacity of the Leydig cells with age and/or reduction in Leydig cell number.

3.1 Leydig Cell Changes with Aging

In old rats, Leydig cell number increases, possibly in an attempt to compensate for an age-related deficit in Leydig cell function, but also in response to a decreased testicular blood flow. Binding of LH, and uptake of hCG by old and young testes, did not differ, but the old testis was less responsive to a single injection of gonadotropin. The production of testosterone in response to 1, 5, and 20 IU hCG is lower in the old testis when compared to young, although the percentage increase above baseline may be equal at both ages[26] (Figure 13-2). If multiple injections of hCG are given for a 3- or 7-day period, the response of young, middle-aged, and old rats does not differ significantly although the initial response decreased with age.[27] Daily injections of hCG in old rats returned to normal metabolic enzyme levels and testicular blood flow but could not restore the circulating levels of testosterone. Leydig cells from old rat testes incubated *in vitro* show a reduced response to hCG stimulation, but the *in vitro* secretion of testosterone was unchanged because of the higher Leydig cell number in old rats.[2] It seems likely the gonadotropin binding per Leydig cell declines with age despite the absence of alteration in number or affinity of binding sites for hCG, due to the increased number of Leydig cells in the old rat.

3.2 Gonadotropins and Steroid Feedback Inhibition

Plasma gonadotropins show variability with age in men. Increased LH but normal FSH levels have been reported or increased FSH but normal LH, or increases in both LH and FSH, with the rise in FSH usually greater than that of LH. Levels of FSH and LH were equivalent to those seen after castration or in postmenopausal women. Increased gonadotropin secretion is associated with unchanged plasma levels of testosterone.[28] Thus, a decrease in sex steroid feedback inhibition of the hypothalamic-pituitary axis could result from a decreased Leydig cell function and lead to increased gonadotropin secretion. FSH secretion may be independently regulated by inhibin. As inhibin levels decrease with age, FSH rises, thereby accounting for the greater rise seen in this hormone in some studies.[15,29]

Release by the pituitary of LH and FSH in response to LHRH in elderly men is lower than would be expected in view of their elevated basal gonadotropin levels (the higher the basal gonadotropin level the greater is the magnitude of response of LHRH).[28] A consistent reduction in LH response after blocking hypothalamic estrogen receptors with clomiphene citrate may be the consequence of the freeing of gonadotropin secretion from steroid feedback inhibition in elderly men. As LHRH was not measured in these experiments, it was impossible to separate alterations at the hypothalamic and pituitary level.[15] Some impairment of the pituitary gonadotropin secretory response in aging men follows and the hypothalamic response may be reduced as well. In a study of the basal and LHRH-stimulated serum concentrations of the common alpha-subunit of the glycoproteins (LH and FSH) in 69 healthy men aged 25 to 89 years, a significant age-related delay in the timing of the peak alpha-subunit response was observed.[30] This observation was interpreted as evidence of an age-related alteration in healthy men in the secretion and/or metabolic clearance of pituitary gonadotropins.

Molecular Changes with Aging — LH and FSH

The declining levels of plasma testosterone in old male rats are accompanied by a reduction in plasma levels of FSH and LH and an increase in prolactin.[31] The reduction by 25% in the pituitary content of LH in old male rats and the increase in the molecular size of LH (assessed by gel filtration) in the circulation and pituitaries of 20- to 24-month-old male rats[32] are possibly due to an alteration in sialic acid content of the LH molecule. The LH in old male rats can be restored to the molecular size of LH in young animals by 12 days of testosterone proprionate injections. This indicates that the decreased testosterone levels in old male rats may affect the LH molecule itself. Following castration, old (23 to 30 months) male rats show a smaller rise in LH and FSH than young animals (3 to 6 months). The high LH levels in old castrated males are significantly reduced by testosterone proprionate injections. This indicates an increase in sensitivity to testosterone negative feedback.[33] The relative response of LH and FSH to a single injection of GnRH was equal in old (21-month) and young (4-month) male Wistar rats (Figure 13-3).[34] Multiple injections of GnRH produced significantly smaller increases in LH and FSH in old than in young male rats suggesting that the function of the pituitary changes with age (Figure 13-4).

3.3 Hypothalamic Function

The lower hypothalamic content of GnRH in 26-month-old compared to 4-month-old male rats suggests a lower release of GnRH, hence a reduced LH and FSH secretion. The number of medial preoptic area (MPOA) neurons expressing the GnRH gene, as determined by mRNA levels, and hypothalamic GnRH peptide content was decreased in old compared to young male Fischer 344 rats.[35] However, the number of neurons expressing GnRH in the MPOA and the circulating levels of gonadotropins and testosterone were similar in orchidectomized old and young rats. This suggests that the age-related decrease in GnRH synthetic capacity and the subsequent lowered gonadotropin levels are the result of testicular feedback.[35]

Immunocytochemical identification of luteinizing hormone–releasing hormone in neurons from male, Fischer 344 rat brain, preoptic areas revealed no age-dependent difference either in total numbers or sizes of LHRH neurons nor in their distribution in the brain. Examination of such neurons at the level of the electron microscope revealed significant changes in the synaptic organization with the most significant change being an increased input to LHRH perikarya. This synaptic input to the LHRH neuron perikarya membrane increased threefold in middle-aged rats (12 to 14 months), and tenfold in old rats (20 to 23 months), compared to the young group (2 to 4 months). Density of synaptic input to LHRH neurons at the dendritic membrane did not change with age.[36] This finding has been confirmed in Sprague–Dawley rats which are not so susceptible to testicular tumors as the Fischer 344 strain.[37]

Reduced content and turnover rate of the neurotransmitters dopamine and norepinephrine, with increased turnover of serotonin in the hypothalamus of old males, may be responsible for the decreased LH and FSH and increased prolactin secretion observed. This neurotransmitter imbalance may also be responsible for the smaller postcastration rise of LH seen in the old rat.

Neuropeptide Y (NPY) is involved in the control of hypothalamic luteinizing hormone-releasing hormone (LHRH) and it stimulates the release and potentiates the action of LHRH on LH secretion. Aged male rats exhibit markedly reduced concentrations of NPY in the median eminence

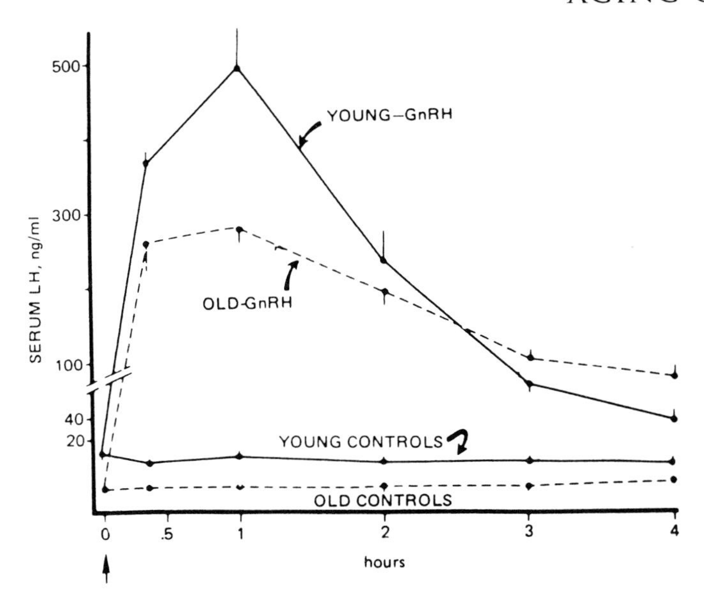

FIGURE 13-3 *Effects of a single injection (arrow) of GnRH on serum LH in young (4 month) and old (21 month) male Wistar rats.* Vertical lines indicate SEM. (From Bruni, J. F., Huang, H. H., Marshall, S., and Meites, J., *Biol. Reprod.*, 17, 309–312, 1977. With permission.)

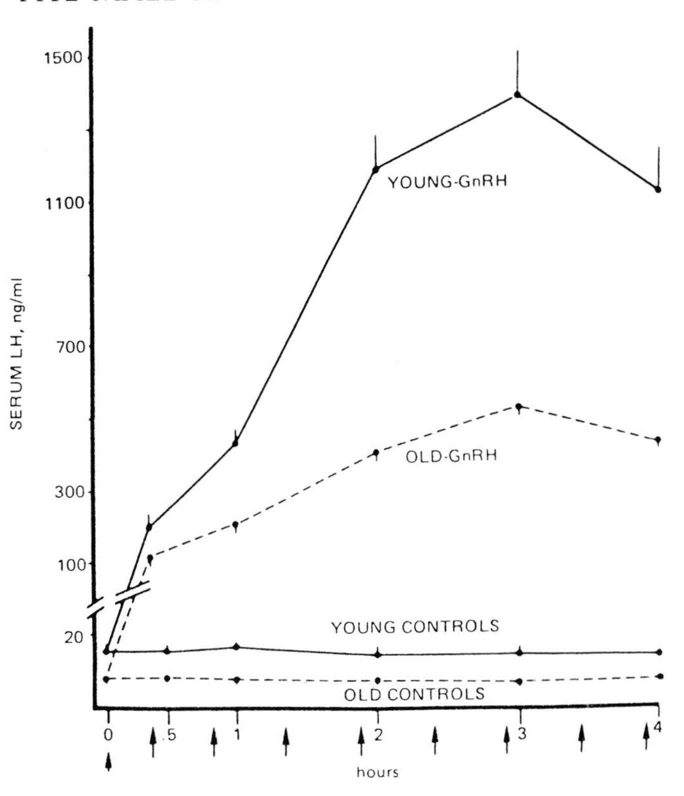

FIGURE 13-4 *Effects of multiple injections (arrows) of GnRH on serum LH in young (4-month) and old (21-month) male Wistar rats.* Vertical lines indicate SEM. (From Bruni, J. F., Huang, H. H., Marshall, S., and Meites, J., *Biol. Reprod.*, 17, 309–312, 1977. With permission.)

and in the arcuate, medial preoptic, suprachiasmatic, paraventricular, dorsomedial, and ventromedial hypothalamic nuclei. The age-related decline in NPY content in these brain regions is associated with a decline in serum testosterone and LH levels. The medial basal hypothalamic tissues of 13-month-old rats release significantly less NPY in response to K^+ depolarization than tissue from 2.5-month-old animals.[38,39] In contrast, these authors found that the K^+ evoked LHRH release from the same tissues was unimpaired. Consequently, it can be demonstrated that the hypothalamus and pituitary of aging male rats show functional defects.[31,40]

3.4 Accessory Sex Organs

The epididymis grows slowly throughout adult life and shows little evidence of cell renewal. Pigment granules (lipofuscin) are laid down in the epithelial cells of the epididymis, seminal vesicles, and prostate with age. In the human seminal vesicle, fluid volume falls from 5 ml at 21 years to 2.25 ml over the age of 60. Lipofuscin is deposited in the columnar but not basal cells of the epithelium, and epithelial cell height decreases. The walls of the seminal vesicles decrease in thickness and the mucosal folds are lost from ages 30 to 40 onward. The weight of the seminal vesicles declines with age in the rat with a gradual decrease in citric acid levels (thought to be an indicator of hormonal stimulation) from 12 to 20 weeks of age. Fructose levels in the seminal vesicles do not change with age, and DHT binding remains unchanged, but specific prolactin-binding sites are decreased with age.[2,3]

Aging changes in the prostate are presented in Chapter 19 and therefore are not discussed here.

4 SEXUAL FUNCTION

In men, events leading to orgasm decrease with age. The curve of total impotence in men is exponential, with 0.4% of men being impotent at age 25, 6.7% at age 50, and 55% at age 75.[41] Erotic responsiveness, speed of attaining erection and its duration, the amount of preorgasmic mucus secreted, nocturnal erections and emissions, and the capacity for multiple climax all decline with age.[41] Although the frequency of sexual activity declines substantially with age, only small changes in hormone levels and spermatogenic indices have been reported and reproductive capacity is maintained in healthy old men.[42]

Testosterone is important for the maintenance of male sex drive and potency in humans. Between 60 to 80 years, men with high testosterone tend to be more sexually active than those with low testosterone (Figure 13-5).[43] Muscle mass, smoking, or coronary heart disease do not significantly influence either sexual activity or testosterone levels. Alcohol intake over 4 oz per day reduces sexual activity but not testosterone. Increased percentage of body fat does the reverse. Thus, serum testosterone only slightly coincides with male vigor. Fall in testosterone levels in men did not appear to be correlated with total testosterone and sexual activity, potency, or libido;[44] thus, declining androgen levels and reduced sexual potency are separate and only partly related aspects of the aging process.

The declining level of testosterone is only one of the many factors contributing to the age-related decline in sexual activity. Men with high levels of sexual activity in the first 1 to 2 years of marriage or, generally, between the ages of 20 to 39, are likely to sustain higher levels into their 60s and 70s.[45] However, age itself appears to be the most important determinant of sexual frequency so far investigated, and further study of this complex area is needed.

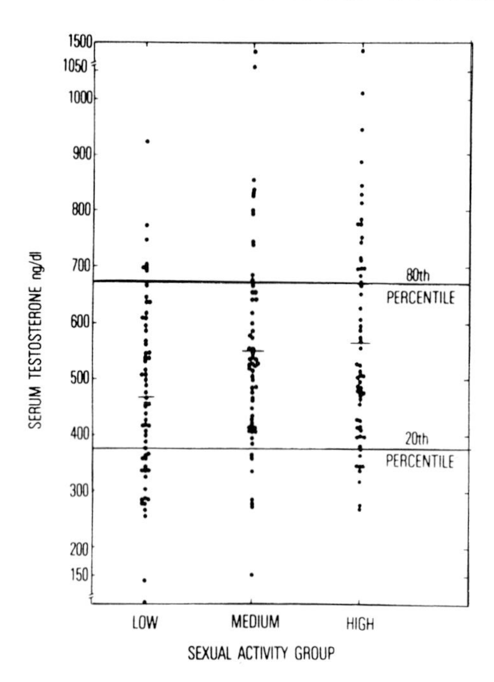

FIGURE 13-5 *Individual and mean serum testosterone concentrations in 183 men from the Baltimore Longitudinal Study aged 60 to 79 years, divided into groups according to sexual activity.* (From Tsitouras, P. D., Martin, C. E., and Harman, S. M., *J. Gerontol.*, 37, 288–293, 1982. With permission.)

Sexual Function in Aged Animals

Rats of 10, 27, and 44 weeks were given the opportunity to mate with a proestrus female. During a 2-h period, standard measures of copulatory behavior, e.g., mounting, intromission, and ejaculation, were documented by video recording. It was observed that the number of ejaculations in 10-week-old rats over the 2-h period was 6 to 8, while in 27-week-old males it was 3 to 4 during the first hour, after which they refrained from sexual activity. A further decline in copulatory behavior was observed in the 44-week-old males.[46] In middle-aged, sexually experienced rats, the probability of mating is significantly reduced and, when it does occur, the time to intromission and ejaculation is lengthened. Sexual motivation is also decreased in middle-aged (13- to 15-month) rats as compared to the young. In aged rats, erection frequency is reduced, but the number of spontaneous ejaculations is increased[47] as is the number of ejaculations required to reach sexual exhaustion.[48,49] The amount of anogenital investigation directed to receptive females by old rats was significantly reduced compared to that of young and middle-aged rats.[50]

In aged male CBF_1 mice, the three component parts of copulation (mounting, intromission, and ejaculation) show independent rates of deterioration with age (ejaculatory failure in these animals being associated with urogenital pathology).[51] The decrease in mounting and intromission behavior observed with age is not due to muscular debilitation, loss of the capacity to arouse, or a general reproductive/endocrine lesion, but may be attributable to physiologic changes in the specific neural or neuromuscular tissues responsible for this behavior. A comparison of the reproductive capabilities of 6- and 24-month-old C57BL/6NNia male mice after being paired for 1 month with a 4-month-old proven fertile female revealed that while all the younger males mated, only 46% of aged males did so. Subsequent to the matings, 96% of young males compared with only 42% of the aged males sired a litter. No statistical differences in litter size or congenital defects for the offspring could be demonstrated between the groups. The old males that failed to mate exhibited atrophied testes with fewer, less-mobile sperm, hypertrophied seminal vesicles, and a higher proportion of degenerating epithelium lining in the seminiferous tubules.[52]

In rats, restoring plasma testosterone levels in middle-aged rats to those found in young adults restores the capacity for erection, but sexual motivation and the latency periods for copulatory behavior were not restored to young levels.[47,53] In contrast, 15-month-old rats receiving 1 mg/kg yohimbine, a compound which strongly stimulates sexual motivation and arousal in young animals, were restored to nontreated, 3-month-old rat levels in terms of intromission and ejaculation in mating tests.[54] Previously impotent old male rats receiving 0.25 mg/kg of (–) deprenyl, the selective inhibitor of monamine oxidase B, exhibited restored sexual function as determined by intromissions and ejaculations. This restored sexual function persisted for several weeks subsequent to the single injection of (–) deprenyl.[55] An alternative strategy to drug therapy has been adopted in an attempt to restore full reproductive function to impotent 18- to 20-month male rats. Each aged male received an anterior hypothalamic graft removed from a 17- to 19-day-old fetus and placed into the anterior third ventricle. Control animals either did not undergo the surgical procedure or received a graft of cerebral cortical tissue. Before and subsequent to the transplantation surgery, reproductive function of each aged animal was assessed by placing them overnight with four 11- to 12-week-old proestrus female rats. Vaginal smears of all female rats were monitored the following morning with the presence of sperm indicative of copulation and ejaculation. Seven of the ten impotent aged males that received hypothalamic grafts were restored to full sexual function, with 106 pups being fathered from nine females. None of the control rats and only one of the animals receiving a cerebral cortex implant had sexual function restored, with six pups being delivered from one female. Serum testosterone, LH, and pituitary LH in the rats receiving a hypothalamic implant and restoration of sexual function were significantly higher than those of control animals.[56]

The capacity to sustain multiple ejaculations in rhesus macaque monkeys declines from 9 to 30 years of age. Each male was given a test of sexual exhaustion, defined as a 45-min period without a mount when paired with a female, with each of five ovariectomized, estrogen-treated females. Males in the age range 25 to 30 years did not achieve more than two ejaculations before reaching sexual exhaustion, while animals in the age ranges 19 to 20 years and 9 to 15 years displayed five to six ejaculations before meeting the criteria of sexual exhaustion. Although very old males (25 to 30 years) mounted as often as younger males, they achieved fewer intromissions, and the latency to intromission and ejaculation was longer.[57]

Sexual motivation (probability of mating to ejaculation and the rate of mating) decreases in male macaques over the age of 20 years when compared to those about 10 years old. Circulating testosterone levels are not decreased in old rhesus monkeys, but steroid hormone–binding globulin levels are significantly higher in the older group and showed high negative correlation with ejaculation, intromissions per minute, and contacts per minute.[58] Restoration of testosterone levels in the old males had no effect on sexual behavior. Therefore, at present, it appears that neither an alteration in testosterone levels nor decreased sensitivity to testosterone accounts for the changes shown in sexual behavior in old animals.

These findings have been confirmed in old male rhesus macaques following treatment with GnRH. GnRH has been reported to enhance sexual performance in a number of species. Measures of sexual behavior and serum testosterone and LH serum levels were determined in old intact and old testosterone-implanted castrated monkeys in response to two doses of GnRH. The mean intromission rate of old intact males was significantly lower following treatment with 100 mμ/g of GnRH compared with control animals. This failure to enhance sexual performance was also observed in GnRH-injected young rhesus males. In both young and old monkeys, serum LH and testosterone levels were significantly increased in response to GnRH doses.[59]

REFERENCES

1. Steger, R. W., Huang, H. H., and Meites, J., Reproduction, in *CRC Handbook of Physiology of Aging,* Masoro, E. J., Ed., CRC Press, Boca Raton, FL, 1981, 333.
2. Steger, R. W. and Huang, H. H., The reproductive decline in male rats, in *Neuroendocrinology of Aging,* Meites, J., Ed., Plenum Press, New York, 1983, 123.
3. Talbert, G. B., Aging of the reproductive system, in *Handbook of the Biology of Aging,* Finch, C. E. and Hayflick, L., Eds., Van Nostrand Reinhold, New York, 1977, 318.
4. Parkening, T. A., Fertilizing ability of spermatozoa from aged C57BL/6NNia mice, *J. Reprod. Fertil.,* 87(2), 727, 1989.
5. Taylor, G. T., Weiss, J., and Pitha, J., Epididymal sperm profiles in young adult, middle-aged, and testosterone-supplemented old rats, *Gamete Res.,* 19(4), 401, 1988.
6. Paniagua, R., Nistal, M., Saez, F. J., and Fraile, B., Ultrastructure of the aging human testis, *J. Electron. Microsc. Tech.,* 19, 241, 1991.
7. vom Saal, F. S. and Finch, C. E., Reproductive senescence; phenomena and mechanisms in mammals and selected vertebrates, in *The Physiology of Reproduction,* Vol. 1, Knobil, E. and Neill, J., Eds., Raven Press, New York, 1988.
8. Humphreys, P. N., The histology of the testis in aging and senile rats, *Exp. Gerontol.,* 12, 27, 1977.
9. Johnson, L., Zane, R. S., Petty, C. S., and Neaves, W. B., Quantification of the human Sertoli cell population: its distribution, relation to germ cell numbers, and age-related decline., *Biol. Reprod.,* 31, 785, 1984.
10. Wright, W. W., Fiore, C., and Zirkin, B. R., The effect of aging on the seminiferous epithelium of the brown Norway rat, *J. Androl.,* 14(2), 110, 1993.
11. Johnson, L. and Neaves, W. B., Age-related changes in the Leydig cell population, seminiferous tubules, and sperm production in stallions, *Biol. Reprod.,* 24, 703, 1981.
12. Seymour, F. I., Duffy, C., and Koerner, A., A case of authenticated fertility in a man of 94, *JAMA,* 105, 1423, 1935.
13. Merry, B. J. and Holehan, A. M., Serum profiles of LH, FSH, testosterone and 5α-DHT from 21 to 1000 days in *ad libitum* fed and dietary restricted long-lived rats., *Exp. Gerontol.,* 16, 431, 1981.
14. Harman, S. M. and Tsitouras, P. D., Reproductive hormones in aging men. I. Measurement of sex steroids, basal LH, and Leydig cell response to hCG, *J. Clin. Endocrinol. Metab.,* 51, 35, 1980.
15. Harman, S. M., Reproductive decline in men, in *Neuroendocrinology of Aging,* Meites, J., Ed., Plenum Press, New York, 1983, 203.
16. Stearns, E. L., MacDonald, J. A., Kauffman, B. J., Lucman, T. S., Winters, J. S., and Fairman, C., Declining testis function with age: hormonal and clinical correlates, *Am. J. Med.,* 57, 761, 1974.
17. Fingscheidt, U. and Nieschlag, E., The response of inhibin to human chorionic gonadotrophin is decreased in senescent men compared with young men, *J. Endocrinol.,* 123(2), R9, 1989.
18. Korenman, S. G. and Stanik, S., The male reproductive system and aging, in *Altered Endocrine Status During Aging,* Cristofalo, V. J., Baker, G. T., Adelman, R. C., and Roberts, J., Eds., Alan R. Liss, New York, 1984.
19. Steger, R. W., Huang, H. H., Hodson, C. A., Leung, F., Meites, J., and Sacher, G. A., Effects of advancing age on the hypothalamic-hypophysial-testicular functions in the male white-footed mouse (Peromyscus leucopus), *Biol. Reprod.,* 22, 805, 1980.
20. Swanson, L. J., Desjardins, C., and Turek, F. W., Aging of the reproductive system in the male hamster: behavioral and endocrine patterns, *Biol. Reprod.,* 26(5), 791, 1982.
21. Simpkins, J. W., Kalra, P. S., and Kalra, S. P., Alterations in the daily rhythms of testosterone and progesterone in old male rats, *Exp. Aging Res.,* 7, 25, 1981.
22. Saksena, S. K. and Lau, I. F., Variations in serum androgens, estrogens, progestins, gonadotropins and prolactin level in male rats from prepubertal to advanced age, *Exp. Aging Res.,* 5, 179, 1979.
23. Gruenewald, D. A., Hess, D. L., Wilkinson, C. W., and Matsumoto, A. M., Excessive testicular progesterone secretion in aged male Fischer 344 rats: a potential cause of age-related gonadotropin suppression and confounding variable in aging studies, *J. Gerontol.,* 47(5), B164, 1992.
24. Zirkin, B. R., Santulli, R., Strandberg, J. D., Wright, W. W., and Ewing, L. L., Testicular steroidogenesis in the aging brown Norway rat, *J. Androl.,* 14(2), 118, 1993.
25. Vermeulen, A., Leydig cell function in old age, in *Hypothalamus. Pituitary and Aging,* Everitt, A. V. and Burgess, J. A., Eds., Charles C Thomas, Springfield, IL, 1976, 458.
26. Riegle, G. D. and Miller, A. E., Aging effects on the hypothalamic-hypophyseal-gonadal control system in the rat, in *The Aging Reproductive System,* Schneider, E. L., Ed., Raven Press, New York, 1978.
27. Harman, S. M., Danner, R. L., and Roth, G. S., Testosterone secretion in the rat in response to chorionic gonadotropin: alterations with age, *Endocrinology,* 102, 540, 1978.
28. Harman, S. M., Tsitouras, P. D., Costa, P. T., and Blackman, M. R., Reproductive hormones in aging men. II. Basal pituitary gonadotropins and gonadotropin responses to luteinizing hormone releasing hormone, *J. Clin. Endocrinol. Metab.,* 54, 547, 1982.
29. Baker, H. W. G., Bremner, W. J., Burger, H. G., deKretser, D. M., Dulmanis, A., Eddie, L. W., Hudson, B., Keogh, E. J., Lee, V. W. K., and Rennie, G. C., Testicular control of follicle stimulating hormone secretion, *Rec. Prog. Horm. Res.,* 32, 429, 1976.
30. Blackman, M. R., Tsitouras, P. D., and Harman, S. M., Reproductive hormones in aging men. III. Basal and LHRH-stimulated serum concentrations of the common alpha-subunit of the glycoprotein hormones, *J. Gerontol.,* 42(5), 476, 1987.
31. Smith, W. A. and Conn, P. M., Causes and consequences of altered gonadotropin secretion in the aging rat, in *Experimental and Clinical Interventions in Aging,* Walker, R. F. and Cooper, R. L., Eds., Marcel Dekker, New York, 1983.
32. Conn, P. M., Cooper, R., McNamara, C., Rogers, D. C., and Shoenhardt, L., Qualitative change in gonadotropin during normal aging in the male rat, *Endocrinology,* 106, 1549, 1980.
33. Shaar, C. J., Euker, J. S., Riegle, G. D., and Meites, J., Effects of castration and gonadal steroids on serum LH and prolactin in old and young rats, *J. Endocrinol.,* 66, 45, 1975.
34. Bruni, J. F., Huang, H. H., Marshall, S., and Meites, J., Effects of single and multiple injections of synthetic GnRH on serum LH, FSH and testosterone in young and old male rats, *Biol. Reprod.,* 17, 309, 1977.
35. Gruenewald, D. A. and Matsumoto, A. M., Age-related decreases in serum gonadotropin levels and gonadotropin-releasing hormone gene expression in the medial preoptic area of the male rat are dependent upon testicular feedback, *Endocrinology,* 129(5), 2442, 1991.
36. Witkin, J. W., Aging changes in synaptology of luteinizing hormone-releasing hormone neurons in male rat preoptic area, *Neuroscience,* 22(3), 1003, 1987.

37. Witkin, J. W., Increased synaptic input to gonadotropin-releasing hormone neurons in aged, virgin, male Sprague-Dawley rats, *Neurobiol. Aging,* 13(6), 681, 1992.
38. Sahu, A., Kalra, P. S., Crowley, W. R., and Kalra, S. P., Evidence that hypothalamic neuropeptide Y secretion decreases in aged male rats: implications for reproductive aging, *Endocrinology,* 122(5), 2199, 1988.
39. Sahu, A., Kalra, S. P., Crowley, W. R., and Kalra, P. S., Aging in male rats modifies castration and testosterone-induced neuropeptide Y response in various microdissected brain nuclei, *Brain Res.,* 515, 1, 1990.
40. Meites, J., Steger, R. W., and Huang, H. H., Relation of the neuroendocrine system to the reproductive decline in aging rats and human subjects, *Fed. Proc.,* 39, 3168, 1980.
41. Kinsey, A. C., Pomeroy, W. B., and Martin, C. E., *Sexual Behavior in the Human Male,* W.B. Saunders, Philadelphia, 1948.
42. Tsitouras, P. D., Effects of age on testicular function, *Endocrinol. Metab. Clin. N. Am.,* 16(4), 1045, 1987.
43. Tsitouras, P. D., Martin, C. E., and Harman, S. M., Relationship of serum testosterone to sexual activity in healthy elderly men, *J. Gerontol.,* 37, 288, 1982.
44. Davidson, J. M., Chen, J. J., Crapo, L., Gray, G. D., Greenleaf, W. J., and Catania, J. A., Hormonal changes and sexual function in aging men, *J. Clin. Endocrinol. Metab.,* 57, 71, 1983.
45. Martin, C. E., Factors affecting sexual functioning in 60–79 year old married males, *Arch. Sex. Behav.,* 10, 399, 1981.
46. Hokao, R., Saito, T. R., Wakafuji, Y., Takahashi, K. W., and Imamichi, T., The change with age of the copulatory behavior of the male rats aged 67 and 104 weeks, *Exp. Anim.,* 42(1), 75, 1993.
47. Gray, G. D., Smith, E. R., Dorsa, D. M., and Davidson, J. M., Sexual behavior and testosterone in middle-aged rats, *Endocrinology,* 109, 1597, 1981.
48. Larsson, K., Age differences in the diurnal periodicity of male sexual behavior, *Gerontologia,* 2, 64, 1958.
49. Larsson, K. and Essberg, L., Effect of age on the sexual behavior of the male rat, *Gerontologia,* 6, 133, 1962.
50. Mencio-Wszalek, T., Ramirez, V. D., and Dluzen, D. E., Age-dependent changes in olfactory-mediated behavioral investigations in the male rat, *Behav. Neural. Biol.,* 57(3), 205, 1992.
51. Huber, M. R. H., Bronson, F. H., and Desjardins, C., Sexual activity of aged male mice: Correlation with level of arousal, physical endurance, pathological status and ejaculatory capacity, *Biol. Reprod.,* 23, 305, 1980.
52. Parkening, T. A., Collins, T. J., and Au, W. W., Paternal age and its effects on reproduction in C57BL/6NNia mice, *J. Gerontol.,* 43(3), B79, 1988.
53. Chambers, K. C. and Phoenix, C. H., Testosterone is more effective than dihydrotestosterone plus estradiol in activating sexual behavior in old male rats, *Neurobiol. Aging,* 7(2), 127, 1986.
54. Smith, E. R. and Davidson, J. M., Yohimbine attenuates aging-induced sexual deficiencies in male rats, *Physiol. Behav.,* 47(4), 631, 1990.
55. Dallo, J., Yen, T. T., Farago, I., and Knoll, J., The aphrodisiac effect of (-)deprenyl in non-copulator male rats, *Pharmacol. Res. Commun.,* 20(Suppl.), 25, 1988.
56. Huang, H. H., Kissane, J. Q., and Hawrylewicz, E. J., Restoration of sexual function and fertility by fetal hypothalamic transplant in impotent aged male rats, *Neurobiol. Aging,* 8(5), 465, 1987.
57. Phoenix, C. H. and Chambers, K. C., Old age and sexual exhaustion in male rhesus macaques, *Physiol. Behav.,* 44(2), 157, 1988.
58. Chambers, K. C. and Phoenix, C. H., Diurnal patterns of testosterone, dihydrotestosterone, estradiol and cortisol in serum of rhesus males: relationship to sexual behavior in aging males, *Horm. Behav.,* 15, 416, 1981.
59. Phoenix, C. H. and Chambers, K. C., Sexual performance of old and young male rhesus macaques following treatment with GnRH, *Physiol. Behav.,* 47(3), 513, 1990.

III
SYSTEMIC AND ORGANISMIC AGING

14 AGING OF THE THYROID GLAND AND BASAL METABOLISM

Paola S. Timiras

1 THYROID FUNCTION AND REJUVENATION?

The hormones of the thyroid gland, primarily thyroxine (T4) and triiodothyronine (T3), have traditionally sparked the interest of investigators seeking hormonal determinants of aging. Early studies in humans suggested that certain signs of aging resemble those of thyroid insufficiency or hypothyroidism. Individuals affected by hypothyroidism develop a number of signs that could be interpreted as "precocious senility" including a reduced metabolic rate, hyperlipidemia and accelerated atherosclerosis, early aging of skin and hair, slow reflexes, and slow mental performance. Since these patients improve markedly after hormonal replacement therapy, it has been argued that similar symptoms in normally aging individuals may represent effects secondary to thyroid involution with age. The well-known action of thyroid hormones in controlling development made it logical to suspect that these hormones might also control the rate or site of aging.

As early as the turn of this century, experimenters optimistically attempted rejuvenation or prolongation of life through hormone administration. While thyroid hormone administration to some animals (rats) seemed to shorten rather than prolong life,[1] in others (fowl) it caused an apparent dramatic rejuvenation.[2] Delayed and impaired growth and maturation associated with a significant prolongation of the lifespan were also reported in rats made hypothyroid at an early (first postnatal week) age,[3,4] resembling the effects reported after food restriction (Chapter 24). However, the aging process in euthyroid elderly humans was never significantly slowed or altered by the administration of these hormones nor the lifespan affected by hypo- or hyperthyroidism.

The capacity to maintain an euthyroid (normal) state continues during aging despite a number of changes in various aspects of thyroid hormone production, secretion, and action. As for all other endocrines, normal thyroid function in the elderly is marginal and easily endangered by repeated challenging demands and stress; the ensuing dysthyroid (abnormal) state may lead to decreased overall physiologic competence, disease, and aging.[5-10] In the present chapter, we will consider first the aging of the thyroid gland and the hypothalamo–pituitary–thyroid axis, and, second, the possible consequences of this aging on the whole organism and on specific functions — basal metabolism, thermoregulation — regulated by thyroid hormones. Also included by virtue of their intimate anatomic association with the thyroid gland is aging of the parathyroids and of the thyroid C cells secreting calcitonin.

2 STRUCTURAL AND FUNCTIONAL CHANGES IN HYPOTHALAMO–PITUITARY–THYROID AXIS

During development, the thyroid gland is necessary for whole-body and organ growth and development, and maturation of the central nervous system. In adulthood, it has an essentially metabolic function, regulating tissue oxygen consumption and thereby maintaining metabolic rate as well as influencing some aspects of behavior. The thyroid gland is not essential for life, but in its absence there is poor resistance to cold, mental and physical slowing, and, in children, mental retardation and dwarfism. Reciprocally, in hyperthyroidism, metabolic and behavioral alterations threaten well-being and survival.

Structure and Regulation

The thyroid gland is a bilateral organ that bridges the lower larynx and upper trachea with a narrow isthmus. A third pyramidal lobe, remnant of the thyroglossal duct, is not unusual. As one of the most vascularized endocrines, it receives blood from the superior thyroid arteries, branches of the external carotid artery, and is drained by corresponding veins into the internal jugular vein (Figure 14-1). In normal individuals, vascularity, size, and microscopic structures vary with the levels of the pituitary tropic hormone, thyroid-stimulating hormone (TSH) or thyrotropin, nutrition, temperature, sex, and age.

The functional units of the thyroid gland are multiple, variable-sized follicles, formed by a single layer of epithelial cells, filled with colloid (a proteinaceous material containing thyroglobulin, a glycoprotein necessary for the synthesis of T3 and T4) (Figure 14-2). At high magnification, the cell surface lining the follicle is rich in microvilli that project into the follicular lumen where the colloid is secreted; hormones are secreted into the blood at the opposite basal cell pole adjoining the rich capillary net.

The control of the thyroid gland in the elderly, as in individuals of all other ages, cannot be viewed in isolation but must be considered in the context of

- Regulation at the hypothalamo–pituitary axis
- Thyroid hormone metabolism
- Interactions with receptors at target cells

These connected levels of integration (Figure 14-3) can be briefly outlined as follows:

1. At the level of the hypothalamus, thyrotropin-releasing hormone (TRH), a tripeptide, is secreted into the portal capillaries relaying secreted TRH to the anterior pituitary. There, TRH stimulates the anterior lobe thyrotropes to synthesize and release thyroid-stimulating hormone (thyrotropin or TSH). Secretion of TSH is, in turn, inhibited by the negative feedback of thyroid hormones and stimulated or inhibited by stimuli from higher brain centers in response to environmental changes.

0-8493-8979-8/94/$0.00+$.50

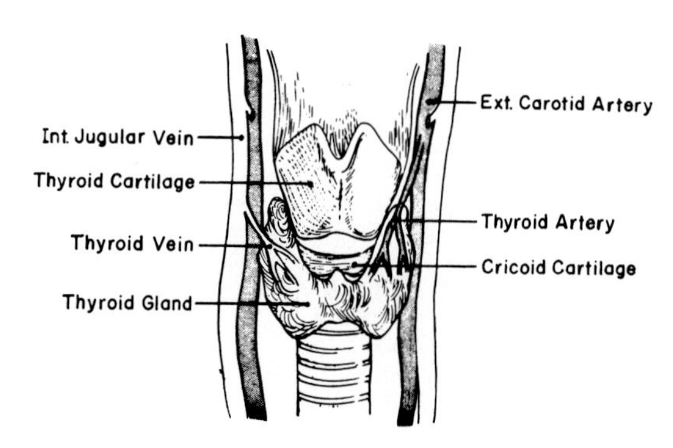

FIGURE 14-1 *Diagrammatic representation of the thyroid gland.* The thyroid and cricoid cartilages and the major blood vessels are included. Note that four parathyroid glands are embedded in the superior and inferior poles of the thyroid gland. C cells secreting calcitonin are dispersed throughout the thyroid gland. (Drawing by S. Oklund.)

2. At the pituitary level, TSH is a glycoprotein secreted by the basophilic thyrotropes. TSH secretion is regulated by negative feedback of thyroid hormones, i.e., the higher the serum levels of these hormones, the lower TSH release and vice versa, and stimulation from TRH. In the absence of TSH (e.g., by hypophysectomy in experimental animals) thyroid function is depressed and the thyroid gland atrophies; administration of TSH stimulates the thyroid gland and increases circulating levels of thyroid hormones.
3. At the thyroid level, thyroid hormones, thyroxine (T4), triiodothyronine (T3), and to a much lesser extent, reverse T3 (rT3), are iodothyronines, iodine-containing derivatives of the amino acid tyrosine. They are synthesized by iodination and condensation of the tyrosyl residues of thyroglobulin molecules stored in the colloid of the thyroid follicle. Iodinated thyroglobulin enters the thyroid cells by endocytosis and is hydrolyzed there to liberate T4 and T3, which are released into the circulation. Cells of the thyroid gland contain TSH receptors. Binding of TSH to its receptors activates the enzyme adenylate cyclase with increase in intracellular cAMP. Most of TSH actions are mediated through this cAMP increase, but some depend on stimulation of cell membrane phospholipids.

The major secreted product of the thyroid gland is T4, while T3 is secreted only in small amounts and derives mainly from the peripheral deiodination of T4. One third of circulating T4 is converted to T3 in peripheral tissues. Both hormones are present in serum either bound to proteins or in the free state. T3 is less tightly bound to plasma proteins than is T4 and is therefore more readily available for cellular uptake. The free hormone is biologically active and interacts with specific receptors localized in the membrane, mitochondria, cytoplasm, and nucleus of responsive cells. T3 binds more avidly than T4 to the nuclear receptors to a much greater extent than T4; hence, T3 is more rapidly and biologically active than T4. T3 and T4 are deiodinated and deaminated in the tissues. In the liver, they are conjugated, (e.g., to glucuronic acid), pass into the bile, and are excreted into the intestine. Conjugated and free hormones are also excreted by the kidney.

This aspect of thyroid function may be summarized by pointing out that

- The major source of circulating T3 is not from thyroid secretion, but from peripheral deiodination of T4.
- The negative feedback at the pituitary anterior lobe is principally through T4 taken from the circulation and converted in the thyrotrope to T3 by thyrotrope deiodinase.
- The peripheral deiodination depends on the physiological state of the organism. It allows an autonomy of response of the tissues to the hormones. Deiodination can convert T4 (a less active hormone) to T3 (a more active hormone) or not. The conversion depends on activities of the various deiodinating enzymes.

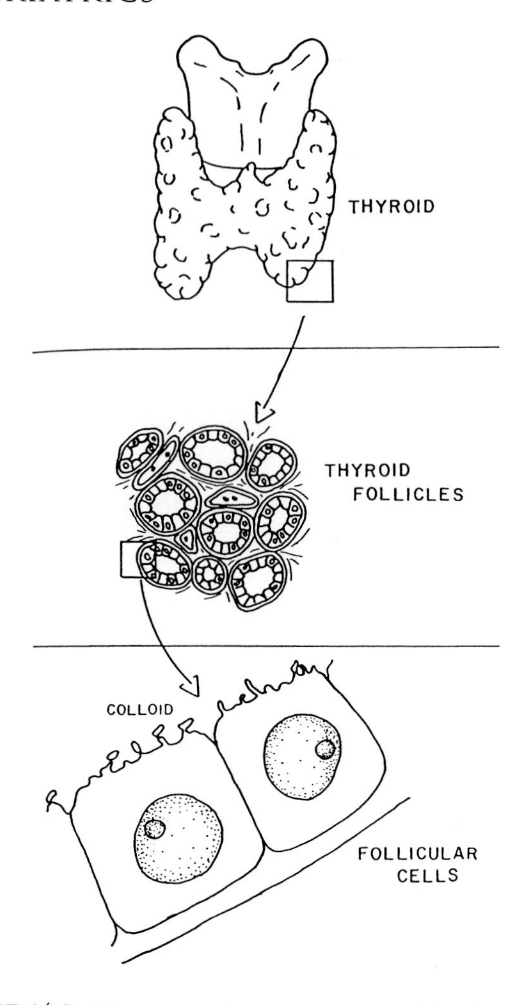

FIGURE 14-2 *Diagrammatic representation of the thyroid* (top), *the thyroid follicles* (middle), and *the follicular cells* (bottom). The microvilli from the follicular cells project into the follicular colloid.

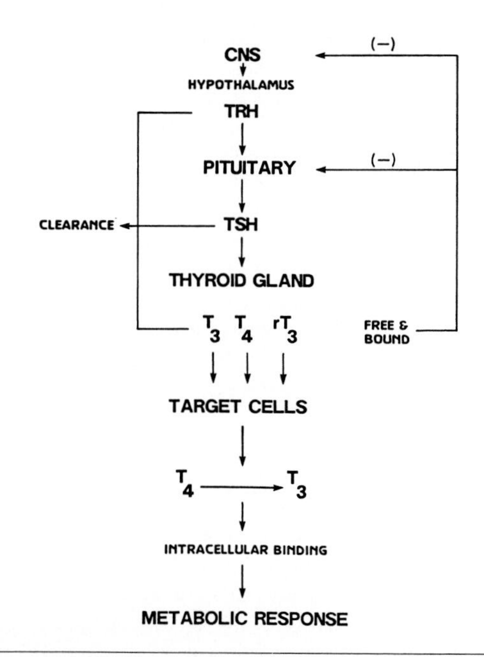

FIGURE 14-3 *Diagram of the interrelations of the hypothalamo–pituitary–thyroid axis.*

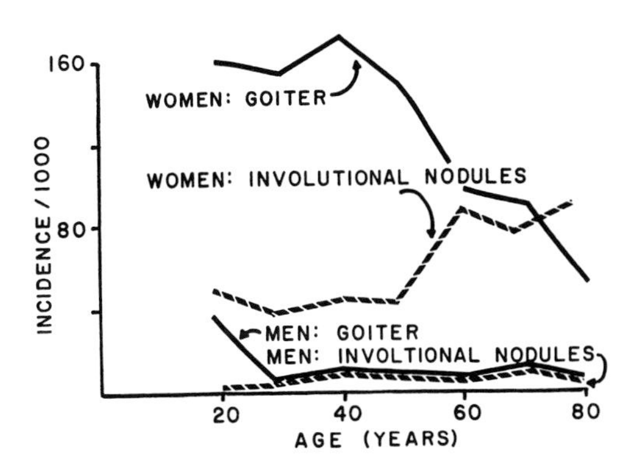

FIGURE 14-4 *Incidence of goiter and involutional nodules in the thyroid gland as a function of age in women and men.*

TABLE 14-1

Comparison of Results From Thyroid Function Tests in Young (n = 40) and Elderly (n = 178)

Test		Mean ± SD
Total T4	Elderly	7.6 μg/dl ± 1.5
	Young	8.8 μg/dl ± 1.7
Free T4	Elderly	1.5 ng/dl ± 0.2
	Young	1.7 ng/dl ± 0.2
T3	Elderly	128 ng/dl ± 21
	Young	170 ng/dl ± 29
TSH	Elderly	3.3 mIU/ml ± 1.9
	Young	1.5 mIU/ml ± 1.0

Modified from Runnels, B. L. et al., *J. Gerontol. Biol. Sci.*, 46, B39, 1991.

2.1 Structural Changes

From maturity to old age, the size of the thyroid gland decreases, although in some cases, the weight of the gland may remain unchanged or even increase.[11] The increase is due most often to the presence of nodules (small lumps or masses of hyperplastic and hypertrophic cells) or to mild endemic goiter (hypothyroidism due to iodine deficiency). The general shape of the gland remains unchanged, although microscopic changes are frequent and include

- Distention of follicles
- Discoloration of colloid
- Flattening of the follicular epithelium suggesting reduced secretory activity
- Decreased number of mitoses, suggesting decreased cell turnover
- Increased fibrosis of interstitial connective tissue
- Parenchyma and vascular changes of an atherosclerotic nature leading to decreased transport between cells, blood, and follicles

All these changes suggest reduced function. When the cells are inactive, as in the absence or deficiency of TSH or in some old individuals, follicles are large, more colloid is accumulated in the follicular cavity due to reduction of endocytosis, and the lining cells are flat and display signs of reduced secretory activity. However, in acute illness, thyroid hormonal output may be greatly increased.[12] When thyroid cells are active, as following TSH stimulation, follicles are small and cells are cuboidal or columnar with signs of active endocytosis, i.e., transport of the colloid into the cell.

Nodules

Nodules, primarily involutional in older individuals, are characterized by localized tissue proliferation with overlapping areas of cell involution or stimulation as occur frequently in the hyperthyroidism of the elderly. The prevalence of micronodules and clinically palpable nodules may both increase[13] and decrease[14] with aging. The variability of this finding reflects geographic differences in the prevalence of endemic goiter (overall enlargement of the thyroid) due essentially to iodine deficiency. Endemic goiter is more prevalent in young women than men and may be caused by a relative iodine deficiency associated with poor dietary habits and nutritional deficiencies that may occur at adolescence or the increased nutritional demands created by pregnancy. With aging, the prevalence of goiter tends to decrease while that of multiple nodules increases (Figure 14-4). Multinodular goiter is also associated with increased antithyroid antibodies. The much lower prevalence of goiter and nodules in men remains essentially unchanged into old age.

2.2 Changes in Thyroid Hormone Serum Levels

Changes with aging in the levels of these hormones are controversial due to fluctuations in response to a variety of stimuli and disease. Total and free T4 serum levels may be unchanged, slightly decreased, or increased depending upon age, sex, and general health. In a study of healthy geriatric individuals, serum levels of T4, free T4, and TSH differed between young and old individuals.[15] Data from this study (Table 14-1) and previous ones permit the following general conclusions:

1. Secretion and metabolic clearance of T4 progressively decrease with advancing age to a level of almost half that of the younger controls. Because both of these factors decline to a similar degree, total T4 levels (bound and free) remain essentially normal.
2. Thyroid-binding globulin levels are unchanged.
3. T3 (and rT3) levels tend to decline with age but generally remain in the normal range.
4. TSH levels are elevated in 10% of the elderly, even in the absence of other manifestations of hypothyroidism.
5. The increase in TSH is associated with an increase in antithyroid antibodies.

These data also reveal a greater individual variability among the elderly than the young subjects. Despite this variability, it may be stated that values in the elderly are in the lower normal range for T4 and T3 and in the higher normal range for TSH; borderline abnormal values are more usual in women than men.

The observation that T4 levels remain unchanged is reminiscent of the case of adrenocortical hormones where a slower metabolism (in the liver and tissues) and excretion (by the kidney) of the hormones compensate for the reduced secretion and maintain the levels essentially constant (Chapter 11).

Serum T3 may be significantly reduced in some aged individuals.[16-18] This age-related decrease of blood T3 concentration presumably may reflect:

- Impaired T4 to T3 conversion in tissues
- Increased T3 degradation (metabolism or excretion)
- Decreased T3 secretion from the thyroid gland due to either failure of stimulation by TSH or intrinsic, primary alteration of the gland

Based on the current data, it is impossible to distinguish among the foregoing possibilities. However, T4 to T3 conversion falls with age, at least in experimental animals (rat).[19] With respect to laboratory animals, not only are there differences from humans but also intraspecies variation. In the Sprague–Dawley rat, one of the most used animals for aging research, serum T4 rather than T3 is markedly reduced with aging (Figure 14-5).[20]

A certain number of elderly show elevated TSH with a T4 which falls within the "normal" range (as established by values obtained in the younger adult population). Under normal feedback conditions, TSH levels are elevated in response to low T4 values. This discrepancy in high TSH/normal T4 levels in a good percentage of elderly may indicate alterations in thyroid–pituitary–hypothalamus feedback. It may also indicate that a significant portion of the elderly are hypothyroid and would benefit from hormone replacement therapy. It has been proposed (but, so far, not adopted) to raise the cutoff of T4 values (indicating hypothyroidism) to reflect the large number of elderly with T4 levels that fall in the low normal range.[21]

Inasmuch as hormonal levels are measured by their immunologic properties, the discrepancy in high TSH normal T4 levels may also be explained by changes in TSH, which would retain immunoreactivity but would be less biologically active, hence the need for the higher TSH levels to maintain normal T4 (see below).

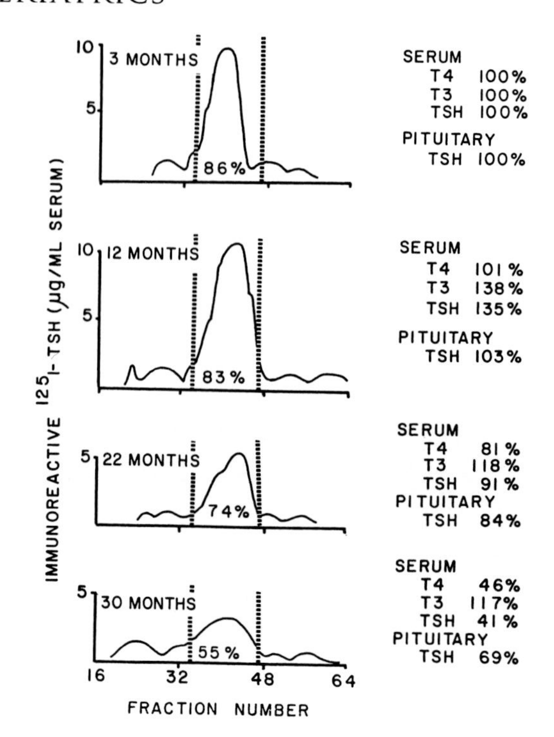

FIGURE 14-5 *Changes with age in TSH immunoreactivity.* On the left side, changes in the fractionation profiles of TSH at different ages (3, 12, 22, and 30 months) in male Fischer 344 rats. TSH extracted from sera was radioimmunoassayed to obtain the illustrated profiles. The amount of immunoreactive TSH in each fraction was corrected for the original volume of serum and for the recovery from the affinity column so that the four plots are directly comparable. The percentages are relative to the total TSH recovered. On the right side, serum T4, T3, and TSH and pituitary TSH values measured at the same corresponding ages, taking 3 months as 100%. (Redrawn from Choy, V. J., et al., *Mech. Ageing Dev.,* 10, 151, 1979.)

2.3 Changes in the Pituitary–Thyroid Axis

TSH secretion from the pituitary is the principal regulator of thyroid gland function and is, in turn, controlled by the hypothalamic releasing hormone (TRH) and by direct inhibitory feedback of high circulating thyroid hormone levels. In this way, the pituitary–thyroid axis is capable of bringing about the appropriate adjustments to internal and external environmental changes. Other regulators of thyroid function include direct autonomic inputs (thyroid follicular cells are innervated by the sympathetic nervous system and are sensitive to sympathetic signals). Another little-known regulatory source may be represented by local control from some still unidentified iodinated compounds.

The major regulatory pathways remains the TRH–TSH axis. When TSH levels are low, the ability of the thyroid gland to secrete thyroid hormones is reduced and, *vice versa,* when TSH levels are high the activity of the gland is increased. TSH action on the thyroid gland involves

- Increase in iodide trapping and binding
- T3 and T4 synthesis
- Thyroglobulin secretion into colloid
- Colloid endocytosis into thyroid cells
- Increase in blood flow

Prolonged stimulation of the thyroid gland by TSH results in cell hypertrophy with overall enlargement of the gland or goiter.

With aging, TSH circulating levels are usually elevated when T4 or T3 circulating levels are lowered. However, in rats, qualitative changes also occur. TSH is a glycoprotein and is present in several forms of different molecular weight and immunoreactivity. In old rats, there is a progressive decrease in the major TSH form, which is also biologically the most active, and an increase in the proportion of high- and low-molecular-weight forms (Figure 14-5).[20,22] The biological significance of this age-related increase in TSH polymorphism still remains speculative. Polymorphism is not limited exclusively to TSH but occurs in other glycoproteic hormones of the pituitary, such as the gonadotropins in which the proportion of polymorphic forms also increases with aging. In rats, not only are TSH levels and TSH polymorphism increased with aging, but the typical circadian cyclicity of the hormone is abolished. The functional significance of TSH rhythmicity is still obscure; however, the loss of specific pulsatile signals may suggest a progressive failure with aging of fine tuning of thyroidal function.[23] Such a failure is supported by the observation that, in some species, the low T3 and T4 serum levels are associated with normal or reduced (rather than elevated) TSH levels. Inasmuch as the major amounts of pituitary T3 (effective in the negative feedback inhibition of TSH) may be derived from intrapituitary deiodination of T4, it is possible that the low thyroid hormone levels are due to decreased systemic deiodination (without affecting the levels of pituitary T3) or that the thyroid hormones–pituitary feedback is impaired.[24]

The majority of studies describe unchanged or slightly elevated circulating TSH levels without marked changes in circadian periodicity in humans and rats.[14,23-25] For TSH as well as for hypothalamic TRH, current results are in conflict particularly in relation to sex differences, a decrease in TRH having been reported exclusively in either males or females, depending on the investigator. Other studies show increased TSH responsiveness to TRH in apparently normal elderly subjects in whom T4 levels are unchanged and basal TSH levels slightly increased.[26]

TSH Antibodies and Thyroid Autoimmune Diseases

The frequency of antithyroid antibodies, primarily directed against the cell-surface receptors for TSH on thyroid gland cells, increases with age. This increased incidence may be due to the decline in the self-recognition ability of the immune system (Chapter 7) as well as to the increased TSH polymorphism.

The incidence of two autoimmune diseases of the thyroid —Graves' disease or toxic diffuse goiter, and Hashimoto's disease or chronic lymphocytic thyroiditis — increases with age. Graves' toxic goiter is associated with hyperfunction or hyperthyroidism (with low TSH, high T3, and T4). Hashimoto's thyroiditis is associated with hypofunction or hypothyroidism (with high TSH, low T3, and T4) (Table 14-2). Both are prevalent in women, in whom the incidence reaches a peak around 40 years of age. TSH levels are not detectable in Graves' disease because of continuous stimulation of the thyroid receptors by TSH antibodies and consequent high T4 and T3 secretion. TSH levels are either normal or elevated in Hashimoto's disease as a consequence of low T3 and T4 levels and TSH blockage by TSH antibodies.

TABLE 14-2

Autoimmune Diseases of the Thyroid

Characteristics	Graves' disease	Hashimoto's thyroiditis
Thyroid status	Hyperthyroid	Hypothyroid
TSH	Generally undetectable	Normal to elevated
T4, T3 (serum)	Above normal	Below normal
Antibodies(ABs)	ABs compete with TSH at receptor sites Loss of TSH control over thyroid function	Some ABs block TSH actions
Autoantibodies against thyroglobulin, T3, T4, thyroid microsomal, and nuclear components	Generally present	Generally present destructive of thyroid
Lymphocytic invasion	Limited	Marked
Female-to-male ratio	As high as 10:1	As high as 10:1

Although the events that trigger the disease are not known, the immune system is strongly implicated. Graves' and Hashimoto's diseases may represent the extremes of a continuum of signs and symptoms resulting from a deranged immune system with hyperthyroidism at one end and hypothyroidism at the other. Antibodies against TSH receptors or thyroid cell components are present in both disorders, although in different proportions. TSH-receptor antibodies compete with TSH at the receptor site on the thyroid cell. Some of these antibodies stimulate T3 and T4 secretion as in Graves'; others block access of TSH at the receptor sites and reduce T3 and T4 secretion.

TSH-receptor antibodies are not the only antithyroid antibodies present in autoimmune disorders. The most common antibodies, important clinically, are those directed against thyroglobulin, the thyroid cell protein, precursor of T3 and T4, antibodies against thyroidal microsomal, and nuclear components and antibodies against T3 and T4. All these antibodies lead to thyroid destruction; they are more frequent in Hashimoto disease.

Another feature of both disorders is invasion of the thyroid tissue by lymphatic tissue. This infiltration is more aggressive in Hashimoto's with extensive destruction of the thyroid gland.

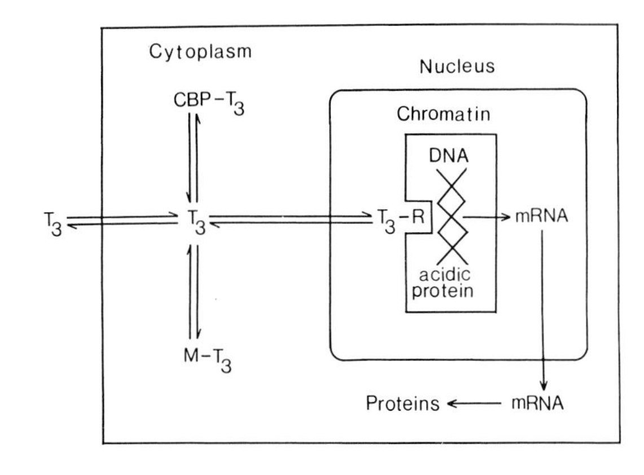

FIGURE 14-6 *Model for T3 interaction with target cell: CBP, cytosol-binding proteins; M, mitochondria; R, nuclear receptor.* As T3 enters the cell, it may be bound to a cytosol-binding protein, in reversible equilibrium with a small pool of free T3 which can interact reversibly with T3 receptors in the nucleus and perhaps also with receptors in the mitochondria.

2.4 Thyroid Hormone Receptors

As it is difficult to reconcile a relatively normal adrenocortical function (at least with respect to glucocorticoid secretion) to a severely decreased capacity to withstand stress with aging, similarly, the relative adequacy of the pituitary–thyroid axis into old age seems to belie the apparent symptoms of altered thyroid state of the elderly. One possible explanation is that age-related alterations occur essentially at the peripheral level,[27] primarily within the cell and involve, to a lesser extent, the hypothalamo–pituitary–thyroid axis. The study, then, of intracellular hormone receptors and their eventual changes with aging and their impact on cell function may prove useful in the elucidation of intracellular alterations.

Thyroid hormone receptors, primarily for T3, have been identified in the nucleus, mitochondria, plasma membrane, and cytosol[28] (Figure 14-6). The biologic actions of the hormone occur primarily through nuclear binding and stimulation of protein synthesis. The number of receptors appears to be inversely related to hormone levels so that receptors increase in number when the hormone level is low as in hypothyroidism (up-regulation) and decrease in number when hormone level is high as in hyperthyroidism (down-regulation). *In vitro* studies of T3 nuclear binding in rat liver and brain show that the receptor number remains unchanged with aging, although low circulating T4 levels *in vivo* might have forecast an upward regulation. *In vivo* the major source of intracellular T3 is local conversion from T4; thus, the nuclear receptor binding is dependent on the availability of T4 and T3 to the cell and hormone availability is decreased with aging.

After *in vivo* administration of radiolabeled T4, cytoplasmic-free T4 and T3 values are lower, but bound values are higher in brain (cerebral hemispheres) of old rats. Furthermore, the amount of T3 derived intracellularly from conversion of T4 is significantly reduced in the old animals, and nuclear binding of both T3 and, to a lesser extent, T4 is also reduced.[29] These data indicate a reduction with aging of intracellular production of T3 (from T4). At the same time, plasma T4 levels decrease with aging in Long–Evans rats. These two factors — reduced secretion of T4 from the thyroid gland and reduced intracellular conversion of T4 to T3 — would be responsible for a reduced intracellular T3 binding and consequent reduced T3 actions (Figure 14-7).

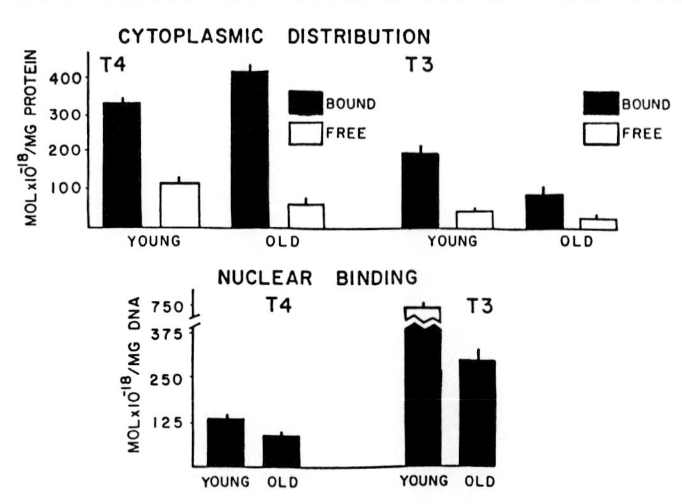

FIGURE 14-7 *Changes with age in T3 and T4 cytoplasmic distribution and nuclear binding.* Bound T4 increases and free T4 decreases in the cerebral hemispheres of young (2-month-old) as compared to old (24-month-old) male Long–Evans rats. Lower T3 levels in older animals suggest depressed T4 deiodination with aging. The depressed free cytoplasmic T4 and T3 levels are also reflected in decreased nuclear binding of these hormones, particularly for T3. Similar results were also reported in the liver. (Drawn from data taken from Reference 29.)

3 EFFECTS OF THYROID HORMONES

3.1 Major Actions

Of the major actions of thyroid hormones listed in Table 14-3, those directing growth and development cease once adulthood has been reached. In some instances, however, thyroid hormones appear to promote growth of adult tissues, even those which reach maturity early in life, such as the nervous tissue. For example, a facilitatory action of thyroid hormones has been proposed —although currently disputed — in the recovery of spinal cord and peripheral nerve injuries in humans and other animals. Catch-up whole-body growth and rehabilitation of some behaviors, impaired by thyroid deficiency at an early age, have been described in adult rats after return to the euthyroid state.[30-32]

Thyroid hormones also modulate such widespread processes as oxygen consumption, regulation of basal metabolic rate, and calorigenesis. Other actions of thyroid hormones that may be relevant to the aging process and affect the lifespan are those involved in

- Intermediary metabolism, particularly cholesterol metabolism
- Regulation of cardiac output due to stimulatory interaction of thyroid hormones and catecholamines on cardiac rate
- Decreased peripheral resistance due to increased cutaneous vasodilation
- Behavioral changes such as rapid mentation, irritability, restlessness, probably mediated through neurotransmitter effects and not through changes in oxygen consumption, which is unaffected by thyroid hormones in adult and aged brain

For a better comprehension of some of these actions, the reader is referred to the corresponding chapters (Chapters 8 to 10 for the Nervous System and Chapters 16 and 17 for the Cardiovascular System). We will consider here metabolic rate, calorigenesis, and cholesterol metabolism.

TABLE 14-3

Thyroid Hormones Regulate
Calorigenesis
Metabolism
Brain maturation
Behavior
Growth and development

3.2 Basal Metabolic Rate

Basal metabolic rate (BMR) is the rate at which oxygen is consumed and carbon dioxide is produced under conditions of physical and mental rest and in comfortable ambient temperature. BMR, significantly influenced by thyroid hormones, decreases consistently with aging, but the magnitude of the fall varies with the study and the criterion of measurement (body weight versus lean body mass). BMR is customarily calculated in terms of oxygen consumption in kilocalories per unit of time and of body surface. With this method, the value for a healthy (euthyroid) male weighing 80 kg and aged 20 to 30 years is 39.5 kcal/m^2 of body surface per hour.

With aging, from adolescence to old age, metabolic rate decreases progressively, in both men and women — in the latter, values are always slightly lower than in men — from 46 kcal in men and 43 in women at 14 to 16 years of age to 35 and 33, respectively, at 70 to 80 years. The cause of this progressive decline in BMR is not definitely known. It may be associated, at least in part, with the age-related declining levels of thyroid hormones, primarily T3, although, as described above, this decline appears generally slight and does not occur in all aged individuals.

The decreased BMR with aging has also been ascribed to a progressive increase in adipose body mass (less metabolically active) relative to lean body mass (more metabolically active) (Chapter 21). The metabolically active cytoplasmic mass would decrease with aging, as shown by an age-related decline in total body potassium (the major intracellular electrolyte) and urinary excretion of creatine (a product of muscle metabolism). However, these metabolic alterations themselves could also result from the changing thyroid hormone levels.

Is Aging Associated With Alterations in Hormone Metabolism and Tissue Demand Rather Than Inadequate Secretion?

As previously discussed in relation to the adrenal, and as will be seen in relation to the pancreas and insulin, the peripheral metabolism of thyroid hormones and tissue demands for these hormones appears to be altered with aging; thus, T4 and corticosteroid turnover rates decline and insulin resistance appears to develop with age. Because the circulating hormone levels are maintained constant, several authors have concluded that peripheral demand for the metabolic hormones declines with increasing age. The most common explanation offered for this decline in demand is the loss of metabolically active tissue mass with aging, as suggested above. However, this explanation has several flaws: endocrine changes begin early in life, before significant changes in body composition; changes in body composition do not occur in all individuals and when they occur they may be secondary to the endocrine changes.

If a smaller lean body mass cannot account for the reduction in peripheral hormone metabolism, perhaps an explanation may be found elsewhere, for example, in some metabolic alteration in the target tissues, such as an overall decline in protein synthesis with aging. Thus, endocrine changes which occur with aging may ultimately be associated with a decline in overall protein synthesis which begins with the cessation of growth and continues to fall throughout the duration of the lifespan. Thyroid hormones with their widespread actions on cell metabolism represent an ideal model to illustrate age-related changes in metabolic-endocrine interrelations.

Hypothyroidism as a "Protective" Response in the Elderly

Thyroid hormones, necessary for growth and development when energy requirements are high, may become detrimental when the only energy needed is for homeostasis. A selective impairment of the general anabolic actions of thyroid hormones may occur with cessation of growth and subsequent aging without a concomitant loss of catabolic effects. For example, injections of high doses of T4 in young animals are tolerated quite well; the animals respond with increased appetite and more rapid growth. The same doses injected to the adult animals result in muscle wasting and weight loss. These and other findings have led to the suggestion that "there is an homeostatic wisdom in the arrangement whereby conversion of T4 to T3 is inhibited when catabolism is already overactive."[17] From this perspective, the age-related decline in thyroid hormone metabolism would reflect not so much a reduced demand for tissue utilization as an increased need for protection against the catabolic actions of these hormones.[33] Therefore, the reduced intracellular availability of T3 for nuclear binding and consequent effects on protein synthesis may represent a compensatory beneficial response to the decreased metabolic needs of the aging individual.

3.3 Calorigenesis

Thyroid hormones stimulate oxygen consumption in almost all tissues with the exception of adult brain, testes, uterus, lymph nodes, spleen, and anterior pituitary. Increased oxygen consumption leads to an increased cellular metabolic rate and this to increased heat production or calorigenesis.

The magnitude of the calorigenic effect of thyroid hormones depends on several factors, such as interaction with catecholamines and initial metabolic rate, the higher the levels of catecholamines and the lower the metabolic rate at the time of T3 administration, the greater the calorigenic effect. The calorigenic effect of thyroid hormones contributes to the maintenance of body temperature together with a number of other metabolic and neural adjustments. With aging, thermoregulation is progressively impaired, and this decline may be due in part to alterations in thyroid function.

3.4 Thermoregulatory Changes

Thermoregulation involves a series of adjustments destined to maintain body temperature constant in homeothermic (warm-blooded) animals, including humans. These adjustments maintain a balance between heat production (stimulated by muscular exercise, assimilation of food, and hormones regulating basal metabolism) and heat loss (induced by conduction and radiation through the skin and mucosae, sweat, respiration, urination, and defecation). The balance depends on a group of reflex and hormone responses that are integrated in the hypothalamus and operate to maintain body temperature within a narrow range in spite of wide fluctuations of environmental temperature.

The well-documented *increased susceptibility of older people to hypothermia and heat stroke* reflects the less-efficient temperature regulation that commonly is associated with aging. In the elderly, thermoregulatory inefficiency to cold or heat is usual and results from

- Decreased heat production
- Decreased body mass
- Reduced muscle activity
- Less efficient shivering
- Reduced sweating response
- Less efficient vasomotor responses
- Decline in temperature perception

Thus, alterations occur at several levels of thermoregulation: in peripheral temperature sensation, at hypothalamic autonomic and neuroendocrine centers, and in higher cerebrocortical centers, which control perception and coordinate the multiple inputs that determine the effectiveness of adaptative adjustments.[34] Healthy older people fail to respond with vasoconstriction to cooling and show diminished or increased sensitivity to cold.[35] The higher mortality of the elderly during "heat" or "cold waves" is well documented and may be attributed not only to physiologic decrements but often to economic (inadequate diet, poor clothing, housing) and emotional and mental (depression, dementia) impairments.

In rodents, body temperature falls lower and takes longer to return to normal in old rats immersed in cold water than in younger animals.[36,37] Dietary restriction (i.e., tryptophan deficiency), initiated at weaning, continued to 1 year of age and followed by a normal diet improves thermoregulatory responses. The fall in temperature after cold exposure is less severe and the return to normal more rapid. This persistence of efficient thermoregulation in old animals has been interpreted as a benefit of the delayed aging induced by dietary tryptophan restriction.

Responses to heat are also impaired with aging. For example, the onset of sweating is slower in the elderly.[38] As a consequence of the impaired thermoregulatory competence, the elderly have also a reduced fever response.[39,40] Fever is the most universal hallmark of disease and depends on a "resetting" of the hypothalamic thermostat in response to a variety of agents such as bacteria responsible for the production of interleukin-1 which acts on lymphocytes (Chapter 7). Interleukin-1 enters the brain where it stimulates the local production of prostaglandins activating the fever response. Fever responses are dampened in old humans and old animals.

3.5 Cholesterol Metabolism

Serum cholesterol levels rise with aging in both the human and the rat despite a fall in hepatic cholesterol synthesis indicating a net reduction in overall turnover. A similar situation occurs in hypothyroidism, prompting several attempts to treat hyperlipedemia with thyroid hormones and their analogs.[41] Several studies suggest that a causal relationship exists between decreased thyroid activity and elevated serum cholesterol. In young rats, thyroidectomy results in reduced cholesterol turnover and elevated serum levels, while in old rats there is no significant postoperative change. Similar results have been observed in a wide variety of mammalian species. The implication is that there is a reduction in thyroid hormone secretion and/or a decrease in the sensitivity of cholesterol metabolism to thyroid hormones. In humans, a hyperbolic fall in serum cholesterol occurs with increasing serum T3 concentrations,[42] and a similar inverse correlation is seen between declining serum T3 and the age-related increase in serum cholesterol levels. Taken together, these results support the inference that age changes in cholesterol metabolism may be secondary to age changes in the thyroid axis.

The specific defect in cholesterol metabolism that develops in the hypothyroid state would result from a reduction in turnover of the serum low-density lipoproteins (LDL), elevated in both hypothyroidism and aging, while metabolism of high-density lipoproteins (HDL) remains unchanged. The significance of altered lipoprotein metabolism in the development of atherosclerosis (Chapter 17) suggests that early attempts by investigators to relate declining thyroid function to factors predisposing to atherosclerotic lesions may not be entirely without foundation.[43]

Thyroid Status and Longevity

As noted previously, the lower circulating T3 in the elderly and T4 in the aged rat, and the reduced conversion of T4 to T3 in target tissues, may represent a beneficial compensation to the catabolic actions of the hormones; a certain "homeostatic wisdom", reflecting the changing metabolic needs of the organism. This is supported by studies in which rats made hypothyroid neonatally outlived the corresponding controls by about 4 months for males and 2 months for females.[3,4] In these experiments, the mortality of hypothyroid rats was similar to that of controls until 24 months old but markedly lower thereafter. Maximum life duration was 35 months for male hypothyroid and 31 for male controls; it was 38 months for female hypothyroid and 36 months for female controls. The sex difference may be related to the higher T3 and T4 levels in males than in females and their greater reduction through neonatal intervention. The life-extending effects of hypothyroidism resemble those found after pituitary ablation.[5] They also resemble effects produced by food restriction, for the body weight of hypothyroid animals is significantly reduced and the inhibition of growth may act as an antiaging factor (Chapter 24). When the thyroid hormone levels are increased through the administration of exogenous T4 over many months (12 and 22 months), the average lifespan is significantly shortened in the treated animals.[3,4] If T4 treatment is initiated at an already senescent age (26 months), the lifespan is not affected. Thus, the life-shortening effects of excess thyroid hormones are due not to the direct action of the hormones to initiate or promote old-age diseases, the direct cause of death, but rather to acceleration of the aging process which may result, in turn, from a more rapid timetable of development. Thyroid hormones seem to act as pacemakers capable not only of controlling certain key events during development (the metamorphosis of tadpoles into frogs) but also of intervening in the initiation of the aging processes.[44]

4 ABNORMAL THYROID STATES IN THE ELDERLY

The precariousness of the euthyroid state in the elderly is translated into a greater frequency of thyroid disorders with age. This age-related increase is often masked by the atypical manifestations of the disease. Thus, although modalities for treatment of thyroid disease are readily available and straightforward, the subtleties of diagnosis in the elderly often provide a challenge for the clinician. A diagnosis of thyroid disease in the elderly can be delayed or missed entirely because the possibility of thyroid disease is overlooked. In the elderly, signs and symptoms of thyroid disease frequently are minimal or atypical and are commonly assumed to be caused by either the normal aging process or other diseases.[8,15,45-49]

At all ages, thyroid disease is 3 to 14 times more frequent in women than in men. Abnormalities of thyroid function occur more often in older patients who are institutionalized or ill than in the healthier elderly who live in the community.

TABLE 14-4

Common Signs and Symptoms of Hyperthyroidism in the Elderly

Cardiovascular abnormalities
Congestive heart failure symptoms
Atrial fibrillation
Angina
Pulmonary edema
Tremor
Nervousness
Weakness
Weight loss and anorexia (poor appetite)
Palpable goiter (may not be present)
Eye findings (may not be present)
Thyroid nodules (nonspecific)

Overall, however, the prevalence of hypothyroidism is said to be between 0.5 to 4.4% in the elderly. Likewise, the prevalence of hyperthyroidism has been reported to be between 0.5 and 3%. As already discussed, the prevalence of thyroid nodules increases with age, reaching 5% by age 60. Nodules are present in 90% of women over the age of 70 and 60% of men over the age of 80. This makes it impractical to routinely perform a workup on all elderly patients with thyroid nodules.

4.1 Hyperthyroidism

Thyroid hormone excess or hyperthyroidism is caused by either Graves' disease (i.e., diffuse toxic goiter) or nodular toxic goiter. Because the signs and symptoms that occur in older patients are often attributed to another illness, the term "masked hyperthyroidism" has been introduced (Table 14-4). Cardiovascular abnormalities are quite common in older patients with hyperthyroidism. In one study, 79% of subjects had an abnormal cardiovascular examination, 67% had symptoms of congestive heart failure, 39% were in atrial fibrillation, 20% had symptoms of angina, and 8% presented in pulmonary edema. Tachycardia, or fast heart beat, even in the presence of atrial fibrillation, is less impressive in older than in younger hyperthyroid patients. Typical signs and symptoms such as tremor, nervousness, and muscular weakness may be ignored because they are considered common in the elderly. Weight loss, another common symptom in the elderly, may lead to an evaluation for malignant gastrointestinal lesion. A goiter may or may not be palpable and the eye findings of Graves' disease may be absent. Since thyroid nodules are common in the elderly, it is difficult to arouse suspicion of significant thyroid disease solely on the basis of their presence.

4.2 Apathetic Hyperthyroidism

Apathetic hyperthyroidism is another term used for a thyroid disease in the elderly, the characteristics of which could potentially be confused with a hypothyroid state (Table 14-5). Laboratory diagnosis of hyperthyroidism includes assay of serum T4 and T3 determination, and a radionuclide thyroid scan can help differentiate Graves' disease from toxic multinodular goiter.

4.3 Treatment

Treatment of hyperthyroidism includes medical, radiological, and surgical approaches single or combined. In the medical treatment, the goal is to reduce thyroid hormone production. The most commonly used antithyroid medication

TABLE 14-5

Characteristics of Apathetic Hyperthyroidism

Blunted effect, i.e., withdrawal behavior with apathy and depression
Absence of hyperkinetic motor activity
Slowed mentation
Proximal muscle weakness
Edema of the lower extremity
Droopy eyelids
Cardiovascular abnormalities
Diarrhea

is oral propylthiouracil. The major disadvantage is the recurrence of hyperthyroidism after treatment is stopped. Administration of radioiodide is usually the preferred treatment in the elderly. The isotope is actively collected by the thyroid gland, particularly in the regions of active proliferation, and the fast-growing and secreting tissue is destroyed. Radioiodide can be administered in conjunction with medical treatment since the effect of the former is gradual and not usually complete until 3 months after treatment. Close follow-up is necessary in view of the possibility of postradioiodide hypothyroidism.[50] Surgery of the thyroid gland, an uncommon intervention in the elderly, may be necessary for large multinodular glands resistant to radioiodide.

4.4 Hypothyroidism

Thyroid hormone deficiency, or hypothyroidism, is caused by autoimmune thyroiditis or as a consequence of treatment for hyperthyroidism. Just as "masked hyperthyroidism" exists in the elderly, so does "masked hypothyroidism" (Table 14-6). Typical symptoms of hypothyroidism such as fatigue, weakness, dry skin, hair loss, constipation, mental confusion, depression, and cold intolerance may be attributed to old age instead of thyroid disease. In addition, the insidious onset and slow progression of hypothyroidism makes it more difficult to diagnose. An elevated TSH is the most reliable symptom of hypothyroidism since T4 may or may not be decreased in mild cases.

Common signs and symptoms in elderly hypothyroidism are shown in Table 14-6.

Treatment of hypothyroidism is thyroid replacement in the form of oral levothyroxine.[51] Follow-up determination that TSH levels have returned to normal is the best indication that adequate replacement has been achieved. Over-replacement is dangerous in older patients and especially in those with cardiac disease. Safe replacement should be started with low initial doses of 0.025 to 0.05 mg of levothyroxine per day. Maintenance dose in the elderly is generally lower than in younger patients and, once established, usually remains constant (around 0.1 mg per day).

In view of the difficulties in clinically diagnosing thyroid disease in the elderly, the role of laboratory screening becomes very important.[52] Although arguments have been made that such screening, including a serum TSH, in the healthy ambulatory population of elderly is not worthwhile, certain guidelines can be established (Table 14-7).

Thyroid disease states, although potentially difficult to diagnose clinically in the elderly, can be detected if the possibility is entertained. The relative ease and success of treatment makes such detection well worthwhile.

TABLE 14-6

Frequently Missed Common Signs and Symptoms of Hypothyroidism in Elderly Patients

1. Cardiovascular abnormalities such as dyspnea (a sensation of shortness of breath) in over one half of patients, chest pain is present in up to one quarter, an enlarged heart in most, and bradycardia (or a slow heart beat).
2. Anorexia (poor appetite), and constipation are common.
3. Muscular weakness is noted by one-half of patients.
4. Mild anemia is present in roughly one half of patients.
5. Depression is present in up to 60%.
6. Joint pain has also been described as a symptom.

5 THE PARATHYROID AND THYROID C CELLS

In humans, the four parathyroid glands, embedded in the thyroid gland, secrete the parathyroid hormone (PTH). So-called C cells dispersed throughout the thyroid gland, the parathyroids, and thymus secrete another hormone, calcitonin (CT). Both hormones play a significant role in the maintenance of calcium homeostasis (Chapter 21). They do not appear to be consistently altered with aging.

5.1 Major Actions of Parathyroid Hormone and Calcitonin

PTH, a polypeptide secreted in response to hypocalcemia, raises the concentration of plasma calcium by

- Increasing renal calcium reabsorption
- Mobilizing calcium from bones by stimulating osteoclastic activity (i.e., destruction of bone cells)
- In the presence of adequate amounts of vitamin D, stimulating the absorption of calcium from the small intestine
- Lowering levels of plasma inorganic phosphate by inhibiting renal reabsorption of phosphate.

Calcitonin, also a polypeptide, is secreted not only in response to calcium levels but also to gastrointestinal hormones such as glucagon. CT receptors are found in bones and the kidneys, and the major actions of this hormone involve regulation of plasma calcium levels. They include (1) lowering of plasma calcium and phosphate levels by inhibiting bone resorption and (2) increasing calcium excretion in urine. CT has also some minor action on water and electrolytes and decreases gastric acid secretion.

5.2 Changes with Aging

Studies of structural changes in the aging human parathyroids are few, and those studies from laboratory animals reveal only minor changes, such as the presence of degenerating cells containing colloid and mitochondria showing bizarre

TABLE 14-7

Conditions for Thyroid Laboratory Screening in Elderly Patients (Over 60 Years Old)

Admission to an acute care hospital, nursing home, or psychiatric ward
Dementia or altered mental status
Proximal myopathy
Any tachyarrhythmia including atrial fibrillation

patterns. Structural changes of C cells are rare, although, in old rats, the ratio of C cells to thyroid cells appears to increase. A study of immunoassayable PTH shows a decline after 60 years of age in white men but not in white women, with variable effects depending on the race, while in other studies the hormone levels increase with aging.[53] In these experiments, some of the biologically inactive fragments of the hormone molecule may be responsible for the high immunoassayable levels.

Both PTH and CT are derived by the action of proteases from larger pre-prohormones, which are less active biologically but immunologically similar; with aging, the processing of the precursor hormones may be altered, resulting in the secretion of the less-active pre-prohormones.[54] Little is known of the changes with aging in CT and whether and how they affect the aging of bone. A decrease in CT has been reported in humans and is greater in men than women. Since CT decreases bone resorption, its potential usefulness in the therapy of age-related demineralization and osteoporosis has been explored but without beneficial results. Further discussion of the potential role of these and other hormones in aging of bone is presented in Chapter 21.

REFERENCES

1. Robertson, T. B.. The influence of thyroid alone and of thyroid administered together with nucleic acids upon the growth and longevity of the white mouse, *Aust. J. Exp. Biol. Med. Sci.*, 5, 69, 1928.
2. Crewe, F. A. E., Rejuvenation of the aged fowl through thyroid medication, *Proc. R. Soc.*, 45, 252, 1924–1925.
3. Ooka, H., Fujita, S., and Yoshimoto, E., Pituitary-thyroid activity and longevity in neonatally thyroxine-treated rats, *Mech. Ageing Dev.*, 22, 113, 1983.
4. Ooka, H. and Shinkai, T., Effects of chronic hyperthyroidism on the lifespan of the rat, *Mech. Ageing Dev.*, 33, 275, 1986.
5. Everitt, A. V., The thyroid gland, metabolic rate and aging, in *Hypothalamus, Pituitary and Aging,* Everitt, A. V. and Burgess, J. A., Eds., Charles C Thomas, Springfield, IL, 1976, 511.
6. Gregerman, R. I. and Davis, P. J., Effects of intrinsic and extrinsic variables on thyroid hormone economy. Intrinsic physiologic variables and non-thyroidal illness, in *The Thyroid,* Werner, S. C. and Ingbar, S. H., Eds., Harper and Row, New York, 1978, 223.
7. Ingbar, S. H., The influence of aging on the human thyroid hormone economy, in *Geriatric Endocrinology,* Greenblatt, R. B., Ed., Raven Press, New York, 1978, 13.
8. Cole, G. M., Segall, P. E., and Timiras, P. S., Hormones during aging in *Hormones in Development and Aging,* Vernadakis, A. and Timiras, P. S., Eds., SP Medical & Scientific Books, New York, 1982, 477.
9. Green, M. F., The endocrine system, in *Principles and Practice of Geriatric Medicine,* Pathy, M. S. J., Ed., John Wiley & Sons, Chichester, England, 1991, 1061.
10. Blumenthal, H. T. and Perlstein, I. B., The aging thyroid. I. A description of lesions and an analysis of their age and sex distribution, *J. Am. Geriatr. Soc.*, 35, 843, 1987.
11. Hegedus, L., Perrild, H., Poulsen, L. R., Anderson, J. R., Holm, B., Schnohr, P., Jensen, G., and Hansen, J. M., The determination of thyroid volume by ultrasound and its relationship to body weight, age, and sex in normal subjects, *J. Clin. Endocrinol. Metab.*, 56, 260, 1983.
12. Gregerman, R. I. and Solomon, N., Acceleration of thyroxine and triiodothyronine turnover during bacterial pulmonary infections and fever, implications for the functional state of the thyroid during stress and in senescence, *J. Clin. Endocrinol. Metab.*, 27, 93, 1967.
13. Studer, H., Riek, M. M., and Greer, M. A., Multi-nodular goiter, in *Endocrinology,* Grune & Stratton, De Groot, L. J., Cahill, G. F., and Odell, W. D., Eds., New York, 1979, 489.
14. Tunbridge, W. M., Evered, D. C., Hall, R., Appleton, D., Brewis, M., Clark, F., Evans, J. G., Young, E., Brid, T., and Smith, P. A., The spectrum of thyroid disease in a community, the Whickham survey, *Clin. Endocrinol.*, 7, 481, 1977.
15. Runnels, B. L., Garry, P. J., Hunt, W. C., and Standefer, J. C., Thyroid function in a healthy elderly population, implications for clinical evaluation, *J. Gerontol. Biol. Sci.*, 46, B39, 1991.
16. Bermudez, F., Surks, M. I., and Oppenheimer, J. H., High incidence of decreased serum T3 concentration in patients with nonthyroidal disease, *J. Clin. Endocrinol. Metab.*, 41, 27, 1975.
17. Chopra, I. J., *Triiodothyronines in Health and Disease,* Monographs in Endocrinology, Vol. 18, Springer-Verlag, New York, 1981.
18. Herrmann, J., Heinen, E., Kroll, H. J., Rudorff, K. H., and Kruskemper, H. L., Thyroid function and thyroid hormone metabolism in elderly people, low T3 syndrome in old age?, *Klin. Wochenschr.*, 59, 315, 1974.
19. Ooka, H., Changes in extrathyroidal conversion of thyroxine (T4) to triiodothyronine (T3) in vitro during development and aging of the rat, *Mech. Ageing Dev.*, 10, 151, 1979.
20. Choy, V. J., Klemme, W. R., and Timiras, P. S., Variant forms of immunoreactive thyrotropin in aged rats, *Mech. Ageing Dev.*, 19, 273, 1982.
21. Rock, R. C., Interpreting thyroid tests in the elderly: updated guidelines, *Geriatrics,* 40, 61, 1985.
22. Klug, T. L. and Adelman, R. C., Evidence for a large thyrotropin and its accumulation during aging in rats, *Biochem. Biophys. Res. Commun.*, 77, 1431, 1977.
23. Murialdo, G., Costelli, P., Fonzi, S., Parodi, C., Torre, F., Cenacchi, T., and Polleri, A., Circadian secretion of melatonin and thyrotrohpin in hospitalized aged patients, *Aging,* 5, 39, 1993.
24. Klug, T. L. and Adelman, R.C., Altered hypothalamic-pituitary regulation of thyrotrophin in male rats during aging, *Endocrinology,* 104, 1136, 1979.
25. Blichert-Toft, M., Hummer, L., and Dige-Petersen, H., Human thyrotrophin level and response to thyrotrophin-releasing hormone in the aged, *Gerontol. Clin.*, 17, 191, 1975.
26. Ohara, H., Kobayahi, T., Shiraishi, M., and Wada, T., Thyroid function of the aged as viewed from the pituitary-thyroid system, *Endocrinol. Jpn.*, 21, 377, 1974.
27. Naidoo, S. and Timiras, P. S., Effects of age on the metabolism of thyroid hormones by rat brain tissue in vitro, *Dev. Neurosci.*, 2, 213, 1979.
28. Eberhardt, N. L., Valcana, T., and Timiras, P. S., Triiodothyronine nuclear receptors. An in vitro comparison of the binding of triiodothyronine to nuclei of adult rat liver, cerebral hemisphere and anterior pituitary, *Endocrinology,* 102, 556, 1978.
29. Margarity, M., Valcana, T., and Timiras, P. S., Thyroxine deiodination, cytoplasmic distribution and nuclear binding of thyroxine and triiodothyronine in liver and brain of young rats, *Mech. Ageing Dev.*, 29, 181, 1985.
30. Meisami, E., Complete recovery of growth deficits after reversal of PTU-induced postnatal hypothyroidism in the female rat: a model for catch-up growth, *Life Sci.*, 34, 1487, 1984.

31. Tamasy, V., Meisami, E., Vallerga, A., and Timiras, P. S., Rehabilitation from neonatal hyperthyroidism: spontaneous motor activity, exploratory behavior, avoidance learning and responses of pituitary-thyroid axis to stress in male rats, *Psychoneuroendocrinology,* 11, 91, 1986.
32. Tamasy, V., Meisami, E., Du, J.-Z., and Timiras, P. S., Exploratory behavior, learning ability, and thyroid hormonal responses to stress in female rats rehabilitating from postnatal hyperthyroidism, *Dev. Psychobiol.,* 19, 537, 1986.
33. Utiger, R. D., Decreased extrathyroidal triiodothyronine production in nonthyroidal illness: benefit or harm?, *Am. J. Med.,* 69, 807, 1980.
34. Collins, K. J., Autonomic failure and the elderly, in *Autonomic Failure,* Bannister, R., Ed., Oxford University Press, New York, 1983, 489, 33.
35. Collins, K. J., Effects of cold on old people, *Br. J. Hosp. Med.,* 38, 506, 1987.
36. Segall, P. E. and Timiras, P. S., Age-related changes in thermoregulatory capacity of tryptophan-deficient rats, *Fed. Proc.,* 34, 83, 1975.
37. Segall, P. E. and Timiras, P. S., Pathophysiologic findings after chronic tryptophan deficiency in rats: a model for delayed growth and aging, *Mech. Ageing Dev.,* 5, 109, 1976.
38. Foster, K. G., Ellis, F. P., Dore, C., Exton-Smith, A. N., and Wiener, J. S., Sweat responses in the aged, *Age Aging,* 5, 91, 1976.
39. Petersdorf, R. G., Disturbances of heat regulation, and chills and fever, in *Harrison's Principles of Internal Medicine,* Isselbacker, K. J., Adams, R. D., Braunwald, E., Petersdorf, R. G., and Wildon, J. D., Eds., McGraw-Hill, New York, 1980, 43.
40. Norman, D. C., Grahn, D., and Yoshikawa, T. T., Fever and aging, *J. Am. Geriatr. Soc.,* 33, 859, 1985.
41. Burrow, G. N., Thyroid hormone therapy in nonthyroid disorders, in *The Thyroid, A Fundamental and Clinical Text,* Werner, S. C. and Ingbar, S. H., Eds., Harper and Row, Hagerstown, MD, 1978, 974.
42. Bantle, J. P., Dillmann, W. H., Oppenheimer, J. H., Bingham, C., and Runger, G. C., Common clinical indices of thyroid hormone action: relationships to serum free 3,5,3′-triiodothyronine concentration and estimated nuclear occupancy, *J. Clin. Endocrinol. Metab.,* 50, 286, 1980.
43. Wren, J. C., Thyroid function and coronary atherosclerosis, *J. Am. Geriatr. Soc.,* 16, 696, 1968.
44. Walker, R. F. and Timiras, P. S., Pacemaker insufficiency and the onset of aging, in *Cellular Pacemakers,* Carpenter, D., Ed., John Wiley & Sons, New York, 1982, 345.
45. Gambert, S. R., Atypical presentation of thyroid disease in the elderly, *Geriatrics,* 40, 63, 1985.
46. Hurley, J. R., Thyroid disease in the elderly, *Med. Clin. N. Am.,* 67, 497, 1983.
47. Klein, I. and Levey, G. S., Unusual manifestations of hypothyroidism, *Arch. Intern. Med.,* 144, 123, 1984.
48. Heikoff, L. E., Luxenberg, J., and Feigenbaum, L. Z., Low yield of screening for hypothyroidism in healthy elderly, *J. Am. Geriatr. Soc.,* 32, 616, 1984.
49. Sawin, C. T., Castelli, W. P., Hershman, J. M., McNamara, P., and Bacharach, P., The aging thyroid. Thyroid deficiency in the Framingham Study, *Arch. Intern. Med.,* 145, 1386, 1985.
50. Young, R. E., Jones, S. J., Bewsher, P. D., and Hedley, A. J., Age and the daily dose of thyroxine replacement therapy for hypothyroidism, *Age Aging,* 13, 293, 1984.
51. Rosenbaum, R. L. and Barzel, U. S., Levothyroxine replacement dose for primary hypothyroidism decreases with age, *Ann. Intern. Med.,* 96, 53, 1982.
52. Walfish, P. G. and Gryfe, C. I., Testing your older patient's thyroid function, *Geriatrics,* 37, 135, 1982.
53. Endres, D. B., Morgan, C. H., Garry, P. J., and Omdahl, J. L., Age-related changes in serum immunoreactive parathyroid hormone and its biologic action in healthy men and women, *J. Clin. Endocrinol. Metab.,* 65, 724, 1987.
54. Wongsurawat, N. and Armbrecht, H. J., Comparison of calcium effect on in vitro calcitonin and parathyroid hormone release by young and aged thyroparathyroidal glands, *Exp. Gerontol.,* 22, 263, 1987.

15 THE ENDOCRINE PANCREAS AND CARBOHYDRATE METABOLISM

Paola S. Timiras

The age-reduced ability to maintain glucose homeostasis after a glucose challenge has long been realized.[1] The mechanisms of this age-related impairment in carbohydrate economy with aging have been extensively investigated and their elucidation is now emerging despite several confounding factors. These confounding factors are nonspecific and are shared by many other body functions. They include

- Difficulty of differentiating the effects of growth and development from those of aging
- Composition of the diet with regard both to caloric and carbohydrate content
- Exercise
- Body composition with particular reference to body fat content
- Presence of pathologic conditions
- Role of medications and drugs

These factors are also applicable to animal models, especially rodents which may present multiple, but not all, signs of age-related carbohydrate alterations.

In this chapter, we will examine first the aging of the endocrine function of the pancreas and then the effects of this aging on carbohydrate metabolism and lifespan.

1 AGING OF THE ENDOCRINE PANCREAS

Several hormones participate in the regulation of carbohydrate metabolism. Four of them are secreted by the cells of the islets of Langerhans in the pancreas: two, insulin and glucagon, with major actions on glucose metabolism and two — somatostatin and pancreatic polypeptide — with modulating actions on insulin and glucagon secretion. Other hormones affecting carbohydrate metabolism include epinephrine, thyroid hormones, glucocorticoids, and growth hormone. Some of their actions have been discussed in the respective chapters (11 and 14).

Structure and Function of the Pancreas

The pancreas lies inferior to the stomach, in a bend of the duodenum. It is both an endocrine and an exocrine gland. The exocrine functions are concerned with digestion (Chapter 20). The endocrine function consists primarily of the secretion of the two major hormones, insulin and glucagon. Four cell types have been identified in the islets, each producing a different hormone with specific actions:

- A cells produce glucagon.
- B cells produce insulin.
- D cells produce somatostatin.
- F or D1 cells produce pancreatic polypeptide.

These hormones are all polypeptides. Insulin is secreted only by the B cells whereas the other hormones are also secreted by the gastrointestinal mucosa, and somatostatin is also found in the brain.

Both insulin and glucagon are important in the regulation of carbohydrate, protein and lipid metabolism:

- *Insulin is an anabolic hormone; that is, it increases the storage of glucose, fatty acids and amino acids in cells and tissues;*
- *Glucagon is a catabolic hormone, that is, it mobilizes glucose, fatty acids, and amino acids from stores into the blood.*

Somatostatin may regulate, locally, the secretion of the other pancreatic hormones; in brain (hypothalamus) and spinal cord it may act as a neurohormone and neurotransmitter. The function and origin of the pancreatic polypeptide are still uncertain although the hormone may influence gastrointestinal function and promote intra-islet homeostasis. A diagram of an islet of Langerhans is presented in Figure 15-1 and a list of the pancreatic hormones in Table 15-1.

1.1 Changes with Aging

With aging, few morphologic changes have been reported in the endocrine pancreas in humans.[2,3] Among these are

- A certain degree of atrophy
- An increased incidence of tumors
- Deposition of amyloid material and lipofuscin granules

In rats, a decrease in the proportion of large to small islets and an increase in B cells has been interpreted as a possible compensatory mechanism for the decreased responsiveness of tissues to insulin.[4] Somatostatin levels also increase with aging, and treatment of islets with antisomatostatin antibodies may partially reverse impairments of the glucose-stimulated insulin response.[5] The totality of these changes is, in most cases, minor and does not produce significant cellular alterations, except perhaps for decreased cAMP activity.[6]

Secretion and Actions of Insulin

Insulin is synthesized in B cells as part of a larger preprohormone — preproinsulin — which includes a 23-amino acid leader sequence attached to proinsulin; this leader sequence is lost upon entrance of the molecule into the endoplasmic reticulum leaving the proinsulin molecule. Kallikrein, an enzyme present in the islets, aids in the conversion of proinsulin to insulin. In this conversion, a C peptide chain is removed from the proinsulin molecule producing the disulfide-connected A and B chains that are insulin.

Insulin secretion is pulsatile and is regulated by a variety of stimulatory and inhibitory factors, most of them related to glucose metabolism and the effects of cAMP. Insulin secretion is stimulated by high blood glucose levels and reduced when blood glucose is low. Other stimulatory factors include several amino acids, intestinal hormones, acetylcholine (parasympathetic stimulation), and others. Inhibitory factors include somatostatin, norepinephrine (sympathetic stimulation), and others.

Once in the circulation, insulin is degraded within minutes in the liver and kidneys. C peptide and Kallikrein are also present in the circulation, having been secreted with the insulin. Antibodies to com-

0-8493-8979-8/94/$0.00+$.50

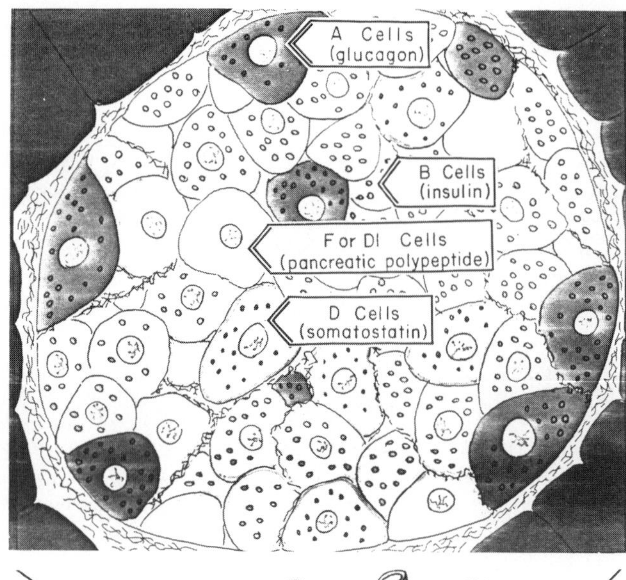

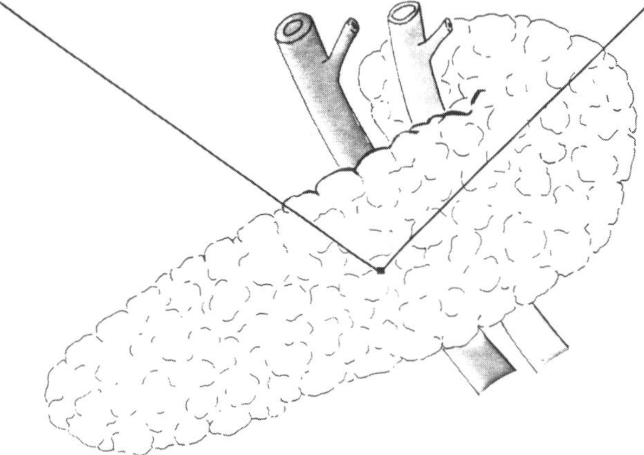

FIGURE 15-1 *Diagrammatic representation of the pancreas with pancreatic cells.*

ponents of islet cells have been detected in a high proportion of patients with insulin-dependent diabetes, that is, diabetes due to insulin deficiency. Antibody attack on B cells leads to extensive loss of these cells, characteristic of insulin-dependent diabetes and initiated by genetic mechanisms.

Insulin binds with specific membrane receptors forming an insulin–receptor complex, which is taken into the cell by endocytosis. Insulin receptors are found in almost all cells of the body. The insulin–receptor, a tetramer, is made up of two alpha and two beta glycoprotein subunits. The beta subunit is a protein kinase that catalyzes the phosphorylation of proteins, an activity resulting in a change in the number of "transporters", i.e., protein carriers of glucose. Intracellular free glucose concentration is low (due to rapid, efficient phosphorylation of glucose); therefore, a certain amount of glucose moves into the cell even in the absence of insulin. With insulin, however, the rate of glucose entry is much increased due to facilitated diffusion as mediated by transporters.

The insulin–receptor complex enters the lysosomes where it is cleaved, the hormone internalized, and the receptor recycled. Increased circulating levels of insulin reduce the number of receptors — down-regulation of receptors — and decreased insulin levels increase — up-regulation — the number of receptors. The number of receptors per cell is increased in starvation and decreased in obesity and acromegaly; receptor affinity is decreased by excess glucocorticoids.

The actions of insulin, listed in Table 15-2, are mediated by the binding of the hormone to membrane receptors to trigger several

TABLE 15-1

Major Pancreatic Hormones

Pancreatic site	Hormone	Alternate source
B Cells	Pre-proinsulin Proinsulin Insulin (+ connecting C peptide)	—
A Cells	Proglucagon Glucagon (+ glicentin)	GI mucosa
D Cells	Somatostatin	GI mucosa CNS
F or D Cells	Pancreatic Polypeptide	GI mucosa

Note: GI, gastrointestinal system; CNS, central nervous system.

TABLE 15-2

Major Actions of Insulin

- Facilitation of glucose transport through certain membranes (e.g., adipose and muscle cells)
- Stimulation of the enzyme system for conversion of glucose to glycogen (liver and muscle cells)
- Slow-down of gluconeogenesis (liver and muscle cells)
- Regulation of lipogenesis (liver and adipose cells)
- Promotion of protein synthesis and growth (general effect)

simultaneous actions. A major effect of insulin is to promote the entrance of glucose and amino acids in cells of muscle, adipose tissue, and connective tissue. Glucose enters the cell by facilitated diffusion along an inward gradient created by low intracellular free glucose and by the availability of a specific carrier called transporter. In the presence of insulin, the rate of movement of glucose into the cell is greatly stimulated in a selective fashion.

In the liver, insulin does not affect the movement of glucose across membranes directly but facilitates glycogen deposition and decreases glucose output. Consequently, there is a net increase in glucose uptake. Insulin induces or represses the activity of many enzymes; however, if these actions are direct or indirect is not known. For example, insulin suppresses the synthesis of key gluconeogenic enzymes and induces the synthesis of key glycolytic enzymes such as glucokinase. Glycogen synthetase activity is also increased. Insulin likewise increases the activity of enzymes involved in lipogenesis.

1.2 Glucose Tolerance Test

Despite the paucity of the morphologic changes, a number of functional tests of glucose metabolism suggest impaired competence with aging.[7-9] Many clinical studies indicate a very slight (about 1 mg/dl per decade) age-related increase in fasting blood glucose levels in healthy individuals — not significantly affected by sex or adrenal steroids. This aging-related increase in blood glucose levels may be less marked or may not occur in nonobese, physically active elderly.

This modest age-related increase in fasting plasma glucose levels is often associated with striking changes in response to a glucose challenge. Indeed, more than 50% of randomly selected subjects over 60 years of age have significantly altered glucose tolerance tests when compared with younger individuals.

A typical oral glucose tolerance test measures the response of plasma glucose to a standard oral test dose of glucose; the variables are represented by

- Fasting (preglucose) plasma glucose levels
- Time and magnitude of the peak plasma glucose after administration
- Time required for plasma glucose levels to return to normal (preglucose) levels

In young adults, fasting plasma glucose levels range from 76 to 110 mg/dl; after oral administration of glucose, glucose levels rise to a peak of about 120 mg/dl after 30 min but return to preglucose levels within 1 h. In diabetes mellitus, fasting glucose is greater than 115 mg/dl, and the 1, 1.5, and 2 h postglucose values are greater than 185, 165, and 140 mg/dl, respectively. In the elderly, blood glucose levels are higher and take longer to return to normal compared to the younger individual. These findings have led to hundreds of publications and considerable controversy has resulted from the apparent necessity of labeling more than half of the older population as diabetic. This clinical dilemma may be neatly resolved by introducing age-corrected nomograms for the glucose tolerance test or by substituting higher fasting glucose, rather than decreased glucose tolerance as the principal criterion for diagnosis. The progressive increase in glucose intolerance with age cannot, however, be denied.

1.3 Insulin Resistance in the Elderly

What is the mechanism of the loss of glucose tolerance with age? The answer has been sought at

- The pancreas level, where insulin secretion may be depressed
- The peripheral level, where resistance of target tissues to insulin may be increased[10]

The most accepted of these two hypotheses is that normal aging is associated with failure of the ability of insulin to stimulate glucose uptake by peripheral tissue.

With regard to a possible reduced secretion of insulin, some investigators have shown that the initial insulin rise in response to glucose is depressed, but the late response is not.[11] Even if an altered responsiveness to insulin does exist, it is minor and insufficient to contribute significantly to the observed impairment in glucose tolerance. Indeed, insulin secretion itself and its metabolism, hepatic extraction of insulin, insulin half-life, insulin clearance, etc., are not significantly changed in the elderly. In fact, in some cases, the higher blood glucose levels prevailing in the elderly are associated with higher insulin levels than in younger subjects.

A corollary to the decreased-insulin-secretion hypothesis is a proportional increase in the secretion of proinsulin, which has considerably less biologic activity than insulin. Proinsulin levels in the basal state do not differ significantly in the elderly as compared to younger controls, but, after glucose loading, the amount of proinsulin relative to that of insulin is greater in some older individuals.

Considerable attention has been directed to the enzymes which process prohormones.[12] Enzymes for cleaving proinsulin to insulin and C peptides would become less efficient with aging, hence the high circulating levels of the prohormone.[13] Alternatively, the increased resistance of peripheral tissues to insulin with aging may create a greater demand for insulin with consequent insufficient time before secretion for cleaving the prohormone and, therefore, the increased release of the prohormone rather than the hormone. Irrespective of its causes, the elevation of the proinsulin-to-insulin ratio with aging is too small to account for the alterations in glucose tolerance.

One of the best arguments in support of the hypothesis that peripheral tissues become more resistant to the actions of insulin with aging is the very fact that insulin secretion and metabolism remain normal even in advanced age. The peripheral glucose uptake in the elderly is only one third of that in the young, and this low uptake had been associated by some investigators with a decreased number of insulin receptors. However, most investigators agree that the number and affinity of insulin receptors are similar in young and old.[14]

If receptors are not altered, then the defect or defects in peripheral glucose uptake are likely to reside at the postreceptor level and may involve a number of cytoplasmic processes such as receptor-mediated phosphorylation/dephosphorylation events or the generation of intracellular mediators following receptor-ligand interactions. The possibility that postreceptor reactions may be responsible for the apparent "resistance to insulin" is further supported by the observation that minimal receptor occupancy is required for insulin action. Therefore, even reduced number/affinity of receptors would be sufficient for carrying on insulin actions.

In conclusion, the mechanisms for glucose intolerance with aging are still a subject of debate; the prevalent view is that they reflect a major reduction in peripheral tissue responsiveness to glucose and insulin. Current interpretation of the data remains controversial, but the possibility of defects in the cascade of postreceptor reactions is being actively investigated (Table 15-3).[15,16]

1.4 Other Mechanisms Responsible for Glucose Intolerance in the Elderly

Other factors may be responsible or related to glucose intolerance with aging (Table 15-3). They are

- Loss of hepatic sensitivity to insulin and reduced glycogenesis
- Increased glucagon levels
- Changes in diet and exercise regimen
- Loss of lean muscle mass and increase in adipose tissue

The last factor — that is, increase in adipose tissue — may contribute to the decreased ability of insulin to facilitate cell glucose uptake. With aging, adipose tissue cells (adipocytes) increase in number and in size. Enlargement reduces the concentration of receptors on the cell surface.[17] This relative reduction, coupled with a possible reduction in the absolute number of receptors suggested by some investigators, would lead to reduced insulin binding and decreased cell response to the hormone.[18]

However, whether adipocyte size or adipocyte aging is primarily responsible for the increased insulin resistance remains unresolved. Attempts to clarify this controversy have managed to further confuse the issue: for example, caloric (food) restriction in rats reduces adipose tissue and adipocyte size and maintains responsiveness to insulin and glucagon into old age, suggesting a cause — small adipocyte size, and effect — high insulin sensitivity — relationship. However, caloric restriction retards the aging process, in general, and persistence of normal insulin response may merely reflect this antiaging affect, also reflected in the maintenance of small size adipocytes.

TABLE 15-3

Mechanisms of Loss of Glucose Tolerance with Aging

	Pancreatic level
Reduced insulin secretion	Depressed initial response to glucose (normal late response to glucose) Increased ratio of circulating proinsulin to insulin
	Peripheral level
Tissue insulin resistance	Reduced tissue uptake of glucose Reduced postreceptor response Reduced hepatic glucogenesis Loss of lean body mass with increase in adipose tissue

Despite the close relationship between aging, and obesity and insulin resistance, it must be realized that adipose tissue is responsible for the removal of less than 5% of injected glucose in both humans and rats. The primary sites of glucose uptake are the liver and muscles. It is in these tissues that the age-related defect in transport/intracellular metabolism must develop, and several factors must intervene simultaneously to explain the increased insulin resistance of the elderly.

1.5 Multiple Endocrine Causes of Altered Carbohydrate Metabolism with Aging

It must be kept in mind that, besides alterations of the endocrine pancreas, obesity and hyperglycemia may be caused by a number of endocrine disorders (e.g., Cushing's syndrome, acromegaly, insulin receptor abnormalities). This multifactorial etiology may be profitably studied in animal models such as the ob/ob mouse, which is obese and shows hyperinsulinemia and insulin resistance. The etiology of the syndrome has been related to a number of neuroendocrine defects involving the hypothalamic satiety center, the neurotransmitter serotonin, temperature regulation, and alterations in glucocorticoid and thyroid hormones. Despite these metabolic and endocrine disorders, the lifespan is little affected, and, as the animal ages, there is a remission of symptoms. Spontaneous remission with aging of diabetic or prediabetic symptoms is also a common finding in humans.

2 GLUCAGON CHANGES IN THE ELDERLY

Glucagon, secreted by the A cell, derives, like insulin, from a larger polypeptide precursor (proglucagon) and, like insulin, is degraded in the liver. The A cell also produces a large molecule, glicentin, with some glucagon activity. In the A-cell granule, glucagon is located in the center and glicentin in the periphery. Both glucagon and glicentin can also be found in cells of the intestinal mucosa. Little is known of the exact function of glicentin or its relation to glucagon.

The actions of glucagon are, in general, antagonistic to those of insulin. Glucagon promotes glycogen and lipids breakdown and conversion of nonglucose molecules to glucose and ketone bodies. It raises blood sugar because it stimulates adenylate cyclase in liver cells. While insulin serves as a hormone of fuel storage, glucagon serves as a hormone of fuel mobilization through glycogenolysis and gluconeogenesis. Following a meal, stimulation of B-cell secretion of insulin and suppression of A-cell secretion of glucagon serve to store fuels in liver, muscle, and adipose tissue. Conversely, during starvation, stimulation of glucagon secretion and suppression of insulin secretion direct the breakdown of fuels stored intracellularly — as reflected in lipolytic and ketogenic actions — to meet the energy needs of the brain and other tissues (Table 15-4). A related role for glucagon as the hormone of injury and insult has been proposed. For example, impaired glucose tolerance and hyperglycemia noted with infections, trauma, burns, and myocardial infarctions are all associated with increased plasma glucagon. The significance of this increase is still unclear but may have some relevance to glucagon secretion in aging.

TABLE 15-4

Glucagon in Aging

Anti-insulin effect

- Gluconeogenesis
- Glycogenolysis
- Lipolysis
- Ketogenesis

Anti-injury effect

- Gluconeogenesis

In aging

- Paradoxical increase despite glucose intolerance and hypoglycemia
- Anti-injury effect

Few detailed studies have been conducted on the secretion and metabolism of glucagon in the elderly, and those available do not indicate any abnormality. However, after glucose administration, an apparently paradoxical rise in glucagon has been reported in experimental animals as well as in patients with insulin-resistant diabetes.[19] This paradoxical increase may be triggered in response to injury, one of the actions of the hormone, and may contribute to the defect in glucose metabolism. Other still little known actions of glucagon on motor, vascular, and secretory function in the intestinal tract are relatively unstudied in the elderly.

3 ACCELERATED AGING AND DIABETES MELLITUS TYPES I AND II

Diabetes mellitus is a health problem widespread in humans. It affects almost 5% of the population in the U.S., and the number of those affected is rapidly rising with the increasing elderly population. In the U.S. and Western Europe, the disease affects about 16% of the aged 65 and over, and the size of the diabetic population is likely to continue to increase. In those countries with more limited food availability the number of diabetics is much lower (as low as 2%); the low rates have been attributed to a variety of factors among which the diet is most important, but also exercise, lifestyle, and heredity may all contribute to the low risk and reduced severity of the disease. In all cases, when the disease is present, it has a tremendous impact not only on the immediate health but also on the long-term viability of the individuals who are often disabled by it.[20-22]

There are essentially two main categories of diabetes mellitus:

- Type I, insulin-dependent diabetes, primarily found in children (juvenile diabetes) but also known to occur occasionally at other ages
- Type II, non–insulin-dependent diabetes, or late-onset diabetes, occurring after age 40 and older.

In both cases, the metabolic disturbances are essentially similar. However, type I is due to insulin deficiency and type II is due to insulin resistance (insulin levels are normal or even elevated).

A variety of "diseases of old age" (e.g., coronary heart disease, glomerulonephrosis, retinopathy, limb gangrene, stroke, cataract) are consequences or complications of diabetes and major cause of ill health and mortality. While some diabetics live long lives with little indisposition, in general the rate of disability of diabetics is two to three times greater than in nondiabetics. For example, in diabetics, blindness is about 10 times and gangrene about 20 times more common and 14% of diabetics (usually the elderly) are bedridden for an average 6 weeks per year.

3.1 Insulin Deficiency, Diabetes Mellitus (Type I)

The impact of insulin action is best evaluated by observing the metabolic disturbances that occur in the event of insulin deficiency, Type I diabetes (Table 15-5). Major signs of this disease characterized by severe metabolic alterations are

1. Hyperglycemia or high blood sugar due to reduced glucose uptake by cells, together with increased hepatic glucogenesis
2. Glycosuria (glucose in urine) and diuresis (increased urinary excretion, polyuria); as a consequence of hyperglycemia, the threshold for glucose reabsorption by the kidney is exceeded and glucose with accompanying water are excreted
3. Hunger symptoms, due to low intracellular glucose, leading to increased food intake or polyphagia; despite polyphagia, body weight is lost due to inability of assimilating glucose and lack of growth due to reduced cellular uptake of amino acids
4. Decreased entry of amino acids into muscle and decreased protein synthesis in liver with consequent loss of weight and impairment of growth in children
5. Increased lipolysis together with reduced lipogenesis; insulin is necessary for glucose uptake in adipose tissue and for lipogenesis, and it controls the release of free fatty acids into the circulation, which is increased
6. Increased ketone body production in the liver due to increased circulating fatty acids; as a consequence of the ketone bodies, the pH of body fluid is decreased
7. Acidosis, following pH fall, and dehydration, due to polyuria, will combine to lead to loss of consciousness or coma
8. Vascular complications superimposed on aging-induced vascular changes

Management of this type of diabetes revolves primarily on the administration of insulin in appropriate doses and delivered

TABLE 15-5

Characteristics of Diabetes Mellitus

Decreased glucose uptake	Hyperglycemia
	Decreased glycogenesis
	Increased hepatoglucogenesis
	Glycosuria
	Polyuria
	Polydipsia
	Polyphagia
Increased protein catabolism	Increased plasma amino acid
	Increased gluconeogenesis
	Weight loss, growth inhibition
	Negative nitrogen balance
Increased lipolysis	Increased free fatty acids
	Ketosis
	Acidosis
Vascular changes	Microangiopathies

at appropriate time intervals (e.g., after meals) to mimic the endogenous secretion.

3.2 Non–Insulin-Dependent Diabetes Mellitus (Type II)

This type of diabetes is associated with old age as it occurs at later ages (40 years and older). It is characterized by two abnormalities:

1. Decreased tissue and cellular response to insulin, or insulin resistance
2. Some abnormalities of insulin secretion, such as a delayed response to elevated glucose levels (e.g., altered glucose tolerance test, decreases pulsatility of insulin secretion)

As already discussed, insulin action is a complex process involving multiple steps, and most of these are affected in type II diabetes. Decreases in insulin binding, receptor tyrosine kinase activity, insulin–receptor internalization, generation of mediators, and glucose transporter translation and activity have been reported. Management of this type of diabetes does not depend exclusively on hormone replacement therapy by the administration of insulin. Currently, it relies on four modalities:

1. A hypocaloric weight reduction diet
2. The establishment of a regular physical exercise program
3. The administration of drugs, e.g., sylfonylureas capable of stimulating insulin secretion, cell sensitivity to insulin, and glucose utilization
4. If the above procedures are insufficient, the administration of insulin, if given in sufficient (high) doses, will control the hyperglycemia; however, high levels of insulin may have toxic effects and may induce obesity, which, in turn, may aggravate the diabetes

Irrespective of the causes of the disease, the consequences — that is, hyperglycemia and its derived effects — will have an extremely serious impact on aging and survival and health at late ages.

3.3 Pathological versus Physiological Considerations

While historically the majority of investigators have concerned themselves with the increased incidence of diabetes with aging, in recent years a growing number of them have reversed this focus to ask whether there might not be an acceleration

TABLE 15-6

Diabetes and Accelerated Aging

Diabetes	Aging
Microangiopathy	—
Cataracts	Cataracts
Neuropathy	Neuropathy
Accelerated atherosclerosis	Atherosclerosis
Early decreased fibroblast proliferation	Decreased fibroblast proliferation
Autoimmune involvement	Autoimmune involvement
Skin changes	Skin changes

of aging in diabetes (Table 15-6). Patients with diabetes display an increased incidence of several features commonly associated with aging: cataracts (Chapter 10), microangiopathy,[23] neuropathy,[24] dystrophic skin changes,[22] and accelerated atherosclerosis.[25] Accelerated atherosclerosis is a major feature of the various genetic syndromes reported to resemble premature aging, and all of these syndromes include abnormal glucose tolerance.[26] Further, in normal aging, in patients with progeria and in diabetics, the proliferative capacity of cultured fibroblasts is reduced, perhaps due, in part, to a reduced response to insulin and to growth factors. Insulin resistance has been reported in cells from patients with Werner's syndrome and in normal aging and progeria. In addition, in both juvenile-onset and maturity-onset diabetics, the rate of collagen aging (the aging of collagen having been represented as the fundamental aging process) is accelerated (Chapter 22). And, finally, the putative autoimmune etiology of juvenile-onset diabetes, observations of immune dysfunction in aging and in diabetes, and the reports of increased pancreatic amyloidosis in senile humans and animals have excited the interest of proponents of an immunogenesis of aging.[31] Collectively, these studies point to the possibility of an intriguing relationship between diabetes and aging.

The great deal of genetic variability in human populations with respect to the lifespan also applies to predisposition to diabetes.[26] Twin studies have demonstrated a high degree of concordance in late-onset diabetics which would make the incidence of a putative recessive diabetic gene greater than 40%. However, genetic factors in diabetes are at present poorly understood, probably owing to the considerable heterogeneity in this difficult-to-define disorder. Juvenile-onset diabetes has been recently linked to several specific major histocompatibility complex phenotypes (HLA) which may predispose selected individuals to viral infection or autoimmune reactions (Chapter 7), but the basis of genetic factors in late-onset diabetes remains obscure. The apparent high heritability of late-onset diabetes may indicate a pathology complicated by the effects of "normal aging", or, alternatively, "diabetes" may represent an acceleration of basic aging processes in a large, genetically predisposed percentage of the population.

Not all members of a population or even all populations develop coronary heart disease, yet this pathology is clearly age related, and remains the largest single cause of death in the elderly in the industrialized nations. Similarly, while diabetes (or abnormal glucose tolerance) may not be demonstrable in all aging individuals or populations, it remains unequivocally age related, and ranks as the sixth-largest killer in the U.S. Whatever genetic and pathological components ultimately prove to underlie selected specific manifestations of aging or the evident heterogeneity in diabetes and atherosclerosis, it is perhaps significant in this regard to note that the currently recommended treatment for late-onset diabetes mellitus is carefully restricted diet and regular exercise,[27] and both are commonly considered to be the normal healthy individual's best defenses against atherosclerosis and senility, while the best assurance of a long lifespan remains the thoughtful choice of long-lived parents.

REFERENCES

1. Spence, J. C., Some observations on sugar tolerance with special reference to variations found at different ages, *Q. J. Med.*, 4, 314, 1920–1921.
2. Andres, R., Aging and diabetes, *Med. Clin. N. Am.*, 55, 835, 1981.
3. Sugawara, K., Kobayashi, T., Nakanishi, K., Kajio, H., Ohkubo, M., Sugimoto, T., Murase, T., Itoh, T., Hara, M., and Kosaka, K., Marked islet amyloid polypeptide-positive amyloid deposition: a possible cause of severely insulin-deficient diabetes mellitus with atrophied exocrine pancreas, *Pancreas*, 8, 312, 1993.
4. Reaven, E. P., Wright, D., Mondon, C. E., Solomon, R., Ho, H., and Reaven, G. M., Effect of age and diet on insulin secretion and insulin action in the rat, *Diabetes*, 32, 175, 1983.
5. Chaudhuri, M., Sartin, J. L., and Adelman, R. C., A role for somatostatin in the impaired insulin secretory response to glucose by islets from aging rats, *J. Steroid Biochem.*, 38, 431, 1983.
6. Lipson, L. G., Bobrycki, V. A., Bush, M. J., Tietjen, G. E., and Yoon, A., Insulin values in aging: studies on adenylate cyclase, phosphodiesterase, and protein kinase in isolated islets of Langerhans in rats, *Endocrinology*, 108, 620, 1981.
7. Davidson, M. B., The effect of aging on carbohydrate metabolism: a review of the English literature and a practical approach to the diagnosis of diabetes mellitus in the elderly, *Metabolism*, 28, 688, 1979.
8. Bennett, P. H., diabetes in the elderly: diagnosis and epidemiology, *Geriatrics*, 39, 37, 1984.
9. Taylor, R. and Agius, L., The biochemistry of diabetes, *Biochem. J.*, 250, 625, 1988.
10. Reaven, G. M., Insulin resistance in non-insulin dependent diabetes mellitus. Does it exist and can it be measured?, *Am. J. Med.*, 74, 3, 1983.
11. DeFronzo, R. A., Glucose intolerance and aging, *Diabetes Care*, 4, 493, 1981.
12. Marx, J. L., A new marker for diabetes, *Science*, 215, 651, 1982.
13. Gold, G., Reaven, G. M., and Reaven, E. P., Effect of age on pro-insulin and insulin secretory patterns in isolated rat islets, *Diabetes*, 30, 77, 1981.
14. Goldfine, I. D., The insulin receptor: molecular biology and transmembrane signaling, *Endocrine Rev.*, 8, 235, 1987.
15. Caro, J. F., et al., Insulin receptor kinase in human skeletal muscle from obese subjects with and without non-insulin dependent diabetes, *J. Clin. Invest.*, 79, 1330, 1987.
16. Fink, R. I., Kolterman, O. G., Griffin, J., and Olefsky, J. M., Mechanisms of insulin resistance in aging, *J. Clin. Invest.*, 71, 1523, 1983.
17. Olefsky, J. M. and Reaven, G. M., Effects of age and obesity on insulin binding to isolated adipocytes, *Endocrinology*, 96, 1486, 1975.

18. Bolinder, J., Kager, L., Ostamn, J., and Arner, P., Differences at the receptor and postreceptor levels between human omental and subcutaneous adipose tissue in the action of insulin on lipolysis, *Diabetes,* 32, 117, 1983.
19. Simonson, D. C. and DeFronzo, R. A., Glucagon physiology and aging: evidence for enhanced hepatic sensitivity, *Diabetologia,* 25, 1, 1983.
20. Green, D. R., Acute and chronic complications of diabetes mellitus in older patients, *Am. J. Med.,* 80 (Suppl.) 39, 1986.
21. Jackson, R. A. and Finucane, P., Diabetes mellitus, in *Principles and Practice of Geriatric Medicine,* Pathy, M. S. J., Ed., John Wiley & Sons, New York, 1991, 1123.
22. Funnell, M. M. and Merritt, J. H., The challenges of diabetes and older adults, *Nursing Clin. N. Am.,* 28, 45, 1993.
23. Siperstein, M. D., Unger, R. H., and Madison, L. L., Studies of muscle capillary basement membranes in normal subjects, diabetic, and prediabetic patients, *J. Clin. Invest.,* 47, 1973, 1968.
24. Harati, Y., Diabetic peripheral neuropathies, *Ann. Intern. Med.,* 107, 546, 1987.
25. Pyorala, K., Diabetes and atherosclerosis: an epidemiologic view, *Diabetes Metab. Rev.,* 3, 463, 1987.
26. Goldstein, S., Human genetic disorders that feature premature onset and accelerated progression of biological aging, in *The Genetics of Aging,* Schneider, E. L., Ed., Plenum Press, New York, 1978, 171.
27. Laws, A. and Reaven, G. M., Physical activity, glucose tolerance, and diabetes in older adults, *Ann. Behav. Med.,* 13, 125, 1991.

III
SYSTEMIC AND ORGANISMIC AGING

16 CARDIOVASCULAR ALTERATIONS WITH AGE: ATHEROSCLEROSIS, CORONARY HEART DISEASE, HYPERTENSION

Paola S. Timiras

Cardiovascular disease continues to be the major cause of death in the U.S. and other industrialized, socioeconomically advanced countries. Indeed, cardiovascular disease associated with advancing age remains the most important single worldwide cause of death in old age in both sexes. This is, in part, because people are living longer and are more susceptible to the occurrence of degenerative diseases. It is also due to some unknown aspects of modern life which are increasing the incidence of atherosclerosis. This chapter will attempt to define different types of arterial pathology and illustrate highlights of arterial structure. It will review some of the major lesions of atherosclerosis and their etiopathogenesis. Taking coronary heart disease and hypertension as examples, this chapter will also very briefly examine some of the clinical consequences of atherosclerosis, the chronic arterial disorder that is the major cause of heart attacks and strokes.

1 DEFINITIONS

1.1 Arteriosclerosis

Arteriosclerosis is a generic term for any vascular degeneration that leads to progressive thickening and loss of resiliency of the arterial wall. One type of arteriosclerosis is *atherosclerosis,* which refers to specific vascular alterations, such as atheromas or plaques characterized by a combination of fatty accumulation in the intima and an increase in connective tissue in the subintimal layers of the arterial wall. It is this form of arteriosclerosis that is the most widespread and, at the same time, the most threatening, inasmuch as it affects those vessels such as the aorta, the coronary, and the cerebral arteries that are extremely important in providing the necessary blood supply for the heart, brain, and other vital organs. Atherosclerosis, then, is the vascular disorder that underlies most arteriosclerotic heart disease or coronary heart disease and also plays a major role in cerebrovascular disease. Atherosclerosis dwarfs all other single causes of mortality in the U.S. and represents by far the major cause of death from cardiovascular diseases.

Arterial diseases may also arise from congenital structural defects, from inflammatory diseases (e.g., syphilitic aortitis), from hypersensitivity or autoimmune diseases which principally affect the smaller vessels and may lead to thromboangiitis obliterans, and from specific capillary lesions as in diabetic angiopathy.

1.2 Progressiveness and Universality of Atherosclerosis

Atherosclerosis, characterized by its *progressive* onset culminating in overt manifestations in old age, its widespread distribution throughout the arterial tree, and its consequences leading to severe disability or death, has been studied as a prototype of cardiovascular changes with age.[1-3] In this respect, atherosclerosis must be viewed as a disease that, sooner or later, affects all of us. Working silently over the years from early childhood, it gradually destroys the arteries, ultimately preventing the exchanges of gases and nutrients necessary to keep organs, tissues, and cells alive and functioning normally.

Although a scourge of modern civilization and often discussed in relation to the pressures of an urban technocratic society, atherosclerosis has been with us from ancient times; the disease has been detected in Egyptian mummies and described in early Greek writings. There are several ways in which atherosclerosis impairs the normal function of the arteries:

- It may corrode the arterial walls to such a degree that they suddenly yield to the pressure of the blood inside and explode in a massive hemorrhage.
- In reaction to its destructive processes, it may set off a secondary proliferation of the tissues, thereby leading to a gradual blockage of the arterial lumen.
- The changes induced in the arterial wall may induce clotting of the blood within the diseased artery and, in this way, obstruct blood flow.

Thus, while atherosclerosis is a progressive disease that develops slowly over years or decades, the final "accidents" for which it is responsible (hemorrhage, thrombosis) may be initiated within only a few seconds.

One of the characteristics of atherosclerosis is its *universality* in almost all animal species and throughout all populations within a species. So insistent and progressive is its onslaught with advancing age that it is generally considered to be an inevitable manifestation of aging — a "wearing out" of the arteries. In statistical terms, the accidents induced by atherosclerotic arterial disease, associated with high blood pressure, have been found to be responsible for more deaths in middle-aged and elderly persons than all other diseases put together — at least in North America and Europe. Indeed, death by heart attack or stroke is now so common that we have come to regard it as a natural end of the lifespan; at the same time, however, much of the gravity of atherosclerosis is the accepted conviction that it kills us prematurely, in the sense that alterations in the arteries are capable of irreparably damaging such vital organs as the heart and brain at a time when the functional competence of these structures is otherwise sound. From the analogy of the heart or brain as a motor and the arteries as the pipes that convey the fuel to the motor comes the idea that if the motor is deprived of fuel because of a breakdown in the pipe system,

0-8493-8979-8/94/$0.00+$.50

it will stop working, even though the motor itself is without defect.[4]

Although atherosclerosis is universal and its etiopathogenesis similar in all organisms, there are some individual differences, probably related to the preponderance of one risk factor over the others. Thus, the first effects (early lesions) on the arterial lining may differ whether the risk factor is hyperlipidemia, or diabetes, or hypertension or smoking, the major events being, respectively, increased lipid deposition, or particular arterial pathology, or concomitant endocrine (angiotensin–aldosterone) alterations or accumulation of free radicals.

The question of how long we might live were it possible to prevent atherosclerosis remains unanswered; it would seem that complete prevention of atherosclerosis would add several years to life.[5,6] Scientists throughout the world representing many disciplines are now attacking the problem both to understand the nature of the disease and to find means of preventing or curing it. It has been suggested that we are now, in the 1990s, entering a "new era" of atherosclerosis research in which it is possible not only to prevent but also to arrest or to induce regression of both the lesions and their clinical manifestations.

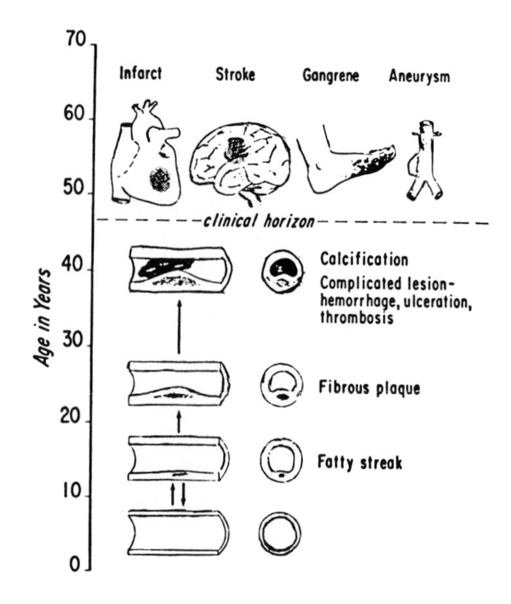

FIGURE 16-1 *Natural history of atherosclerosis shown in this diagrammatic concept of the pathogenesis of human atherosclerotic lesions and their clinical manifestations.* (From McGill et al., *Atherosclerosis and Its Origin,* Sandler, M. and Bourne, G. H., Eds., Academic Press, New York, 1963, 39. With permission.)

2 CAUSES AND PATHOLOGY

2.1 Course of Atherosclerosis

Although the consequence of atherosclerotic lesions become manifest clinically in the fourth decade of life and thereafter, atherosclerosis is not exclusive to advanced age but, rather, represents the culmination of progressive changes in the arterial wall from childhood.[7] Vascular changes which occur in infancy and childhood are clearly identifiable microscopically. They consist of intimal thickening, cell proliferation, accumulation of proteoglycans, matrix, and formation of fibrous matrix. Some, but not all, contain lipids and, in this case, can be readily observed as fatty streaks. *Atherosclerosis in humans may be defined as a chronic inflammatory, smooth muscle, proliferative response to an initially damaging agent (e.g., high blood lipids, altered carbohydrate metabolism or immune function, accumulation of free radicals).*

The general pattern of the lesions varies from one individual to another and among different populations throughout the world. The approximate time sequence involved in the development of atherosclerotic lesions with respect to specific pathologic changes has been generally established (at least in North America) to proceed in the following order (Figure 16-1): the fatty streaks appear in the arteries during the first decade of life (as early as the first years or even months or days of life) and continue into the second decade; they appear in the cerebral arteries in the third decade. The fibrous or "pearly" plaques appear from the second decade on; clinical consequences (e.g., cardiac infarct, stroke, gangrene, and aneurysm) occur from the fourth decade on.[8]

Atherosclerosis seems to develop in several "waves" throughout the lifespan, inasmuch as early and late stages of the lesions can be found side by side in the same vessel or in different vessels of the same person. However, this view assumes that the early stage is represented by the fatty streaks and advanced stages of the same lesion by the appearance of plaques, a concept that is negated by the observation that the two types of lesions are often found in different locations and that some populations with a fair amount of fatty streaks do not subsequently develop a commensurate number of plaques. That the lesions are distinct and independent is supported by the fact that all lesions do not progress to the same degree in all individuals and that is possible for the two types of lesions, early and late, to coexist.

From a morphologic and biochemical point of view, the natural history of the disease remains a subject of controversy. It has been traditionally accepted, for example, that atherosclerosis is primarily an alteration of the intima and that atherosclerotic lesions start in the innermost layers where they can be confined for a long time, and eventually spread to the entire arterial wall. The question of the primary site of the lesions, however, loses some of its importance in the light of the current concept that the entire arterial wall represents a functioning unit.

So far, the study of atherosclerosis has been approached from two main directions:

1. The study of atherosclerotic lesions in man, i.e., the analysis of the various morphologic and biochemical components of the lesions, the timetable of their appearance, their location, their consequences, and their relationship to the normal and external environment of the body.
2. The attempts to reproduce the human disease in animals to discover the factors that may be capable of preventing or curing such lesions. However, it is recognized that, in general, animal observations are not always referable to man and that atherosclerotic lesions, in particular, differ in different species.

2.2 Types and Structure of Arteries

In its journey from the heart to the tissues, the blood passes through channels of six principal types: elastic arteries, muscular arteries, arterioles, capillaries, venules, and veins. In this system, the arteries show a progressive diminution in diameter as they recede from the heart, from about 25 mm in the aorta to 0.3 mm in some arterioles. Likewise for the veins, the diameter is small in the venules and progressively increases as the veins approach the heart. All arteries are comprised of three distinct layers — intima, media, and adventitia — but the proportion and structure of each varies with the size and function of the particular artery. The morphology of the arteries is illustrated in Figure 16-2 and represented diagramatically in Figure 16-3. A large artery, like the aorta, is comprised of the following layers, going from the lumen to the most external layers:

1. The intima, or innermost layer, consists of a layer of endothelial cells separated from the inner layer by a narrow layer of connective tissues that anchors the cells to the arterial wall.
2. A large layer of elastic fibers forms the *elastica interna* layer.
3. Below this layer are concentric waves of *smooth muscle cells* intermixed with elastic fibers. Elastic lamellae and smooth muscle cells are embedded in a ground substance rich in proteoglycans. Proteoglycans are formed of disaccharides bound to protein and serve as binding or "cement" material in the interstitial spaces. The outer layer of the media is penetrated by branches of the vasa vasorum.
4. Between the smooth muscle layer and the adventitia, there is again another layer of elastic fibers, the *elastica externa*. Layers 2, 3, and 4 form the media.
5. The outer layer or *adventitia* is formed of irregularly arranged collagen bundles, scattered fibroblasts, a few elastic fibers, and blood vessels that, because of their location, are called *vasa vasorum* or vessels of the vessels.

This structure of the aorta and large arteries corresponds well to their function as blood reservoirs and their need to stretch or recoil with the pumping action of the heart. The wall of the arterioles contains fewer elastic fibers but more smooth muscle cells than that of the aorta. The arterioles represent the major site of the resistance to blood flow and small changes in their caliber cause large changes in total peripheral resistance. Muscle cells are innervated by noradrenergic nerve fibers which are constrictor in function, and, in some cases, by cholinergic nerve fibers which dilate the vessels.

The *capillary structure* shows a diameter just large enough to permit the red blood cells to squeeze through in single file. In the same manner as the intima of the arteries, the capillary wall is formed of a layer of endothelial cells resting on a basement membrane. The major function of the capillaries is to promote exchange of nutrients and metabolic end products between the blood and the interstitial tissues. Such exchanges are facilitated by the presence of specialized junctions, gaps, or fenestrations.

2.3 Some Characteristics of Blood Flow and Arterial Function

The arterial system provides not only for circulation of the blood as a whole but, when necessary, for the special needs or functions of a particular organ. Certain organs — brain, heart, kidney — receive a larger proportion of blood than others and, within the same organ, blood flow varies considerably depending on the degree of activity, as dramatically evidenced in the 30-fold increase in blood flow to the exercising muscle. The velocity of blood flow declines gradually, from approximately 8 cm per second in the medium-sized arteries to 0.3 cm per second in the arterioles. Pressure, on the other hand, remains high in the large and medium-sized arteries but falls rapidly in the small arteries to low levels of 30 to 40 mmHg; the magnitude of the drop varies depending on the degree of arteriolar constriction.

TABLE 16-1

Summary of Factors Regulating Arteriolar Diameter

Vasodilator	Vasoconstrictor
Reduced oxygen tension	Norepinephrine
Reduced pH	Epinephrine
Increased CO_2	Angiotensin II
Increased temperature	Vasopressin
Lactic acid	
Histamine release	
Potassium ions	
Adenosine and nucleotide	
Kinins	

With aging, reduced capacity for arteriolar dilation or constriction because of

- Reduced elasticity
- Collagen crosslinkage
- Calcification
- Adrenergic receptor sensitivity
- Atherosclerosis

The mechanisms that regulate arteriolar diameter and hence blood flow through the arteries involve both local (e.g., release of chemical substances into tissue) and systemic factors (e.g., stimulation of the baroreceptors and the vasomotor center) (Table 16-1). The arterial system as a whole is never static but continuously undergoes structural changes and adaptations that permit the organism to respond to changing requirements for blood supply.

Even at birth, arteries vary in structure and in distribution depending upon the hemodynamic conditions under which they operate. Arteries continue to change with maturation and advancing age of the individual, changes that are continuously affected by extrinsic as well as intrinsic factors; in the process of adapting to environmental stimuli, structural changes occur at particular sites, and the gross pattern of vascular distribution to an organ or body part may undergo considerable change, as in the development of collateral circulation.

Evidence from human and experimental observations suggests that the rigid classification of the various elements that form the arterial wall into definite "species" of cells has to be abandoned in favor of a far more versatile view of the cellular configuration of the arteries. It seems that these cells are not irrevocably specialized but, rather, that they can assume more than one function when the need arises. For example, a muscle cell will not only contract upon stimulation but, under certain conditions, it could also phagocytize lipids and even produce collagen or elastic fibers — a potential that might have a bearing on the formation of atherosclerotic lesions.

2.4 Blood Supply to the Arterial Wall and Metabolic Exchanges

Blood supply of a nonfunctioning organ or part of an organ can be diminished and, conversely, blood supply of an actively functioning part can be increased by three basic mechanisms:

1. Arteriovenous anastomoses, i.e., short channels that connect arterioles to venules, bypassing the capillaries
2. Specialized muscular arrangements in the walls of arteries, as in the sphincters of hepatic and splenic arteries

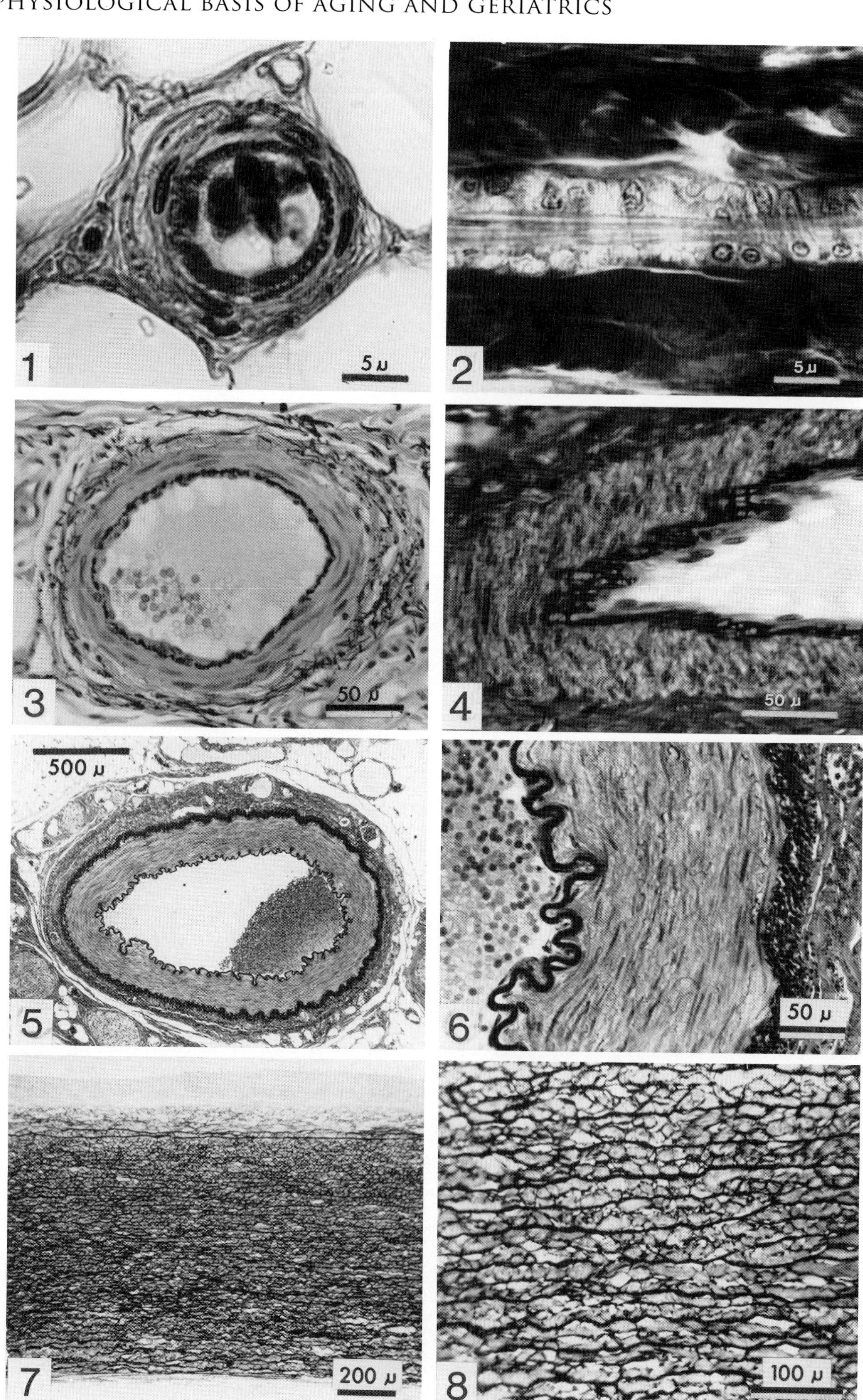
1
5 μ
2
5 μ
3
50 μ
4
50 μ
500 μ
5
6
50 μ
7
200 μ
8
100 μ

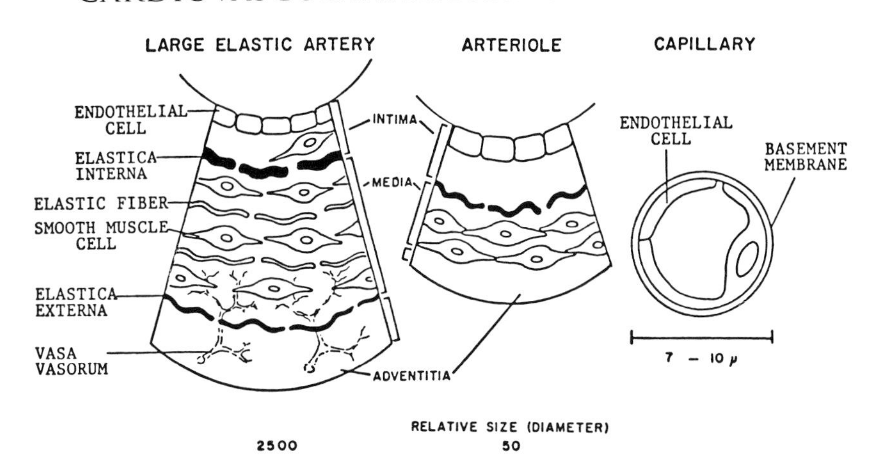

FIGURE 16-3 *Diagrammatic representation of the arterial wall illustrating a large (elastic) artery, an arteriole, and a capillary.* (Drawing by S. Oklund.)

TABLE 16-2

Localized Factors Contributing to Atherosclerotic Lesions

Marginal vascularization of arterial wall
Relative ischemia
Limited metabolic exchange
Blood turbulence and mechanical stress

3. Arrangement of the capillaries in the capillary bed in such a manner that there is a preferential capillary channel from arteriole to venule (a controversial assumption that presupposes contractile structures in the capillaries)

One of the best examples of how blood circulation adjusts to hemodynamic requirements is represented by the establishment of a *collateral circulation* (i.e., circulation that is carried on through secondary channels) when the main arterial supply to a specific organ or tissue is cut off. Adequate blood supply to the heart is generally assured at all times by the presence of a rich system of anastomoses not only between the two main coronary arteries but also between the coronary arteries and arteries from the pericardium, lungs, thorax, and diaphragm. Thus, in the case of thrombosis of one of the coronaries, provided that the process of occlusion is sufficiently slow, competent collateral circulation can be established.

The extent of *vascularization of the arterial* wall varies from species to species; for example, vascularity is less well developed in man, rabbit, and chicken (species highly susceptible to atherosclerosis) than in the horse, goat, and cow (species less susceptible to the disease).[9] The type and amount of the blood supply to the vascular wall may represent a central problem of angiopathy;[10,11] that is, the lack of adequate vascularization would represent a major cause of arterial disease (Table 16-2). As stated above, the intima and inner media obtain their nutrition by diffusion from the lumen; thus, any process that causes a thickening of the intima, as well as any process that damages the vasa vasorum, might be expected to cause an ischemic type of injury to the arterial tissue (i.e., due to blood insufficiency). Conversely, when the atherosclerotic lesion has reached the stage of plaque, regardless of the causes of plaque formation (e.g., lipid infiltration, thrombosis), the lesion becomes vascularized. Thus, although an inadequacy of blood supply to the arterial walls has been implicated in the initial stages of atherosclerosis, the increased vascularization that is associated with further development of the lesion has been viewed as an aggravating factor in advanced stages of the disease.[12] Increased vascularization may also be responsible, at least in part, for the sudden accidents, such as hemorrhage, with or without thrombosis, that are characteristic of these advanced stages.

Connected with the problem of vascularization and likewise involved with the special structure of the arterial wall are the exchanges of metabolic products between the vessel wall and lumen and *vice versa*. Diffusion, which normally regulates nutritional exchanges between blood and tissues, varies with the layer of the arterial wall. In the intima, for example, because of its proximity to the blood stream, nutrients

FIGURE 16-2 (facing page)

Morphology of normal arterial vessels.
(1) Arteriole — cross section (human mesentery). Stain: iron hematoxylin–aniline blue. Note: Red blood cells in lumen, endothelial nucleus, internal elastic membrane, smooth muscle cells with elongate nuclei.
(2) Arteriole — longitudinal section (cat ileum submucosa). Stain: Mallory–Azan. Note: Smooth muscle cells coiling around endothelial tube which contains the nucleus of an endothelial cell.
(3) Small artery — cross section (human external ear, subcutaneous tissue). Stain: Verhoeff and Van Gieson. Note: Distinct internal elastic membrane and smooth muscle of media; elastic fibers are beginning to accumulate to form an external layer.
(4) Small artery — tangential section showing fenestrated internal elastic membrane (human external ear, subcutaneous tissue). Stain: as in (1). Note: All elastic membranes in the arterial tree bear fenestrae (window-like openings).
(5) Medium artery — cross section, low power (human mesenteric artery). Stain: as in (1). Note all layers of the wall, intima (with internal elastica), media, and adventitia (with externa elastica) are distinct.
(6) Medium artery — cross section, high power. Stain: as in (1). Note: Internal elastic membrane, well-developed muscular media, adventitia with external elastic tissue disposed as coarse fibers in helices, hence, cut tangentially.
(7) Large artery — cross section, low power (human aorta). Stain: as in (3). Note: thick intima and high content of elastic tissue (appearing as black lines).
(8) Large artery — cross section, high power (human aorta). Stain as in (3). Note: multiple thick membranes forming concentric tubes interconnected by finer cross membranes. The interstices are filled mainly with collagenous connective tissue and sparse smooth muscle.
(Courtesy of Dr. E. S. Evans.)

diffuse from the blood, and products of arterial tissue metabolism are discharged into the lumen in the reverse direction. The other layers are too thick to be nourished by diffusion; in the adventia and outer media, metabolic needs are supplied by the vasa vasorum, leaving the inner portion of the media metabolically undersupplied and, therefore, at risk. Hence, any change in thickness of the tissue layer that the metabolic substances must cross to and from the lumen or any alterations in the circulation of the vasa vasorum will lead to alterations in metabolic exchanges and accumulation of metabolic by-products. Once the lesion has been established, metabolic injury will aggravate the lesion and impair eventual recovery processes.[13]

2.5 Localization of Lesions

It must be kept in mind that atherosclerotic lesions are focal; that is, they have preferred sites of occurrence and sites where they are seldom found. Thus, other factors to be considered in evaluating the etiology and progression of the atherosclerotic lesion are their location along the arterial wall and the influence of *blood turbulence*. Preferred sites are around the orifices of arteries branching off a major artery, such as the intercostal arteries from the descending aorta. Another preferred site is at the bifurcation of a large artery into two smaller ones, such as the abdominal aorta bifurcating into the two iliac arteries. It is at these preferred sites that the lesion begins and progresses rapidly. One of the reasons for these preferential locations is that in these regions — *orifices* and *bifurcations* — blood flow increases in velocity, exceeds critical velocity, and become turbulent. It is this turbulence, that is, increased velocity and more erratic flow, that creates a mechanical stress at the orifice or bifurcation, favoring the onset and progression of the lesion.[13]

3 STRUCTURAL CHANGES IN ATHEROSCLEROTIC LESIONS

3.1 Timetable of Lesions

As mentioned above, atherosclerotic lesions begin at an early age and progress continuously throughout life. The localization, rate of progression, or type of lesion varies widely depending on several factors related to the structure and function of the vessel considered, specific hemodynamic physiologic requirements, and a number of associated pathologic conditions — either local (e.g., hemorrhages, thrombi) or systemic (e.g., hypertension, diabetes) (Table 16-3).[14,15] In all cases, atherosclerosis occurs fundamentally as a localized lesion of the wall. Although its diffusion — in the vessel itself, in the area of the vessel, and throughout the whole body — may be considerable, the steadily progressing pathologic process is always confined to one focal point, the form and size of which depends on both local and generalized conditions. The atherosclerotic plaque or atheroma represents the characteristic site at which the histogenesis of the disease can best be analyzed.[16] Although it has not yet been possible to detect the precise beginning of an atherosclerotic lesion, there are a sufficient number of characterizing features in those lesions to assign them the descriptive terms "early" and "advanced" (Figures 16-4 and 16-5).

3.2 Early Lesions

For large elastic arteries like the aorta, lesions usually begin in the form of scattered foci in which the innermost layers of the vascular wall show signs of damage accompanied by the growth of repair tissue. In the smaller arteries, the progression of events is less easily identifiable, but damage and repair processes still represent the most important characteristics of the early lesion in these structures.

TABLE 16-3

General Characteristics of Atherosclerotic Lesions

Early onset — progressive
Focal lesions
Early lesions
Advanced lesions
Damage
Repair
Regression?
Localized-type lesions, progression influenced by
- Local factors — vessel structure and metabolism, blood turbulence
- Systemic factors — diabetes, hypertension, stress

In many cases, the endothelial cells of the intima undergo changes of the membrane — they become "sticky"; simultaneously, monocytes, a type of leukocyte (from the immune system) circulating in the blood, also become "sticky". These changes facilitate the attraction between these two cells and the invasion of the arterial wall by the monocyte. In the arterial wall, the monocyte is transformed into a macrophage and begins to engorge with lipids to form the so-called "foam cells".

As the intima thickens as a result of an increase in tissue fluid in the intimal ground substance, there is a disruption and disintegration of the innermost elastic lamellae, followed first by moderate influx and then swelling and flooding of the area with amorphous materials, primarily proteins and sulfated proteoglycans. The proteins seem to derive from the blood, perhaps as a consequence of the increased permeability of the damaged endothelium (as in all inflammatory edemas) and are often coupled with lipids, which become visible when uncoupled. The origin of proteoglycans is uncertain; they may derive from the blood or they may be formed *in situ;* in any case, they are very similar to the materials that accumulate in most young repair tissues of the body where they help to build collagen fibers, the principal component of scars (Chapter 21). At this point, early lesions are essentially proliferative, with endothelial and smooth muscle cells as well as macrophages releasing growth factors and cytokines, which stimulate cell proliferation. At this time the presence of T lymphocytes is also to be noted, probably the expression of some immune response of still unidentified origin.

Other studies of experimentally induced atherosclerosis suggest that the initial lesion consists of proliferation of muscle cells in the media where their multiplication gives rise to a cell mass impinging into the intima. These muscle cells degenerate and accumulate lipids, and the resulting mass will be invaded by macrophages and connective tissue in an attempt to circumscribe and encapsulate. This view of the early lesions forms the basis of the muscle myoclonal hypothesis[17,18] which identifies the abnormal proliferation of muscle cell clones as the first event in atherosclerosis pathogenesis; this proliferation would be triggered by a variety of toxic substances and mediated through the release of local growth factors.

The next phases of the atherosclerotic lesion involve the *repair and protective processes* that characterize any inflammation, i.e., the attempt to circumscribe or eliminate the production of masses of new fibers — mostly collagen but also elastic fibers — within and above the area of injury. One of the consequences of these processes is the further aggravation of intimal hyperplasia or thickening. These initial lesions contain very little abnormal fat, if any, and, appropriately, some

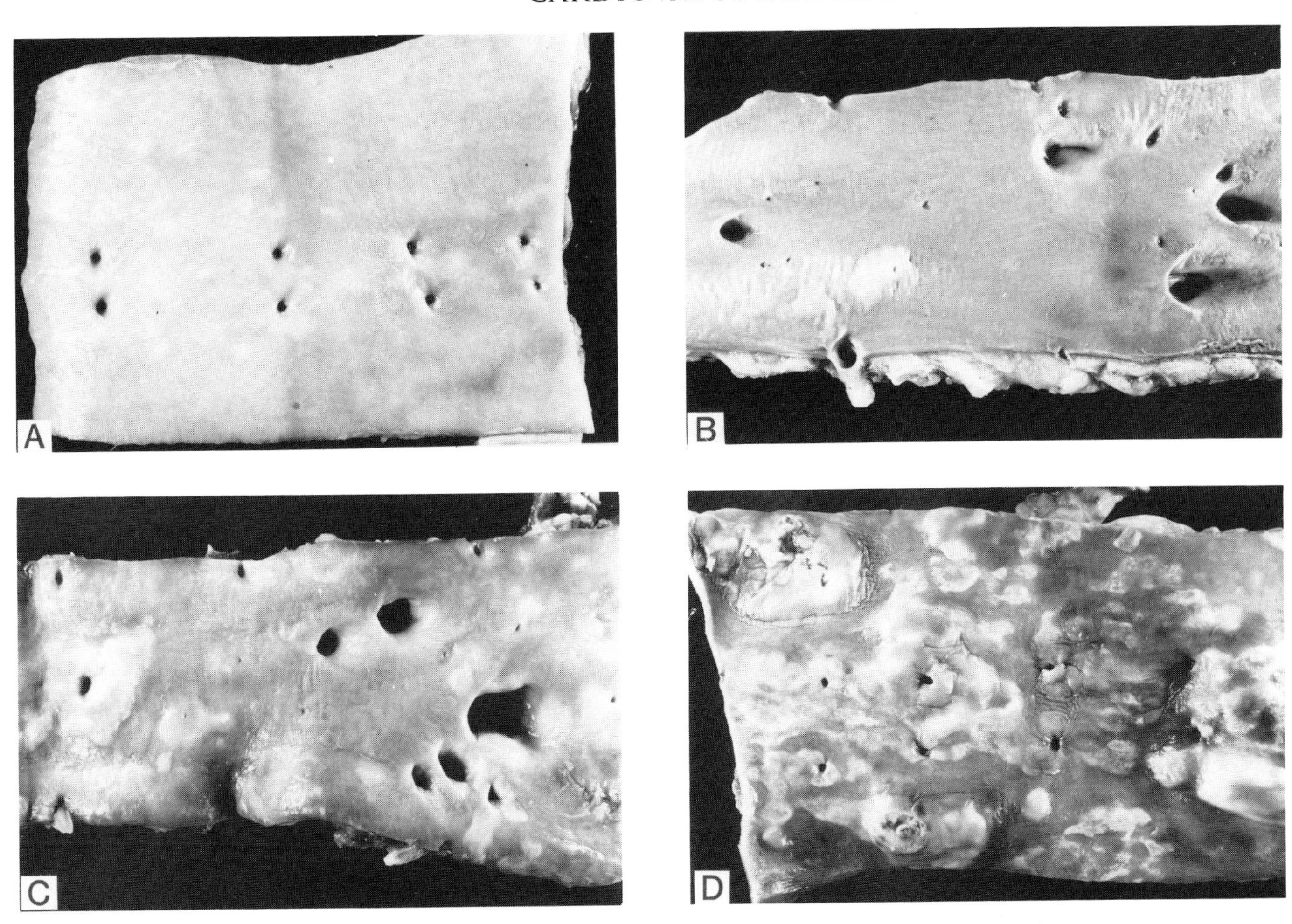

FIGURE 16-4 *Progression of morphologic changes in human aorta from early to advanced atherosclerosis.* Aortas were split open and the intima exposed for photography. (A) Aorta (thoracic) of a 32-year-old male showing early fatty plaques (represented by lighter coloration) localized mainly around the orifices of the intercostal arteries. As discussed in the text, because of the contribution of hemodynamic factors in the genesis of the atherosclerotic lesion, orifices of collateral arteries are frequently the site at which lesions first appear. (B) Aorta of a 24-year-old female showing early fatty plaques, also around the orifices of collateral arteries. (C) Aorta of a 55-year-old male showing advanced plaques characterized not only by a greater amount of fatty material but also by fibrotic thickening of the wall. (D) Aorta of a 65-year-old male showing large, complicated, and calcified plaques. (Courtesy of Dr. O. N. Rambo.)

investigators have classified this period as the "prelipid" stage of atherosclerosis. Sooner or later, lipid does appear in some of these lesions, mostly in their basal portions and not only within the cells themselves (local smooth muscle cells and invading and proliferating macrophages) but also in the intercellular spaces and on the disintegrating elastic lamellae. The lipid material, first in the form of small droplets, gradually fills up the cells imparting a "foamy" appearance in the histologic section. With the increase in the number of foam cells and the amount of extracellular lipid, this fat accumulation becomes visible to the naked eye as tiny yellow spots or streaks in the inner lining of the arteries, the so-called "fatty spots" or "fatty streaks". Taken as evidence of lesion, these fatty streaks are most commonly found in the aorta of children and younger individuals, although similar foci have been described in octogenarians and centenarians who showed a so-called "juvenile" atherogenic index.

3.3 Ground Substance

Studies of the ground substance (extracellular matrix) and its major constituents, the proteoglycans, show progressive quantitative and qualitative changes of these substances as the atherosclerotic lesion progresses. Alone, as part of the larger group of glucosaminoglycans, and in combination with other components of the ground substance, such as hyaluronic acid and embedded substances such as collagen and elastin, the proteoglycans regulate some important viscoelastic and water-binding properties of tissues, including those of the arterial wall (Table 16-4). Impairment of these properties with aging would result in alterations of the ground substance with reduction of its functions of mechanical support and hydration. Such alterations would translate into a weakening of the arterial wall with an inability to support the compressive load of normal blood pressure; it would also result in chemical alterations of transport and binding of water-soluble substances.

3.4 Advanced Complicated Lesions

These lesions, found mainly in adults and elderly persons in whom autopsic examinations are conducted more frequently, have been studied more exhaustively than the early lesions typical of the first decade of life. Detailed morphologic descriptions of advanced human atherosclerosis can be found in most textbooks of pathology and in specific texts dealing with atherosclerosis. Only a very brief summary will be presented here.

FIGURE 16-5 *Progression of microscopic changes in human arteries from early to advanced atherosclerosis.* (A) Earliest fatty plaque in fibrous intima (aorta from 24-year-old male: hematoxylin–eosin stain; ×152). (B) Large, atheromatous cystic plaque (carotid artery from a 65-year-old male; Verhoeff and Van Gieson stain; ×24). (C) Atheromatous plaque showing both fatty infiltration and alterations of elastic tissue (carotidendarterectomy in 68-year-old male; stain as in (B); ×95). (D) Large ulcerated calcified plaque with metaplastic bone (carotid artery with occlusion in 79-year-old male; stain as in (B); ×24). (Courtesy of Dr. O. N. Rambo.)

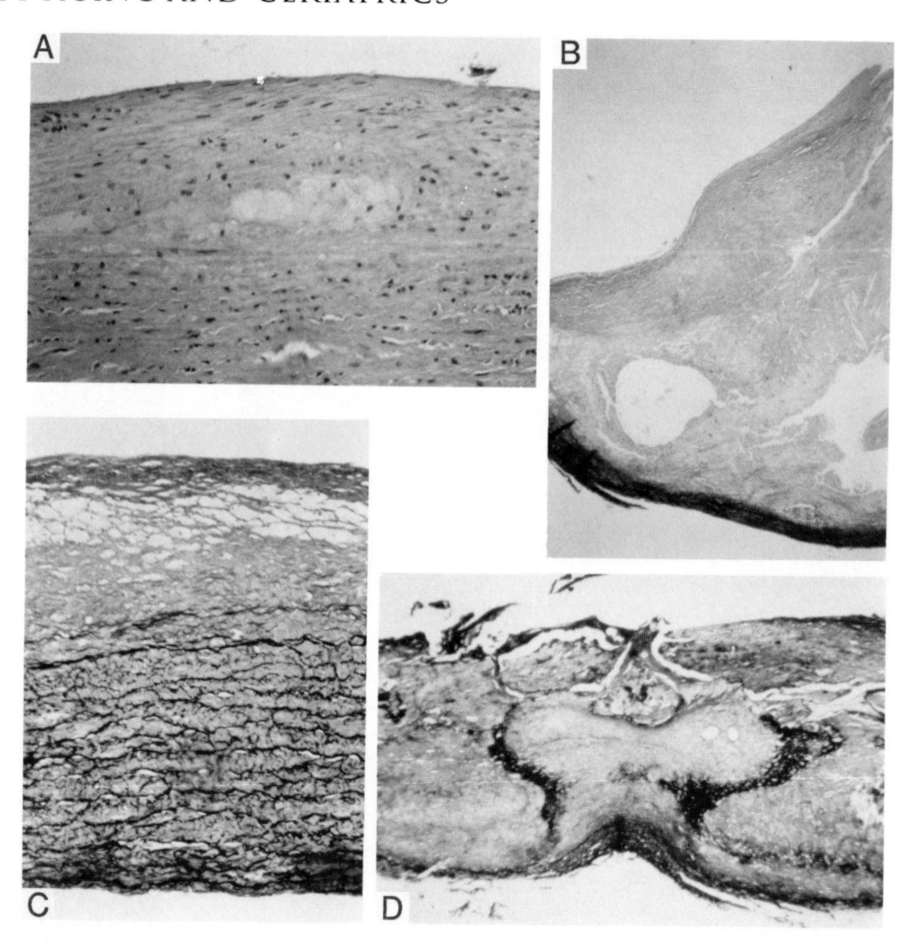

With the passing of years, more and more lipids accumulate into the fatty streaks of the established lesion (Table 16-5); the foam cells increase in number to the extent that the cells in the center of the arterial wall die — probably because of lack of oxygen supply and because the enormous amount of fat in their cytoplasm displaces or alters the organelles concerned with normal cellular function. The lipid released from the disintegrating foam cells, along with extracellular lipid, assembles in large pools, partly in the form of cholesterol crystals and partly as an amorphous mixture of triglycerides, phospholipids, and sterols. The whole material has the consistency of a soft paste or gruel (hence the term "atheroma" from the Greek "ather" indicating a gruel-like substance). An aorta that is riddled with atheromatous plaques may contain as many as 100 or more times its normal content of lipids.

The mass of extracellular lipid acts as an irritant to the arterial wall and provokes a proliferative reaction in the surrounding vascular tissue, similar to the inflammatory reaction that occurs in response to any foreign body encountered by the organism. To be noted here is that the major lipid component is the *low-density lipoprotein* (LDL), particularly its *oxidized form* (Chapters 6 and 17). Newly formed LDL is relatively benign, but it undergoes oxidation in the arterial wall. Oxidized LDL acquires a new configuration, binds to new receptors, and becomes less susceptible to removal by the high-density lipoprotein (HDL). Oxidized LDL would also be able to attract monocytes from the circulating blood; once the monocytes are trapped in the arterial wall and transformed in macrophages, oxidized LDL would inhibit their motility, thereby trapping them in the arterial wall.

The resulting atheroma develops like a sac, encapsulating the gruel, but remaining much thicker on the lumen side of the arterial wall where it forms a thick barrier between the blood stream and the gruel. At this stage, the atherosclerotic lesion has progressed considerably from its earlier manifestation as a fatty streak; not only is it larger and thicker, but it seems to rise above the inner surface of the artery like a cushion, resembling an encapsulated abscess. Because of its appearance, the lesion has been given the name of "atheromatous abscess", but it is also called a "raised plaque" or a "fibrous plaque", and its characteristic pearly white color

TABLE 16-4

Probable Role of Ground Substance in Early Atherosclerotic Lesions

Major components	Properties
Glucosaminoglycans (proteoglycans)	Viscoelastic (impaired with aging, reduced mechanical support)
Hyaluronic acid	Water-binding (with aging, reduced hydration, altered transport)
Collagen	
Elastin	

TABLE 16-5

Percentage of Total Lipids in Human Aortic Intima at Different Ages and in Different Types of Lesions

	Age (years)		Type of lesion		
Lipid group	Normal 15	Intima 65	Fatty streak	Fibrous plaque	Calcified fibrous plaque
Total lipid mg/100mg dry tissue	4.4	10.9	28.2	47.3	50.0
Cholesterol ester	12.5	47.0	59.7	54.1	56.3
Free cholesterol	20.8	12.2	12.7	18.4	22.4
Triglycerides	24.8	16.6	10.0	11.1	6.5
Phospholipids	41.9	24.2	17.6	16.6	14.8

Adapted from Smith, E. B., *J. Atherosci. Res.*, 5, 224, 1965. With permission.

resulting from the high content of collagen fibers in the capsule has given it also the name of "pearly plaque".

One of the main characteristics of the advanced atheroma is its progressiveness. As the plaque grows with fat, it consumes more and more of the arterial wall underneath it, transforming the cells into foam cells and disintegrating one elastic lamella after another. In this process, the entire media is destroyed and the atheroma invades the adventitia, which then reacts by setting up a series of inflammatory-like responses such as hyperemia (due to vascular invasion) and lymphocytic infiltration. Simultaneously, the capsule of the advanced atheroma, perhaps in a compensatory effort, thickens considerably, building a new arterial wall; however, containing neither muscle nor elastic fibers but rather almost exclusively scar (connective) tissue, this wall is functionally less efficient. As the pearly plaque becomes established, calcium deposits precipitate on the gruel, the capsule, or both in the form of either fine granules, thin strips, or huge masses. In the coronaries, for example, with the accumulation of large amounts of calcium over the years, the arteries become exceedingly hard and brittle, hence the term, "hardening" or "sclerosis" of the arteries.

Up to this stage, the changes in the atherosclerotic lesion have been counterbalanced by repair processes and, consequently, there has been no loss of tissue; indeed, in the sense that the lesion continues to form scar tissue, it can be viewed as "productive". In this respect, the function of the vessel, although impaired, is not drastically altered inasmuch as there is still a lumen and an intact, though thickened, wall with a relatively smooth lining permitting blood supply to the tissues. Some lesions can remain in this stage for an indefinite period of time, whereas others eventually undergo changes that cause the breakdown of the vessel, inviting the perils that have made atherosclerosis a deadly disease.

3.5 Complications of Advanced Lesions

When the capsule of the atheroma breaks away, the plaque is transformed into an ulcer, part of the exposed gruel is carried away by the bloodstream, and the rest forms an uneven, abnormal surface over which the blood clots. In some cases, the ulceration is accompanied by varying degrees of hemorrhage into the gruel or under the lips of the ulcer; before the original arterial wall has been destroyed by the atheromatous process, the ulceration breaks through the remnant of the wall causing rupture of the artery and massive hemorrhage into the space outside. Although little is yet known of the precise factors that promote ulceration and hemorrhage in plaques, it has been suggested that they might be favored by certain local changes in the lesions (e.g., extensive cell necrosis in the capsule) as well as hemodynamic events (e.g., sudden rise in blood pressure).

Other complications of advanced atherosclerosis include narrowing or widening of the arterial lumen, *thrombosis,* and *rupture* of the arterial wall. The widespread view that atherosclerosis narrows and tends to shut down the arterial lumen is valid mainly for some arteries, such as the coronaries, which are embedded into an unyielding environment. Arterial occlusion by stenosis (narrowing) of the lumen is usually a slow process, and collateral circulation, as previously described, frequently has time to establish itself so that a sufficient blood supply reaches the area normally serviced by the stenotic vessel.

Large arteries, such as the aorta, are widened by the atherosclerotic process as a result of the progressive weakening of the wall by the formation of scar tissue, causing it to give way to the mounting pressure within. In these cases, the arteries not only widen but also tend to lengthen, bending and twisting in the process. *Aneurysms,* balloon-like bulges that press upon neighboring structures and often burst (with subsequent hemorrhage), frequently occur in a given spot in the wall that is much weaker than the rest.

3.6 Thrombosis

Thrombosis represents the process by which a *plug of clot,* or *thrombus,* is formed in a blood vessel (or in one of the heart cavities) by coagulation of the blood, either remaining at the point of its formation or propagating along the arterial wall. It is distinguished from the *embolus* which is also a clot of coagulated blood but is carried in the blood current, and an *"embolism,"* which represents the plugging of an artery by a clot (embolus) that has been brought to its place by the blood current. For reasons not yet fully understood, thrombi develop more frequently in atherosclerotic than in normal arteries, appearing particularly on the ulcerated plaques,[19] on the arterial wall, or wherever there is a crack or fissure in the plaque. "Mural" thrombi, which are small and flat and develop over the surface of the wall of the large arteries, are relatively harmless inasmuch as they do not seriously impede the flow of blood through the vessel. When, however, the thrombus is large or develops in a small artery (such as a coronary artery or a cerebral vessel), it can fill its entire lumen and block all flow of blood, with disastrous results for the tissues that were supplied by the plugged vessel. Such an *occlusion* generally occurs suddenly, within minutes or hours, leaving little or no time for a collateral circulation to become established. In addition to its coincidence with atherosclerosis, the occurrence of thrombosis also has been related to blood chemistry (e.g., high blood lipids) and hemodynamic changes (e.g., hypertension).

Rupture of the arterial wall that has been weakened by atherosclerosis also can be triggered by hypertension. When the rupture occurs in the relatively small cerebral arteries, the result is a "stroke". When it occurs in the aorta, especially in the descending portion, the usual result is massive bleeding in the abdominal cavity (a consequence surgeons try to avoid by reconstruction).

3.7 Role of Platelets

In the previous discussion, we emphasized the involvement of the arterial wall in the formation of the atherosclerotic lesion. Equally important is the participation of some of the blood components. Cholesterol and lipoproteins will be discussed in Chapter 17. Other blood constituents include the platelets, a type of blood cells, and the blood coagulation system. An example of this association — arterial wall and blood constituents — is represented by thrombus formation, one of the complications of the atherosclerotic process.[19,20-22]

As already mentioned, fatty and fibrous plaques are more frequent and severe at the bifurcation of an artery. In this region, endothelial denudation, that is, ulceration or even minor loss of continuity of the endothelial surface overlying the atheroma, usually initiates the formation of the thrombus within the arterial wall.

Alterations in the endothelial lining cells and in the collagen from the subintimal layers cause platelets to adhere to the roughened surface. This surface adhesion is a normal function of platelets. Hydrolysis of some of the constituents (phosphatidylinositol) of the platelet inner membrane induces the release of the contents of the granules which are abundant in these cells. The release is associated with calcium flux and the formation of thromboxane A2, a membrane lipid derivative which further activates the release process. The substances released by the platelets are *growth factors* which promote further aggregation of platelets and initiation of clotting mechanisms. It is to be noted here that the universally used analgesic and antipyretic drug aspirin causes a moderate inhibition of thromboxane formation, an action that would be of value in reducing the incidence of recurrences in patients with some forms of stroke.

Growth Factors in Atherosclerosis

An increasing long list of growth factors suggests that these local secretions actively participate in the formation, or attempt to repair, the atherosclerotic process. Among these, fibroblast growth factor (FGF) and platelet-derived growth factor (PDGF) stimulate proliferation of cells from the arterial wall. Other growth factors derive from the immune system, such as platelet-activating factor (PAF), a cytokine secreted by monocytes as well as platelets and capable of stimulating platelet aggregation of the site of arterial injury.

The plug or thrombus formed by platelet aggregation and release and by clotting stimulation superimposed on the alterations of the arterial wall is friable, and the fibrino–platelet material may break off and enter the circulation, as an embolus. Emboli circulate in the blood and may occlude some smaller artery, for example, in the eye or brain. Alternatively, the thrombocytic process may completely occlude the artery at the level of the original plaque.

4 EXPERIMENTAL ATHEROSCLEROSIS AND ANGIOGRAPHY

As indicated before, atherosclerotic lesions have been found in all animal species investigated. In addition, lesions resembling those occurring in humans can be produced in animals, particularly by manipulation of the diet, although other methods (e.g., trauma of the arterial wall) are also successfully used for these purposes (Table 16-6). The various aspects of atherosclerosis from onset to potential therapy are being actively studied by a variety of experimental *in vivo* and *in vitro* approaches. *In vivo* experiments consist of studies of naturally occurring and experimentally induced atherosclerosis.[4,23,24] Atherosclerotic lesions have been found in all animal species investigated but with specific characteristics for each species. Comparison of lesions among species may provide some interesting information, which is marred, however, by the difficulty of extrapolating from one species to another and especially to humans.

Induction of lesions resembling those occurring in humans have been produced in animals by various methods based on the etiopathogenesis of the lesions:

1. By manipulation of the diet (e.g., high-fat [cholesterol] diets, high-carbohydrate diets)
2. By administration of hormones, such as adrenocorticoids (to mimic stimulation of the adrenal cortex by stress)
3. By alterations of the arterial wall, e.g., by mechanical injury or bacterial or viral infections

TABLE 16-6
Experimental Atherosclerosis

In Vivo

Interspecies studies
Dietary manipulation (lipids)
Trauma of arterial wall (mechanical, viral)
Alteration of clotting system
Hormones, stress

In Vitro

Culture of individual cell types
Effects of various factors (toxic substances) on cultured cells
Molecular studies of early lesion
Action of local growth-promoting factors

4. By alterations of the clotting system and of the platelets
5. By the administration of oxidants or exposure to oxidation; generating conditions (to induce accumulation of free radicals)
6. By the administration of local growth-promoting or growth inhibitory factors

Progress in the study of atherosclerosis in humans has been greatly facilitated by the availability of imaging techniques for the arterial wall *in vivo,* such as *angiography,* associated with computer analysis. This technique, which involves the X-ray visualization of blood vessels after the injection of radio-opaque, contrast material, is extremely useful both to identify and to quantify the lesions and to follow their progression (or regression) under natural conditions or in response to various treatments. The arteries most frequently examined by angiography are the carotid and coronary arteries as well as the femoral artery. The greatest drawbacks to angiography include its cost and potential for complications (allergic reaction to contrast agents, renal failure consequent to contrast administration, bleeding, thrombus formation, pain during the procedure). All of these potential problems may be minimized by observing proper techniques and medication.

5 THEORIES OF ATHEROSCLEROSIS

From the description of atherosclerosis presented so far, it emerges clearly that the atherosclerotic lesion is a complex, multifactorial process. However, at the present state of our knowledge, our ability for understanding the lesions may be better advanced by concentrating on studying simple parts of the process rather than trying to tie everything together into a single theory. In fact, this is just what is current today with four major theories, each dealing with different processes in the arterial wall. There are also several minor theories but we will address here only the major ones. They include the following:

1. **Lipid accumulation theory.** Alterations of blood lipoproteins such as accumulation of low-density lipoprotein (LDL) or altered ability of the cell to interact with LDL, especially oxidized LDL, would result in the accumulation of lipids in the arterial wall. Lipid accumulation would result from an imbalance between the input of cholesterol into the cell and its removal by HDL as discussed in Chapter 17. An

example of this accumulation is represented by the foam cells mentioned above. Lipid infiltration involves not only muscle cells but also migrating monocytes and macrophages and T lymphocytes; thus, our understanding should cover not only how lipid infiltrates the arterial wall but also how immune reactions participate in this accumulation. While, at least in humans, lipid infiltration appears to be an early event, the open question is what effect does that lipid have on the subsequent stages leading to the advanced lesion and on the clinical effects of the disease?

2. **Smooth muscle cell proliferation.** According to the original theory, the atherosclerotic lesion would begin as a mutagenic event in muscle cells triggered by mechanical, chemical, or viral injury. Such an event would result in the transformation of the muscle cell from a normal to a replicating cell, and the proliferative state would lead to the formation of a benign tumor acting as a stimulus to inflammation and capable of inducing both damage and repair mechanisms of the lesion.
3. **The thrombogenic theory.** This theory states that the onset of the lesion may be related to early alterations of the endothelial cells of the intima (or the basement membrane) leading to a small hemorrhage and thrombus formation. Current research in this area investigates the surface proteins of the platelets, the mechanisms of interaction of platelets with the connective tissue of the arterial wall, the series of events leading to the activation of the platelet to release its granule content (representing growth factors), and the role of the blood clotting system in contributing to thrombosis.
4. **The free radical theory,** discussed in Chapter 6.

The presence of at least four major theories of atherosclerosis indicates that none can explain satisfactorily the atherosclerotic process. However, they have opened the way to intensive experimentation *in vivo* and *in vitro* which, with the current advances in molecular biology, are beginning to provide the necessary answers for effective prevention and therapy.

6 CORONARY HEART DISEASE

One of the major life-threatening consequences of atherosclerosis involves the coronary arteries, that is, the arteries that supply blood to the heart. Atherosclerosis of the coronaries leads to coronary heart disease (CHD), also called ischemic heart disease, which implies reduced blood flow to the heart and, consequently, both angina pectoris and myocardial infarction. Other possible consequences include arrhythmias due to defects of impulse conduction and electrocardiographic (ECG) changes. All of these conditions lead to heart failure, severe disability, and death.

CHD continues to be the major cause of disability and death in the U.S. Its prevalence increases with age and this prevalence shows a significant sex difference, with women having a much lower incidence than men; this sex difference disappears after 70 years of age. Stress is another significant contributing factor to the high CHD incidence in the elderly (Figure 16-6).

6.1 CHD Mortality Declining?

The prevalence of CHD is still high despite a steady decrease in mortality from this cause since the late 1960s. This decline includes all sectors of the adult population. In the 35- to 74-year age group, the rate of mortality due to CHD has fallen by 30%, and this is all the more remarkable because it turned around after a sustained period of rising CHD as far back as at least 1940. The current striking decline in the U.S. is greater than that observed in any other country, but do not forget that the incidence there is among the highest.

Apparently we have been doing the "right things" during the past 25 years in regard to CHD and the major adult cardiovascular diseases in general.[25-27] Two major factors may account for this:

- The first, and most important, is the improvement in lifestyle, amelioration of the diet, increased physical exercise (see also Chapters 17 and 21), and cessation of cigarette smoking.
- The second factor is therapeutic improvement, including better control of hypertension, better medical care in the hospital for patients with CHD, and more widespread use of coronary by-pass surgery and other methods of coronary revascularization.

The right and left coronary arteries arising from the aorta as it emerges from the left ventricle are the major vessels supplying blood to the myocardium (the muscle layer of the heart). Venous drainage is via a superficial system ending primarily in the coronary sinus and via a deep system that drains the remainder of the heart and empties directly into the heart chambers.

Coronary flow at rest (250 ml/min) is 5% of the cardiac output. The heart extracts 70 to 80% of the oxygen, close to the maximum amount possible from this blood. Therefore, oxygen consumption can be significantly increased on demand

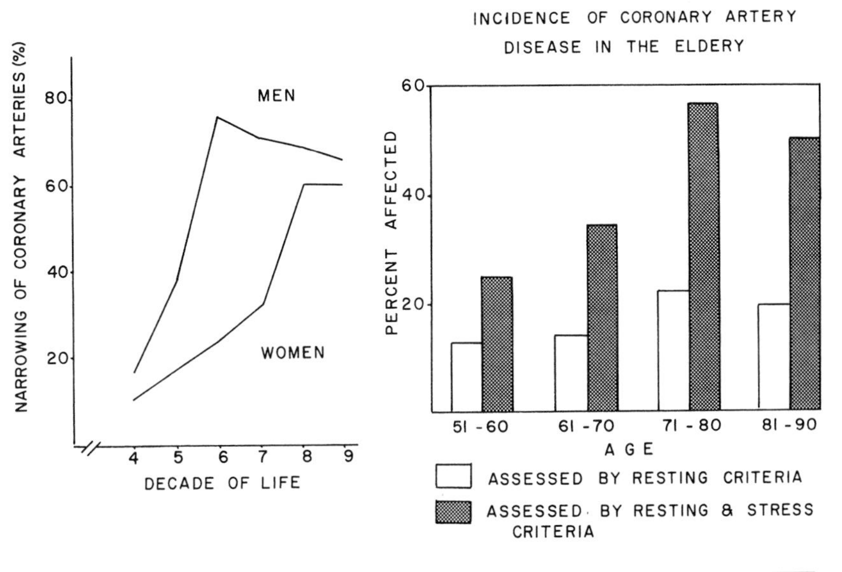

FIGURE 16-6 *Percentage narrowing of coronary artery lumen with aging (on the left).* Note sex differences with narrowing coronary artery occurring earlier and being more severe in men than women. *Incidence of coronary artery disease with aging (on the right).* Note the increase in incidence with age, particularly under conditions of stress. (Drawing by S. Oklund.)

only by increasing blood flow. Coronary flow is influenced by several local factors which regulate flow through vasodilation or vasoconstriction (Table 16-7).

Major vasodilators include reduced oxygen and increased carbon dioxide concentration, increased products of metabolism (hydrogen ions, potassium, lactic acid), local agents (prostaglandins, adenine nucleotides, and adenosine), and neural stimulation (stimulation of parasympathetic innervation). Major vasoconstrictors include angiotensin II (Chapter 11) and neural stimulation of the sympathetic innervation. As for the brain, the cardiac circulation is preferentially preserved when that of other organs is compromised.

TABLE 16-7

Local Regulation of Coronary Blood Flow

Vasodilation	Vasoconstriction
Low Oxygen	Angiotensin II
High CO_2	Sympathetic stimulation (direct)
High H^+	
High K^+	
High lactic acid	
High prostaglandins	
High adenine nucleotides and adenosine	
Vagal stimulation	

6.2 CHD Consequences

When progressive narrowing of the coronary lumen reduces flow through the coronary artery to the point that the myocardium becomes ischemic, angina pectoris develops. If the myocardial ischemia is very severe and prolonged, irreversible changes occur in the cardiac muscle and the result is myocardial infarction. The heart stops functioning and death ensues within a few minutes. If ischemia is less severe, death may not occur, but it may generate permanent functional impairment.

6.3 Signs and Symptoms

The major sign of angina pectoris is squeezing or pressure-like pain, retrosternally, radiating to the left shoulder, arm, hand, neck, and jaw. The pain often appears with exertion, emotion, or a large meal. Anginal pain in the elderly is often less marked, possibly due to reduced activity or altered pain perception; it may be effort-induced or may occur at rest or in bed and may present as headache or epigastric pain relieved by antacids (Table 16-8). Major signs and symptoms of myocardial infarction are variable in the elderly, although chest pain remains the commanding feature in patients admitted to the coronary care units. Other symptoms include breathlessness, confusion, behavior change, fainting, palpitations, vomiting, sweating, abdominal pain, and hypotension. The pain may last over 30 min and is not relieved by multiple doses of nitroglycerine. Diagnosis is confirmed by ECG changes (Table 16-8). With the establishment of the infarct, irreversible changes occur in the myocardium; muscle cells first become leaky and the rise in serum enzymes and isoenzymes is a biochemical diagnostic sign of the infarct. The first enzyme to be elevated is serum glutamic oxaloacetic transaminase (SGOT) followed by creatinine phosphokinase (CPK) and lactic dehydrogenase (LDH).

6.4 Risk Factors in CHD

CHD epidemiology, so important in identifying the causes and subsequent prevention and treatment, can best be studied by focusing on the consequences. Indeed, the decreased mortality from CHD may be ascribed primarily to identification of risk factors and to their prevention. Studies in various populations have identified the following risk factors that predispose to the development of CHD: age, genetic predisposition, hypertension, diabetes mellitus, hypercholesterolemia, and cigarette smoking. Other risk factors include obesity, poor physical fitness and lack of exercise, and personality type (Table 16-9).

In a 1964 to 1972 study on a cohort of 7500 white Kaiser Permanente Medical Care Program members, initially free of CHD and ranging in age from 60 to 79 years of age, the mortality risk factors until 1980 were increasing age, high systolic blood pressure and moderate to high serum cholesterol, cigarette smoking, alcohol consumption, increased body weight, alterations of pulse and white blood count, and high serum uric acid. For males, the most important risk factors were age, blood pressure, and body weight; and for females, age, blood pressure, cigarette smoking, cholesterol, uric acid, and nonconsumption of alcohol (no drinks *versus* one to two drinks per day).[28]

TABLE 16-8

Symptoms of Angina Pectoris and Acute Myocardial Infarction in the Elderly

Angina Pectoris

Pain, less marked than in adult; may present as headache or epigastric distress

Myocardial Infarction

Variable presentation with chest pain, including breathlessness, confusion, fainting, GI symptoms, sweating, hypotension, etc.

These lists dictate specific preventive measures such as treatment of hypertension and diabetes, elimination of cigarette smoking, amelioration of dietary habits towards an optimal body weight, and encouragement of measures to improve physical fitness.

6.5 "New Era" of Management

The reduction of mortality due to CHD reported in the 1980s had arisen from the awareness of contributing factors and their alleviation or amelioration (Table 16-10). New developments in this area demonstrate that it is possible to induce regression/reversal or at least to arrest the progression of atherosclerotic lesions by appropriate dietary and drug interventions. Reduction of LDL (the atherogenic lipoprotein) blood levels, by administration of drugs or manipulation of the diet, results in a significant regression of the atherosclerotic lesions. This regression occurred in cases of hyperlipoproteinemia or hypercholesterolemia (familial or not), in men and women, and in younger as well as older (65+ years) individuals.[29-36] Individuals who have had a myocardial infarction, coronary by-pass surgery, or angioplasty should be treated promptly and aggressively, to lower levels of LDL and stabilize plaques.

Current studies show that the administration of antioxidants, by reducing the level of oxidized LDL, may also be beneficial. So far promising results with antioxidant therapy have been reported in nonhuman primates and swine. The effectiveness of these treatments supports the view that the atherosclerotic process is not, as previously thought, an

TABLE 16-9

Major Risk Factors in Coronary Heart Disease

Age
Genetic predisposition
Hypertension
Diabetes mellitus
Hypercholesterolemia
Cigarette smoking
Obesity
Poor physical fitness and lack of exercise
Personality type (?)

TABLE 16-10

Major Types of Coronary Heart Disease Treatment

Medical treatment
Diet
Exercise
No smoking
Pharmacologic agents
Surgical treatment
Aortocoronary bypass graft
Percutaneous coronary angioplasty with streptokinase/TPA[a] anticoagulant therapy

[a] TPA, tissue plasminogen activator.

inexorably progressive condition; rather, the new research emphasizes the similarities between atherosclerotic and inflammatory processes — once the cause of the inflammation has been removed, the arterial wall would "remodel" and repair its structure. Thus, medical treatment includes

1. Treatment of the underlying disease, if any (hypertension, diabetes mellitus, hyperlipidemias)
2. Behavioral therapy: low cholesterol, low fat diets, cessation of smoking, reduced stress and increased physical exercise, especially for cardiac rehabilitation
3. Administration of pharmacologic agents with the intention of
   - reducing LDL and cholesterol blood levels
   - decreasing free radical levels (antioxidants)
   - increasing cardiac blood flow, reducing cardiac work, and preventing clotting

Surgery has also been successful with aortocoronary bypass grafts and transluminal coronary angioplasty (mechanical dilation of the area of constriction) with anticoagulant therapy by the intracoronary injection of the enzyme streptokinase or even better of tissue plasminogen activator (TPA), a recombinant protease.[37-40]

7 HYPERTENSION

7.1 Definition and Role as CHD Risk Factor

Blood pressure depends on cardiac output and peripheral vascular resistance. The World Health Organization defines normal blood pressure as below 140 mmHg for the systolic pressure and 90 mmHg for the diastolic, and hypertension as a sustained elevation of the arterial pressure with a systolic blood pressure of 160 mmHg or more, and a diastolic of 95 mmHg or more, or both. The condition is present in over 60% of people over 60 years of age. This is an often silent disease, that is, presenting no symptoms; therefore, the true incidence of hypertension is probably much higher.

Hypertension is an important risk factor for cardiovascular morbidity and mortality in males and females at any age. The interaction of atherosclerosis and hypertension is complex but crucial in the progression of cardiac and cerebrovascular disease. Although it is unclear whether hypertension plays a greater role than any other of the risk factors discussed above in the etiology of CHD, or whether atherosclerosis is the major cause of hypertension, no one questions the existence of an interaction between the two. The importance of hypertension as the most prominent reversible risk factor for cardiovascular and cerebral complications in the elderly cannot be overemphasized.

Blood pressure increases with age, but the systolic and diastolic pressures behave differently over time. Systolic pressure rises slowly commencing in the early adult years and continuing on to old age. Diastolic pressure rises steadily in early adulthood but begins to decline at about age 60. This accounts for the high prevalence of isolated systolic hypertension in the elderly (Figure 16-7). The elderly with combined systolic and diastolic hypertension have disproportionately higher systolic pressures.[42]

7.2 Essential Hypertension

Most elderly patients with elevated blood pressure have essential hypertension, which, in fact, comprises 90% of the cases (Tables 16-11 and 16-12). Its etiology is unknown, but all of the affected elderly have increased peripheral resistance, probably related to atherosclerosis. Resistance to blood flow depends to a minor degree on blood viscosity but to a greater degree on the diameter of the vessels, primarily the arterioles. Reduction of the arteriolar diameter increases resistance, and increased resistance results in increased diastolic pressure. Increased rigidity of the larger arteries contributes to the increase in systolic pressure.

The remaining 10% of elevated blood pressure cases are due to other factors, including *chronic renal and endocrine diseases* and *polypharmacy*. Renal hypertension may be due to kidney damage secondary to infections, glomerulonephritis, and pyelonephritis, to polycystic disease, or to narrowing of the renal artery, primarily due to atherosclerosis. Hypertension of endocrine origin may be due to alterations of thyroid function or to increased secretion of adrenal steroids, as in Conn's syndrome (with aldosterone hypersecretion), virilizing hyperplasia (with increased mineralocorticoids), or Cushing's syndrome (with glucocorticoid hypersecretion).[43] Tumors of the adrenal medullary tissue, generally outside the adrenal, are called pheochromocytoma. They induce a type of hypertension which is often episodic (but may be sustained) and is characterized by very high systolic pressure. Coarctation of the aorta is a congenital narrowing of the thoracic aorta, producing severe hypertension in the upper part of the body.

Numerous drugs, alone, or in combination, may also induce hypertension as a side effect. Singling out steroids, one can speak of "pill hypertension". This refers to hypertension in women undergoing chronic treatment with oral contraceptives; however, there is evidence that these women are predisposed to hypertension in any case.

7.3 Hypertension in the Elderly

Hypertension is characterized in the elderly by marked lability (Table 16-13).[44-47] Therefore, casual measurements of blood pressure can be particularly unreliable, and several readings should be taken over an interval of time. The measurement

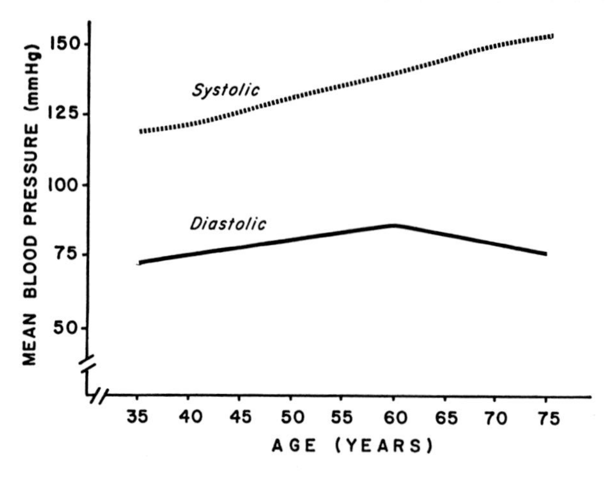

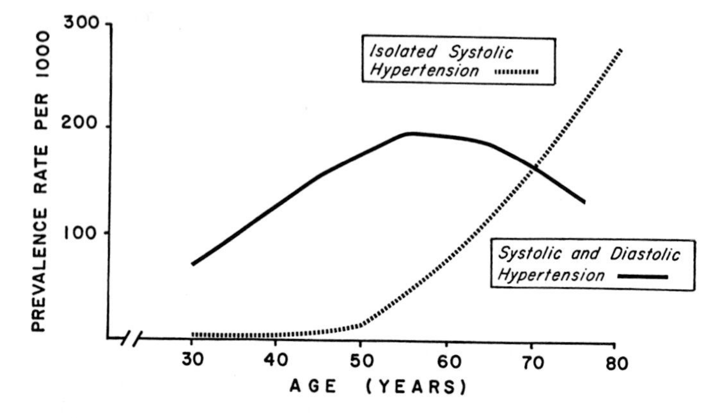

FIGURE 16-7 (A) *Increase with aging in mean systolic blood pressure but not in diastolic pressure;* (B) *Prevalence of isolated systolic hypertension continues to increase with aging, whereas combined systolic and diastolic hypertension increases at a younger age but levels off after 60 years of age.*

should also be taken with the individual in different positions, sitting, reclining, standing, for orthostatic hypotension (occurs upon standing and produces syncope) due to reduced autonomic responsiveness is prevalent in the elderly.

Orthostatic (postural) hypotension consists in a fall in blood pressure upon passing suddenly from the supine to the standing position. It occurs in some young and adult individuals and in many elderly. The hypotension causes a transitory cerebral ischemia (reduced blood flow) with dizziness, dimness of vision, and even fainting or syncope (loss of consciousness). The fall in blood pressure is due to inability to compensate rapidly for the gravitational changes in blood pools that occur with standing, depriving the brain of the required amounts of blood. The major compensation on assuming the upright position is triggered by the drop in blood pressure in the baroreceptors (carotid sinus and aortic arch) with consequent increase in cardiac rate, venoconstriction, and increased levels of renin and aldosterone. All these responses induce peripheral vasoconstriction and increased cardiac output with normalization of the blood pressure and decreased cerebral vascular resistance with increased blood flow. Failure or slowing down of all of these compensatory mechanisms results in hypotension. This abnormality is quite frequent in the elderly because of autonomic insufficiency, abnormal baroreceptors, myocardial arrhythmias, and impaired cerebral vascular adjustments.[48] It is also often the consequence of various medications to which the elderly may be particularly sensitive.

The blood pressure must be taken in both arms and peripheral pulses examined to exclude coarctation of the aorta. General physical examination must exclude endocrine disorders, and an abdominal examination may reveal an enlarged polycystic kidney. Auscultation of abdominal bruits are of little diagnostic help in the elderly. With the auscultatory method of measuring blood pressure, the first sounds heard (after releasing pressure in the inflatable cuff) represent the systolic pressure (as read in mmHg on the sphygmomanometer); these are followed by the so-called sounds of Kortkow, and the disappearance of these sounds corresponds to the diastolic pressure (as read in mmHg). In some elderly patients, there is an auscultatory gap which consists of a silent interval below the beginning of the sounds of Korotkow at systole. In this case, blood pressure must first be measured by palpation to avoid missing a very elevated systolic blood pressure, as may occur by auscultation alone.

Signs of hypertensive end organ damage include left ventricular enlargement, cardiac failure, and retinal changes.

TABLE 16-11

Causes of Hypertension

Essential hypertension	Unknown
Renal hypertension	Glomerulonephritis
	Pyelonephritis
	Polycystic disease
	Renal artery narrowing
Endocrine	Hypo/hyperthyroidism
	Hypersecretion of aldosterone (Conn's syndrome)
	Hypersecretion of glucocorticoids (Cushing's syndrome)
	Congenital virilizing adrenal hyperplasia
	Adrenal medullary tumor (pheochromocytoma)
Coarctation of the aorta (narrowing)	
Polypharmacy	Sympathomimetic amines
	Steroids
	Nonsteroid analgesics
	Tricyclic antidepressants, etc.

TABLE 16-12

Factors that Contribute to Hypertension in the Elderly

Atherosclerosis
Renal impairment
Hormonal defects
Blunted baroreceptor responses

TABLE 16-13

Clinical Manifestations of Hypertension in the Elderly

Lability of blood pressure
Prevalence of orthostatic hypotension
Presence of an auscultatory gap

TABLE 16-14

General Guidelines for Treatment of Hypertension in the Elderly

Use nonpharmacologic maneuvers when possible
Use pharmacologic therapy cautiously
Avoid sudden changes in blood pressure
Watch for orthostatic hypotension

Malignant hypertension — that is, a type of severe hypertension with an accelerated course of 1 to 2 years — is rare but not unknown in the elderly, and this disorder may or may not be associated with very high levels of blood pressure. Fundoscopy, the examination of the blood vessels of the fundus of the eye, will show evidence of hemorrhages and soft exudates with or without edema of the papilla. Proteinuria is often present.

7.4 Consequences of Hypertension

The majority of hypertensive elderly are asymptomatic, i.e., have no apparent subjective symptoms. Among the factors that contribute to the development of hypertension at all ages, those pertaining specifically to old age, include atherosclerosis, which decreases arterial compliance and narrows blood vessels, renal impairment, hormonal defects, as well as blunted responsiveness of baroreceptors, the peripheral sensors of blood pressure.

7.5 Management

Hypertension of specific origin is curable by eliminating the cause; therefore, the cause must be sought out and the appropriate treatment initiated whether by surgery, hormonal suppression, or drug administration (Table 16-14). However, the likelihood of detecting a treatable cause of hypertension in the elderly patient is small, and more specialized investigation is only justified if there is any clear evidence of a primary underlying cause. In the absence of such specific cause, treatment consists primarily in keeping the blood pressure within normal limits with the help of single or combined antihypertensive drug treatment (see Chapter 23).

REFERENCES

1. Bierman, E. L., Arteriosclerosis and aging, in *Handbook of the Biology of Aging,* Finch, C. E. and Schneider, E. L., Eds., Van Nostrand Reinhold, New York, 1985.
2. Lee, K. T., Ed., *Atherosclerosis,* Annals of the New York Academy of Sciences, Vol. 454, New York Academy of Sciences, New York, 1985.
3. Kottke, B. A. and Rooke, T. W., Disorders of the blood vessels, in *Principles and Practice of Geriatric Medicine,* Pathy, M. S. J., Ed., John Wiley & Sons, New York, 1991, 625.
4. Constantinides, P., *Experimental Atherosclerosis,* Elsevier, Amsterdam, 1965.
5. Gresham, G. A., *Reversing Atherosclerosis,* Charles C Thomas, Springfield, IL, 1980.
6. Morrison, L. M. and Schjeide, O. A., *Arteriosclerosis: Prevention, Treatment and Regression,* Charles C Thomas, Springfield, IL, 1984.
7. Ross, R., The pathogenesis of atherosclerosis, *N. Engl. J. Med.,* 314, 488, 1986.
8. McGill, H. C., Geer, J. C., and Strong, J. P., Natural history of human athero-sclerotic lesions, in *Atherosclerosis and Its Origin,* Sandler, M. and Bourne, G. H., Eds., Academic Press, New York, 1963, 39.
9. Schlichter, J. and Harris, R., The vascularization of the aorta. A comparative study of the aortic vascularization of several species in health and disease, *Am. J. Med. Sci.,* 218, 610, 1949.
10. Gozna, E. R., Marble, A. E., Shaw, A., and Holland, J. G., Age-related changes in the mechanics of the aorta and pulmonary artery of man, *J. Appl. Physiol.,* 36, 407, 1974.
11. Ross, R., George Lyman Duff Memorial Lecture. Atherosclerosis: a problem of the biology of the arterial wall cells and their interactions with blood components, *Arteriosclerosis,* 1, 293, 1981.
12. Jellinek, H., Detre, Z., and Veress, B., *Transmural Plasma Flow in Atherogenesis,* Akademiai Kiado, Budapest, 1983.
13. Patel, D. J. and Vaishnav, R. N., *Basic Hemodynamics and its Role in Disease Processes,* University Park Press, Baltimore, 1980.
14. Moore, S., Ed., *Injury Mechanisms in Atherogenesis,* The Biochemistry of Disease, Vol. 9, Marcel Decker, New York, 1981.
15. Vikhert, A. M. and Zhdanov, V. S., *The Effects of Various Diseases on the Development of Atherosclerosis,* Pergamon Press, New York, 1981.
16. Woolf, N., *Pathology of Atherosclerosis,* Butterworth Scientific, London, 1982.
17. Benditt, E. P., The origin of atherosclerosis, *Sci. Am.,* 236, 74, 1977.
18. Ross, R. and Glomset, J. A., Atherosclerosis and the arterial smooth muscle cell: proliferation of smooth muscle is a key in the genesis of lesions of atherosclerosis, *Science,* 180, 1332, 1973.
19. Bang, N. U., et al., *Thrombosis and Atherosclerosis,* Year Book Medical Publishers, Chicago, 1982.
20. Schwartz, S. M., Gojdusek, C. M., and Selden, S. C., Vascular wall growth control, the role of the endothelium, *Arteriosclerosis,* 1, 107, 1981.
21. George, J. N. and Shattil, S. J., The clinical importance of acquired abnormalities of platelet function, *New Engl. J. Med.,* 324, 27, 1991.
22. Widhalm, K. and Sinzinger, H., Current aspects of atherosclerosis lipids, lipo-proteins, platelets, prostaglandins and experimental findings, *Atherogenesis,* Vol. 5, M. Maudrich, Wein, 1983.
23. Sandler, M. and Bourne, G. H., Histochemistry of atherosclerosis in the rat, dog and man, in *Atherosclerosis and Its Origin,* Sandler, M. and Bourne, G. H., Eds., Academic Press, New York, 1963.
24. Likar, I. N. and Robinson, R. W., Atherosclerosis: Cattle as a model for study in man, *Monographs on Atherosclerosis,* Vol. 12, S. Karger, Basel, 1985, 1.
25. Pell, S. and Fayerweather, W. E., Trends in the incidence of myocardial infarction and in associated mortality and morbidity in a large employed population, 1957–1983, *N. Engl. J. Med.,* 312, 1005, 1985.
26. Stamler, J., Coronary heart disease: doing the "right things", *N. Engl. J. Med.,* 312, 1053, 1985.
27. Ornish, D., Brown, S. E., Schwerwitz, L. W., et al., Can lifestyle changes reverse coronary heart disease?, The Lifestyle Heart Trail, *Lancet,* 336, 129, 1990.
28. Sidney, S., *Risk factors for coronary heart disease in the elderly,* Am. Heart Assoc. 26th Annual Conference on Cardiovascular Disease Epidemiology, San Francisco, March 3 to 5, 1986.
29. Kane, J. P., Malloy, M. J., Ports, T. A., Phillips, N. R., Diehl, J. C., and Havel, R. J., Regression of coronary atherosclerosis during treatment of familial hypercholesterolemia with combined drug regimens, *JAMA,* 264, 3007, 1990.

30. Blankenhorn, D. H., Nessim, S. A., Johnson, R. L., et al., Beneficial effects of combined colestipol-niacin therapy on coronary atherosclerosis and coronary venous bypass grafts, *JAMA,* 257, 3233, 1987.
31. Brown, G., Albers, J. J., Fisher, L. D., et al., Regression of coronary artery disease as a result of intensive lipid-lowering therapy in men with high levels of apolipoprotein B, *N. Engl. J. Med.,* 323, 1289, 1990.
32. Buchwald, H., Varco, R. L., Matts, J. P., et al., Effects of partial ideal bypass surgery on mortality and morbidity from coronary heart disease in patients with hypercholesterolemia. Report of the Program on the Surgical Control of the Hyperlipidemias (POSCH), *N. Engl. J. Med.,* 323, 946, 1989.
33. Cashin-Hemphill, J., Mack, W. J., Pogoda, J. M., et al., Beneficial effects of colestipol-niacin on coronary atherosclerosis. A 4-year follow-up, *JAMA,* 264, 3013, 1990.
34. Malloy, M. J., Kane, J. P., Kunitake, S. T., and Tun, P., Complementarity of colestipol, niacin, and lovastatin in treatment of severe familial hypercholesterolemia, *Ann. Intern. Med.,* 107, 616, 1987.
35. Steinberg, D. and Witztum, J. L., Lipoproteins and atherogenesis. Current Concepts, *JAMA,* 264, 3047, 1990.
36. Watts, G. F., Lewis, B., Brunt, J. N. H., et al., Effects on coronary artery disease of lipid-lowering diet, or diet plus cholestyramine, in the St. Thomas Atherosclerosis Regression Study, *Lancet,* 339, 563, 1992.
37. Gersh, B. J., Kronmal, R. A., Schaff, H. V., Frye, R. L, Ryan, T. J., et al., Comparison of coronary artery bypass surgery and medical therapy in patients 65 years of age or older. A randomized study from the Coronary Artery Surgery Study (CASS) registry, *N. Engl. J. Med.,* 313, 217, 1985.
38. Kennedy, J. W., Ritchie, J. L., Davis, K. B., Stadius, M. L., Maynard, C., and Fritz, J. K., The Western Washington randomized trial of intracoronary streptokinase in acute myocardial infarction. A 12-month follow-up report, *N. Engl. J. Med.,* 312, 1073, 1985.
39. Koren, G., Weiss, A. T., Hasin, Y., Appelbaum, D., Welber, S., et al., Prevention of myocardial damage in acute myocardial ischemia by early treatment with intravenous streptokinase, *N. Engl. J. Med.,* 313, 1384, 1985.
40. Mock, M. B., Reeder, G. S., Schaff, H. V., Holmes, D. R., Jr., Vlietstra, R. E., Smith, H. C., and Gersh, B. J., Percutaneous transluminal coronary angioplasty versus coronary artery bypass. Isn't it time for a randomized trial?, *N. Engl. J. Med.,* 312, 916, 1985.
41. Sun, M., The coming competition among clot-busting drugs, *Science,* 240, 1267, 1988.
42. Rowe, J. W., Systolic hypertension in the elderly, *N. Engl. J. Med.,* 309, 1246, 1983.
43. Laragh, J. H., Atrial natriuretic hormone, the renin-aldosterone axis, and blood pressure-electrolyte homeostasis, *N. Engl. J. Med.,* 313, 1330, 1985.
44. Niarchos, A. P. and Laragh, J. H., Hypertension in the elderly. II. Diagnosis and treatment, *Mod. Geriatr. Cardiovasc. Dis.,* 49, 49, 1980.
45. Tuck, M. and Sowers, J., Hypertension and aging, in *Endocrine Aspects of Aging,* Korenman, S. G., Ed., Elsevier Biomedical, New York, 1982, 81.
46. Franklin, S. S., Geriatric hypertension, *Med. Clin. N. Am.,* 67, 395, 1983.
47. Williams, B. O., The cardiovascular system, in *Principles and Practice of Geriatric Medicine,* Pathy, M. S. J., Ed., John Wiley & Sons, New York, 1991, 573.
48. Weisfeldt, M. L., Ed., *The Aging Heart — Its Function and Response to Stress,* Raven Press, New York, 1980.

III
SYSTEMIC AND ORGANISMIC AGING

17 PLASMA LIPOPROTEINS: THEIR METABOLISM AND ROLE IN ATHEROSCLEROSIS

Trudy M. Forte

As discussed in an earlier chapter, the cause of cardiovascular disease, primarily atherosclerosis, is complex, and many factors can contribute to the development of premature atherosclerosis. Factors that singly or in combination are contributors to the premature manifestation of the disease and early death include hypercholesterolemia, hypertension, diabetes mellitus, smoking, genetic predisposition, obesity, and stress. As this list of offenders implies, plasma cholesterol and, by extension, plasma lipoproteins have a significant role in the process of atherogenesis. The structure, synthesis, and metabolism of plasma lipoproteins are reviewed in this chapter, and their role as either positive or negative risk factors in premature atherosclerosis is presented.

1 LIPOPROTEIN NOMENCLATURE, STRUCTURE, AND COMPOSITION

Lipoproteins, specific lipid-containing macromolecules in the plasma, have an important role in the development of the atherosclerotic lesion.[1] Certain lipoproteins predispose to premature atherosclerosis, while others have a protective role. Lipids circulating in the plasma include free fatty acids, which are bound to albumin, and cholesterol, triglyceride, and phospholipids, which are bound to specific proteins, the apolipoproteins, to form lipoprotein complexes. There are five major classes of lipoproteins, which are recognizable on the basis of size and lipid content. The density of these lipoproteins is inversely proportional to their lipid content so that high-density lipoproteins have the lowest total lipid and the highest protein content. The major lipoproteins from least dense to most dense are chylomicrons, very-low-density, intermediate-density, low-density, and high-density lipoproteins.

Generally, one thinks of lipoproteins in terms of pathologic conditions such as coronary artery disease, but in fact they are molecules that are necessary for the maintenance of overall body functions. They are extremely important in the body since the lipids they carry, especially the triglycerides, are a major source of energy for cells. Fat-soluble vitamins are also transported by lipoproteins. Cholesterol transported by these macromolecules is utilized by the cells for cell division, cell growth, and membrane repair; cholesterol is also essential for the production of steroid hormones, both adrenocortical hormones, and sex hormones. An overabundance of lipoproteins, however, particularly those carrying cholesterol, can be deleterious, particularly by predisposing to premature cardiovascular disease.

As mentioned previously, lipoproteins are protein–lipid complexes that transport insoluble lipids, principally cholesterol and triglyceride, in the blood and interstitial fluids. A schematic of a generic lipoprotein is illustrated in Figure 17-1. The lipoprotein particle is essentially an oil droplet (hence, globular in shape) stabilized by a surface coat of hydrophilic molecules, including proteins and phospholipids. The core of the particle consists of the highly insoluble lipids, cholesteryl ester and triglyceride. Free cholesterol may also be associated with the surface components, as indicated in the illustration. As protein (the apolipoproteins) content on the particles increases relative to the lipid content, the particles become smaller and denser. The difference in densities of the particles is the fundamental basis for the nomenclature used for defining lipoproteins, i.e., very-low-density (VLDL), intermediate-density (IDL), low-density (LDL), and high-density (HDL) lipoproteins.

Major Classes of Lipoproteins

The major lipoprotein classes and their composition are summarized in Table 17-1. Essentially, lipoproteins are grouped into three major categories:

1. *Chylomicron (CM) and Very-Low-Density Lipoprotein (VLDL).* These are relatively low in protein, phospholipid, and cholesterol, but high (55 to 99%) in triglyceride. In more general terms, these particles are referred to as triglyceride-rich lipoproteins.
2. *Intermediate-Density Lipoproteins (IDL) and Low-Density Lipoproteins (LDL).* These are characterized by high levels of cholesterol, mainly in the form of cholesteryl esters. The latter form of cholesterol is highly insoluble. Since up to 50% of the LDL mass is cholesterol, it is not surprising that LDL has a significant role in the development of atherosclerotic disease.
3. *High-Density Lipoprotein (HDL).* The hallmarks of these particles are their high protein content (50%) and relatively high phospholipid content (30%). HDL are generally divided into two subclasses, HDL_2 and HDL_3; of the two, HDL_2 are large and less dense, and HDL_3 are smaller and more dense.

2 PLASMA LIPOPROTEIN CONCENTRATIONS

The most commonly used clinical indicator for measuring potential risk of premature cardiovascular disease is the level of plasma lipids. Fasting levels of triglyceride, cholesterol, and HDL cholesterol can often be used to identify possible abnormalities. The expected normal adult plasma lipid levels are shown in Table 17-2. Females characteristically have lower triglyceride concentrations (80 mg/dl) than males (120 mg/dl) and have higher HDL cholesterol (55 mg/dl versus 43 mg/dl for males). For comparison, the lipid levels from cordblood of normal, full-term newborns are also provided. The newborn infant has triglyceride and total cholesterol levels one half to one third those of the adult. The HDL cholesterol levels are relatively high (35 mg/dl) in the newborn, where the ratio of total cholesterol to HDL cholesterol is 2 compared with the adult values of 3.5 for females and 4.6 for males. HDL are considered beneficial; that is, they are protective against atherosclerosis, whereas LDL are positive risk factors. Therefore, the lipid levels in infants are perhaps the most "ideal", for at

0-8493-8979-8/94/$0.00+$.50

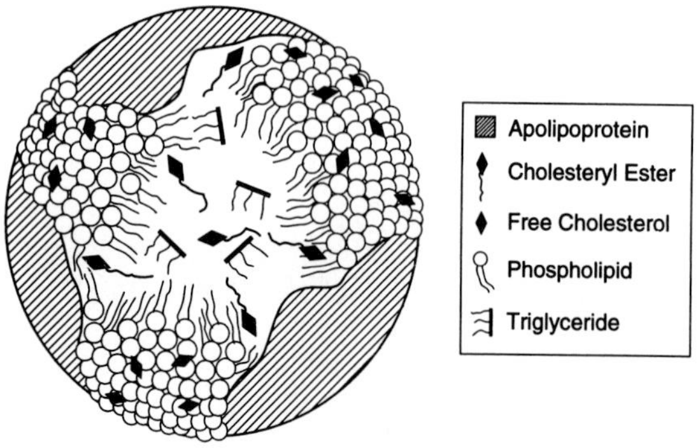

FIGURE 17-1 *Generalized organization of mature plasma lipoproteins.* Particles are overall globular in morphology with polar (water-soluble) components on the surface to stabilize the particle in the aqueous plasma environment. The polar constituents are primarily protein (apolipoprotein) and phospholipid. Some free cholesterol is also on the surface. The core of the particle consists of the nonpolar (highly insoluble in an aqueous environment) components, cholesteryl ester and triglyceride.

birth, plasma total cholesterol is low while HDL cholesterol is relatively high. Except for genetic abnormalities (such as homozygous familial hypercholesterolemia, to be discussed later), the vascular walls of neonates are free of fatty streaks. Fat accumulation appears in the first years of life, indicating that dietary input and environmental factors probably influence the initiation and progression of atherosclerosis. At birth, no distinction can be seen between male and female infants since sex hormone concentrations are low in levels and apparently have little metabolic influence at this stage of development.

3 FACTORS INFLUENCING LIPOPROTEIN LEVELS

In the adult, gender differences have a definite effect on plasma lipid levels, as shown by the higher levels of triglyceride and total cholesterol and the lower levels of HDL cholesterol in the males as compared to the females (Table 17-2). Indeed, females, before menopause, tend to have lower total cholesterol than males, but the most functionally significant difference is the higher HDL cholesterol in females, i.e., 55 mg/dl in females versus 43 mg/dl in males. This increase in HDL cholesterol is in a specific subclass of HDL, the HDL_2. Epidemiologic studies suggest that elevated HDL_2 plays a protective role in atherogenesis. In women, the hormone estrogen has an important role in elevating HDL_2 and, thus, tends to protect premenopausal women from an early onset of atherosclerosis. Indeed, the incidence of cardiovascular disease increases after menopause (Chapter 12).

TABLE 17-2

Normal Plasma Lipid Levels (mg/dl)

	Triglyceride	Total cholesterol	HDL cholesterol
Adult female	80	190	55
Adult male	120	200	43
Neonate	38	70	35

HDL_2 levels can also be increased by physical exercise. It is known that male marathon runners have HDL patterns similar to those of females. An obvious conclusion is that exercise is beneficial in maintaining healthy HDL levels. Cigarette smoking, on the other hand, decreases HDL levels, while cessation of smoking reverses the effect. Other factors that contribute to abnormal lipoprotein concentrations include: genetic disorders that increase VLDL and LDL and/or decrease HDL levels, including diabetes, obesity, and hypertension.[2] Regulation of plasma lipid levels is clearly complex and involves both genetic and environmental components.

4 DIFFERENTIAL ROLE OF LIPOPROTEINS IN ATHEROSCLEROSIS

As already mentioned, HDL have a beneficial effect inasmuch as high plasma HDL levels correlate with decreased risk of cardiovascular disease. Conversely, elevated LDL cholesterol is directly linked with an increased risk of cardiovascular disease. However, within the LDL class, one can distinguish specific subclasses, LDL subclass pattern A and LDL subclass pattern B, which correlate differently with regard to their contribution to cardiovascular disease.[3] Pattern A LDL are large (>25.5 nm diameter) buoyant particles while pattern B LDL are less buoyant and smaller in size (<25.5 nm). The latter LDL pattern (but not pattern A) is associated with an increased risk of cardiovascular disease; pattern B is also associated with increased plasma triglyceride, elevated apolipoprotein B concentrations, and decreased HDL, particularly HDL_2, concentrations; all are additional risk factors in cardiovascular disease. LDL subclasses are genetically influenced; pattern B appears to be associated with an autosomal dominant allele(s) that has a rather high population frequency of 25 to 30%. Interestingly, expression of the pattern B phenotype is age dependent. This phenotype is not expressed in males until approximately 20 years of age and in females is not expressed until menopause. LDL pattern B expression is likely to be influenced by hormones as well as other factors.

The plasma triglyceride and cholesterol levels shown in Table 17-2 are those of young adults; these levels tend to increase with advancing age. The effect of aging on plasma cholesterol concentrations is demonstrated in Figure 17-2. The age-related increase in plasma cholesterol is less marked between 20 and 40 years in women than in men; however, females show a distinct upward

TABLE 17-1

Nomenclature and Composition of Plasma Lipoproteins

	Weight (percent)			
Nomenclature	Protein	Phospholipid	Cholesterol	Triglyceride
Chylomicron (CM)	1–2	3–6	2–7	80–95
Very-low-density lipoprotein	5–10	15–20	10–15	55–65
Intermediate-density lipoprotein	19	19	38	23
Low-density lipoprotein	20–25	22	45	10
High-density lipoprotein	45–50	30	20	5

FIGURE 17-2 *Plasma cholesterol levels in women (——) versus men (- - -) from age 20 to 70.* White women during the premenopausal period have lower total cholesterol levels than corresponding males. After menopause (between 50 to 60 years), there is a distinct increase in total cholesterol in women. Figure derived from data published by NIH.[19] (Drawing by S. Oklund.)

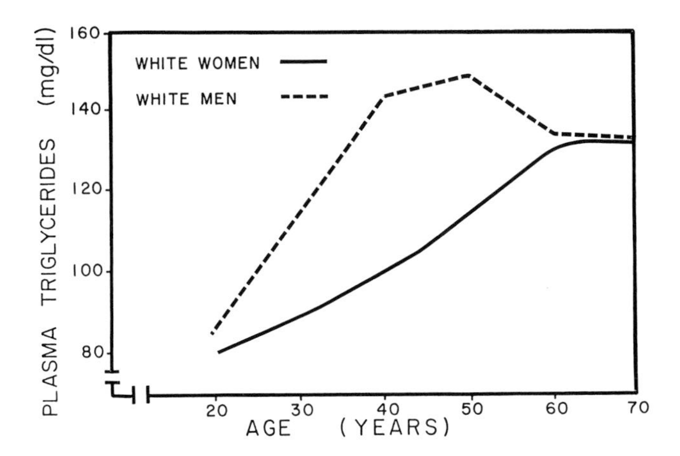

FIGURE 17-3 *Plasma triglyceride levels in women (——) versus men (- - -).* White males have higher levels between 20 to 50 years of age, while both sexes are quite comparable after age 50. Figure derived from data in Reference 19. (Drawing by S. Oklund.)

trend at approximately 55 years, an age that coincides with menopausal events such as decreased estrogen levels.

The increase in levels of plasma lipids with aging is also evident with respect to triglycerides (Figure 17-3). This lipid class increases noticeably with age where, in men, the increase is significant at 40 years, while in women the levels increase significantly between 50 and 60 years.

5 SYNTHESIS OF LIPOPROTEINS

The vital effects of the lipoproteins in well-being and survival and their implication in the etiopathology of cardiovascular disease necessitate a close consideration of the metabolism of these macromolecules. Indeed, in recent years, a wealth of research is providing a clearer understanding of their synthesis, distribution, and catabolism, as well as their binding to cell receptors and intracellular metabolism.

There are two major sites of synthesis of lipoproteins: the liver and the small intestine (Figure 17-4). The liver is

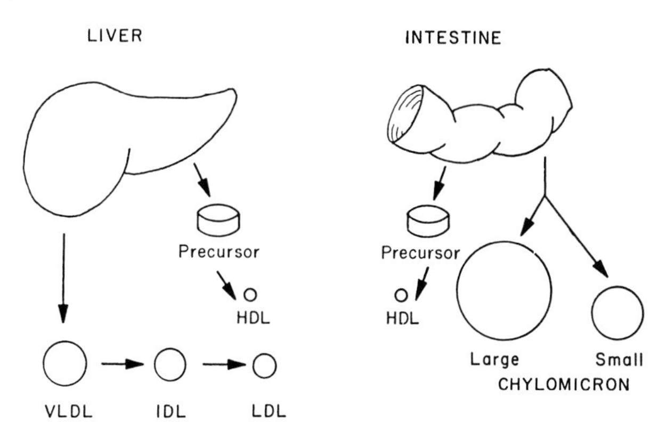

FIGURE 17-4 *Sites of synthesis of the major plasma lipoproteins.* Liver secretes VLDL which, following lipolysis, generates smaller, denser particles including intermediate density (IDL) and low density (LDL) lipoprotein particles. High density lipoproteins are also secreted by the liver; however, as indicated on the schematic, the nascent or precursor particle is discoidal in shape rather than spherical. These nascent particles are subsequently transformed in the plasma to the mature, spherical forms. The small intestine secretes mainly triglyceride-rich particles in the form of large and small chylomicrons. The intestinal cells also secrete nascent HDL, which are similar in their chemical characteristics to those secreted by the liver. (Drawing by S. Oklund.)

the major organ regulating cholesterol homeostasis. Parenchymal liver cells secrete VLDL, which are large particles rich in triglycerides. VLDL loses core lipids, mainly triglyceride, through lipolysis and gives rise to IDL and LDL. The generally accepted mechanism for the formation of LDL is by way of this precursor-product relationship wherein large triglyceride-rich VLDL is the precursor and the smaller, cholesteryl ester–rich particle, LDL, is the final product. The liver is also the site of synthesis for HDL. Newly secreted HDL, however, are chemically and structurally different from circulating HDL. The newly secreted particles, like membranes, possess mainly phospholipid and free cholesterol organized as a bilayer, are discoidal in shape, and contain little or no core lipids. Such discoidal HDL are "immature" forms of HDL, whereas the spherical ones are mature HDL. Discoidal HDL have also been termed "nascent" or precursor HDL, and normally they are rapidly converted to mature HDL by the plasma enzyme lecithin: cholesterol acyltransferase (LCAT) discussed below.

Lipoprotein secretion by the intestine is regulated to a great extent by what we eat. Dietary lipids are secreted in the form of large and small chylomicrons (Figure 17-4). The intestine, like the liver, also secretes an HDL, which is discoidal in shape. This intestinal precursor of HDL must also undergo transformation via the LCAT enzyme reaction in order to assume the morphology and chemistry of normal circulating plasma HDL.

6 APOLIPOPROTEINS: THEIR ROLE IN LIPOPROTEIN METABOLISM

Although the lipid moiety in lipoproteins is involved in processes of growth and survival as well as development of disease, it is the proteins associated with the lipids that direct the metabolism and ultimate fate of lipoproteins. Lipoprotein proteins, or apolipoproteins, are important constituents that impart specific properties and functions to the molecule. The major apolipoproteins (apo) are apoAI, apoAII, apoB, apo(a), apoCI, apoCII, apoCIII, and apoE. Their molecular weights vary from 8000 Da for apoCI to 800,000 for apo(a), as shown in Table 17-3. The function and origin of apolipoproteins are summarized in Table 17-4.

TABLE 17-3

Major Apolipoproteins

Apolipoprotein	Molecular wt. (daltons)	Lipoprotein class in which found
ApoAI	28,000	HDL
ApoAII	17,000	HDL
ApoB-100	540,000	VLDL, IDL, LDL
ApoB-48	260,000	CM
Apo(a)	300,000–800,000	LDL
ApoCI	8,000	VLDL, HDL
ApoCII	10,000	VLDL, HDL
ApoCIII	12,000	VLDL, HDL
ApoE	34,000	VLDL, IDL

Apolipoprotein AI

ApoAI is the major protein of HDL and is synthesized in both the liver and the intestine. The levels of plasma apoAI may turn out to be a better indicator of atherosclerotic risk than plasma cholesterol or even HDL cholesterol. This apolipoprotein is the major activator of the enzyme LCAT, which, as indicated in Figure 17-5, is necessary for transformation of the discoidal precursor HDL to the spherical mature HDL. ApoAI is also required to convert free cholesterol transported on mature HDL to cholesteryl ester as illustrated in Figure 17-6.

Epidemiological studies have abundantly shown an inverse relationship between plasma concentrations of apoAI, HDL, and risk of atherosclerosis. Only recently have experimental models been developed to directly test the hypothesis that increases in HDL and its associated apoAI do indeed protect against the development of atherosclerotic lesions. The C57Bl/6 mouse is susceptible to dietary saturated fat and cholesterol and on a high fat diet develops aortic fatty streaks in 3 to 4 months. Rubin and associates[4] developed a C57Bl/6 transgenic mouse model containing the human apoAI gene. This transgenic mouse expresses high levels of human apoAI and HDL. When challenged with a high-fat, high-cholesterol diet, the transgenic mice did not develop fatty streaks in the aorta as shown in Figure 17-7B; in comparison, nontransgenic control mice developed typical lesions,[5] Figure 17-7A. These studies are the best experimental evidence, to date, that HDL (and its apoAI) play a direct role in prevention of atherosclerosis.

TABLE 17-4

Properties of Major Apolipoproteins

Apolipoprotein	Function	Origin
ApoAI	Activator of LCAT[a]	Intestine, Liver
ApoAII	Inhibitor of LCAT	Liver
ApoB-100	Recognition of LDL receptor, triglyceride transport from liver cell	Liver
ApoB-48	Triglyceride transport from intestinal cell	Intestine
Apo(a)	Inhibits fibrinolysis	Liver
ApoE	Recognition of LDL receptor	Liver
ApoCI	Activator of LCAT	Liver
ApoCII	Activator of lipoprotein lipase	Liver
ApoCIII	Modulate apoE uptake; lipoprotein lipase inhibitor	Liver

[a] LCAT = lecithin:cholesterol acyltransferase.

Apolipoprotein AII

ApoAII is the second most abundant apolipoprotein on HDL and is produced primarily by the liver. Its functional significance is not completely understood, although it is thought that it may inhibit LCAT and thereby modulate HDL metabolism by indirectly influencing the conversion of free cholesterol to cholesteryl ester.

Apolipoprotein B-100

Apo B-100 is an important protein synthesized by the liver and associated with the less dense lipoprotein classes including VLDL, IDL, and LDL. It is the sole protein of LDL which possesses one molecule of apoB-100. ApoB-100 has two important functions:

1. It is necessary for the assembly and secretion of triglyceride-rich particles by the liver. In a rare genetic disease, abetalipoproteinemia, where apoB-100 is not secreted by the liver, this organ becomes fatty because of the intracellular accumulation of triglycerides. The lack of apoB-100 has serious metabolic implications since hydrolysis of liver-derived VLDL normally produces LDL. The latter are important transporters of cholesterol, required for both normal growth and development but are lacking in patients with abetalipoproteinemia.
2. ApoB-100 is also a ligand, i.e., a substance that binds to specific membrane receptors. The receptor that recognizes apoB-100 is called the LDL receptor or the apoB-E receptor; its function is to internalize LDL into the cell, thereby delivering cholesterol for various cellular functions. All cells possess LDL receptors to a greater or lesser degree. Clearly, receptors are functionally necessary to mediate normal cell metabolism; however, an overabundance of LDL and apoB-100 can lead to saturation of the receptors with a consequent accumulation of excess cholesterol in the plasma and initiation of the atherosclerotic process. The concentration of apoB-100 in plasma is a good indicator of atherosclerotic risk, for elevated apoB levels correlate with elevated levels of circulating cholesterol.

Apolipoprotein B-48

ApoB-48 is a truncated form of apoB-100 synthesized in the human intestine but not in the liver; it is required for the assembly and secretion of

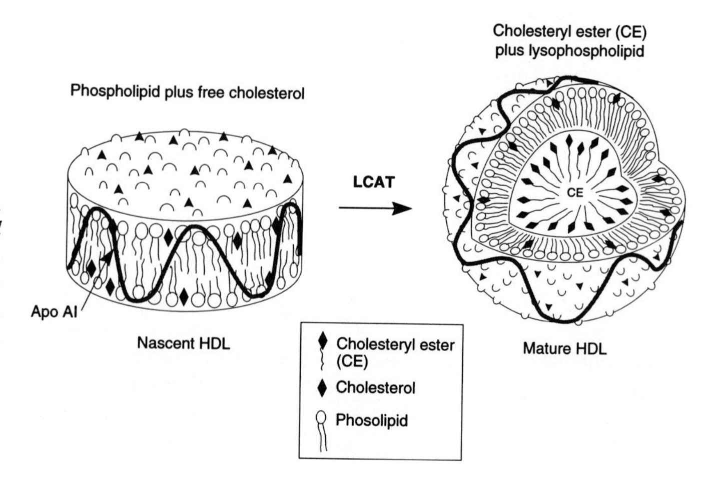

FIGURE 17-5 *Lecithin:cholesterol acyltransferase (LCAT) activity and transformation of nascent discoidal HDL to mature plasma HDL.* Nascent HDL secreted either by the intestine or the liver consist of bilayers of phospholipid and free cholesterol stabilized by apolipoprotein AI. In the presence of the enzyme LCAT, free cholesterol is esterified to cholesteryl ester with the concomitant production of lysolecithin. Cholesteryl ester is highly nonpolar and hence moves to the interior of the HDL, thus forming a core and converting a disk to a sphere. Lysolecithin, the other product of the reaction, is bound to albumin.

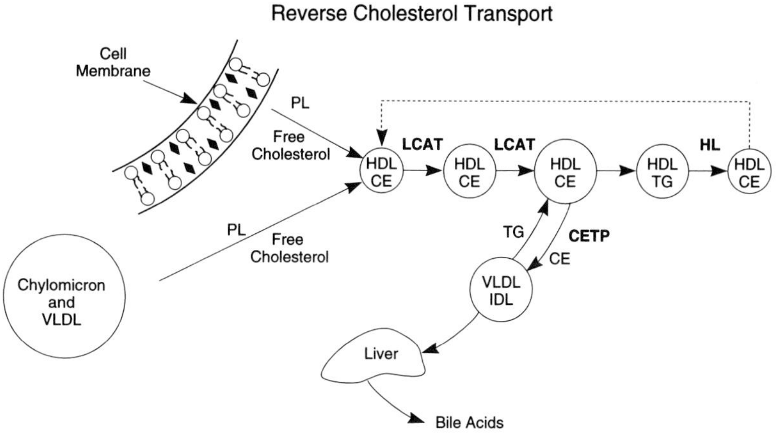

FIGURE 17-6 *The role of HDL in removal of excess cholesterol from cell membranes and lipoprotein surfaces.* Small HDL can remove excess free cholesterol and phospholipid from membranes and chylomicron/VLDL surfaces. In the presence of LCAT, the cholesterol is converted to cholesteryl ester (CE), thereby generating a larger HDL particle. The cholesteryl ester in the core of this large HDL is exchanged for triglyceride (TG) from VLDL and IDL; this exchange is facilitated by cholesteryl ester transfer protein (CETP). The HDL core now becomes triglyceride enriched; however, hepatic lipase (HL) hydrolyzes the triglyceride, thus regenerating a small HDL which can recycle. The cholesteryl ester transferred to VLDL/IDL is taken up by the liver through the apoB-E receptor and the excess cholesterol is degraded to bile acids and excreted. This pathway, involving LCAT, through which excess cell cholesterol is returned to the liver for excretion is termed reverse cholesterol transport. (Drawing by S. Oklund.)

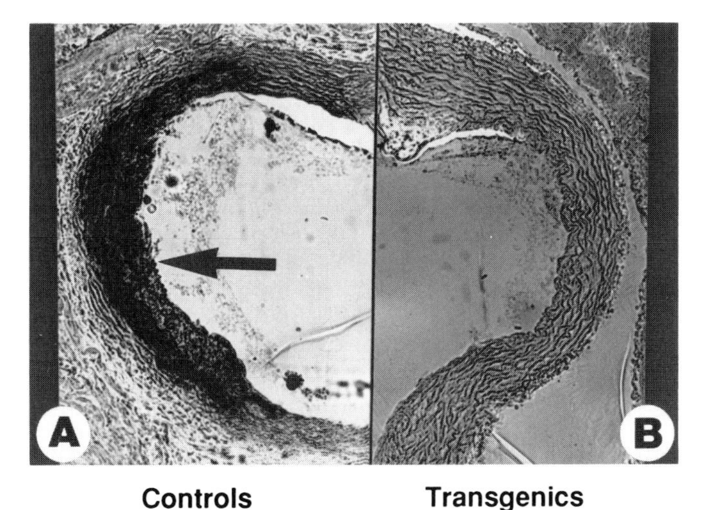

FIGURE 17-7 *Micrographs showing protection against atherosclerosis in transgenic mice expressing the human apoAI gene.* (A) Control, nontransgenic mouse; section through the aorta of a mouse maintained on a high fat and cholesterol diet for 3 to 4 months. The mouse shows the accumulation of lipid on the aortic wall as evidenced by the accumulation of dark staining material (arrow). (B) Transgenic mouse; section through the aorta of a transgenic mouse expressing high levels of human apoAI. The animal was maintained on a high fat and cholesterol diet but shows no evidence of atheroma formation. These studies suggest a direct role for apoAI and HDL in protection against premature atherosclerosis. (Photos kindly provided by Dr. Edward Rubin.)

chylomicrons which transport dietary lipids into the bloodstream. ApoB-48 is a product of the apoB-100 gene, but editing of the mRNA in intestinal cells leads to a stop codon which signals premature termination of apo-B translation with the end result that the molecular weight of the apoB determined on sodium dodecylsulfate polyacrylamide gels is only 48% that of the apoB-100 protein; hence the designation, apoB-48 (reviewed in Reference 6). The cellular process of apoB editing in humans is developmentally regulated and apoB-48 is not produced by enterocytes until the 13th week of gestation.[7] Prior to this point only apoB-100 is synthesized by intestinal cells. Since apoB-48 is required for the transport of chylomicrons from the intestinal cell, in abetalipoproteinemia where apoB is not secreted, the intestinal cells become lipid laden. In addition such individuals have steatorrhea (presence of excess lipids in stools) and diarrhea ensues along with malnutrition.

Reverse Cholesterol Transport

Cell Membrane
PL
Free Cholesterol
Chylomicron and VLDL
PL
Free Cholesterol
HDL CE
LCAT
HDL CE
LCAT
HDL CE
HDL TG
HL
HDL CE
TG
CETP
CE
VLDL IDL
Liver
Bile Acids

Apolipoprotein (a)

More than 25 years ago, Berg[8] discovered an unknown antigen associated with LDL in some subjects. He named this component lipoprotein antigen or lipoprotein (a). The metabolic significance of lipoprotein (a), abbreviated Lp(a), was not appreciated until recently; it is now known that Lp(a) is an important risk factor in coronary artery disease.[9] Lipoprotein (a) is a unique subset of LDL; hence, its lipid composition is like LDL. The protein associated with Lp(a) consists of the apoB-100 molecule bonded to the apo(a) molecule by a disulfide bridge. The apo(a) protein is a highly glycosylated protein of variable molecular weight (300,000 to 800,000 Da). The apo(a) protein has high homology with plasminogen, which hydrolyzes fibrin and aids in the dissolution of clots. Plasminogen possesses five pretzel-shaped protein units named "kringels". Apo(a) contains kringel 5 of plasminogen and variable numbers of the kringel 4 unit; the number of kringel 4 units in the apo(a) structure is responsible for the variation in molecular weight of the protein. Plasma Lp(a) concentrations are genetically controlled; in addition, there is an inverse relationship between apo(a) size and Lp(a) concentration. Individuals with elevated Lp(a) (greater than 30 mg/dl) have increased risk of coronary heart diseases. Lp(a), however, is not correlated with other known risk factors such as concentrations of LDL cholesterol, HDL cholesterol, apoAI, and apoB.[10] It is an independent risk factor for atherosclerosis. Lp(a) is not responsive to dietary and environmental factors; it also is not readily controlled by pharmacological agents, although it has been suggested that niacin, neomycin, and anabolic steroids may decrease Lp(a) levels. The physiological role of Lp(a) in coronary artery disease is not completely understood. However, the mechanisms may involve (1) inhibition of fibrinolysis since the molecule may interfere with plasminogen function and (2) the binding of Lp(a) particles to the extracellular matrix in the subendothelial space of the artery wall.[11] Bound Lp(a) can be modified and taken up by the scavenger receptor on macrophages, thus contributing to foam cell formation in fatty streaks.

Apolipoprotein E

ApoE is another apolipoprotein that is recognized by the LDL receptor (apoB-E receptor). It plays a particularly important role in targeting VLDL remnants to the liver where they are catabolized. VLDL remnants are the smaller, denser particles enriched in cholesteryl ester, which are formed following hydrolysis of triglyceride-rich particles. It is thought that apoE may also play a role in promoting the efflux of excess cholesterol from peripheral cells and tissues and its return to the liver. This pathway for cholesterol transport is known as reverse cholesterol transport.[12] Reverse cholesterol transport is probably important in decreasing the burden to cholesterol for the cell, particularly the cell membrane. A family has recently been described that has a genetic defect whereby apoE is not synthesized. Homozygous family members have premature coronary artery disease and elevated plasma cholesterol and triglyceride levels, thus suggesting that apoE plays a significant role in human lipid metabolism.

Apolipoprotein Cs

The apoC proteins are often referred to as the "small" proteins. They are synthesized in the liver and function principally as cofactors in the enzyme systems that hydrolyze lipoproteins (Table 17-4). ApoCI acts as an activator of LCAT and, in this respect, is functionally similar to apoAI. ApoCI levels in plasma are extremely low compared with apoAI, which suggests that apoCI may have other as yet unidentified function(s).

ApoCII is a cofactor in lipoprotein lipase activation; the protein is essential for degradation of chylomicrons and VLDL whereby core triglycerides are hydrolyzed to free fatty acids (Figure 17-8). The absence of plasma apoCII leads to gross elevation of plasma triglyceride levels.

ApoCIII has a functional role in two areas of lipoprotein metabolism: it acts as an inhibitor of lipoprotein lipase activity and hence modulates triglycerides hydrolysis, and it also modulates the cellular uptake of apoE-containing lipoprotein particles.

7 ENZYMES IN LIPOPROTEIN METABOLISM

7.1 Lecithin: Cholesterol Acyltransferase (LCAT)

As indicated earlier, this enzyme, synthesized by the liver and secreted into the plasma, is involved in the maturation of HDL. It plays a crucial, albeit indirect, role in removal of excess cholesterol from cells and tissues (the so-called reverse cholesterol-transport pathway). The proposed role of LCAT in the transformation of immature or nascent HDL into mature HDL is illustrated in Figure 17-5. Newly formed or nascent precursor HDL consist of small, disk-shaped bilayers of phospholipids stabilized on their rim by HDL apolipoprotein AI. Free cholesterol is inserted between the phospholipids in a fashion similar to that in cell membranes. In the plasma compartment, where LCAT is present, the disks are rapidly converted into spheres upon the formation of cholesteryl ester by the following reaction:

$$\text{phospholipid} + \text{free cholesterol} \xrightarrow{\text{LCAT}} \text{cholesteryl ester} + \text{lypophospholipid}$$

The major phospholipid in lipoproteins is lecithin. The fatty acid in the second position on lecithin is removed and added to cholesterol at the position of the hydroxyl group; this then gives rise to cholesteryl ester (highly insoluble) and lysolecithin. Lysolecithin formed in the reaction rapidly binds to plasma albumin. Cholesteryl ester is extremely nonpolar and moves into the core, thus inducing the structural change from disk to sphere.

LCAT is also responsible for enlarging the size of mature, circulating HDL (Figure 17-6). Small HDL can pick up excess cholesterol from cells, and, in the presence of LCAT, the added cholesterol is esterified. This mechanism of loading circulating HDL with cholesterol is extremely important, for it allows for efflux and removal of excess cholesterol from peripheral cells. The cholesteryl ester–enriched HDL is a large buoyant HDL_2 particle. The cholesteryl ester in the large HDL is exchanged for triglyceride from VLDL or IDL, a process mediated by a cholesteryl ester transfer protein (CETP)[13] as indicated in the diagram in Figure 17-6. Triglyceride in the large HDL is hydrolyzed by a specific lipase, hepatic lipase (HL), which does not require a cofactor. The net result of triglyceride hydrolysis is the regeneration of a smaller, denser HDL_3. This allows HDL to recycle and continue its function in transporting cholesterol out of cells. The cholesterol ester exchanged into VLDL and/or IDL is returned to the liver, as shown in Figure 17-6, and taken up by the apoB-E receptors in the cells. Excess cholesterol is excreted as bile. The mechanism of unloading excess cholesterol from peripheral cells and returning it to the liver, a process involving HDL, is called the reverse cholesterol-transport pathway. Reverse cholesterol transport may represent an important step in preventing atherosclerosis.

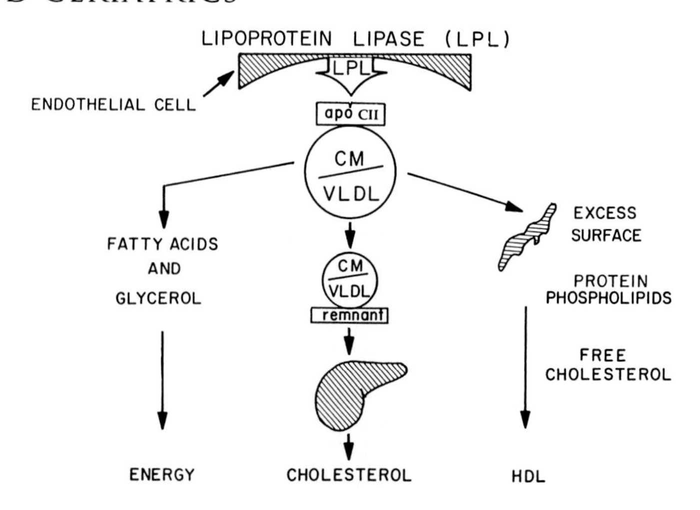

FIGURE 17-8 *Schematic diagram of the physiological function of lipoprotein lipase.* The enzyme found on the lumenal surfaces of capillary endothelium requires apolipoprotein (apo) CII as an activator. Triglyceride-rich particles including chylomicrons (CM) and very-low-density lipoproteins (VLDL) carry apoCII and activate the enzyme; in so doing, the triglyceride core is hydrolyzed, and the constitutive parts of triglyceride, free fatty acids and glycerol, are liberated. Removal of core triglyceride creates a remnant particle which is smaller and denser and contains a high proportion of cholesterol. Most of the remnants are removed by the liver. Excess surface components (protein), phospholipid, and free cholesterol generated by shrinking of the core are taken up by HDL. This enzyme is a key one in clearing triglyceride-rich lipoproteins. (Drawing by S. Oklund.)

HDL also has an integral role in the normal catabolism of chylomicrons and VLDL. During lipolysis of these triglyceride-rich particles, excess surface material, including protein, phospholipid, and free cholesterol, is generated. Such excess surface material is removed by uptake into HDL, as suggested in Figures 17-6 and 17-8.

The critical role of LCAT in maintaining HDL function and structure is dramatically illustrated by a rare disease, familial LCAT deficiency. In these patients, LCAT is not synthesized and secreted by the liver; as a consequence, HDL levels are extremely low, and the circulating form of HDL is discoidal in shape. Free cholesterol levels in almost all cells are elevated, and life-threatening sequelae include fatty liver, degenerative kidney disease, and premature coronary artery disease. This disease demonstrates without a doubt that HDL cholesterol esterification is necessary for reverse cholesterol transport and the prevention of premature atherosclerosis.

7.2 Lipoprotein Lipase (LPL)

This enzyme is important for the metabolism of triglyceride-rich lipoproteins. LPL is synthesized by adipocytes, heart, and kidney, but migrates from the sites of synthesis to the capillary endothelium, where it is bound to the cell surface. LPL is responsible for catabolism of the large, triglyceride-containing lipoproteins, principally chylomicrons and VLDL. It catalyzes the hydrolysis of triglyceride to free fatty acids and glycerol and requires the presence of apoCII, which is an activator of the enzyme. The action of LPL in the degradation of chylomicrons and VLDL is illustrated in Figure 17-8. VLDL and chylomicrons interact with LPL at the endothelial cell surface where hydrolysis occurs. The triglyceride core of these large particles is removed, thus generating fatty acids, excess surface material, and lipoprotein remnants. The fatty acids are utilized by cells for energy. Chylomicrons and VLDL remnants that are generated during lipolysis are rich in cholesteryl ester and contain apoE on their surfaces. Remnants are rapidly cleared from the plasma through apoB-E receptor–mediated uptake and subsequent degradation by the liver. This degradation step is vital for normal cholesterol metabolism. Some individuals possess an apoE protein that has a single amino acid substitution which leads to loss of recognition of the protein by the apoB-E receptor. In this disease, the remnants accumulate in the plasma and are directly linked with an increased risk to premature atherosclerosis.

Excess surface comp‹ nents will also be produced when the triglyceride core in chyıomicrons and VLDL shrinks. These components, especially phospholipids and free cholesterol, are incorporated into HDL as shown in Figure 17-6.

8 INTERACTION OF LIPOPROTEINS WITH CELLS

8.1 The LDL Receptor

As indicated previously, lipoproteins interact with cells and are responsible for delivery of cholesterol to both hepatic and nonhepatic tissues. LDL are the major transporters of cholesterol in the body; cholesterol in the form of cholesteryl ester is delivered to cells by uptake of LDL by a specific receptor mechanism, the LDL receptor (or apoB-E receptor). Although all cells possess apoB-E receptors to a greater or lesser degree, the liver, based on total mass, contains the largest number of apoB-E receptors. This receptor is functionally extremely important in the regulation of intracellular cholesterol synthesis and flux and, in addition, it regulates further synthesis of the receptor. The function of the high affinity apoB-E receptor is outlined in Figure 17-9 and is based on the work of Brown and Goldstein.[14,15] Essentially, liver and peripheral cells possess high affinity specific receptors that recognize LDL (or apoE-containing remnants). LDL (or remnant) binds to the receptor and is internalized in a structure referred to as the endosome. The endosome is converted into a lysosomal body wherein hydrolysis and degradation of the internalized lipoprotein takes place. The LDL is broken down into its molecular constituents, amino acids and cholesterol.

In the cell, catabolism of LDL results in the activation of several self-regulatory processes. Accumulation of free cholesterol up-regulates, that is, stimulates, the enzyme acyl-coenzyme A:cholesterol acyltransferase (ACAT), which reesterifies cholesterol. The accumulation of cellular cholesterol, then down-regulates, i.e., decreases, the enzyme 3-hydroxy-3-methylglutaryl coenzyme A (HMG CoA) reductase, which is the rate-limiting step in cellular cholesterol synthesis. In other words, the cell slows down its own machinery for synthesizing cholesterol when adequate amounts are being delivered to it by the

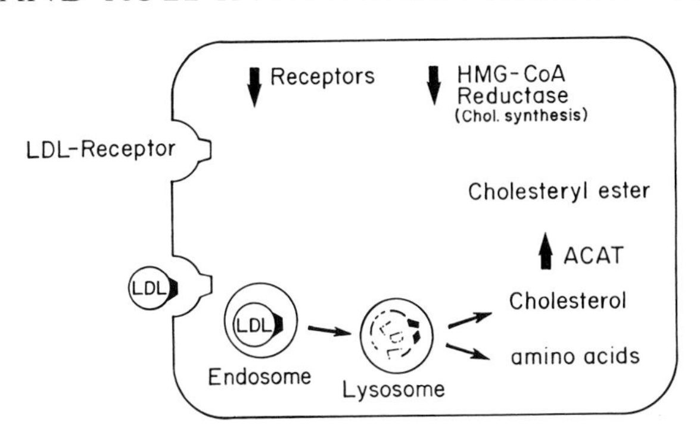

FIGURE 17-9 *Schematic outline of the function of the LDL (also called the apoB-E) receptor.* This receptor plays an important role in catabolism of cholesterol-containing lipoproteins. Lipoproteins carrying apolipoproteins (apo) B or E are recognized by the receptor and bound. The bound particles are internalized in a membrane-bound sac, the endosome. Hydrolytic enzymes invade the vesicle which becomes a lysosome in which proteins and lipids are broken down to amino acid and free cholesterol. In the cytosol, cholesterol is esterified to cholesteryl ester by the cellular enzyme, acyl coenzyme A:cholesterol acyltransferase (ACAT). Cholesterol coming into the cells down-regulates, or decreases, the activity of the cell's own cholesterol-making machinery. The rate-limiting step in *de novo* synthesis of cell cholesterol is 3-hydroxy-3-methylglutaryl coenzyme A (HMG CoA) reductase; therefore, this enzyme is the one regulated by receptor-mediated cholesterol accumulation. In addition to a decrease in HMG CoA-reductase, accumulation of cellular cholesterol also decreases the number of LDL receptors on the cell surface. Overall, the receptor-mediated process is a finely tuned system in the regulation of cholesterol metabolism. (Drawing by S. Oklund.)

lipoproteins. In addition to down-regulating the HMG CoA-reductase, degraded LDL down-regulates synthesis of the apoB-E receptor, and thus the cell reduces its uptake of cholesterol. The entire cell is finely tuned for an elegant maintenance of cholesterol homeostasis.

Receptor-mediated uptake and degradation of apoB- and apoE-containing lipoproteins is important in maintaining normal plasma cholesterol levels. Elevated levels of cholesterol-containing particles such as VLDL remnants and LDL could result in saturation of the apoB-E receptors and accumulation of excess cholesterol in the plasma, with the increased risk to cardiovascular disease. The risk to cardiovascular disease is also enormously increased when apoB-E receptors are defective or deficient. This does happen in certain genetic defects where either the receptor protein is not synthesized or the synthesized protein is defective. Patients with such receptor defects are hypercholesterolemic, and their condition is called familial hypercholesterolemia. In the homozygous form, patients can have staggeringly high plasma cholesterol levels, as high as 1000 mg/dl as compared to the normal 150 to 200 mg/dl. These patients have precocious atherosclerosis and, unless managed extremely carefully, will not survive the second decade of life. The disease state is a clear case of overproduction and underutilization of LDL. Heterozygotes for the disease have decreased numbers of functional receptors and, therefore, have elevated plasma cholesterol and premature coronary artery disease.

8.2 The Scavenger Receptor

In addition to the apoB-E receptor, another receptor, the scavenger receptor, is functionally important; the scavenger pathway

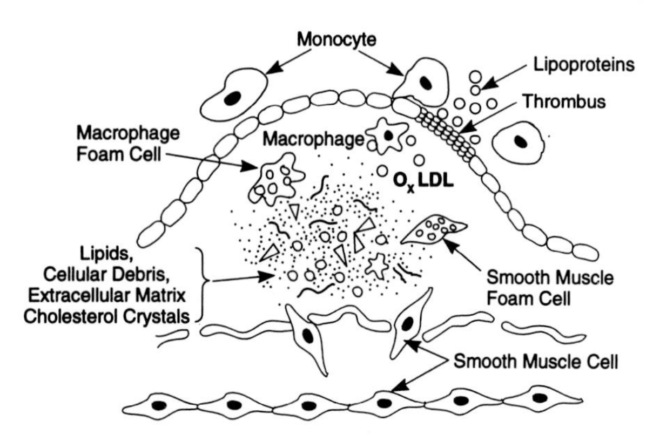

FIGURE 17-10 *Schematic of the role of macrophages and lipoproteins in foam cell and atheroma formation.* Early steps in atherosclerotic lesion formation involve the appearance of foam cells in the subendothelial space. Current research suggests that oxidatively modified LDL (OxLDL) play an important role in production of the foam cells. OxLDL are taken up by the scavenger receptor of macrophages in the artery wall; however, the internalized OxLDL cholesterol is not reutilized, and it accumulates within the cell in the form of lipid droplets. Macrophages with large quantities of accumulated lipid droplets are foam cells. The macrophage/foam cell in the artery wall ultimately dies and releases its residue of cholesterol which accumulates in the lesion. The macrophage also releases factors which stimulate smooth muscle cells to proliferate and accumulate lipids, thus further augmenting the atherosclerotic process.

TABLE 17-5

Diseases Related to Lipoprotein Abnormalities

Disorder	Clinical Findings
Secondary Hyperlipoproteinemias	
Diabetes mellitus	Atheromas
	Pancreatitis
	Xanthomas
Hypothyroidism	Atheromas
Estrogen excess	Pancreatitis
(oral contraceptives)	Xanthomas
Primary Hyperlipoproteinemias	
Single gene	Atheromas
	Pancreatitis
	Xanthomas
Multiple genes	Atheromas

for lipoprotein uptake is functional in macrophages. This receptor is a high-affinity, non-LDL receptor that recognizes modified LDL and actively removes it. Modified lipoproteins may accumulate in the plasma because of altered apoB structure or because normal apoB-E receptors are defective. In the latter case, lipoprotein residence time in the plasma is increased, thus increasing the likelihood of protein alterations, particularly by oxidative processes. Modified lipoproteins are bound and internalized by the macrophage scavenger receptor. A current theory on the development of the atherosclerotic lesion involves this pathway (Figure 17-10).[16,17] During the early events of lesion development, monocytes enter the arterial intima in regions of endothelial damage. The cytokines, specialized cell signals, trigger the differentiation of monocytes into macrophages and stimulate smooth muscle cells to proliferate. Macrophages accumulate lipids by scavenging modified lipoproteins, particularly oxidized LDL, present at the injury site. Scavenging LDL is probably a protective function; however, if the insult continues, more macrophages accumulate at the site and modified LDL are accumulated intracellularly in great abundance. The macrophage cannot, however, utilize the extra burden of cholesterol, which accumulates within the cell in the form of lipid droplets, the so-called foam cells. These cells eventually die. It is believed that the lipids released from the cells may be taken up by the smooth muscle cells, which also become lipid-laden, the smooth muscle foam cells.

9 ■ HYPERLIPOPROTEINEMIA

When the concentration of plasma lipoproteins exceeds an arbitrary normal value — which is typically defined as the 95th percentile of a random population — the condition is designated hyperlipoproteinemia. This value varies with age and sex. Clinical concern arises because an elevated concentration of lipoproteins can accelerate the development of atherosclerosis. Not only may atheromas form, but deposition of cholesterol may also occur in tendons and skin, producing raised nodules called xanthomas; these are overt manifestations of severe hypercholesterolemia.

Primary hyperlipoproteinemias are due to a single gene defect or to a combination of genetic factors; in addition, genetic factors may be exacerbated by environmental or dietary factors. The incidence of the different primary hyperlipoproteinemias ranges from 1 in 250 in the population to 1 in 1 million. Hyperlipoproteinemias are designated either primary or secondary (Table 17-5). Secondary hyperlipoproteinemias are complications of more generalized disturbances such as diabetes mellitus, hypothyroidism, excessive intake of alcohol, or the intake of high doses of estrogen. Knowledge of the plasma concentrations of cholesterol and triglycerides usually reveals the class of lipoprotein that is high, and this is useful in making a diagnosis and in designing proper drug and/or diet therapy.

Treatment protocols for hyperlipoproteinemias are summarized in Table 17-6. Basic to the treatment of all hyperlipidemias is a diet that maintains normal body weight and that minimizes the plasma lipid concentration. If a patient is overweight, weight loss should be attempted and then maintained on a diet low in cholesterol and saturated animal fats and relatively high in polyunsaturated vegetable oils. These polyunsaturated fats improve the palatability when saturated fats are restricted and also reduce the concentrations of plasma LDL cholesterol. Only rare individuals require a diet severely reduced in total fat.

Patients with secondary hyperlipidemia require treatment of the underlying disorder (diabetes, hypothyroidism, excess alcohol consumption, estrogen excess, etc.) and should reduce all other risk factors, such as smoking and hypertension and maintain physical fitness.

Primary hyperlipoproteinemia requires more aggressive treatment; in addition to a proper diet, drugs that lower plasma lipoprotein concentrations are used.[18] These drugs function either by diminishing the production of lipoproteins or by increasing the efficiency of their removal.

Nicotinic acid (niacin) has long been used to reduce the production of VLDL and, in so doing, it also lowers LDL. It also increases HDL concentrations by decreasing HDL clearance from the plasma. Nicotinic acid does produce cutaneous flush and pruritus (itching) involving the face and upper body, but this appears to subside with continued use. It may, however, interfere with compliance, that is, the patient's willingness to continue with the drug.

Fibric acid derivates such as clofibrate and gemfibrozil are effective in decreasing the synthesis of VLDL and increasing VLDL catabolism (by increasing the activity of LPL).

The most encouraging new drugs in treating hypercholesteremia are the HMG CoA-reductase inhibitors, including lovastatin, pravastatin, simvastatin, and fluvastatin. To compensate for a reduction in cholesterol synthesis in the liver, the number of hepatic receptors for LDL increases, and this brings about a reduction in plasma LDL.

Normally, cholesterol in the liver is converted to bile acids, which are delivered to the small intestine lumen where some of the cholesterol in the form of bile acid is ultimately excreted and some is reabsorbed. Normally, formation of bile acids has a negative-feedback effect on further production of bile acids. Removing the bile acids so that they no longer exert negative feedback speeds up the conversion of cholesterol to bile acids and reduces body pools of cholesterol, including cholesterol sequestered in xanthomas. Resins such as cholestyramine and colestipol have such an effect; they readily bind bile acids in the intestinal lumen and increase the flux of cholesterol from the liver.

Both nicotinic acid and HMG CoA-reductase inhibitors may be used in combination with one of the bile acid–binding resins. These combinations are usually synergistic, allowing the doses of both substances to be lowered. It is thus apparent that drugs utilized in treating dysfunctions of lipid metabolism must be selected and tailored according to the individual condition.

TABLE 17-6

Treatment of Hyperlipoproteinemias

Diet: low cholesterol and animal fat; relatively high polyunsaturated fats

Drug Therapy:

Class	Metabolic effect
HMG CoA-reductase inhibitors (lovastatin, pravastatin, simvastatin, fluvastatin)	Inhibits cholesterol biosynthesis; enhances LDL clearance; lowers plasma cholesterol
Bile acid sequestrants (cholestyramine, colestipol)	Binds and removes bile acids in intestine; increases cholesterol conversion to bile; increases LDL clearance
Fibric acid derivatives (clofibrate, gemfibrozil)	Reduces synthesis and increases catabolism of VLDL (raises LPL activity)
Nicotinic acid	Reduces synthesis of VLDL; lowers LDL; elevates HDL by reducing its clearance

ADDENDUM

Since the original writing of this chapter, studies have shown that one of the isoforms of apolipoprotein E, apoE4, is associated with increased risk for familial late-onset Alzheimer's disease (AD). As the number of apoE4 alleles increases, there is an increase in risk for AD and also a decrease in the average age of onset.

REFERENCES

1. Steinberg, D., Lipoproteins and atherosclerosis: a look back and a look ahead, *Arteriosclerosis,* 3, 28–301, 1983.
2. Breslow, Jan L., Genetics of lipoprotein disorders, *Circulation,* 87(Suppl. III), III-16–III-21, 1993.
3. Austin, M. A., King, M. C., Vranizan, K. M., and Krauss, R. M., Atherogenic lipoprotein phenotype: a proposed genetic marker for coronary heart disease risk, *Circulation,* 82, 495–506, 1990.
4. Rubin, E. M., Ishida, B. Y., Clift, S. M., and Krauss, R. M., Expression of human apolipoprotein A-I in transgenic mice results in reduced plasma levels of murine apolipoprotein A-I and the appearance of two new high density lipoprotein size subclasses, *Proc. Natl. Acad. Sci. U.S.A.,* 88, 434–438, 1991.
5. Rubin, E. M., Krauss, R. M., Spangler, E. A., Verstuyft, J. G., and Clift, S. M., Inhibition of early atherogenesis in transgenic mice by human apolipoprotein AI, *Nature,* 353, 265–267, 1991.
6. Young, S. G., Recent progress in understanding lipoprotein B, *Circulation,* 82, 1574–1594, 1990.
7. Teng, B., Verp, M., Salomon, J., and Davidson, N. O., Fetal editing is regulated and widely expressed in human tissues, *J. Biol. Chem.,* 265, 20616–20620, 1990.
8. Berg, K., A new serum type system in man: the LP system, *Acta Pathol. Microbiol. Scand.,* 59, 369–382, 1963.
9. Scanu, A. M. and Fless, G. M., Lipoprotein (a) heterogeneity and biological relevance, *J. Clin. Invest.,* 85, 1709–1715, 1990.
10. Sandkamp, M., Funke, H., Schulte, H., Kähler, E., and Assman, G., Lipoprotein (a) is an independent risk factor for myocardial infarction at a young age, *Clin. Chem.,* 36, 20–23, 1990.
11. Kostner, G. M. and Krempler, F., Lipoprotein (a). *Curr. Opinion Lipidology,* 3, 279–284, 1992.
12. Roheim, P. S., Atherosclerosis and lipoprotein metabolism: role of reverse cholesterol transport, *Am. J. Cardiol.,* 57, 3C–10C, 1986.
13. Tall, A. R., Plasma lipid transfer proteins, *J. Lipid Res.,* 27, 361–367, 1986.
14. Brown, M. S. and Goldstein, J. S., How LDL receptors influence cholesterol and atherosclerosis, *Sci. Am.,* 251, 58–66, 1984.
15. Brown, M. S. and Goldstein, J. S., A receptor-mediated pathway for cholesterol homeostasis, *Science,* 232, 341–47, 1986.
16. Ross, R., The pathogenesis of atherosclerosis — an update, *N. Engl. J. Med.,* 314, 488–500, 1986.
17. Steinberg, D. M., Parthasarathy, S., Carew, T. E., and Witztum, J. L., Beyond cholesterol: modifications of low-density lipoprotein that increase its atherogenicity, *N. Engl. J. Med.,* 320, 915–924, 1989.
18. Levy, R. L., Troendle, A. J., and Fattu, J. M., A quarter century of drug treatment of dyslipoproteinemia, with a focus on the new HMG CoA reductase inhibitor fluvastatin, *Circulation,* 87(Suppl. III) III-45–III-53, 1993.
19. The Lipid Research Clinics Population Studies Data Book. The Prevalence Study, Vol. 1, 1980 DHEW Publication no. (NIH) 80-1527, Government Printing Office, Washington, D.C., 1980.
20. Corder, E. H., Saunders, A. M., Strittmatter, W. J., Schmechel, D. E., Gaskell, P. C., Small, G. W., Roses, A. D., Haines, J. L., and Pericak-Vance, M. A., Gene dose of apolipoprotein E type 4 allele and the risk of Alzheimer's disease in late onset families, *Science,* 261, 921, 1993.

III
SYSTEMIC AND ORGANISMIC AGING

18 AGING OF RESPIRATION, ERYTHROCYTES, AND THE HEMATOPOIETIC SYSTEM

Paola S. Timiras

1 THE LUNG: AN ORGAN "BATTERED" FROM WITHIN AND WITHOUT

Respiratory function includes both an external process — oxygen (O_2) absorption and carbon dioxide (CO_2) removal from the body — and an internal metabolic process — cellular gaseous exchanges. This chapter is concerned with the former, that is, the uptake of O_2 and excretion of CO_2 through the lungs and their transport by erythrocytes to and from the tissues.

In the respiratory system,

- *The gas-exchange organ* is represented by the two lungs.
- *The pump that ventilates the lungs* is represented by
  - *the chest wall*
  - *the respiratory muscles,* which increase or decrease the size of the thoracic cavity
  - *the brain centers and nerve tracts,* which control the muscles

In addition to gaseous exchange, the lungs perform other functions. They participate in

- *Immunologic defenses* of the body (by phagocytizing particles from the inspired air and from the blood)
- *Metabolic functions* (by synthesizing, storing, or releasing into the blood such substances as surfactant and prostaglandins)
- *Endocrine functions* (by transforming angiotensin I into angiotensin II, a powerful vasoconstrictor and stimulus for aldosterone secretion, Chapter 11)
- *The actions of a few biologically active peptides,* some with pressor (e.g., VIP, vasoactive intestinal peptide) and some neuronal (e.g., opioid peptides) activity

The respiratory system in humans is mature by age 20.[1] Pulmonary function begins to gradually decline in healthy subjects after the age of 25.[2] This decline is linked to progressive deleterious changes that occur in respiratory structures including the lung, the thoracic cage, and respiratory muscles as well as the respiratory centers in the central nervous system (CNS). These changes, however, are minor compared to the constant effects of the environment and other insults to the respiratory system — infections, pollution, cigarette smoking, disordered immune responses, unfavorable working conditions — to which the organism is exposed throughout the lifespan.[3-5]

The major function of the lungs is to ensure the efficient exchange of air (oxygen) with the environment. However, this very function is performed at the peril of contamination from the many toxic substances transported in the air. The degree and the rate of age-related changes in structure and function of the lungs are variable and dependent on the habits of the individual (particularly smoking, physical exercise), the environment in which he/she lives (urban versus rural), and disease (infections, industrial diseases). The lungs are "battered" not only by external insults but also by formation of oxygen radicals, so deleterious to tissues, in the pulmonary cells in immediate contact with the gaseous environment rich in O_2. Toxic effects of O_2 are discussed in Chapter 6.

Pulmonary alterations with aging have been described extensively. These alterations often result in a variety of symptoms such as

- Reduced maximum breathing capacity
- Weakening of respiratory muscles
- Decreased elasticity of thoracic cage and chest wall
- Less efficient emptying of the lungs
- Increased rigidity of internal lung structures
- Earlier and easier fatigability

However, in the absence of disease, none of these functional decrements, singly or in combination, is sufficient to severely incapacitate the old individual. The majority of the elderly are capable of maintaining their life-style and a satisfactory respiratory function under resting (steady state) conditions. Some of the impairments become manifest when ambient conditions worsen or when pathology ensues. Given continuing exposure to external and internal insults, respiratory diseases are more prevalent in older individuals than in the general population. Among these diseases, incidence and severity of infections, chronic obstructive disease (emphysema), and cancer increase with aging.

In this chapter, some of the age-related changes that may occur in the lungs as a consequence of environmental insults are considered first, together with some respiratory diseases prevalent in the elderly. In a second part, the aging of the hematopoietic system and, particularly, of the erythrocytes (red blood cells), is discussed in terms of the respiratory function of these cells in carrying O_2 and CO_2 to and from the lungs and tissues. Erythrocytes have a relatively short lifespan of about 120 days and may serve as a model to study cellular aging.

2 AGING-ASSOCIATED CHANGES IN THE LUNG

Air passes through the respiratory system as follows:

1. First, it passes through the nasal passages (nares) where it is filtered of the larger contaminants.
2. Next, it enters the pharynx where it is warmed and absorbs water vapor.

0-8493-8979-8/94/$0.00+$.50

3. Then, it flows down the trachea and through the bronchi and bronchioles.
4. Finally, it proceeds through the respiratory bronchioles and the alveolar ducts to the alveoli. *The alveoli are the functional units of the lungs.*

Alveolar Structure and Function

The major structural characteristic of the alveolus is the close proximity of the capillary blood and the alveolar air, separated only by the capillary endothelium and the thin basement membrane supporting the alveolar cells. This arrangement facilitates gas exchanges between blood and air.

Two types of epithelial cells (pneumocytes) line the alveolus. Type I cells are extremely thin with few intracellular organelles and are designated agranular pneumocytes. Type II cells contain many organelles and lipid droplets and are designated granular pneumocytes. These type II cells produce surfactant, a proteolipid that coats the alveolar cells and lowers the surface tension at the air–fluid interface. Another and less frequent cell type in the lung is the alveolar macrophage. These cells are loaded with digestive enzymes and can ingest foreign materials, as do white blood cells. They migrate from the blood stream and patrol the tissues of the lung on the alveolar side, gliding on the surfactant.

2.1 Structural Changes

The architecture of the lung is altered in aging. The lungs are more voluminous, the alveolar ducts and respiratory bronchioles are enlarged, while the alveoli become shallower and flatter with loss of septal tissue (Table 18-1). These changes do not affect appreciably total lung capacity (the maximum volume of air in the lungs and airways) when this volume is corrected for the age-related decrease in height (Figure 18-1). However, air distribution is altered: duct air is increased but alvedar air is decreased. As air transport occurs primarily through the alveolar surface, overall air transport is reduced. The amount of elastic tissue, abundant in the lung and partly responsible for the stretchability of this organ, is decreased with age while fibrous tissue is increased. The decrease in elastic tissue is associated with an increase in elastin, a major protein of this tissue. This apparently paradoxical increase has been explained as the consequence of structural changes in elastin with age; the "old" elastin undergoes a number of cross-linkages (as do other connective tissue fibers, Chapter 22) which decrease its extensibility and increase its resistance to solution during purification. The importance of lung elasticity is illustrated by the condition of emphysema, in which the lungs lose their elasticity due to disruption of elastic tissue. The nature of the exact changes in the elastic fibers during aging is unclear, but it is likely that alteration in the distribution of elastic tissue is more functionally significant than changes in amounts. With aging, abnormal location or structure of the fibers may be responsible for impairments of ventilation and perfusion of the lungs.

The net effect of these structural alterations is a decreased alveolar surface area, which is 75 m^2 at 30 years and decreases by 4% per decade thereafter. The configuration and mechanical properties of the chest wall also change with aging, due to increased curvature of the spine and calcification of the intercostal cartilage. All these changes resemble those found in emphysema (see below).

TABLE 18-1

Morphologic Changes in the Thorax and Lung with Age

Morphological change	Functional significance
	Thorax
Calcification of bronchial and costal cartilage	↑ Resistance to deformation of chest wall (A elastic work)
↑ Stiffness in costovertebral	↑ Use of diaphragm in ventilation
↑ Rigidity of chest wall	
↑ Anterior-posterior diameter (kyphosis)	↓ Tidal volume Response to exercise hyperpnea
Wasting of respiratory muscles	↓ Maximal voluntary ventilation
	Lung
Enlarged alveolar ducts ↓ Supporting duct framework, enlarged alveoli	↓ Surface area for gas exchange
Thinning, separation of alveolar membrane	↑ Physiologic dead space
↑ Mucous gland ↓ Number, thickness of elastic fibers(?) ↓ Tissue extensibility (alveolar wall)	↓ Lung elastic recoil VC, RV/TLC, and[a] ↓ ventilatory flow rate ↓ ventilation distribution ↑ resistance to flow in small airways
↓ Pulmonary capillary network ↑ Fibrosis of pulmonary capillary intima	↓ Ventilation: blood flow equality

[a] VC = vital capacity; RV = residual volume; TLC = total lung capacity.
Modified from Reddan, W. G., *Exercise and Aging*, Smith, E. L. and Serfass, R. C., Eds., Enslow Publishing, Hillside, NJ, 1981. With permission.

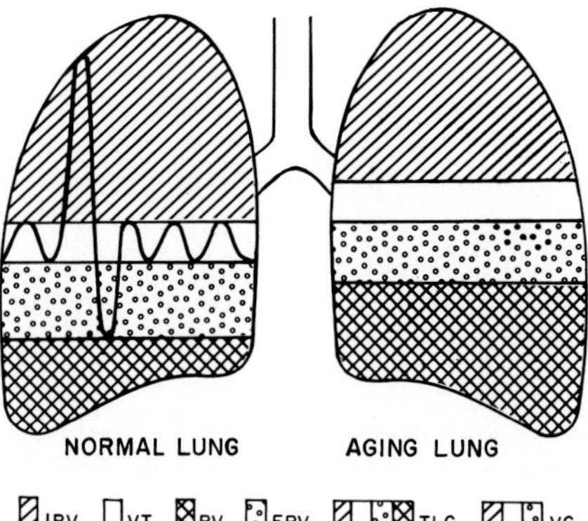

FIGURE 18-1 *Changes in lung volumes with aging.* With aging, note particularly the decrease in VC and the increase in RV. IRV, inspiratory reserve volume; TV, tidal volume; RV, residual volume; ERV, expiratory reserve volume; TLC, total lung capacity; VC, vital capacity.

2.2 Changes in Lung Volumes

Lung volume and pressure characteristics change dramatically from birth to death, with major and rapid changes during childhood and adolescence and slower but progressive alterations with increasing age. These changes involve both the lungs and chest wall, and are often divergent (Table 18-1).[6] The chest becomes stiffer because of calcification of costal cartilages, while the lungs become more distended because of a slightly increased compliance (i.e., stretchability) and decreased recoil. Consequently, lung volumes are altered and ventilation rate at rest, and, especially, during physical exercise is decreased.

Lung Volumes and Measurements

Total lung capacity (TLC) is 6 l in men (4.2 l in women) distributed in the following volumes (in men):

- *Tidal volume,* amount of air that moves into and out of the lungs with each quiet inspiration and expiration (= 0.5 l)
- *Inspiratory reserve volume* (= 3.3 l) and *expiratory reserve volume* (1 l), amount of air that moves into or out of the lungs following maximal inspiration or expiration
- *Residual volume,* air left in the lungs after maximal expiration (= 1.2 l)
- *Dead space,* air in the airways (= 150 ml)
- *Vital capacity,* the greatest amount of air that can be expired after a maximal inspiration. Vital capacity is frequently taken as an index of pulmonary function and ranges from 3 to 4 l in adult females to 4.5 to 5.5 l in adult males. A more precise index is to measure vital capacity per unit of time; for example, in asthma, vital capacity appears normal but when timed, it shows a significantly prolonged time because of bronchial constriction
- *Pulmonary ventilation rate* (or respiratory minute volume) is normally 6 l/breath × 12 breaths/min
- *Maximal voluntary ventilation* is the largest volume of air that can be moved into and out of the lungs in 1 min by voluntary effort; it may be as high as 125 to 180 l/min

TABLE 18-2

Lung Volumes and Ventilation Variables in 20- and 60-Year-Old Individuals Matched for Height and Weight

	Men		Women	
Variable	20 Years	60 Years	20 Years	60 Years
Total lung capacity (TLC) (l)[a]	7.20	6.90	5.10	4.70
Vital capacity (l)	5.20	4.00	4.17	3.29
Functional residual capacity (l)	2.20	3.50	2.40	2.50
Residual volume (% TLC)	25	40	28	40
Forced expiratory volume in 1 s				
l/BTPS	4.45	3.17	3.26	2.26
% VC	81	71	80	70
Maximum voluntary ventilation (l/min)	150	99	110	77
Maximal expiratory flow at 50% VC (liters/s^{-1})	5.00	3.80	4.40	2.70
Closing volume (% VC)	8	25	8	25
Recoil pressure of lung at 60% TLC (cm H_2O)	7.8	4.4	7.8	4.4
Recoil pressure of chest wall at 60% TLC (cm H_2O)	–6.0	–4.0	–6.0	–4.0

[a] l = liter; BTPS = body temperature and pressure saturated with water vapor.
Modified from Reddan, W. G., *Exercise and Aging,* Smith, E. L. and Serfass, R. C., Eds., Enslow Publishing, Hillside, NJ, 1981. With permission.

In some elderly, vital capacity may decrease to approximately 75% of its value in the 7th decade as compared to that at the age of 17 years (taken as men/women average of 4.8 l). Residual volume increases nearly 50% during this time. However, total lung capacity remains unchanged (Table 18-2 and Figure 18-1). Other measures of ventilatory mechanics decline as well; these include forced expiratory volume per second, which decreases by 32 ml/year in males and 25 ml/year in females starting from the age of 25 years, and specific airway conductance, which also progressively slows with aging.

2.3 Changes in Arterial PO_2

As with many other functions, not all parameters of pulmonary function follow a similar timetable of decline. For example, expiratory flow rates begin to decline at an age when vital capacity is still intact.[7] This early decrease in expiratory rate may reflect the reduced elastic recoil in the elderly; it may cause a premature closure, during expiration of some regions of the lungs with trapping of air in sites distal to the closure and, consequently, poor mixing of inspired air. This abnormality is responsible, at least in part, for the age-related changes in arterial PO_2 (i.e., partial pressure of O_2, which at sea level is 160 mmHg in the air and 149 in the lungs) and arterial–alveolar PO_2 differences (alveolar PO_2 is 100 mmHg, arterial PO_2 is 40 mmHgG).[8]

Among the most important physiological changes with aging are

- The progressive reduction in arterial PO_2 due to premature airway closure
- Loss of elastic recoil

These conditions result in an imbalance of lung ventilation and gas perfusion, and these, combined with decreased cardiac output, lead to lowered O_2 uptake and pressure.[9] As a nomogram has been adopted in the case of the age-related increase in fasting blood glucose (Chapter 15), similarly a nomogram has been proposed to calculate the expected arterial O_2 tension in relation to age with the arterial O_2 tension falling to 75 mmHg in the 7th decade. In contrast, PCO_2 remains remarkably constant throughout life despite a reduction in CO_2 sensitivity in most tissues, possibly including the brain.

2.4 Control of Ventilation

Control of ventilation by *brain centers in medulla and pons* and by peripheral, *carotid and aortic bodies chemoreceptors,* is markedly altered in the elderly.[10] It is unclear whether altered ventilation is due to

- Intrinsic alteration of neural control, such as decreased sensory perception of PCO_2, pH, and PO_2
- Loss of synchrony of higher CNS inputs
- Alterations of mechanical factors such as stiffness of chest wall
- Reduced neuromuscular competence/responsiveness to neural inputs

None of these hypotheses is fully accepted and more research is needed in this area.[11] Responses to hypercapnia (increased PCO_2) and hypoxia (reduced PO_2) are reduced by 50% in the aged as compared to the young.[12]

2.5 Responses to Exercise

Physical exercise stimulates the active tissue to utilize more O_2 and to eliminate more CO_2 which requires *coordinated cardiovascular and respiratory adjustments.* Circulatory changes increase blood flow to the exercising muscle (Chapter 21). Respiratory adjustments include

- Increased ventilation, to provide more O_2, eliminate more CO_2, and dissipate heat

- Increased extraction of O_2 from the blood in the exercising muscle
- Increased blood flow and shift in the O_2 dissociation curve

With aging, lung ventilation — already impaired under quiet conditions — is further altered during exercise:

- It becomes inadequate to muster the necessary adjustments to meet the increased demands of exercise
- Loss of elastic recoil and decreased functional residual capacity (i.e., gas volume remaining in lungs at the end of quiet expiration) inhibit the effective range of tidal volume
- The early closure of the airways similarly inhibits the expiratory flow
- Dyspnea, or shortness of breath, ensues and necessitates the early cessation of exercise.

The reduction of vital capacity with aging restricts the potential tidal volume that may be reached during maximal exercise. At rest, only a minor fraction of the potential lung volume and flow changes is used, and even during maximum exercise, the maximal inspiratory and expiratory flow rates and volumes are not usually reached. With increasing age, during moderate to heavy exercise, the ability to reach maximal rates and volumes is severely curtailed, and, therefore, ventilation cannot increase sufficiently to provide for the increased metabolic demands.[13]

In addition to difficulty in adjusting ventilatory responses to exercise, the elderly have an earlier onset of the shift from aerobic (requiring O_2) to anaerobic (independent of O_2) metabolism. Both the time required to reach steady state level at the onset of exercise and the time required to return to pre-exercise resting levels is prolonged with age after moderate to heavy exercise. Alveolar–capillary gas exchange is reduced and alveolar–arterial PO_2 differences are increased with exercise in middle-aged men. Nevertheless, gas diffusion capacity at 65 years is comparable to that at younger ages.[14,15]

2.6 Respiratory Muscles

The impaired efficiency of the lungs in some aged individuals to shift from resting to maximal function also depends on the decline in the strength and endurance of the respiratory muscles (Chapter 21). Although the diaphragm is the predominant muscle to drive ventilation at rest, during increasing exercise, the rib cage and abdominal muscles assist the diaphragm in augmenting ventilation rates. Because of the increased stiffness with aging of the rib cage, the diaphragm takes over a higher proportion of the mechanical effort needed for increasing ventilation. The diaphragm is a muscle not easily fatigued; therefore, whether it becomes fatigued in old age together with the other accessory respiratory muscles (abdominal and intercostal) during increasing ventilatory activity needs further investigation. Exertional dyspnea (i.e., shortness of breath after exercise) is common among the elderly when exercising. Yet the respective roles of decrements in muscle strength, increase in thoracic stiffness, and loss of lung compliance remain to be elucidated.

Inasmuch as fatigue after muscular exercise has a strong CNS component, other factors besides muscular decrements such as impaired coordination, insufficient motivation, arthritic involvement of the joints, and others may also play an important role in limiting the adaptive competence of the elderly in undertaking physical exercise.

2.7 Benefits of Physical Exercise

Despite the overall decline in performance during exercise, the elderly can and should undertake a regimen of physical exercise adequate to their capabilities and needs.[14-16] (Discussion of such regimens is presented in Chapter 25.) Other factors influencing performance of physical exercise are discussed in relation to skeletal and muscular changes with aging (Chapter 21). Age-related respiratory changes during sleep are discussed in Chapter 9.

3 SOME PULMONARY DISORDERS WITH AGING

The major respiratory disorders of the elderly include: chronic bronchitis (i.e., chronic infection or inflammation of the bronchi), often coexisting with emphysema (see below), neoplasia, and lung infections, particularly pneumonia. While some of these disorders, e.g., bronchitis and emphysema, are not necessarily life-threatening, they represent a considerable "burden" for the well-being of the elderly in terms of days of hospitalization, physician consultation, days of sickness, and almost continuous discomfort. From the perspective of the entire lifespan, the proportion of deaths due to respiratory diseases (in Western countries) is highest (approximately 30%) in the first year of life, falling to lower values (approximately 5%) in late adolescent and early adulthood. From the 5th decade, the incidence of respiratory disease rises steadily and, in those over 85 years of age, accounts for 25% of all deaths. In the last 20 years, the death rate from respiratory disease has fallen with the exception of bronchiogenic carcinoma (still rising especially in young women) and pneumonia (remaining constantly high).[3-5] Of the respiratory diseases, emphysema, pneumonia, and tuberculosis will be considered briefly.

3.1 Oxygen Toxicity in Lungs

The overall effects of oxygen free radicals in tissues and cells have been presented in Chapter 6. In the lungs, these molecules may contribute both to the acute oxygen toxicity that occurs when individuals breathe higher than normal concentration of oxygen and to the chronic damage to elastic tissue that leads to emphysema.

The acute toxicity is of particular significance for critically ill patients who require respirators. Although the lowest possible doses of oxygen are given, the potential damage to the lung must always be considered. Sometimes it may be necessary to choose between "allowing a patient to die immediately and giving pure oxygen which may kill in days".[17] Experiments in rats show that breathing oxygen increases the production of free radicals in pulmonary epithelial cells and macrophages, with consequent death of the animals within 3 days. If, however, animals are protected by the administration of the enzymes, superoxide dismutase and catalase, death of the animal is prevented. The chronic damage to lung tissue by oxygen is based on the inflammatory reaction that represents the first event in emphysema. The response to the phagocytic cells that infiltrate the inflammed tissue includes not only the liberation of free radicals but also the induction of enzymes such as elastase, which breaks down the elastic tissue of the lungs. Under normal conditions, elastase activity is controlled by inhibitors such as α_1-protease inhibitor. Thus, emphysema would result from inactivation of this enzyme. Cigarette smoking would also stimulate free radical formation, thereby further aggravating the cell damage.

3.2 Emphysema and Its Physiopathology

This disorder, together with chronic bronchitis, represents an often intolerable burden for the elderly.[18-20] Thousands of individuals 60 years of age and older require help with some aspects of daily living as a result of this disease. In the U.S., 3.3% of disabilities are ascribed to this disease. These numbers are only estimations, as age alone is associated with a significant increase in the prevalence of morning cough, even in nonsmokers, and a decline in respiratory competence.

This respiratory disease is characterized by

1. Diffuse distention and overaeration of the alveoli
2. Disruption of intra-alveolar septa
3. Loss of pulmonary elasticity
4. Increased lung volume
5. Associated impairment of pulmonary function

Functional impairment leads to

1. Disturbed ventilation
2. Altered air and blood flow
3. Frequently partial obstruction of bronchi (hence the often used name of obstructive disease).

Causes of emphysema range from an inherent defect of elastic tissue to association with fibrotic pulmonary diseases, such as silicosis, to the consequence of chronic diffused bronchitis due to age and aggravated by cigarette smoking.[21,22]

3.3 Emphysema and Cigarette Smoking

Heavy cigarette smoking is the most common cause of emphysema. The smoking increases the number of pulmonary alveolar macrophages, which release a chemical substance that attracts leukocytes to the lungs. Leukocytes in turn release proteases as elastase, which attacks the elastic tissue in the lungs. Normally, a plasma protein α_1-antitrypsin inactivates elastase and other proteases. In emphysema, the activity of this enzyme is decreased. This inactivation of the enzyme may be promoted by oxygen radicals which are released by the leukocytes (Chapters 6 and 7). Thus, with smoking, there is both an increased production of elastase and a decreased activity of the inactivating enzyme with the resulting destruction of elastic fibers.

An hereditary association has been noted with the affected individuals characterized by an α_1-antitrypsin deficiency. In the homozygous individual, emphysema develops early in life (about 20 years of age) even in the absence of cigarette smoking.

The most frequent manifestations of emphysema are

1. Exertional dyspnea and increase in lung compliance
2. Chronic productive cough (with mucus)
3. Labored inspiration and expiration (wheezing) and more work required for breathing
4. Enlarged and barrel-shaped chest, as the chest wall expands because of increased lung volume and increased use of accessory (shoulder and abdominal) muscles
5. Minor respiratory infections (of no consequence in young individuals with normal lungs) producing fatal and near-fatal disturbances of respiratory function
6. Loss of elasticity and restructuring of alveoli by large air sacs, resulting in poor and uneven alveolar ventilation and inadequate perfusion of the underventilated alveoli
7. *Hypoxia* (low O_2 levels) and *hypercapnia* (high CO_2 levels).

Hypercapnia induces acidosis, at first compensated by urinary retention of bicarbonate. When this compensatory mechanism fails, especially in the aged in whom renal competence is marginal (Chapter 19), the ensuing *respiratory acidosis* represents a medical emergency and must be treated accordingly. Hypoxia stimulates production of red blood cells which are increased in number (polycythemia). This contributes to *hypertension,* which causes the right side of the heart to enlarge, and then to cardiac failure (the so-called "cor pulmonale" or congestive right heart failure).[23]

Management

Given the burden inflicted by the disease and its frequency, treatment and prevention are still inadequate. Emphasis on prevention with cessation of cigarette smoking continues to yield only slow responses despite extensive efforts to counsel and educate. Treatment is essentially symptomatic and includes (1) the use of pharmacological agents such as bronchodilators (to relieve bronchial spasm), mucus liquefiers (to thin the mucous secretions) and antibiotics (to control potential infections); (2) administration of O_2 to be used cautiously to prevent acidosis; and (3) exercise to strengthen abdominal muscles and diaphragm to aid in lung ventilation. The use of respiratory aids in the form of aerosols, sprays, etc. may also be of benefit.[24]

3.4 Pneumonia

Pneumonia is an inflammatory process of the lung parenchyma most commonly caused by infection. The infectious agent (often the *Bacillus pneumococcus*) is frequently present among the normal flora of the respiratory tract. The development of pneumonia must, therefore, be usually attributed to an impairment of natural resistance. Indeed, the decline of immune competence with aging (Chapter 7) may explain why pneumonia remains, despite the availability of antibiotics, a serious, life-threatening problem with the elderly.

In most cases of community-based epidemics of pneumonia, the elderly are more susceptible with respect to severity and complications (e.g., lung abscess, bacteremia) as well as mortality (as high as 80% in those aged 60 and older).[25,26] The rise in pneumonia after the 7th decade registered in several Western countries including the U.S. may simply reflect the type of death certification, pneumonia being the terminal expression of other diseases. However, studies of hospitalized elderly suggest a true increase.[27] While the true cause may be failure of immunologic competence, the immediate cause has been ascribed to aspiration in the lungs of oropharyngeal flora during sleep. This is a common occurrence in the elderly in whom the mucobronchial defense barrier is impaired due to deficient ciliary activity, mucus production, and mechanical reflexes.

Presentation and Management

The diagnosis of pneumonia in the elderly is more difficult than in the young because of the atypical presentation. The classic features of chest pain, cough, and purulent or blood-stained sputum (spit) are uncommon. There is a lack of cough, toxic confusion predominates, and dehydration occurs early. Progression of the disease may produce further lung damage (e.g., abscess), or aggravate extrapulmonary manifestations such as the confusional state or induce additional damage such as pericarditis (inflammation of the pericardium, the sac surrounding the heart), ischemic heart disease, and meningitis (the membranes surrounding the brain). Treatment involves the use of antimicrobial agents, primarily antibiotics, the correction of dehydration, and ultimately the treatment of the underlying disease. Prevention includes the use of anti-influenza vaccination and perhaps pneumococcal vaccination (still under investigation). Rehabilitation after recovery is important both physiologically and psychologically and requires the use of an individually tailored program, based on exercise, oxygen therapy, and coordinated support of family and health providers.

3.5 Tuberculosis Revisited

Tuberculosis, induced by *Mycobacterium tuberculosis,* was considered for the earlier part of this century an infectious disease definitively eradicated by chemotherapy and improved socioeconomic and hygienic conditions. Indeed, tuberculosis is generally viewed as one of the most easily treatable serious infectious diseases likely to occur among older adults. The disease is chronic in nature, and the agent that causes it may remain dormant for many years but may be reactivated as immune defenses are reduced in old age (Chapter 7). The disease may be more widespread and severe in the elderly consequent to other unfavorable conditions such as malnutrition, alcoholism, and superimposed diseases.

In the 1980s, tuberculosis resulted in several outbreaks in nursing homes, where both reactivation and primary contact infection may occur. Nursing home residents are at greater risk for tuberculosis than elderly persons living in the community because of easier transmission of the disease under more crowded living conditions. It is important that all nursing home residents be tested for tuberculosis upon admission to the facility and that vigorous preventive measures be taken to stop the spread of the infection. Even in the elderly living in the community, the geriatrician must be alert to the possibility of the infection, often masked by atypic (simulating cold, influenza, pneumonia) symptoms.[28,29]

4 LIFE CYCLE OF ERYTHROCYTES*

4.1 Erythrocyte Function and Lifespan

The major function of red blood cells (erythrocytes) is to carry oxygen (O_2) and carbon dioxide (CO_2) to and from the tissues. This function is made possible by the presence in the cell of the pigment hemoglobin which binds to O_2 and CO_2. The erythrocytes are biconcave disk-like cells produced in the bone marrow. In mammals they lose their nuclei before leaving the bone marrow; they survive 120 days in the circulation. Because of their gas-carrying capacity, they can be viewed as an extension of the lungs.

The study of blood and blood cells is the object of continuing research, with a wealth of literature available on the contributing role of the hematopoietic system to the functional integrity of the organism. There are notable *species differences;* for example, geriatric anemia does not occur in mice. Specifically, in old mice, a reduced erythrocyte lifespan occurs, but the hematocrit of the young and old animal is the same. Enzyme analysis confirmed that older animals have a chronologically younger population of erythrocytes than young animals.[30] In contrast, geriatric anemia in humans is often associated with disease and may not be recognized if controlled for health.

Aging studies of the hematopoietic system are easy as compared to the collection of other tissues *in vivo,* and alterations in the entire organism or in specific organs or tissues are often reflected in alterations of the blood. The preservation and storage of blood and plasma for purposes of transfusion has indirectly provided a substantial amount of knowledge on the aging of blood and blood cells.

The erythrocyte represents one of the most thoroughly studied "aged" cells, with its particular life cycle well delineated. In addition, as a cell that undergoes continual renewal in the life of the organism, the erythrocyte has been studied as an example of how aging in general may affect cell turnover. Dietary restriction increases the mean and maximal lifespan in numerous species. However, mice fed a diet 40% restricted in calories exhibited erythrocytes with shortened lifespan, while controls never exhibited anemia. Given the beneficial effects of dietary restriction (Chapter 24), an enhanced erythropoiesis should have been expected, but, on the contrary, the mouse erythrocyte lifespan was shortened by some yet unknown biophysical event.[31] Similar studies undertaken in epithelial cells of the intestinal mucosa, cells that are also continuously replaced during the lifespan of the organism, have demonstrated (by tritiated thymidine autoradiography) that the variability from cell to cell within a cell population increases gradually with aging and that cell turnover is accomplished in a progressively longer period of time (Chapter 20). This shortening of the cell lifespan may be related mainly to the lengthening of the "interphase" (G1) portion of the DNA cycle, that is, the phase that precedes DNA synthesis and is presumably related to the accumulation of necessary precursors.[32]

Although the cause of the interphase lengthening is not known, it may make a cell more susceptible to damage. For example, DNA damage occurs in older cells with an increase in single-strand breaks in erythrocytes from aged chickens.[33] However, it is not clear, as the authors note, if these changes were truly reflective of *in vivo* aging or a result of the preparative alkaline sucrose sedimentation technique employed.

Since changes in the circulating erythrocyte reflect studies of an *already aged cell,* it may be more important to investigate age-dependent changes in *erythropoietic stem cells.* After old rodents (mice and rats) were exposed to hypoxia or bleeding, their erythropoietic stem cells served as donor tissue to genetically anemic host mice. Donors, irrespective of age, cured the anemia. Furthermore, anemic mice with old donor erythropoietic grafts responded as well as young mice to erythropoietic stimuli.[34] These studies are of particular relevance to geriatric medicine. For example, in a comprehensive mouse study analyzing 16 hematologic parameters, only the hematocrit (i.e., the percentage of the volume of blood occupied by erythrocytes) decreased as a function of age and not of disease.[35] When evaluating the human literature, it is not uncommon to identify an elderly patient with "old-age anemia". However, in the "Framingham study" no decrease in hematocrit was observed in elderly individuals when controlled for health.[36] Renal anemia is caused in part by a reduced lifespan of erythrocytes and decreased erythropoietin synthesis. Erythropoitein, a glycoprotein, is produced mainly in the kidney, is inactivated in the liver, and regulates (by stimulation of RNA synthesis) the production and release of of erythrocytes from the bone marrow in reponse to hypoxia, cobalt salts, andragens, and possibly other hormones. To aid in treatment, human recombinant erythropoietin administration with parathyroid hormone offers a valuable therapy in the hemolytic patient.[37] An explanation for the success of this therapy could be that erythropoietin normalizes the circulating erythrocyte volume.[38] Thus, what appears to be a disease of old age should be reevaluated

* This section (pages 230–231) contributed by R. W. Atherton, Ph.D., Department of Zoology and Physiology, University of Wyoming, Laramie, WY.

to determine if other factors, e.g., endocrine, could be the cause of anemia in older patients.

4.2 The Erythrocyte as a Model of Cellular Aging

The following discussion will deal primarily with the erythrocyte as a prototype of a cell in which the life cycle is telescoped — developing, acquiring specific functional competence, aging, and dying within a relatively short period of time, an average of 120 days in man — a process that is repeated in countless millions of such cells throughout the human lifespan. Indeed, when one studies the changes in erythrocyte structure and function with age, one must consider not only the life cycle of each cell but also the modifications that are associated with the development and aging of the total organism.[39]

In most vertebrates, the primitive and embryonic erythrocyte is characteristically different from the adult type in its morphologic, functional, and biochemical properties. Typical of these cells at all age periods is that their lifespan is set; even during embryogenesis, erythrocytes are studied as models of "programmed cell death" or "apoptosis" (Chapter 5). Erythropoietin, in the process of regulating erythropoiesis, reduces DNA damage in erythroid progenitor cells and thus diminishes apoptosis.[40] Molecular biology studies of the erythropoietin receptor demonstrate that a modified form (truncated by alternative splicing) exists in the early progenitor cells. The truncated form, when transfected into cells, makes them more prone to cell death than those with the full-length receptor. It is proposed that the full-length receptor may transduce a signal to prevent programmed cell death, thus expanding the role for erythropoietin in the regulation of erythropoiesis.[41]

The changes characteristic of the aging erythrocyte include[42-45]

- Diminished cell size
- Increased cell density (probably due to decreased lipid content)
- Relative decrease of intracellular potassium
- Increase of sodium (probably due to permeability changes of the cell membrane)
- Reduction in the activity of certain enzymes (e.g., glucose-6-phosphate dehydrogenase)
- Reduction of surface charge and increase in fragility
- Less reversible deformatility
- Increased agglutinability

Indeed, it appears that in the older erythrocytes, "loss of elasticity" and "deformity" may represent two of the major causes of cell removal and cell death.[45] A senescent cell antigen may appear on aged cells such that binding of IgG autoantibodies aids in the removal of erythrocyte by macrophages. Hemoglobinopathies may serve as a genetic marker for erythrocyte removal from the circulation as well as glycosylation[46] and methylation.[47]

4.3 Changes in Hemoglobin

Whether alterations also occur in the content, structure, and function of hemoglobin within the aging erythrocyte remains to be clarified. In general, gross alterations have not been observed with the aging of the cell, but it is entirely possible that subtle functional modifications do occur, especially in response to stress. Studies of the incorporation of ^{51}Cr in young and old erythrocytes have shown a preferential labeling of the younger cell, but whether this is due to differences in cell size, permeability, metabolism, or aging of the hemoglobin itself has not been established. Hemoglobin, like collagen (Chapter 22), may undergo specific changes with age, the "old hemoglobin" differing from the "young hemoglobin" in a number of physicochemical properties (in the same manner as embryonic and fetal hemoglobin differs from the adult form). Inasmuch as the main function of hemoglobin is to carry O_2, oxygen dissociation curves of young and old erythrocytes generally reveal that older erythrocytes exhibit a significantly greater oxyhemoglobin saturation than younger erythrocytes — a finding that perhaps may be attributed to macromolecular changes with age in the structure of the hemoglobin.[48]

5 AGING OF THE HEMATOPOIETIC SYSTEM

Inasmuch as the erythrocytes are formed in the hematopoietic system and primarily in the bone marrow, the function of this system has been studied to determine whether it undergoes functional impairment with aging. In general, few changes have been reported with respect to either stem-cell kinetics, bone blood flow, bone marrow cellularity, or erythropoiesis *per se*.[49] However, when particular strains of mice are studied, there are demonstrable differences in the hematopoietic system of aged mice.[50,51]

One of the primary observations drawn from studies of the aging erythrocyte is that, despite the specific changes that characterize its life cycle, the effects of aging, in general, on cell turnover, cell renewal, and cell function appear to be minimal, at least under steady-state conditions. In addition, it seems that the capacity of the hematopoietic system to respond to increased functional demand, such as imposed by hypoxia, is not impaired by aging. For example, both young and old rats when exposed to low oxygen pressure respond with a comparable increase in hematocrit value, total circulating hemoglobin, and number of erythrocytes as well as total red cell volume and total blood volume.[52] Thus, it has been concluded that the homeostatic function of the blood, vital to the regulation of bodily function, is not significantly affected by aging. The regenerative potential of the erythrocyte, seemingly inexhaustible *in vivo,* stands in contrast with studies *in vitro* that suggest, rather, that some cells (fibroblasts), despite their proliferative capacity, ultimately lose the ability to regenerate. The erythrocyte as a product of the erythron with a specific protein of paramount importance remains a most suitable model for aging; it may provide direction for some most interesting and profitable avenues for the study of aging.[53,54]

REFERENCES

1. Meisami, E. and Timiras, P. S., Eds., Respiratory development, in *Handbook of Human Growth and Developmental Biology,* Volume 3, Part B, CRC Press, Boca Raton, FL, 1990, 131.
2. Masoro, E. J., Ed., *CRC Handbook of Physiology in Aging,* CRC Press, Boca Raton, FL, 1981.
3. Davies, B. H., The respiratory system, in *Principles and Practice of Geriatric Medicine,* Pathy, M. S. J., Ed., John Wiley & Sons, Chichester, England, 1990, 663.

4. Mahler, D. A., Ed., *Pulmonary Disease in the Elderly Patient,* Marcel Dekker, New York, 1993.
5. Kauffmann, F. and Frette, C., The aging lung: an epidemiological perspective, *Resp. Med.,* 87, 5, 1993.
6. Reddan, W. G., Respiratory system and aging, in *Exercise and Aging,* Smith, E. L. and Serfass, R. C., Eds., Enslow Publishing, Hillside, NJ, 1981.
7. Hurwitz, S. A., Allen, J., Liben, A., and Becklake, M. R., Lung function in young adults — evidence for differences in the chronological age at which various functions start to decline, *Thorax,* 35, 615, 1980.
8. Davis, C., Campbell, E. J. M., Openshaw, P., Pride, N. B., and Woodroof, G., Importance of airway closure in limiting maximal expiration in normal man, *J. Appl. Physiol.,* 48, 695, 1980.
9. Åstrand, I., Åstrand, P.-O., Hallback, I., and Kilbom, A., Reduction in maximal oxygen uptake with age, *J. Appl. Physiol.,* 35, 649, 1973.
10. Levitsky, M. G., Effects of aging on the respiratory system, *Physiologist,* 26, 102, 1985.
11. Peterson, D. D., Pack, A. I., Silage, D. A., and Fishman, A. P., Effects of aging on the ventilatory and occlusion pressure responses to hypoxia and hypercapnia, *Am. Rev. Resp. Dis.,* 124, 387, 1981.
12. Dill, D. B., Hillyard, S. D., and Miller, J., Vital capacity, exercise performance and blood gases at altitude as related to age, *J. Appl. Physiol.,* 48, 6, 1980.
13. Shephard, R. J., Ed., *Physical Activity and Aging,* Croom Helm, London, 1978.
14. Shephard, R. J., Ed., *Physiology and Biochemistry of Exercise,* Praeger, New York, 1982.
15. McConnell, A. K., Semple, E. S., and Davies, C. T., Ventilatory responses to exercise and carbon dioxide in elderly and younger humans, *Eur. J. Appl. Physiol. Occup. Physiol.,* 66, 332, 1993.
16. Warren, B. J., Nieman, D. C., Dotson, R. G., Adkins, C. H., O'Donnell, K. A., et al., Cardiorespiratory responses to exercise training in septuagenarian women, *Int. J. Sports Med.,* 14, 60, 1993.
17. Marx, J. L., Oxygen free radicals linked to many diseases, *Science,* 235, 529, 1987.
18. Bignon, J. and Scarpa, G. L., Eds., *Biochemistry, Pathology, and Genetics of Pulmonary Emphysema,* Pergamon Press, New York, 1981.
19. Snider, G. L., Symposium on emphysema, *Clinics in Medicine,* W. B. Saunders, Philadelphia, 1983.
20. Sterling, G. M., Ed., *Respiratory Disease,* William Heinesman Medical Books, London, 1983.
21. Sherrill, O. L., Lebowitz, M. D., Knudson, R. J., and Burrows, B., Longitudinal methods for describing the relationship between pulmonary function, respiratory symptoms and smoking in elderly subjects: the Tucson Study, *Eur. Resp. J.,* 6, 342, 1993.
22. Sparrow, D., O'Connor, G. T., Rosner, B., De Molles, D., and Weiss, S. T., A longitudinal study of plasma cortisol concentration and pulmonary function decline in men. The Normative Aging Study, *Am. Rev. Resp. Dis.,* 147, 1345, 1993.
23. Murphy, M. L. and Bone, R. C., *Cor Pulmonale in Chronic Bronchitis and Emphysema,* Futura Publ., Mount Kisco, NY, 1984.
24. Petty, T. L. and Nett, L. M., Eds., *Enjoying Life with Emphysema,* Lea & Febiger, Philadelphia, 1984.
25. Macfarlane, J. T., Finch, R. G., Ward, M. J., and Macrae, A. D., Hospital study of adult community-acquired pneumonia, *Lancet,* 2, 255, 1982.
26. Fox, R. A., Treatment recommendations for respiratory tract infections associated with aging, *Drugs Aging,* 3, 40, 1993.
27. Mylotte, J. M. and Beam, T. R., Jr., Comparison of community-acquired and nosocomial pneumococcal bacteraemia, *Am. Rev. Resp. Dis.,* 123, 265, 1981.
28. Stead, W., Tuberculosis among the elderly: forgotten but not gone, *Health Lett.,* 5, 7, 1989.
29. Stead, W. W., Lofgren J., P., Warren, E., and Thomas, C., Tuberculosis as an epidemic and nosocomial infection among the elderly in nursing homes, *N. Engl. J. Med.,* 312, 1483, 1985.
30. Magnani, M., Rossi, L., Stocchi, V., Cucchiarini, L., Piacentini, G., and Fornaini, G., Effect of age on some properties of mice erythrocytes, *Mech. Ageing Dev.,* 42, 37, 1988.
31. Hishinuma, K. and Kimura, S., Dietary restriction induces microcytic change and shortened life span of erythrocytes without anemia in mice, *Int. J. Vitam. Nutr. Res.,* 59, 406, 1989.
32. Thrasher, J. D. and Greulich, R. C., The duodenal progenitor population. I. Age related increase in the duration of the cryptal progenitor cell, *J. Exp. Zool.,* 159, 39, 1965.
33. Karran, P. and Ormerod, M. G., Is the ability to repair damage to DNA related to the proliferative capacity of a cell? The rejoining of X-ray-produced strand breaks, *Biochim. Biophys. Acta,* 229, 54, 1973.
34. Harrison, D. E., Normal function of transplanted marrow cell lines from aged mice, *J. Gerontol.,* 30, 279, 1975.
35. Finch, C. E. and Foster, J. R., Hematologic and serum electrolyte values of the C57BL/6J male mouse in maturity and senescence, *Lab. Anim. Sci.,* 23, 339, 1973.
36. Gordon, T. and Shurtleff, D., Means at each examination and interexamination variation of specific characteristics: Framingham study exams 1–10, in *The Framingham Study: An Epidemiological Investigation of Cardiovascular Disease,* Kannel, W. B., Ed., Washington, D.C., 1973.
37. Hampl, J., Riedel, E., Scigalla, P., Stabell, U., and Wendel, G., Erythropoiesis and erythrocyte age distribution in hemodialysis patients undergoing erythropoietin therapy, *Blood Purif.,* 8, 117, 1990.
38. Najean, Y., Moynot, A., Deschryver, F., Zins, B., Naret, C., Jacquot, C., and Drueke, T., Kinetics of erythropoiesis in dialysis patients receiving recombinant erythropoietin treatment, *Nephrol. Dial. Transplant.,* 4, 350, 1989.
39. Landaw, S. A., Factors that accelerate or retard red blood cell senescence, *Blood Cells,* 14, 47, 1988.
40. Koury, M. J. and Bondurant, M. C., Erythropoietin retards DNA breakdown and prevents programmed death in erythroid progenitor cells, *Science,* 248, 378, 1990.
41. Nakamura, Y., Komatsu, N., and Nakauchi, H., A truncated erythropoietin receptor that fails to prevent programmed cell death of erythroid cells, *Science,* 257, 1138, 1992.
42. Bishop, C. and Van Gastel, C., Changes in enzyme activity during reticulocyte maturation and red cell aging, *Haematologia,* 3, 29, 1969.
43. Luque, J., Delgado, M. D., Rodriguez-Horche, Company, M. T., and Pinilla, M., Bisphosphoglycerate mutase and pyruvate kinase activities during maturation of reticulocytes and ageing of erythrocytes, *Biosci. Rep.,* 7, 113, 1987.
44. Waugh, R. E., Narla, M., Jackson, C. W., Mueller, T. J., Suzuki, T., and Dale, G. L., Rheologic properties of senescent erythrocytes: loss of surface area and volume with red blood cell age, *Blood,* 79, 1351, 1992.
45. Linderkamp, O., Friederich, E., Boehler, T., and Ludwing, A., Age dependency of red blood cell deformability and density: studies in transient erythroblastopenia of childhood, *Br. J. Haematol.,* 83, 125, 1993.

46. Winterhalter, K. H., Nonenzymatic glycosylation of proteins, in *Cellular and Molecular Aspects of Aging. The Red Cell as a Model,* Eaton, J. W., Konzen, D. K., and White, J. G., Eds., Alan R. Liss, New York, 1985.
47. Clark, S., The role of aspartic acid and asparagine residues in the aging of erythrocyte membranes: cellular metabolism of racemized and isomerized forms by methylation reactions, in *Cellular and Molecular Aspects of Aging. The Red Cell as a Model,* Eaton, J. W., Konzen, D. K., and White, J. G., Eds., Alan R. Liss, New York, 1985.
48. Edwards, M. J., Koler, R. D., Rigas, D. A., and Pitcairn, D. M., The effect of in vivo aging of normal human erythrocytes and erythrocyte macromolecules upon oxyhemoglobin dissociation, *J. Clin. Invest.,* 40, 636, 1961.
49. Little, K., Bone marrow and aging, *Gerontologia,* 15, 155, 1969.
50. Hellebostad, M., Sanengen, T., and Halvorsen, S., Variations in erythrocytes throughout a lifetime. Studies in a high-leukaemic mouse strain, the AKR/O strain, and a non-leukaemic strain, the WLO strain, *Blut,* 61, 358, 1990.
51. Boggs, S. S., Patrene, K. D., Austin, C. A., Vecchini, F., and Tollerud, D. J., Latent deficiency of the hematopoietic microenvironment of aged mice as revealed in W/Wv mice given +/+ cells, *Exper. Hematol.,* 19, 683, 1991.
52. Garcia, J. F., Erythropoietic response to hypoxia as a function of age in the normal male rat, *Am. J. Physiol.,* 190, 25, 1957.
53. Swislocki, N. I., Impromptu discussion of comparative aspects of red cell aging, in *Cellular and Molecular Aspects of Aging. The Red Cell as a Model,* Eaton, J. W., Konzen, D. K., and White, J. G., Eds., Alan R. Liss, New York, 1985.
54. Platt, D., *Blood cells, Rheology, and Aging,* Springer-Verlag, New York, 1988.

III
SYSTEMIC AND ORGANISMIC AGING

19 THE KIDNEY, THE LOWER URINARY TRACT, THE PROSTATE, AND BODY FLUIDS

Mary L. Timiras

As with other systems, the urinary tract is both directly and indirectly affected by aging. Direct effects are exemplified by intrinsic cellular changes involving the nephron as well as the muscles of the bladder or the prostate. Indirect effects may be secondary to cardiovascular, endocrine, or metabolic alterations occurring with aging that have vital repercussions on urine formation and excretion. Reciprocally, physiologic decrements in renal function and disturbances of the lower urinary tract may not only alter the elimination of end-products of metabolism but also may impair other functions such as regulation of body fluids, acid-base balance, and blood pressure. Finally, given the metabolic and excretory functions of the kidney with respect to drugs of environmental, recreational, and therapeutic origin, the decline of renal function with aging can modify the renal handling of drugs and thereby alter drug pharmacokinetics, including toxicity and therapeutic effectiveness. Such drug-related considerations are important at any age but are particularly so in the elderly to whom multiple drugs are simultaneously prescribed, increasing dangers of polypharmacy and iatrogenicity.

In the present chapter, the following aging-related changes will be reviewed:

1. Changes in renal function, including considerations of drug excretion
2. Changes in the lower urinary tract, including the problems of urinary incontinence and prostatic pathology
3. Changes in body fluids and acid–base balance

1 AGING-RELATED CHANGES OF RENAL FUNCTION

All parameters of renal function are affected by aging, but the age of onset, rate and course of changes, and consequences vary. Thus, glomerular filtration, closely dependent on the efficiency of the renal blood flow and the integrity of the glomerular basement membrane, appears affected at an earlier age and more severely than tubular reabsorption and secretion. After the age of 30, in humans, renal function as measured by several functional tests (see below) gradually decreases and by 60 years has been further reduced to half. This decline has been ascribed to a gradual loss of nephrons and diminished enzymatic and metabolic activity of tubular cells as well as an increased incidence of pathologic processes (primarily atherosclerosis, which involves the renal blood circulation, an essential factor in determining renal competence). Likewise, a number of experimental animal studies show an overall decline of renal competence involving glomerular and tubular functions. A discussion of declining renal function is preceded by a brief summary of renal structure and function.

Some Gross Structural Characteristics and Major Functions of the Kidney

The kidneys are paired organs lying on either side of the descending (abdominal) aorta and receive their blood via the renal arteries, two main collaterals of the aorta. A smooth outer capsule covers the cortex, which contains the majority of the glomerular portion of the individual nephron (Figure 19-1). The nephron represents the functional unit and grossly is formed of the glomerulus, in the cortex, and the renal tubule which dips down and occupies the medulla, the inner portion of the renal parenchyma. Running parallel to the tubules in the medulla is a network of blood vessels, the vasa recta, which participate, with the tubular loop of Henle, in controlling the osmolality of the medulla through a counter-current mechanism. The urine formed in the nephron flows into collecting ducts and out through the calyces and the pelvis to the ureter.

The glomerulus is formed by a tuft of capillaries between the entering, afferent, and the exiting, efferent, arterioles. Filtration is through a fenestrated glomerular endothelium which is separated by a basal lamina from the interdigitated epithelial cells, the so-called podocytes, of the tubular epithelium. The filtrate has the same composition as plasma, except for the absence of proteins. The epithelial cells lining the various segments (proximal tubule, loop of Henle, distal tubule, collecting duct) of the renal tubule have been divided into types and subtypes on the basis of minor differences in histologic structure; there is some evidence that these differences correlate with differences in function, which is either reabsorption or secretion. Cells of the afferent arteriolar wall and the abutting distal tubule form the juxtaglomerular apparatus, site of the formation and release of renin. Cells of the distal convoluted tubule and the collecting duct are sensitive to hormones, aldosterone for sodium reabsorption, and antidiuretic hormone (ADH) for water reabsorption.

The kidneys adjust the amount of water and electrolytes excreted in order to keep the body fluid relatively constant in volume and composition as well as maintain blood pH through the secretion of H^+, that is, maintain fluid and acid-base balance and homeostasis. Kidneys also excrete end products of metabolism. They have certain endocrine functions: they are responsible for the secretion of renin, hence regulating the renin-angiotensin system and, indirectly, blood pressure and aldosterone secretion; they activate vitamin D and thereby play a role in Ca^{2+} metabolism; they secrete erythropoietin and maintain the hemoglobin level (Table 19-1).

Tests of Renal Function

At all ages, the competence of renal function is assessed by a number of tests among which the most routinely used are urine volume (per 24 hours), analysis of urine constituents and urine concentration, and dilution tests. Other measures of renal function include clearance tests, that is, clearance of a substance from the plasma during passage through the kidney (for example, inulin to measure glomerular filtration rate and *para*-aminohippuric acid, PHA, to measure renal blood flow). Radiologic, isotopic, ultrasonic, magnetic resonance spectroscopy and imaging, and tomographic imaging as well as biopsies and other less common tests represent more laborious and expensive means of assessing specific conditions.

1.1 Glomerular Function

The function of the glomerulus is to selectively filter plasma to produce a glomerular filtrate which, under normal healthy

0-8493-8979-8/94/$0.00+$.50

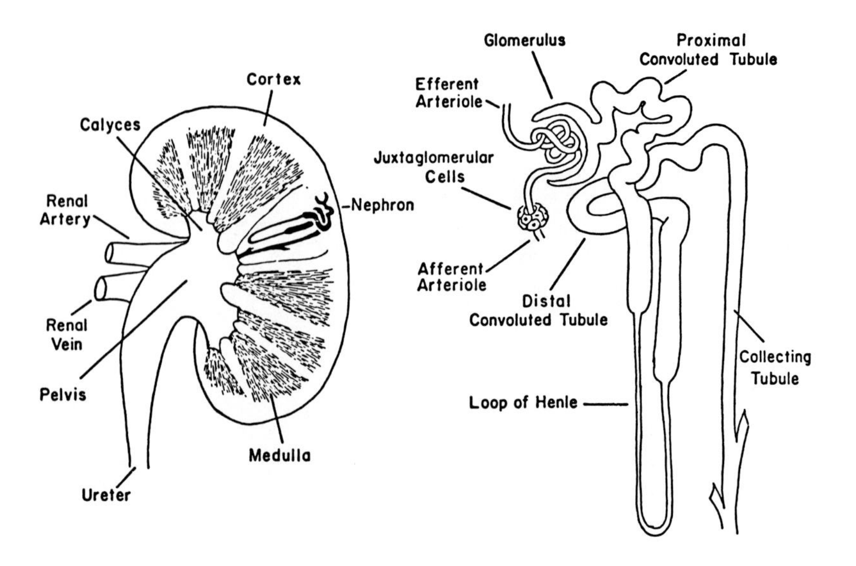

FIGURE 19-1 *Diagram of the kidney* (A) *and of a nephron* (B). (Drawing by S. Oklund.)

TABLE 19-1

Major Functions of the Kidney
Water and electrolyte regulation
Metabolic products excretion
Hydrogen ion excretion and maintenance of blood pH
Endocrine functions
Renin–angiotensin (blood pressure)
Vitamin D activation (Ca^{2+} metabolism)
Erythropoietin (hemoglobin levels)

conditions, is practically free of proteins. Glomerular filtration is determined by measuring the clearance of plasma and the excretion in the urine of a substance, such as inulin or creatinine, that is freely filtered through the glomeruli but not secreted or reabsorbed by the tubule. This technique establishes the filtration rate in a normal average young man as approximately 125 ml/min. With aging, commencing as early as 30 years, the glomerular filtration rate decreases progressively and significantly (Figure 19-2).[1-3]

Creatinine Clearance

Both inulin, a polysaccharide administered intravenously, and creatinine, a normal endogenous product of protein metabolism, are usually utilized to assess glomerular function. Creatinine clearance, which does not involve any drug administration, is most frequently used and is calculated by the following formula:

$$\text{Creatinine clearance} = \frac{(140 - \text{age}) \times \text{body weight (in kg)}}{72 \times \text{serum creatinine (in mg\%)}}$$

A 20-year-old individual, with a serum creatinine of 1 mg% and a body weight of 72 kg, has a creatinine clearance of 120 mg% whereas a 90-year-old with the same body weight and serum creatinine has a creatinine clearance of 50 mg%, a greater than 50% reduction. The reduction with aging in the urinary output of creatinine, a muscle-specific metabolite, has been interpreted as reflecting a reduction in lean body mass as demonstrated to occur in old age by loss of radioactively labeled potassium, the major intracellular electrolyte.

The above equation is valid only with the assumption that renal function decreases with age at a rate of 1 ml/min/year after age 40. Although this provides a better estimate than no correction at all, recent studies have shown great variation in the rate of decline of renal function. In some individuals, no decrease in renal clearance can be detected with advancing age.[4]

Creatinine clearance is often measured in elderly individuals, not merely as a test of renal function but also as a guide to drug administration, so that dosage can be adjusted depending on age, serum creatinine, and creatinine clearance, as well as a measure of efficiency of renal excretion.

Of the factors that affect the glomerular filtration rate, the most important are characteristics of the glomerular wall and renal blood flow. Studies in humans show that the decrease in glomerular function may be due primarily to a loss or alteration of glomeruli and secondarily to alterations in blood flow, although the reverse sequence may also be true. Irrespective of the primary site of the lesion, studies in rats show a 30% incidence of necrotic glomeruli with age while, after food restriction, the incidence is reduced to 2%.[5]

Microscopic and ultramicroscopic studies of the glomerulus in rats reveal a progressive thickening (from 1300 Å neonatally to 4800 Å at the old age of 2 years) of the basement membrane in some areas (focal or segmental thickening) and collapse (with loss of distinct layers) in others. Podocytes are lost or undergo swelling or atrophy. Proliferation of interstitial collagen leads to progressive sclerosis. As a consequence of these alterations in the glomerular structure, abnormal proteinuria appears in about 25% of male rats at a young age and its incidence and severity increases progressively with age, although less so in female animals.[6]

In humans, widespread glomerular necrosis is rare. However, even in humans, histologic studies show a thickening of the glomerular basement membrane with concomitant biochemical alterations such as a progressive decrease (e.g., 3.7% reduction per decade) in some amino acids,

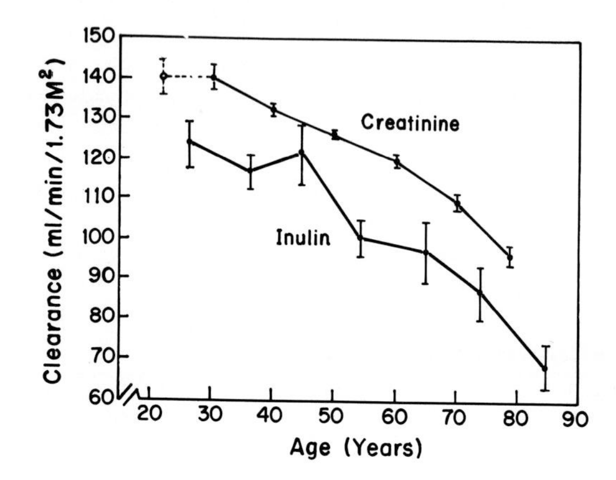

FIGURE 19-2 *Glomerular filtration rate with age.* Decrease in glomerular filtration rate with age as measured by creatinine and inulin clearance. Creatinine is a normal endogenous metabolic product and inulin a polymer of fructose.

suggesting a diminution of the collagenous component.[7] In humans, however, alterations of renal blood flow appear to be primarily responsible for the glomerular alterations and age-related decrease in blood flow have been reported;[3,7] moreover, these changes are potentially reversible.[9]

Aging-Related Changes in Day/Night Urine Excretion

In the normal adult, urine flow and electrolyte excretion follow a day/night pattern with higher levels in the daytime. This pattern may have evolved to permit undisturbed sleep. In the elderly, the rhythm is shifted, with increased water and electrolyte excretion during the night. This shift may involve the glomerular filtration along with hormonal control of urine concentration and may be regulated by extrarenal, hypothalamic factors.[10]

1.2 Glomerular Function and Dietary Proteins

Changes with aging in glomerular function do not pose any threat by themselves to well-being; likewise, renal function is not seriously compromised with aging, even in octagenerians. If, however, intrinsic renal disease or surgical loss of renal tissue or other factors, such as diet, add to the glomerular burden of age, then the course of glomerular sclerosis and reduced glomerular filtration rate may be hastened appreciably. Under normal conditions, very little protein is filtered in the glomeruli and excreted in the urine. One manifestation of impairment in glomerular filtration is altered permeability with consequent proteinuria, i.e., more than the usual trace amounts of protein in the urine. In addition, high protein diets make it more difficult to prevent filtration of protein. Experiments show that dietary proteins may induce glomerular damage and that reduction of dietary protein intake reduces this damage. For example, feeding rats a diet low in protein or restricted in total calories reduces age-related proteinuria. A similar beneficial effect has been demonstrated after long-term administration of the adrenal steroid, dehydroepiandrosterone, which would act by reducing weight gain and/or oxidative processes or by some still unclear hormonal action.[11] Dietary restriction seems to extend the life of laboratory animals and such extension may involve the beneficial effects of a low-protein diet on renal function.[12] Thus, rats consuming 75 kcal of food per day, a really large quantity, show high protein excretion throughout life from 4 mg/day at 70 days, just after sexual maturation, to 88 mg/day at the old age of 900 days. The normal food consumption is more in the range of 50 to 65 kcal per day and, with this diet, protein excretion is halved. If food is severely restricted to 12.5 kcal/day, the rise in proteinuria with age is abolished. The same modification is apparent even with short-time food restriction in old animals. When the diet is reduced from 60 to 25 kcal per day, protein excretion decreases by 40% in 1 week and continues at this level for the next week, even after discontinuation of the restricted diet (Figure 19-3).

In humans, the current diet offers sustained (three meals a day) rather than intermittent (feast or famine of animals in the wild) intake of protein. Such sustained "excess" would impose rigorous demands on the glomerular filtration rate and renal blood flow, thereby contributing to the decline of these functions with aging. It may also be responsible for the inexorable progressiveness of the renal deficits and the greater incidence of renal pathology in the elderly. Calorie or protein restriction fairly early in the course of renal disease slows the rate of decline in glomerular filtration. This is true not only for aging but also for metabolic and renal disorders; this is the case, for example, of obesity, in which diet modification leading to body weight loss reduces the often-present proteinuria associated with glomerular sclerosis.[13]

1.3 Tubular Function

The aged kidney is unable to concentrate urine as does the young kidney, and, because of this, water and electrolyte metabolism are potentially critical in the elderly, even those who have no overt signs or laboratory tests indicating renal dysfunction.

The ability to concentrate or dilute urine is gradually lost, with the result that the elderly individual is unable to cope optimally with either dehydration or water load.[14] This inability is manifested both after stimulation and inhibition of antidiuretic hormone (ADH) secretion (Chapter 11).[15] Administration of ethanol, a drug which is a known ADH inhibitor, reduces circulating ADH levels up to 120 min after administration in young individuals. In old subjects, ADH levels are decreased immediately after alcohol ingestion but increase thereafter, as shown at 120 min in Figure 19-4. In the young individual, low ADH levels are associated with the expected diuresis or increased water clearance. In old individuals, ADH inhibition occurs early after alcohol administration and then disappears; concomitantly diuresis does not occur or is only minimal.

As for ADH inhibition, ADH stimulation induces different responses in the elderly as compared to the young. In response to the same osmotic challenge, ADH levels increase, more so in old subjects than in young. Thus, administration of an hypertonic sodium chloride solution increases ADH levels much more in the elderly than in the young, but this

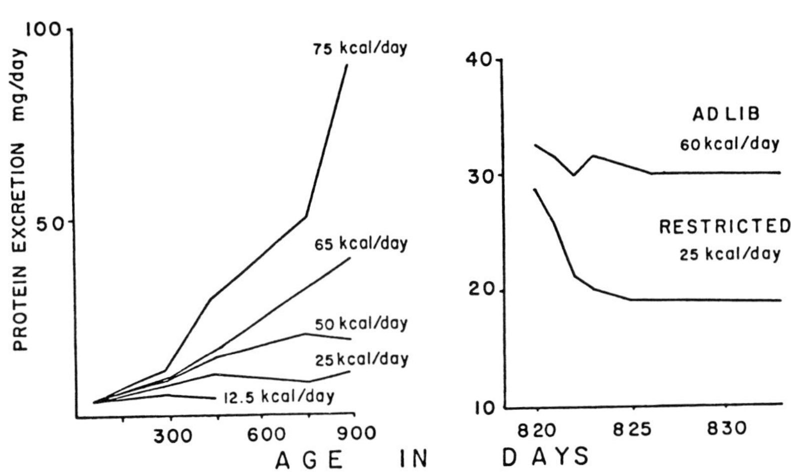

FIGURE 19-3 *Dietary caloric restriction reduces aging-related proteinuria in male rats.* (A) Proteinuria is reduced proportionately to the severity of the caloric restriction; severe food restriction (12.5 kcal/day) initiated at a young age (70 days) abolishes the steep rise in protein excretion with aging (but is not compatible with long life), and a less severe restriction (25 or 50 kcal/day) markedly inhibits it. (B) This effect of caloric restriction is also observed in old (820 days) rats in which about 1 week restriction was sufficient to reduce proteinuria significantly.

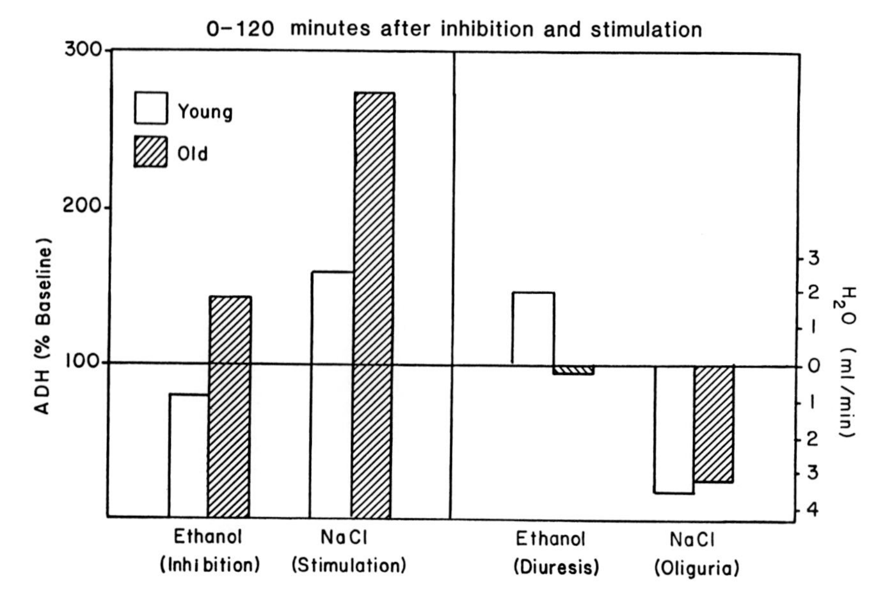

FIGURE 19-4 *Age-related changes in tubular responses to antidiuretic hormone (ADH).* Ethanol inhibits ADH and is less efficient in this inhibition in the elderly as reflected in the diuresis which ensues. Hypertonic sodium chloride increases ADH more in the elderly than the young; however, the expected increase in water retention does not occur.

increase in ADH levels is not accompanied in the elderly by the expected increase in water retention observed in the young (Figure 19-4). Attempts to identify the site(s) of the decreasing function, either renal or extrarenal, have not yet led to definitive conclusions. Administration of a standardized dose of ADH to young and old individuals shows a decline in the ability of the tubules to perform osmotic work and increase water retention in old individuals.[16] However, further studies suggest that decrements in the ability of cellular membranes in the collecting duct to become more water permeable under ADH influence occur primarily in old individuals with renal infections or hypertension; these decrements should not, therefore, be viewed as the usual accompaniment of aging.[17] Further experiments show that, in aged rats, decreased responsiveness of collecting tubular epithelium to ADH is the most likely explanation for the impairment of urine-concentrating ability, [18] while reduced secretion and levels of ADH would not play an important role in this impairment. Whether the defect in renal regulation of water excretion lies in the hypothalamus, the pituitary, or the kidney, there is no doubt about the relative inability of the aged kidney to concentrate or dilute urine and the consequent ease with which the elderly person may develop acute or chronic renal failure. As long as the total water intake is adequate — 2.5 to 3.0 l/day — renal function will be adequate. Crises will arise when water intake is diminished due to loneliness, immobility, confusion, fright of incontinence, etc. This may also occur with the unnecessary use of diuretics.

Blood urea and blood creatinine are usually unchanged in aged, healthy individuals, even though, as in the case of creatinine, their urinary excretion may be altered. Normal serum creatinine, despite decreased excretion, may reflect low production, perhaps related to reduced muscle mass and function. Blood urea may be reduced due to diminished protein intake, often present in the elderly or, more rarely, to liver disease and insufficient urea production.

1.4 Renal Disease in the Elderly

Compounding the relatively minor impairment of renal function in the healthy elderly is the increased incidence of renal diseases and the reduced ability to handle the excretion of drugs. The major renal disorders of the young and middle aged — acute nephritis, collagen disorders, and malignant hypertension — are uncommon in old age. If they occur, they present and are treated, as in the young, with caution to protect water and electrolyte balance.

In the aged population, the common problems affecting renal function are related to damage induced by infections or drugs or by hypertension, or they are consequent to miscellaneous disorders such as tuberculosis, nephritis, diabetes mellitus, amyloidosis, collagen disorders, etc. (Table 19-2). If untreated, these disorders may lead, perhaps more easily than in the young and with more life-threatening consequences, to impairment and, finally, to failure of renal function.[21]

1.5 Failure of Renal Function: Pathogenesis and Management

Renal failure refers to the inability of the kidneys to handle electrolytes and other substances that must be excreted. Normally, one third of the nephrons can eliminate all of the normal waste products from the body and prevent their accumulation in body fluids. When the number of functioning nephrons fall to 10 to 20% of normal, urinary retention and death follow.

Urinary failure is designated clinically as either acute or chronic. The acute form is the most frequent in the elderly. In both cases, the consequences depend to a great extent on the food and water intake of the individual. With moderate intake, the most important signs are

- Generalized edema, resulting from salt and water retention
- Acidosis, resulting from failure to excrete normal acidic products
- High concentration of nonprotein nitrogens, especially urea, resulting from failure to excrete metabolic end-products
- High concentration of other urinary products, including creatinine, uric acid, phenols, etc.

This condition of failing excretion is called *uremia* because of the high concentrations of normal urinary excretory products that collect in the body fluids (Table 19-3).

TABLE 19-2

Common Renal Problems in the Aged
Renal failure
Impaired drug excretion
Urinary tract infections
Hypertension
Miscellaneous disorders
Tuberculosis
Nephritis
Diabetes
Etc.

Acute renal failure refers to a sudden cessation of renal function following various insults to the normal kidney. The causes of acute renal failure are numerous and are divided into prerenal, renal, and postrenal depending on whether they are traced to alterations of the kidney itself or to extrarenal alterations (Table 19-4). Prerenal causes, most apt to be found in the aged, include: loss of body fluids (e.g., vomiting, diarrhea), inadequate fluid intake (often associated with the overuse of diuretics and laxatives), and surgical shock or myocardial infarction.

Renal causes are relatively rare in the elderly; they include drug toxicity due to certain antibiotics (sulfonamides, aminoglycosides, amphoteracin B), X-ray contrast materials (in a rather dehydrated individual), drug-induced immunologic reactions,[22] infectious diseases, Gram-negative bacteremia with shock or peritonitis, thrombosis and other circulatory alterations due to atherosclerosis, and intravascular hemolysis such as may follow transurethral resection of the prostate.

Postrenal causes are due primarily to urinary tract obstructions, such as occurs with prostatic enlargement.

Chronic Renal Failure

Chronic renal failure results from slow, progressive renal dysfunction. It is more often a disease of young and middle-aged individuals, as survival is reduced in the major instances such as in glomerulonephritis and polycystic disease. The most common causes in the elderly are (1) progressive renal sclerosis (due to atherosclerosis), (2) chronic pyelonephritis, and (3) slow but progressive enlargement of the prostate.

When urine output is acutely reduced to 20 to 200 ml/day (oliguria) following any of the foregoing events, the patient is in acute renal failure. Renal tubular necrosis is the characteristic finding and scattered basement membrane disruption occurs. The urine contains protein, red cells, epithelial cells, and characteristic dirty brown casts. Rate of protein catabolism in the body governs the rate of accumulation of metabolic products in body fluids — and the signs of uremia appear with nausea, vomiting, diarrhea, lethargy, and hypertension.

Acute renal failure may be resolved, spontaneously, in a few days or up to 6 weeks.[23] Management depends on the mechanism that is responsible for the failure. Therefore, the initial step involves the differentiation between a prerenal oliguric state versus an intrinsic renal state. This is important because the treatment for a hypovolemic — reduced blood volume — prerenal state consists of fluid and intravascular volume replacement; such treatment can lead to a dangerous fluid overload in a patient with intrinsic renal disease such as acute tubular necrosis. Examination of the urine for the fractional excretion of sodium is the safest and most reliable method for making this differentiation.

Fractional Sodium Excretion

The fractional excretion of sodium measures the kidney capacity to conserve sodium and excrete creatinine. Fractional excretion of sodium (FE_{Na}) is determined according to the following formula:

$$FE_{Na} = \frac{U_{Na} \div P_{Na}}{U_{cr} \div P_{cr}} \times 100 \quad \begin{cases} < 1 \rightarrow \text{prerenal} \\ > 1 \rightarrow \text{intrinsic renal} \end{cases}$$

TABLE 19-3

Signs of Renal Failure

Signs of Renal Failure
Generalized edema
Acidosis
Increased circulating nonprotein nitrogens (urea)
Increased circulating urinary retention products

TABLE 19-4

Selected Causes of Acute Renal Failure

Selected Causes of Acute Renal Failure
Prerenal
Loss of body fluids
Inadequate fluid intake
Surgical shock or myocardial infarction
Renal
Drug toxicity
Immunologic reactions
Infectious diseases
Thrombosis
Postrenal
Intravascular hemolysis
Urinary tract obstruction

The urine sodium concentration is divided by the plasma sodium concentration and this is then divided by the urine creatinine concentration over the plasma creatinine concentration. When this value, multiplied by 100, is less than 1, then prerenal azotemia — high blood nitrogen — is responsible for the renal failure. When the fractional excretion of sodium is greater than 1, this indicates intrinsic renal failure.

Another method — simpler but more dangerous, especially for the elderly — is to administer a fluid challenge of either isotonic saline or mannitol together with a potent diuretic such as furosimide. A response to these maneuvers may indicate that one is dealing with a prerenal azotemia.

Management of acute renal failure includes

- Treatment of the underlying cause; for example, the discontinuation of nephrotoxins, should any be implicated
- Careful fluid monitoring to avoid fluid overload and congestive failure
- Electrolyte monitoring, such as hyperkalemia and acidosis
- Prevention of infection, as this is the major cause of mortality in patients with acute tubular necrosis
- The alteration of the diet to limit sources of nitrogen, potassium, phosphorus, and sulfate

This involves limiting protein to no more than 40 g/day and supplying sufficient calories (at least 3000) to prevent endogenous protein catabolism.

Dialysis Treatment and Kidney Transplantation in the Elderly

The question often arises as to what extent "heroic interventions" are justifiable and advisable in the elderly. In the case of renal failure, dialysis and kidney transplantation represent interventive measures that are widely utilized with considerable success in the young and adult. Evidence for their rational use with a favorable outcome in the elderly has been shown from the relatively few cases treated so far and from animal experimentation. Clearly, while all contraindications and immediate and long-term risks of these measures must be taken carefully into consideration, as they may be magnified in the elderly, age alone should not deter from their appropriate use.[24] Dialysis is indicated on an emergency basis in the following situations:

- Blood volume overload leading to congestive heart failure
- Hyperkalemia, i.e., high blood potassium, which is refractory to medical management
- Severe acidosis
- Neurologic abnormalities resulting from a rapid increase in azotemia
- Severe hyponatremia, i.e., low sodium blood levels

Dialysis should not be withheld from an elderly patient in acute renal failure merely on the basis of his/her age. It is not usually easy to accurately determine the patient's prognosis at the onset of the acute illness and, therefore, major decisions in management, such as the use of dialysis, should be initially made with the assumption that a positive outcome will occur. Thus, aggressive dialysis in acute renal failure should be the rule rather than the exception.

Indication and success of transplantation of organs such as the kidney depend on meeting several criteria such as normal function and competency of the organ to be transplanted and the age (young) and health (good) of the donor. With respect to age, although the use of kidneys from older donors is controversial, a number of data suggest that age alone should not eliminate using older kidney donors when their renal function and tissue matches are good.[25] Pregnancy puts a strain on kidney function, but a report of a successful pregnancy in a renal transplant recipient with a kidney from a 75-year-old donor supports the view that an old kidney may function normally.[26] Systematic studies in young adult rats — bilaterally nephrectomized and receiving kidneys transplanted from animals of progressive ages, until quite advanced age — show a good recipient survival rate, but old kidneys functioned less well than young on the average.[27] It is possible that the kidneys from old donors are more susceptible to the hypoxia induced by surgery or age-related renal metabolic changes. Reciprocally, kidney transplants in older individuals (with young kidney donors) seem to fare as well as in the young once the surgical and pharmacologic measures appropriate to the age of the individuals are taken into account (Chapters 3 and 5).

1.6 Kidney Susceptibility to Drugs

The kidneys are particularly susceptible to the toxic effects of drugs and other chemical agents because

1. Blood flow to the kidney is high: 20% of cardiac output.
2. Drugs tend to accumulate in the renal medulla as water is removed from the glomerular filtrate.
3. Drug accumulation increases when renal function is impaired.
4. Reduced hepatic enzyme activity in the elderly increases circulating drug levels and renal toxicity.
5. Incidence of autoimmune disorders increases with aging with consequent hypersensitivity reactions in the kidney (Table 19-5).

The manifestations of renal drug intoxication are not unique to the aged kidney; they are, however, more frequent and, when they occur, they may be more severe than in the adult. In the aged kidney, toxic effects are seen with lower doses than in the adult, and the consequences are more dangerous, taking into account the multiplicity of drugs taken by the elderly, the generally long duration of treatment, and the often impaired conditions of the kidney (Table 19-6).

Examples of drugs and the mechanisms whereby they induce renal damage include

- Dehydration-induced uremia due to use of diuretics and laxatives
- Obstruction of urinary tract due to deposition of crystalline matters such as calcium from excessive administration of vitamin D
- Vascular lesions produced by thiazide diuretics
- Glomerular damage by penicillin-like antibiotics
- Interstitial and tubular damage from radiological contrast media
- Papillary necrosis due to analgesics

2 AGING-RELATED CHANGES IN THE FUNCTION OF THE LOWER URINARY TRACT

As already stated in Chapter 2, it has been written that the major health problems of the elderly can be easily recalled by listing them under five words: Instability, Immobility, Incontinence, Impaired cognition, and Iatrogenic diseases, all starting with the letter "I".[28] Of these, one of the most embarrassing and distressing is urinary incontinence. Failure of urinary continence often results from alterations in the function of some of the structures of the lower urinary tract, and, in fact, it may sometimes involve all of these structures; it will, therefore, represent one of the major topics to be discussed in this section.

Recent surveys show that incontinence occurs in 10 to 30% of community-dwelling elderly and in 50 to 60% of those living in institutions.[29] However, statistics are not accurate inasmuch as people often fail to disclose this condition.

Many consider it inevitable and many refuse to admit to it. Normal control of bladder and urethral functions are taken for granted by the majority of individuals and, in fact, remain efficient in many individuals well into old age. However, when it occurs, failure of this control is often considered a main threat to the welfare of those affected; it conjures up fears of rejection which are often real and further restrict the social interactions of the elderly. Urinary incontinence may not only result from aging of the urinary tract but also may reflect disorders of locomotion and of cognitive function and, as such, may be viewed as an indicator of overall decline of physiologic competence.

Structure and Function of the Lower Urinary Tract

The structures involved in transfer, storage, and excretion of urine are the ureter, the bladder, and the urethra (Figure 19-5). These retroperitoneal and pelvic organs are similar in both sexes except for the urethra and associated structures, such as the prostate, which are sex-specific.

All these structures consist essentially of smooth muscle lined by mucosa. The largest is the bladder which is a smooth muscle chamber comprised of a body, formed of the detrusor muscle with a weblike structure, a neck, and a trigone near the neck through which the ureters and urethra pass. The detrusor muscle, by contraction, is mainly responsible for emptying the bladder. Release of urine from the bladder is controlled by an internal and external sphincter. The internal sphincter responds to pressure built up in the bladder by stimu-

TABLE 19-5

Etiopathology of Renal Drug Toxicity

High renal blood flow
Increased drug concentration and accumulation in kidney
Increased hepatic enzyme inhibition in the elderly
Increased autoimmune disorders in the elderly

TABLE 19-6

Some Examples Of Drug Toxicity in Elderly Individuals with Diminished Renal Function

Drugs	Response
Aminoglycosides	Deafness and kidney toxicity
Nitrofurantoin	Neuropathy
Oral antidiabetics	Hypoglycemia
Digoxin	Vomiting and irregular heartbeat
Nonsteroidal anti-inflammatory agents	Renal failure
Anticholinergics	Confusional states

lation of stretch receptors; the external sphincter, composed of striated muscle, is under voluntary control (Figure 19-6). The main function of these structures is micturition, that is, the process by which the bladder empties when it becomes filled with urine. Basically, micturition may be considered a special reflex, stimulated or inhibited by higher brain centers. Like defecation, it is subject to voluntary facilitation or inhibition. The events of micturition consist of progressive filling of the bladder with urine from the kidneys through the ureters. Once the bladder is filled, there is a tension threshold which initiates a reflex to micturate or, at least, a conscious desire to urinate.

The innervation of the bladder consists of three components, the two branches of the autonomic nervous system and the somatic nerves which align with the type of muscle, smooth or striated. The autonomic fibers, both sympathetic and parasympathetic, act on smooth muscles (detrusor, trigone, and internal sphincter), and the somatic fibers travel to the skeletal (striated) muscle (external sphincter). The act of micturition involves a complex coordination of neural and muscular responses leading to (1) relaxation of the internal sphincters under parasympathetic control and of the external sphincter, under somatic control, (2) constriction of the ureteral sphincters under sympathetic control, and (3) contraction of the detrusor muscle, under parasympathetic control. The entire sequence is under voluntary control and can be initiated or inhibited at will. In the absence of inputs from the cerebral cortex and other high brain centers, spinal reflexes may take over control and induce automatic emptying of the bladder whenever the retained volume of urine reaches a critical level. When the voluntary system is active, the desire to urinate becomes apparent after approximately 150 to 300 ml of fluid have collected.

2.1 Urinary Continence

In view of the complexity of the mechanisms regulating micturition, the high prevalence of incontinence in the elderly should not be surprising.[30] Continence depends on a long list of physiologic requirements, several of which undergo changes with aging (Table 19-7). These include motivation to be continent, adequate cognitive function to understand the need for continence, and mobility and dexterity to find and utilize the appropriate sanitary facilities. Normal urinary tract function, another necessary requirement for continence, depends on appropriate storage and emptying of the bladder; urine storage in the bladder depends, in turn, on its relaxation (absence of involuntary contractions, accommodation of increased urine volume under low pressure, closure of sphincters, etc.) as well as its contraction (normal contractile mechanisms, absence of anatomic obstruction, sphincter relaxation, etc.) Physical and environmental barriers (e.g., bed rails, distant toilet location) must also be considered as possible causes of incontinence. Finally, the administration of drugs for a variety of ills and their effects, particularly on cognitive functions, may also contribute to the incontinence.

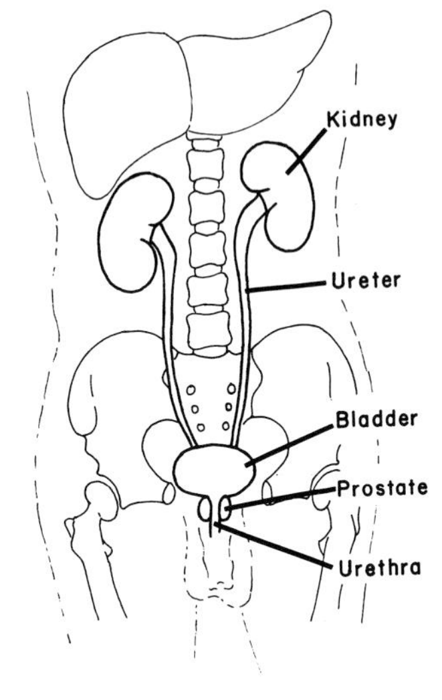

FIGURE 19-5 *Diagram of the major components of the urinary tract in man.* (Drawing by S. Oklund.)

2.2 Causes and Course of Urinary Incontinence in the Elderly

Urinary incontinence can be defined as the involuntary loss of urine sufficient in amount and frequency to be a social/health problem. It is most frequent in the elderly; it is stressful not only for those affected but also for their family and caregivers and can play a decisive role in the decision to place an elderly subject in a nursing home.[31-33]

Although urinary incontinence is not an inevitable consequence of aging, certain aging-related changes may contribute to the etiology of incontinence (Table 19-8). For example, in postmenopausal women, the reduction in estrogen levels weakens the tissues around the pelvic floor and bladder outlet, decreases the tone of urethral smooth muscle, and causes atrophic vaginitis, all of which contribute to the decrease in urethral contractility and pressure with aging. In elderly men, hypertrophy of the prostate represents one of the major causes of incontinence as it leads to a decline in urinary out-flow rate, increased risk of urine retention, and increased instability of the detrusor muscle. Other causes include delirium, effects of drugs such as anticholinergics (inducing urinary retention) and diuretics (inducing polyuria), and diseases such as infections and diabetes. A mnemonic device to remember the major causes of acute incontinence is the word DRIP, where D stands for delirium, R for urinary retention due to anticholinergic drugs and restricted motility, I for infection, and P for polyuria due to diuretics or diabetes (Figure 19-7).

Many of the causes of acute incontinence may be easily diagnosed and reversed. However, when they have been addressed without any significant improvement, the incontinence is considered persistent and its management is considerably more difficult than the acute form. Persistent urinary incontinence is categorized into four types to which correspond specific etiologies and interventions:

- Stress incontinence, due to increased intra-abdominal pressure as in coughing and laughing, is caused by weakness of the pelvic floor musculature and is common in multiparous women.
- Urge incontinence, due to inability to delay voiding after perception of bladder fullness, may be caused by mild

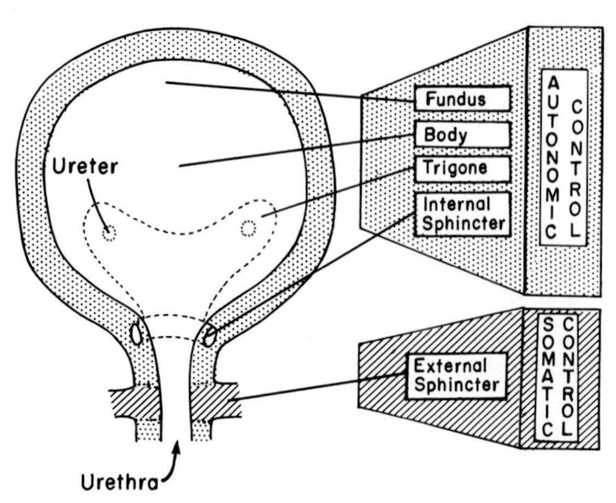

FIGURE 19-6 *Schematic of urinary bladder with structures under autonomic and somatic nervous control.* (Drawing by S. Oklund.)

out-flow obstruction by a central nervous system disorder such as stroke.

- Overflow incontinence, due to a leakage of urine resulting from mechanical forces of an overdistended acontractile bladder, may be due to prostatic obstruction or to neurogenic disturbances such as a neurogenic acontractile bladder
- Functional incontinence, due to inability or psychologic unwillingness to get to the toilet, is caused by cognitive deficits, psychologic conditions, or unavailability of caretakers.

Differential Diagnosis of Incontinence Types

It is based not only on the urinary symptoms but also on a complete history and physical examination including evaluation of the mental status, physical activity, and ambulation. An abdominal exam should look for signs of bladder distension; a rectal exam for prostate size, rectal sphincter tone, and the presence of fecal impaction; a pelvic exam for uterine prolapse or atrophic vaginitis. Laboratory tests should include a urinalysis, a urine culture, and postvoid residual. This is a simple test to determine how much urine remains in the bladder after voiding. A catheter is placed in the bladder after the subject has voided and the volume of urine is measured. If the volume of urine is greater than 100 ml, then a more complete urologic or gynecologic evaluation is warranted.

2.3 Management of Incontinence

Management varies with the type of incontinence (Table 19-9).[34,35] Strengthening of abdominal and pelvic muscles by exercise (Kegel's exercises),[36,37] administration of estrogen, and surgical bladder neck suspension are needed for stress incontinence; administration of bladder relaxants, for urge incontinence; intermittent catheterization or removal of the prostate for overflow incontinence; and for functional incontinence, habit training, scheduled toileting, undergarment devices, and indwelling catheterization (taking into account the risks involved, such as danger of infection) are needed.

As indicated above, urinary incontinence is a widespread problem among the elderly. And yet, despite its prevalence and its serious consequences for those affected, it receives little to no attention because of social taboos and ignorance of its physiologic and pathologic causes. Few of the health care providers are aware and knowledgeable of the problem. The assumption that once urinary incontinence occurs in an elderly individual it is to be accepted as an inevitable consequence of aging is false.

TABLE 19-7

Physiologic Requirements for Continence

Motivation to be continent
Adequate cognitive function
Adequate mobility and dexterity
Normal lower urinary tract function
  Storage
    No involuntary bladder contractions
    Appropriate bladder sensation
    Closed bladder outlet
    Low pressure accommodation of urine
  Emptying
    Normal bladder contraction
    Lack of anatomic obstruction
    Coordinated sphincter relaxation and bladder contraction
Absence of environmental or iatrogenic barriers

TABLE 19-8

Age-Related Changes Contributing to Incontinence

In Females
Estrogen deficiency
  Weak pelvic floor and bladder outlet
  Decreased urethral muscle tone
  Atrophic vaginitis
In Males
Increased prostate size
  Impaired urinary flow
  Urinary retention
  Detrusor instability

Urinary incontinence in the elderly can and must be treated in each and every case. Specific questioning as to urinary continence must be carried out in all routine evaluations of elderly patients, and, in this, as in all disorders, an understanding of the physiopathology is essential for a rational diagnostic evaluation and treatment.

2.4 The Lower Urinary Tract and the Excretion and Action of Drugs

Dysfunction of the lower urinary tract amenable to pharmacologic interventions is rare, except for infections. Urinary retention, secondary to surgery or spinal cord injury, may also be relieved pharmacologically, but this and infections, although quite frequent in the elderly, are not exclusive of this age group. Infections are treated with antimicrobial agents and urinary retention with cholinergic agonists. The characteristics of neural control of micturition are presented in Table 19-10. Both parasympathetic and sympathetic impulses control bladder emptying, with cholinergic fibers inducing contraction of the detrusor muscle and relaxation of the internal sphincter muscle and thereby promoting emptying, and adrenergic fibers having the opposite effects. Somatic nerves also influence micturition by regulating the contraction/relaxation of the external sphincter. Cholinergic agonists (such as the drugs bethanecol and oxybutinin) are used to facilitate bladder emptying. While these drugs may act on other parasympathetic targets as well and, therefore, induce side effects, they are preferred to bladder catheterization with its attendant danger of infection.

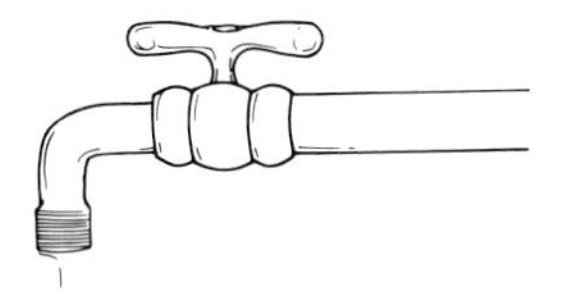

- D elirium
- R etention, Restricted mobility
- I nfection, Inflammation, Impaction
- P olyuria

FIGURE 19-7 *Mnemonic device for the major causes of acute urinary incontinence.*

In addition to being the target of specific drugs administered for therapeutic effects, the lower urinary tract, particularly the bladder, may bear the consequences or side effects of many prescribed, as well as over-the-counter, drugs. Thus, the bladder may become a victim of polypharmacy. Drugs in this category include decongestants, antihistamines, antidiarrheals, antipsychotics (such as phenothiazines), and tricyclic antidepressants. Even if taken as prescribed, their usage may add up to increasing toxicity. Most of these drugs have some autonomic activity. The greater number have anticholinergic actions. They block the parasympathetic responses; for example, over-the-counter sleeping, asthma, and antidiarrheal medicines contain the belladonna alkaloids such as atropine or scopolamine or synthetic substitutes. They block the parasympathetic responses, inhibit micturition, and induce a degree of urinary retention (Table 19-11).

Acting on the other branch of the autonomic nervous system are the sympathetic or adrenergic drugs. Sympathomimetics, such as some decongestants and α-adrenergic blockers, relax the detrusor muscle, constrict the sphincters, and promote urine retention. Some of the α-adrenergic blockers, extensively utilized as antihypertensive drugs, have also some additional side effects, distressing in the older male for they also prevent contraction of the vas deferens and inhibit ejaculation (Table 19-12).

3 AGING OF THE PROSTATE

3.1 Structure and Functions

The prostate is a secretory organ that is part of the male reproductive system. Its main function is the production and secretion of a specific fluid which adds to sperm and the secretory products of other structures of the male reproductive tract to form the semen. Inasmuch as the prostrate surrounds the urethra, and its enlargement produces alterations of the lower urinary tract, it is often described with the urinary system. Thus, the changes considered here are those affecting the function of the lower urinary tract, particularly the bladder — the neck of which is encircled by the prostate — and the urethra (Figure 19-5).

In the adult, the prostate, weighing about 20 g, consists of three to five lobes, which have distinct significance in the development of benign enlargement or tumors. Histologically, it is a tubulo–alveolar gland with epithelial secretory cells distributed along a basement membrane and separated by an abundant fibromuscular stroma. The secretory product of the gland contributes not only fluid to the semen but also several enzymes and constituents that stimulate the motility and viability of sperm.

Aging changes can be detected after the age of 40 years. Between 40 and 60 years, a series of alterations have been described that are classified as presenile. They consist essentially of atrophy of smooth muscle and proliferation of fibrous tissue and are associated with a flattening of the secretory epithelium. However, these changes are not uniform nor distributed throughout the gland; in fact, these presenile atrophic changes appear prominent in the outer regions while the inner mass undergoes hyperplasia. After the age of 60, the entire prostate undergoes a slower but more uniform atrophy with the accumulation of small concretions that represent residual secretory products.

While with aging, there is an overall atrophy of the prostate, one third or more of males over the age of 60 have a so-called benign prostatic hyperplasia (also referred to as hypertrophy), characterized by a nodular enlargement of prostatic tissue, particularly the inner portion of the gland and leading to significant obstruction of the urethra and urine outflow from the bladder.[38,39] The etiology of this enlargement is still obscure but has been ascribed, in addition to hormonal factors, to heredity, infection, atherosclerosis, sexual activity, etc.[40] The incidence of the hypertrophy is extremely low in some Oriental races, reflecting, perhaps, an hereditary factor. The presence of functioning testes is necessary for the development of hyperplasia, as no cases have been found in eunuchs.

The prostatic enlargement is most frequently adjacent to the neck of the bladder and the urethra, where it obstructs urine outflow. This obstruction results in profound changes in the bladder, ureters, and kidneys. Micturition requires increased pressure of the detrusor muscle, producing hypertrophy and thickening of the bladder wall.

As the prostate continues to enlarge, the bladder finally reaches the limits of compensatory hypertrophy, after which

TABLE 19-9

Management of Urinary Incontinence

Type	Management
Stress	Exercises
	α-Adrenergic agonists
	Estrogen
	Surgery
Urge	Bladder relaxants
	Surgery
Overflow	Surgery
	Catheterization
Functional	Habit training
	Scheduled toileting
	Hygienic devices

TABLE 19-10

Neural Control of Micturition

Muscle (type)	Parasympathetic nerves (cholinergic)	Sympathetic nerves (adrenergic)	Somatic nerves
Detrusor (smooth)	Contraction	Relaxation	No effect
Trigone			
Internal sphincter (smooth)	Relaxation	Contraction	No effect
External sphincter (striated)	No effect	No effect	Relaxation

TABLE 19-11

Drugs Secondarily Affecting the Lower Urinary Tract

Autonomic action	Drug type
Anticholinergic	Antidiarrheals
	Sedatives
	Antiasthmatics
	Antihistamines
	Antipsychotics (phenothiazines)
	Tricyclic antidepressants
	Belladonna
	Alkaloids
	(Over-the-counter)
Adrenergic	Decongestants
Adrenergic block	Antihypertensives

there is a gradual decompensation and, eventually, complete failure to contract and to void with urinary retention.

The most serious consequence of prostatic enlargement is progression of obstruction to the point of retrograde filling of the ureters and renal pelvis with eventual hydronephrosis and infection (pyelonephritis). If the obstruction is not relieved, renal failure with uremia and death follow (Figure 19-8).

3.2 Diagnosis and Management of Prostatic Enlargement

Benign prostatic enlargement can be detected by palpation on rectal examination and by urinary signs including hesitancy and straining to initiate micturition, and reduced force and caliber of urinary stream. In most serious cases, the above signs are associated with hematuria, hydronephrosis, and pyelonephritis with attending signs and symptoms. In the absence of appropriate treatment, there is anuria and uremia. All of these signs and symptoms may be related to failure of emptying the bladder. They can be relieved by the administration of α-adrenoreceptor antagonists[41] and catheterization of the bladder. Treatment of prostate disease includes not only surgical removal, irradiation, and chemotherapy, but alteration of the sex hormone environment as well (Table 19-12). While surgical resection of the prostate has been the traditional approach, alternative modalities are now gaining popularity. This is because transurethral surgery for prostatic hypertrophy carries significant morbidity in patients over 80 years of age.[42] In addition to postoperative complications of up to 27% in this population, 20% of men develop symptoms requiring a repeat operation within 8 years.[43] Newer surgical techniques, such as insertion of urethral stents, may decrease morbidity; however, the long-term effects are not yet known.[44] Another new modality is the use of finasteride, a 5-α-reductase inhibitor, that prevents the conversion of testosterone to its tissue active form, dihydrotestosterone.[45] As with all new drugs, its long-term efficacy and safety have yet to be determined.

Hormonal Involvement in Prostatic Enlargement and Prostate Cancer

The prostate is a target for androgens, the male sex hormones; as such, it contains hormone receptors, particularly for the testosterone metabolite, dihydrotestosterone, the most functionally active androgen. The testes secrete primarily androgens. In addition, the testes secrete small amounts of estrogens which, contrary to the ovary, increase with aging.

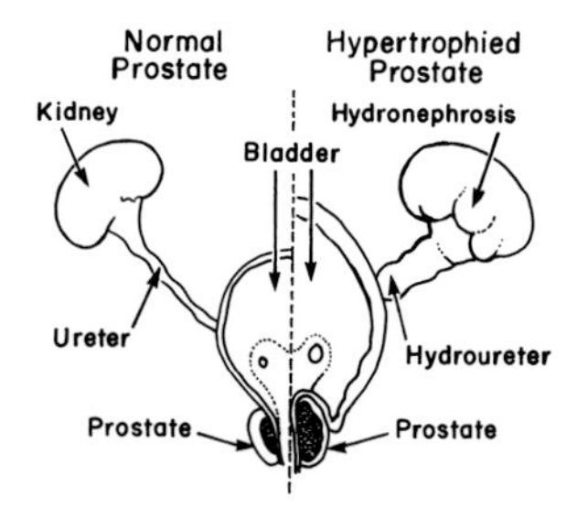

FIGURE 19-8 *Consequences of prostate hypertrophy.* The enlarged prostate constricts the urethra with consequent difficulties in bladder emptying and the occurrence of retrograde filling of the ureters and of the renal pelvis (hydronephrosis). (Drawing by S. Oklund.)

It is on this endocrinologic basis that prostatic carcinoma is treated either by decreasing the levels of androgens by castration or by administration of antiandrogens GnRH agonists, or estrogens. Such a hormonal treatment, while not particularly effective in benign prostatic hyperplasia, controls prostatic cancer in 70 to 80% of cases.

Prostate cancer is now the most frequently diagnosed cancer in males, and is the second leading cause of mortality due to cancer in the U.S.[46] The challenge of managing this disease is in differentiating between latent prostate cancer, which will not lead to clinical illness, versus the aggressive and eventual terminal disease, which it has the potential to become. While the prevalence of histologic prostate cancer is high (up to 80% of octogenarians),[47] its clinical manifestations are found in only a fraction of these individuals. A diagnosis at an earlier age tends to coincide with a poorer prognosis. The issue of screening for this disease is one of the more controversial in urology. This has in part been brought on by the advent of a new screening test, the prostate-specific antigen (PSA), which many primary care providers are obtaining on a routine basis. PSA levels between 4 ng/ml and 10 ng/ml in one study are associated with a 22% risk of prostate cancer and serum values over 10 ng/ml with a 67% risk.[48] The American Cancer Society recommends an annual rectal examination for all men over age 40. As with all cancers, treatment is tailored to the condition of the patient and the stage of the disease.[49] The current trend to treat this disease conservatively in frail elderly patients with hormonal therapy represents a striking example of how changes in the hormonal environment may intervene in the course of malignant changes in a hormone target organ.

4 WATER AND ELECTROLYTE DISTRIBUTION AND ACID–BASE BALANCE

The kidney of the normal, healthy, older individual is well capable of maintaining water and electrolyte distribution and acid–base balance within homeostatic limits. This is remarkable as some changes do occur in body composition. With aging, there is generally a loss of fat-free body weight and tissue mass, a reduction in body mineral, and a gain in body fat.[50-52] These age differences carry over to sex differences, with women showing a greater increase in total body weight due to increased body fat and men maintaining their body weight. This maintained body weight is due to reciprocal changes in lean body mass (decreased) and body fat (increased) (Figure 19-9). Some gross comparisons in "reference men" between 25 and 70 years show the following changes, at age 25: fat, 14%, water, 61%, cell solids, 19%, and bone mineral 6%; and at age 70: 30%, 53%, 12%, and 5%, respectively.

These changes in body composition may vary based on several factors related either to the individual or the methodology. For example, gain in body fat often increases into the 60s but declines thereafter and, at all ages, depends on a number of variables among which the degree of physical activity is most consistent. In contrast, lean body mass or body cell mass (as measured by body density through water

TABLE 19-12

Therapeutic Interventions for Prostatic Enlargement

Benign prostatic hyperplasia
- Bladder catheterization
- Surgical resection
- Finasteride
- Urethral stent

Prostatic carcinoma
- Castration
- Estrogen administration
- Antiandrogenic progestins
- GnRH agonists — leuprolide
- Surgery, irradiation

displacement or helium dilution) continues to decrease with aging, more in males than in females.[51] Some studies show a 3.6% decrease per decade from age 30 to 70 and, thereafter, 9% per decade. If nonfat mass is measured by total body potassium, one observes that potassium decreases with aging as does lean body mass early, but, with advancing age, after 70 years, potassium decrease is more pronounced than lean body mass.[53] It has been suggested that at older ages the degenerative loss of lean tissue is in part replaced by other tissue that is low in potassium[54] and has a lower metabolic rate.[50] An alternative explanation with respect to regression of metabolic rate with aging is that the oxygen uptake of cells in old individuals is not significantly different than in young; rather, there are fewer functioning cells.[55]

For example, while total body water is diminished with age, extracellular water remains unchanged and intracellular water decreases. The reduction in total body and intracellular water in the absence of change in extracellular water can be taken as further support for a loss of functioning cells (lean body mass) with increasing age (Chapter 21).

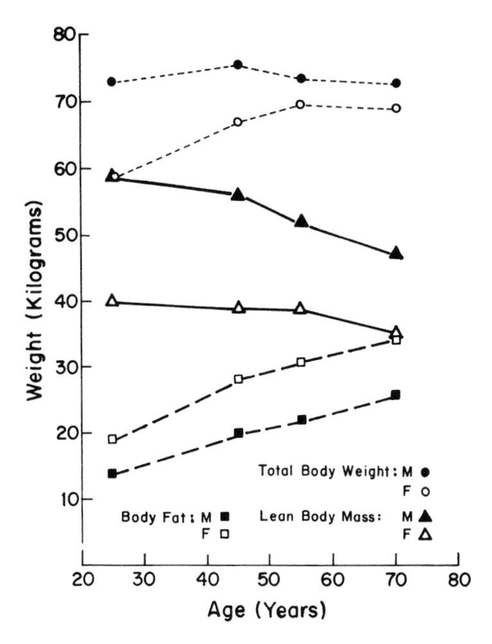

FIGURE 19-9 *Age-related weight changes in body composition of males and females.* Males maintain overall body weight despite a decrease in lean body mass and an increase in body fat. Females not only have an increase in body weight, but this is mostly due to the increase of body fat.

REFERENCES

1. Shock, N. W., Kidney function tests in aged males, *Geriatrics,* 1, 232, 1946.
2. Davies, D. F. and Shock, N. W., Age changes in glomerular filtration rate, effective renal plasma flow, and tubular excretory capacity in adult males, *J. Clin. Invest.,* 29, 496, 1950.
3. Rowe, J. W., Andres, R., Tobin, J. D., Norris, A. H., and Shock, N. W., The effect of age on creatinine clearance in men: a cross-sectional and longitudinal study, *J. Gerontol.,* 31, 155, 1976.
4. Lindeman, R. D., Tobin, J., and Shock, N. W., Longitudinal studies on the rate of decline in renal function with age, *J. Am. Geriatr. Soc.,* 33, 278, 1985.
5. Durakovic, Z. and Mimica, M., Proteinuria in the elderly, *Gerontology,* 29, 121, 1983.
6. Bolton, W. K. and Sturgill, B. C., Ultrastructure of the aging kidney, in *Aging and Cell Structure,* Vol. 1, Johnson, J. E., Jr., Ed., Plenum Press, New York, 1981.
7. Smalley, J. W., Age related changes in the amino acid composition of human glomerular basement membrane, *Exp. Gerontol.,* 15, 43, 1980.
8. Brod, J., Changes of renal function with age, *Scripta Medica,* 223, 1968.
9. McDonald, R. K, Solomon, D. H., and Shock, N. W., Ageing as factor in renal hemodynamic changes induced by standardized pyrogen, *J. Clin. Invest.,* 30, 457, 1951.
10. Kirkland, J. L., Lye, M., Levy, D. W., and Banerjee, A. K., Patterns of urine flow and electrolyte excretion in healthy elderly people, *Br. Med. J.,* 287, 1665, 1983.
11. Pashko, L. L., Fairman, D. K., and Schwartz A. G., Inhibition of proteinuria development in aging Sprague-Dawley rats and C57BL/6 mice by long-term treatment with dehydroepiandrosterone, *J. Gerontol.,* 41, 433, 1986.
12. Everitt, A. V., Porter, B. D., and Wyndham, J. R., Effects of caloric intake and dietary composition on the development of proteinuria, age-associated renal disease longevity in the male rat, *Gerontology,* 28, 168, 1982.
13. Brenner, B. M., Meyer, T. W., and Hostetter, T. H., Dietary protein intake and the progressive nature of kidney disease: the role of hemodynamically mediated glomerular injury in the pathogenesis of progressive glomerular sclerosis in aging, renal ablation and intrinsic renal disease, *N. Engl. J. Med.,* 307, 652, 1982.
14. Phillips, T. L., Rolls, B. J., Ledingham, J. G. G., et al., Reduced thirst after water deprivation in healthy elderly men, *N. Engl. J. Med.,* 311, 753, 1984.
15. Helderman, J. H., Vestal, R. E., Rowe, J. W., Tobin, J. D., Andres, R., and Robertson, G. L., The response of arginine vasopressin to intravenous ethanol and hypertonic saline in man: the impact of aging, *J. Gerontol.,* 33, 39, 1978.
16. Miller, J. H. and Shock, N. W., Age differences in renal tubular response to antidiuretic hormone, *J. Gerontol.,* 8, 446, 1953.
17. Lindeman, R. D., Lee, T. D., Yiengst, M. J., and Shock, N. W., Influence of age, renal diseases, hypertension, diuretics and calcium on antidiuretic responses to suboptimal infusion of vasopressin, *J. Lab. Clin. Med.,* 68, 206, 1966.
18. Bengele, H. H., Mathias, R. S., Perkins, J. H., and Alexander, E. A., Urinary concentrating defect in the aged rat, *Am. J. Physiol.,* 240, 147, 1981.
19. Nunez, M., Iglesias, C. G., Roman, A. B., Commes, J. L. K., Becerra, L. C., Rome, J. M. T., and Del Pozo, S. D., Renal handling of sodium in old people. A functional study, *Age Ageing,* 7, 178, 1978.

20. Cox, J. R. and Shalaby, W. A., Renal disease, in *Principles and Practice of Geriatric Medicine,* Pathy, M. S. J., Ed., John Wiley & Sons, Chichester, England, 1985.
21. Frocht, H. and Fillit, H., Renal disease in the geriatric patient, *J. Am. Geriatr. Soc.,* 32, 28, 1984.
22. Bennett, W. M., Luft, F., and Porter, G. A., Pathogenesis of renal failure due to aminoglycosides and contrast media used in roentgenography, *Am. J. Med.,* 69, 767, 1980.
23. Oliveria, D. B. G. and Winearls, C. G., Acute renal failure in the elderly can have a good prognosis, *Age Ageing,* 13, 304, 1984.
24. Nicholls, A. J., Waldek, S., Platts, M. M., Moorhead, P. J., and Brown, C. B., Impact of continuous ambulatory peritoneal dialysis on treatment of renal failure in patients aged over 60, *Br. Med. J.,* 13, 304, 1984.
25. Matus, A. J., Simmons, R. L., Kjellstrand, C. M., Buselmeier, T. J., and Najarian, J. S., Transplantation of the aging kidney, *Transplantation,* 21, 160, 1976.
26. Coulam, C. B. and Zincke, H., Successful pregnancy in a renal transplant patient with a 75-year-old kidney, *Surg. Forum,* 32, 457, 1981.
27. Van Bezooijen, K. F. A., deLeeuw-Israel, F. R., and Hollander, C. F., Long-term functional aspects of syngeneic orthotopic rat kidney grafts of different ages, *J. Gerontol.,* 29, 11, 1974.
28. Feigenbaum, L., Geriatric medicine and the elderly patient, in *Current Medical Diagnosis and Treatment,* Schroeder, S. A. and Krupp, M. A., Eds., Lange, Los Altos, CA, 1991, 21.
29. Herzog, A. R. and Fultz, N. H., Prevalence and incidence of urinary incontinence in community dwelling populations, *J. Am. Geriatr. Soc.,* 38, 273, 1990.
30. Williams, M. E. and Pannill, F. C., Urinary incontinence in the elderly, *Ann. Intern Med.,* 97, 895, 1982.
31. Williams, M. E., A critical evaluation of the assessment technology for urinary continence in older persons, *J. Am. Geriatr. Soc.,* 31, 657, 1983.
32. Freed, S. Z., Urinary incontinence in the elderly, *Hosp. Pract.,* 17, 81, 1982.
33. Brocklehurst, J. C., Incontinence, in *Geriatric Medicine Annual,* Ham, R. J., Ed., Medical Economics Books, Oradell, NJ, 1982.
34. Resnick, N. M. and, Yalla, S. V., Management of urinary incontinence in the elderly, *N. Engl. J. Med.,* 313, 800, 1985.
35. Wein, A. J., Pharmacologic treatment of incontinence, *J. Am. Geriatr. Soc.,* 38, 317, 1990.
36. Kegel, A. H., Progressive resistance exercise in the functional restoration of the perineal muscles, *Am. J. Obstet. Gynecol.,* 56, 238, 1948.
37. Kegel, A. H. and Powell, T. O., The physiologic treatment of urinary stress incontinence, *J. Urol.,* 63, 808, 1950.
38. Glynn, R. J., Campion, E. W, Bouchard, G. R., and Silbert, J. E., The development of benign prostatic hyperplasia among volunteers in the normative aging study, *Am. J. Epidemiol.,* 121, 78, 1985.
39. Hieble, J. P. and Caine, M., Etiology of benign prostatic hyperplasia and approaches to its pharmacological management, *Fed. Proc.,* 45, 2601, 1985.
40. Mawhinney, M. G., Etiological considerations for the growth of stroma in benign prostatic hyperplasia, *Fed. Proc.,* 45, 2615, 1985.
41. Caine, M., Clinical experience with adrenoceptor antagonists in benign prostatic hypertrophy, *Fed. Proc.,* 45, 2604, 1985.
42. Mebust, W. K., Holtgrewe, H. L., Cockett, A. T. K., and Peters, P. C., Transurethral prostatectomy: immediate and post-operative complications: a cooperative study of thirteen participating institutions evaluating 3885 patients, *J. Urol.,* 141, 243, 1989.
43. Roos, N. P., Wennberg, J. E., Malenka, D. J., et al., Mortality and re-operation after open and transurethral resection of the prostate for benign prostatic hyperplasia, *N. Engl. J. Med.,* 320, 1120, 1989.
44. DuBeau, T. E. and Resnick, N. M., Controversies in the diagnosis and management of benign prostatic hypertrophy, *Adv. Intern. Med.,* 37, 55, 1991.
45. Gormley, G. J., Stoner, E., Bruskewitz, R. C., et al., The effect of finasteride in men with benign prostatic hyperplasia, *N. Engl. J. Med.,* 327, 1185, 1992.
46. Boring, C. C., Squires, T. S., and Tong, T., Cancer statistics 1991, *CA Cancer J. Clin.,* 41, 6, 1991.
47. Moon, T. D., The aging prostate: prostate cancer, in *Endocrinology and Metabolism in the Elderly,* Morley, J. E. and Korenman, S. G., Eds., Blackwell Scientific, Cambridge, MA, 1992, 273.
48. Catalona, W. J., Smith, D. S., Ratliff, T. L., et al., Measurement of prostate-specific antigen in serum as a screening test for prostate cancer, *N. Engl. J. Med.,* 324, 1156, 1991.
49. Moon, T. D., Prostate cancer, *J. Am. Geriatr. Soc.,* 40, 622, 1992.
50. Allen, T. H., Anderson, E. C., and Langham, W. H., Total body potassium and gross body composition in relation to age, *J. Gerontol.,* 15, 348, 1960.
51. Forbes, G. B. and Reina, J. C., Adult lean body mass declines with age: some longitudinal observations, *Metabolism,* 19, 653, 1970.
52. Steen, B., Bruce, A., Isaksson, B., Lewin, T., and Svanborg, A., Body composition in 70-year-old males and females in Gothenburg, Sweden. A population study, *Acta Med. Scand. Suppl.,* 611, 87, 1977.
53. Cox, J. R. and Shalaby, W. A., Potassium changes with age, *Gerontology,* 27, 340, 1981.
54. Behnke, A. R. and Myhre, L. G., Body composition and aging: a longitudinal study spanning four decades, in *Oxygen Transport to Human Tissues,* Loeppky, J. A. and Riedesel, M. L., Eds., Elsevier, New York, 1982.
55. Shock, N. W., Watkin, D. M., Yiengst, M. J., Norris, A. H., Gaffney, G. W., Gregerman, R. I., and Falzone, M. D., Age differences in the water content of the body as related to basal oxygen consumption in males, *J. Gerontol.,* 18, 1, 1963.

20 AGING OF THE GASTROINTESTINAL TRACT AND LIVER

Paola S. Timiras

1 HETEROGENEITY OF GASTROINTESTINAL FUNCTIONS AND EFFECTS OF DIET

The study of the aging of the gastrointestinal (GI) system discloses few alterations in gastrointestinal function, although cellular changes occur and involve secretory activity and motility of the major structures. Such changes are similar to those elsewhere in the body, but, in the absence of localized disease, function is usually maintained in line with or even above requirements. Disorders and diseases, however, become more common with advancing age and involve all levels of the gastrointestinal tract, starting with the mouth and extending to the rectum, anus, and pelvic floor musculature (Figure 20-1). In geriatric clinics, about 20% of all patients have significant gastrointestinal symptoms, and morbidity from gastrointestinal diseases, such as cancer of the colon, is second only to that from lung cancer.[1-4]

Of the two major appendices of the gastrointestinal tract — the liver, the largest gland in the body, and the pancreas, with both endocrine and exocrine functions — it is the liver that appears to be severely affected by aging. Given the numerous functions of this organ, alterations with aging have widespread repercussions for the well-being of the entire organism. However, in the liver as in the whole organism, not all functions are affected simultaneously or with equal severity. The bile, produced in the liver for optimal digestion and absorption of lipids, is important at all ages and is particularly crucial in the elderly whose poor diet may be deficient in several essential elements (e.g., lipid-soluble vitamins); bile formation remains quite stable in healthy individuals well into old age.

Another important hepatic function, the detoxification of many (therapeutic and recreational) drugs is progressively restricted with advancing age and contributes, with the increased use of drugs (polypharmacy), to the greater susceptibility of the elderly to the potential toxicity of excessive or incorrect medication (Chapter 23). In the pancreas, the changes with aging of endocrine functions have been discussed in Chapter 15. Changes with aging of the exocrine functions, which involve powerful protein-splitting (proteolytic) enzymes, may lead to impaired digestion and absorption.

In this chapter, aging of the gastrointestinal tract will be presented first, followed by a brief discussion of the aging pancreas and liver. While the focus will be on how late and how well digestion and absorption are retained, it must be kept in mind that limiting food intake may have beneficial effects. Since early studies, more than 40 years ago, evidence has been accumulating steadily that a dietary reduction of caloric intake or of specific food constituents will prolong life significantly in rodents. The earlier the restriction begins, the more successful is the subsequent retardation of growth and prolongation of life. Such dietary effects have also been demonstrated in other species. It will always remain unthinkable to perform such restrictive experiments in humans. However, encouragement to limit caloric intake and to prevent or treat obesity is easily justified. Not only is life prolonged, but frequency and severity of a variety of pathological changes are reduced in experimental animals (Chapter 24).

The major function of the gastrointestinal system is to provide the organism with nutritive substances, vitamins, minerals, and fluids. The elderly are an heterogeneous group of individuals in many important aspects of nutrition. This is true in relation not only to age but also to degree of health and disease, degree of physical activity, as well as psychological and socioeconomic characteristics. Thus, "young elderly" (aged 65 to 75 years) are distinguished from "old elderly" (over 75 years). However, the latest edition of the American Recommended Dietary Allowances (Food and Nutrition Board, 1980) groups individuals aged 51 years and older together into a single age group. Nevertheless, the elderly display a wide variety of individual differences, none better described than the aging-related changes in the mouth.

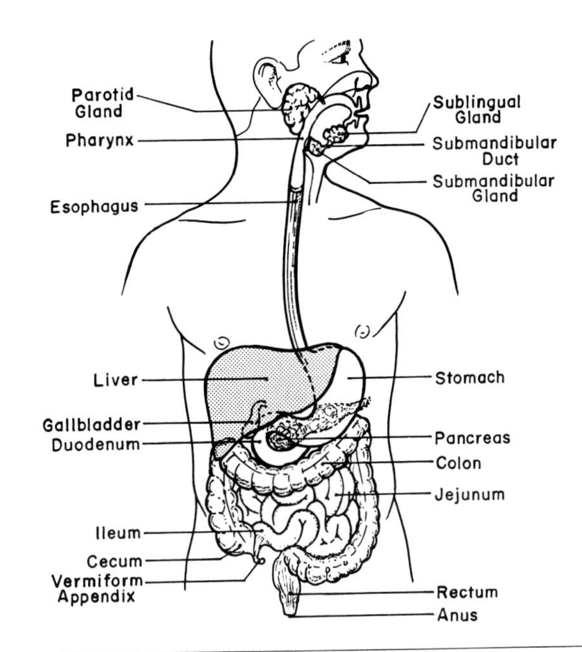

FIGURE 20-1 *Schematic representation of the gastrointestinal tract, with the liver and pancreas.*

0-8493-8979-8/94/$0.00+$.50

2 ■ TEETH, GUMS, AND ORAL MUCOSA

In the mouth, food is mixed with saliva, chewed, and propelled into the esophagus. Chewing (mastication), a matter of breaking down large food particles, is a function of the teeth. The saliva contains the digestive enzyme *ptyalin,* which plays a minor role in starch digestion. It also contains the glycoprotein *mucin,* which lubricates the food and facilitates its passage through the esophagus on the way to the stomach.

With aging, the *teeth* undergo characteristic changes:[5,6]

- They acquire a yellowish-brown discoloration from staining by extrinsic pigments from beverages, tobacco, and oral bacteria.
- The pulp recedes from the crown and the root canal becomes narrow and thread-like.
- The roots become brittle and fracture easily during extractions.
- The odontoblast layer lining the pulp chamber becomes irregular and discontinuous.
- The pulp undergoes fibrosis and calcification.
- Concomitantly, with faster destruction than reconstruction of the dentin, the mandibular and maxillary bones in which the teeth are embedded undergo the same aging processes as all other bones.
- Osteoporotic changes (i.e., increased bone loss) result in looser teeth, contributing ultimately to tooth loss.

The surfaces of the teeth involved in chewing become progressively worn down throughout life. This attrition is a consequence not only of chewing but also, in some individuals, of the habit of grinding or clenching the teeth together (so-called bruxism), often during REM sleep (Chapter 9). Abrasion (sometimes due to improper brushing) and erosion (often aggravated by the demineralizing action of soft drinks) are frequent. Although new caries (i.e., cavities with decay) are uncommon in the elderly, loss of interest in dental hygiene and a decline in dexterity needed for tooth brushing may lead to plaque accumulation and caries.[7] About 50% of elderly in the U.S. have lost the majority of their teeth by age 65 and about 75% by age 75. Teeth loss can be prevented by restoring appropriate hygienic measures self-managed or provided by a dental hygienist.[8]

Recession of the gingivae (gums) occurs in all elderly. The epithelial attachment which forms a cuff around the tooth at the interface with the gums recedes and opens the way to accumulation of particulate material with bacteria (i.e., plaque), swelling, inflammatory hyperplasia, or low-grade infection. Whether such a gum recession is a physiologic process or the result of chronic peridontitis (i.e., inflammation of the peridontal membrane) due to a variety of local irritative factors (e.g., ill-fitting dentures) remains to be clarified. Indeed, after the age of 40, chronic peridontitis is the major cause of tooth loss.[7] Peridontal disease is not only common in aging humans but is also usually found in aging experimental animals (e.g., mice, rats, dogs, monkeys, baboons). In both humans and experimental animals, the disease is due to local factors as well as some systemic predisposing disease (e.g., diabetes) or stress (e.g., cold exposure).[9] Indeed, in germ-free animals, recession of gums does not occur with age.

With aging, the epithelium of the oral mucosa becomes relatively thin and atrophic. Specialized structures, such as the papillae of the tongue also become atrophic and this atrophy is associated with loss of taste (Chapter 10). Other structures, like the palatal mucosa, undergo edema and keratinization (i.e., accumulation of a highly insoluble protein, keratin), a condition which seems to be delayed or prevented by the wearing of dentures. Keratin, also a normal component of the skin, is the product of epidermal cells and undergoes characteristic changes with development as well as extensive crosslinking with age and a number of conditions. The oral epithelium also exhibits increasing amounts of glycogen and alterations in collagen (Chapter 22).

Oral Diseases in the Elderly

While the above changes in oral structures with aging are gradual and relatively benign, they may predispose the involved tissues to a variety of pathologic conditions. Although very few oral diseases are characteristic of old age, many pathologic states are seen with greater frequency than in younger individuals. Among these, chronic periodontal disease (discussed above), xerostomia, mucositis and mucosal atrophy, leukoplakia, and malignant neoplasia (these latter beyond the scope of the present discussion) are the most common.[5]

Xerostomia or *dry mouth* may be due to a large variety of etiologic factors. In the elderly, the major causes are

- Primarily, atrophy of the salivary glands
- Decline of salivary secretion
- Systemic disease (e.g., diabetes)
- Heavy cigarette smoking
- Anxiety and depression
- Several medications (antihypertensives, antidepressants, antihistamines) depending on the dose

In this condition, the oral cavity is extremely dry, the mucosa appears red, dry, fissured, and often coated with food particles and sloughed-off cells. Therapy involves cessation of the underlying cause (e.g., smoking) or treatment of the underlying systemic disease (diabetes) or the use of artificial saliva. The decrease in salivary volume is associated with enzymatic changes, such as reduction in amylase activity and in electrolytes.[10-12]

Mucositis and *mucosal atrophy* are frequent occurrences in elderly individuals in whom the oral mucosa has become atrophic and less resistant to the irritation of oral noxious stimuli such as trauma, hot foods, smoking, to infections, and to chemotherapeutic agents or radiation therapy. These stimuli result in a chronic inflammatory process (mucositis) and, in more severe cases, in ulceration with pain.

Leukoplakia, or *keratosis,* represents an hyperplasia of the mucosa with accumulation of keratin, hence, the name of "white patch". It is rarely seen in young individuals but is frequent after 60 years. It may be caused by pipe or cigarette smoking, by ill-fitting dentures, or by infections (e.g., candidiasis). It is associated with precancerous histologic alterations, and, in this case, it must be treated as if it were a carcinoma.[13]

3 ■ SWALLOWING AND PHARYNGOESOPHAGEAL FUNCTION IN THE ELDERLY

Dysphagia, or *difficulty in swallowing,* is a common complaint of elderly individuals. It can result from alteration of any of the components of deglutition, a very complex motor activity, involving the mouth, the esophagus, and several levels of nervous control.

Deglutition or swallowing is a reflex response that pushes the contents of the mouth into the esophagus. The afferent stimuli are generated by the voluntary collection of the oral contents on the tongue and their propulsion backward into the pharynx and are carried by several nerves to the medulla oblongata where they are integrated. The efferent fibers are carried, also by several nerves to the pharyngeal musculature and the tongue. Inhibition of respiration and closure of the glottis are part of the reflex. Swallowing is impossible when the mouth is open; it is very rapid during eating but continues at a slower rate between meals. Upon swallowing, the upper portion of the esophagus relaxes to permit entrance of the swallowed material, which then progresses through the esophagus to the stomach by peristaltic movements (circular waves). In the standing position, liquids and semisolid foods may fall by gravity to the lower esophagus where the musculature relaxes upon swallowing and permits the passage of food in the stomach.

The act of swallowing is divided into three stages, all of which are affected by aging.[14,15] The first stage, in which the material to be swallowed is passed from the mouth to the pharynx, is a voluntary act, mediated through stimulation of skeletal muscles. These undergo aging-related changes with atrophy and increasing weakness common to all skeletal muscles (Chapter 21). The second stage, reflexive in nature, is very short in duration but very complex in its neural control and involves the relaxation of the sphincter between the pharynx and the esophagus. In the third stage, reflex transport sweeps the contents onward through smooth muscle peristalsis. All these stages that require a precisely timed contraction/relaxation sequence are affected; they may become desynchronized and result in less efficient deglutition. Dysphagia of varying degrees of severity is a common complaint. In the most severe cases, dysphagia is associated with symptoms of choking and drowning with aspiration or regurgitation of food while mild dysphagia may be found in otherwise healthy elderly. Severe dysphagia is always a symptom of a systemic disease, either of the muscle or the nervous system. In addition to being an uncomfortable and unpleasant symptom, dysphagia leads to reduced and altered nutritional intake, particularly in the elderly. Malnutrition, weight loss; and dehydration are common features of this condition.[16]

Presbyesophagus, or *old esophagus,* is the most common disorder of the esophageal motility associated with old age.[17,18] Radiologic and manometric studies reveal an increased incidence of nonperistaltic contractions (with failure of the lower esophageal sphincter to relax), a reduced amplitude of peristaltic contractions, and a decreased responsiveness to cholinergic stimulation.

Other motor disorders of the esophagus in addition to those mentioned above and occurring at all ages, including old age, are achalasia, in which food accumulates in the esophagus and the organ becomes extremely dilated; sphincter incompetence, which permits reflux of acid from the stomach; and aerophagia, or ingestion of air, which can be regurgitated or absorbed in the intestine or expelled as flatus.

4 THE STOMACH AND DUODENUM: AGE-RELATED PHYSIOLOGIC AND PATHOLOGIC CHANGES

The stomach serves as a food reservoir and, by its churning movements (due to the presence of three smooth muscle layers), breaks down ingested food. It secretes the gastric juice which contains a variety of substances: digestive *enzymes* (e.g., pepsin), *mucus* (which lubricates the food),

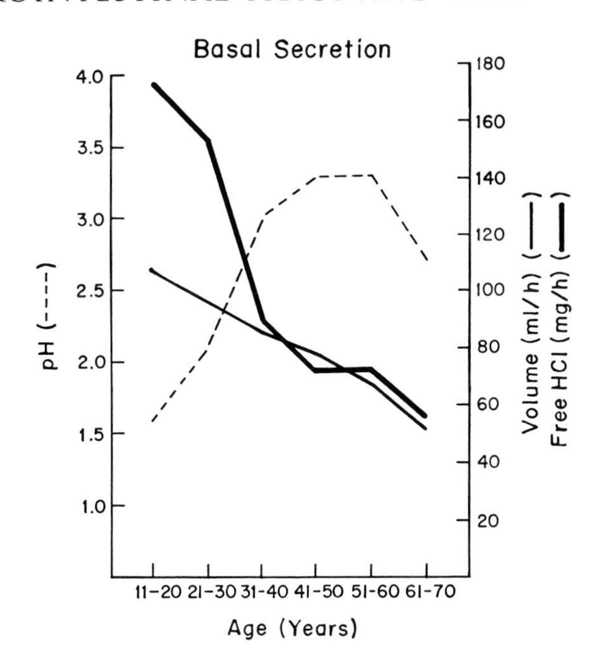

FIGURE 20-2 *Changes with age, under basal (preprandial) conditions of gastric secretion.* Note, with advancing years, the moderate decrease in total volume and the greater decrease in free hydrochloric acid (HCl). As the free HCl decreases, pH increases.

hydrochloric acid (which destroys the ingested bacteria, aids in protein digestion, and is necessary for the transformation of iron for the synthesis of hemoglobin), *hormones* (gastrin, glucagon, somatostatin), and special *peptides* (VIP, substance P).

In addition, the cells of the gastric mucosa secrete an intrinsic factor (a glycoprotein) necessary for the absorption of Vitamin B_{12} from the small intestine. Vitamin B_{12} is necessary for the maturation of red blood cells, and deficiency of the vitamin or an intrinsic factor lead to a severe type of anemia (pernicious anemia).

With aging, some of the major changes in the stomach and neighboring duodenum involve the hydrochloric acid and pepsin secretions which are decreased, under basal conditions, in the normal elderly (Figure 20-2). Changes with aging in some of the digestive enzymes may be consequent directly to changes in the enzyme secreting cells and organs or indirectly to hormonal and neural regulatory alterations (Table 20-1). The direct changes may involve cellular or enzymatic reduction (Chapter 5). More likely, the failure lies in alterations in the regulatory mechanism of enzyme synthesis and release due either to a change in the hormonal or neural stimuli or to a change in response (altered receptors).

However, with increasing age, the incidence of *peptic ulcers* is a common condition of Western society. Increased acid secretion is the major feature of gastric and duodenal ulcers. The number of parietal cells is increased perhaps because of genetic predisposition or as a result of increased stimulation. Vagal overactivity has also been demonstrated. Acid production is inhibited in normal aged subjects by low intragastric pH, but, in duodenal ulcer, this inhibition fails.

4.1 Regulation of Gastric and Duodenal Acidity

The hydrochloric acid secreted by the parietal cells in the body of the stomach is concentrated enough to produce tissue

TABLE 20-1

Possible Mechanisms of Aging-Related Changes in Digestive Enzyme Secretion

Enzyme secreting organs
Reduction in
Number of cells
Enzyme concentration
Enzyme synthesis and release
Hormonal and neural regulation of enzyme-secreting organs
Reduction in
Number of gastrointestinal endocrine cells
Hormone concentration
Impairment of
Sensitivity of endocrine cells to digestive stimuli
Alteration of
Distribution and metabolism of gastrointestinal hormones
Number and affinity of endocrine or neural receptors

damage. However, in the normal adult, the mucosa of the stomach and the duodenum (the first part of the small intestine in proximity of the stomach) does not become irritated or digested. The gastric mucus was once thought to serve as a protective layer, but now the surface membranes of the mucosal cells and the tight junctions between the cells are considered the main barrier to acid damage. It appears that the stomach also has an alkaline secretion, which may play a role in protecting the stomach. In the duodenum, cells from the Brunner's glands secrete a thick alkaline mucus that probably helps protect the duodenal mucosa from the gastric acid. Independent of the Brunner's cells, duodenal cells secrete appreciable amounts of bicarbonate which contribute to this protection.

A number of substances (aspirin, ethanol, caffeine, bile salts) tend to disrupt this barrier and cause gastric irritation. Disruption of the barrier may be severe enough to induce ulcers, i.e., loss of cells with bleeding and necrosis of tissues. These ulcers called *peptic ulcers* (for their relation to digestion) may be situated in the stomach (usually near the pylorus, the region close to the duodenum) or the duodenum and will then be designated as gastric and duodenal ulcers.

Excessive acid production plays a major role in the pathogenesis of these ulcers. *Gastrin,* one of the gastrointestinal hormones, secreted by the gastric mucosal cells as well as the pituitary, the pancreas (fetal), and the brain, also stimulates gastric acid and pepsin secretion. In this manner, increased secretion of gastrin would contribute to the genesis of peptic ulcers. Gastrin may also participate to some degree in regenerating processes inasmuch as it stimulates mucosal growth and gastric motility. The growth-promoting action may occur through stimulation of the epidermal growth hormone, secreted by the duodenal Brunner's glands.

The incidence of peptic ulcer is on the rise in developing countries, particularly in India and Africa. In Western countries, including the U.S., it is decreasing in the overall population (perhaps reflecting advances in treatment) but increasing in the elderly group. In this group not only is the incidence on the rise, but the severity of the disease and its consequences are also greater than in younger individuals.[19]

Elderly patients tend to predominate among those admitted to hospitals for peptic disease, and patients over 60 years of age account for nearly 50% of those with gastric and 40% of those with duodenal ulcers.[20] In these individuals, age is the major mortality risk. While, in many cases, the causes of the disease are unknown, drugs (aspirin, alcohol) and cigarette smoking may be important factors in the pathogenesis of the disease.

Management of Peptic Ulcer

Procedures used in the treatment of ulcers include (1) dietary, (2) pharmacologic, and (3) surgical interventions. Pharmacologic interventions are aimed at inhibiting acid secretion and enhancing mucosal resistance to acid.

A variety of antacids, most of which contain aluminum and magnesium hydroxide or calcium carbonate, are available. Inhibition of parasympathetic inputs (which stimulate acid secretion) by atropin gives variable responses and has many undesirable side effects. Histamine (an amine derived from the amino acid histidine and a powerful stimulator of gastric secretion) receptor blockers, such as cimetidine and ranitidine (some of the most commonly prescribed drugs in the U.S.), are also often used. Drugs capable of inhibiting H^+K^+ ATPase and epidermal growth hormone are at the experimental stage. A number of substances increase the resistance of mucosal cells to acid by forming adherent protein complexes at the ulcer site. Usually the pharmacologic treatment is associated with special diets and cessation of cigarette smoking, inasmuch as healing rates of duodenal ulcer are probably adversely affected by cigarette smoking and the incidence of ulcers is higher in smokers than in nonsmokers.[20]

If pharmacologic, dietary, and hygienic measures fail, surgical intervention is advisable, with the caution necessary for the older patient. In addition to ulcers, disruption of the protective mucosal barrier may frequently occur with aging as a consequence of a variety of factors and contribute to gastric pathology. Whether alterations in the mucosal cells are spontaneous or secondary to chronic toxicity of drugs and alcohol or to mucosal ischemia due to hypotension remains unanswered. Disruption of the barrier allows luminal hydrogen ions to diffuse into mucosal cells causing damage. Gastrointestinal bleeding is often a consequence of duodenal and gastric ulcers;[20] it may also be associated with less severe mucosal alterations such as *gastritis* and gastric erosions, a form of gastric irritation and inflammation. The incidence of these alterations increases with aging. The incidence of chronic and acute gastritis increases with age so that evidence of it is found in 50% of individuals over the age of 60 years. In view of the overall declining physiologic competence of the elderly and the debilitating nature of the disease, mortality rises rapidly with advancing age.

4.2 Vascular Alterations

As indicated above, gastrointestinal *bleeding* in the elderly presents a special problem — both incidence and mortality are high after the age of 60.[20] Another cause of bleeding in the elderly is vascular malformations (ectasias) of the intestinal vessels, which are easily subject to bleeding. These ectasias occur throughout the GI tract, particularly the small bowels, 40%; stomach, 30%; and colon, 25%. They are associated with aortic stenosis, age-related degenerative changes in tissues, inherited collagen, or ground substance defects.[21]

Circulatory alterations are not confined to bleeding but also encompass *ischemia* (i.e., local and temporary deficiency of blood), which can lead to mesenteric infarction. Ischemia may produce symptoms of varying severity from transient intestinal discomfort, to abdominal angina, infarction, or, in cases of segmental ischemia, to ischemic colitis.

Carcinoma of the Stomach

For unknown reasons (perhaps a change in dietary and other habits?), the incidence of carcinoma of the stomach, once one of the most frequent cancers in men, has been declining in the last 20 years. It is still relatively frequent and is situated primarily in the lower regions of the stomach (antrum and pylorus). It has a very unfavorable prognosis. Peak incidence is reached in 80- to 90-year-old men and may have a familial occurrence. Diagnosis is made on the basis of gastroscopy with biopsy. Unfortunately treatment, either by surgery, radiation, and/or chemotherapy shows a low (5 to 10%) 5-year survival rate.

5 THE SMALL AND LARGE INTESTINES

Structure of the Small Intestine Mucosa

Throughout the length of the small intestine, the mucosa displays many folds and is covered with villi formed of a single layer of columnar epithelium and containing a network of capillaries and one lymphatic vessel. The free edges of the mucosal cells are divided into microvilli, which form a "brush border". Folds, villi, and microvilli augment considerably the absorptive surface. The mucosal cells are formed from mitotically active undifferentiated cells located at the bottom of the villi (in the so-called crypts of Lieberkuhn). They migrate up to the tips of the villi where they are sloughed into the intestinal lumen. The average life of these cells lasts 2 to 5 days, thereby representing a potential model (still little utilized) for the study of cellular aging. The crypts are also the site of active secretion of water and electrolytes. Studies in humans have reported, with aging, a reduced height and a frequent convoluted pattern of villi; changes were not observed in villus width, cell height, or mucosal thickness.[22-24]

5.1 Changes in Intestinal Absorption

In a number of animal species, advancing age is accompanied by alterations involving one or several of the following changes:

- Overall shape of the villus
- Increase in collagen
- Mitochondrial changes
- Lengthening of the crypts
- Prolonged replication time of the crypt stem cells

All of these changes are minor and none appears sufficient to explain the impaired absorption often found in the elderly. Other factors may intervene:

- Altered villus motility limiting functional surface area
- Inadequate intestinal blood supply (due to atherosclerotic involvement of major intestinal vessels)
- Impaired water "barrier" restricting diffusion and transport

Research concerning absorption of nutrients and vitamins is sparse in humans and even in laboratory animals. There are, however, with increasing age, reports of reduced absorption of several substances (e.g., sugar, calcium, iron) while digestion and motility remain relatively unchanged.[25,26] Among the substances for which absorption has been studied in the elderly, calcium probably presents the best evidence for a gradual reduction with increasing age (Table 20-2). In addition to changes in bone calcium, with aging, calcium absorption and transport are altered.

Calcium balance studies in humans reveal a significant drop in absorption with aging, starting at about age 60. When comparing young and old subjects, the young individuals are capable of responding to a low-calcium diet by increasing intestinal calcium absorption, a response no longer found in the elderly. Mechanisms responsible for this reduced calcium absorption include

- Decreased intake or malabsorption of Vitamin D
- Lack of sunlight and conversion (hydroxylation) in the skin
- Altered hepatic and/or renal hydroxylation of vitamin D
- Impaired cellular binding of calcium.

TABLE 20-2

Mechanisms of Decreased Intestinal Calcium Absorption with Aging
Decreased intake of vitamin D (poor nutrition)
Decreased vitamin D conversion in skin (reduced sunlight exposure)
Decreased intestinal absorption
Decreased vitamin D metabolism (hepatic) and activation (renal)
Decreased cellular calcium binding (decreased receptors)

There is conflicting evidence for other substances, such as dextrose, xylose, iron, vitamin B_{12}, or fats, which may be reduced or may be absorbed normally at all ages.

Increased general frailty and weight loss may occur in the old, without evidence of any specific underlying cause, in the presence of a well-maintained appetite and a balanced diet. However, a relatively high percentage of old individuals suffer from malabsorption; at least 7% of residents in nursing homes are likely to have impaired absorptive ability.

5.2 Absorption of Nutrients and Malabsorption

Adequate nutrition is indispensable at all ages but especially in the young who must provide extra calories for growth and in the old whose gastrointestinal function is only marginal. Indeed, dietary interventions to ensure a long and vigorous life have been popular for many centuries. Once achieved, old age has been recognized as a period requiring special attention to dietary habits.[27]

In the intestine, primarily the small intestine, the intestinal contents are mixed with mucus, pancreatic juice, and bile. Digestion, which begins in the mouth and stomach, is completed in the lumen and mucosal cells of the small intestine. Digestion depends upon a number of enzymatic processes under neural and hormonal stimuli which can be affected in the elderly (Table 20-1). The products of digestion are then absorbed, along with vitamins and fluid in the small intestine (water also in the colon) and carried to the liver by the portal blood. The small intestine, with its many folds, finger-shaped villi, and array of microvilli on the luminal side of the cells, is particularly well designed for this function of absorption. The rate of nutrient transport by the intestine is related primarily to the surface area that is functionally exposed to the luminal contents. Several factors are important for the maintenance of optimal function of the small intestine (Table 20-3) and may be affected by aging. Thus, any change in intestinal architecture or diffusion barrier of the mucosal cells greatly affects transport. With aging some anatomic changes have been described in animals and in apparently healthy aged individuals.[22,23,28,29]

Malabsorption and Disease

Many generalized conditions, such as *rheumatoid arthritis* (Chapter 21), afflicting the elderly may have some detrimental effects on absorption.[30] A number of small bowel disorders may cause malabsorption, but their incidence is low and symptomatology vague. Other frequent causes of malabsorption include

TABLE 20-3

Important Factors for the Maintenance of Optimal Small Intestinal Function

Anatomic integrity of the absorbing small intestine and normal intestinal mucosal cell replication
Normal gastrointestinal secretions. Including basal and postprandial secretions from salivary, gastric, pancreatic, and hepatic cells
Coordinated gastrointestinal motility
Normal intestinal uptake and transintestinal transport. Including adaptation to intraluminal substrates
Adequate intestinal blood supply
 to maintain cell oxygenation and
 to maintain cell nutrient supply
Normal defense mechanisms against toxic injurious agents from the intestinal lumen
 normal clearance of bacteria in the intestinal lumen
 immunologic response to injury
 mucosal cell wall integrity to prevent macromolecular uptake
 mucosal cell detoxification of toxic absorbed materials

- Infections (e.g., after gastrointestinal surgery, diarrhea)
- Small intestine diverticula (i.e., small dilations or pockets leading off from the intestinal tube)
- Pancreatic insufficiency (rare)
- Celiac disease (alterations of mucosa and cell transport)
- Mental disorders (e.g., dementia)[31]

Many old individuals with malabsorption are severely undernourished, weak, and debilitated. Management, therefore, includes both treatment of the specific underlying disease and appropriate diet. With energetic treatment, even severely ill patients have a good prognosis.

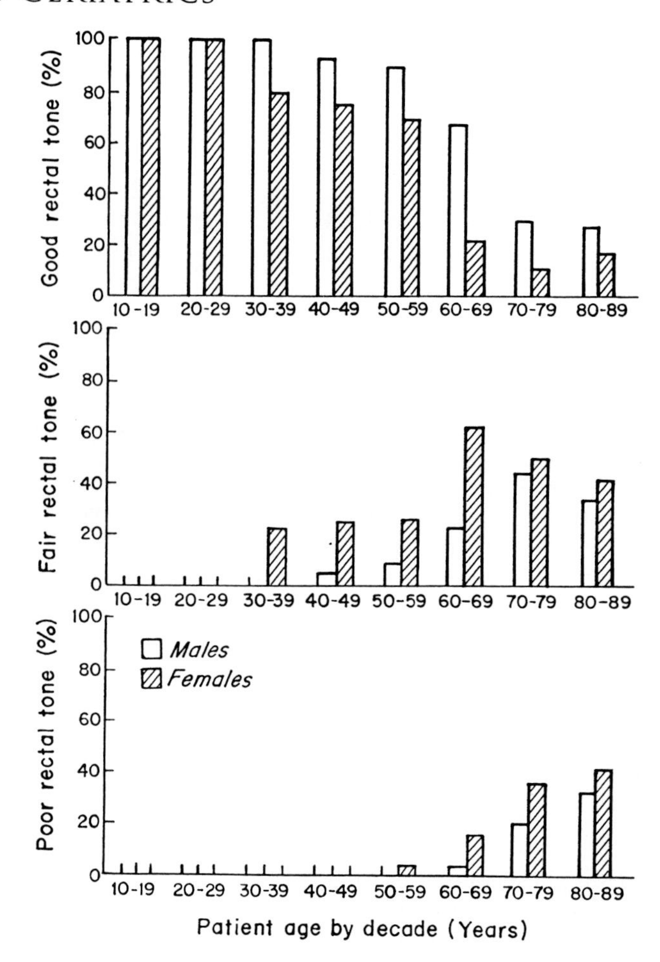

FIGURE 20-3 *Relationship of sphincter tone to age.* (From Stewart, E. T. and Dodds, W. J., *Am. J. Roentgenol.*, 132, 197, 1979. With permission.)

5.3 Aging of the Large Intestine

Disorders of the large bowel are almost exclusive to the elderly. The physiopathology of such disorders is still little known, but given the multiple clinical problems, more attention is currently being dedicated to the study of physiologic changes with aging in this intestinal segment.

Anatomical changes are similar to those of the small intestine and include (1) atrophy of the mucosa, (2) proliferation of connective tissue, and (3) vascular changes, mostly of an atherosclerotic nature.

The large intestine's major functions are storage, propulsion, and evacuation of the intestinal content (feces). Especially important are those conditions associated with bowel motility such as constipation or diarrhea.[32]

In the colon (the last portion of the large intestine before the rectum), the most obvious aging-related change is the increased prevalence of *diverticula,* small, pocket-like mucosal herniations through the muscular wall. They vary in diameter from 3 mm to more than 3 cm and are present in 30 to 40% of persons over the age of 50, with increasing incidence thereafter. Diverticula are often responsible for severe bleeding from the rectum and often become inflamed, causing diverticulitis. A highly refined, low-residue diet, as often consumed today, may be responsible for the formation of diverticula. The lack of dietary fiber and bulk is associated with spasm of the colon. The intraluminal pressure builds up and the mucosa eventually pushes through the muscular coat at weak points, usually where the colonic blood vessels pierce the muscle to supply the mucosa. Diverticula become filled with packed feces and may ulcerate into the thinned mucosa causing infection and inflammation.

The presence of diverticula may induce nonspecific abdominal pain, diarrhea, or constipation. A diet to increase the fiber content may alleviate these symptoms. Major complications include diverticulitis, hemorrhage, and colonic obstruction or perforation requiring surgery and rigorous pharmacologic and dietary treatment.[33]

5.4 Incontinence

As discussed in Chapter 19, urinary incontinence is one of the major afflictions of old age. The same tragic consideration applies to fecal incontinence (Figure 20-3). The maintenance of normal control on fecal evacuation is regulated by complex neuromuscular functions (Figure 20-4). Should any physiologic decrement occur in the activity of the intestine itself, or in the muscles of the pelvic floor, or in the neural inputs, then the control of defecation may break down. A person with efficient sphincters may find control impossible during an attack of severe diarrhea. In the elderly, loss of sphincter muscle strength (Figure 20-5), merely as a consequence of aging, creates a more difficult problem when confronted with diarrhea. Certain neurologic conditions affect the pelvic floor muscles and these may be so severe that, even with a normal stool, continence cannot be preserved. At all ages, there may be organic deficiencies in the muscle ring due to trauma, and such deficiencies are more likely to occur in the elderly person.

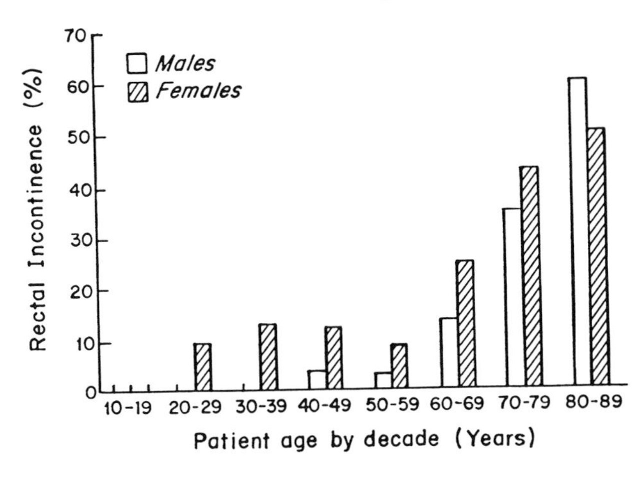

FIGURE 20-4 *Relationship of age and sex to fecal incontinence.* (From Stewart, E. T. and Dodds, W. J., *Am. J. Roentgenol.*, 132, 197, 1979. With permission.)

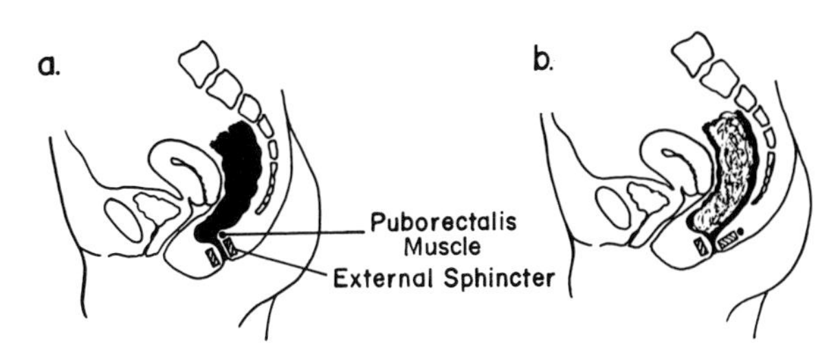

FIGURE 20-5 *Diagrammatic representation of some of the structures involved in defecation (here, in women).* Note that the external sphincter and the puborectalis muscle play an important role in maintaining normal fecal continence: (a) when the rectum is empty and the sphincter is closed, and (b) when it is full, and the sphincter is relaxed and displaced. Note that one important action of the puborectalis muscle is to maintain angulation (bend) between the lower rectum and anal canal, upon which continence is largely dependent. Also note that with increased rectal pressure, the activity of the sphincter increases, thereby protecting the individual from involuntary defecation. However, above a certain pressure (fecal volume) this protection is lost.

Pelvic Structures and Defecation

Distension of the rectum, the last portion of the intestine, initiates reflex contractions of its musculature. In humans, the internal involuntary sphincter is excited by the sympathetic but inhibited by the parasympathetic nerve supply. The external sphincter (voluntary) is innervated by somatic nerves. The urge to defecate begins when the rectal pressure increases to a certain level (about 55 mmHg) at which time both sphincters relax and the rectal contents are expelled. At a lower rectal pressure, defecation can be initiated by voluntary relaxation of the external sphincter and contraction of the abdominal muscles. Thus, defecation is a spinal reflex that can be voluntarily inhibited by contraction of the external sphincter or facilitated by its relaxation.

With aging, the rectal muscle mass is decreased in size and the sphincter is weakened. The external sphincter is always the most affected of the pelvic floor muscles. The high incidence of incontinence in the elderly makes it essential to exclude any possible underlying gastrointestinal infection or systemic disease. One classification is shown in Table 20-4 and may be compared with the causes of urinary incontinence (Chapter 19).

Fecal incontinence may be caused by

1. Neurogenic alterations involving the cortex (e.g., dementia) or the spinal cord (e.g., failure to inhibit defecation upon entrance of feces into rectum)
2. Muscle atrophy (due to trauma as in prolonged and difficult labor), a direct analogy to stress incontinence of urine (Chapter 19)
3. Constipation (due to immobility, poor reflexes, difficulty in reaching the toilet)
4. Diarrhea, which may cause incontinence at all ages but more frequently at older ages[34-36]

Knowledge of the etiology of incontinence leads to some practical interventions. A first step is to rule out constipation or diarrhea and, if present, to treat them.[32] If fecal incontinence persists, then other causes must be sought and appropriate treatment established.

Constipation

This is considered one of the most common gastrointestinal complaints of the elderly.[37] Its prevalence seems to be greater in women, although this sex difference may not be real but rather due to the larger number of old women and their overall greater degree of disability (Chapter 3). The major cause is decreased motility of the large intestine, but diet (unbalanced with respect to bulk) and lack of exercise may also be implicated in constipation. Treatment involves redressing the latter two factors.

TABLE 20-4
Classification of Fecal Incontinence

Cause	Consequence
Neurogenic	
Cortical	Loss of inhibition
Spinal	Reduced reflex activity
Muscle atrophy	"Stress" incontinence
Retention	Constipation
Overflow (bacteria, virus, allergies)	Diarrhea

Carcinoma of the Large Intestine

Carcinoma of the large bowel is the second (after lung carcinoma) most common malignancy in individuals over 70 years of age. Cancer of the colon would be more frequent in women and cancer of the rectum in men, *Polyps* resulting from hypertrophy of the intestinal mucosa and extending into the intestinal cavity are also frequent. They may be benign tumors or possible precursors of carcinoma.

6 AGING OF EXOCRINE PANCREAS

Besides its hormonal secretions, the pancreas produces a pancreatic juice containing enzymes (amylase, lipase, proteases) important for digestion. The enzymes are discharged by exocytosis and their secretion is controlled by a reflex mechanism and by the hormones, secretin and cholecystokinin. The major enzyme is trypsin which is secreted as an inactive proenzyme, trypsinogen. Some uncertainty exists regarding the effects of advancing age upon pancreatic secretion.[38] The senile gland is smaller, harder than normal (due to increasing fibrosis), and yellow-brown (due to accumulation of lipofuscin). Of the major enzymes, some (amylase) remain constant whereas others (lipase, trypsin) decrease dramatically.[39] Secretin-stimulated pancreatic juice and bicarbonate concentration remain unchanged.[40]. Little is known so far about age-related changes in the hormones that regulate pancreatic function. While a decline in some functions of the pancreas occurs with aging, the genesis of this decline is unknown but has been related to

- Diet
- Drugs (e.g., alcoholism)
- Vascular sclerosis
- General fibrosis
- Lack of cell regeneration

However, as only one tenth of pancreatic secretion is needed for normal digestion, it is not probable that age alone could lead to a significant pancreatic insufficiency capable of inducing severe digestive disorders.

In general, these age-related changes do not seriously compromise pancreatic function, but their presence may increase the incidence of pancreatic disease (acute and chronic pancreatitis, cancer) in the elderly.

7 AGING OF THE LIVER

The liver is an organ with many functions, some of which are

- Bile formation
- Carbohydrate storage and metabolism
- Ketone body formation
- Reduction and conjugation of steroid hormones
- Inactivation of polypeptide hormones
- Detoxification of many drugs and toxins
- Manufacture of plasma proteins
- Urea formation
- Regulation of lipid metabolism

Little is known, however, of the changes that may occur with "normal" aging.[41] As is the case with other multifunctional organs, not all functions age at the same pace. In this section, we will consider some changes in morphology and function, particularly bile excretion. Changes in enzyme activity with aging and hepatotoxicity of various drugs, especially in the elderly, are considered in Chapters 5 and 23, respectively.

7.1 Structural Changes

Atrophy and decreased weight of the liver are common in elderly individuals and are often independent of organ sclerosis and fibrosis. The reduced weight is due to a decrease in cell number rather than in cell size in contrast to other instances such as malnutrition in which reduction of cell size is the cause of decreased liver weight. In fact, cell size is often increased. Cell loss begins in the 60s, slowly at first and then more rapidly in the 80s and thereafter.

Hepatic Cells

The liver is organized in lobules formed of hepatic cells, hepatocytes, lined up in rows irradiating from the center of the lobule to the periphery. The hepatic veins are situated in the center of the lobule and the biliary ducts, the portal veins, and hepatic arteries at the periphery. The specialized capillaries, the sinusoids, are lined with phagocytic cells (Kupffer's cells) that engulf bacteria or other foreign particles (see Chapter 7). The lobules are separated by a small amount of interlobular connective tissue in which are found the blood vessels and the beginnings of the biliary ducts and lymphatic vessels (Figure 20-6).

With advancing age, binucleate cells appear and, after 70 years, hypertrophy of the cells and polyploidization (increased nucleus size with more than two full sets of homologous chromosomes).[42] The enlarged cells are interspersed with "dark", degenerating cells, probably products of the usual cycle of cell degeneration/regeneration of the hepatic parenchyma.

With aging, cell regeneration would be slowed down perhaps due to the absence or deficit of growth factors for cell replication or to the excess of growth inhibitory factors. Hence, cell hypertrophy may be interpreted as a compensatory reaction to cell loss. That this is the case and that aged hepatic cells are active is supported by the increased activity of some enzymes (e.g., succinic dehydrogenase).

The number of mitochondria is decreased whereas their size is increased, mitochondrial hypertrophy being a compensatory response to decreased number. The increased mitochondrial volume and the presence of "giant" mitochondria may be ascribed to the incompleteness of their division following synthesis of new mitochondrial material in the presence of hypertrophy. However, regeneration of liver after partial hepatectomy is slower in older rats, even after hormonal (cortisol, thyroxine) administration. Indeed, the very compensatory increase in volume when regenerative power is more limited would accelerate cell loss and thereby result in a vicious cycle of cell destruction and compensatory hypertrophy.

Smooth endoplasmic reticulum is decreased in the rat hepatocytes, a morphologic correlate underlining the age-related reduction in the hepatic capacity to metabolize drugs (Chapter 23). Other aspects of hepatic structure are not markedly altered in the old rat, and maintenance of normal morphology agrees with minimal or no alteration of hepatic metabolic functions such as protein synthesis.[43]

Only a few studies have investigated whether collagen changes in the liver with aging; those conducted so far suggest a moderate increase[44] ascribed, perhaps, to hepatic injury.[45]

7.2 Functional Changes

Alteration of hepatic structure and enzymatic functions with aging is moderate. In the healthy elderly, routine tests of liver function involving the metabolism and elimination of specific dyes (e.g., bromosulphalein) and radioisotopes, and protein synthesis do not show significant differences between individuals aged 50 to 69 and 70 to 89 years (Table 20-5).[46,47] Similarly, tests of liver dysfunction, including cholestasis (blockage of bile flow), cell necrosis, inflammation and impaired detoxification, are usually negative in the absence of specific liver damage and systemic disease.

In contrast to humans, studies in experimental animals (rats) show consistent reduced removal from the blood of many of the dyes used to measure hepatic function, storage, and transport.[48,49] These changes have been ascribed to both circulatory alterations and increased collagen deposition (and hence impaired transport) in the liver.

7.3 Bile Formation

Bile production is a major function of the liver. Bile, considered as both a secretion and an excretion, is formed continuously by the hepatic cells. It is a greenish-yellow fluid composed of water, bile salts, bilirubin, cholesterol, and various inorganic salts.

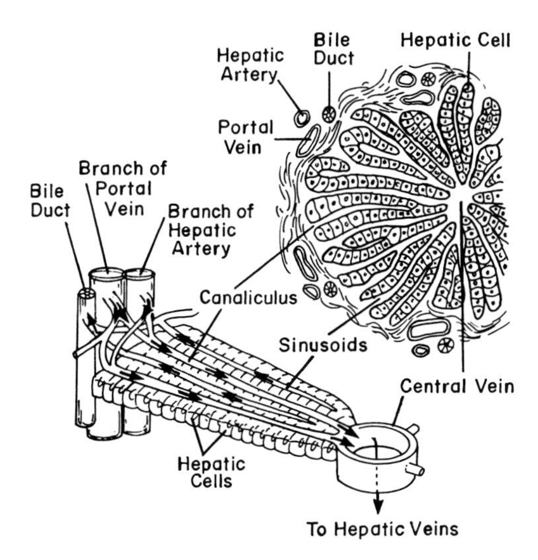

FIGURE 20-6 *Diagrammatic representation of the liver, which is organized in lobules formed of hepatic cells, hepatocytes, lined up in rows irradiating from the center of the lobule to the periphery, shown in sagittal plane, top, and a closer view, bottom.* The hepatic veins are situated in the center of the lobules and the biliary ducts, the portal veins, and hepatic arteries at the periphery. The specialized capillaries, the sinusoids, are lined with phagocytic cells (Kupffer's cells) that engulf bacteria or other foreign particles. The lobules are separated by a small amount of interlobular connective tissue in which are found the blood vessels and the beginnings of the biliary ducts and lymphatic vessels.

TABLE 20-5

Liver Function Tests

Function	Test
Hepatic clearance	Serum bilirubin levels
	Bromosulphophtalein and other dyes
	Radioisotope scans
Synthesis	Plasma albumin
	Prothrombin time
	Blood urea
Cell necrosis	Transaminases
	Lactic acid dehydrogenase
Cholestasis	Conjugated bilirubin levels
	Alkaline phosphatase
	5′-Nucleotidase
	Gammaglutamyltranspeptidase
Inflammation	Gammaglobulin level
Detoxification	Ammonia
	Mercaptans

The Biliary System

Bile formed of the hepatic cells is carried through several bile caniculi to the right and left hepatic ducts, which join to form, outside the liver, the common hepatic duct. This duct joins the cystic duct that drains in the gallbladder, the reservoir for bile located on the undersurface of the liver. The hepatic duct unites with the cystic duct to form the common bile duct, which enters the duodenum, usually united with the pancreatic duct, and pours the bile into the duodenum.

The bile, primarily through the action of its salts, has several important functions in the digestion by emulsifying lipids and activating the lipid enzymes, lipases. Bilirubin, the major pigment of the bile, results from the breakdown of hemoglobin, myoglobin, and respiratory enzymes in the reticuloendothelial system of the liver and spleen. Unconjugated bilirubin thus formed is carried to the liver cell where it is conjugated and bound to proteins; this conjugated form is water soluble. In the colon, conjugated bilirubin is hydrolyzed to urobilin and excreted.

FIGURE 20-7 *Incidence of bile duct stones at cholecystectomy in relation to age.* Note the sharp increase from 60 years on.

Little is known of changes with aging in bilirubin metabolism. However, biliary disease is common and its incidence increases steadily with age (Figure 20-7).[50] Biliary disease (gallstones), a common problem of most Western societies, affects 15 to 20% of adults of all ages and 30 to 50% of elderly persons by age 75 with a ratio of 2:1 for females to males.[51] As people live longer, the incidence of biliary disease has increased in the last 20 years. Of the patients with gallstones, 40 to 60% show no symptoms, and it is possible that the condition starts this way in most individuals.

In the gallbladder, the bile is concentrated by absorption of water (water is 97% in the liver bile and 89% in the gallbladder bile). Stones form in the gallbladder or bile ducts when a substance that is not normally present appears in the bile or the composition of the bile changes so that a normal constituent precipitates. For example, cholesterol stones form when the proportion of cholesterol, lecithin, and bile salts in the bile are altered.

Several aspects of biliary disease are characteristic of the elderly. These include

- A greater incidence of acute versus chronic cholicystitis (i.e., inflammation of the gallbladder)
- The presence of stones in the bile duct
- The recurrence of the disease after a previous operation
- The greater severity of the disease and the higher mortality

An increasing incidence of stones in bile ducts and gallbladder with aging has been well documented. With this increase, the incidence of related complications (e.g., jaundice, pancreatitis, cholecystitis, liver abscesses, and systemic sepsis) also rises. The treatment of choice is surgical and consists in the removal of the stone(s); it entails, in the elderly, more complex operations and more rigorous follow-up to control or correct the disease.

Bilirubin is an Antioxidant

Bilirubin, formed in the tissues by the breakdown of hemoglobin, is the major pigment of the bile. It is conjugated in the liver, and, except for a small portion that escapes in the blood, it is excreted via the bile ducts in the intestine. It is generally regarded as a potentially cytotoxic, lipid-soluble waste product that needs to be excreted. However, bilirubin is also capable of efficiently scavaging peroxyl radicals.[52] In this function, it would join other metabolic products such as uric acid (a product of purine metabolism) and taurine (a product of cysteine metabolism) with antioxidant properties. Thus, bilirubin would be "beneficial as a physiological, chain-breaking antioxidant". Therefore, the study of alterations in bilirubin metabolism with aging, hitherto little investigated, is clearly worth pursuing.

REFERENCES

1. Levitan, R., G.I. problems in the elderly. I. Aging-related considerations, *Geriatrics,* 44, 53, 1989.
2. Levitan, R., G.I. problems in the elderly. II. Prevalent diseases and disorders, *Geriatrics,* 44, 80, 1989.
3. Shamburek, R. D. and Farrar, J. T., Disorders of the digestive system in the elderly, *N. Engl. J. Med.,* 322, 438, 1989.
4. Morris, J. S., Dew, M. J., Gelb, A. M., and Clements, D. G., Age and gastrointestinal disease, in *Principles and Practice of Geriatric Medicine,* Pathy, M. S. J., Ed., John Wiley & Sons, New York, 1991, 417.
5. Walker, D. M., Oral disease, in *Principles and Practice of Geriatric Medicine,* Pathy, M. S. J., Ed., John Wiley & Sons, New York, 1991, 381.
6. Toga, C. J., Nandy, K., and Chauncey, H. H., Eds., *Geriatric Dentistry: Clincal Application of Selected Biomedical and Psychosocial Topics,* Lexington Books, Lexington, MA, 1981.
7. Baum, B. J., Characteristics of participants in the oral physiology component of the Baltimore longitudinal study of aging, *Community Dent. Oral. Epidemiol.,* 9, 128, 1981.
8. Griffiths, J. E., The dental need of the elderly and the delivery of care, in *Principles and Practice of Geriatric Medicine,* Pathy, M. S. J., Ed., John Wiley & Sons, New York, 1991, 353.
9. Allen, E. F., Statistical study of primary causes of extractions, *J. Dent. Res.,* 23, 453, 1944.
10. Grad, B., Diurnal, age, and sex changes in the sodium and potassium concentrations of human saliva, *J. Gerontol.,* 9, 276, 1954.
11. Garrett, J. R., Some observations on human submandibular salivary glands, *Proc. R. Soc. Med.,* 55, 488, 1962.
12. Shannon, I. L., A saliva substitute for dry mouth relief, in *Geriatric Dentistry: Clincal Application of Selected Biomedical and Psychosocial Topics,* Toga, C. J., Nandy, K., and Chauncey, H. H., Eds., Lexington Books, Lexington, MA, 1981, 161.
13. Waldron, C. A. and Shafer, W. G., Leukoplakia revisited. A clinicopathological study of 3,256 oral leukoplakia, *Cancer,* 36, 1386, 1975.
14. Patel, G. K., Diner, W. C., and Texter, E. C., Swallowing and pharyngoesophageal function in the aging patient, in *The Aging Gut. Pathophysiology, Diagnosis, and Management,* Texter, E. C., Ed., Masson Publ., New York, 1983, 83.
15. Sonies, B. C., Parent, L. J., Morrish, K., and Baum, B. J., Durational aspects of the oral-pharyngeal phase of swallow in normal adults, *Dysphagia,* 3, 1, 1988.
16. Hellemans, J., Vantrappen, G., and Pelemans, W., Oesophageal problems, in *Gastrointestinal Tract Disorders in the Elderly,* Hellemans, J. and Vantrappen, G., Eds., Churchill Livingstone, New York, 1984, 17.
17. Soergel, K. H., Zboralske, F. A., and Amberg, J. R., Presbyesophagus, esophageal motility in nonagenarians, *J. Clin. Invest.,* 43, 1472, 1964.
18. Hollis, J. B. and Castell, D. O., Esophageal function in elderly men — a new look at "presbyesophagus", *Ann. Intern. Med.,* 80, 371, 1974.
19. Sonnenberg, A., Schmid, P., Muller-Lissner, S. A., Vogel, E., and Blum, A. L., What makes duodenal ulcer heal and relapse, *Gastroenterology,* 78, 1266, 1980.
20. Kumpuris, D., Gastrointestinal bleeding in the older patient, in *The Aging Gut. Pathophysiology, Diagnosis, and Management,* Texter, E. C., Ed., Masson Publ., New York, 1983, 57.
21. McGinty, D. P., Bowles, M. H., Patel, G. K., and Texter, E. C., Vascular ectasias of the gastrointestinal tract: a new problem in our aging population, in *The Aging Gut,* Texter, E. C., Ed., Masson Publ., New York, 1983, 75.
22. Webster, S. G. P., Small bowel morphology and function. in *Gastrointestinal Tract Disorders in the Elderly,* Hellemans, J. and Vantrappen, G., Eds., Churchill Livingstone, New York, 1984, 85.
23. Webster, S. G. P. and Leeming, J. T., The appearance of the small bowel mucosa in old age, *Age Ageing,* 4, 168, 1975.
24. Warren, P. M., Pepperman, M. A., and Montgomery, R. D., Age changes in small-intestinal mucosa, *Lancet,* 2, 849, 1978.
25. Montgomery, F. D., Haboubi, N., Mike, N., Chesner, I., and Asquith, P., Causes of malabsorption in the elderly, *Age Aging,* 15, 235, 1986.
26. Nordin, B. E. C., Wilkinson, R., Marshall, D. H., Gallagher, J. C., Williams, A., and Peacock, M., Calcium absorption in the elderly, *Calcif. Tissue Res.,* 21, 442, 1976.
27. Beaumont, D. M. and James, O. F. W., Aspects of nutrition in the elderly, *Clin. Gastroenterol.,* 14, 811, 1985.
28. Webster, S. G. P., Absorption of nutrients in old age, in *Principles and Practice of Geriatric Medicine,* Pathy, M. S. J., Ed., John Wiley & Sons, New York, 1991, 407.
29. Holt, P. R., Digestive disease and aging. Past neglect and future promise, *Gastroenterology,* 85, 1434, 1983.
30. Pettersson, T., Wegelius, O., and Skrifuars, B., Gastrointestinal disturbances in patients with severe rheumatoid arthritis, *Acta Med. Scand.,* 188, 139, 1970.
31. McEvoy, A. and James, O. F. W., Malabsorption syndromes, in *Gastrointestinal Tract Disorders in the Elderly,* Hellemans, J. and Vantrappen, G., Eds., Churchill Livingstone, London, 1984, 96.
32. Wienbeck, M. and Erckenbrecht, J., Motor function of the large intestine. Constipation and diarrhoea, in *Gastrointestinal Tract Disorders in the Elderly,* Hellemans, J. and Vantrappen, G., Eds., Churchill Livingstone, London, 1984, 134.
33. Painter, N. S. and Burkitt, D. P., Diverticular disease of the sigmoid colon: twentieth century problem, *Clin. Gastroenterol.,* 4, 3, 1985.
34. Stewart, E. T. and Dodds, W. J., Predictability of rectal incontinence on barium enema examination, *Am. J. Roentgenol.,* 132, 197, 1979.
35. Brocklehurst, J. C., The problem of faecal incontinence, in *Gastrointestinal Tract Disorders in the Elderly,* Hellemans, J. and Vantrappen, G., Eds., Churchill Livingstone, New York, 1984, 147.
36. Mathers, S. E. and Swash, M., Physiology and pathophysiology of sphincter function, in *Principles and Practice of Geriatric Medicine,* Pathy, M. S. J., Ed., John Wiley & Sons, New York, 1991, 487.
37. Castle, S., Constipation: endemic in the elderly?, *Med. Clin. N. Am.,* 73, 1497, 1989.
38. Laugier, R. and Sarles, H., Pancreatic function and diseases. in *Gastrointestinal Tract Disorders in the Elderly,* Hellemans, J. and Vantrappen, G., Eds., Churchill Livingstone, London, 1984, 243.

39. Meyer, J. and Necheles, H., Studies in old age. IV. The clinical significance of salivary, gastric, and pancreatic secretion in the aged, *JAMA,* 115, 2050, 1940.
40. Rosenberg, I. R., Friedland, N., Janowitz, H. D., and Dreiling, D. A., The effect of age and sex upon human pancreatic secretion of fluid and bicarbonate, *Gastroenterology,* 50, 191, 1966.
41. Kitani, K., Ed., *Liver and Aging,* Elsevier/North-Holland, Amsterdam, 1978.
42. Tauchi, H. and Sato, T., Age change in size and number of mitochondria of human hepatic cells, *J. Gerontol.,* 23, 454, 1968.
43. Schmucker, D. L., A quantitative morphological evaluation of hepatocytes in young, mature and senescent Fischer 344 male rats, in *Liver and Aging,* Kitani, K., Ed., Elsevier/North-Holland, Amsterdam, 1978, 21.
44. Barrows, G. H., Schrodt, G. R., Greenberg, R. A., and Tamburro, C. H., Changes in stainable collagen in the aging normal human liver, *Gastroenterology,* 79, 1099, 1980.
45. Popper, H., Ed., *Collagen Metabolism in the Liver,* Stratton Intercontinental Book Corp., New York, 1975.
46. Koff, R. S., Garvey, A. J., Burney, S. W., and Bell, B., Absence of an age effect on sulfobromophthalein retention in healthy men, *Gastroenterology,* 65, 300, 1973.
47. Kampmann, J. P., Sinding, J., and Moller-Jorgensen, I., Effect of age on liver function, *Geriatrics,* 30, 91, 1975.
48. van Bezooijen, C. F. A. and Knook, D. L., A comparison of age-related changes in bromsulfophthalein metabolism of the liver and isolated hepatocytes, in *Liver and Aging,* Kitani, K., Ed., Elsevier/North-Holland, Amsterdam, 1978, 131.
49. Skaunic, V., Hulek, P., and Martinkova, J., Changes in kinetics of exogenous dyes in the ageing process, in *Liver and Aging,* Kitani, K., Ed., Elsevier/North-Holland, Amsterdam, 1978, 115.
50. Harness, J. K., Styrodel, W. E., and Talsma, S. E., Symptomatic biliary tract disease in the elderly patient, *Am. Surg.,* 52, 442, 1986.
51. Hermann, R. E., Biliary disease in the aging patient, in *The Aging Gut,* Texter, E. C., Ed., Masson Publ., New York, 1983, 27.
52. Stocker, R., Yamamoto, Y., McDonagh, A. F., Glazer, A. N., and Ames, B. N., Bilirubin is an antioxidant of possible physiological importance, *Science,* 235, 1043, 1987.

III
SYSTEMIC AND
ORGANISMIC
AGING

21 AGING OF THE SKELETON, JOINTS, AND MUSCLES

Paola S. Timiras

1 GENERAL CHARACTERISTICS OF MUSCULOSKELETAL AGING

As with other systems of the body, the musculoskeletal system — including the skeleton, joints, and muscles — passes through phases of growth, maturation, and decline. The first two phases have been studied extensively in pediatric medicine. The postmature phase, characterized by a number of involutional processes, is less well understood. It is this latter phase that is addressed in the present chapter.

Compared with other systems, the skeleton is rugged and durable; it usually carries on its tasks into advanced years, and normal and/or abnormal aging of the skeleton is seldom the cause of death. However, it is subject, as are other parts of the body, to various hazards, primarily trauma, deficient metabolism and nutrition, and multiple degenerative changes. The skeleton generally resists damage well and has an efficient self-repair capability.

While aging of the skeleton usually occurs without conscious awareness of the changes on the part of the individual, aging of the articulations (joints) induces considerable physical pain and causes severe disability. Arthritic diseases, some of the most common expressions of joint aging, are among the most frequent and debilitating diseases of old age. The functional impairment and pain resulting from normal or pathologic aging of the joints limit the movement of elderly individuals, thus hindering their ability to care for themselves, eroding their independence, and, by forcing them to varying degrees of immobility, contributing to the decline in competence of other systems (such as the circulatory system). Some pathology of the joints begins at a relatively young adult age (rheumatoid arthritis may begin in adolescence) but with increasing prevalence with advancing age. Thus, prevalence of rheumatoid arthritis is less than 1% before the age of 30 years and thereafter rises in each decade to 1 to 3% in the late 50s and 8 to 11% in the late 60s. Other manifestations of articular aging occur at later ages; osteoarthritis that affects 85% of persons 70 to 79 years of age is one of the major causes of invalidism, confining the affected individuals to bed or to the wheelchair.

The aging process in muscle is greatly complicated by the fact that muscle fibers do not constitute an homogenous tissue (e.g., differences between skeletal and cardiac muscle, striated and smooth muscle). Furthermore, the state of the tissue at any time depends on the extent and nature of many influences (e.g., nutritional, neural, hormonal, exertional). The characteristic decline of muscular performance associated with advancing age is, therefore, variable and may be caused not only by primary aging changes in the muscle fibers but also by aging of other body systems — nervous, vascular, and endocrine.

2 AGING OF THE SKELETON

The skeleton, the heaviest and most durable part of the body, provides the body framework and derives its properties from the unique characteristics of bone. In addition to this function of body support, the bone provides storage for calcium and other minerals, thereby aiding in mineral homeostasis; it also contributes, in association with the lungs and kidneys, to the maintenance of acid–base balance by providing additional phosphate and carbonate for buffers.

2.1 Structure and Function of Bone

Bone is a hard form of connective tissue consisting of bone cells or osteocytes embedded in a collagenous protein matrix that has been impregnated with mineral salts, especially phosphates of calcium. The matrix occurs in two phases: an organic phase that consists of collagen, proteins, and glucosaminoglycans (Chapter 22) and an inorganic phase that consists mainly of hydroxyapatites (calcium phosphate) and minor amounts of other minerals. The collagen fibers provide resilience and the minerals, hardness. Given individual variations, it can be stated that in childhood about two thirds of the bone substance is composed of connective tissue, whereas in aged individuals about two thirds consists of minerals. This transposition in content results in decreased flexibility and increased brittleness with advancing age.[1]

Bone Histology

Bone is distinguished histologically as "compact bone", found in shafts of long bones and outer surfaces (periosteum) of flat bones; "cancellous bone", making up the trabecular space containing the bone marrow; and the "woven bone", an immature form of bone also involved in fracture repair. The cells in bone are primarily concerned with bone formation and resorption. Osteocytes, mainly in cancellous bone, maintain bone structure; osteoblasts in the periosteum and endosteum form the bone matrix; and osteoclasts in the same regions as the osteoblasts resorb bone by phagocytosis and digestion in their cytoplasm. Bone is well vascularized and contains an abundant nervous network.

2.2 Bone Remodeling

Maintenance of bone structure and function is a dynamic process regulated by a number of factors, as illustrated in Figure 21-1. Calcium in bone is derived from circulating free calcium, which is dependent upon adequate calcium absorption, mainly through the gastrointestinal tract. Circulating calcium also depends on its excretion, mainly through the kidneys. A number of osteogenic factors promote bone formation through stimulation of osteoblastic activity, or bone resorption through osteoclastic activity.

Bone is in a continuous state of flux. Although net bone mass does not change throughout much of adult life, bone is

0-8493-8979-8/94/$0.00+$.50

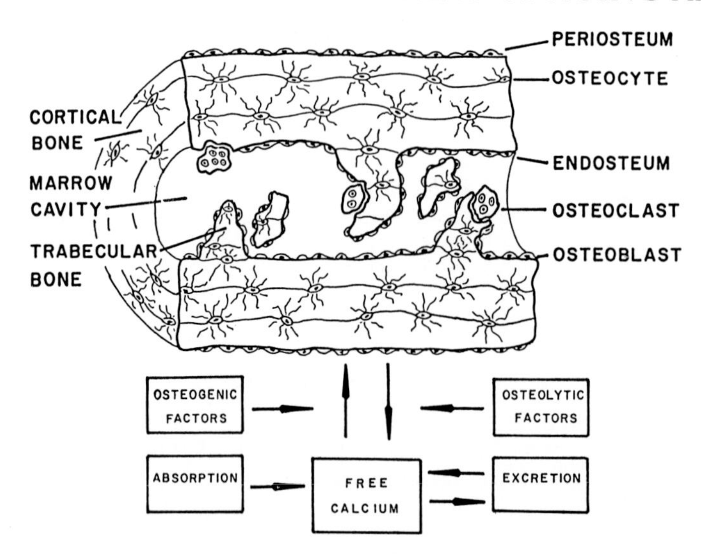

FIGURE 21-1 *Factors responsible for maintenance of bone structure.* Calcium in bone is derived from circulating free calcium which depends on calcium absorption and excretion. Formation of new bone (osteogenesis) depends on osteoblasts, and destruction of all bone (osteolysis) depends on osteoclasts. New bone is continuously formed in the periosteum and endosteum. Active bone is formed of osteocytes, which are part of the cortical and trabecular bone. Blood cells are continuously produced in the bone marrow.

never metabolically at rest and constantly remodels and reappropriates its mineral stores along lines of mechanical stress. The major site of remodelling is the cancellous bone, lining the bone marrow cavities. Thus, bone maintenance includes formation of new bone by the osteoblasts, resorption of old bone by osteoclasts, and carrying out of mature bone functions by osteocytes. The mechanisms responsible for the calcification of newly formed bone matrix are uncertain despite intensive investigation; precipitation of calcium phosphate may depend on some critical concentration and is associated with the activity of certain enzymes (e.g., alkaline phosphatase) and specific proteins capable of binding calcium.

With aging, the balance between rates of bone formation and bone resorption is disturbed, and the ensuing changes lead to a decrease in bone mass. After age 40, formation rates remain constant while resorption rates increase. Over several decades, the skeletal mass may be reduced to half the value it had at 30 years. Progression of this loss may be measured by counting the number of Haversian canals or osteons, which refer to a feature of bone structure represented by cylinders of bone containing a central blood vessel and nerve fibers. With advancing age, the number of these osteons increases together with an increase in the bone shaft beginning at the ends of long bones. Between the ages of 42 and 52 years, this medullary cavity extends to the neck of the femur; between 61 and 74 years it reaches the epiphyseal line close to the articular end. Similar changes occur in the humerus. After 61 years, the outer surface becomes rough and the cortex thinner; the medullary cavity reaches the epiphyseal line; and after 75 years, there is little spongy tissue left and the cortex is very thin. Articular surfaces become very thin and may collapse.

The major factor in these age-related changes is a loss of bone matrix. However, this loss is confined primarily to the inner bone core. In the elderly, the periosteal tissue at the outer surface of the bones tends to remain constant or even increases in some bones (e.g., metacarpal bone of the hand),[2] but the endosteal tissue at the interior of bones is increasingly resorbed.[3]

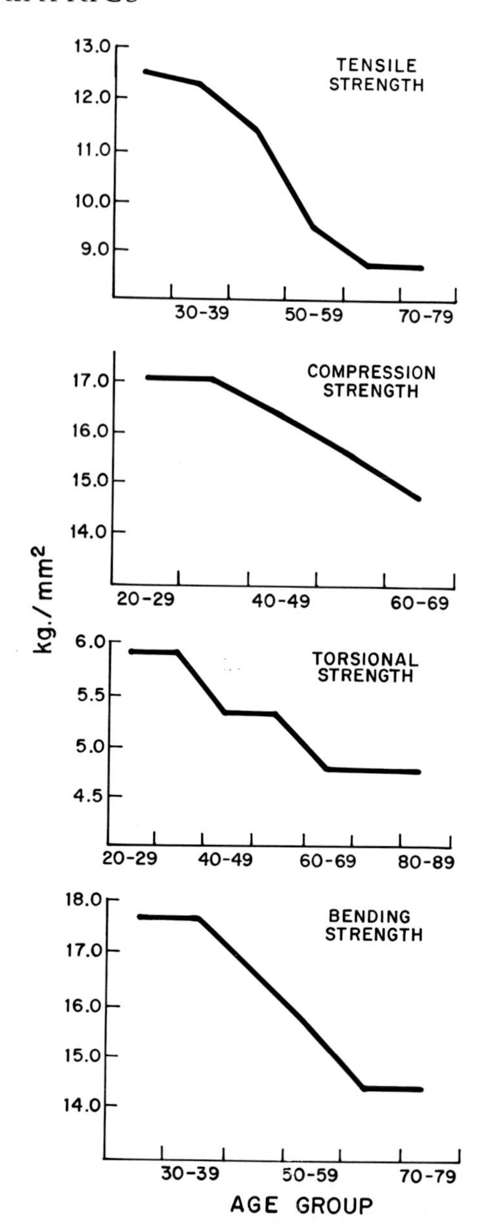

FIGURE 21-2 *Changes in bone strength with aging.* Bone strength provides the ability of the bone to undergo force applied in (bottom to top) bending, twisting (torsional strength), compressing, and stretching (tensile strength).

2.3 Bone Strength

Bone strength is an important property which allows bones to withstand the forces applied in the various movements of daily life. Several tests of bone strength have been studied and the results show life cycle trends from youth through maturity into old age (Figure 21-2) and disclose a consistent decline of strength with aging.[4] Comparison of the timetable of aging of several musculoskeletal components reveals the fastest decrease in strength of the cartilage followed by muscle, bone, and tendon last. Also, the comparison of bone with other tissues shows a much slower rate of decline for bone than for intestine and muscle but faster than for kidney.[5]

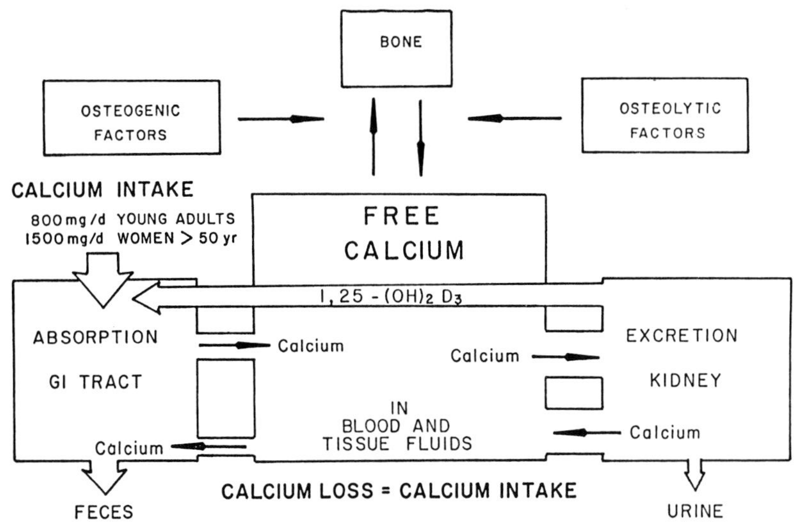

FIGURE 21-3 *Calcium metabolism.* Note the different daily requirements of calcium with age: almost double in women after 50, as compared to young adults.

2.4 Factors Affecting Bone Aging: Changes in Calcium Metabolism

While factors that affect bone remodeling are not completely known, adequate circulating calcium is vital in maintaining bone mass. Indeed, the body regulates few parameters with greater fidelity than the concentration of extracellular calcium. The constancy of these levels depends in part on its absorption in the intestinal mucosa and its excretion from the kidneys (Figure 21-3).

The human body contains 1100 g of calcium or 1.5% of body weight; 99% of this calcium is in the skeleton. The need for calcium increases with aging. For example, while for young adults the recommended daily allowance is 800 mg, for women over 50 years of age it is increased to over 1500 mg. This increased need for calcium may be explained by a progressively less efficient absorption from the upper intestinal tract (Chapter 23).[6] Thus, in the elderly more dietary calcium is needed to maintain an adequate calcium balance. In addition to dietary calcium, other dietary components, physical exercise, and gender influence bone growth and aging.[7-9]

2.5 Calcium Metabolism

Three hormones, parathyroid hormone, calcitonin (Chapter 14), and calcitriol, are primarily concerned with calcium metabolism (Figure 21-4):

- *Parathyroid hormone* directly increases bone resorption and mobilizes calcium, thereby elevating plasma calcium. It also depresses plasma phosphate by increasing its urinary excretion.
- *Calcitonin* lowers body calcium by inhibiting bone resorption.
- *Calcitriol* (dehydroxycholecalciferol, vitamin D_3), a sterol derivative, increases calcium and phosphate absorption from the intestine and decreases their renal excretion; it also enhances bone resorption.

Calcitriol derives through a series of steps either from the diet through absorption in the intestine, or through transformation of cholesterol in the skin by the action of sunlight. In the skin the first step is the formation of previtamin D_3, which is converted to vitamin D_3 or cholecalciferol. Vitamin D_3 from the skin and intestine is carried to the liver where it is hydroxylated to calcidiol, which, in turn, is carried to the kidney where it is again hydroxylated to form the active product calcitriol.

The regulation of the relationship between calcium in blood and in bone depends primarily on parathyroid hormone, but it also involves other factors acting on calcium in tissues and cells. Indeed, calcium has many functions and is known to be a "universal regulator".[10,11] A summary of the many actions of calcium groups them into six categories: cell movement, cell excitability, cell secretion, phagocytosis, intermediary metabolism and respiration, and cell reproduction (Table 21-1).

Despite the universal role of calcium in cell function and metabolism, few studies have inquired into the potential changes in calcium activity with aging. The most prominent involvements of calcium related to aging are listed in Table

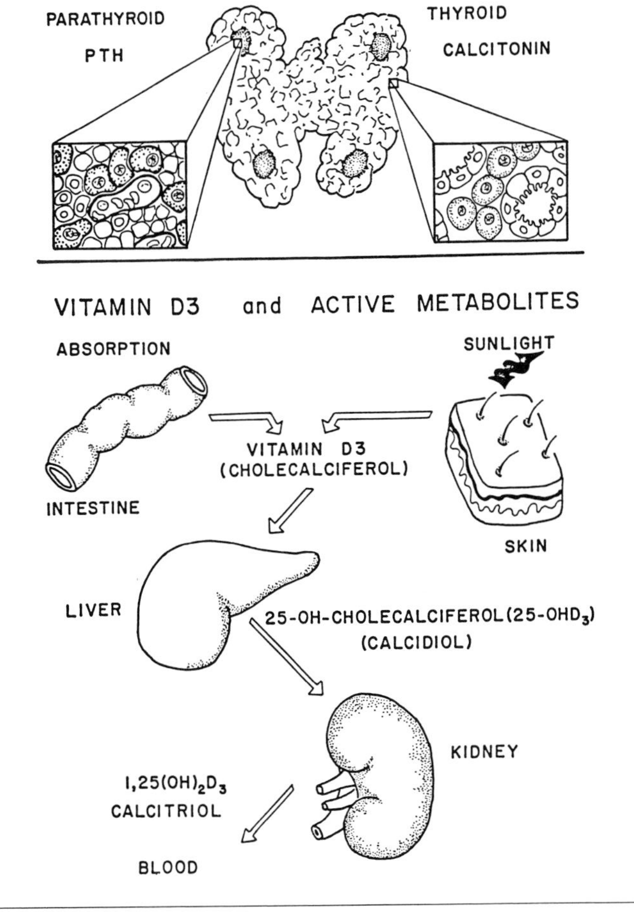

FIGURE 21-4 *Major hormones that regulate calcium metabolism.* (Top) The parathyroid gland secretes the parathyroid hormone (PTH) which acts directly on bone to increase its resorption and to mobilize calcium, thereby increasing plasma calcium. Calcitonin from the thyroid gland inhibits bone resorption and lowers serum calcium. (Lower part) Calcitriol, a sterol derivative, increases calcium absorption from the intestine and decreases renal excretion. It also enhances bone resorption. Vitamin D_3 derives from dietary sources and is absorbed in the intestine or is formed in the skin under the influence of ultraviolet radiation. In the liver, vitamin D_3 is converted to calcidiol which, in turn, is activated in the kidney to calcitriol, the most active form in calcium regulation.

TABLE 21-1

Intracellular Calcium Actions

Action	Examples of change of state
(1) Cell movement	Muscle contraction
	Ciliate and flagellate movement
	Chemotaxis
(2) Cell excitability	Muscle action potential
	Myocardial action potential
	Response of the eye and other photoreceptor types to light
(3) Cell secretion	Neurotransmitter and hormone secretion
	Exocrine secretion gland
(4) Cell phagocytosis	Vesiculation of particles or soluble substances
(5) Intermediary metabolism and respiration	Glucose production
	Lipolysis
	Blood coagulation
(6) Cell reproduction	Lymphocyte transformation
	Ovum fertilization and sperm capacitation
	Ovum maturation and meiosis

Note: In each category, changes in intracellular calcium translate into specific actions. For example, in (1), by promoting cell movement, calcium causes muscle contraction and ciliate and flagellate movement and chemotaxis. In (2), the production of an action potential in appropriate cells causes muscle contraction or photoreceptor response. In (3), another pertinent example related to neural and endocrine functions is the role of calcium in cell secretion of neurotransmitters and hormones. In (4), calcium is involved in the uptake of particles by phagocytosis or uptake of soluble substances by vesiculation. (5) refers to the role of calcium in glucogenesis, lipolysis, and prostaglandin synthesis. And, finally, (6) intracellular calcium acts in the regulation of various phases of cell reproduction.

21-2. Considering the many actions of calcium in cell metabolism and function, it would seem reasonable to hypothesize that alterations in these actions have an important bearing on the causes and course of the aging process. Research in this direction is still limited and should be pursued actively.

2.6 Hormonal Regulation of Bone Metabolism

Bone changes with aging apparently occur without marked alterations in the parathyroid gland or in the levels of parathyroid hormone (Chapter 14). However, in some old women, parathyroid hormone levels increase with aging, but the contribution of this increase to bone loss may be minimal.[12] There seem to be some racial differences with black and Oriental postmenopausal women having lower levels of parathyroid hormone and higher levels of calcium well into advanced age than white women. There seems to be a gender difference as well, with men maintaining lower levels coincident with a lower incidence of osteoporosis than women (Chapter 12).

The increase in parathyroid hormone levels with aging may represent a compensatory response to reduced intestinal absorption of calcium, hence lower plasma calcium levels. Reduced exposure to sunlight in the immobile and housebound elderly will also impair vitamin D manufacture in the skin and accentuate any dietary deficiencies. (Table 21-3) Reduced renal degradation and excretion of parathyroid hormone may be another important factor; however, it is still controversial whether such a reduction occurs and how serious it is.

TABLE 21-2

Intracellular Calcium-Dependent Changes Relating to Aging

Actions	Relation to aging
(1) Intermediary metabolism and respiration	Activation of oxygen radicals
	Obesity and diabetes
	Arthritis and other diseases
(2) Tissue calcification	Loss of bone calcium (osteoporosis)
	Abnormal deposition on normal and injured tissues (e.g., atherosclerotic lesions)
(3) Cell excitability	Changes in cell potentials
	Changes in response to drugs
(4) Cell secretion	Alterations in neurotransmitter and hormone production
(5) Phagocytosis	Alterations in immune responses

Note: The most prominent involvements of calcium related to aging appear to be (1) triggering the intracellular production of oxygen radicals which, as discussed in Chapter 6, would lead to the accumulation of potentially toxic substances; (2) to affect normal and abnormal bone metabolism; a possible role in the pathogenesis of a number of diseases such as arthritis, obesity, diabetes, cystic fibrosis, etc.; calcification of injured and necrotic tissues, such as occurs in atherosclerotic lesions (Chapter 16); (3) a possible role of calcium in age-related changes in cell excitability and in the action of several classes of drugs such as anesthetics, analgesics, or cardiovascular drugs; (4) by its involvement on ion movement, calcium may affect neurotransmitter and hormone production and secretion; and (5) its participation in phagocytosis may relate to aging of the immune system (Chapter 7).

Diseases of the Parathyroid Gland

Diseases of this gland are infrequent at all ages; however, they bear mentioning here for the symptoms associated with aging may mask parathyroid pathology. Indeed, if the disorder is recognized as a possible parathyroid dysfunction and is corrected, the symptoms described as aging changes may be ameliorated. Hyperparathyroidism presents with a variety of symptoms such as increased plasma calcium (hypercalcemia), renal calculi, peptic ulceration, and, in a few individuals, mental aberrations with psychotic components. The latter have been sometimes erroneously identified as senile dementia of the Alzheimer type (Chapter 9). In advanced stages, characteristic bone lesions are present. Hypoparathyroidism is quite rare at all ages and is easily recognizable for it generally follows ablation of the glands during thyroid gland surgery.

Regulation of bone metabolism depends on several hormones and in aging may depend on a number of endocrine changes:

1. In women, the decrease in *estrogen* levels at menopause plays a key role in the induction of a common age-related degenerative bone disorder known as osteoporosis[13] and discussed already in Chapter 12. Replacement therapy with estrogens prevents osteoporosis (Chapter 12).
2. *Glucocorticoids* lower plasma calcium levels and over long periods of time may cause osteoporosis by decreasing bone formation (due to inhibition of cellular replication, protein synthesis, vitamin D, absorption of calcium from the intestine, and function of osteoblasts) and increasing bone resorption (due to stimulation of parathyroid hormone secretion).
3. *Growth hormone* increases calcium excretion in urine, but it also increases intestinal absorption of calcium and this effect is greater than that on excretion and hence produces

TABLE 21-3

Parathyroid Hormone Changes with Aging

Increased parathyroid hormone plasma levels
Causes
Decreased calcium intestinal absorption
Decreased production of Vitamin D (skin)
Parathyroid tumors
Consequences
Increased bone resorption (osteoporosis)
Symptoms resembling SDAT
Experimental progeria-like syndrome (rats)
Lesser increase of parathyroid hormone levels
Men
Black and Oriental women
Lower incidence and severity of osteoporosis

a positive calcium balance. Somatomedins, induced by growth hormone, stimulate protein synthesis in bone.

4. *Thyroid hormones* may induce hypercalcemia and hypercalciuria and in some cases may also produce osteoporosis, but the mechanisms of these actions are unclear.
5. *Insulin* promotes bone formation, and there is significant bone loss in diabetes.
6. *Local growth or inhibitory factors* also influence bone formation and resorption, but their significance is uncertain although they may be involved in the continuing remodeling of bone, perhaps in response to stress. Most of these factors (e.g., epidermal growth factor, fibroblast growth factor, platelet-derived growth factor) promote bone growth. Prostaglandins E secreted by certain tumors increase plasma calcium levels. Osteoclast activating factor, produced by the lymphocytes, induces bone loss in tumors of the bone marrow. It has been suggested, but not proven, that this factor may be involved in the formation of marrow cavities in normal bones.
7. *Physical activity* also affects bone in several ways: it increases stress and strain on the skeleton due to muscular contraction and gravity, it improves blood flow to exercising muscles and, indirectly, increases venous return, and it stimulates bone mineral accretion. These effects are most marked in the young but are also affected, although by a lesser degree, in the old.[14]

2.7 Aging-Related Fractures

Fracture patterns in the elderly differ markedly from those in the younger adult. Whereas in the younger adult (20 to 50 years) considerable violence is required to break the bones, in the elderly fractures result from minimal or moderate trauma. This is due to the progressive loss of absolute bone volume, both compact and spongy bone, with aging. While the consequences of bone loss become manifest at 40 to 45 years of age in women and 50 to 60 years in men, bone loss may in fact begin much earlier, at the end of the growth period.

In younger adults, fracture incidence is lower in females than in males; in the elderly, however, fracture incidence is much higher in females than in males, especially for fracture of the vertebral bodies, the lower end of the forearm, and the proximal femur. The relation between menopause and accelerated bone loss (osteoporosis) in females is discussed in Chapter 12.

Not only is the incidence of fractures higher in the elderly than young, but the sites of fracture are often different. In the elderly, fractures occur through cancellous bone, usually next to a joint, rarely through the shaft of the bones, the most frequent site of fracture in the young. Orthopedic interventions to prevent or repair fractures are more difficult and recovery is slower in the elderly than in the young, due to bone fragility and the overall less robust condition of the elderly.[15,16]

In addition to osteoporosis, other age-related bone disorders are associated with fractures. One of the most frequent is *Paget's disease* or *Osteitis Deformans,* characterized by pain, deformity, and fractures of the bones. It affects men more than women. The early lesions are localized osteolytic lesions and are followed by active remodeling, leading to excessive and somewhat disorganized ("mosaic" or "woven" aspect) bone tissue balance. The resulting bone, although more dense than normal bone, has lost part of its elastic properties and is more susceptible to fractures.[15] Complications include neoplastic transformation, neurologic signs due to compression of CNS structures, and heart failure. Management is based on symptomatic administration of aspirin for pain relief and of calcitonin and diphosphonates for decrease of bone resorption and bone formation.

Ethnic Differences in Bone Mass and Consequences in Aging

The rates of fractures due to osteoporosis and other bone diseases associated with old age are substantially lower among black persons than white persons (three times lower among black women and five times lower among black men).[17] These differences in rates of fractures are generally attributed to the 10 to 20% greater mass and density of adult bone in blacks and may begin before birth.[18] In blacks, bone remodeling also proceeds at a lower rate than in whites.[19] Taken together, these observations suggest that hormonal, metabolic, and genetic factors play a role in these black/white differences. Age-matched bone density and mass among other ethnic groups in the U.S. population, although less studied than those among blacks and whites, show significant differences; for example, a survey of proximal femur fracture rates show that incidence rates are higher among Asian women than white women.

3 AGING OF THE JOINTS

The pattern and prevalence of joint changes in the elderly reflect the following:

- Certain disorders arise more frequently with increasing age.
- There is a steady cumulative effect with age inasmuch as many joint disorders are chronic.

By age 65, 80% of the population has some articular disorder. Beyond the aging changes affecting the articulations themselves, the problems of added illness, frailty, diminished motivation, and social isolation affect the problems of management and outcome of the disorder.[20,21]

Osteoarthritis and rheumatoid arthritis are examples of articular disorders occurring with increasing frequency from middle to old age. In fact, osteoarthritis is so common among the elderly that it is often assumed to be a normal accompaniment of aging rather than a disease. Radiological signs of osteoarthritis are universal in the later decades of life and, in many cases, the changes are not associated with significant pain or disability. The dividing line between disease and normal aging is particularly difficult to draw with respect to degenerative changes of the cartilage.[20]

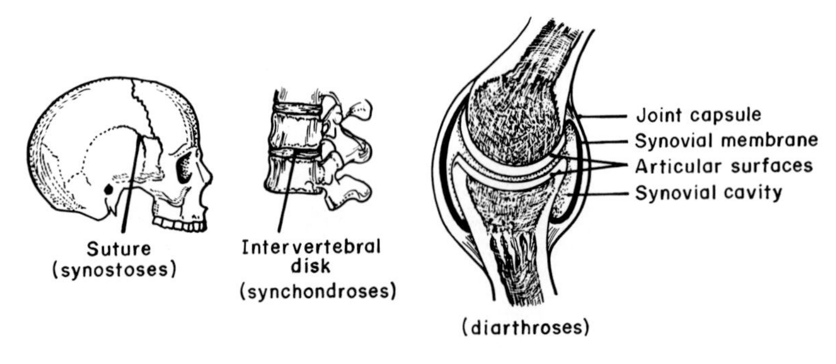

FIGURE 21-5 *The three major types of joints.*

Structure of the Joints

The articular system is comprised of simple and complex joints associated with the skeletal system and with their blood and nerve supplies. They represent junctions between two or more bones or cartilages (Figure 21-5). In the skull, the joints (synarthroses) are immovable, and the connected bones are separated by a very thin layer of connective tissue. The articulations between the vertebrae (amphiarthroses) are a little more movable and the bones, in this case, are united by dense fibrous tissue and intervening cartilage. Most bones are freely movable (diarthroses), as the adjoining ends are coated with smooth cartilage separated by a short tube of strong fibrous tissue containing *synovial fluid.* Joint cartilage is a special type of connective tissue with an extracellular matrix of proteoglycans and collagen, synthesized by the cartilage cells or chondrocytes.

3.1 Aging-Related Changes in Joints

With aging, *bony excrescences or osseous outgrowths (called osteophytes or bone spurs)* occur on the heads of long bones, e.g., on the head of the femur in 33% of individuals beyond 50 years of age: they are not progressive and are viewed as a purely age-related phenomenon. Areas where smooth cartilage has been replaced by a rough surface occur in all joints and are often seen as early as the second decade and spread to the periphery of the joint with age.

The *undulations and hollows* that can be seen on the cartilage surface by electron microscopy increase both in depth and diameter with increasing age, and the outer-surface irregularities become more common. Whether these consistent surface changes may be related to the development of osteoarthritis is not known. Osteoarthritis is associated with *cartilage thinning.*

It is puzzling to note that, in general, cartilage actually becomes thicker with increasing age,[22] with the exception of the patella (kneecap) which becomes thinner, especially in women.[23] As cartilage ages, it loses some of its elasticity and becomes *more easily stretched.* Such changes lead to *easier fatigability and higher susceptibility to osteoarthritis.* There is a progressive *pigmentation* of the cartilage cells due to deposition of amino acid derivatives (probably from the protein of the matrix), and a gradual reduction in collagen but no change in water content; this latter change is invariably associated with the early stages of osteoarthritis. The lack of changes or the presence of relatively minor biochemical changes with aging negates a generalized time-dependent deterioration of chondrocyte function in articular cartilage.[24]

* This section contributed by Mary Letitia Timiras, M.D., Newark Beth Israel Medical Center, Newark, NJ.

3.2 Pathological and Clinical Aspects of Musculoskeletal Aging*

Normal and pathological changes of the musculoskeletal system in the elderly are frequent and severely curtail the well-being of this population. They contribute significantly to Immobility, one of the dreaded "five I's" of geriatrics (Chapter 3). It is difficult at best to encourage an elderly individual to maintain a program of physical fitness or activity if mobility is limited by joint pain. Chronic pain and disability of whatever cause or nature inflict psychological pain and dampen the morale of any individual, especially the elderly already prone to depression. The loss of the ability to participate in certain physical activities can be quite devastating to an elderly individual who may be suffering from other concomitant losses.

The list of musculoskeletal disorders of the elderly is quite long, and an all-inclusive description of these disorders is beyond the scope of this text. A few of the more common entities will be described briefly in this section as a comparison and elucidation of "normal" aging changes.

Musculoskeletal disorders can essentially be divided into two major types (Table 21-4):

- Those which affect joints without involvement of other organ systems
- Those which affect the musculoskeletal system as a manifestation of systemic disease involving several organ systems

Most of the systemic disorders mentioned in this section are often referred to as "collagen–vascular" as they are frequently manifested by changes in connective and/or vascular tissues. Several of these disorders are also referred to as "autoimmune" because they are associated with the presence of antibodies which attack and damage the host's own tissue (Chapter 7).[25] Some of the characteristics of the most frequent of these disorders are summarized here as examples of articular pathology with aging.

3.3 Osteoarthritis and Gout

As already indicated, the most common articular disorder of the elderly is *osteoarthritis* or *degenerative joint disease.* The former term is preferable because it carries less negative connotation of decay and degeneration. Although it affects over three quarters of the elderly, only 30% have any symptoms thereof and only 10% suffer significant disability therefrom. In a large proportion of the elderly, osteoarthritic lesions are discovered incidentally (for example, at the occasion of a chest X-ray). Risk factors include heredity as well as a history of trauma or injury to a certain body part or bone. Little is known of the pathogenesis of osteoarthritis. Why the process occurs at an accelerated pace in some individuals and why it affects certain joints and not others remain a mystery.

The early changes occur in the articular cartilage that lines the joint surfaces of the bones and are due to a loss of proteoglycans.[26] Eventually the joint space narrows. The bone underlying the articular cartilage is itself undergoing aging changes and becomes more susceptible to damage and microfractures. In the attempt to repair the damage, cysts and osteophytes form in addition to sclerosis.

Osteoarthritis is not a systemic disease; it affects only individual joints. Susceptible locations include the hands, hips, knees, feet, and spine. Symptoms usually begin with aches

TABLE 21-4

Types of Musculoskeletal Disorders

	Nonsystemic	Systemic
Noninflammatory	Osteoarthritis	
Inflammatory	Gout	Nonautoimmune Polymyalgia rheumatica Giant cell arteritis Autoimmune Rheumatoid arthritis Systemic Lupus Erythematosus

TABLE 21-5

Management of Osteoarthritis

Physical Interventions	Medical Treatment
Weight loss	Acetaminophen (Tylenol)
Cane	Aspirin
Exercise to increase muscle strength	Nonsteroidal anti-inflammatory drugs
Orthotics	Intra-articular steroid injection
Intra-articular lavage	
Arthroplasty (joint replacement)	

and pain in the involved joints. Initially pain occurs with motion or weight bearing. In later stages, pain can occur at rest. Hand involvement occurs more frequently in women than in men and usually affects the distal interphalangeal joints. Small paired dorsal cysts of these affected joints are called Heberden's nodes. The first carpometacarpal joint at the base of the thumb is also commonly affected and gives the thumbs a squared-off appearance. Hip involvement occurs more frequently in men and can present with knee or groin pain.

Diagnosis and Treatment of Osteoarthritis

Diagnosis is made from clinical presentation and X-rays. It is important to differentiate the disease from other much more serious and life-threatening illnesses. For example, an elderly person with back pain should be carefully evaluated to ensure that no serious pathology such as metastatic malignancy exists before attributing the symptoms to osteoarthritis.

While significant advances have been made in the treatment of old age-associated diseases (e.g., atherosclerosis, Chapter 16), less progress has occurred with respect to osteoarthritis. There are no measures known that prevent or reverse the progressive joint damage. Therefore, management revolves around symptomatic relief of pain (Table 21-5). In view of the lack of inflammatory activity in the affected joints, some argue that acetaminophen (Tylenol) is a good first-line medication, particularly in those susceptible to the gastrointestinal effects of aspirin, which is just as effective. The nonsteroidal anti-inflammatory agents (such as ibuprofen) are more potent but potentially have more side effects.

Avoidance of heavy weight bearing can be helpful especially in osteoarthritis of the knee. Weight loss, use of a cane, and orthotics (for the feet) can aid in diminishing some of the weight bearing. Muscle strengthening is also to be encouraged especially in knee osteoarthritis. Intraarticular steroid injection and lavage have been used with temporary amelioration of the symptoms. For osteoarthritis of the hip and knee, joint replacement or arthroplasty are available to individuals who present low operative risk and to whom continued activity is important.[16]

Another localized joint disorder is *gout*. Unlike osteoarthritis, however, gout induces, in the articulation involved, an inflammatory response which is manifested by joint swelling (due to fluid accumulation), warmth, and the presence of white blood cells in the joint fluid. *Urate crystals in the fluid are diagnostic of gout.* This disease can occur in persons younger than 60 years of age who have a metabolic defect in the breakdown of uric acid, a product of protein metabolism formed from the breakdown of purines and synthesized from glutamine. Gout occurs commonly in the elderly because of *decreased urate excretion by the kidney*. This diminished urinary excretion can be further exacerbated by the use of diuretics (for reducing hypertension) which further decrease urate excretion (Chapters 19 and 23). The most frequently affected joint is the great toe (podagra), but other joints such as ankle, knee, elbow, and wrist can also be involved.

An articular alteration similar to gout is *pseudogout* or *chondrocalcinosis* or *calcium pyrophosphate deposition disease*. This condition is distinguished from gout by the presence in the joint fluid or in the calcifications in joint cartilage of calcium pyrophosphate crystals. In both conditions, the management involves the use of nonsteroidal anti-inflammatory drugs (e.g., aspirin, indomethecin). The routine use of substances that either decrease production of uric acid (naproxen, colchicine) or increase its excretion (probenecid, sulfinpyrazone) is recommended only when serum uric acid levels are moderately increased and gouty attacks occur frequently.

3.4 Autoimmune Arthritides

Cartilage is an *avascular tissue,* which is usually protected from immune recognition and, in this sense, is truly an *immunologically privileged structure.* However, both proteoglycans of the cartilage matrix and chondrocytes manifest antigenic properties that are usually sequestered in the intact tissue. It is possible to envision that alteration of the integrity of cartilage, irrespective of the cause, may facilitate a leaching out into the circulation or may promote exposure of antigenic molecules of the tissue. An ensuing humoral or cell-mediated immune response would induce not only inflammation but also cartilage degradation. Furthermore, because of the antigenic cross-reactivity between matrix constituents and tissues elsewhere in the body, sensitization initiated at one particular articulation may be responsible for the genesis of autoimmunicity throughout the body, thereby creating a systemic autoimmune disease. This is the case of *rheumatoid arthritis,* which is a systemic disease involving several organ systems in addition to the musculoskeleton.[27,28] It occurs in all age groups but is more frequent in the elderly, of whom 75% undergo a clinical course similar to that of young patients (e.g., weight loss, malaise, vasomotor disturbances, symmetric joint swelling with stiffness, redness, warmth and pain, subcutaneous nodules, sometimes enlarged spleen and lymph nodes). However, 25% can exhibit atypical symptoms such as more severe proximal muscle weakness, more marked morning stiffness, higher fever, and more severe weight loss and malaise, while the inflammatory involvement of the joints may be less evident.

3.5 Pathogenesis of Rheumatoid Arthritis

Although rheumatoid arthritis is considered an autoimmune disease, multiple factors may contribute to its pathogenesis including genetic predisposition and neural involvement besides impaired immunological control of virus infection and autoantibody formation. Involvement of the nervous system is suggested by the release at the sensory nerve endings to the joint tissues of the neuropeptide, substance P, involved in the transmission of pain signals.[29] Substance P would stimulate the release of prostaglandin and activate the enzyme collagenase from the synovial cells (synoviocytes). Additionally, substance P would stimulate proliferation of these cells. The symmetrical distribution of the joints affected and the role of psychologic trauma in the onset and exacerbation of the disease are clinical observations which support a role for the nervous system in the disease. These observations point to a pathway by which the nervous system might be directly involved in diseases of the joints.[28]

Differentiation from osteoarthritis is based on the different preferential localization of the joint lesions and the presence in rheumatoid arthritis of some markers such as rheumatoid factor, increased antinuclear antibodies, elevation of erythrocyte sedimentation rate, and gamma globulins. The management of rheumatoid arthritis is complex and includes both steroidal (corticosteroids) and nonsteroidal anti-inflammatory drugs. Gold injections — and more recently, oral gold — and agents used in chemotherapy are also effective.

Another autoimmune disease also involving the articulations is *systemic Lupus Erythematosus*. The disease can occur in all age groups; however, it is less common and generally exhibits a clinically milder course in the elderly than in younger individuals.

3.6 Polymylagia Rheumatica

This syndrome (abbreviated PMR) rarely occurs in persons less than 50 years of age. The average annual incidence in a studied population increased from about 20 per 100,000 persons aged 50 to 59 years to 112 per 100,000 persons aged 70 to 79 years.[30] It occurs in females more than in males and virtually never occurs in blacks and Orientals.

PMR is characterized by pain and stiffness in pelvic and shoulder girdles, which is worse in the morning (making it difficult to get out of bed) and after periods of inactivity. This reflects the inflammatory involvement of the synovial membranes of these joints.[31] This is a systemic disease and, in some elderly individuals, there may be only general symptoms such as weight loss, fever, anorexia, malaise, and apathy. Anemia is commonly a complication. An elevated erythrocyte sedimentation rate is one of the nonspecific diagnostic findings. A prompt response to moderate doses of corticosteroids is also characteristic of this disorder.

3.7 Giant-Cell Arteritis

This condition, also abbreviated GCA or called temporal arteritis, often overlaps with PMR, 60% of patients with GCA having PMR as well. It involves an inflammatory response in medium-sized arteries incorporating giant cells; hence, the diagnosis is made by biopsy of arteries. The temporal arteries are most commonly affected (hence the name) and are more readily accessible for biopsy. Headache is the most common symptom; however, painful chewing caused by inflammation of the facial artery, transient double vision due to ischemia of the extraocular muscles, and sudden unilateral blindness due to occlusion of the terminal branches of the ophthalmic artery may also occur. This illness can lead to very serious problems stemming from the ischemia that results when an artery is affected. High doses of corticosteroids are required and patients may have to remain on lower doses to prevent relapse. It is important to exclude the possibility of GCA complications in patients with PMR although obtaining temporal artery biopsies in all PMR patients is controversial.

Pattern and Management of Joint Disorders in the Elderly

Most of the disorders involving the articulations and described here have these qualities:

- They become manifest for the first time beyond the age of 60 years.
- They began at a much younger age.
- They are often generalized to the entire organism.
- At the level of the articulations, they may be superimposed on normal aging processes.

Establishing the diagnosis of an arthritis, especially when it is acute, tends to be more problematic in the aged than in young individuals, but, in view of the high incidence of these disorders in the elderly, it is important that the diagnosis and treatment be carefully conducted to keep the symptoms under control and to maintain mobility and independence of living as long as possible.

4 AGING OF SKELETAL MUSCLE

Muscle strength declines with increasing age irrespective of the muscle group considered. Peak of muscle strength occurs between 20 and 30 years of age and the decline is continuous thereafter, accelerating progressively with aging, although the rate of aging may vary depending on the muscle groups studied and the type of test used for measurement of muscle function. For example, the diaphragm remains active throughout life and undergoes little change with aging (Chapter 18). In contrast, the soleus muscle of the leg may be relatively inactive in the less mobile elderly and shows decreased strength with aging.[32] Many of the age-related changes identified in muscle are similar to those of other tissues, but those occurring in muscle are particularly prominent because of the proportionally high distribution of muscle throughout the body and its relationship to lean body mass (body mass minus bone, mineral, fat, and water).[33] However, physical training, even in the elderly, can increase muscle power.[34]

Skeletal Muscle Structure and Function

The major function of muscle is to contract with utilization of energy and production of work and heat. Muscle is formed of individual muscle fibers containing fibrils which are divisible in filaments made up of contractile proteins, *myosin, actin, tropomyosin, and troponin*. Myosin forms the thick muscle filaments, and the three other proteins form the thin filaments. Cross-linkages are formed between myosin and actin molecules. During contraction, by breaking and reforming of cross-linkages between myosin and actin, the thin filaments slide over the thick filaments, thereby shortening muscle length and utilizing ATP for energy. Muscle fibrils are surrounded by the sarcotubular system formed of vesicles and tubules distinguished into a T system and a sarcoplasmic reticulum. The function of the T system is the rapid transmission of the action potential from the cell membrane to all myofibrils. The sarcoplasmic reticulum regulates calcium movement and muscle metabolism.

Muscle contraction is initiated by calcium release through depolarization at the neuromuscular junction; the action potential is transmitted to all fibrils via the T system and triggers the release of calcium from the sarcoplasmic reticulum. This process is called excitation–contraction coupling.

4.1 Aging-Related Changes in Skeletal Muscle

Macroscopically, as muscles age they

- Become smaller in size (i.e., become atrophic)
- Lose the usual red-brown color of normal muscle due to the myoglobin pigment

- Become yellow due to the deposition of lipofuscin pigment and increased fat cells, or gray due to increased amounts of connective (fibrous) tissue

Microscopically,

- Muscle fibers decrease in number and increase in size variability.
- The T system and sarcoplasmic reticulum proliferate.
- The synthesis of contractile protein is decreased.
- The number of mitochondria is reduced.

Which of these changes represents the primary aging phenomenon remains controversial.[35,36] Together with the muscular changes, the following changes occur at the neuromuscular junction:[37]

- The capability to sustain transmission of the nerve impulse from the neuronal axon to the muscle fiber decreases, as muscle and nerve fibers show increased refractoriness.
- The amount of acetylcholine (ACh), the neurotransmitter at the neuromuscular junction, declines.
- Motor nerve conduction velocity is reduced.
- The balance between nerve terminal growth and degeneration becomes less stable.
- Membrane alterations are manifested in reduced membrane potentials, lowered uptake of choline (the ACh precursor), and less uniform distribution of ACh receptors.

These changes, which are characterized by their high variability (i.e., some neuromuscular junctions remain unaffected), may be due to

- Intrinsic muscle changes
- Alterations of the neuromuscular junctions and/or
- Mental impairment, observed in demented individuals (Chapter 9)

4.2 Muscle Energy Sources and Metabolism

Muscle activity requires energy; the muscle has been called a machine for converting chemicals to mechanical energy. In the case of the heart, the requirement is for continued sustained work with limited resting time when compared to skeletal muscle activity. Activity in skeletal muscle is on demand with intervening periods of contractile inactivity. To provide for the energy requirements, muscle has an abundant blood supply, numerous mitochondria, and a high content of myoglobin, the latter an iron-containing pigment, resembling hemoglobin. Myoglobin serves as an oxygen supply during contractions which cut off blood flow, and it also facilitates the diffusion of oxygen to mitochondria where the oxidative reactions occur.

With aging, generation and supply of energy to the muscles may be decreased. However, few data are available, as methods of measurement of maximum oxygen intake under maximum effort or even at rest may not be well tolerated, especially by the frail elderly.[38] The safest approach is to examine metabolic activity in the context of daily routine such as walking (which normally demands 35 to 40% of maximum oxygen intake) or after muscle activity of such intensity to induce difficulty in speaking due to breathlessness (usually the anaerobic threshold, 60 to 70% of maximum oxygen intake). Under both conditions, metabolic activity is decreased and cardiac rate increased thereby reducing muscle efficiency and accelerating fatigability. However, even in very old individuals, physical activity performance may be considerably improved with exercise. In addition, physical training may increase the ability to utilize lipids as an energy source for exercise. Long-term heavy exercise may decrease LDL and have a beneficial effect on cardiovascular function (Chapter 17), but it may also increase the production of free radicals with unfavorable consequences on the same cardiovascular function (Chapter 6).

4.3 Responsiveness of Aging Skeletal Muscle to Exercise

Muscle responsiveness to appropriate stimuli is well demonstrated by the improvement induced by physical exercise in muscle strength, ambulatory ability, and endurance in the young and adult as well as in the elderly, including the "oldest old" (80 years and older).[34] For example, high-resistance weight training in nonagenarians leads to significant gains in muscle strength, size, and motility.[39,40] Exercise training in 60- to 70-year-old men (but not women) increases bone density as well.[14]

Notwithstanding the alterations at the muscle fibers and myoneural junction, reinnervation by motor nerve sprouting may occur.[41] Such "sprouting effects" have provided evidence for the concept of a continual, dynamic turnover of synaptic sites in muscle. Despite some conflicting reports, the prevalent view is that remodeling of aging neuromuscular junctions would replace those that degenerate after a limited lifetime.[42] Thus, adaptation to training may remain operative in some motor units well into old age.[43] In old rats, treadmill running and ablation of one of two synergistic muscles significantly increases (by 45 to 75%) muscle mass and force-generating capacity. The increase in old animals is comparable to that occurring in adults and young rats.[44]

4.4 Responsiveness of Aging Skeletal Muscle to Growth Hormone

The declining activity with advancing age of growth hormone (GH) from the pituitary (Chapter 11) may contribute to the decrease in lean body mass and the increase in mass of adipose tissue that occur in old age. GH acts on the liver and other tissues to stimulate the production of growth factors (somatomedins) such as insulin-like growth factor I (IGF-I) also known as somatomedin C. IGF-I is responsible for most of the metabolic and growth-promoting actions of GH. On the premise that GH administration may improve muscle and bone mass and strength, healthy old men aged 61 to 81 years with low IGF-I blood levels were injected with biosynthetic human GH for 6 months. During the 6 months after treatment, these men showed a 9% increase in lean body mass, a 15% decrease in adipose tissue, and a 2% increase in density of some (but not all) bones.[45]

While encouraging and supportive of the beneficial effects of GH in adult and older individuals, these results should be considered as preliminary. GH and growth factors have a variety of actions in addition to those on muscle and adipose tissues. Several of these actions may lead to conditions reminiscent of acromegaly and to complications in the elderly such as hyperinsulinemia, arthritis, hypertension, edema, and congestive heart failure. Other aspects of this treatment remain to be elucidated: What are the effects of long-term treatment of GH on muscle function (strength, motility) and quality of life? When (at which age) should the treatment begin? What is the minimal effective dose? What are the treatment costs? Further studies should be performed before adopting the use of GH treatment for the elderly.[46]

4.5 Physical Fitness: Programs of Exercise for the Aged

Physical activity programming for the older adult requires more individualized care and more knowledge of the musculoskeletal system than with any other age group. Several factors must be recognized and critically evaluated before establishing a useful program, one which provides the most benefits without endangering the individual. These factors include

1. The wide spectrum of physical abilities displayed by the older individuals and the considerable variability among individuals
2. The difficulty of detecting and assessing these abilities by medical examination, including electrocardiogram under exercise (e.g., treadmill) (stress ECG), and forestalling any potential dangers to safety[37]
3. The necessity of distinguishing between the "young" old (still preserving some life-long activity habits) and the "old" old (with mental and/or physical disorders restricting activity)

While many older individuals are limited in their ability to pursue regular physical activity, the need for a minimum level of daily exercise to prevent cardiovascular and musculoskeletal complications and further age-related deterioration is well established. A number of exercise programs are being tested and implemented to improve mechanical functioning, including amelioration of performance of skeletal muscle and cardiovascular function.[47] In addition, enhancement of body image, prolongation of independent living, and medical prevention through physical exercise and improved fitness are other important reasons for encouraging further exploration and implementation of physical fitness programs in the elderly population.[48]

5 AGING OF CARDIAC MUSCLE

The major function of the heart is to pump blood through the systemic and pulmonary circulations and thereby transport respiratory gases, nutrients, and metabolic products to and from tissues. It is not surprising, therefore, that any alteration in cardiac function will have life-endangering consequences. Structural changes with aging involve all components of the heart, including the myocardium, the cardiac muscle (closely resembling the skeletal muscle with the difference that myocardial fibers act as a syncytium), the cardiac conduction system (the specialized pacemaker tissue that can initiate repetitive action potentials), and the endocardium (the endothelial layer lining the internal surface of the cardiac cavities) (Figure 21-6).

Heart disease, associated with advancing age and atherosclerosis, remains the most important single cause of death worldwide in old age in both sexes (Chapter 16). In individuals over the age of 65, heart disease accounts for approximately 70% or more of all cardiovascular deaths in many countries including the U.S. Usually, pathologic changes in the heart are superimposed on physiologic changes which may be relatively minor but which, with the consistent and marked vascular atherosclerotic lesions, predispose the heart to a variety of pathologic conditions, most of them lethal.

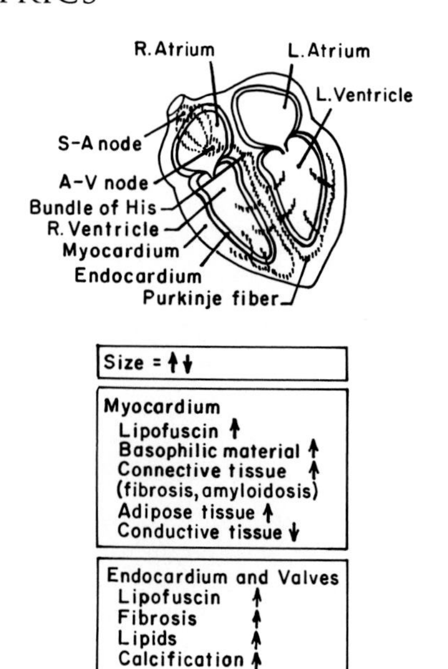

FIGURE 21-6 *Some age-related changes in the heart.*

5.1 Aging-Related Changes in Cardiac Structure

Cardiac structural changes may be related to the aging process itself or may be secondary to disease. They may be

- Primary to the heart
- Secondary to vascular lesions or to pulmonary disease
- So severe as to be present at rest
- Manifest only under conditions of increased demand, such as physical exercise

They involve, in varying degrees, all cardiac elements:

1. Muscle
2. Connective tissue
3. Conduction tissue
4. Endocardium
5. Valves
6. Cardiac vasculature

When present, aging changes usually result in a general decline of all aspects of organ and integrated functions. This decline, associated with concomitant atherosclerosis of the coronary arteries, the major arteries of the heart, leads to severe pathology such as coronary heart disease, the major cause of cardiovascular disease in the elderly (Chapter 16). Another consequence of age-related changes in heart and vasculature is hypertension, which today is amenable to successful treatment (Chapter 16).

Cardiac size and weight remain essentially unchanged with aging, although some studies suggest enlargement, particularly of the left ventricle, due to increased muscle mass or adipose tissue accumulation, or atrophy, due to loss of muscle mass. A constant change in cardiac muscle cells is the accumulation of lipofuscin (Chapter 5) often associated with accumulation of basophilic material, perhaps a possible by-product of glycogen metabolism. Connective tissue represents 20 to 35% of cardiac mass, and this proportion either remains unchanged or increases slightly with aging.

Adipose tissue accumulates as fat deposits both in the ventricles and the interatrial septum. In the latter, such deposits may displace conduction tissue in the sino-atrial node and lead to disturbances of conduction in some severe cases. Other extracellular deposits include some degree of fibrosis and amyloidosis. Fibrosis such as occurs in myocardial infarction (Chapter 16) reduces the contractile efficiency of cardiac muscle; amyloid degeneration and accumulation is also found in the neuritic plaques and amyloid deposits around cerebral vessels in Alzheimer's disease (Chapters 5 and 8). Mitochondrial DNA damage, perhaps due to free radical damage (Chapter 6) and to lipid alterations,[49] may play a role in decreased metabolic activity of human hearts and deterioration of function.[50]

Specialized conducting cells (forming the sino-atrial, and atrioventricular nodes and the intraventricular bundle of His) may be lost with aging, although this loss is usually moderate and may be associated with an increase in connective tissue elements, particularly collagen and elastin. It is not clear, however, whether and to what extent these changes interfere with conduction.

Endocardial changes — lipofuscin deposition and varying degrees of fibrosis — are probably influenced by mechanical factors such as blood flow. These mechanical factors may also be responsible for thickening of the atria and valves, which also show lipid deposition and calcification. The timetable of these valvular lesions varies with each valve.

5.2 Aging-Related Changes in Cardiac Function: Cardiac Output

The overall expression of cardiac function is *cardiac output, i.e., the amount of blood pumped by the heart into the circulation per unit of time (in the healthy adult, cardiac output is 5.5 l/min).* Cardiac output depends on stroke volume and cardiac rate as well as on venous return (Figure 21-7). Stroke volume depends on strength of muscle contractility, cardiac rate on autonomic innervation, and venous return (amount of blood returning to the heart) on the competence of the veins.

With aging, cardiac output may be reduced in a large number of elderly, although it remains efficient well into old age in others. This reduction involves alterations in one or all three components of cardiac output. For example, the venous compartment is altered probably due to changes in the elastic and smooth muscle components as represented by the *varicose veins,* tortuous and enlarged veins, most frequently found in the lower limbs, or *hemorrhoids,* enlargement of the rectal veins (Table 21-6). Both varicose veins and hemorrhoids occur at all ages but are more frequent at older ages. Another condition related to the aging of the venous tree is *thrombophlebitis,* an inflammation of the veins due to the presence of a thrombus induced by slow blood flow in the unusually dilated veins.

Postural hypotension is common in the elderly and its incidence increases with age. This condition, characterized by a fall in blood pressure rising from the supine to the standing position, has been ascribed to autonomic, particularly sympathetic, insufficiency, to low circulating levels of epinephrine, or to decreased activity of the baroreceptors, which sense blood pressure.

Strength of ventricular contraction may be altered by a number of factors, some of which are illustrated in Figure 21-8; some are related to changes in the muscle itself (loss of myocardial tissue, as described for skeletal muscle), others to alterations in autonomic nervous control particularly with respect to levels of catecholamines, or to effects of hormones and selected drugs, and alterations in blood oxygen, electrolytes, and pH.

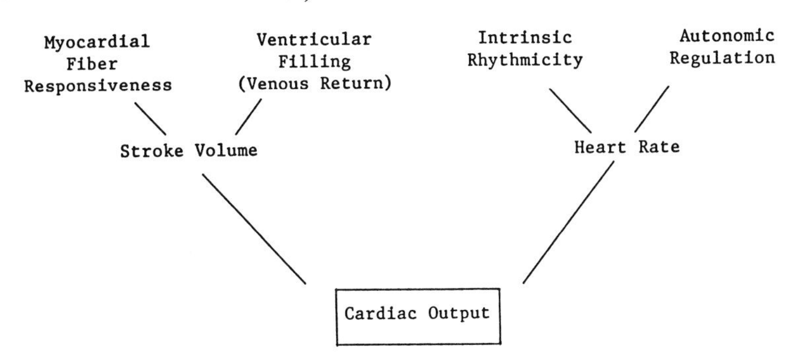

FIGURE 21-7 *Factors regulating cardiac output.*

5.3 Cardiac Output During Physical Exercise and Increased Peripheral Resistance

The decreasing cardiac output with age may not have significant impact on the circulation at rest, but it severely curtails the ability of the heart to respond to increased demand such as physical exercise. This potential cardiac insufficiency is particularly reflected in the cardiac rate which remains essentially unchanged at rest but cannot adjust efficiently to physical exercise. During exercise, the expected rise in cardiac rate necessary to increase cardiac output in order to provide for the increased oxygen consumption by the exercising muscles is much lower in the elderly than in the young. The maximum achievable heart rate with exercise decreases linearly with age and may be calculated empirically taking 220 beats per minute as the maximum in the adult. The age changes can be calculated by subtracting the age of the individual from the 220 value (Figure 21-9). Thus, in an 80-year-old the maximal heart rate that can be achieved while exercising is 220 – 80 = 140 beats per min. Although the decline is progressive, the change is more precipitous after 50 years of age.

During exercise in the young adult, cardiac output is also increased by increasing stroke volume. Here, the most important factor is the vasodilation in exercising muscles; vasodilation leads to a fall in vascular resistance, thereby augmenting venous return to the heart, increasing ventricular blood-filling in the diastole with a resulting stronger stroke volume. With aging, peripheral vasodilation is less efficient or absent and muscle mass is decreased; vascular resistance is increased due to atherosclerosis and cannot be overcome by the strength of cardiac contraction.

Normally, less than 1% of the energy liberated in cardiac tissue is due to anaerobic metabolism; the heart is highly dependent on aerobic processes. Under basal (at rest) conditions, 35% of the caloric needs of the heart are provided by carbohydrates, 5% by ketones and amino acids, and 60% by fat. The proportion of these substrates does vary with the nutritional state. Oxygen consumption of the heart is primarily dependent on the heart rate and the contractile state of the myocardium among other factors. Increasing the heart rate increases oxygen consumption, at least temporarily until the appropriate compensatory adjustments enter into effect, and so, too, does increased peripheral resistance, which causes the heart to work harder. This explains why in angina pectoris (Chapter 16) there is a relative deficiency of oxygen due to the greater oxygen demand to expel blood against increased peripheral pressure.

TABLE 21-6

Effects of Aging on Venous Return
Venous competence depends on
Venous diameter
Intrathoracic pressure
Total blood volume
With aging
Altered venous smooth muscle and elastin (varicose veins, hemorrhoids)
Thrombophlebitis
Orthostatic hypotension (sympathetic tone, baroreceptor activity, polypharmacy)

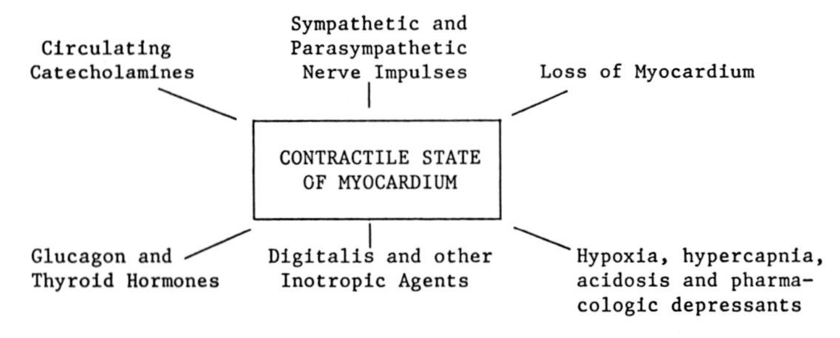

FIGURE 21-8 *Factors that influence the contractile state of the myocardium.*

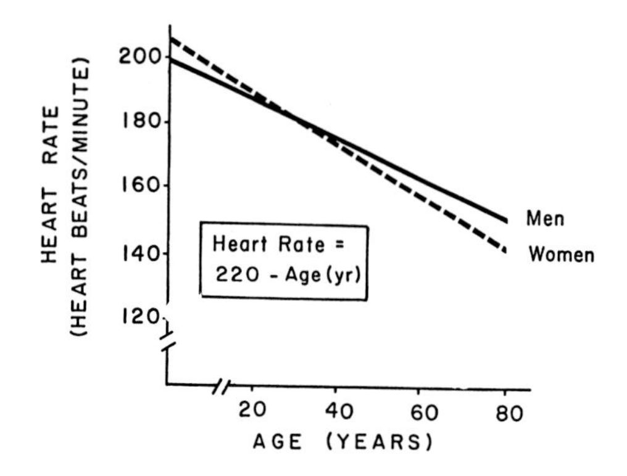

FIGURE 21-9 *Changes in cardiac rate with age.*

Cardiac Changes with Disuse

Decreased cardiac output similar to that described for the elderly also occurs in young individuals after as little as a few weeks of muscle disuse, as found with bed rest or in individuals exposed to low gravity, e.g., astronauts (Chapter 3). Typical in the older individual after retirement is reduced physical activity; therefore, the reduced physical capacity of the elderly may be ascribed to aging *per se* and may have superimposed thereon the consequences of disuse. If, truly, a part of the reduced physical capacity of the elderly is the consequence of disuse, then, indeed, exercise may be of benefit in conditioning and rehabilitating cardiac function, in particular, and in the overall functional competence of the elderly.

Cardiac Innervation and Some Disorders with Aging

Intrinsic conduction tissue is characterized by an unstable potential related to an unstable permeability to potassium. With aging, intracellular accumulation of lipofuscin, extracellular amyloidosis, loss of conduction cells, and fatty and fibrotic infiltrates lead to a decline in function and an increase in instability. Thus, *cardiac dysrhythmias* or *arrhythmias* (alterations of cardiac rhythmic contractions) are frequent in elderly subjects whose hearts are more vulnerable to such biochemical insults as hypoxia (decreased oxygen), hypercapnia (increased carbon dioxide), acidosis (decreased pH), and hypokalemia (decreased blood potassium). All these conditions increase cardiac irritability, particularly of the atria. This, coupled with reduced cardiac output and coronary blood flow, leads (in the elderly, more than in young adults) to cardiac failure and death.

Another conduction defect in the elderly is the *interruption or block of conduction,* most often between atria and ventricles. Cardiac blocks reflect potential cardiac disorders but are themselves usually asymptomatic and non–life-threatening.

The extrinsic innervation of the heart involves the two branches of the autonomic nervous system: the parasympathetic vagal innervation slows heart rate, the sympathetic increases cardiac rate and strength of contraction (Table 21-7). Both undergo changes with aging. Some constant findings are decreased inotropic (force of cardiac contraction) responses to catecholamines (primarily, epinephrine and norepinephrine) and decreased sensitivity to a variety of drugs (e.g., sympathetic agonists or antagonists) or hormones (e.g., thyroid hormones). Decreased sympathetic responses may be due to loss of neural cells resulting in reduced catecholamine content. In the rat, for example, cardiac norepinephrine is halved between the age of 1 month (young) and 28 months (old). Another contributing factor may be the reduction in number and/or affinity of beta receptors. Although little is known in man, in experimental animals the number of receptors appears to be markedly decreased.

5.4 Adaptive Adjustments in the Aging Heart

Despite some of the cardiovascular alterations described with aging in this and previous chapters (Chapters 16 and 17), the heart is capable of mustering those compensatory responses necessary to maintain adequate function. For example, while the rate at which the left ventricle fills with blood during early diastole declines markedly between the ages of 20 and 80, the enhanced filling in during later diastole keeps ventricular filling adequate in elderly individuals. The myocardium remains capable of adaptive hypertrophy to maintain normal heart volume and pump function in the presence of a moderate increase in systolic pressure. Such hypertrophy may be mediated through myocyte (myocardial cells) hyperplasia[51] and may be due to persistent activation of cardiac gene expression and its hormonal regulation.[52]

Although cardiac rate increases less during severe exercise in many elderly as compared to young individuals, stroke volume may increase to compensate for the smaller increase in heart rate.[53] The smaller increase in heart rate during exercise has been attributed to a reduced efficiency of beta-adrenergic modulation[54] rather than to intrinsic alterations of the senescent myocardium.[55,56] Current evidence shows that overall cardiac function in most elderly people in good health is adequate to meet the body's needs for pressure and flow at rest and to sustain physical exercise.

Excitation–Contraction Coupling

In skeletal muscle, the nerve impulse triggers the release of calcium from its stores in the sarcoplasmic reticulum and initiates contraction. Calcium binds to troponin C, and in so doing, uncovers several myosin binding sites on actin. These molecular changes result in a decrease in the number of cross-linkages that bind myosin to actin and facilitate the sliding of the actin and myosin filaments along each other and induce shortening of the muscle during contraction. In cardiac muscle during the action potential, calcium ions diffuse into the myofibrils from stores in the sarcoplasmic

TABLE 21-7
Autonomic Cardiac Regulation

Parasympathetic stimulation
Decreased heart rate
Decreased conduction
Increased refractory period
Sympathetic stimulation
Increased heart rate
Increased strength of contraction

reticulum and also from the T system. Cardiac T tubules are much larger and contain many times more calcium than those in skeletal muscle. This extra supply of calcium from the T tubules is at least one factor responsible for prolonging the cardiac muscle action potential for as long as one third of a second, 10 times longer than in skeletal muscle. Once calcium concentration is lowered, chemical interaction between myosin and actin ceases and the muscle relaxes. ATP provides the energy necessary for the active transport of calcium.

With aging, the excitation–contraction coupling in the heart appears altered. In experimental animals or using isolated heart models, excitation appears insufficient — or less sufficient than in the young — to trigger the release of calcium from its stores; the calcium stores themselves would be reduced and the accumulation of calcium in the sarcoplasmic reticulum would be slowed. Likewise, the active transport of calcium appears inhibited and the duration of contraction and relaxation is prolonged.[57,58] Alterations with aging in calcium stores and movements are also reflected in diminished cardiac responses to inotropic agents such as catecholamines. These agents exert their stimulatory action on cardiac contraction by loading calcium stores and by facilitating calcium transport in the cells. With aging, both calcium storage and movements would be altered, hence the reduced effectiveness of the inotropic factors to activate cardiac contraction.

5.5 Hormonal and Chemical Influences on the Aging Heart

Some hormones, such as glucagon and thyroid hormones, influence myocardial contractility. Glucagon (Chapter 15) increases the formation of cAMP through its binding to receptors other than β-receptors. Its inotropic action therefore can be beneficial to individuals suffering from toxicity due to the administration of β-adrenergic blockers. Thyroid hormones increase the number of β-adrenergic receptors in the heart, and their effects resemble those of sympathetic stimulation (Chapter 14).

Various drugs also affect myocardial contractility. Xanthines, such as caffeine, exert their inotropic effect by inhibiting the breakdown of cAMP. The inotropic effect of digitalis and related drugs is due to their inhibitory effects on Na^+K^+ ATPase; this inhibition increases intracellular sodium which, in turn, increases calcium availability to the cell and initiates contraction (Chapter 23). Depressants such as barbiturates depress myocardial contractility.

To what extent age-related changes in myocardial function will influence and be influenced by hormones and drugs in elderly individuals is still not certain. Responses to inotropic agents are altered in the aged under some conditions. Evaluation of these data should take into consideration that cardiac responses depend in large measure on the condition of the vascular tree and that the agents considered exert both cardiac and extracardiac effects, and the latter may, indirectly, influence the myocardium.

REFERENCES

1. Antich, P. P., Pak, C. Y., Gonzales, J., Anderson, J., Sahkaee, K., and Rubin, C., Measurement of intrinsic bone quality in vivo by reflection ultrasound: correction of impaired quality with slow-release sodium fluoride and calcium citrate, *J. Bone Mineral Res.*, 8, 301, 1993.
2. Garn, S. M., Rohmann, C. G., Wagner, B., and Ascoli, W., Continuing bone growth throughout life: a general phenomenon, *Am. J. Phys. Anthropol.*, 26, 313, 1967.
3. Roche, A. F., Aging in the human skeleton, *Med. J. Aust.*, 2, 943, 1966.
4. Smith, E. L., Sempos, C. T., and Purvis, R. W., Bone mass and strength decline with age, in *Exercise and Aging: The Scientific Basis,* Smith, E. L. and Serfass, R. C., Eds., Enslow Publishers, Hillside, NJ, 1981, 59.
5. Yamada, H., *Strength of Biological Materials,* Williams & Wilkins, Baltimore, 1970.
6. Francis, R. M., Peacock, M., Storer, J. H., Davies, A. E., Brown, W. B., and Nordin, B. E., Calcium malabsorption in the elderly: the effect of treatment with oral 25-hydroxyvitamin-D3, *Eur. J. Clin. Invest.*, 13, 391, 1983.
7. Grynpas, M. D., Hancock, R. G., Greenwood, C., Turnquist, J., and Kessler, M. J., The effects of diet, age and sex on the mineral content of primate bones, *Calcif. Tissue Int.*, 52, 399, 1993.
8. Nilas, L., Nutrition and fitness in the prophylaxis for age-related bone loss in women, *World Rev. Nutr. Diet.*, 72, 102, 1993.
9. Holloszy, J. O., Exercise, health and aging: a need for more information, *Med. Sci. Sports Exerc.*, 25, 538, 1993.
10. Anghileri, L. J. and Tuffet-Anghileri, A. M., *The Role of Calcium in Biological Systems,* CRC Press, Boca Raton, FL, 1982.
11. Campbell, A. K., *Intracellular Calcium, its Universal Role as Regulator,* John Wiley & Sons, New York, 1983.
12. Flicker, L., Lichtenstein, M., Coman, P., et al., The effect of aging on intact PTH and bone density in women, *J. Am. Geriatr. Soc.*, 40, 1135, 1992.
13. Nuti, R. and Martini, G., Effects of age and menopause on bone density of entire skeleton in healthy and osteoporotic women, *Osteoporosis Int.*, 3, 59, 1993.
14. Blumenthal, J. A., Emery, C. F., Madden, D. J., et al., Effects of exercise training on bone density in older men and women, *J. Am. Geriatr. Soc.*, 39, 1065, 1991.
15. Courpron, P., Bone disorders, in *Principles and Practice of Geriatric Medicine,* Pathy, M. S. J., Ed., John Wiley & Sons, Chichester, England, 1991, 1217.
16. Wallace, W. A. and Prince, H. G., Orthopaedic management of the elderly, in *Principles and Practice of Geriatric Medicine,* Pathy, M. S. J., Ed., John Wiley & Sons, Chichester, England, 1991, 1275.
17. Pollitzer, W. S. and Anderson, J. J., Ethnic and genetic differences in bone mass: a review with a hereditary vs environmental perspective, *Am. J. Clin. Nutr.*, 50, 1244, 1989.
18. Gilsanz, V., Roe, T. F., Mora, S., Costin, G., and Goodman, W. G., Changes in vertebral bone density in black girls and white girls during childhood and puberty, *N. Engl. J. Med.*, 325, 1597, 1991.
19. Weinstein, R. S. and Bell, N. H., Diminished rates of bone formation in normal black adults, *N. Engl. J. Med.*, 319, 1698, 1988.
20. Pullar, T. and Wright, V., Diseases of the joints, in *Principles and Practice of Geriatric Medicine,* Pathy, M. S. J., Ed., John Wiley & Sons, Chichester, England, 1991, 1237.
21. Roth, R. D., Joint diseases associated with aging, *Clin. Podiatr. Med. Surg.*, 10, 137, 1993.

22. Armstrong, G. G. and Gardner, D. L., Thickness and distribution of human femoral head articular cartilage. Changes with age, *Ann. Rheum. Dis.*, 36, 407, 1977.
23. Meachim, G., Bentley, G., and Baker, R., Effect of age on thickness of adult patellar articular cartilage, *Ann. Rheum. Dis.*, 36, 563, 1977.
24. Ball, J. and Sharp, J., Osteoarthrosis, in *Copeman's Textbook of the Rheumatic Diseases,* 5th ed., Scott, J. T., Ed., Churchill Livingstone, London, 1978, 595.
25. Kay, M. M. B., Galpin, J., and Makinodan, T., Eds., *Aging, Immunity, and Arthritic Disease,* Raven Press, New York, 1980.
26. Hamerman, D. and Klagsbrun, M., Osteoarthritis. Emerging evidence for cell interactions in the breakdown and remodeling of cartilage, *Am. J. Med.*, 78, 495, 1985.
27. van Schaardenburg, D., Hazes, J. M., de Boer, A., Zwinderman, A. H., Meijers, K. A., and Breedveld, F. C., Outcome of rheumatoid arthritis in relation to age and rheumatoid factor at diagnosis, *J. Rheumatol.*, 20, 45, 1993.
28. Cash, J. M. and Wilder, R. L., Neurobiology and inflammatory arthritis, *Bull. Rheum. Dis.*, 41, 1, 1992.
29. Lotz, M., Carson, D. A., and Vaughan, J. H., Substance P activation of rheumatoid synoviocytes: neural pathway in pathogenesis of arthritis, *Science,* 235, 893, 1987.
30. Chuang, T.-Y., Hunder, G. G., Ilstrup, D. M., and Kurland, L. T., Polymyalgia rheumatica: a 10-year epidemiologic and clinical study, *Ann. Intern. Med.*, 97, 672, 1982.
31. Chou, C.-T. and Schumacher, H. R., Jr., Clinical and pathologic studies of synovitis in polymyalgia rheumatica, *Arthritis Rheum.*, 27, 1107, 1984.
32. Pearson, M. B., Bassey, E. J., and Bendall, M. J., Muscle strength and anthropometric indices in elderly men and women, *Age Ageing,* 14, 49, 1985.
33. Cunningham, D. A., Rechnitzer, P. A., Howard, J. H., and Donner, A. P., Exercise training of men at retirement: a clinical trial, *J. Gerontol.*, 42, 17, 1987.
34. Fiatarone, M. A., Marks, E. C., Ryan, N. D., et al., High-intensity strength training in nonagenarians. Effects on skeletal muscle, *JAMA,* 263, 3029, 1990.
35. Lexell, J., Ageing and human muscle: observation from Sweden, *Can. J. Appl. Physiol.*, 18, 2, 1993.
36. Shafiq, S. A., Lewis, S., Leung, B., and Schutta, H. S., Fine structure of aging skeletal muscle, in *Aging and Cell Structure,* Vol. 1, Johnson, J. E., Jr., Ed., Plenum Press, New York, 1984, 333.
37. Smith, D. O. and Rosenheimer, J. L., Aging at the neuromuscular junction, in *Aging and Cell Structure,* Vol. 2, Johnson, J. E., Jr., Ed., Plenum Press, New York, 1984, 113.
38. Shephard, R. J., *Physical Activity and Aging,* 2nd ed., Aspen Publishers, Rockville, MD, 1987.
39. Fiatarone, M. A. and Evans, W. J., Exercise in the oldest old, *Geriatr. Rehabil.*, 5, 63, 1990.
40. Fiatarone, M. A., O'Neill, E. F., Doyle, N., Clements, K. M., Roberts, S. B., Kehayias, J. J., Lipsitz, L. A., and Evans, W. J., The Boston FICSIT study: the effects of resistance training and nutritional supplementation on physical frailty in the oldest old, *J. Am. Geriatr. Soc.*, 41, 333, 1993.
41. Andonian, M. H. and Fahim, M. A., Effects of endurance exercise on the morphology of mouse neuromuscular junctions during ageing, *J. Neurocytol.*, 16, 589, 1987.
42. Wernig, A. and Herrera, A. A., Sprouting and remodeling at the nerve-muscle junction, *Prog. Neurobiol.*, 27, 251, 1986.
43. Stanley, S. N. and Taylor, N. A., Isokinematic muscle mechanics in four groups of women of increasing age, *Eur. J. Appl. Physiol. Occup. Physiol.*, 66, 178, 1993.
44. White, T. P., personal communication.
45. Rudman, D., et al., Effects of human growth hormone in men over 60 years old, *N. Engl. J. Med.*, 323, 1, 1990.
46. Vance, M. L., Growth hormone for the elderly?, *N. Engl. J. Med.*, 323, 52, 1990.
47. Shephard, R. J., Physical fitness. Exercise and aging, in *Principles and Practice of Geriatric Medicine,* Pathy, M. S. J., Ed., John Wiley & Sons, Chichester, England, 1991, 279.
48. Bortz, W. M., *We Live Too Short and Die Too Long,* Bantam Trade Paperback, New York, 1992.
49. Lewin, M. B. and Timiras, P. S., Lipid changes with aging in cardiac mitochondrial membranes, *Mech. Ageing Dev.*, 24, 343, 1984.
50. Hayakawa, M., Sugiyama, S., Hattori, K., Takasawa, M., and Ozawa, T., Age-associated damage in mitochondrial DNA in human hearts, *Mol. Cell. Biochem.*, 119, 95, 1993.
51. Anversa, P., Palackal, T., Sonnenblick, E. H., Olivetti, G., Meggs, L. G., and Capasso, G. M., Myocyte cell loss and myocyte cellular hyperplasia in the hyperplasia aging heart, *Circulation Res.*, 67, 871, 1990.
52. Chien, K. R., Knowlton, K. U., Zhu, H., and Chien, S., Regulation of cardiac gene expression during myocardial growth and hypertrophy: molecular studies of an adaptive physiologic response, *FASEB J.*, 5, 3037, 1991.
53. Rodeheffer, R. J., Gersternblith, G., Becker, L. C., Fleg, J. L., Weisfeldt, M. L., and Lakatta, E. G., Exercise cardiac output is maintained with advancing age in healthy human subjects: cardiac dilatation and increased stroke volume compensate for a diminished heart rate, *Circulation,* 69, 203, 1984.
54. Rowe, J. W. and Troen, B. R., Sympathetic nervous system and aging in man, *Endocrinol. Rev.*, 1, 167, 1980.
55. Lakatta, E. G., Altered autonomic modulation of cardiovascular function with adult aging, perspectives from studies ranging from man to cell, in *Pathobiology of Cardiovascular Injury,* Stone, H. L. and Weglocki, W. B., Eds., Martinius Nijhoff, Boston, 1985, 441.
56. Lakatta, E. G., The aging heart, lifestyle and disease, *Ann. Intern. Med.*, 113, 455, 1991.
57. Lakatta, E. G. and Yin, F. C., Myocardial aging: functional alterations and related cellular mechanisms, *Am. J. Physiol.*, 242, H927, 1982.
58. Lakatta, E. G., Gersternblith, G., Angell, C. S., Shock, N. W., and Weisfeldt, M. L., Prolonged contraction duration in aged myocardium, *J. Clin. Invest.*, 55, 61, 1975.

III
SYSTEMIC AND ORGANISMIC AGING

22 THE SKIN AND CONNECTIVE TISSUE: AGING CHANGES

Mary L. Timiras

Perhaps no aspect of aging is as dramatic or readily obvious as that which occurs in the skin and its appendages. The development of either gray hair or facial wrinkles represents irrefutable evidence of the passage of time and the aging process. Many descriptions of aging skin fail to distinguish between intrinsic aging changes and changes caused by environmental insults that accumulate with exposure time. For example, chronic solar damage changes are far more prevalent in the elderly since they result from cumulative exposure over time.[1] Intrinsic changes in the structure and function of skin that occur with aging also make the skin more vulnerable to external insults. It is not surprising, then, that dermatologic problems are very common in the elderly.[2] Several studies describe how almost one half of persons over 65 years of age have at least one dermatologic disease requiring medical attention.[3,4] About one third of these have more than one skin problem. Multiple skin conditions are characteristic of the very old, and the common ones are different from those affecting the young.[5,6]

A connective tissue component particularly involved in skin changes with aging is collagen. Because it is ubiquitous throughout the body and undergoes identifiable changes with age, collagen has been intensively studied as a possible primary source for the onset of aging processes. It has been proposed, for example, that the striking changes that take place with age in the structure and chemistry of collagen fibers and the surrounding extracellular matrix (previously designed as ground substance) may derive from metabolic alterations in tissues — as manifested by the occurrence of cross-linkages — and, thus, represent a fundamental mechanism by which overall functional impairment is induced in the aged.[7] In order to provide an understanding of the nature and origin of age-related changes in the skin and of how these changes lead to dermatologic problems in the aged, this chapter will review some of the anatomic and physiologic changes that occur in the skin and collagen with normal senescence. Pressure sores will be discussed as a clinical example of skin dysfunction in the elderly.

1 AGING OF THE SKIN

The skin is one of the largest organs of the body and accounts for approximately 16% of the total body weight. It is part of the integument, a covering of the entire body, which, in addition to the skin, includes the nails, hair, and various types of glands, all accessory organs derived from the skin. The following are the major functions of the skin:

1. It provides a barrier to exclude harmful substances and prevents dessication.
2. It plays a role in the control of body temperature.
3. It readily repairs itself.
4. It receives sensory stimuli: touch, pressure, temperature, and pain.
5. It excretes waste products through sweat glands and secretes special products such as milk from mammary glands.

The skin consists of an epithelium, the epidermis, and the dermis, considered as part of the connective tissue.

Epidermis and Dermis

The main layers of the skin consist of the surface layer, known as the epidermis, and a deeper connective tissue layer, known as the dermis. The interface between the epidermis and dermis is normally uneven and forms wavy interdigitations. The projections of the dermis into the epidermal layer are called dermal papillae.

The epidermal cells produce keratin, a fibrous protein essential for protection, and form the dead superficial layers of the skin (stratum corneum). The superficial keratinized cells are continuously exfoliated and are replaced by cells that arise from the basal layer of the epidermis. The sequence of changes of an epidermal cell from the basal layer of the epidermis to keratinization and exfoliation normally takes from 15 to 30 days depending on the region of the body.

The epidermis is locally specialized to form various skin appendages such as the hair, nails, sweat glands, and oil or sebaceous glands. Other types of cells that are found in the epidermis include melanocytes, which produce the pigment melanin that protects against ultraviolet irradiation, and Langerhans cells that are believed to be responsible for recognition of foreign antigens.

Since the epidermis has no blood vessels, the dermis, which has abundant blood vessels, plays an important role in supplying the epidermis with nutrients. Although the epidermis does contain nerve endings, there are more in the dermis because it also contains the nerves that lead to the sensory nerve endings. In addition, epidermal appendages, such as the hair follicles, sebaceous glands, and sweat glands, extend down into the dermis. Below the dermis is looser connective tissue which usually consists of subcutaneous adipose tissue that plays a role in fat metabolism.

1.1 Epidermal Changes

Dryness and/or roughness of the aging skin is one of the most readily appreciated changes that occur. This could be due either to a decrease in the moisture content of the stratum corneum or to a decrease in vertical height and increase in overall surface area of epidermal cells. The increased dryness results in a surface likened to "shingles on an old roof" (Figure 22-1).[8] The exact pathogenesis of the drying process is still unclear but probably represents a combination of several contributing factors.

Another well-recognized change that occurs with aging is the increased incidence of both benign and malignant epidermal neoplasms. Although there is no question that chronic exposure to ultraviolet light contributes to this problem, some intrinsic cellular alterations of basal cells may also contribute to the increased incidence of skin cancer in the elderly.

One of the histologic changes that occurs most consistently in the elderly is the flattening of the dermo–epidermal interface and effacement of the dermal papillae. This reduces the total area of dermo–epidermal junction per area of external body surface area. This change predisposes older persons to blister formation and shear-type injuries and easy

0-8493-8979-8/94/$0.00+$.50

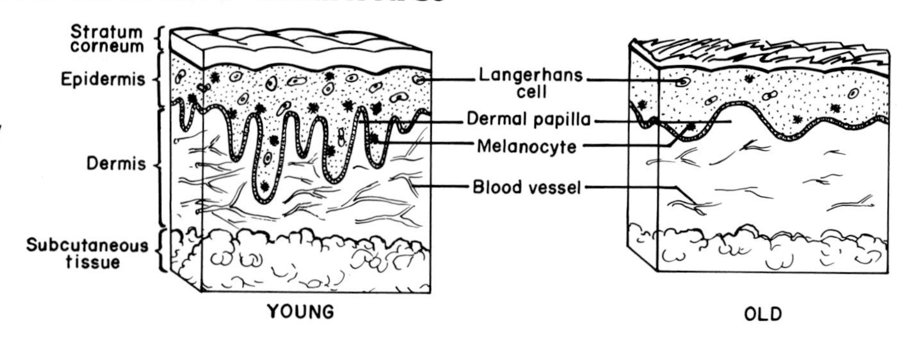

FIGURE 22-1 *Histologic changes in normal aging skin.* Changes from top to bottom include (1) rougher stratum corneum, (2) fewer melanocytes and Langerhans cells, (3) flattening of the dermo-epidermal junction, (4) fewer blood vessels in the dermis, and (5) less subcutaneous tissue.

abrasions. Another observation is an age-associated decrease in epidermal turnover rate of about 50% between the 3rd and 7th decades of life. This means fewer basal cells are replaced and it takes longer for a basal cell to reach the stratum corneum and be exfoliated. The slower movement prolongs the exposure of epidermal cells to potential carcinogens and contributes to an increased incidence of skin cancer; in addition, it is responsible for the slowing of wound healing.

The number of melanocytes decreases with age by approximately 8 to 20% per decade after the age of 30 in both exposed and unexposed areas.[9] This reduction leads to irregular pigmentation especially in sun-exposed areas (hence the "age spots" frequently seen on the back of the hands), and to the inability to tan as deeply as when younger. However, other studies seem to negate or at least to minimize this reduction in melanocytes.[5] The number of Langerhans cells decreases as well. This change is expected to contribute to a decline in the skin's cell-mediated immune response. The reduction of melanin and its protective action, the reduced inflammatory warning signs, and the reduced immune capacity combine to increase the risk for tumorigenesis. In fact, elderly patients require longer ultraviolet exposure to develop erythema and edema (sunburn) than younger patients. Thus, the body's warning system as well as its defense system in relation to skin cancer becomes blunted with age. In addition, ultraviolet light further reduces the quantity of Langerhans cells. These changes together with the increase in cumulative irradiation exposure possibly explain why non-melanomatous skin cancers are prone to occur on sun-exposed skin.

1.2 Dermal Changes

The dermis in elderly individuals has a decreased density, with fewer cells and blood vessels. The total amount of collagen decreases 1% per year in adulthood; therefore, skin thickness decreases linearly with age after 20. Collagen itself becomes thicker, less soluble, and more resistant to digestion by collagenase with age. Changes in the number and types of cohesive bonds make collagen stronger and more stable. Collagen alteration predisposes the dermis to tear-type injury since there is less "give" in the tissue. Architectural rearrangements of fibers may also be responsible for changes in dermal tissue properties.

The total amount of hyaluronic acid and dermatan sulfate, components of the extracellular matrix, decreases in the dermis with aging, affecting the viscosity of the dermis which, in turn, may alter the rate of dermal clearance of substances. Changes in the elastic fibers of the dermis also result in loss of stretch and resilience. A consequence is skin sagging and wrinkling and predisposition to injury of the underlying tissues following trauma. While even the very old (beyond 85 years) can effectively repair extensive wounds, elderly individuals, in general, lag behind younger controls at every stage of wound healing.[10,11]

Pale skin results from the decrease in dermal blood vessels. Skin surface temperature is also decreased due to the diminished vascularity. These changes, together with a decrease in the thickness of the subcutaneous tissue, make thermoregulation more difficult in the elderly. The vascular changes described above also result in a decrease in dermal clearance of foreign materials which can prolong or exacerbate cases of contact dermatitis.

2 AGING OF SKIN APPENDAGES

Older individuals produce less sweat because sweat glands either decrease in number or functional efficiency, a decrease that interferes further with thermoregulation. Although the number of sebaceous glands remains constant with age, their size increases while the sebum output as well as wax production decline with age. The diminished production of sweat and oil no doubt contributes to skin dryness and roughness in the elderly.

The rate of linear nail growth decreases with aging. Nail plates usually become thinner, more brittle, and fragile.

Hair graying occurs because of a progressive loss of functional melanocytes from hair bulbs. By the age of 50, it is said that 50% or more of the population have at least 50% of their body hair gray regardless of sex or hair color. Heredity does play an important role in hair graying as well. A decrease in the number of hair follicles in the scalp and consequent increased balding have also been described.[12]

Although no changes in the free nerve endings in aged skin have been found, the number of Pacinian and Meissner's corpuscles (end organs responsible for the sensation of pressure and light touch) decreases with age (Chapter 10). This results in decreased sensation, which predisposes the elderly to injury. It also decreases the ability to perform fine maneuvers with the hands. A summary of structural and functional changes with aging in the main skin components, epidermis, dermis, and appendages, is presented in Tables 22-1 and 22-2, respectively.

2.1 Projecting the Image

Upon a first encounter, the older individual, like any other, is judged by the image he projects. The skin is a major signpost of time. We estimate age, personality, status, race, and health merely by looking. A psychological study shows that aged persons who are physically attractive are more optimistic and have a better personality and health than the unattractive.[13] One must always have a proper regard for the skin and bear in mind that a number of the skin lesions afflicting the elderly are preventable. As noted previously,

TABLE 22-1

Changes in Normal Aged Skin

Decrease	Increase or other changes
Epidermis	
Epidermal turnover rate	Severe dryness and roughness
Number of melanocytes	Flattening of dermoepidermal junction
Langerhans cells	
Dermis	
Density	Stiffer collagen
Cells	Stiffer elastic fibers
Blood vessels	
Clearance of foreign substances	
Appendages	
Sweat production	Gray hair
Sebaceous glands	Thinner nails
Hair follicles	
Rate of nail growth	
Sensory end organs	

some of the lesions are due to the passage of time, but most are due to cumulative environmental insults and should, and can, be avoided.

2.2 Pressure Sores

The pathophysiology of pressure sores illustrates several principles in geriatric management.[14] Although not the most commonly seen dermatologic diagnosis in the elderly (skin cancer is), the problem of pressure sores is frequent enough and produces serious enough consequences to deserve discussion.[15] Pressure sores were formerly seen more often in younger patients with chronic diseases or spinal cord injuries.[16] Now they are becoming more of a problem in the elderly. The prevalence of pressure ulcerations increases with age such that patients over 70 years of age account for 70% of those affected. In this age group, 70% of patients develop pressure sores within 2 weeks of hospital admission.[17] As with urinary incontinence, the presence of pressure ulcers and their status frequently play a prominent role in the decisions made regarding the ultimate management of a patient.

Pressure sores represent a dreaded complication of one of the "five I's" of geriatrics: Immobility. The most commonly affected sites are the sacrum, ischial, trochanteric, and calcaneal tuberosities, as well as the lateral malleolus. They arise from four different mechanisms: pressure, shear, friction, and moisture leading to maceration. The role of pressure is the most critical in the development of the pressure ulcer. The average period of time necessary to produce pressure-associated changes varies but can be as little as 2 h. Hence, a patient who does not move him/herself every 2 h, for whatever reason, is at risk for developing sores. Shearing forces are produced by an improper position, friction occurs with improper handling of the patients, and moisture occurs due to perspiration or urinary and fecal incontinence. Due to all these factors, the skin changes previously described increase the risk of developing pressure sores.

A uniform classification system of pressure sores has been proposed and is routinely utilized in their management.[17,18]

TABLE 22-2

Functional Changes in Aging Skin

Decreased function	Increased function
Wound healing capability	Blister formation
Cell-mediated immune response	Incidence of infection
Thermoregulation	Incidence of cancer
Clearance of foreign substances	Dryness
Tanning	Roughness
Elasticity	Fragility
Sweat and oil production	Sensory deprivation
Thickness	

Stage I occurs when there is nonblanchable erythema of intact skin. Stage II occurs when there is partial-thickness skin loss involving epidermis and/or dermis. The ulcer is superficial and presents clinically as an abrasion, blister, or shallow crater. Full thickness skin loss involving damage or necrosis of subcutaneous tissue, which may extend down to, but not through, underlying fascia represents Stage III. Finally, Stage IV occurs when there is full thickness skin loss with extensive destruction, tissue necrosis, or damage to muscle, bone, or supporting structures (i.e., tendon).

There is no cure for a pressure sore once it develops. Even if an ulcer heals, there is always a significant chance that it will recur.[19] Therefore, prevention of the development of pressure sores, as with many other geriatric problems, is the most important aspect of management. Risk factors for the development of pressure sores include: incontinence, edema, obesity, diabetes with neuropathy, sepsis, vascular disease, immobility due to fractures, dementia or restraints, and finally systemic factors related to malnutrition such as hypoalbuminemia, anemia, and vitamin deficiency. Once patients at risk are identified, specific measures need to be taken to avoid the causative factors mentioned earlier. This is best accomplished through the use of a multidisciplinary team approach. The medical primary care provider optimizes physiologic function and treats any underlying illnesses. The nursing staff ensures feeding, turning, positioning, and initiates a bowel and bladder program. The nutritionist develops an aggressive nutritional strategy, and finally physical therapy can institute a mobilization plan, encourage strengthening exercises, and provide special pressure-sparing equipment.

Relief from pressure is the most important factor in preventing as well as treating pressure sores. In patients at risk, this may necessitate turning the patient in bed every 2 h. Padding for the bed or chair as well as special air and water mattresses have also been recommended.[20] Providing a clean and moist environment for tissue to heal is important for Stages II to IV. This is usually accomplished with sterile gauze moistened with normal saline or synthetic colloid dressings. In addition, for Stages III and IV, it is necessary to remove necrotic debris. This is accomplished through surgical or mechanical dressing change debridement, as well as enzymatic products. Avoiding irritating substances such as betadine or hydrogen peroxide also helps with wound healing. Finally, treating local infection with frequent dressing changes or debridement, and systemic infections with empirical antibiotics, will also help wounds heal faster. If the wound is large enough, plastic surgery utilizing tissue flaps may be necessary to fill in large gaps. The management of pressure sores is much more successful and satisfying when underlying pathophysiologic principles are utilized in conjunction with a comprehensive and multidisciplinary approach.

3 ■ AGING OF COLLAGEN

Collagen, with elastin and ground substance, represents one of the components of connective tissue, generally found around the structural elements that constitute organs and tissues but ubiquitous throughout the body. Connective tissue itself, formed of cells, fibers, and extracellular matrix, functions in a number of important ways — serving as mechanical support, means of exchange of metabolites between blood and tissues, storage of fuel in its adipose cells, protection against infection, and repair of injury. Collagenous fibers are present in all types of connective tissue but vary greatly in their abundance as well as in the importance of their function and the range of aging effects.[21] Collagen undergoes continuous changes with age. It forms a large portion (30 to 40%) of all proteins of the body. It is easily available and its structure and chemistry are fairly well known. Therefore, it offers a valuable model for gerontologic tissue studies. Indeed, the demonstration that cross-linkages do occur biologically and that highly cross-linked collagen structures increase proportionally during the course of natural aging has been largely responsible for the cross-linking theory of aging.

3.1 Structure and Chemistry of Collagen

Collagens are a family of macromolecules deposited in the form of fibers. They occur almost chemically pure in the tendons, as a convolute of fibers in an extracellular matrix in the skin, as the substance of bones on which calcium is deposited, or as the interstitial fibrous tissue between muscle fibers, in membranes of articulations, in the arterial wall, and so on. The central feature of all collagen molecules is their stiff, triple-stranded helical structure. Three collagen polypeptide chains, called α chains, are wound around each other in a regular helix to generate a rope-like collagen molecule conventionally referred to as tropocollagen, about 300 nm long and 1.5 nm in diameter. In unstained preparations of loose connective tissue, the collagen fibers appear as colorless strands. When collagen fibers are treated with a special stain and viewed at high magnification, several transversal bands can be resolved at intervals of 64.9 nm. The explanation of this striation has been based on the generally accepted view that tropocollagen molecules (see below) come together in a parallel arrangement and overlap each other by about one quarter of their length to produce a staggered array resulting in cross-striations (Figure 22-2).

Although flexible, collagen fibers offer great resistance to a pulling force. For example, the breaking point of human collagenous fibers (as in tendon) is only reached with a force of several hundred kilograms per square centimeter, and their elongation at this point is only a few percentage points. In this respect, the flexibility and stability of the collagen fiber have been related to that of a cable that ties a ship to shore; it is sufficiently flexible to be curled when not in use and, yet, does not allow movement of the ship at anchor. Similarly, blood vessels, skin, and tendon are capable of transmitting tension and compression as they are simultaneously restrained from deformation by their collagen (and elastin) fibers.[7]

Of the amino acids comprising the polypeptide chains, two —hydroxyproline and hydroxylysine — do not occur in significant amounts in other animal proteins, and their content in a tissue, therefore, can be taken as an index of its collagen content. Collagen molecules are held together by electrostatic bonds, and the different amino acid chains are

FIGURE 22-2 *Electron micrograph of a negatively stained microfibril of collagen isolated from rat tendon.* One dark and one light segment represent a period produced by the arrangement of tropocollagen molecules. Magnification 96,541×. (Courtesy of Dr. N. B. Gilula.)

connected by intermolecular cross-links to form fibrils, the strength of these bonds increasing progressively with age. From the moment the collagen is formed, the completed fiber does not take part in metabolism and is regarded as a metabolically inert component of the extracellular space.

Tropocollagen fibers are usually embedded in a matrix of amorphous ground substance, a very viscous solution formed mainly of proteoglycans. Originally, the polysaccharides are modified hexose polymers usually bound to proteins; some contain a sulfate group (chondroitin-6-sulfate), whereas others carry a large number of negatively charged ions (hyaluronate) and are therefore described as polyanions. Ground substance also contains interstitial fluid filtered from the capillaries, and water is loosely absorbed to the ground-substance components as water of hydration. Plasma proteins may also penetrate the ground substance and be bound to the polysaccharides. Both water and proteins are returned to the circulation — in the case of water, through the lymphatic system, and in the case of proteins, either directly (for the free proteins) or after release from the polysaccharides by the action of proteases (for the bound proteins). Thus, ground substance may serve not only as a lubricant (to facilitate the movement of the joints), as a barrier to the spread of bacteria (that have gained access to the tissues), and as a "plasticizer" (to diminish friction and wear between collagen fibers), but also as a storage site for water, ions, and proteins.[22]

Collagen is produced by fibrocytes, the cellular elements of the connective tissue, from amino acids that are extruded into the ground substance where they then undergo maturational changes leading to the typical structure of the adult tropocollagen (Figure 22-3). The collagen chains are originally formed from larger precursors which contain both the amino acids, which give rise to collagen, and some extra amino acids, called extension peptides, which do not form collagen. These extension peptides have at least two important functions:

1. They guide the intracellular formation of the triple-stranded collagen molecule.
2. They prevent the intracellular formation of large collagen fibrils, which would be catastrophic for the cell.

With aging, either of these functions could be altered as well as the proteolysis of this material.

3.2 Growth and Aging of Collagen

Growth of the collagen fibers after their birth proceeds in two phases: the tropocollagen fibers aggregate in the axial direction and they continue to grow in thickness by apposition, i.e.,

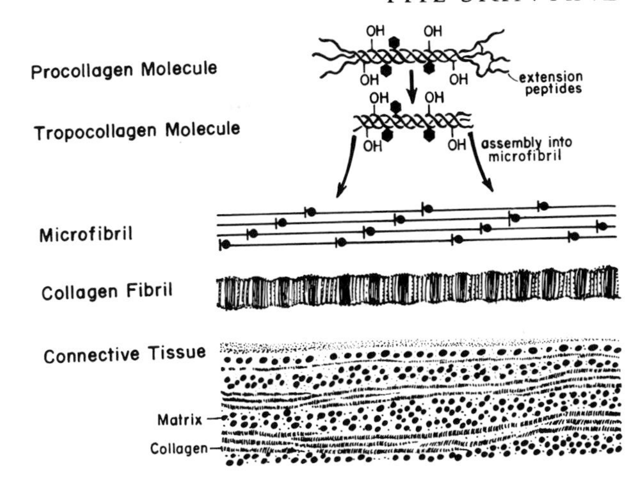

FIGURE 22-3 *Schematic representation of the composition of a collagen fibril and the distribution of tropocollagen molecules in the connective tissue.* Tropocollagen molecules, 2800 Å in length, are aligned in a staggering fashion, overlapping by one quarter of their length (640 Å).

by accretion of more soluble particles. Extracellular matrix is also produced by fibrocytes, and, when newly formed, it appears under the electron microscope as vacuolated, consisting of a water-rich phase (the vacuoles) and a less water-rich phase (the walls of the vacuoles). Some time after secretion of the ground substance, the tropocollagen begins the end-to-end alignment process, with the fibrils orienting themselves parallel to each other in the denser (less water-rich) areas of the ground substance.[23] The formation and maturation of tropocollagen and extracellular matrix are closely interrelated; for example, proteoglycans from the ground substance seem to regulate fibril formation, as shown by the observations of a decrease in the quantity of these proteins in tendons when the tropocollagen fibers have grown to their ultimate size.

With the passage of time, both tropocollagen and extracellular matrix undergo molecular changes leading to structural, biochemical, and functional alterations, including decreased chemical extractability, decreased effectiveness of enzymatic degradation, increased cross-linkage, and increased hydrophobicity. As a result of these changes, the collagen becomes tougher, more crystalline, and more difficult to dissolve, its tensile strength is reduced, and its plasticizing function is impaired or lost. In the case of the extracellular matrix, age-related changes in its physicochemical composition and a decrease in turnover of some of its constituents induce an increase in density and aggregation, thereby rendering the ground substance less permeable to the flow of substances moving through it; as a consequence, cellular nutrition becomes altered.

Collagen solubility decreases with age and this decrease has been related to increased cross-linking.[24] The appearance of cross-links between collagen fibers is well documented and is generally regarded as a normal index of maturational change. With increasing age, however, as the collagen becomes less degradable, the number of cross-linkages and especially of intra- and intermolecular ester bonds increases.[25] Other properties of collagen, such as its extractability and solubility, are also affected by age. In young rats, for example, the chemically insoluble fraction of collagen can be effectively degraded when treated with cortisol or other hormones of the adrenal cortex, whereas in older animals hormonal treatment does not restore the reduced capacity of the animal to degrade collagen.[24] The greater resistance of older collagen to enzymatic degradation may be ascribed to either or both of two factors: (1) the decrease in amount and activity of collagenase or (2) the increased resistance of the more tightly held cross-linked fibers to collagenase activity. In relation to the first factor, cortisol is capable of stimulating collagenase activity in young but not in old animals. In this regard, it should be recalled that glucocorticoids secreted by the adrenal cortex have a gluconeogenic activity; that is, they are capable of converting amino acids to glucose. Some of the amino acid components of collagen are strongly gluconeogenic, and it has been suggested that in case of stress, the increased levels of glucocorticoids would be capable of stimulating activity, thereby releasing the amino acids from collagen converting them to glucose, thus producing the energy necessary to withstand the stressful situation. In older animals, however, such an energy conversion is prevented because of the metabolic inertia of senile collagen with the result that resistance to stress is decreased in advancing age.[24]

Heating of collagen (e.g., tendon fibers from the rat tail) to approximately 60°C induces shrinkage of the fibers to about one fourth of their original length. Shrinkage is greater and the contraction and subsequent relaxation periods are more prolonged in older than in younger animals. These differences in thermal responses are accompanied by chemical differences. For example, hyroxyproline, the amino acid characteristic of collagen, is liberated during contraction; in the young animal, the contracting tendon fibers release hydroxyproline quickly and in large quantities, whereas in the old animal, hydroxyproline is released more slowly and to a much smaller extent. Progressive changes in mechanical tension and chemical composition occur throughout life. The liberation of hydroxyproline complexes from collagen during thermal contraction is quantitatively related to the age of the animal. This observation has facilitated the study of aging of collagen in such tissues as skin, skeletal muscles, and arterial walls, in which mechanical shortening of the fibers cannot be measured. Regardless of the index chosen, however, the quantity of labile collagen was shown to decrease with age in all tissues studied.[7]

3.3 Other Changes in Connective Tissue Elements

It is difficult to envision that changes in collagen would occur independently from alterations in the other components of the connective tissue. Changes with age in elastin structure and function have been thoroughly investigated.[26] A marked reduction in the volume of the extracellular matrix or in relation to the collagen content of a tissue is also a typical consequence of aging and is often accompanied by a decrease in interstitial water content. Indeed, a progressive reduction in the relative tissular content of water is consistently observed throughout the lifespan in many organs and species. The decrease in extracellular matrix has been explained by its faster turnover as compared to that of collagen, for extracellular matrix must be renewed within days or weeks, whereas the more inert collagen persists for a considerably longer period of time. Changes in the composition of the extracellular matrix with age have also been described and appear to differ from tissue to tissue with a progressively increasing occurrence of cross-links.

REFERENCES

1. Forbes, P. D., Davies, R. E., and Urbach, F., Aging, environmental influences and photocarcinogenesis, *J. Invest. Dermatol.,* 73, 13, 1979.
2. Shenefelt, P. D. and Fenske, N. A., Aging and the skin. Recognizing and managing common disorders, *Geriatrics,* 45, 57, 1990.
3. Johnson, M. L. T. and Roberts, J., Prevalence of dermatologic disease among persons 1–74 years of age. Advance Data, U.S. Department of Health, Education and Welfare, Washington, D.C., 1977.
4. Gilchrest, B. A., Age-associated changes in the skin, *J. Am. Geriatr. Soc.,* 30, 139, 1982.
5. Kligman, A. M., Grove, G. L., and Balin, A. K., Aging of human skin, in *Handbook of the Biology of Aging,* 2nd ed., Finch, C. E. and Schneider, E. L., Eds., Van Nostrand Reinhold, New York, 1985, 820.
6. Marks, R., Skin disorders, in *Principles and Practice of Geriatric Medicine,* Pathy, M. S. J., Ed., John Wiley & Sons, Chichester, England, 1990, 1011.
7. Verzar, F., Intrinsic and extrinsic factors of molecular aging, *Exp. Gerontol.,* 3, 69, 1968.
8. Fenske, N. A. and Lober, C. W., Structural and functional changes of normal aging skin, *J. Am. Acad. Dermatol.,* 15, 571, 1986.
9. Quevedo, W. C., Szabo, G., and Vicks, J., Influence of age and ultraviolet on the populations of dopa-positive melanocytes in human skin, *J. Infect. Dis.,* 52, 287, 1969.
10. Grove, G. L., Age-related differences in healing of superficial skin wounds in humans, *Arch. Dermatol. Res.,* 272, 381, 1982.
11. Grove, G. L. and Kligman, A. M., Age-associated changes in human epidermal cell renewal, *J. Gerontol.,* 38, 137, 1983.
12. Selmanowitz, V. J., Rizer, R. L., and Orentreich, N., Aging of the skin and its appendages, in *Handbook of the Biology of Aging,* Finch, C. E. and Hayflick, L., Eds., Van Nostrand Reinhold, New York, 1977, 496.
13. Graham, J. A., The psychotherapeutic value of cosmetics, *Cosmetic Technol.,* 5, 25, 1983.
14. Bar, C. A. and Pathy, M. S. J., Pressure sores, in *Principles and Practice of Geriatric Medicine,* Pathy, M. S. J., Ed., John Wiley & Sons, Chichester, England, 1990, 1037.
15. Allman, R., Epidemiology of pressure sores in different populations, *Decubitus,* 2, 30, 1989.
16. Reuler, J. B. and Cooney, T. G., The pressure sore. Pathophysiology and principles of management, *Ann. Intern. Med.,* 94, 661, 1981.
17. Cooney, T. G. and Reuler, J. B., Pressure sores, *West J. Med.,* 140, 622, 1984.
18. Perez, E. D., Pressure ulcers, Updated guidelines for treatment and prevention, *Geriatrics,* 48, 39, 1993.
19. Constantian, M. B. and Jackson, H. S., Biology and care of the pressure ulcer wound, in *Pressure Ulcers. Principles and Techniques of Management,* Constantian, M. B., Ed., Little, Brown and Co., Boston, 1980, 69.
20. Ferrell, B. A., Osterweil, D., and Christenson, P., A randomized trial of low-air-loss beds for treatment of pressure ulcers, *JAMA,* 269, 494, 1993.
21. Hall, D. A., *The Ageing of Connective Tissue,* Academic Press, London, 1976.
22. Sobel, H., Aging of ground substance in connective tissue, in *Advances in Gerontological Research,* Vol. 2, Strehler, B. L., Ed., Academic Press, New York, 1967, 205.
23. Bondareff, W., Submicroscopic morphology of connective tissue ground substance with particular regard to fibrilcollogenesis and aging, *Gerontologia,* 1, 222, 1957.
24. Houck, J. C., DeHesse, C., and Jacob, R., The effect of aging upon collagen catabolism, *Symp. Soc. Exp. Biol.,* 21, 403, 1967.
25. Seifter, S., Gallop, P. M., and Franzblau, C., Some aspects of collagenolytic action, *Trans. N.Y. Acad. Sci.,* 23, 540, 1961.
26. Sinex, F., Cross-linkage and aging, in *Advances in Gerontological Research,* Vol. 1, Strehler, B. L., Ed., Academic Press, New York, 1964, 165.

23 PHARMACOLOGY AND DRUG MANAGEMENT IN THE ELDERLY

Judith L. Beizer and Mary L. Timiras

The use of medications by elderly patients has always been a focus of concern for health professionals caring for the aging patient. The elderly make up 13% of the U.S. population; yet, it is estimated that they consume 30% of prescribed medications and 40 to 50% of over-the-counter (OTC) medications (Figure 23-1).[1] Physiologic changes, coupled with increased use of medications, place older patients at risk for adverse effects and drug interactions. The concept of "polypharmacy", originally meaning "many drugs", has acquired a derogatory connotation of excessive and unnecessary use of medication (Table 23-1). Studies have shown that 9 to 31% of hospital admissions in elderly patients may be medication related.[2-4] In addition, the elderly seem to be two to three times as likely to experience adverse drug reactions as compared to younger adults.[5]

Despite the fact that the elderly use so much medication, there are relatively few data on specific dosing of medications for older patients. The practitioner must be familiar with the physiologic changes of aging in order to predict how these will affect the pharmacokinetics and pharmacodynamics of a specific drug.

1 PHYSIOLOGIC CHANGES AFFECTING PHARMACOKINETICS

Pharmacokinetics is defined as the handling of a drug within the body, including its absorption, distribution, metabolism, and elimination. There are various physiologic changes that occur with aging that may affect drug disposition in the body. A summary of pharmacokinetic changes with aging is illustrated diagrammatically in Figure 23-2.

1.1 Absorption

While the extent of gastrointestinal (GI) absorption of most drugs is not significantly changed in the elderly, there are several changes in the GI tract which may affect the pattern of absorption (Chapter 20). Gastric emptying time may be prolonged, causing a delay in absorption from the GI tract. This would only be clinically significant for acutely administered medications, such as analgesics. GI motility may be decreased, thereby prolonging the absorption phase of a medication. Decreased mucosal cells have been noted in the aging GI tract, and intestinal blood flow may be diminished, particularly if the patient has congestive heart failure. Another important factor is that approximately 40% of the elderly have decreased gastric acid secretion resulting in achlorhydria or hypochlorhydria.[6] This may impair the disintegration and/or dissolution of orally administered medications. Absorption by active transport is reportedly decreased, but since the majority of medications are absorbed by passive diffusion, changes in absorption are not usually clinically significant.

Absorption of medications via the intramuscular route (IM) has not been well studied in the elderly, but some predictions about absorption can be made. Absorption of IM injections may be impaired in the elderly because of decreased peripheral blood flow, particularly in patients with peripheral vascular disease. Increased connective tissue in aging muscle (Chapters 21 and 22) can impair permeability, decreasing systemic absorption of an IM injection. Intramuscular injections may also be difficult to administer and painful to the patient due to the decreased muscle mass in many elderly patients.

Systemic absorption must also be considered when using a transdermal product in older patients. Transdermal absorption of medications has not been extensively studied in the elderly, but there are changes in the aging skin which may affect transdermal absorption (Chapter 22). Older patients tend to have decreased skin hydration and decreased surface lipid content, factors important for transdermal penetration. They may also have increased keratinization, further impairing absorption. Decreased peripheral blood flow and compromised microcirculation may impair systemic absorption from transdermal products.[7] Of the currently marketed transdermal products, most (nitroglycerine, estradiol, clonidine, fentanyl) are commonly prescribed for older patients and should therefore be studied in this population.

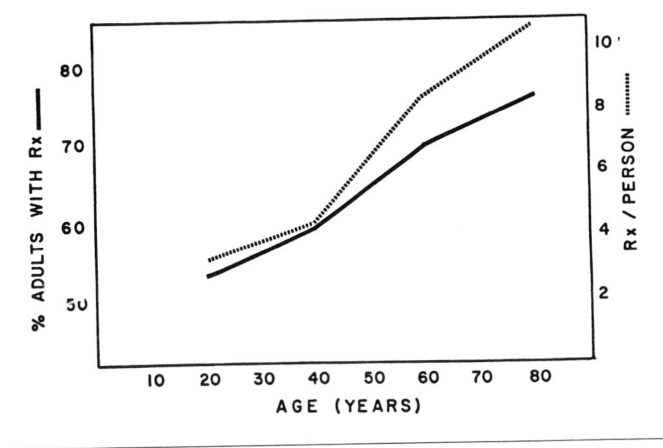

FIGURE 23-1 *Increase with age in the percentage of adults receiving prescriptions (——) and in the number of medications prescribed per person (• • • •).*

1.2 Distribution

The distribution of medications within the body is dependent on whether the drug is lipid soluble or water soluble and the extent of protein binding. *Age-related changes in body composition*

TABLE 23-1

Features of Polypharmacy
Medication not indicated
Duplicate medications
Concurrent interacting medications
Contraindicated medications
Inappropriate dosage
Drug treatment of adverse drug reaction
Improvement following discontinuance

0-8493-8979-8/94/$0.00+$.50

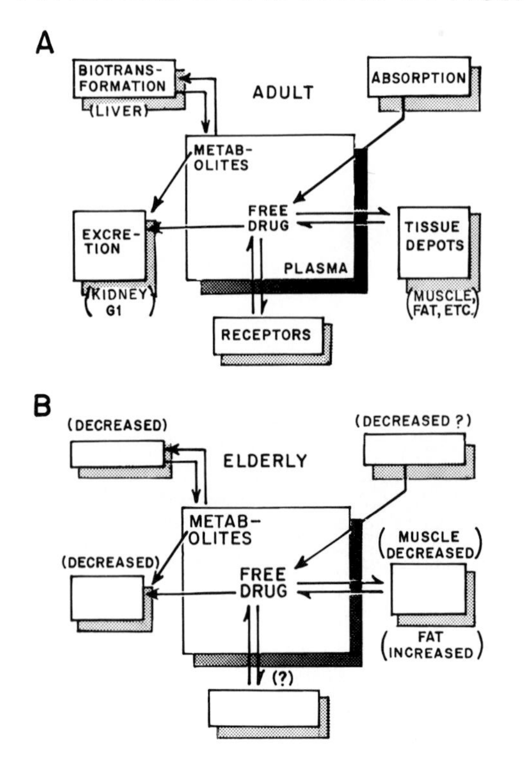

FIGURE 23-2 *Drug kinetics in adults and elderly.* (A) In the adult, active free drug depends upon absorption, metabolism, and excretion. It also depends on type of tissue for the position and duration of time in body. The cellular action is dependent on receptor binding. (B) In the elderly, persistence and elevated active free drug is due to decreased metabolism and excretion. Tissue deposits change and the net result is prolonged duration of action. Receptor changes are as yet little known. (Drawing by S. Oklund.)

may significantly affect drug distribution. With aging, lean body mass and total body water decrease, while fat content increases (Chapter 21).[8] This can significantly increase the volume of distribution of fat-soluble drugs such as the benzodiazepines. Clinically, it may take longer both to reach steady state levels and for the drug to be eliminated from the body.

The extent to which *drugs bind to proteins* can significantly alter the volume of distribution. Many elderly, particularly those living alone in the community, have inadequate protein intake and therefore are hypoproteinemic. Since only the unbound portion of a drug is pharmacologically active, a reduction of plasma protein, specifically albumin, can result in higher free drug levels and increased effects. This is clinically significant for highly protein-bound drugs such as phenytoin (an antiepileptic drug) and warfarin (an anticoagulant drug). Practitioners should be especially careful when prescribing two or more highly protein-bound drugs. It is also postulated that the affinity of albumin for drugs is diminished in the elderly, meaning that drugs may be more easily displaced from their binding sites. Basic drugs such as lidocaine and disopyramide are bound to α_1 glycoprotein, which is increased with aging, though the changes in binding are rarely clinically significant.

Cardiac output may be decreased in the elderly, altering regional blood flow. Decreased splanchnic, renal, and peripheral blood flow affects the distribution and, therefore, the effectiveness of some medications. Tissue barriers may be altered with age, also affecting the outcome of the drug.

1.3 Metabolism

The liver is the main site of metabolism or biotransformation of drugs and with increased age there is a decrease in both hepatic mass and blood flow (Chapter 20).[9] The first-pass metabolism of some medications is decreased with aging, thus increasing the relative bioavailability of the medicine.

Hepatic metabolism of drugs occurs mainly by either phase I oxidative reactions or phase II conjugation reactions. Various factors besides age — gender, genetics, smoking, alcohol, and medications — affect metabolism. With age, phase I reactions are decreased, particularly in older men. Medications affected include theophylline, propranolol, and diazepam, to name a few. Phase II reactions are minimally affected by age. Examples here are lorazepam, oxazepam (antianxiety drugs), acetaminophen (an analgesic), and isoniazid (an antibacterial drug, particularly useful in the therapy of tuberculosis, an infection currently on the rise among some elderly, Chapter 18).

The inducibility of hepatic enzymes by smoking, alcohol, and drugs appears to be diminished, though results from various studies are conflicting.[10] Age-related changes in cimetidine-induced enzyme inhibition have not been seen. It appears that enzyme activity is decreased by the same amount in both elderly and younger subjects.[10]

1.4 Elimination

Renal elimination is considered the most significant pharmacokinetic change in the elderly (Chapter 19). Between the 4th and 8th decades of life, renal mass decreases on average by 20% and renal blood flow decreases by 10 ml/min per decade after age 30.[11] Glomerular filtration rate, as expressed by the creatinine clearance (CrCl), also declines linearly with age beginning in the fourth decade. While measuring creatinine clearance via a 24-h urine collection is the best way to obtain an accurate level, these collections are often difficult, if not impossible, in the elderly patient and usually entail inserting a Foley catheter. A recent study found that the creatinine clearance measured from an 8-h urine collection was not significantly different from the clearance measured via a 24-h collection.[12] Of the patients, 84% would have received the correct dose of a renally eliminated drug if the 8-h value was used.

Alternatives to urine collections rely upon empirical estimations of creatinine clearance. A common approach relates renal function to the serum creatinine alone. However, in the elderly, serum creatinine is an inadequate measure of renal function because creatinine production decreases as muscle mass decreases. Therefore, it is more appropriate to estimate creatinine clearance using an equation which takes into account age, weight, serum creatinine (Scr), and gender. A commonly used equation is that developed by Cockcroft and Gault (Equation 1) where the weight used is the patient's lean body weight (LBW) (Equation 2) if the patient is overweight.[13,14]

$$\mathrm{CrCL}_{\mathrm{men}} = \frac{(140 - \mathrm{Age})\mathrm{LBW}(\mathrm{kg})}{72 \times \mathrm{Scr}(\mathrm{mg/dl})}$$
$$\mathrm{CrCl}_{\mathrm{women}} = \mathrm{CrCl}_{\mathrm{men}} \times 0.85 \qquad (1)$$

$$\mathrm{LBW}_{\mathrm{men}} = 50\mathrm{kg} + 2.3\ \mathrm{kg/in\ above\ 5\ feet}$$
$$\mathrm{LBW}_{\mathrm{women}} = 45.5\ \mathrm{kg} + 2.3\ \mathrm{kg/in\ above\ 5\ feet} \qquad (2)$$

The Cockcroft–Gault equation is useful when calculating doses for drugs which are eliminated by glomerular filtration such as aminoglycosides, vancomycin, digoxin, lithium, and H_2 antagonists. In most cases, when renal function is decreased, the dose of the medication remains the same, but the interval between doses will be increased.

Drugs that are eliminated by tubular secretion also exhibit decreased excretion with age. This is clinically significant for a medication such as nitrofurantoin which must be secreted in the urine to treat urinary tract infections. It is ineffective in patients with a creatinine clearance <40 ml/min and therefore is rarely useful in the elderly.

2 PHYSIOLOGIC CHANGES AFFECTING PHARMACODYNAMICS

Pharmacodynamics refers to the processes involved in the interaction between a drug and an effector organ that results in a response, either therapeutic or adverse. Pharmacodynamics measures the intensity, peak, and duration of action of a medication. With aging, physiologic changes may affect the body's response to medications. Some changes increase the older patient's sensitivity to a drug, some may decrease the effect of the medication, and some may make the older person more susceptible to the adverse effects of a drug. The predictability of a drug's response is decreased in the elderly, and the practitioner cannot only rely on the pharmacokinetic changes but must include pharmacodynamic factors as well.

Pharmacodynamic changes in the elderly can best be understood by reviewing examples of medications in which pharmacodynamic changes have been described. In the following section, several medications that illustrate pharmacodynamic changes are discussed.

Postural hypotension appears to be more of a problem in older patients as compared to younger patients. This is due to a decrease in baroreceptor function and decreased peripheral venous tone (Chapter 21).[15,16] When a younger person taking a vasodilator stands up abruptly, the body responds to the immediate hypotension with a reflex tachycardia. This tachycardia does not always occur in older patients, putting them at risk for dizziness, syncope, and falls. For this reason, vasodilators should be used cautiously in the elderly. Older patients who are treated with hypotensive or vasodilatory drugs should be instructed to change positions slowly when rising from a lying or sitting position. Increased postural (orthostatic) hypotension is seen with nitrates, nifedipine, tricyclic antidepressants, antipsychotics, and diuretics.

The *hypnotic–sedative benzodiazepines* are (unfortunately) widely used in the geriatric population. Based on pharmacokinetic factors, we can choose a benzodiazepine that would be relatively well tolerated in the elderly (short half-life and conjugated metabolism), but pharmacodynamic factors must also be considered. Reidenberg and colleagues studied the relationship between diazepam dose, plasma level, age, and central nervous system (CNS) depression.[17] Their results demonstrated that older patients require a lower dose and plasma level of diazepam than younger patients to reach the same level of sedation. An increased sensitivity was also demonstrated by Castleden and colleagues who found increased psychomotor impairment in older subjects as compared to younger subjects while on nitrazepam.[18]

In general, the elderly will be more sensitive to the effects of a medication which depresses or excites the CNS. The elderly have less CNS reserve and therefore are more sensitive to an "insult" by a medication (Chapters 8 to 10).

Though the data are conflicting, most studies on *adrenergic β-receptors* have demonstrated that the elderly have a decreased β-receptor sensitivity as evidenced by a decreased cardiac response to the β-agonist isoproterenol and the β-antagonist (blocker) propranolol.[19] It was originally thought that this decreased response was due to a decrease in the number of β-receptors. Instead, no overall decrease was demonstrated, but rather a decrease in the number of receptors that have a high affinity for these drugs.[20] The elderly have a decreased ability to form high affinity binding complexes. Desensitization of the receptors would occur, since norepinephrine levels increase with age (Chapter 11), yet response to β agonists decreases.[21] The clinical implications of all of these changes are that the elderly may have a decreased response to β blockers and agonists, including the β_2 bronchodilators.

Elderly patients seem to be more susceptible to the side effects of *antipsychotic medications*. Extrapyramidal symptoms, orthostatic hypotension, and anticholinergic effects occur more frequently than in younger patients and are less well tolerated.[22] Elderly patients on antipsychotics are more likely to experience *Parkinsonism,* probably because of an already depleted dopamine reserve (Chapter 8). Adding a dopamine antagonist, such as a phenothiazine or haloperidol, can "tip" the older patient into Parkinsonism or unmask latent Parkinson's disease. In one series of new Parkinson's cases, 51% of the patients had drug-induced Parkinsonism.[23] Clinical features cleared completely in only 66% of the cases, taking up to 36 weeks to resolve. Elderly patients experience *tardive dyskinesia* (i.e., abnormal involuntary movements) more frequently and earlier in treatment, even when on low doses of neuroleptics.[24] This may be due to a "hypersensitivity" of the dopamine receptors in the nigrostriatum. Tardive dyskinesia is more likely to be persistent and severe in the elderly, and women seem to be particularly at risk.

Physiologic changes in the geriatric patient may exaggerate the effects of *anticholinergic agents* (Chapters 8 and 9). Slowed GI motility increases the risk of constipation. Urinary retention is enhanced in patients with an outflow obstruction such as an enlarged prostate. Anticholinergic-induced CNS effects such as delirium and memory impairment may be more pronounced in the elderly because of decreased CNS reserve. Dry eyes and mouth (signs of parasympathetic block) may be more pronounced in the elderly and are more bothersome than in younger patients.

Elderly patients seem to be more sensitive to the effects of *anticoagulants* such as warfarin, and decreased doses are needed to adequately anticoagulate the older patient.[25,26] Just why this sensitivity occurs is not entirely clear. There may be a relative deficiency of vitamin K or vitamin K-dependent clotting factors in the elderly. There may also be increased concentrations of inhibitors of coagulation. In addition, warfarin is highly protein-bound to albumin, so alterations in binding must be considered. Whether age itself is an independent risk factor for complications of warfarin therapy has been debated.[27,28] Either way, it is prudent for the practitioner to closely monitor all elderly patients on warfarin therapy.

3 ADVERSE DRUG REACTIONS IN THE ELDERLY

As stated earlier, studies have demonstrated that the elderly are two to three times more at risk for adverse drug reactions as compared to younger adults.[5] This is due to a number of factors summarized in Table 23-2, the most important of which include

- Increased number of medications taken by the elderly
- Increased sensitivity to medications as described by the pharmacokinetic and pharmacodynamic changes that occur

Many adverse reactions are iatrogenic in nature. For example, they may be due to

- Choice of an inappropriate medication or dosage
- Inadequate monitoring of the patient
- Failure to recognize or notice adverse effects by both the patient and prescriber
- Drug–drug interactions or drug–disease interactions
- Noncompliance

Subtle effects such as GI complaints, dizziness, mental status changes, change in libido, instability and falls, and bowel or bladder habits may be attributed to "old age" or may be treated as a new disease state. Health professionals should inquire about specific adverse effects and should encourage the patient to report any unusual occurrences while on medication.

Noncompliance is the failure of a patient to follow instructions regarding medications. It represents another factor often contributing to adverse drug reactions. Psychosocial complications such as poverty, dementia, and loneliness exacerbate this problem. Effective communication between the prescriber and patient and/or caregiver can help eliminate this factor.

TABLE 23-2

Factors Associated with Increased Incidence of Adverse Drug Reactions in the Elderly

Reduced (small) stature
Reduced renal and hepatic function
Cumulative insults to body
Disease
Faulty diet
Drug abuse
Medications, multiple and potent
Altered pharmacokinetics
Noncompliance

3.1 Specific Principles

Although a comprehensive review of all drug classes is impractical, certain groups of medications are problematic enough in the elderly to warrant separate consideration. For example, *psychotropic medications* are responsible for the most adverse drug reactions.[29] Such reactions include worsening mental status, falls, dehydration, orthostasis, extrapyramidal signs, and tardive dyskinesia. Therefore, their use should be reserved only for circumstances in which they are needed to enable the patient to remain functional.[30] Hypnotics (i.e., drugs inducing sleep) should not be used in chronic conditions or without an evaluation for depression.[31] If benzodiazepines are utilized, only those with inactive metabolites should be considered.[32] A side-effect profile that minimizes anticholinergic effects should be chosen when prescribing antidepressants or antipsychotics.[33,34]

Analgesics (drugs relieving pain) are another group of medications frequently prescribed and therefore also warrant special consideration. Nonsteroidal anti-inflammatory drugs (NSAIDs) are currently ubiquitous despite the plethora of adverse drug reactions that may arise from their use.[35] The most common side effects are gastrointestinal symptoms ranging from gastritis to life-threatening hemorrhage.[36,37] Renal effects are usually reversible as are photosensitivity and urticaria. Pulmonary edema arises from congestive heart failure due to fluid and sodium retention. NSAIDs also interfere with hypertension treatment. CNS effects range from headache to altered mental status to frank psychosis. Finally, genitourinary effects include ejaculatory dysfunction. In view of the aforementioned problems, geriatricians are conservative in the use of NSAIDs. The nonpharmacologic management of osteoarthritis (i.e., physical therapy) is utilized whenever possible, and, finally, acetaminophen is recommended for analgesia (Chapter 21). The use of narcotics is also discouraged except in extreme cases because of their adverse affect on mental status as well as their propensity to exacerbate constipation.

In the interest of preventive medicine, the use of *antihypertensive agents* is on the rise.[38] Studies have been carried out in the frail elderly treated with these medications and certain trends have evolved as to their use in this population. Calcium channel blockers as well as inhibitors of the angiotensin-converting enzyme (ACE) (i.e., conversion of angiotensin I to angiotensin II, the most effective hypertensor) (Chapter 11) have become standard first-line drugs because of their effectiveness and their favorable side-effect profile in elderly patients.[39,40] Agents that work directly on the CNS (such as methyldopa) are generally avoided, as are the β blockers for reasons previously described. Diuretics are still utilized but require frequent monitoring for electrolyte imbalance. Checking for orthostatic changes remains important for all antihypertensive medications in the elderly.

Finally, management of an elderly patient on *digoxin* illustrates all of the principles already mentioned. Digoxin is one of the most used preparations of digitalis from the leaf of the foxglove plant. The main action of digoxin is its *ability to increase the force of myocardial contraction.* Beneficial consequences of this action include increased cardiac output, decreased cardiac size, venous pressure and blood volume, slowdown of cardiac rate, promotion of diuresis, and relief of edema. Many of these actions will benefit alterations in cardiac functions at all ages, including old age (Chapter 21). Administration of digoxin to an older individual with cardiac alterations depends on several considerations. First, the indications for the use of digoxin have been scrutinized and digoxin is no longer utilized in an indiscriminate manner to everyone with congestive heart failure.[41] Rather, the indications for its use have been narrowed to specific clinical situations.[42,43] Second, because of its reduced elimination, it is usually given in lower doses in the elderly. Digoxin serum levels are monitored to prevent toxicity. Even at normal therapeutic levels, it can produce a range of adverse effects including anorexia and altered mental status. Studies have been performed to examine the effects of withdrawing patients from this drug after chronic administration, showing that many do well.[43] There also have been reported some sensitivity changes at the digoxin receptor level in elderly patients, showing different end organ reaction to the drug.[44] Thus, this classic medication serves as a good example of a drug that undergoes altered pharmacokinetics as well as altered pharmacodynamics in the elderly.

4 GENERAL GUIDELINES

In the interest of avoiding adverse drug reactions, a number of principles, partly based on physiologic considerations, should be followed:

1. Nonpharmacologic management should be used whenever possible.
2. The number of drugs prescribed should be kept to a minimum.
3. The drug regimen should be simplified to aid in compliance.
4. A diagnosis as specific as possible should be made.
5. Treatment for that diagnosis should be prescribed only with clear goals or end points in mind.
6. An appropriate dosage should be chosen (i.e., start low and go slow).
7. Drug levels should be consistently monitored.
8. Clinical progress should be assessed as indicated.

As during development, during aging as well, physiologic responses are dynamically related to the changing functional competence. Therefore, the drug regimen needs to be regularly reviewed and reassessed for possible changes. This includes an "obsessive" (i.e., as detailed as possible!) drug history which may necessitate the patient bringing in all medications as well as over-the-counter drugs for scrutiny by a health professional. Finally, one must suspect a drug reaction when any otherwise unexplained symptoms occur such as a change in mental status.

One of the most common interventions a geriatrician makes is to discontinue medications. It has been shown that, when done judiciously, most patients benefit from this maneuver.[45] However, more research is needed in patients whose demographic characteristics match those of the patients treated. Equipped with a basic knowledge of physiologic changes that occur with aging, pharmacologic principles, and common sense, health care professionals should be able to prescribe medications for elderly patients in a safe and effective manner.

REFERENCES

1. National Medical Expenditure Survey. Prescribed medicines: a summary of use and expenditures by medicare beneficiaries: research findings, U.S. Department of Health and Human Services Publication 89-3448, National Center for Health Services Research and Health Care Technology Assessment, Rockville, MD, 1989.
2. Grymonpre, R. E., Mitenko, P. A., Sitar, D. S., Aoki, F. Y., and Montgomery, P. R., Drug-associated hospital admissions in older medical patients, *J. Am. Geriatr. Soc.*, 36, 1092, 1988.
3. Colt, H. G. and Shapiro, A. P., Drug-induced illness as a cause for admission to a community hospital, *J. Am. Geriatr. Soc.*, 37, 323, 1989.
4. Williamson, J. and Chopin, J. M., Adverse reactions to prescribed drugs in the elderly: a multicentre investigation, *Age Ageing*, 9, 73, 1980.
5. Nolan, L. and O'Malley, K., Prescribing for the elderly. I. Sensitivity of the elderly to adverse drug reactions, *J. Am. Geriatr. Soc.*, 36, 142, 1988.
6. Baron, J. H., Studies of basal peak acid output with an augmented histamine test, *Gut*, 4, 136, 1963.
7. Roskos, K. V., Maibach, H. I., and Guy, R. H., The effect of aging on percutaneous absorption in man, *J. Pharmacokinet. Biopharmaceut.*, 17, 617, 1989.
8. Yuen, G. J., Altered pharmacokinetics in the elderly, *Clin. Geriatr. Med.*, 6, 257, 1990.
9. James, O. F. W., Drugs and the ageing liver, *J. Hepatology*, 1, 431, 1985.
10. Durnas, C., Loi, C. M., and Cusack, B. J., Hepatic drug metabolism and aging, *Clin. Pharmacokinet.*, 19, 359, 1990.
11. Bennett, W. M., Geriatric pharmacokinetics and the kidney, *Am. J. Kidney Dis.*, 16, 283, 1990.
12. O'Connell, M. B., Wong, M. O., Bannick-Mohrland, S. D., and Dwinell, A. M., Accuracy of 2- and 8-hour urine collections for measuring creatinine clearance in the hospitalized elderly, *Pharmacotherapy*, 13, 135, 1993.
13. Cockcroft, D. W. and Gault, M. H., Prediction of creatinine clearance from serum creatinine, *Nephron*, 16, 31, 1976.
14. Lott, R. S. and Hayton, W. L., Estimate of creatinine clearance from serum creatinine concentration — a review, *Drug Intell. Clin. Pharm.*, 12, 140, 1978.
15. Caird, F. I., Andrews, G. R., and Kennedy, R. D., Effect of posture on blood pressure in the elderly, *Br. Heart J.*, 35, 527, 1973.
16. Gribbon, B., Pickering, T. G., Sleight, P., and Peto, R., Effect of age and high blood pressure on baroreflex sensitivity in man, *Circ. Res.*, 29, 424, 1971.
17. Reidenberg, M. M., Levy, M., Warner, H., Coutinho, C.B., Schwartz, M. A., Yu, G., and Cheripko, J., Relationship between diazepam dose, plasma level, age and central nervous system depression, *Clin. Pharmacol. Ther.*, 23, 371, 1978.
18. Castleden, C. M., Kay, C. M., and Parsons, R. L., Increased sensitivity to nitrazepam in old age, *Br. Med. J.*, 1, 10, 1977.
19. Vestal, R. E., Wood, A. J. J., and Shand, D. G., Reduced β-adrenoreceptor sensitivity in the elderly, *Clin. Pharmacol. Ther.*, 26, 181, 1979.
20. Feldman, R. D., Limbird, L. E., Nadeau, J., Robertson, D., and Wood, A. J. J., Alterations in leukocyte β-receptor affinity with aging. A potential explanation for altered β-adrenergic sensitivity in the elderly, *N. Engl. J. Med.*, 310, 815, 1984.
21. Scarpace, P. J., Decreased receptor activation with age. Can it be explained by desensitization?, *J. Am. Geriatr. Soc.*, 36, 1067, 1988.
22. Peabody, C. A., Warner, M. D., Whiteford, H. A., and Hollister, L. E., Neuroleptics and the elderly, *J. Am. Geriatr. Soc.*, 35, 233, 1987.
23. Stephen, P. J. and Williamson, J., Drug-induced parkinsonism in the elderly, *Lancet*, 2, 1082, 1984.
24. Morton, M. R., Tardive dyskinesia: Detection, prevention and treatment, *J. Geriatr. Drug. Therapy*, 2, 21, 1987.
25. Shepherd, A. M. M., Hewick, D. S., Moreland, T. A., and Stevenson, I. H., Age as a determinant of sensitivity to warfarin, *Br. J. Clin. Pharmacol.*, 4, 315, 1977.
26. Redwood, M., Taylor, C., Bain, B. J., and Matthews, J. H., The association of age with dosage requirement for warfarin, *Age Ageing*, 20, 217, 1991.
27. Gurwitz, J. H., Goldberg, R. J., Holden, A., Knapic, N., and Ansell, J., Age-related risks of long-term oral anticoagulant therapy, *Arch. Intern. Med.*, 148, 1733, 1988.
28. Landefeld, C. S. and Goldman, L., Major bleeding in outpatients treated with warfarin: incidence and prediction by factors known at the start of outpatient therapy, *Am. J. Med.*, 87, 144, 1989.

29. Thompson, T. L., Moran, M. G., and Nies, A. S., Psychotropic drug use in the elderly, *N. Engl. J. Med.*, 308, 194, 1983.
30. Maletta, G., Mattox, K. M., and Dysken, M., Guidelines for prescribing psychoactive drugs in the elderly. I, *Geriatrics*, 46, 40, 1991.
31. Kales, A., Soldatos, C. R., and Kales, J. D., Sleep disorders: insomnia, sleepwalking, night tremors, nightmares, and enuresis, *Ann. Intern. Med.*, 106, 582, 1987.
32. Salzman, C., et al., Long vs. short half-life benzodiazepines in the elderly, *Arch. Gen. Psychiatry*, 40, 293, 1983.
33. Jenike, M. A., Psychoactive drugs in the elderly: antidepressants, *Geriatrics*, 43, 43, 1988.
34. Jenike, M. A., Psychoactive drugs in the elderly: antipsychotics and anxiolytics, *Geriatrics*, 43, 53, 1988.
35. Sack, K. E., Update on NSAIDS in the elderly, *Geriatrics*, 44, 71, 1989.
36. Fries, J. F., et al., Toward an epidemiology of gastropathy associated with nonsteroidal antiinflammatory drug use, *Gastroenterology*, 96, 647, 1989.
37. Griffin, M. R., Ray, W. A., and Schaffner, W., Nonsteroidal anti-inflammatory drug use and death from peptic ulcer in elderly patients, *Ann. Intern. Med.*, 109, 359, 1988.
38. Applegate, W. B., Hypertension in elderly patients, *Ann. Intern. Med.*, 110, 901, 1989.
39. Jenkins, A. C., Knill, J. R., and Dreslinski, G. R., Captropril in the treatment of elderly hypertensive patients, *Arch. Intern. Med.*, 145, 2029, 1985.
40. Reid, J. L., First-line and combination treatment for hypertension, *Am. J. Med.*, 86, 2, 1989.
41. Fleg, J. L., Gottlieb, S. H., and Lakatta, E. G., Is digoxin really important in treatment of compensated heart failure?, *Am. J. Med.*, 73, 244, 1982.
42. Papadakis, M. A. and Massie, B. M., Appropriateness of digoxin use in medical outpatients, *Am. J. Med.*, 85, 365, 1988.
43. Sueta, C. A., Carey, T. S., and Burnett, C. K., Reassessment of indications for digoxin, *Arch. Intern. Med.*, 149, 609, 1989.
44. Kelly, J. G., Copeland, S., and McDevitt, D. G., Erythrocyte cation transport and age: effects of digoxin and furosemide, *Clin. Pharmacol. Ther.*, 34, 159, 1983.
45. Avorn, J., Soumerai, S. B., Everitt, D. E., et al., A randomized trial of a program to reduce the use of psychoactive drugs in nursing homes, *N. Engl. J. Med.*, 327, 168, 1992.

IV
PREVENTION AND
REHABILITATION

24 EFFECTS OF DIET ON AGING

Brian J. Merry and Anne M. Holehan

1 STUDIES IN LABORATORY ANIMALS

In seeking to gain insight into the biochemical mechanisms that underlie aging, particularly in mammals, it has to be recognized that merely cataloging physiological changes with age will give little understanding of such a complex process. Aging is both multifaceted and hierarchic in its expression, with subtle changes occurring simultaneously at the molecular, cellular, tissue, and organ levels. Further complexity arises at each of these levels as metabolic and homeostatic compensations are made in an attempt to accommodate these age-induced changes in the normal functioning of the animal. To distinguish "cause" and "effect" in the assessment of the biologic mechanisms of aging is therefore extremely difficult, a problem frequently exacerbated by the existence of underlying chronic pathology.

Recognizing these difficulties, attempts have been made to develop experimental models in which the rate of aging can be manipulated in a predictable manner, thus facilitating the distinction between "cause" and "effect". It has, however, proven extremely difficult to devise experimental strategies to retard the rate of aging in mammalian species, and those strategies that have been successful have incorporated some form of restricted feeding. It is evident from the rapid growth of publications over the last 5 to 7 years that utilize diet to retard physiological aging that this animal model is now being exploited as an important experimental tool in aging research.

2 THE DIETARY RESTRICTION MODEL

That restricted feeding has a beneficial effect on survival in animal populations has been recognized for many years (see Reference 1 for an historical account). The effect is both reproducible and robust, being observed in spite of widely differing experimental designs, dietary composition, and severity of restricted feeding. Restricted feeding regimes which successfully extend survival are designed to avoid malnutrition and to provide essential nutrients and vitamins, while restricting calorie intake by 30 to 70% of the *ad libitum* level. One of the simplest but most effective methods of extending lifespan and therefore, by implication, slowing the rate of aging, has been to limit access to the normal diet in rodents so that the body weight of experimental animals is maintained at about 50% that of age-matched, fully fed control animals. This is approximately equivalent to maintaining age-matched animals on half-rations throughout postweaning life. This allows for a continuous slow rate of growth to occur throughout life and results in a 36 to 42% extension in mean and maximum lifespan (see Figure 24-1). Calculation from survival data of the time required to double the rate of mortality in fully fed control rats gives a period of about 100 days, while in animals diet-restricted throughout postweaning life the figure is approximately 200 days.

2.1 Composition of the Diet

Numerous studies have been reported that attempt to elucidate the component in the diet responsible for the effect on the rate of aging. Most of the studies have concentrated on the protein and carbohydrate content of the diet, and conflicting conclusions have been reported. In one detailed investigation into whether the beneficial effects of underfeeding were derived mainly from either protein or caloric restriction, 1600 male rats were divided into five dietary groups.[2] One group was fed a commercial diet *ad libitum* and the other four were fed restricted diets designated low protein–low carbohydrate, low protein–high carbohydrate, high protein–low carbohydrate, and high protein–high carbohydrate. The results from this study were again equivocal in their interpretation with the exception that all four diet-restricted groups had a greater life expectancy than those fed *ad libitum*. Animals fed high-protein diets had markedly superior survival in the first half of life, but feeding high carbohydrate either with high or low protein intake resulted in a steep late mortality. The low-protein–low-carbohydrate diet yielded a high early mortality resulting from cecal impaction but showed the best late-life survival. It was further concluded that the caloric intake of these five dietary groups was linearly related to the lifespan with a slope of approximately −3.9 days kcal/day.

Subsequent attempts to resolve this confused aspect of the model have utilized experimental designs that provide restricted rats with the same amount of protein as *ad libitum* controls but with varying amounts of carbohydrate.[3] It was concluded that rats fed calorie-restricted diets survive longer than those fed *ad libitum* even when both groups have a similar protein intake. Low levels of dietary protein in both calorie-restricted and *ad libitum*–fed animals were associated with a decrease in survival rates, and caloric intake appeared to have a greater effect on survival than protein intake. It is, however, difficult to separate the effects of calorie versus protein restriction on subsequent survival because of the metabolic compensations that occur when energy intake in the diet is a limiting factor. The effect on survival in mammalian species appears to depend on restricting the total energy intake, through the diet, for restriction of fat, protein, or carbohydrate alone, without an overall reduction in total energy intake, is not effective in prolonging maximum survival.[4-6]

2.2 Initiation of Dietary Restriction: Effect of Age

It had been concluded in early studies using rats, mice, and hamsters that dietary restriction for 1 year followed by a return

0-8493-8979-8/94/$0.00+$.50

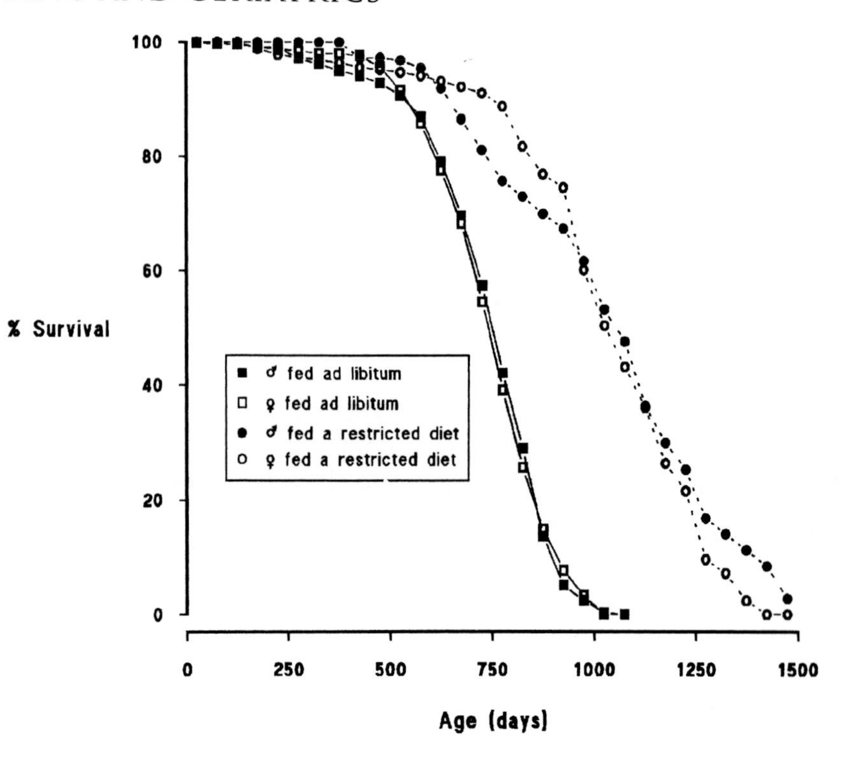

FIGURE 24-1 *Survival profiles of CFY Sprague–Dawley rats fed* ad libitum (■ ♂, □ ♀) *or a restricted diet* (● ♂, ○ ♀) *such that body weight was maintained at 50% that of the fully fed animals.* Group numbers: *ad libitum* = 1094 ♂, 200 ♀; dietary restricted = 450 ♂, 100 ♀.

to full feeding was the most effective nutritional regimen for delaying aging in these species.[7] These conclusions, however, were based not on maximum lifespans or the survival of the last decile (last 10% of the populations) but on mean lifespan data. Changes in the shape of the survival curve will radically alter the average survival age without affecting maximum lifespan and the rate of aging. In the three dietary regimes in which dietary restriction was imposed for (1) the total postweaning lifespan, (2) the first year of life, or (3) the second year of life, little variation in maximum lifespan occurred. It is now accepted that the greatest effect of dietary restriction regimes on subsequent survival is recorded for animals underfed throughout the majority of the postweaning period of life. Weindruch[8] has drawn attention to three studies which have compared survival in control, early-onset, and adult-onset diet-restricted B10C3F1 mice, F344, and Wistar rats. Comparison of either the average lifespan, or the average survival of the longest-lived 10% of the population, in the two diet-restricted groups resulted in lifespans for the adult-onset restricted feeding group which were 89 to 95% as great as those animals experiencing early-onset restricted feeding. The consistency of these observations is intriguing in view of the variation between the three studies in the period separating early- and adult-onset restricted feeding. Only 4.5 months separated the two groups when using F344[9] rats, whereas in the other studies 11 to 13 months separated the two groups.[10,11]

The question therefore arises as to whether there are periods in the life of an animal that are more susceptible to the effects of undernutrition in extending lifespan. Early dietary restriction followed by *ad libitum* feeding has been found not to extend maximum lifespan,[12,13] suggesting that restricted feeding acts through a dynamic process to retard physiological aging. In contrast, feeding a restricted diet (60% *ad libitum*) to 24.5-month-old rats for 70 days resulted in a significant decrease in liver and brain heat-labile aminoacyl–tRNA synthetases.[14] These authors had previously reported an age-associated accumulation of the heat-labile form of aminoacyl–tRNA synthetases in fully fed animals.[15] A potentially important age-sensitive biochemical parameter is therefore susceptible to a short period of restricted feeding very late in the life of the animal.

Successful food restriction of adult rodents appears dependent on the gradual adaptation of the animals to the reduced rations.[16] Adult mice from two long-lived strains were food-restricted at 12 months by restricting food intake for the 1st month of underfeeding to 72% and then in the 2nd month to 56% that of *ad libitum* intake. An increase in maximum and mean lifespan of 10 to 20% was achieved by this procedure (see Figure 24-2).

While the greatest effect on slowing the rate of aging is achieved by restricting the diets of animals throughout most of the postweaning life, the immediate postweaning period is clearly not the only phase of the lifespan sensitive to the effects of underfeeding. Observations from a number of groups indicate that either refeeding previously diet-restricted animals or underfeeding rodents during the preweaning period is detrimental to subsequent survival.

3 EFFECT ON AGE-RELATED PATHOLOGY

It has been recognized from many studies on nutrition and aging that age-related pathologies in laboratory rodents are exquisitely sensitive to the dietary history of the individual animals. A beneficial effect of undernutrition in maintaining tissue integrity (Figure 24-3 and 24-4) and reducing the incidence of chronic glomerulonephritis, myocardial fibrosis, peribranchial lymphocytosis, periarteritis, prostatitis, and endocrine hyperplasias in rats has been recorded,[2,17,18] and aged mice, which have been maintained on a restricted feeding regime, show improved immunity to influenza A virus.[19]

3.1 Tumor Incidence

The majority of studies on chronic underfeeding and pathology have concentrated on the incidence of neoplastic lesions, which is a linearly increasing function of caloric intake or body weight and the protein component of the diet.[20-22] The protein content of the diet has a complex action, with some types of tumor being commonest at high protein intake and others at low protein intake. Dietary restriction has been observed to reduce the frequency of tumors of the pituitary gland, lung, and pancreatic islet cells, while in tumors of soft tissue origin and those of the thyroid gland and the bladder, the frequency of appearance was unaffected, although a delay in onset was observed. Conversely, malignant epithelial tumors, adrenal, parathyroid gland tumors, and reticulum cell sarcomas of the lymphoid glands were significantly increased.[23] These observations have been confirmed in mice

FIGURE 24-2 *Body weights and survival of B10C3F1 mice (A and B) and B6 mice (C and D) fed on control and restricted diets.* Weights are plotted as means ± standard error for all mice alive at the indicated ages. Each point in the survival curves represents one mouse. (From Weindruch, R. and Walford, R. L., *Science,* 215, 1415, 1982. © AAAS. With permission.)

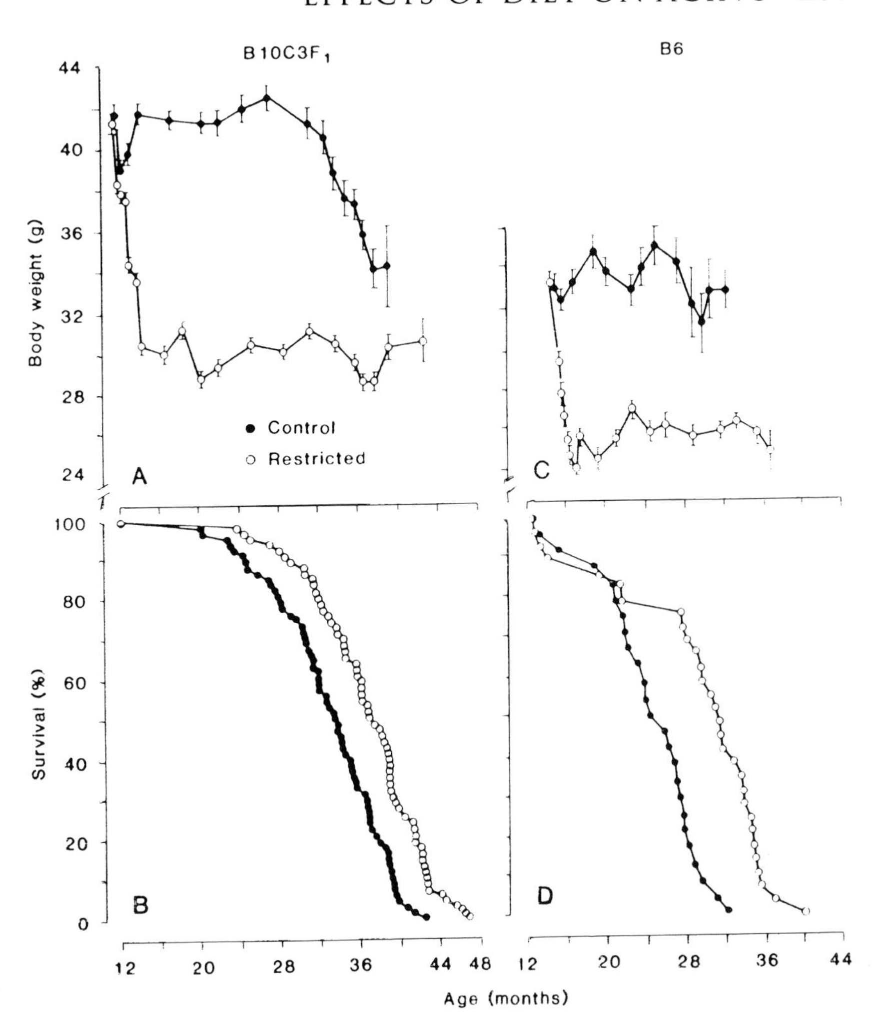

and rats where a reduction of only 20% of the normal food intake resulted in a significant reduction of the most common tumor for each species (mice, liver; rats, pituitary, mammary gland, skin)[24] (see Figure 24-5). The formation of spontaneous lymphoma is inhibited in mice underfed from 12 months of age, but no effect of diet on hepatoma incidence was observed.[16] Thus, a selective effect of diet on the inhibition of tumor formation is again evident.

Chronic calorie restriction reduces mouse mammary tumor virus (MMTV)-induced mammary tumors in C3H/Ou mice regardless of the calorie source (fat versus carbohydrate). Mammary tumorigenesis in C3H mice is associated with integration of MMTV proviral DNA which is thought to act as a putative mammary tumor protooncogene int-1. Calorie restriction appears to decrease the frequency of viral reintegration adjacent to the int-1 gene, thus decreasing its expression. Expression of other putative protooncogenes, int-2, and *ras* in liver tissue were also reduced. The authors considered that both initiation and promotion of mammary tumorigenesis were decreased by calorie restriction in C3H/Ou mice.[25] This study was extended to observe the effects of calorie restriction on thymic lymphoma–prone AKR mice. Thymic expression of murine leukemia virus was uniformly suppressed in 6- and 8-week-old calorie-restricted mice. The latency to median tumor incidence was extended by greater than 3 months and median lifespan was increased by approximately 50%. These observations indicate that retroviral mechanisms involved in the generation of lymphoid malignancy are vulnerable to underfeeding regimes.[26]

In agreement with the observations on extending survival in rats, only a reduction of energy in the diet, rather than a reduction of mineral, fat, or protein intake, was found to depress mortality rates resulting from neoplastic disease in Fischer 344 rats.[27]

These findings support observations in B6/1pr, C3H/1pr, and MRL/1pr mice and Fischer 344 rats that restricted feeding to 50% of control levels results in significantly diminished levels of mRNAs for the oncogenes, c-*myc*, v-*myc*, v-*fos*, v-*abl*, and v-*raf*.[28] The expression of c-*fos* and c-Ki-*ras* mRNAs, activated during hepatic cellular proliferation following partial hepatectomy, was reduced in Fischer 344 rats maintained on a diet containing 40% fewer calories. Mean hepatic levels of [^{3}H]thymidine incorporation were greater in the calorie-restricted group at 18, 24, 28, and 36 h after partial hepatectomy. These findings demonstrate that chronic calorie restriction preserved inducible cellular responses but lowered oncogene expression during cell division.[29]

Response to Carcinogens

Chronic dietary restriction confers a degree of protection against the highly mutagenic action of exogenous carcinogens. The polycyclic hydrocarbon carcinogen dimethylbenz(*a*)anthracene, which binds covalently to DNA after conversion to a reactive epoxide, demonstrates significantly reduced binding to skin DNA from mice maintained on 60% of *ad libitum* food intake.[30] It is suggested that a reduction in the binding of the carcinogen to DNA would retard the initiation of tumor formation. This prediction has been confirmed using a similar experimental design in which reduced calorie intake was achieved by restricting either fat or carbohydrate in the diet. Both strategies were observed to delay the rate and reduce the incidence of carcinoma development in response to dimethylbenz(*a*)anthracene.[31]

Similarly, specific protection against the initiation of intestinal tumors by methylazoxymethanol (MAM) in the highly susceptible Lobund Sprague–Dawley (SD) rat has been demonstrated when such animals are maintained on restricted feeding.[32] Fully fed male Lobund (SD) rats produce a high percentage of visible tumors in the colon and small intestine 20 weeks after a single subcutaneous injection of MAM. When rats were fed 75% of the *ad libitum* food intake 10 days after exposure to MAM, significantly fewer intestinal tumors were observed 20 weeks after injection of the carcinogen. The moderate degree of dietary restriction utilized in this experiment appeared to be acting in a suppressive rather than a chemopreventive mode in which the timing of initiation of restricted feeding was critical. Delayed onset of underfeeding

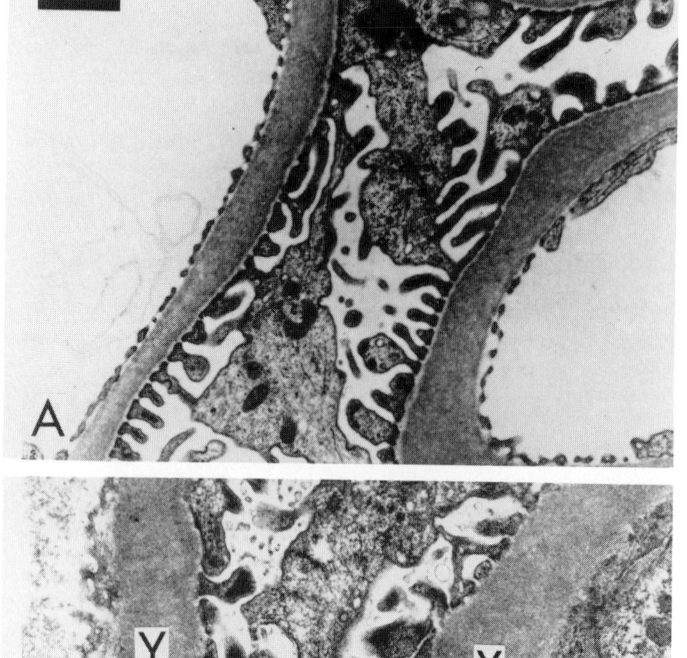

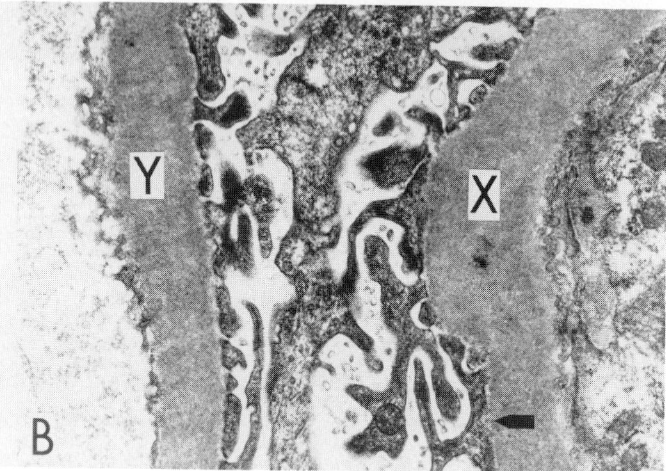

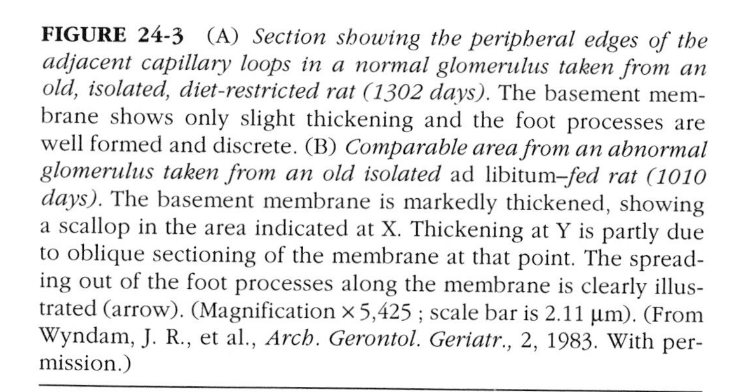

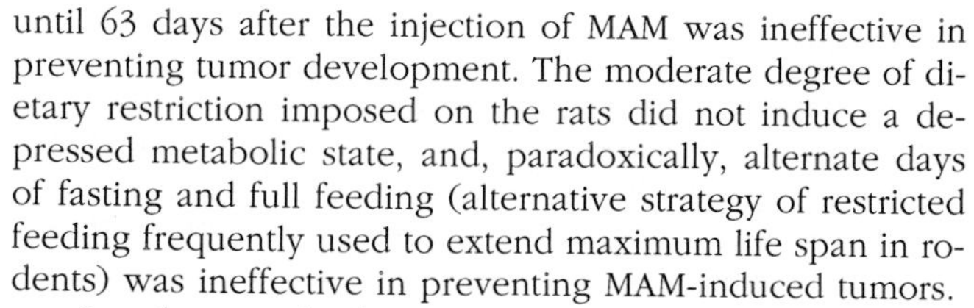

FIGURE 24-3 (A) *Section showing the peripheral edges of the adjacent capillary loops in a normal glomerulus taken from an old, isolated, diet-restricted rat (1302 days).* The basement membrane shows only slight thickening and the foot processes are well formed and discrete. (B) *Comparable area from an abnormal glomerulus taken from an old isolated* ad libitum*–fed rat (1010 days).* The basement membrane is markedly thickened, showing a scallop in the area indicated at X. Thickening at Y is partly due to oblique sectioning of the membrane at that point. The spreading out of the foot processes along the membrane is clearly illustrated (arrow). (Magnification × 5,425 ; scale bar is 2.11 μm). (From Wyndam, J. R., et al., *Arch. Gerontol. Geriatr.*, 2, 1983. With permission.)

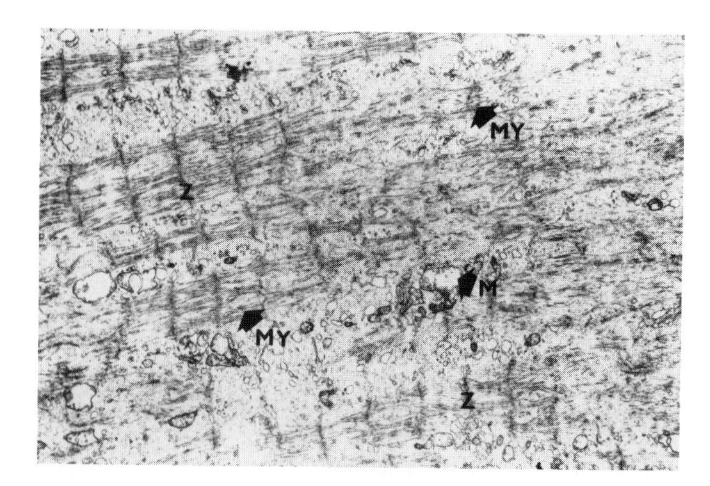

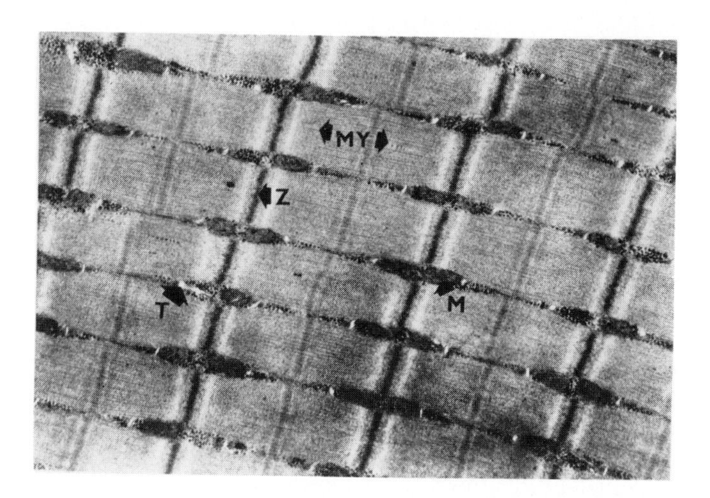

FIGURE 24-4 (A) *Longitudinal section through the gastrocnemius muscle of a control male Wistar rat aged 1010 days.* Myofibrillar breakdown is significant with only thin diffuse Z-bands (arrow) remaining to support the sparse, degenerated myofibrils. The sarcoplasm contains few mitochondria, vesicles, and fine filamentous remnants. (B) *Electron micrograph of a longitudinal section through the gastrocnemius muscle of a food-restricted male Wistar rat aged 1284 days.* There is no evidence of myofibrillar breakdown or structural abnormalities in mitochondria or t-tubules. Abnormal amounts of lipid were not detected. Magnification ×5,248; Z = Z band, M = mitochondrion, MY = myofilaments, T = tubule. (From Everitt, A. V., et al., *Arch. Gerontol. Geriatr.*, 4, 1985. With permission.)

until 63 days after the injection of MAM was ineffective in preventing tumor development. The moderate degree of dietary restriction imposed on the rats did not induce a depressed metabolic state, and, paradoxically, alternate days of fasting and full feeding (alternative strategy of restricted feeding frequently used to extend maximum life span in rodents) was ineffective in preventing MAM-induced tumors.

In a later study this group contrasted the induction of tumors in Lobund calorie-restricted rats using directly and indirectly acting chemical carcinogens, *N*-methyl nitrosourea (MNU) and MAM, respectively.[33] Tumors induced by MAM develop mainly in the small intestine whereas those induced by MNU predominate in the colon (Figure 24-6). Restricting food intake to 75% of *ad libitum* levels did not modify the production of tumors by MNU. It was concluded that the activation step of MAM to methylazoxyformaldehyde, an oxidation dependent on NAD+ dehydrogenases, is impaired in calorie-restricted rats. This interpretation is supported by the observation that azoxymethane (an indirect carcinogen) did not induce adenocarcinomas in Fischer 344 rats maintained on 70% of *ad libitum* food intake[34] and by observations on aflatoxin-B1 (AFB1) binding to DNA in control and diet-restricted rats. Utilizing both an *in vivo* and an *in vitro* approach, restricted feeding (60% of *ad libitum* consumption) was observed to decrease metabolic activation of this model carcinogen and subsequent AFB1–DNA binding.[35]

In an attempt to clarify the details of hepatic metabolism of polycyclic aromatic hydrocarbons in calorie-restricted rodents, Wall et al.[36] employed a liver transplantation technique. Livers from control and calorie-restricted rats were

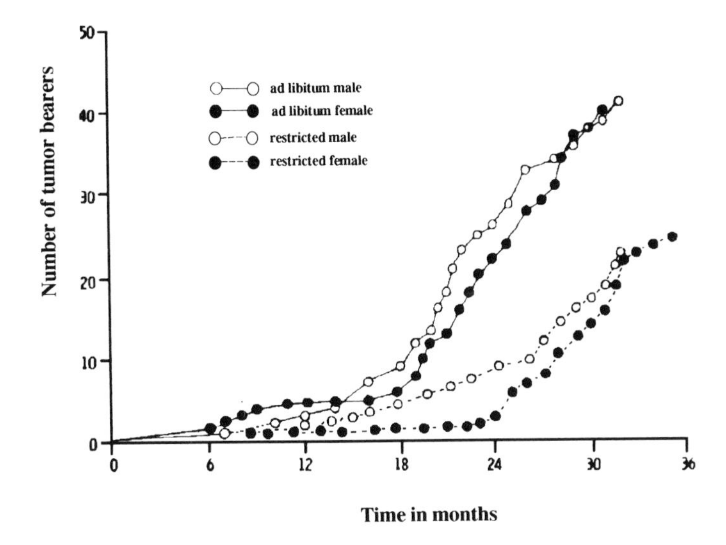

FIGURE 24-5 *Cumulative totals of tumor-bearing specific-pathogen–free Swiss albino mice of the Alderley Park strain.* Control animals were fed *ad libitum* (from 5 g/mouse/day) whereas diet-restricted animals were fed 4 g/mouse/day. Each experimental group consisted of 50 animals. (From Tucker, M. J., *Int. J. Cancer,* 23, 803, 1979. With permission.)

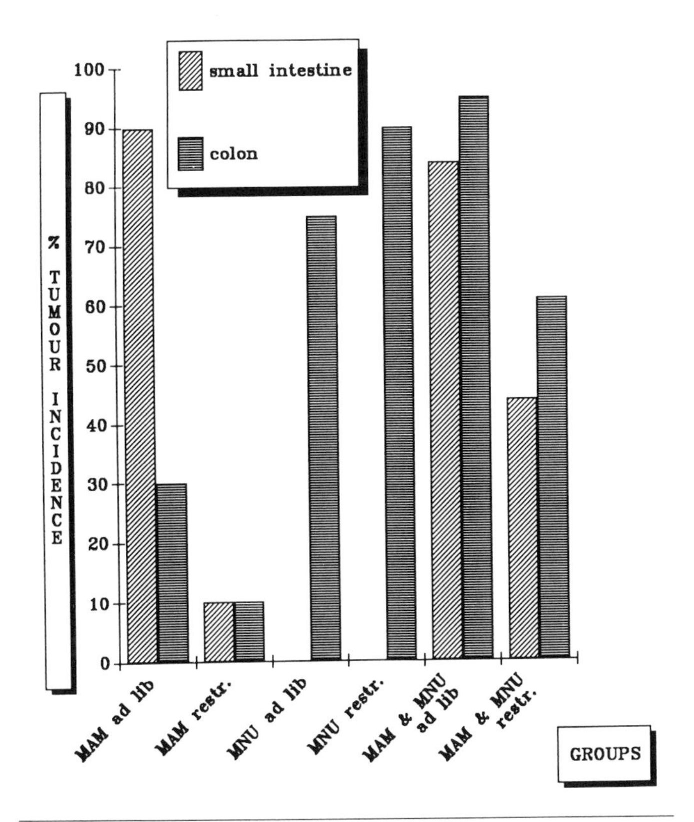

FIGURE 24-6 *The effect of the direct-acting carcinogen N-methylnitrosourea (MNU) and the indirect-acting carcinogen methylazoxymethanol (MAM) on the induction of tumors in the small intestine and colon of male Lobund strain, Sprague–Dawley rats fed either* ad libitum *or at 75% of the* ad libitum *intake.* MAM = 1 dose subcutaneously (30 mg/g); MNU = 3 doses intrarectally at intervals of every other day (0.5 ml of 0.8%). Rats were examined at 20 weeks after onset of the experiment. (Data redrawn from Pollard, M., et al., *J. Natl. Cancer Inst.,* 74, 1347, 1985, reprinted with permission from Merry, B. J., *Rev. Clin. Gerontol.,* 1, 203. Copyright 1991 Edward Arnold, UK.)

transplanted into naive controls. After a 15-min infusion of [^{3}H] benzo[*a*]pyrene, polar metabolites in the liver and blood were measured. Significantly higher levels of polar metabolites were identified in the blood and tissues from the calorie-restricted than for the control group although binding of these metabolites to liver, lung, and kidney DNA was identical in the two dietary groups. To confirm the observation that restricted feeding stimulated the release of polar metabolites, a liver perfusion model was employed. This approach confirmed the earlier observation, and metabolism of the model compound *p*-nitroanisole and glucuronidation of *p*-nitrophenol were observed to be twofold higher in livers from food-restricted rats. Rates of monooxygenation were similar in liver microsomal preparations from the two dietary groups. Wall et al.[36] interpreted these findings as support for the hypothesis that food restriction enhances the supply of cofactors which stimulate metabolism of polycyclic aromatic hydrocarbons.

In a series of studies, Krichevsky and co-workers have reported the effect of restricted feeding regimes on the promotional stage of tumor development and the relationship between the risk of tumorigenesis and the level of obesity. Restricting genetically obese LA/N-cp (corpulent) female rats to 60% *ad libitum* calorie intake after exposure to 7,12-dimethylbenz[*a*]anthracene reduced mammary tumor incidence from 100% at 16 weeks to 27%, in comparison to 21% for phenotypically lean litter mates fed *ad libitum.* Although the energy-restricted obese rats weighed less than *ad libitum*–fed obese rats, the proportion of body fat was not reduced indicating that body fat per se may not be a determinant of tumor promotion.[37] Tumor-bearing rats had higher insulin levels than rats without tumors and *ad libitum*–fed obese animals exhibited marked hyperinsulinemia.

3.2 Effect on Genetic Models of Disease

Chronic restricted feeding has been shown to markedly influence the expression of pathology in rodent species which have been selectively bred as models of human pathology.

The major causes of death in an aging human population are pathologies associated with the cardiovascular system in which hypertension is recognized as a major risk factor. Consequently, animal models have been developed to facilitate a detailed study of the genetic and environmental components of this pathology. Hypertension for which no recognizable cause can be found is called essential or idiopathic. The spontaneously hypertensive rat (SHR), selectively bred from a strain of Wistar rats, has been claimed to be a good model for the study of essential hypertension in humans since the condition is spontaneous, increases in severity with age, is more severe in males, and is often associated with complications of several organs, including heart, brain, and kidneys, similar to those found in human hypertension. The development of hypertension occurs before maturity, followed by a secondary slower increase in blood pressure throughout life with death resulting at about 18 months compared to 24 months in the normotensive WKY control rats (see Figure 24-7). The early death is attributed to six specific lesions: interstitial fibrosis of the kidney, myocardial edema, fatty infiltration of the myocardium, atrial thrombosis, congestion, and vacuolization in the adrenal. Restriction of the normal rat diet to 40% *ad libitum* intake extended mean and maximum lifespan in both normotensive and hypertensive animals.[38,39] Mean lifespan in normotensive animals was increased by 8 months (from 24 to 32 months) while mean lifespan in underfed SH rats was increased by over 12 months (from 18 to over 30 months) (see Figure 24-8). The extension in lifespan in the SH rat was accompanied by a decrease in the specific lesions associated with the hypertensive state to a frequency not significantly different from the diet-restricted normotensive WKY controls. Paradoxically, the development of the hypertension was not affected by the restricted diet, but conversely underfeeding appears to convey protection against end-organ damage and subsequent pathology resulting from the elevated blood pressure. This intriguing observation clearly warrants further study to elucidate the biochemical mechanisms involved.

The KdKd mouse strain exhibits a renal pathology which is inherited as an autosomal recessive and which is similar to nephronophthisis in man (medullary cystic disease). The kidneys appear normal at birth but then develop a progressive renal pathology associated with proteinuria, polydipsia, and autoimmunity directed towards erythrocytes.[40] The KdKd mice die prematurely between 7 and 9 months of age. Restricting dietary energy intake to 50% from the age of 60 days significantly reduced the incidence of renal pathology and the development of the autoimmunity. Survival of 50% was increased to over 16 months. When 8-month-old calorie-restricted mice were returned to a higher energy intake, a rapid onset of renal pathology was observed and death occurred within 8 weeks.[41]

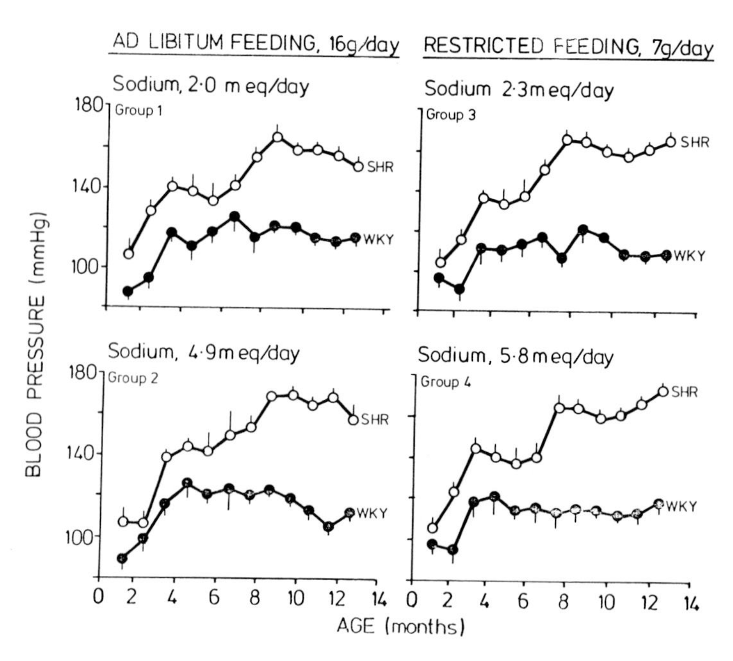

FIGURE 24-7 *The effects of restricted feeding and dietary sodium intake on the development and maintenance of hypertension in the spontaneously hypertensive rat (SHR) compared to the normotensive Wistar Kyoto (WKY) substrain.* (From Lloyd, T. and Boyd, B., in *Function and Regulation of Monoamine Enzymes: Basic and Clinical Aspects,* Usdin, E., Winer, N., Youdin, M. B. H., Eds., Macmillan Publishing, 1981. With permission.)

3.3 Immunologic Response to Dietary Restriction

The delayed appearance and enhanced resistance to many pathologies associated with aging in chronically underfed rodents has inevitably resulted in a detailed evaluation of the immune system in these animals. Most of the studies have been conducted with diet-restricted mice in which the genetics of individual strains are clearly defined. A general response common to mice, rats, and guinea pigs exposed to moderate chronic calorie restriction is a depression in antibody production but, conversely, enhanced cell-mediated immunity.[42] The precise biochemical mechanisms of how calorie restriction induces this adjustment are not understood.

Thymus development is delayed in diet-restricted rodents, and total cell numbers are significantly reduced in the thymus, spleen, lymph nodes, and blood.[43] In spite of a reduction of nucleated white cells in the blood to 20% of control levels, enhanced resistance to pathology and extended survival is observed (see also Chapter 7).[19]

Although dietary restriction will modify longevity in all strains of mice evaluated, the early immunologic effects of underfeeding seem to be strain dependent. In assays of T-cell function, restricted female $B_{10}C_3F_1$ mice outperformed fully fed controls, but when young $C_{57}BL/6J$ mice were assessed, control mice outperformed diet-restricted animals. While dietary influences on the immune system have been shown to be profound, the effects are clearly modified by genetic factors and the associated predisposition to develop particular pathologies rendering the strain short lived. This effect can be illustrated by the two short-lived mouse strains DBA/2F and (NZB × NZW)F_1(B/W). Caloric restriction in B/W mice prolonged lifespan more than protein restriction, whereas DBA/2F mice showed enhanced survival only when the diet was restricted for protein.[44] The diversity of response of these two strains to caloric and protein restriction in terms of subsequent survival appears to result from the influence of diet on the rapidly progressive, highly destructive, autoimmune renal disease that occurs in B/W mice.

Studies on *in vitro* lymphocyte proliferation in response to plant mitogens have shown an enhanced mitotic ability for cells from calorie-restricted animals. A change in the proportion of Lyt-$1^+2^-3^-$ helper T cells has been reported and a population of Lyt-1^+ cells have been observed in spleens from diet-restricted mice which have not been identified in tissue from control animals.[1] Weindruch and Walford have reviewed the effect of calorie restriction on age-related changes in T cell subset ratios in both long- and short-lived strains of mice.[1] These age-related changes appear to be retarded in calorie-restricted animals, particularly the loss of the helper cell type. Retardation of the age-related loss of T-helper cells would explain the increased Con A–induced interleukin-2 production reported for lymphocytes from 19-month, diet-restricted Fischer 344 rats.[45]

Age-related decreases in the proliferative response of lymphoid cells to mitogenic stimuli are associated with changes in fatty acid composition. A decrease in linoleic acid (18:2) and an increase in long-chain unsaturated fatty acids (20:4, 22:4, and 22:5) content is observed. Restricted feeding to 60% of *ad libitum* intake at 6 weeks of age preserved the mitogenic responsiveness and prevented the modification of fatty acid composition.[46] One mechanism through which calorie restriction may delay the loss of age-associated immune functions is through the modulation of the fatty acid composition of phospholipid fractions of spleen cell membranes. It has been proposed that such a modification would enhance binding of interleukin-2 and insulin to their receptors thereby improving T cell proliferation.[47]

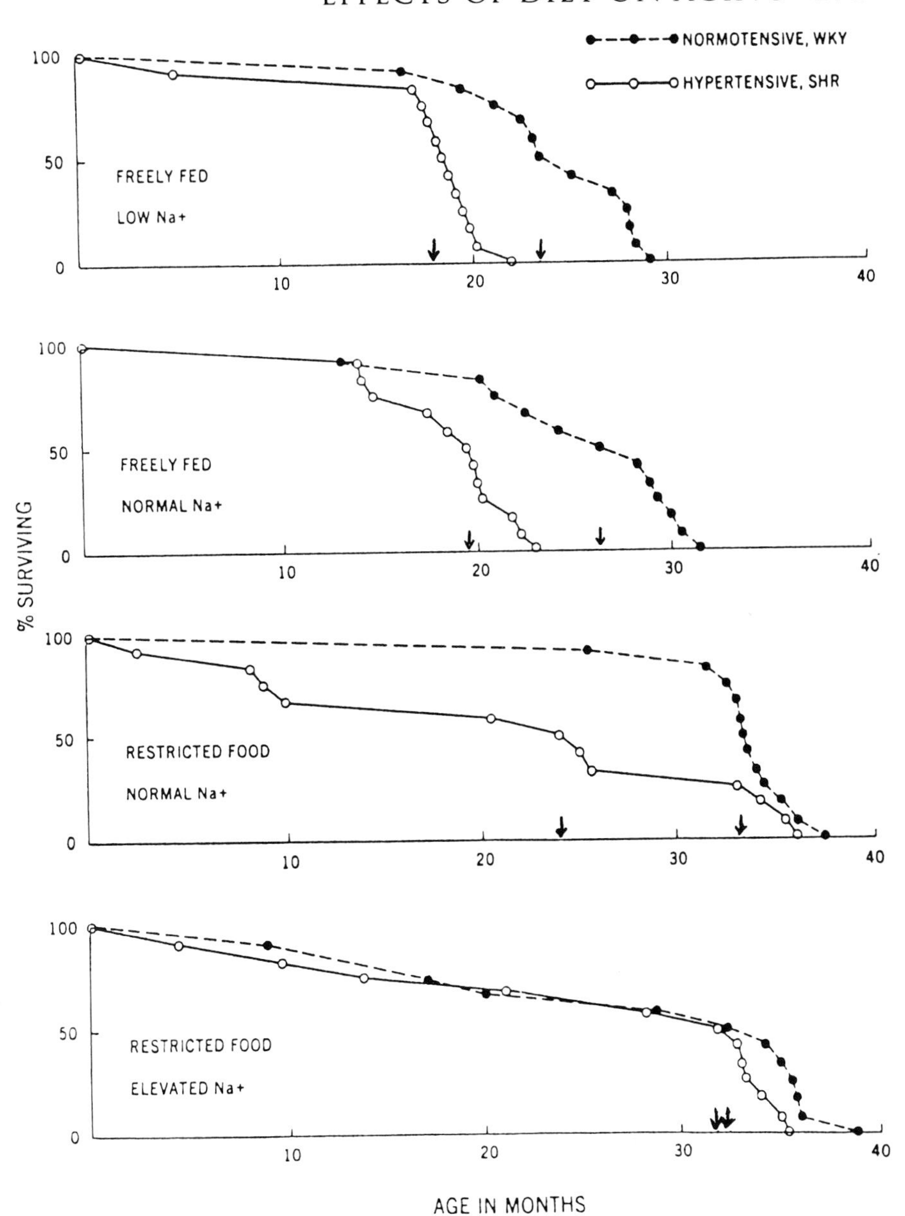

FIGURE 24-8 *Survival curves of* ad libitum*–fed and diet-restricted spontaneously hypertensive rats (SHR) and normotensive controls (WKY).* The arrows indicate the mean lifetime for each strain under the differing dietary conditions. (From Lloyd, T., *Life Sciences,* 34, 401, 1984. With permission.)

Prostaglandin E2 is recognized to have a suppressive effect on cell-mediated immunity. Restricted feeding to reduce the body weight of Emory mice to 66% of control values significantly reduced spleen levels of prostaglandin E2.[48] The importance of this observation in retaining cell-mediated immunity with aging in calorie-restricted rodents is unknown.

The known tumor-suppressive action of calorie-restricted diets has focused attention on the activity of natural killer cells derived from underfed rodents. Basal levels of natural killer cell activity were not increased in rats and were actually decreased in mice by caloric restriction.[49,50] Diet restriction may enhance the responsiveness of natural killer cells to interferon and other T-cell inducer signals for higher cytolytic activity is observed *in vitro* after induction by interferon than for cells from control animals.[50] The antitumor role of macrophages and cytotoxic T lymphocytes in diet-restricted animals has yet to be assessed.

Autoimmune Disease

In an earlier section (see *Response to Carcinogens* in Section 3.1), attention was directed to the protective effect of dietary restriction against tumor formation even when implemented 10 days after exposure to the carcinogen MAM. A somewhat similar effect has been reported in mice susceptible to autoimmune disease. It is a characteristic of the MRL/1 mouse strain that both males and females develop lymphoproliferative diseases, autoimmunity, and a rapidly progressive renal disease. Consequently, MRL/1 mice live approximately 6 months when fed *ad libitum,* but survival can be doubled to over 1 year if mice are underfed, even if the restricted feeding regimen is not implemented until after the onset of the disease. Such delayed dietary restriction very dramatically influenced the appearance of renal pathology and the autoimmune reaction.[51] The ability of calorie-restricted feeding regimes to inhibit the development of autoimmunity in other autoimmune-prone, short-lived mouse strains (BXSB, (NZB × NZW)F_1, NZB) has been confirmed.[52,53] The mechanism of this protective action of underfeeding is again not understood, for the pattern of development of the disease is dissimilar between strains. The development of splenomegaly is prevented or reversed in BXSB mice and the expansion of a non-T, non-B lymphoid cell population prevented. Delaying the imposition of restricted feeding until after the appearance of the autoimmune disease is still effective in reversing the progression of the renal pathology.[53]

In BXSB mice, calorie restriction prevented the formation of anti-DNA antibodies and an increase in circulating immune complexes,[53] whereas in the MRL/1 strain no effect on their formation was recorded.[51] It was observed, however, that diet-restricted MRL/1 and BXSB mice were no longer deficient in interleukin-2 production, particularly in the lymph nodes. This finding is in agreement with the report that in rats maintained on chronic dietary restriction regimes mitogen-induced lymphocyte proliferation, interleukin-2 induction, and protein synthesis were increased in spleen lymphocytes.[45] Autoimmune-prone mice strains characteristically have very deficient production of interleukin-2.[54]

4 METABOLIC RESPONSE TO DIETARY RESTRICTION

Metabolic rate has been widely regarded as an important component of the aging process and it has been central to many hypotheses which seek to explain this phenomenon. The early studies that attempted to measure the basal metabolic rate of diet-restricted rats reported it as lying between normal young and normal old animals.[55] More precisely, when measured at 850 days it was reported to be lower than in control animals if expressed as heat production per unit surface area or, alternatively, higher if expressed against body weight.[56] Indirect support for the idea that the metabolic rate was depressed was presented by Sacher,[57] when he reviewed the data from the earlier study of Ross,[58] in which five dietary regimes were used to observe the effect of calorie intake on survival. It appeared from the calculations that total calorie intake during the lifespan per gram of body weight was the same for each of the five dietary groups, being approximately 102 kcal/g body weight. This conclusion was interpreted as support for the rate-of-living theory of aging.[59] Two subsequent studies have failed to substantiate this conclusion. Male Fischer 344 rats fed 60% of the *ad libitum* intake from 6 weeks consumed a greater number of calories per gram of body weight during their lifetime than did rats fed *ad libitum*.[60] This observation was confirmed in the female CFY strain of Sprague-Dawley rats.[61]

Direct measurement of basal metabolic rate utilizing modern technology has confirmed that it is not depressed in chronically diet-restricted rats. Male and female rats adapted over an 8-week period from weaning to a 50% restriction of energy intake at two levels of protein intake (20% and 10%) showed no change in basal metabolic rate.[62] The growing rats adapted to low energy intake by reducing tissue deposition and increasing the efficiency of energy intake by reducing tissue deposition and increasing the efficiency of energy utilization for tissue maintenance. This observation was confirmed in male Fischer 344 rats maintained at 60% of *ad libitum* food intake. A comparison of oxygen consumption for fully fed and restricted animals at 6 months of age showed no difference in metabolic rate when expressed per kilogram of lean body mass. If the data were normalized to kilograms of body weight, then the metabolic rate was higher in restricted animals, but not when normalized to $kg^{0.67}$ body weight.[63]

A transient reduction in metabolic rate under normal daily living conditions and in basal metabolic rate was identified in Fischer 344 rats fed a restricted diet (40% of *ad libitum* from 6 weeks of age). Within a few weeks of commencing restricted feeding, the metabolic rate of restricted rats was the same as that of rats fed *ad libitum*.[64] A lifetime study of daily metabolic rate was conducted using the same animal model. Metabolic rate was measured indirectly by gas analysis over a 24-h period in control and diet-restricted male rats. Metabolic rate decreased from 6 to 18 months and then increased from 18 to 24 months of age. No significant difference in metabolic rate per unit metabolic mass was observed[65] (Figure 24-9).

4.1 Core Body Temperature

The measurement of metabolic rate and body temperature in female C_{56}BL/6J mice fed *ad libitum* either a 26% protein intake or a 4% protein intake revealed higher oxygen consumption, but with significantly lower rectal temperature in animals maintained on the 4% protein intake.[66] A significantly lower body temperature in mice exposed to early-onset dietary restriction has been confirmed, but late-onset underfeeding (1 year) did not induce a fall in core temperature when compared to age-matched, *ad libitum*–fed controls.[67] Early-onset energy restriction in mice, 201 kJ/week compared to 397 kJ/week, has been observed to induce periods of torpor. This level of energy restriction induces body temperatures of 31°C at ambient air temperatures of 20 to 22°C in C_{57}BL/6 male and female mice and SHN/C_3H F_1 female mice of ages 3 and 13 months, respectively.[68]

Conflicting results have, however, been reported for the effect of chronic dietary restriction on core body temperature in rats. In Sprague–Dawley rats (Cr1:CD[SD]BR strain) the circadian variation of body temperature was similar in restricted and control animals.[69] Technical difficulties, particularly relating to the stress of handling, elevated the core temperature, especially in the younger age groups. It was therefore concluded that the effect of chronic underfeeding on body temperature was different in the two species. Re-examination of core temperature using rectal probes in Sprague–Dawley rats (CFY strain) has revealed a complex response of body temperature to dietary restriction, which may explain in part the conflicting observations previously reported.[13] Male rats maintained at 50% the body weight of age-matched controls showed a fall in core temperature of up to 2°C prior to feeding. Immediately after meal feeding, core temperature rose to control values, but after 9 h at 70 and 120 days, the temperature was again significantly lower in the restricted group. It was noted that with increasing age a subtle adaptation in the body temperature was observed. At 1 year the rhythm in core temperature was similar to that observed in the young animals, with the exception that core temperature after feeding was now significantly higher than in fully fed animals. This significant elevation in core temperature after feeding was again observed in 525-day-old animals, but at this age no significant depression in core temperature prior to feeding could be observed. It was concluded that the response of core temperature in rodents to chronic postweaning dietary restriction involves two components. The metabolic response to the timing and periodicity of feeding are superimposed on the changes in core temperatures that occur with age in both experimental and control animals[13] (see Figure 24-10).

4.2 Mitochondrial Recovery

This was assessed in mice diet-restricted from weaning.[70] Increased state 3 rates but not state 4 rates for respiration supported by glutamate or pyruvate plus malate were observed for liver mitochondria. Dinitrophenol-uncoupled rates were also increased by restricted feeding. The interpretation of these data was that the more efficient uncoupling of mitochondria observed in tissue from diet-restricted mice would result in a reduction of free radical generation in the intact animal and thus retard age-induced mitochondrial damage.

4.3 Enzyme Activities in Response to Dietary Restriction

Studies of the response of specific enzyme activity in tissues from diet-restricted animals have yielded conflicting results. A delay in the maximum activity of liver ATPase, from 200 to 600 days, was observed in rats;[2] however, no effect of restricted feeding on kidney catalase could be detected.[71] Animals maintained on a low-protein diet (4% protein) showed a decrease in the activity of kidney and liver cholinesterase, lactic and malic dehydrogenase, and succinoxidase while heart alkaline phosphatase and cathepsin were unaffected.[72] In a detailed study of the hepatic metabolic response to early- and late-onset dietary restriction, five experimental groups were utilized.[73] Male Fischer 344 rats were divided at 6 weeks of age into

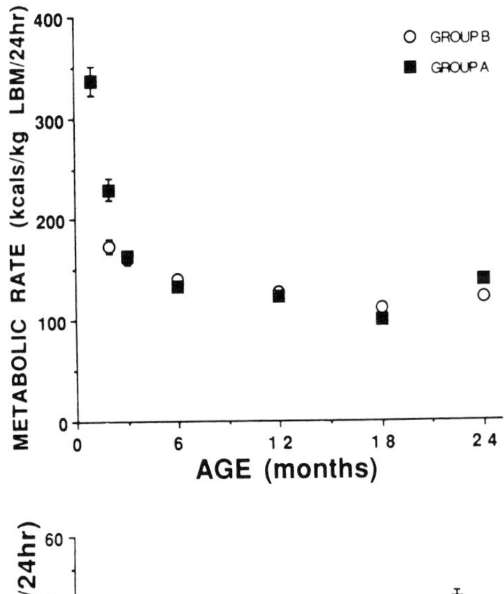

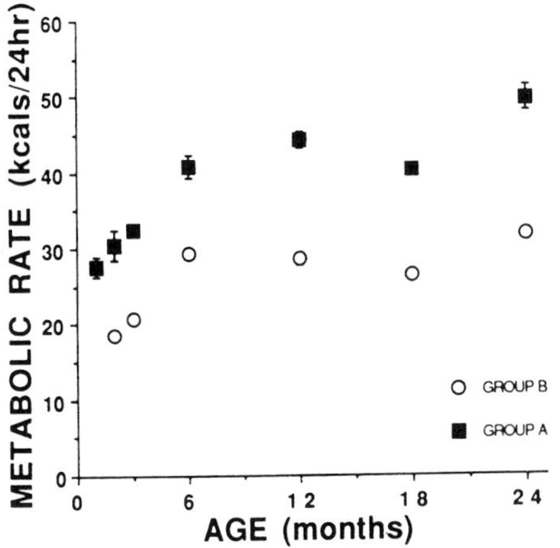

FIGURE 24-9 *Metabolic rate for* ad libitum*–fed and diet-restricted male Fischer 344 rats expressed either over 24 h per rat or as metabolic rate per unit lean body mass (LBM)$^{-1}$ · 24 h^{-1} from rats from 6 weeks to 24 months of age.* Group A (■) fed *ad libitum;* Group B rats (○) fed 60% of *ad libitum* intake. Results are means ± SEM; n = 10 rats. (From McCarter, R. J. and Palmer, J., *Am. J. Physiol.* 263 [*Endocrinol. Metab.,* 26: E448-E452], 1992. With permission.)

Group 1	*Ad libitum* fed
Group 2	Restricted to 60% food intake of Group 1
Group 3	Restricted to 60% food intake of Group 1 until 6 months and then fed *ad libitum*
Group 4	Fed *ad libitum* until 6 months and then restricted to 60% food intake of Group 1
Group 5	Fed the same caloric intake as group 1 but 60% of the protein intake

Hepatic cholesterol concentration increased with age only in Group 1, animals fed *ad libitum*. In Groups 2 and 4, which had the longest periods of restricted feeding, postabsorptive hepatic glycogen concentrations were significantly higher, and enhanced microsomal activity, as indicated by microsomal 3-hydroxy-3-methyl-glutaryl–coenzyme A reductase, was enhanced.

The response of the activity levels of many enzymes is dependent on the type (protein and/or caloric), degree, and duration of the restricted feeding. As the degree of restriction of a 20% protein diet (fed over a period of 6 weeks) increased (75, 50, and 25% of *ad libitum* intake), the activity of liver tryptophan oxygenase activity increased, while conversely quinolinate phosphoribosyl-transferase activity decreased.[74] This trend was reversed for quinolinate phosphoribosyl-transferase activity when the degree of restriction exceeded 50% of *ad libitum* intake. A similar observation has been recorded for the specific activities of liver lysosomal enzymes: acid phosphatase, β-galactosidase, arylsulfatase B, and cathepsin D in rats fed 90, 70, and 50% of *ad libitum* intake from weaning.[75] It was noted that in the early phase of restricted feeding (3 weeks) increased activity of acid phosphatase was observed only in the 50% restricted group while arylsulfatase B activity increased in all restricted groups, with the greatest effects being recorded in the animals restricted to 50% food intake. These biochemical changes were associated with the ultrastructural modification of the hepatocytes and with the appearance of an increased number of autophagic vacuoles and residual bodies. Both the ultrastructural and biochemical changes were interpreted to be a catabolic adaptation to the nutritional stress, and, by 7 weeks of restricted feeding, the specific activity of acid phosphatase and hepatocyte ultrastructure were similar to those in fully fed animals. The activity of arylsulfatase B remained significantly higher in the diet-restricted groups for the duration of the experiment, 24 weeks of restricted feeding.

The effect of restricted feeding on the drug-inducible capacity of liver microsomal cytochrome P450s IA1, IA2 and IIB1, IIB2 was studied in 20-month-old male Fischer 344 rats. Using ELISA and Western Blotting it was possible to demonstrate a significantly higher isosafrole induction of P450-IA1/IA2 and P450-IIB1/IIB2 enzymes in the microsomes from diet-restricted animals compared to *ad libitum*–fed controls. This improved drug-metabolizing capacity induced by restricted feeding appeared to result from increased synthesis of these enzymes, for the mRNAs for P450-IIB1/IIB2 were significantly higher in livers of diet-restricted animals.[76]

While restricted feeding appears to retard the age-associated changes in certain liver enzymes, this effect is not universal. The greatest effect of calorie restriction is on those enzyme systems responsible for detoxification and oxygen metabolism.

4.4 Lipid Peroxidation

The activity of the free radical scavenger superoxide dismutase is inversely related to the protein level of the diet, while lipid peroxidation activity in liver homogenates is positively correlated with the dietary intake of protein.[77] Thus, in both hepatocytes and alveolar macrophages,[78] enhanced superoxide dismutase activity is associated with a reduced capacity for lipid peroxidation in animals fed a low-protein diet. The classical cellular biomarker of aging, lipofuscin, is a fluorescent pigment thought to be derived from lysosomal enzyme–induced lipid peroxidation, which accumulates with time in neurons, heart, testis, adrenals, and other tissues of many mammalian species. It has been confirmed in the heart and brain of male Swiss albino mice examined during the first 12 months of the lifespan that lower levels of such fluorescence age pigments are present when these animals are maintained on a diet containing 4% protein.[79]

Chronic diet restriction in Fischer 344 male rats (60% of *ad libitum* food intake from 6 weeks of age) suppressed the age-associated increase in malondialdehyde production and lipid hydroperoxide formation in liver mitochondrial and

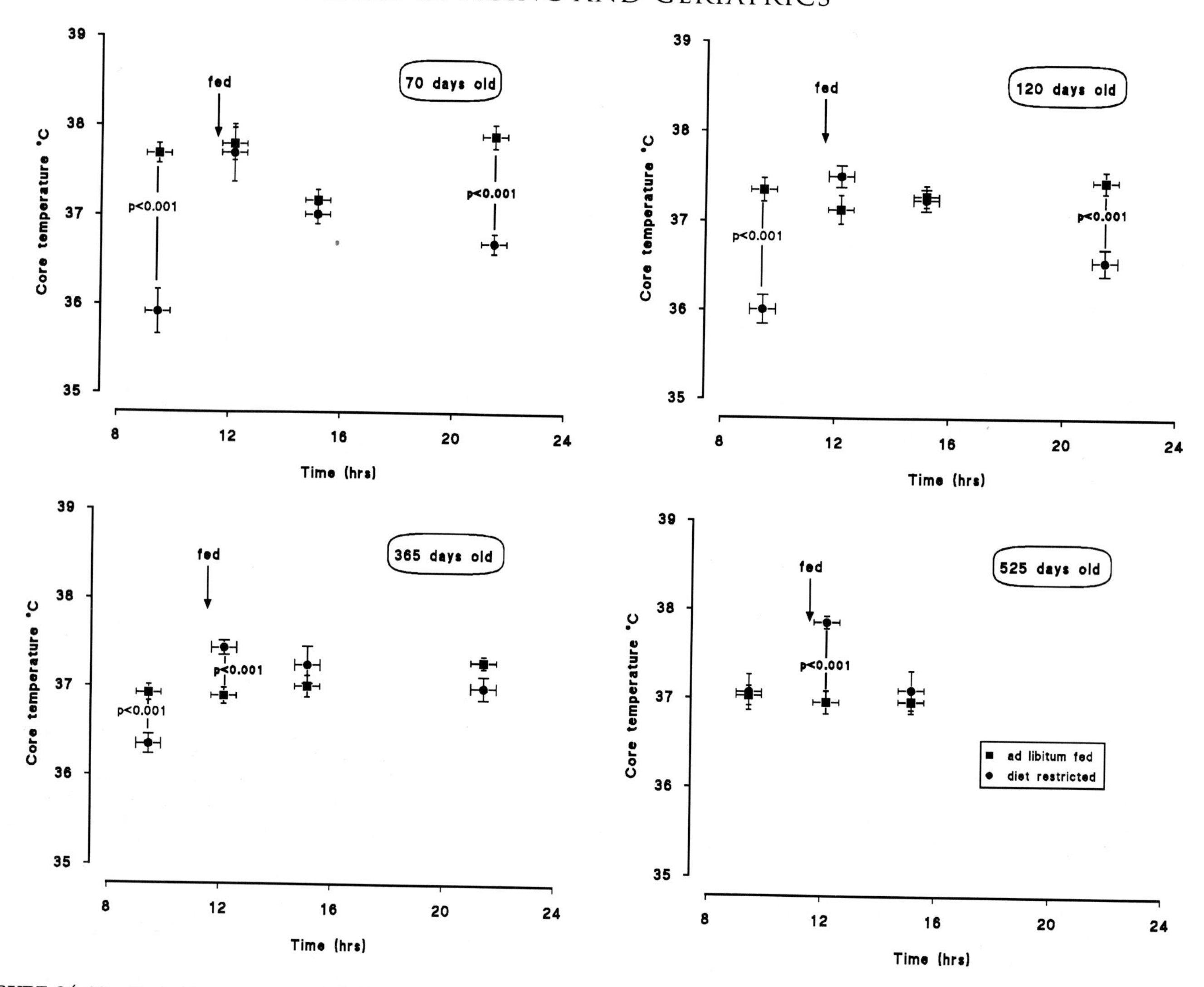

FIGURE 24-10 *Rectal temperature of* ad libitum*-fed and diet-restricted male CFY rats from 70 to 525 days.* Temperatures were recorded during four periods each day, which spanned the period of meal feeding for the restricted animals. Diet-restricted (DR) animals were fed the same diet as control rats but in a limited amount so as to restrict their body weight to 50% that of *ad libitum*–fed, age-matched rats. (Redrawn from Holehan, A. M. and Merry, B. J., *Biol. Reviews,* 61, 329, 1986.)

microsomal membranes[80] (Figures 24-11 and 24-12). In agreement with the observations reported earlier for lymphocyte fatty acid content and membrane composition (see Section 3.3), restricting the calorie intake modified the membrane fatty acid composition in hepatocytes. Linoleic acid content was increased and docosapentaenoic acid content decreased. A higher unsaturation/saturation index was maintained by modulating 18:2 and 22:5 fatty acids. These changes in membrane fatty acid composition were considered to make the membrane more resistant to peroxidation.[80] The fluidity of mitochondrial and microsomal membranes from fully fed control animals show a progressive decline with age whereas those from diet-restricted animals show only slight changes between 6 and 24 months. Age-related peroxidation of membrane lipids may play a significant role in altering membrane fluidity.[81]

In agreement with the findings in liver tissue, diet restriction applied by alternate-day fasting and feeding retarded the age-dependent increase of microviscosity of cerebellar membranes and was able to induce a partial recovery of the age-associated decrease in the β-adrenoreceptor density of cerebellar membranes.[82,83]

4.5 Protein Synthesis

The activity levels of individual enzymes represent a balance between the synthesis and degradation; it is possible that much of the conflicting conclusions derived from studies on specific activities levels may be resolved by observations on protein renewal. *In vivo* protein synthesis and turnover has been studied in a number of tissues from diet-restricted rats.[45,84-87] A significantly higher rate of protein synthesis was recorded in suspensions of freshly isolated kidney cells and cell-free homogenates prepared from testis and spleen lymphocytes from diet-restricted animals (60% of *ad libitum* intake) than from fully fed control animals. In the second year of life in Fischer 344 rats, a significant decline in testicular protein synthesis can be shown *in vitro,* but when rats have been maintained on a restricted diet, no age-related decrease in protein synthesis is observed over the same period. Similarly protein synthesis in isolated hepatocytes from *ad libitum*–fed rats declined 55% between 2.5 and 19 months of age, whereas only a slight decrease with age in protein synthesis was observed in hepatocytes isolated from diet-restricted rats. This had the consequence that at advanced age hepatocytes from diet-restricted animals had a significantly higher rate of protein synthesis than hepatocytes from age-matched, fully fed controls. The higher rate of liver protein synthesis in diet-restricted animals has been confirmed in isolated livers using an *in situ* perfusion technique.[88] The age-related decline in hepatic protein synthesis was not prevented in diet-restricted rats, but beyond 3 months of age significantly higher levels (35%) of ^{14}C-valine incorporation were observed. These findings taken together suggest that a higher rate of protein turnover is maintained in diet-restricted rats.[89]

This observation was confirmed *in vivo* for whole-body protein turnover in Sprague–Dawley rats.[90] Dietary restriction (50% *ad libitum* intake) retarded the developmental decline in protein turnover, resulting in significantly higher fractional synthetic rates (proportion of total protein renewed every 24 h) and degradation rates for restricted animals from 1 year of age. While the data from the whole-animal protein turnover studies confirmed the *in vitro* observations, when the same diet-restricted model was utilized to observe the effect on discrete tissues, the response was found to be more complex. Protein turnover in the ventricular muscle of the heart (see Figure 24-13) and the small and large intestine[91,92] closely followed the response observed for the whole animal, but the general effect on skeletal muscle and lung was a lowering of the rate of protein turnover.[93-95] The general effect on skeletal muscle of chronic dietary restriction was to slow down cellular proliferation and to delay the normal age-related morphologic changes.

Protein synthesis in the liver of diet-restricted rats was depressed when measured *in vivo* in contrast to the findings of the *in vitro* studies with isolated hepatocytes. It was not possible to calculate breakdown and turnover rates for hepatic proteins in this *in vivo* study because of the unknown proportion of the total protein synthesized that was exported from the liver of rats fed a restricted diet. While *in vitro* studies provide for a greater control of experimental conditions, they do have the disadvantage of removing the tissue from the metabolic and endocrine constraints pertaining in the living animal. Both the *in vitro* and *in vivo* approaches provide data only on the rate of total hepatic protein synthesis which may be distorted in old animals by the synthesis of albumin in response to senescent proteinuria.

That the rate of hepatic protein synthesis is directly dependent on the level of circulating thyroid hormones has been clearly demonstrated in surgically thyroidectomized rats.[96] The hormonally sensitive aspect of the translation mechanism is the rate of peptide elongation, and it appears to be independent of any metabolic effect on mitochondrial activity. Surgically thyroidectomized rats that are thyroxine and triiodothyronine deficient have a significantly slower *in vivo* rate of peptide elongation during protein synthesis. Elongation rates can be returned to or above normal by subcutaneous injections of triiodothyronine. It has been established for a number of years that underfeeding in mammals results in a depression of serum levels of thyroxine and triiodothyronine, an observation confirmed in the long-lived diet-restricted rat.[97] It was plausible

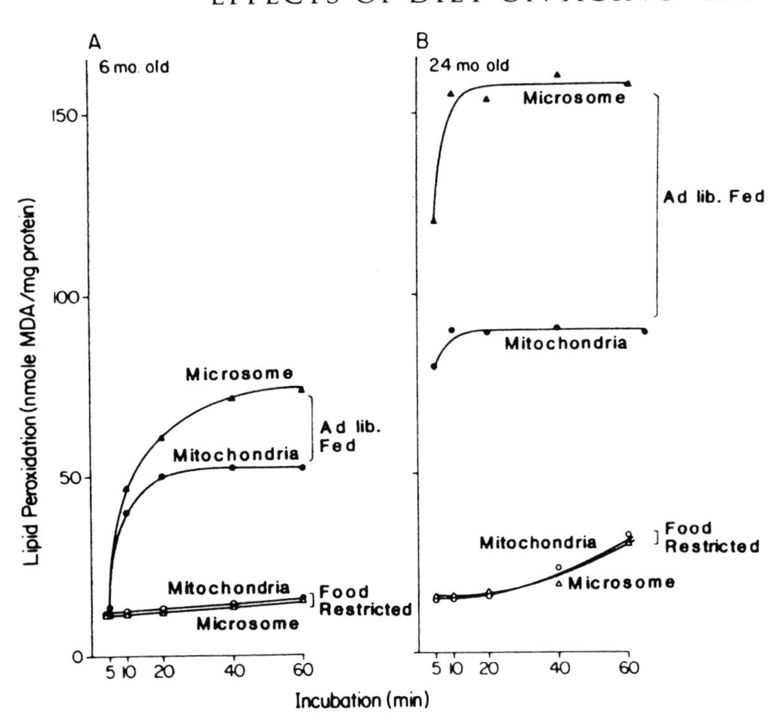

FIGURE 24-11 In vitro *enzyme-dependent lipid peroxidation.* (A) 6-month-old rats; (B) 24-month-old rats. Specific-pathogen–free male Fischer 344 rats were fed a semisynthetic diet either *ad libitum* or at 60% of the *ad libitum* daily caloric allowance starting at 6 weeks of age. Enzyme-dependent lipid peroxidation was carried out in the presence of 1 mmol NADPH and 0.2 mmol $FeSO_4$ complexed with 5 mmol ADP and was measured against malonaldehyde bisdethyl acetal as standard. (From Laganiere, S. and Yu, B. P., *Biochem. Biophys. Res. Commun.*, 145, 1185, 1987. With permission.)

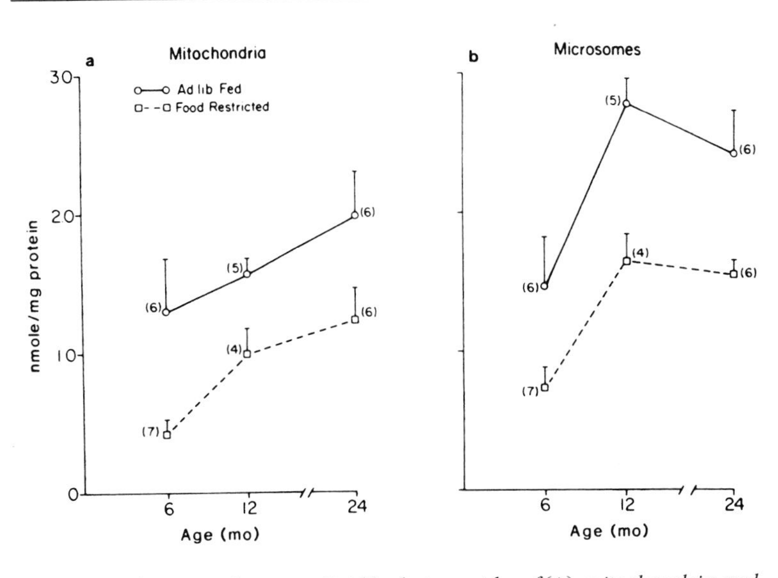

FIGURE 24-12 *Endogenous lipid hydroperoxides of* (A) *mitochondria and* (B) *microsomes.* Statistical differences between *ad libitum*–fed and diet-restricted rats are $0.01 < p < 0.05$. Animal details are the same for Figure 24-11. (From Laganiere, S. and Yu, B. P., *Biochem. Biophys. Res. Commun.*, 145, 1185, 1987. With permission.)

that the depressed rates of hepatic protein synthesis may be explained, at least in part, by changes in the rate of elongation of the nascent polypeptides.

Chronic underfeeding has been found to have two effects on the average transit time for polypeptide synthesis in liver. The initial response observed shortly after the imposition of restricted feeding at weaning was to significantly increase the time required to assemble the average polypeptide from 76.4 to 177.3 s. With increasing age the transit time in fully fed male rats increased to 96.9 s at 2 years, whereas in the underfed animals a decrease in transit time was recorded as the animals recovered from the initial acute response to restricted feeding. By 2 years of age, the diet-restricted animals had an average ribosomal transit time significantly faster than that of age-matched controls. Ribosomal transit times and hepatic protein synthesis in young diet-restricted rats could be returned almost to normal values after 3 days of subcutaneous injections of the hormone triiodothyronine. Administration of this hormone was more effective in restoring transit time values to control levels than was refeeding, although the rate of hepatic protein synthesis rapidly increased on return to refeeding. A period of about 1 month of *ad libitum* feeding was required before ribosome transit times were returned to control values, whereas circulating thyroxine levels returned to control values within a few days.[13,98] One possible explanation for these observations is a persistent impaired peripheral conversion of thyroxine to triiodothyronine in refed diet-restricted rats. Most of the hormonal potency of thyroxine is derived from its conversion to triiodothyronine at peripheral conversion sites, a process impaired during caloric deprivation.[99] It is peripheral conversion of thyroxine to the more active derivative that appears central to the effect on ribosome transit time and, in particular, the delayed recovery observed on return to full feeding (Chapter 14).

A decrease in protein synthesis is a characteristic of many organisms with aging, and it has been suggested from observations recorded in brain, kidney, liver, and skeletal muscle of mice that such a decrease results mainly from an increase in ribosome transit time, i.e., decrease rates of peptide elongation.[100] The increase in the ribosomal transit time for protein synthesis appears to result from impaired binding of aminoacyl–tRNAs to ribosomes, a reaction requiring elongation factor-1α (EF-1α). It has been proposed that the loss of activity of EF-1α with age results from a decline in its synthesis rather than from posttranslational inactivation of normally synthesized molecules.[101] That the decline in protein synthesis with age in cells results primarily from increased ribosomal transit times is still a point of contention. It is of interest within this context that *Drosophila melanogaster* transformed with a P-element vector containing an EF-1α gene under control of hsp70 regulatory sequences have an increase in mean and maximum longevity of 41% when housed at 29.5°C, a temperature at which the hsp70 promoter should enhance transcription of the EF-1α inserts.[102]

4.6 Response of Collagen to Dietary Restriction

The rate of cross-linkage (the formation of stable covalent bonds) between the fibers of the structural protein collagen has been used for many years as a biomarker for the rate of aging. Dietary restriction has been shown in a number of studies to retard the aging of collagen and the deposition of collagen in the kidneys, lung, and liver.[103-105] No effect of diet was found on age-related changes in collagen for the skin.[106] Early studies had reported an increase in the level of calcification for the aorta and kidney in the diet-restricted rat, but this observation may reflect the relatively high mineral content of the early diets rather than a real response to long-term dietary restriction (Chapter 22).

The mammalian eye lens has provided over recent years a model system in which to observe age-related posttranslational

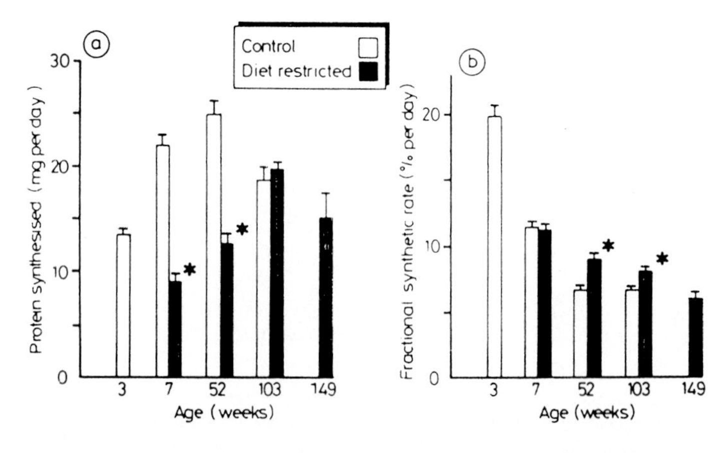

FIGURE 24-13 *Rates of ventricular protein synthesis during postnatal development and long-term dietary restriction.* Values are mean ± SEM of a minimum of five tissues expressed either as the total amount of protein synthesized per day (a) or the fractional rate (b) (k = the percentage of protein mass synthesized per day), $^{*}p < 0.01$. Diet-restricted male rats were fed limited amounts of the normal diet so as to maintain body weight at 50% that of *ad libitum*–fed age-matched controls. (From Goldspink, D. F., et al., *Cardiovas. Res.*, 20, 672, 1986. With permission.)

changes in structural proteins and enzymes.[107] A consistent finding with age in both rodents and humans is a significant decrease in the amount of soluble gamma crystallin, an observation often associated with cataract development (Chapter 10). In a study of a long-lived mouse strain fed either 85 kcal/week (controls) or 50 kcal/week (diet restricted), the high-performance liquid chromatography profiles of the soluble lens proteins of restricted and control mice showed no difference at 2 months, but at 11 months of age significantly different profiles were apparent.[108] The amounts of gamma crystallin in lenses from middle-aged mice maintained on the restricted diet were significantly greater than in lenses from age-matched control mice. This difference was confirmed by sodium dodecylsulfate polyacrylamide gel electrophoresis and persisted in a comparison of 30-month-old animals (see Figure 24-14). In contrast to these observations on the aging rate of the eye lens, no effect of restricted feeding has been found on the rate of retinal aging.[109]

Physiological measurements on vascular smooth muscle have shown that restricted feeding prevented the normal age-related loss in tension observed in late life for *ad libitum*-fed rats.[110] Individual muscles respond in different ways to chronic underfeeding. In the gastrocnemius muscle, a delay in the age-related decline in mass and an associated accumulation of collagen was reported.[111] Examination of the lateral omohyoideus, a fast-twitch muscle, at 6 months of age in underfed rats revealed no effect on the mechanical properties, resting metabolism, ultrastructure, or number of muscle fibers although muscle mass and fiber diameter were decreased.[112]

5 THE PHYSIOLOGIC RESPONSE TO CHRONIC UNDERFEEDING

5.1 Reproduction

The original work utilized such a severe degree of restricted feeding that rats were maintained for prolonged periods (up to 900 days) in a prepubertal condition.[113] This gave rise to the idea that arrested development and enhanced survival

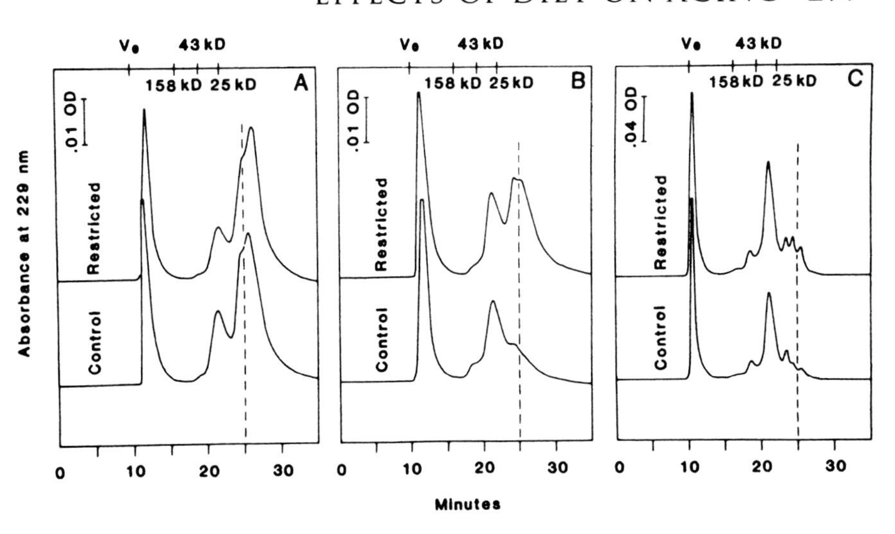

FIGURE 24-14 *High-performance liquid-chromatography (HPLC) elution profiles of soluble proteins from lenses of diet-restricted and control mice aged* (A) *2 months,* (B) *11 months,* and (C) *30 months.* The proteins were monitored at 229 nm and each profile is a representative of an individual lens (N = 4, by age and by diet). Within each age and diet group the chromatograms were highly reproducible and superimposable under identical conditions. It was consistently found that the third peak, gamma crystallins (identified by the dotted line), was significantly greater in 11- and 30-month-old mice on the restricted diet. Mice were fed a semipurified diet providing either 85 kcal per week (controls) or 50 kcal per week (restricted). Abbreviations are V_0, void volume, and OD, optical density. (From Leveille, P. J., et al., *Science,* 224, 1247, 1984. © AAAS. With permission.)

were inextricably linked. Refeeding such animals resulted in rapid sexual maturation, which was followed by a much foreshortened adult phase of the lifespan although total survival was increased. It has subsequently been shown for the rat that increased longevity can be achieved with less severe dietary restriction, which does not inhibit sexual maturation or result in the loss of fertility.[61,114] In male Sprague–Dawley rats maintained at 50% the body weight of age-matched *ad libitum*–fed animals, puberty was delayed by approximately 20 days, as shown by breeding trials, and the circulating levels of the hormones, luteinizing hormone, follicle-stimulating hormone, testosterone, and its peripheral metabolite 5α-dihydrotestosterone.[115] The decline with age in fertility for male rats was similar in both fully fed and diet-restricted animals, decreasing to between 30 and 50% by 2 years. The most notable endocrine disturbance resulting from underfeeding initiated at weaning was a premature fall in serum follicle-stimulating hormone following the initial prepubertal rise, and this may explain the slight delay in the timing of puberty for these animals.

In contrast, puberty in similarly restricted female Sprague–Dawley rats was delayed until 63 to 189 days (range in fully fed rats 34 to 39 days). The delay in the onset of puberty was again associated with decreased circulating levels of follicle-stimulating hormone. Serum progesterone values were significantly depressed in the prepubertal diet-restricted rats while estradiol-17β levels were significantly elevated.[116] The decrease in serum progesterone appears to result from a failure to release pituitary gonadotropins, for there was no significant difference in the elevation of circulating progesterone in response to exogenous gonadotropin stimulation of the ovarian follicles. After sexual maturation more than 70% of diet-restricted female rats demonstrate a normal 5-day cycle characteristic of the CFY strain.

In contrast with the rat, even under very mild restricted feeding regimes, mice are unable to sustain normal estrous cycles.[117] Mice (B6) maintained on 80% normal food intake between 3.5 and 10.5 months were acyclic, being arrested mainly in diestrus.[118] At 12.5 months it was found that the follicular reserves of such mice were twice those of fully fed control animals, and normal estrous cycling could be induced at 10.5 months by a return to full feeding. At this age approximately 80% of control mice are acyclic. Therefore, restricted feeding will delay reproductive aging in mice as judged by the rate of follicular depletion and estrous cycling.

With increasing age, the estrous cycle in fully fed females increases in length and is associated with characteristic irregularities. The most common perturbations of the normal cycle with age are an extension of the period of cornification (persistent estrus) or an increase in the frequency of recurrent pseudopregnancy. Cycle length extension in fully fed female rats is observed as early as 148 to 168 days, while persistent estrus or recurrent pseudopregnancy is typical of the second year of life. Chronic underfeeding abolishes the age-related increase in cycle length and results in a much later appearance of the cycle irregularities. The age-related irregularities of the estrous cycle in fully fed rats are dependent on changes in ovarian endocrine secretion and in the threshold sensitivity of the hypothalamic–pituitary system.[119] Comparison of the serum profiles for luteinizing hormone, progesterone, and estradiol-17β at 3-h intervals across the 5-day cycle in fully fed and diet-restricted rats of 180 to 200 days supports the contention of retarded endocrine aging in underfed animals. At this age, fully fed females are beginning to show the first age-dependent changes in the normal estrous cycles, and clear differences are evident in the temporal relationships of the individual hormones when a comparison is made with diet-restricted animals. The preovulatory peak of luteinizing hormone was reduced in height by underfeeding but occurred approximately 6.6 h earlier in the cycle, whereas a delay of 6 h was observed in the rise of serum estradiol-17β. Significantly greater amounts of follicle-stimulating hormone were released over the cycle of diet-restricted rats, but the total amounts of estradiol-17β and progesterone were reduced.[120]

A series of *in vitro* studies have been undertaken to gain greater insight into the effect of age and diet on the steroid pathways within individual follicles. The follicular content and the release over 4 h of estradiol-17β, progesterone, testosterone, androstenedione, and 20α-dihydroprogesterone were measured during unstimulated steroid synthesis and after luteinizing hormone- or testosterone-stimulated synthesis.[121] It was observed that with increasing age in control follicles there was increased release of progesterone after luteinizing hormone stimulation and enhanced aromatization of available androgen, resulting in a greater activity of the delta-4 pathway to estradiol-17β and reduced activity of the pathway of progesterone metabolism. These age changes were retarded by chronic underfeeding and correlated with the observations on serum hormone levels. A comparison of follicular steroid synthesis in diet-restricted animals of 1 year with that from fully fed animals of 3 months showed no differences. When, in contrast, follicular steroid synthesis was compared in 3-month-old and 1-year-old fully fed animals, a significant age effect was seen. Estradiol-17β synthesis in response to luteinizing hormone or testosterone (a precursor of estradiol-17β) in the incubation media was increased over all stages of the cycle while the release from the follicle of testosterone and 20α-dihydroprogesterone was reduced.

It is possible therefore to demonstrate retarded aging of the endocrine component of reproduction at the level of the individual follicle. In order to distinguish whether the retarded rate of age changes is intrinsic to the individual follicles or is a secondary response to changes induced in the central neuroendocrine system in response to chronic underfeeding, the technique of heterochronic orthotopic ovarian transplantation has been used.[122] This transplantation study demonstrated quite clearly that, in the rat, the age-associated changes in estradiol-17β synthesis and release are reversible and dependent upon changes external to the ovary, i.e., neuroendocrine age changes. Reversal of age changes in follicular steroidogenesis was seen when ovaries were transplanted into either physiologically or chronologically younger animals. The release of estradiol-17β at estrus for follicles from the ovaries of 12-month-old fully fed and 22-month-old diet-restricted animals transplanted into 4-month-old fully fed or 12-month-old diet-restricted animals was significantly reduced.

Extended fertility has been reported in a number of studies where previously diet-restricted animals have been returned to full feeding before being allowed to breed. Female rats diet-restricted from 45 days and then returned to *ad libitum* feeding at 6 to 17 months all successfully produced three to six litters when bred at 16 to 23 months, an age when control females were infertile.[123] Maintaining A strain mice on a calorie-deficient diet induced a state of near sterility, but a return to *ad libitum* feeding at 240 days allowed animals to produce 13 times as many litters after 240 days as their fully fed controls.[124] This finding was confirmed in a later study using A strain mice, but it was observed that although fertility was enhanced on a return to full feeding, the animals were unable to successfully wean their litters.[125] Growth in Long–Evans rats can be suspended or retarded by feeding a diet deficient to varying degrees in the amino acid tryptophan. If growth is reinitiated or accelerated at 17 to 33 months by a return to a complete diet, such animals are then able to reproduce at an age when all control animals are infertile.[126,127] Female Sprague–Dawley rats (CFY strain) diet-restricted from weaning at 21 days are fertile, but overall fertility and litter size are reduced compared to control animals. These females do, however, retain their fertility to an age far exceeding that of the fully fed group (as late as 937 days compared to approximately 550 days).[61,114] In these studies the animals were fertile while retained on the restricted diet and were not returned to full feeding prior to the breeding trial.

5.2 Nonreproductive Endocrinology

In view of the early studies on the diet-restricted rat in which animals were sexually immature and the findings that hypophysectomy linked with hormone replacement therapy could mimic many of the aspects of the restricted feeding model,[128,129] the endocrine response of the long-lived rat is of importance and has been reviewed in some detail elsewhere.[97,130]

The release of growth hormone in rats in which the growth rate is retarded by dietary means is very dependent on the duration and degree of underfeeding. Starvation of adult animals for 2 to 5 days will elevate circulating growth hormone, but chronic, less severe underfeeding results in a depression of both circulating and pituitary growth hormone.[131,132] Mice maintained on a calorie-restricted diet (50 kcal/week compared to 95 kcal/week for control mice) showed decreased cellular content of growth hormone as determined by histochemistry (53% of the mean control value), but no reduction in the number of somatotrophs in the pituitary.[133]

Growth hormone release and plasma values in the rat are characterized by an episodic rhythm of approximately 3.3 h.[134] In 4-month-old male diet-restricted rats fitted with indwelling cannulae and maintained at 50% the normal growth rate, the frequency of release and the plasma concentrations of growth hormone were unaffected, although a restriction in peak duration was observed. That such animals are not growth-hormone deficient has been confirmed by the administration of exogenous ovine growth hormone injected subcutaneously each day over a 3-week period. In fully fed rats, daily injection of ovine growth hormone significantly accelerated the growth rate when compared with saline-injected fully fed controls. No difference in growth rate was evident between growth-hormone- or saline-injected diet-restricted rats.[97] The action of growth hormone on growth promotion is postulated to occur through a family of peptides, the somatomedins, which are depressed in undernutrition. It has been suggested that the somatomedins play a role in the maintenance of body weight and nutritional homeostasis at the level of the central nervous system. Whether specific changes in the somatomedins are important in the diet-restricted long-lived rat has yet to be determined.

It has been shown that maintaining rats on 60% of the average food intake of *ad libitum*–fed animals will increase mean lifespan by 50% and this is associated with delayed skeletal maturation. In such animals the peak of immunoreactive serum parathyroid hormone observed at advanced age in the fully fed animals is not seen in the diet-restricted rats.[135] This age-related hyperparathyroidism in control rats was linked to enhanced bone resorption and bone loss, a process markedly attenuated in the diet-restricted animals. The detailed mechanism by which food restriction prevents age-induced hyperparathyroidism is not understood, but a concomitant decrease in the incidence of renal lesions has been reported.[135] Utilizing the same degree of restricted feeding, it has been shown that the normal age-associated increase in circulating and thyroidal calcitonin is significantly retarded.[136]

A depression in the circulating levels of thyroxine and triiodothyronine is observed within 3 to 7 days of the initiation of restricted feeding, but thyroxine levels could be returned to or above age-matched control values by a return to *ad libitum* feeding for 7 days. The peaks of plasma thyroxine and triiodothyronine (T3) associated with the onset of puberty were delayed by approximately 20 days in the timing of their appearance in diet-restricted rats, which is in agreement with the observations for plasma testosterone.[97,115] The greater effect of underfeeding was recorded for plasma triiodothyronine, the peripheral conversion product of thyroxine. During caloric restriction in the rat, total-body thyroxine-to-triiodothyronine conversion is significantly reduced.[99] Thus, moderate restricted feeding sufficient to prolong lifespan has a greater effect on the peripheral conversion of the thyroid hormones than on their central neuroendocrine control through the release of thyroid-stimulating hormone. The effects of restricted feeding to 60% of *ad libitum* levels on the diurnal variation of serum thyroid hormones has been studied in 6-month-old male Fischer 344 rats. Food restriction was observed to abolish the diurnal variation for T3 and reduced the 24-h mean from 95 ± 1 to 87 ± 3 ng/dl.[137]

In contrast to the thyroid hormones, but in agreement with the effect on progesterone synthesis and release, the changes in plasma glucocorticoids in underfed rats originate primarily through changes in hypothalamic–pituitary control. Elevation in the content of corticotrophin-releasing factor of the median eminence, plasma adrenocorticotropic hormone, plasma corticosterone levels, and adrenal hypertrophy have been reported as the initial response to dietary restriction in rats.[128,138] Prolonged underfeeding is associated with depressed adrenocortical function while calorie restriction is acknowledged to desynchronize the circadian corticoid rhythm between individual animals.[139] It has been shown that 14 days of dietary restriction commenced at weaning (21 days) was sufficient to eliminate the rise in plasma corticosterone normally observed in response to an environmental stress in rats.[97] The stress response was gradually recovered throughout the first year of life and at 1 year of age had returned to that of fully fed animals. In agreement with the observations on thyroxine, 7 days of full feeding were sufficient to return the stress elevation of plasma corticosterone to normal. This inability to elevate the plasma corticosterone to a perceived stress appears to originate at the hypothalamic–pituitary level, for the adrenal cortex in young diet-restricted rats will respond with an elevation in plasma corticosterone to exogenous adrenocorticotropic hormone when administered to conscious animals through indwelling cannulae. A longitudinal lifespan study of the daily concentration pattern of plasma corticosterone in male Fischer 344 rats restricted to 60% of the mean food intake of *ad libitum*–fed controls did not show the age-associated increase observed in *ad libitum*–fed animals. Indeed peak concentrations of total plasma corticosterone were greater in the food restricted rats. There was, however, a marked age-related decrease in the plasma concentration of corticosterone binding globulin (CBG), which was not seen in the *ad libitum*–fed animals. It was calculated from the CBG concentrations that calorie-restricted animals had higher daily mean, plasma free corticosterone concentrations than control animals. Therefore, food-restricted animals are exposed to higher levels of biologically active corticosterone throughout most of their lifespan[140] (Figure 24-15).

Diet restriction to 70% of the intake of control animals in Lobund–Wistar rats had little effect in modulating the age-associated changes in adrenal catecholamines, dopamine, norepinephrine, epinephrine, and dihydroxymandelic acid.[141]

It is recognized that there is a synergistic induction of growth-hormone mRNA biosynthesis by glucocorticoid and thyroid hormones. These hormones can modify the rate of transcription of the growth hormone gene in the normal rat.[142] Such an interplay of two hormones to influence the synthesis of a third has not been studied in the diet-restricted animal.

The morphologic appearance of islet cells in the pancreas of underfed rats is characteristic of those of a chronologically younger animal.[143] A longitudinal study of plasma glucose and insulin concentrations in *ad libitum*–fed and diet-restricted Fischer 344 male rats has revealed that diet-restricted rats have significantly lower plasma glucose levels than control animals. The 24-h plasma glucose concentration is about 15% below that of *ad libitum*–fed animals while plasma insulin levels are maintained in diet-restricted animals at about 50% those of control rats.[144] Although plasma glucose and insulin levels are depressed in diet-restricted rats, the use of fuel per metabolic mass per day is the same as for control animals (see Section 4). The lower levels of plasma insulin may result from a decreased content of pancreatic spermine.[145] The lower plasma glucose level in dietary restricted rats is associated with significantly lower percentage glycosylation of hemoglobin compared to age-matched *ad libitum*–fed rats.[146]

Female C3B10RF1 mice maintained on a 52% reduction in calorie intake had an increase in insulin receptor mRNA of between 15 to 25% over mice fed *ad libitum*. Both insulin receptor and glucocorticoid receptor mRNA increased with age in both diet-restricted and fully fed animals, while neither aging nor calorie restriction modified the hepatic levels of mRNA for insulin-like growth factor-I, RNA polymerase II, elongation factor s-II, or transcription factors Sp1, CCAAT, and enhancer-binding protein or protooncogene c-*jun*.[147]

The mRNAs for glucose-regulated proteins 78 and 94 (GRP78 and GRP94) were reduced approximately 50 and 40%, respectively, in the liver of female C3B10RF1 mice maintained on 40% energy restriction. These changes in transcription were specific and do not represent a general lowering of tissue polyadenylated RNA in response to calorie restriction.[148] The authors comment that elevated tissue levels of GRP mRNA are induced by agents which increase the level of malfolded proteins in the endoplasmic reticulum, and their lowering in the energy-restricted mice may indicate a reduction in malfolded proteins in hepatic cells from these animals.

The age-dependent loss in glucagon-promoted lipolysis in adipocytes is prevented by restricted feeding, and a study of fat cell metabolism with age in response to insulin and epinephrine has shown a reduction in basal glyceride synthesis.[149] Underfeeding preserved the response to insulin and completely prevented the age-related decline in the lipolytic response; however, in both control and restricted rats, the basal conversion of glucose to fatty acids declined rapidly with age. In an *in vitro* culture of vascular smooth muscle cells that retained the age-dependent decrease in β-adrenergic responsiveness characteristic of whole blood vessels, dietary restriction retarded the loss of this response.[150]

5.3 Neurology

In a study on the formation of lipid-peroxidation fluorescence products (age pigments) in the brain and heart of mice that had been maintained on a low-protein diet, significantly lower fluorescence was recorded in the brain by 3 months and in the heart by 12 months of age. Chronic dietary restriction delayed age-related loss of pineal cells in CD rats and resulted in the retention of a more juvenile type of pineal gland.[151] The loss of striatal dopamine receptors that play an integral role in the altered dopaminergic control of physiologic and behavioral functions in the old fully fed rat was delayed by dietary restriction (by feeding every other day).[152] This type of restricted feeding increased mean survival by 40%, and further studies have established that the sparing effect on striatal dopamine receptors is a chronic effect rather than an acute response to restricted food intake.[153]

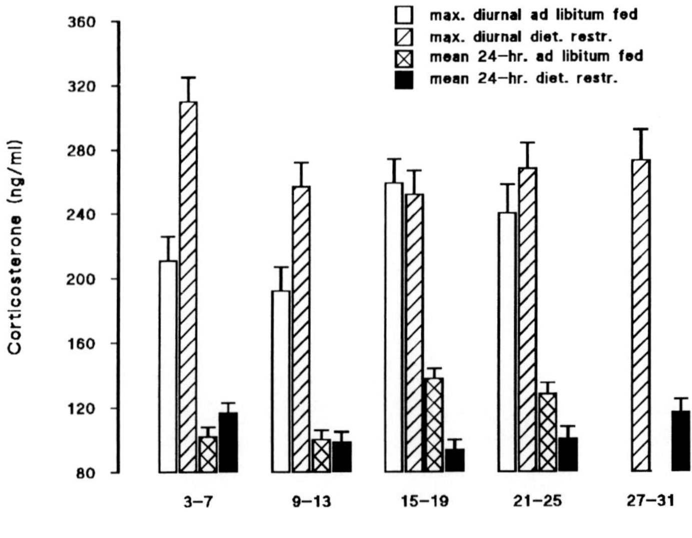

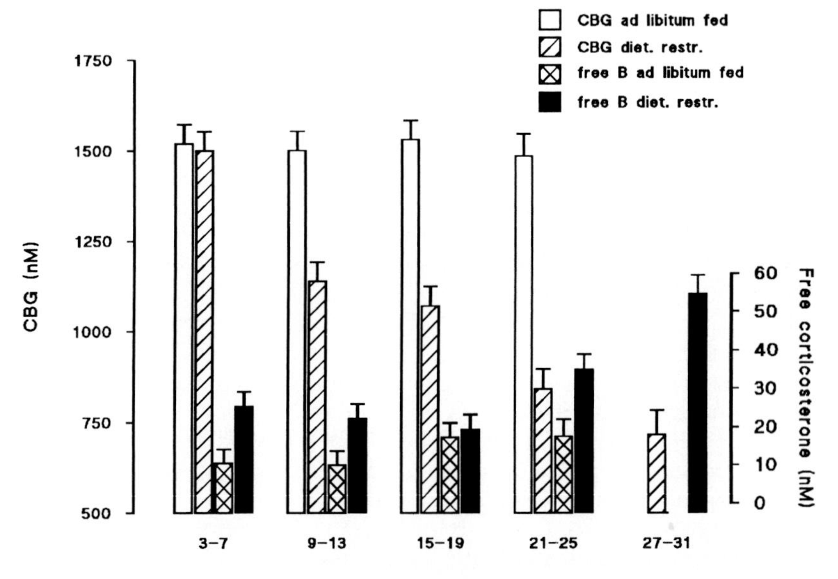

FIGURE 24-15 Top panel: *Maximum diurnal total plasma corticosterone and mean 24-h total plasma corticosterone with age in* ad libitum*-fed and diet-restricted Fischer 344 rats.* Bottom panel: *Plasma concentration of corticosterone-binding globulin (CBG) and calculated mean 24-h plasma free corticosterone concentrations.* Diet-restricted animals were fed 60% of the mean *ad libitum* intake from 6 weeks of age. Data are presented as means values ± SEM. Number of rats measured: *ad libitum* fed — 21 animals, except at 21 to 25 months when 15 animals were measured; diet restricted — 21 animals, except at 21 to 25 months (19 animals) and 27 to 31 months (13 animals). (Data from Sabatino, F., et al., *J. Gerontol.*, 46, B171, 1991.)

Diet restriction (60% of *ad libitum* intake) in 24-month-old Fischer 344 rats significantly increased dopamine levels in the frontal cortex and reduced the age-related increase of serotonin in the occipital cortex. The age-related reduction of cortical tryptophan and plasma ratio of tryptophan to large neutral amino acids in old animals were attenuated by chronic restricted feeding.[154]

These biochemical responses to underfeeding are reflected in the retention of behavioral stereotypes more characteristic of younger control animals when 2-year-old diet-restricted rats are challenged with dopaminergic agonists.[155] A selective effect has so far been seen for the effect of chronic underfeeding on brain aging in rodents. Choline acetyl transferase activity was increased in the striatum, hippocampus, and cerebellum of diet-restricted 2-year-old rats, but little effect was observed in the same brain regions for the activity of glutamic acid decarboxylase, the synthetic enzyme for the neurotransmitter γ-aminobutyric acid.[156]

6 PROPOSED MECHANISMS OF ACTION OF RESTRICTED FEEDING

Although in recent years studies have proliferated detailing the effect of restricted feeding on the physiology and pathology of aging, the molecular mechanism by which these effects can be satisfactorily explained has remained elusive.

6.1 Metabolic

Attempts have been made to explain aging and, more recently, the effects of restricted feeding on the basis of a limited total lifetime energy expenditure. An inverse correlation can be demonstrated within the Mammalia between lifespan and lifetime energy expenditure (approximately 200 kcal/g) in animals ranging in lifespan from 2 years (mouse) to 70 years (elephant). Although most mammalian species have about the same lifetime energy expenditure, in primates it is higher (an average expenditure of 488 kcal/g), and data based on 77 mammalian species indicate that there are three major classes of lifetime energy expenditure.[157] Attempts have been made to calculate the total lifetime energy expenditure for varying degrees of restricted feeding in rats.[57] Two subsequent studies have not substantiated the idea that total energy expenditure per gram of tissue was similar in fully fed and diet-restricted rats. Direct measurement of the metabolic rate has confirmed that total lifetime energy expenditure per gram of tissue in diet-restricted rats is significantly greater than for *ad libitum*-fed controls.[60,61,63,65] Such observations, taken together with protein turnover in underfed rats, suggest that a simple metabolic explanation is extremely unlikely.

6.2 Protein Turnover

It is a general observation that the rate of protein synthesis in many organisms and cells declines with age, and this appears to be associated with a similar decrease in the rate of protein degradation.[86,158] It can therefore be predicted that with age an increase in the half-life of proteins will occur with a decrease in the rate of protein turnover although this interpretation is still a matter of contention.[159] A significant decrease with age in the turnover of two liver enzymes, ornithine decarboxylase and aldolase, can be shown in C_{57}BL/6J mice.[160,161] Further support for a decrease in protein turnover with age has been published, and it is suggested that this increase in protein half-life is an important contributory factor in the appearance of altered proteins in the tissues of older organisms. While providing for a mechanism for the removal of defective proteins, turnover is an important component in enzyme regulation, endowing the cell with the versatility to respond to metabolic change.[162]

As previously stated, in rodents maintained on chronic dietary restriction for all their postweaning life, whole-body protein turnover at advanced chronological age is enhanced while individual tissue response *in vivo* and *in vitro* is more variable. It has been proposed that the decrease in protein turnover with age is the molecular basis for many of the physiologic changes associated with aging, and a retention of a higher rate of protein turnover into advanced chronological age in underfed rats would be predicted. Comparison of the labile component of glucose-6-phosphate dehydrogenase in young and old fully fed and diet-restricted mice revealed a 50% decrease in the proportion of labile enzyme in tissue from underfed animals.[163]

Mice fed a restricted diet (60% of *ad libitum*) for 70 days from as late as 23.5 months of age show a significant decrease in the proportion of heat-labile aminoacyl–tRNA synthetases in liver and brain.[14] Mouse liver parenchymal cells maintained in culture from animals of different ages have been used to investigate the degradation rates for microinjected proteins (horseradish peroxidase, ovalbumin, and pulse-labeled proteins). The half-life of protein degradation in cells from old r. ice is about 50% longer than that in cells from young or middle-aged ones. Dietary restriction initiated at 23 months of age for a total period of 70 days decreased the half-lives of the injected proteins to about 40% that observed in cells from age-matched fully fed mice.[164] The control of protein degradation rates is little understood and the molecular details whereby diet can modify the rate of protein turnover are still to be determined.

6.3 Free Radicals

The free radical theory of aging as proposed by Harman[165] states that aging results from the cumulative deleterious effect of free radicals generated during oxidative metabolism. Starting from the premise that the body weight of diet-restricted rodents results mainly from a reduction in cell size, particularly of postmitotic cells, and only to a lesser extent from a reduction in cell numbers, Harman has argued that underfed animals contain more cells per gram body weight than do fully fed animals.[166] Using the data of McCarter[65] (see Section 4), it can be argued that the rate of oxygen consumption per cell will be less under diet-restricted conditions. The total cell usage of oxygen throughout life would be approximately the same in diet-restricted and fully fed animals, but the rate of use would be different and free radical damage should accumulate at a slower rate (Chapter 6).

A number of studies have attempted to assess rates of free radical–induced damage with aging and in diet-restricted animals.[80,167-170] These studies have shown both a suppression in age-related malondialdehyde production and lipid hydroperoxide formation in liver mitochondrial and microsomal membranes, associated with an increase in the activities of several antioxidant enzymes. Similar observations have been recorded for the effect of restricted feeding on antioxidant enzymes in rat erythrocytes and spleen lymphocytes. Alternate-day fasting and feeding, applied at 3.5 months, prevented the age-associated decrease in catalase and glutathione peroxidase but did not influence superoxide dismutase activity.[83,171]

Calorie restriction alters membrane lipid composition making the membrane lipid less readily peroxidized.[82] An increase in linoleic acid and a decrease in docosapentanoic acid content has been reported. Diet-restricted animals sustain a higher unsaturation/saturation index while lowering the content of the more readily peroxidized 22:5 fatty acid.[46,80]

The cytosolic level of antioxidant enzymes was also modified by restricted feeding for it prevented the age-associated decline in hepatic cytosol catalase activity observed in fully fed control rats. It also prevented the age-related decrease in cellular glutathione and glutathione S-transferase activity that occurs in *ad libitum*–fed animals.[167] A correlation between the increase in the activities of the antioxidant enzymes (superoxide dismutase, glutathione peroxidase, and catalase) and a decrease in lipid peroxidation in the liver tissue of diet-restricted rats has been demonstrated. At 28 months of age, the activities of CuZn superoxide dismutase, catalase, and glutathione peroxidase were 40 to 80% greater in liver tissue from diet-restricted than from control animals.[172]

The finding of enhanced expression of antioxidant enzymes in older diet-restricted mice is of interest in view of the effect of restricted feeding on cataract development in this animal model (see Section 4.6). Eye lenses with cataracts exhibit low activities of catalase, superoxide dismutase, and glutathione peroxidase, associated with high levels of H_2O_2 and lipid peroxides.[173] Diet restriction regimes have been shown to retard the age-related loss of gamma crystallins and the formation of cataracts in the Emory cataract-prone mouse strain.[108,174] This may result from a catalase-mediated lowering of free radical–induced oxidative damage in the lens protein.

Direct measurement of the superoxide radical, hydroxyl radical, and hydrogen peroxide by liver microsomes from rats of varying ages has shown that *ad libitum*–fed animals maintain a higher production of superoxide and hydroxyl radicals when compared to animals maintained on a restricted diet. The food-restricted rats showed higher superoxide dismutase activity in both cytosolic and mitochondrial fractions than control animals.[175]

Very high levels of oxidative damage to DNA occurs during normal metabolism with an estimated steady-state level of damage in rat cells of 10^6 oxidative adducts, and about 10^5 new adducts are formed daily.[176] The rate of DNA damage assessed by measuring 8-hydroxydeoxyguanosine (8-OH dG) in nuclear and mitochondrial DNA is significantly reduced in rats on restricted feeding.[145]

Both oxidative lesions in DNA and oxidatively damaged proteins have been shown to accumulate with aging. Both protein restriction (5 or 10% of the diet as compared with the control level of 20%) and calorie restriction (either 25 or 40% of control intake) was observed to significantly reduce the accumulation of oxidatively damaged proteins (protein carbonyls). Rats fed a diet containing either protein at 5 or 20% were irradiated twice weekly (125 rads per exposure). The low-protein diet (5%) reduced the accumulation of oxidatively damaged proteins resulting from the exposure to ionizing radiation.[177] Measurement of 5-hydroxymethyluracil as a measure of oxidative DNA damage has confirmed the protective effect of calorie-restricted diets.[178]

In spite of these observations on free radical damage in underfed rodents, the only study to assess the combined effect of restricted feeding and the oral administration of an antioxidant (ethoxyquin/2-mercaptoethylamine) on survival in mice has yielded a negative result. The diet-restricted animals had a significantly longer survival (41 months) than control animals, or control or diet-restricted animals supplemented with the antioxidant. The antioxidant supplementation increased hepatic degeneration and increased hepatoma in diet-restricted animals suggestive of a cytotoxic effect.[179]

6.4 DNA and Gene Expression

The activities and mRNA levels of superoxide dismutase (Cu–Zn) and catalase were increased by 24 to 38% and 64 to 75%, respectively, when measured in 21- and 28-month-old rats maintained on a 40% restriction of energy intake. Glutathione peroxidase activity in the liver of diet-restricted rats was significantly higher (37%) at 28 months of age in comparison with fully fed control rats.[172] The increase in the activities of antioxidant enzymes in calorie-restricted rodents appears to arise from an increase in the levels of mRNAs and enhanced levels of transcription[180,181] (Figure 24-16).

Caloric restriction has been shown to delay an age-associated change in hepatic gene expression, which is under endocrine control. Synthesis of the α_{2u}-globulin protein declines with age and this is associated with a 80 to 90% decrease in transcription for the α_{2u}-globulin gene between the ages of 5 and 24 months.[182] The synthesis of α_{2u}-globulin ceases when hepatocytes undergo malignant transformation, and it is recognized to be under multihormonal control. In 18-month-old male Fischer 344 rats restricted to 60% of *ad libitum* feeding from 6 weeks of age, mRNA levels and transcription of the α_{2u}-globulin gene was 1.8 to 3 times higher in comparison with fully fed age-matched controls. The decline in α_{2u}-globulin observed in fully fed animals can be reversed by the administration of testosterone. Senescence in the male causes a marked fall in circulating testosterone which is not retarded to a significant degree by restricted feeding.[115] The derepression of the androgen-repressible senescence marker protein (SMP-2) in the liver of rats is also delayed by restricted feeding. In the 27-month-old diet-restricted rat the mRNA for SMP-2 was 45% the level of the fully fed age-matched control.[183] The effect of restricted feeding on age-dependent changes in gene expression correlate with the hepatic level of immunoreactive cytoplasmic androgen-binding (CAB) protein. Androgen receptor mRNA is almost undetectable in prepubertal (less than 35 days old) and senescent male (greater than 750 days old) rats. Only hepatocytes that express α_{2u}-globulin gene contain androgen receptor mRNA, and the retardation of the age-dependent loss of androgen sensitivity by calorie restriction is due to a concomitant delay in the decline of androgen receptor mRNA synthesis.[184]

No effect of restricted feeding has been observed for single-stranded breaks in DNA for brain, liver, and kidney of $(C_3H \times C_{57}BL/10)F_1$ mice,[185] but unscheduled DNA synthesis is significantly higher in hepatocytes and kidney cells isolated from rats on a restricted diet. Thus, as seen for many biochemical parameters, calorie restriction retards the age-related decline in DNA repair.[186] Both alpha and beta hepatic DNA polymerases from 26-month calorie-restricted mice exhibit a higher level of fidelity than polymerases from *ad libitum*–fed animals.[187] Whether this is merely an effect of restricted feeding or a fundamental part of the mechanism by which retarded aging is induced is not known. This difficulty of interpretation does, however, illustrate a basic problem fundamental to much of the data detailing biochemical and molecular changes in response to calorie restriction.

6.5 Neuroendocrine

Retention of the immature state for prolonged periods of time — the length of which was shown to be directly correlated with the extension in lifespan but inversely correlated with length of adult period of life remaining after refeeding — has led to the repeated suggestion that restricted feeding operates through a neuroendocrine mechanism. Early studies had shown, particularly when adult animals were used, that severe prolonged undernutrition could induce a pseudohypophysectomized state.[188] Further support for this interpretation was obtained from studies utilizing early hypophysectomy linked to cortisone replacement therapy.[129] Such treatment, begun at 70 days, greatly retards growth and aging of tail collagen and prevents the development of proteinuria and subsequent renal disease. Tumor incidence is significantly decreased and survival in dietary restricted and hypophysectomized rats is similarly increased compared to *ad libitum*–fed control rats. Maximum lifespans for *ad libitum*–fed, diet-restricted, and hypophysectomized (with hormone replacement) rats were 1120, 1282, and 1342 days, respectively. The relationship between the antiaging actions of hypophysectomy and underfeeding still needs to be resolved since it is clear from the data cited earlier that the induction of a pseudohypophysectomized state is not obligatory for lifespan extension. Moderate degrees of restricted feeding have produced a far greater effect on survival in terms of maximum lifespan achieved (up to 1800 days) in the rat than have been reported for hypophysectomized animals. It is postulated that growth hormone secretion in the adult phase of the lifespan may have an aging-promotion effect, but although hypophysectomy eliminates this hormone, growth hormone release continues in the diet-restricted animal.

The suggestion has been made that hypophysectomy linked with appropriate hormone maintenance will induce a reversal of age-related pathology and a physiological rejuvenation in rats. This conclusion is based on specific physiological changes known to occur with age, designated biomarkers of aging, but not on survival data. It is known that under certain experimental conditions changes in such biomarkers may be retarded without a true retardation of aging in the whole animal. Indeed it has been reported that hypophysectomized mice had increased motor activity and presented a more youthful appearance even though they were shorter lived than control animals.[189] It is therefore possible that surgical hypophysectomy is an alternative means of inducing chronic underfeeding and that the endocrine changes resulting from such a procedure are secondary to the primary response induced by underfeeding.

Over a number of years a specialized form of restricted feeding has been developed and studied in which rats are maintained on a diet deficient only in one amino acid, tryptophan.[127,190] Such a regime severely retards growth and development and delays the onset of reproductive senescence and tumorigenesis in a manner somewhat similar to severe dietary restriction. The hypothesis has been advanced that this form of restricted feeding may operate through a complex neuroendocrine adjustment involving the neurotransmitter serotonin. It is well established that dietary tryptophan is the sole source of serotonin in the rat and that brain serotonin levels decrease in rats maintained on a diet low in tryptophan.[191] Brain serotonin

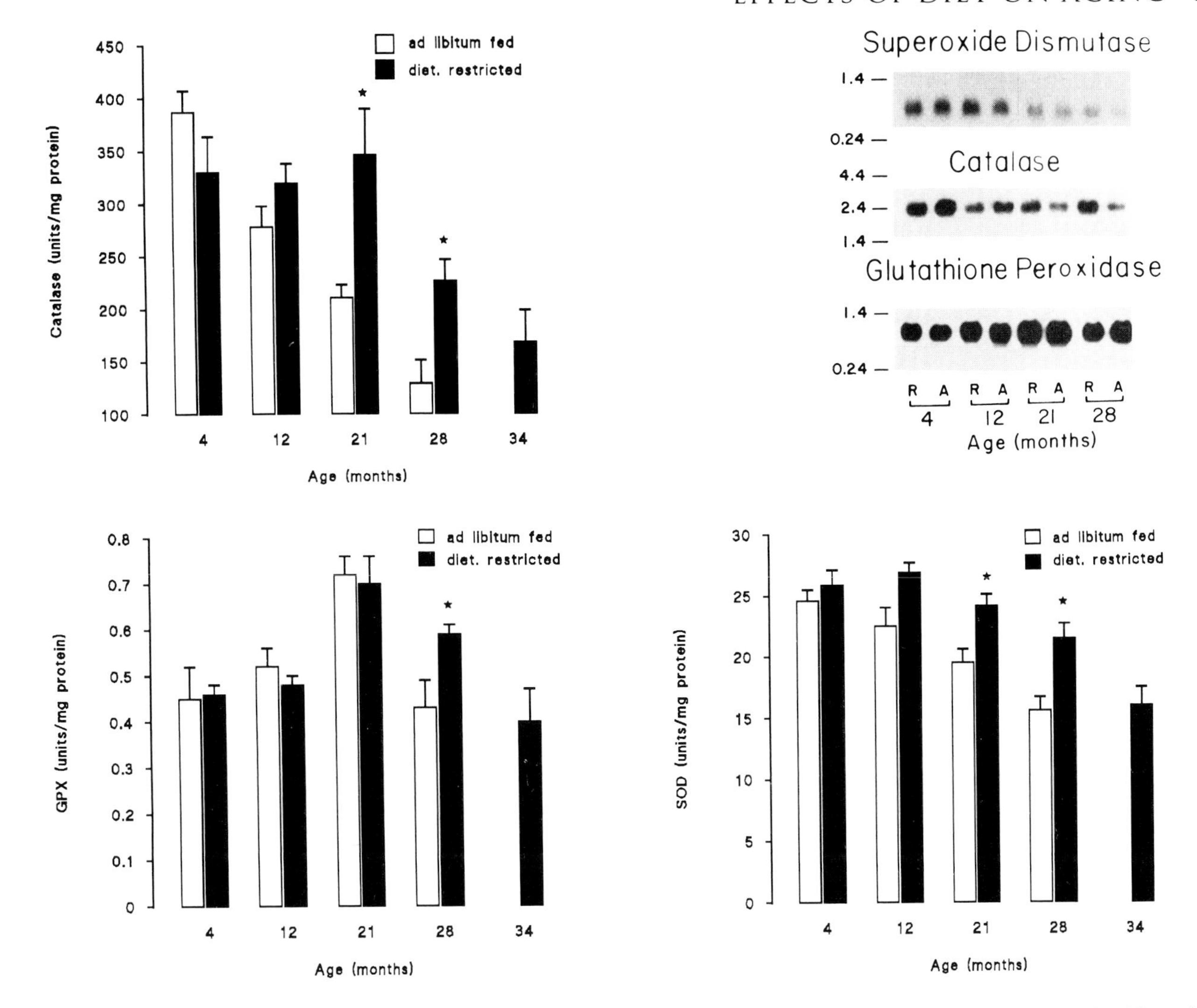

FIGURE 24-16 *Activities of antioxidant enzymes in liver tissue from Fischer 344 male rats fed* ad libitum *or a restricted diet (60% of mean* ad libitum *intake from 6 weeks of age).* Each value is the mean ± SEM for five to eight animals. Effect of age, diet, and their interaction on superoxide dismutase (SOD) activity, effect of age and its interaction with diet on catalase activity, and effect of age on glutathione peroxidase (GPX) activity were significant at $p \leq 0.05$ (2-way ANOVA). Means identified by (*) are significantly different at $p \leq 0.05$ for rats fed *ad libitum* and a restricted diet. *Insert panel shows an autoradiograph of Northern blots for SOD, catalase, and GPX mRNA.* RNA samples were pooled from 5 to 8 animals at 4, 12, 21, and 28 months of age. The RNA (20 µg) was denatured, fractionated on 1.2% agarose gel, transferred to a nylon membrane, and hybridized to ^{32}P-labeled cDNA probes. The autoradiographs are calibrated with 0.24, 1.4, 2.4, and 4.4 kilobase standards. (A) *Ad libitum*–fed rats; (R) diet-restricted animals. (Redrawn from Rao, G., et al., *J. Nutr.*, 120, 602, 1990. With permission.)

concentrations are subject not only to the plasma levels of unbound tryptophan, but also to the concentration of several other large neutral amino acids that share the same transport system at the blood–brain barrier. Since serotonin stimulates growth hormone release while dopamine exerts an inhibitory effect, the possibility exists that diet can modify the monaminergic innervation of the hypothalamus directly and so disturb the endocrine status of the animal. The synthesis of the neurotransmitter acetylcholine and the synthesis and release of dopamine and norepinephrine can be similarly manipulated by altering the dietary levels of their precursor molecules, choline and tyrosine. Thus, it is plausible that dietary restriction can act through a complex neuroendocrine response, which modifies specific gene expression. Since the neuroendocrine system is central to other ontogenetic processes, such as the timing and onset of puberty, it is feasible that the timing and process of senescence may be under similar controls.[192] Thus, dietary restriction could be seen to modify the coordination and timing of ontogenesis.

Tryptophan, however, while being an important precursor for neurotransmitter synthesis, is equally important as a key amino acid in the regulation of the polysome profile in protein synthesis, since it is often the amino acid of lowest concentration in the diet. The interrelationship of endocrine status and protein synthesis and degradation is extremely complex, with at least seven hormones or groups of hormones being known to affect the rate of protein synthesis or protein degradation: insulin, growth hormone, somatomedins, glucagon, thyroid hormones, corticosteroids, and prostaglandins. Modification of the endocrine status of the animal through diet will inevitably have consequences for enzyme induction, free radical generation, gene expression, and protein half-lives, and it is plausible that these effects are

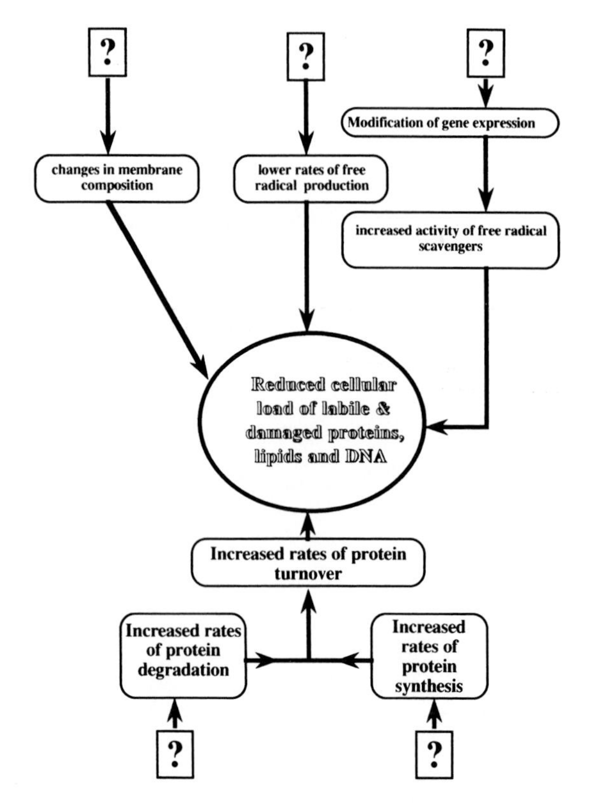

FIGURE 24-17 *Schematic representation of the possible mechanism by which restricted feeding retards aging.* Diet restriction appears to induce a reduction in the cellular load of damaged proteins, lipids, and DNA. This is achieved, in part, by modification of age-related gene expression to increase the activity of free radical scavenger enzymes; to induce changes in membrane composition which makes them more resistant to free radical peroxidation; and to xinduce a metabolic adjustment whereby lower rates of free radicals are produced. In addition there is evidence that protein turnover is enhanced in certain tissues from diet restricted rodents; the effect would be to increase the efficiency of removal of damaged proteins and enhance metabolic flexibility. The metabolic signals linking the lower caloric intake to these observed effects are not known.

the molecular expression of the neuroendocrine response to dietary restriction.

6.6 Immunological Theories

It has been argued that much of observed aging can be explained on the basis of increased autoimmunity and decreased heteroimmunity.[193] Data on the immune function in long-lived diet-restricted rodents support this general contention for a delay in thymic involution, and enhanced T-cell function compared to aged matched controls has clearly been established. The age-related decline in heteroimmunity is retarded in such animals and autoimmune pathologies are suppressed even in strains of mice genetically predisposed to such pathology. Enhanced survival in underfed rodents results in part from the retention of a competent immune system to a greater age than observed in fully fed animals.

Some data suggest that involution and atrophy of the thymus are important in the control of aging in the immune system, and several mechanisms have been advanced to explain this observation.[128,194] It has been suggested that thymic involution and atrophy are a result of clonal exhaustion, the possibility existing that a genetically programmed clock mechanism induces cell death after undergoing a fixed number of cell divisions similar to the fixed *in vitro* cell-doubling potential observed for somatic cells in culture.[195] In diet-restricted rats, mitogen-induced lymphocyte proliferation, interleukin-2 induction, and protein synthesis in spleen and lymph node lymphocytes are characteristic of physiologically younger animals, supporting the idea of delayed clonal exhaustion for such animals. The primary control of immune aging may lie within the endocrine system, a decline in trophic factors with age resulting in a decline in T-cell maturation or an increase in suppressor cell activity. The relationship of enhanced immunological function to altered hormonal control of the immune system in diet-restricted rats has not been studied in detail. The retention of enhanced immunologic surveillance to a greater age in diet-restricted animals clearly facilitates survival by delaying the appearance of and reducing the intensity of age-related pathology. It is far from certain, however, whether this is the primary action of underfeeding since it may well represent another aspect of molecular adaptation common to many cells and tissues.

The molecular mechanism whereby diet can so modify age-related pathology, the response to external carcinogens, and the genetic predisposition to hypertension-related disease while retarding physiologic aging is still far from understood. We have attempted to bring together some of the disparate ideas that have been proposed to explain the underlying mechanism of action of restricted feeding (see Figure 24-17). Although these ideas are presented as discrete hypotheses, they may represent different facets of an overall complex adaptation of the animal to restricted energy intake in order to secure tissue maintenance. While it is now accepted that the rate of physiological aging and the timing of the onset of age-related pathology is susceptible to manipulation through calorie restriction, two fundamental issues remain to be resolved; the first is to determine the biochemical mechanism(s) responsible for these effects, and the second is to determine whether aging can be delayed by calorie restriction in long-lived mammalian species, most notably primate species.

7 THE EFFECTS OF DIET ON AGING IN PRIMATES AND IN THE HUMAN

7.1 Primate Studies

Two studies are now in progress which are designed to directly address the second issue and to determine if biomarkers exist within primate species that can predict the rate of aging. At the National Institute of Aging Gerontology Research Center, rhesus monkeys *(Macaca mulatta)* and squirrel monkeys *(Saimiri sciureus)* are being fed either a diet at or near *ad libitum* levels or 70% of the same diet. The restricted feeding was gradually introduced over a 3-month period. In the initial design, 30 male animals of each species were used, which has subsequently been expanded to 60 male and 60 female rhesus monkeys. Both juvenile (1 year) and adult (3 to 5 years) male rhesus monkeys and juvenile (1 to 4 years) and adult (5 to 10 years) male squirrel monkeys have been used. In addition, *ad libitum*–fed aged rhesus monkeys (>20 years) and squirrel monkeys (>10 years) are included in the study for cross-sectional comparisons. The study has now been running for nearly 5 years. Animals have maintained excellent health status as determined by physical examinations, hematology, and blood chemistry. The relative rates of body weight gain in the diet-restricted animals have been markedly reduced.[196,197]

A second study using restricted feeding to delay aging in primates was begun in 1992. The aim of this project was to determine whether aging processes are retarded by adult-onset diet restriction. The study was begun with 30 adult (8 to 14 years old) male rhesus monkeys, which provided baseline data over the first 3- to 6-month period. After this time 15 animals were assigned to a control group and given free access to a semipurified diet for 6 to 8 h per day. The remaining 15 animals were fed the same diet but at 70% of each individual's baseline intake level. The parameters being studied are body size and composition, physical activity, metabolic rate, glucose tolerance and insulin sensitivity, hematologic indices, immunological function and fingernail growth. After 1 year of study, and in agreement with the NIA study, the animals appear to be in excellent health. A clear effect of restricted feeding on growth rate has been established and these monkey have less body fat than do control animals, whereas the amount of lean body mass is unaltered. The diet-restricted monkeys showed statistically less physical activity than control animals.[198]

It is too early with either of the primate studies to draw any conclusions with respect to an effect of restricted feeding on the rate of aging, but both studies have demonstrated that it is possible to adapt primates to mild diet restriction and retain the animals in good health.

7.2 Human Studies

With respect to human aging little is known of the effect of diet or calorie restriction on the rate of aging. The data presented earlier demonstrate, in rodents, a clear effect of diet on both the timing and intensity of cardiovascular disease, cancer, and brain aging. In the human many diseases associated with aging are susceptible to dietary intervention, and the proposal has been made that nutrition will become a major strategy in reducing the incidence of the major causes of death.[199]

Inferences on the effect of low-calorie diets on human survival can be drawn from such groups as the Okinawan isolate. The Japanese island of Okinawa has an incidence of centenarians which is 2 to 40 times that of other Japanese islands. Detailed study of this population has shown that the dietary energy intake for adults is about 20% less than the national average and for school children it is 62% of the recommended intake for Japan. Death rates from cerebral vascular disease, malignancy, and heart disease is 59, 69, and 59%, respectively, of the average for the rest of Japan. The total death rate for people 60 to 64 years of age was 1280 in Okinawa but 2181 per 100,000 elsewhere in Japan.[200]

Biosphere 2 is a project that has been running from September 1991 which exposes eight subjects (four women and four men) to a diet low in calories. The ecosystem in which the subjects live is a 3.15-acre space that is energetically open (sunlight, electric power, and heat), but is materially closed with air, water, and organic material being recycled. The subjects therefore live on the food crops grown within the ecosystem and the diet produced is low in fat (10% of calories), with an average energy intake of 1780 kcal/day (7748 kJ/day). Over the first 6 months of the project, a significant fall in body weight, mean systolic/diastolic blood pressure, total serum cholesterol, high-density lipoprotein, triglyceride, fasting glucose, and leukocyte count was observed.[201] Such changes are similar to those observed in the calorie-restricted rodent exhibiting extended survival. These observations with those on the Okinawan isolate population suggest that the human may respond to calorie restriction in a manner similar to short-lived rodent species. The survival data from the chronic diet-restriction studies with rhesus and squirrel monkeys should give a strong indication as to whether this is likely to be so.

REFERENCES

1. Weindruch, R. and Walford, R. L., *The Retardation of Aging and Disease by Dietary Restriction,* Vol. 1, Charles C Thomas, Springfield, IL, 1988, 436.
2. Ross, M. H., Protein, calories and life expectancy, *Fed. Proc.,* 18, 1190, 1959.
3. Davies, T. A., Bales, C. W., and Beauchene, R. E., Differential effects of dietary caloric and protein restriction in the aging rat, *Exp. Gerontol.,* 18, 427, 1983.
4. Birt, D. F., Higgenbotham, S. M., Patil, K., and Pour, P., Nutritional effects on the lifespan of Syrian hamsters, *Age,* 5, 11, 1982.
5. Feldman, D. B., McConnell, E. E., and Knapka, J. J., Growth, kidney disease, and longevity of Syrian hamsters *(Mesocricetus auratus)* fed varying levels of protein, *Lab. Anim. Sci.,* 32, 613, 1982.
6. Dalderup, L. M. and Visser, W., Influence of extra sucrose in the daily food on the life-span of Wistar albino rats, *Nature,* 222, 1050, 1969.
7. Stuchlikova, E., Juricova-Horakova, M., and Deyl, Z., New aspects of the dietary effect of life prolongation in rodents. What is the role of obesity in ageing?, *Exp. Gerontol.,* 10, 141, 1975.
8. Weindruch, R., Retardation of aging by caloric restriction in mice, in *The Potential for Nutritional Modulation of Aging Processes,* Ingram, D. K., Baker, G. T., and Shock, N. W., Eds., Food & Nutrition Press, Inc., Trumbull, CT, 1991, 109.
9. Yu, B. P., Masoro, E. J., and McMahan, C. A., Nutritional influences on aging of Fischer 344 rats. I. Physical, metabolic and longevity characteristics, *J. Gerontol.,* 40, 657, 1985.
10. Beauchene, R. E., Bales, C. W., Bragg, C. S., Hawkins, S. T., and Mason, R. L., Effect of age of initiation of feed restriction on growth, body composition, and longevity of rats, *J. Gerontol.,* 41, 13, 1986.
11. Cheney, K. E., Liu, R. K., Smith, G. S., Meredith, P. J., Mickey, M. R., and Walford, R. L., The effect of dietary restriction of varying duration on survival, tumor patterns, immune function, and body temperature in B10C3F1 mice, *J. Gerontol.,* 38, 420, 1983.
12. Nolen, G. A., Effects of various restricted dietary regimes on the growth, health and longevity of albino rats, *J. Nutr.,* 102, 1477, 1972.
13. Holehan, A. M. and Merry, B. J., The experimental manipulation of ageing by diet, *Biol. Rev.,* 61, 329, 1986.
14. Takahashi, R. and Goto, S., Influence of dietary restriction on accumulation of heat-labile enzymes in the liver and brain of mice, *Arch. Biochem. Biophys.,* 257, 200, 1987.
15. Takahashi, R., Mori, M., and Goto, S., Alteration of aminoacyl tRNA synthetases with age: accumulation of heat-labile enzyme molecules in rat liver, kidney and brain, *Mech. Ageing Dev.,* 33, 67, 1985.

16. Weindruch, R. and Walford, R. L., Dietary restriction in mice beginning at 1 year of age: effect on lifespan and spontaneous cancer incidence, *Science,* 215, 1415, 1982.
17. Berg, B. N., Nutrition and longevity in the rat, *J. Nutr.,* 71, 255, 1960.
18. Cornwell, G. G., Thomas, B. P., and Snyder, D. L., Myocardial fibrosis in aging germ-free and conventional Lobund-Wistar rats — The protective effect of diet restriction, *J. Gerontol.,* 46(5), B167, 1991.
19. Effros, R. B., Walford, R. L., Weindruch, R., and Mitcheltree, C., Influences of dietary restriction on immunity to influenza in aged mice, *J. Gerontol.,* 46(4), B142, 1991.
20. Payne, P., Ageing and nutrition, in *Drugs and the Elderly. Perspectives in Geriatric Clinical Pharmacology,* Crooks, J. and Stevenson, I. H., Eds., Macmillan, London, 1979, 38.
21. Ross, M. H. and Bras, G., Tumor incidence patterns and nutrition in the rat, *J. Nutr.,* 87, 245, 1965.
22. Ross, M. H. and Bras, G., Influence of protein under and over nutrition on spontaneous tumor prevalence in the rat, *J. Nutr.,* 103, 944, 1973.
23. Ross, M. H. and Bras, G., Lasting influence of early caloric restriction on prevalence of neoplasms, *J. Natl. Cancer Inst.,* 47, 1095, 1971.
24. Tucker, M. J., The effect of long-term food restriction on tumours in rodents, *Int. J. Cancer,* 23, 803, 1979.
25. Chen, R. F., Good, R. A., Engelman, R. W., Hamada, N., Tanaka, A., Nonoyama, M., and Day, N. K., Suppression of mouse mammary tumor proviral DNA and protooncogene expression: association with nutritional regulation of mammary tumor development, *Proc. Natl. Acad. Sci. U.S.A.,* 87(7), 2385, 1990.
26. Shields, B. A., Engelman, R. W., Fukaura, Y., Good, R. A., and Day, N. K., Calorie restriction suppresses subgenomic mink cytopathic focus-forming murine leukemia virus transcription and frequency of genomic expression while impairing lymphoma formation, *Proc. Natl. Acad. Sci. U.S.A.,* 88(24), 11138, 1991.
27. Shimokawa, I., Yu, B. P., and Masoro, E. J., Influence of diet on fatal neoplastic disease in male Fischer 344 rats, *J. Gerontol.,* 46(6), B228, 1991.
28. Fernandes, G., Khare, A., Laganier, S., Yu, B. P., Sandberg, L., and Friedric, B., Effect of food restriction and aging in immune cell fatty-acids, functions and oncogene expression, *Fed. Proc.,* 46, 567, 1987.
29. Himeno, Y., Engelman, R. W., and Good, R. A., Influence of calorie restriction on oncogene expression and DNA synthesis during liver regeneration, *Proc. Natl. Acad. Sci. U.S.A.,* 89(12), 5497, 1992.
30. Pashko, L. L. and Schwartz, A. G., Effects of food restriction, dehydroepiandrosterone, or obesity on the binding of ^{3}H-7, 12-dimethylbenz(a)anthracene to mouse skin, *J. Gerontol.,* 38, 8, 1983.
31. Birt, D. F., Pinch, H. J., Barnett, T., Phan, A., and Dimitroff, K., Inhibition of skin tumor promotion by restriction of fat and carbohydrate calories in SENCAR mice, *Cancer Res.,* 53(1), 27, 1993.
32. Pollard, M., Luckert, P. H., and Guang-Yan, P., Inhibition of intestinal tumorigenesis in methylazoxymethanol-treated rats by dietary restriction, *Cancer Treat. Rep.,* 68, 405, 1984.
33. Pollard, M. and Luckert, P. H., Tumorigenic effects of direct- and indirect-acting chemical carcinogens in rats on a restricted diet, *J. Natl. Cancer Inst.,* 74, 1347, 1985.
34. Reddy, B. S., Wang, C., and Maruyama, H., Effect of restricted calorie intake on azoxymethane-induced colon tumor incidence in male F344 rats, *Cancer Res.,* 47, 1226, 1987.
35. Chou, M. W., Pegram, R. A., Gao, P., Hansard, S. R., Shaddock, J. G., and Casciano, D. A., The effects of dietary restriction and aging on in vivo and in vitro binding of aflatoxin B1 to cellular DNA, *Biomed. Environ. Sci.,* 4(1–2), 134, 1991.
36. Wall, K. L., Gao, W. S., Qu, W., Kwei, G., Kauffman, F. C., and Thurman, R. G., Food restriction increases detoxification of polycyclic aromatic hydrocarbons in the rat, *Carcinogenesis,* 13(4), 519, 1992.
37. Klurfeld, D. M., Lloyd, L. M., Welch, C. B., Davis, M. J., Tulp, O. L., and Kritchevsky, D., Reduction of enhanced mammary carcinogenesis in LA/N-cp (Corpulent) rats by energy restriction, *Proc. Soc. Exp. Biol. Med.,* 196(4), 381, 1991.
38. Lloyd, T. and Boyd, B., Development and regulation of hypertension in the spontaneously hypertensive rat: enzymatic and nutritional studies, in *Function and Regulation of Monoamine Enzymes: Basic and Clinical Aspects,* Usdin, E., Weiner, N., and Youdin, M. B. H., Eds., Macmillan Press, London, 1981, 843.
39. Lloyd, T., Food restriction increases life span of hypertensive animals, *Life Sci.,* 34, 401, 1984.
40. Lyon, M. F. and Hulse, E. V., An inherited kidney disease of mice resembling human nephronophthisis, *J. Med. Genet.,* 8, 41, 1971.
41. Fernandes, G., Yunis, E. J., Miranda, M., Smith, J., and Good, R. A., Nutritional inhibition of genetically determined renal disease and autoimmunity with prolongation of life in kd/kd mice, *Proc. Natl. Acad. Sci. U.S.A.,* 75, 2888, 1978.
42. Walford, R. L., Liu, R. K., Gerbase-Delima, M., Mathies, M., and Smith, G. S., Long-term dietary restriction and immune function in mice, *Mech. Ageing Dev.,* 2, 447, 1974.
43. Weindruch, R., Kristie, J. A., Naeim, F., Mullen, B., and Walford, R. L., Influence of weaning-initiated dietary restriction on responses to T-cell mitogens and on splenic T-cell levels in a long-lived mouse hybrid, *Exp. Gerontol.,* 17, 49, 1982.
44. Fernandes, G., Yunis, E. J., and Good, R. A., Influence of diet on the survival of mice, *Proc. Natl. Acad. Sci. U.S.A.,* 73, 1279, 1976.
45. Richardson, A. and Cheung, H. T., The relationship between age-related changes in gene expression, protein turnover and the responsiveness of an organism to stimuli, *Life Sci.,* 31, 605, 1982.
46. Laganiere, S. and Fernandes, G., Study on the lipid composition of aging Fischer-344 rat lymphoid cells: effect of long-term calorie restriction, *Lipids,* 26, 472, 1991.
47. Venkatraman, J. and Fernandes, G., Modulation of age-related alterations in membrane composition and receptor-associated immune functions by food restriction in Fischer-344 rats, *Mech. Ageing Dev.,* 63(1), 27, 1992.
48. Meydani, S. N., Lipman, R., Blumberg, J. B., and Taylor, A., Dietary energy restriction decreases Ex Vivio spleen prostaglandin E2 Synthesis in Emory mice, *J. Nutr.,* 120, 112, 1990.
49. Riley, M.-L., Turner, R. J., Evans, P. M., and Merry, B. J., Failure of dietary restriction to influence natural killer activity in old rats, *Mech. Ageing Dev.,* 50, 81, 1989.
50. Weindruch, R., Devens, B. H., Raff, H. V., and Walford, R. L., Influence of dietary restriction and aging on natural killer cell activity in mice, *J. Immunol.,* 130, 993, 1983.
51. Kubo, C., Day, N. K., and Good, R. A., Influence of early or late dietary restriction on lifespan and immunological parameters in MRL/Mp-Ipr/Ipr mice, *Proc. Natl. Acad. Sci. U.S.A.,* 81, 5831, 1984.
52. Kubo, C., Johnson, B. C., Day, N. K., and Good, R. A., Effects of calorie restriction on immunologic functions and development of autoimmune disease in NZB mice, *Proc. Soc. Exp. Biol. Med.,* 201(2), 192, 1992.

53. Kubo, C., Gajar, A., Johnson, B. C., and Good, R. A., The effects of dietary restriction on immune function and development of autoimmune disease in BXSB mice, *Proc. Natl. Acad. Sci. U.S.A.*, 89(7), 3145, 1992.
54. Altman, A., Theofilopoulos, A. N., Weiner, R., Katz, D. H., and Dixon, F. J., Analysis of T cell function in autoimmune murine strains, *J. Exp. Med.*, 154, 791, 1981.
55. Horst, K., Mendel, L. B., and Benedict, F. G., The influence of previous diet, growth and age upon the basal metabolism of the rat, *J. Nutr.*, 8, 139, 1934.
56. Will, L. C. and McCay, C. M., Ageing, basal metabolism and retarded growth, *Arch. Biochem.*, 2, 481, 1943.
57. Sacher, G. A., Life table modification and life prolongation, in *Handbook of the Biology of Aging,* Finch, C. E. and Hayflick, L., Eds., Van Nostrand Reinhold, New York, 1977, 582.
58. Ross, M. H., Aging, nutrition and hepatic enzyme activity patterns in the rat, *J. Nutr.*, 97 (Suppl. 1), 563, 1969.
59. Rubner, M., *Das Problem der Lebensdauer und seine Beziehungen zu Wachstum und Ernahrung,* R. Oldenbourg, Munchen, 1908.
60. Masoro, E. J., Yu, B. P., and Bertrand, H. A., Action of food restriction in delaying the aging process, *Proc. Natl. Acad. Sci. U.S.A.*, 79, 4239, 1982.
61. Holehan, A. M., *The Effect of Ageing and Dietary Restriction upon Reproduction in the Female CFY Sprague-Dawley Rat,* Ph.D. thesis, 1984, University of Hull, Hull, UK.
62. Mohan, P. F. and Narasinga Rao, B. S., Adaptation to underfeeding in growing rats. Effects of energy restriction at two dietary protein levels on growth, feed efficiency, basal metabolism and body composition, *J. Nutr.*, 113, 79, 1983.
63. McCarter, R., Masoro, E. J., and Yu, B. P., Does food restriction retard aging by reducing the metabolic rate?, *Am. J. Physiol.*, 248, E488, 1985.
64. McCarter, R. J., Transient reduction of metabolic rate by food restriction, *Am. J. Physiol.*, 257, E175, 1989.
65. McCarter, R. J. and Palmer, J., Energy metabolism and aging: a lifelong study of Fischer 344 rats, *Am. J. Physiol.*, 263(3 Pt 1), E448, 1992.
66. Leto, S., Kokkonen, G. C., and Barrows, C. H., Dietary protein, life-span and physiological variables in female mice, *J. Gerontol.*, 32, 149, 1976.
67. Weindruch, R. H., Kristie, J. A., Cheney, K. E., and Walford, R. L., Influence of controlled dietary restriction on immunologic function and aging, *Fed. Proc.*, 38, 2007, 1979.
68. Koizumi, A., Tsukada, M., Wada, Y., Masuda, H., and Weindruch, R., Mitotic activity in mice is suppressed by energy restriction-induced torpor, *J. Nutr.*, 122(7), 1446, 1992.
69. Volicer, L., West, C., and Greene, L., Effect of dietary restriction and stress on body temperature in rats, *J. Gerontol.*, 39, 178, 1984.
70. Weindruch, R. L., Cheung, M. K., Verity, M. A., and Walford, R. L., Modification of mitochondrial respiration by aging and dietary restriction, *Mech. Ageing Dev.*, 12, 372, 1980.
71. Stoltzner, G., Effects of life-long dietary protein restriction on mortality, growth, organ weights, blood counts, liver aldolase and kidney catalase in BALB/c mice, *Growth,* 41, 337, 1977.
72. Barrows, C. H. and Kokkonen, G. C., Dietary restriction and life extension — Biological mechanisms, in *Nutritional Approaches to Aging Research,* Moment, G. B., Ed., CRC Press, Boca Raton, FL, 1982, 219.
73. Yu, B. P., Wong, G., Lee, H.-C., Bertrand, H. A., and Masoro, E. J., Age changes in hepatic metabolic characteristics and their modulation by dietary manipulation, *Mech. Ageing Dev.*, 24, 67, 1984.
74. Satyanarayana, U. and Narasinga Rao, B. P., Effect of diet restriction on some key enzymes of tryptophan NAD pathway in rats, *J. Nutr.*, 107, 2213, 1977.
75. Solomon, C., Tuchweber, B., Srivastava, U., and Nadeau, M., Liver lysosomal enzymes in rats during long-term dietary restriction. I. Changes during the developmental period of life, *Mech. Ageing Dev.*, 24, 9, 1984.
76. Horbach, G. J., Venkatraman, J. T., and Fernandes, G., Food restriction prevents the loss of isosafrole inducible cytochrome P-450 mRNA and enzyme levels in aging rats, *Biochem. Int.*, 20(4), 725, 1990.
77. De, A. K., Chipalkatti, S., and Aiyar, A. S., Some biochemical parameters of ageing in relation to dietary protein, *Mech. Ageing Dev.*, 21, 37, 1983.
78. Watson, R. R., Rister, M., and Baehner, R. L., Superoxide dismutase activity in polymorphonuclear leucocytes and alveolar macrophages of protein malnourished rats and guinea pigs, *J. Nutr.*, 106, 1801, 1976.
79. Enesco, H. E. and Kruk, P., Dietary restriction reduces fluorescent age pigment accumulation in mice, *Exp. Gerontol.*, 16, 357, 1981.
80. Laganiere, S. and Yu, B. P., Anti-lipoperoxidation action of food restriction, *Biochem. Biophys. Res. Commun.*, 145, 1185, 1987.
81. Yu, B. P., Suescun, E. A., and Yang, S. Y., Effect of age-related lipid peroxidation on membrane fluidity and phospholipase-A2 — modulation by dietary restriction, *Mech. Ageing Dev.*, 65(1), 17, 1992.
82. Pieri, C., Moroni, F., Falasca, F., Marcheselli, F., and Recchioni, R., Diet restriction decreases the membrane microviscosity of cerebellar membranes of old female Wistar rats, *Boll. Soc. Ital. Biol. Sper.*, 66(10), 915, 1990.
83. Pieri, C., Food restriction slows down age-related changes in cell membrane parameters, *Ann. N.Y. Acad. Sci.*, 621, 353, 1991.
84. Richardson, A. and Birchenall-Sparks, M. C., Age-related changes in protein synthesis, in *Review of Biological Research in Aging,* Rothstein, M., Eds., Alan R. Liss, New York, 1983, 255.
85. Richardson, A., Roberts, M. S., and Birchenall-Sparks, M. C., The possible role of protein synthesis in the aging process, in *Comparative Pathobiology of Major Age-Related Diseases: Current Status and Research Frontiers,* Alan R. Liss, New York, 1984, 47.
86. Richardson, A., The effect of age and nutrition on protein synthesis by cells and tissues from mammals, in *Handbook of Nutrition in the Aged,* Watson, R. R., Ed., CRC Press, Boca Raton, FL, 1985, 31.
87. Birchenall-Sparks, M. C., Roberts, M. S., Staecker, J., Hardwick, J. P., and Richardson, A., Effect of dietary restriction on liver protein synthesis, *J. Nutr.*, 115, 110, 1985.
88. Ward, W. F., Enhancement by food restriction of liver protein synthesis in the aging Fischer 344 rat, *J. Gerontol.*, 43(2), B50, 1988.
89. Ward, W. F., Food restriction enhances the proteolytic capacity of the aging liver, *J. Gerontol.*, 43, B121, 1988.
90. Lewis, S. E. M., Goldspink, D. F., Phillips, J. G., Merry, B. J., and Holehan, A. M., The effects of ageing and chronic dietary restriction on whole body growth and protein turnover in the rat, *Exp. Gerontol.*, 20, 253, 1985.
91. Merry, B. J., Lewis, S. E. M., and Goldspink, D. F., The influence of age and chronic restricted feeding on protein synthesis in the small intestine of the rat, *Exp. Gerontol.*, 27(2), 191, 1992.
92. Merry, B. J., Goldspink, D. F., and Lewis, S. E. M., The effects of age and chronic restricted feeding on protein synthesis and growth of the large intestine of the rat, *Comp. Biochem. Physiol.*, 98A(3/4), 559, 1991.

93. Goldspink, D. F., Lewis, S. E. M., and Merry, B. J., The effects of ageing and chronic dietary intervention on protein turnover and the growth of ventricular muscle in the rat heart, *Cardiovasc. Res.*, XX, 672, 1986.
94. Goldspink, D. F., El Haj, A. J., Lewis, S. E. M., Merry, B. J., and Holehan, A. M., The influence of chronic dietary intervention on protein turnover and growth of the diaphragm and extensor digitorum longus muscles of the rat, *Exp. Gerontol.*, 22, 67, 1987.
95. El Haj, A. J., Lewis, S. E. M., Goldspink, D. F., Merry, B. J., and Holehan, A. M., The effect of chronic and acute dietary restriction on the growth and protein turnover of fast and slow types of skeletal muscle, *Comp. Biochem. Physiol.*, 85A, 281, 1986.
96. Mathews, R. W., Oronsky, A., and Haschemeyer, A. E. V., Effect of thyroid hormone on polypeptide chain assembly kinetics in liver protein synthesis *in vivo, Biol. Chem.*, 248, 1329, 1973.
97. Merry, B. J. and Holehan, A. M., The endocrine response to dietary restriction in the rat, in *The Molecular Biology of Aging*, Vol. 35, Basic Life Sciences, Woodhead, A. D., Blackett, A. D., and Hollaender, A., Eds., Plenum Press, New York, 1985, 117.
98. Merry, B. J. and Holehan, A. M., Effect of age and restricted feeding on polypeptide chain assembly kinetics in liver protein synthesis *in vivo, Mech. Ageing Dev.*, 58, 139, 1991.
99. van Doorn, J., van der Heide, D., and Roelfsema, F., The influence of partial food deprivation on the quantity and source of triiodothyronine in several tissues of athyreotic thyroxine-maintained rats, *Endocrinology,* 15, 705, 1984.
100. Blazejowski, C. A. and Webster, G. C., Effect of age on peptide chain initiation and elongation in preparations from brain, liver, kidney and skeletal muscle of the C57BL/6J mouse, *Mech. Ageing Dev.*, 25, 323, 1984.
101. Webster, S. L. and Webster, G. C., Effect of age on the synthesis of individual cellular proteins, in Proc. XIIIth Int. Congr. Gerontol., New York, July 12–17, 1985, 354.
102. Shepherd, J. C. W., Walldorf, U., Hug, P., and Gehring, W. J., Fruit flies with additional expression of the elongation factor EF-1 live longer, *Proc. Natl. Acad. Sci. U.S.A.*, 86, 7520, 1989.
103. Chvapil, M. and Hruza, Z., The influence of aging and undernutrition on chemical contractility and relaxation of collagen fibres in rats, *Gerontologia,* 3, 241, 1959.
104. Giles, J. S. and Everitt, A. V., The role of the thyroid and of food intake in the aging of collagen fibres, *Gerontologia,* 13, 65, 1967.
105. Deyl, Z., Juricova, M., Rosmus, J., and Adam, M., The effect of food deprivation on collagen accumulation, *Exp. Gerontol.*, 6, 383, 1971.
106. Leto, S., Kokkonen, G. C., and Barrows, C. H., Dietary protein, life-span and biochemical variables in female mice, *J. Gerontol.*, 31, 144, 1976.
107. Bloemendal, H., *Molecular and Cellular Biology of the Eye Lens,* John Wiley & Sons, New York, 1981.
108. Leveille, P. J., Weindruch, R., Walford, R. L., Bok, D., and Horwitz, J., Dietary restriction retards age-related loss of gamma crystallins in the mouse lens, *Science,* 224, 1247, 1984.
109. O'Steen, W. K. and Landfield, P. W., Dietary restriction does not alter retinal aging in the Fischer 344 rat, *Neurobiol. Aging,* 12(5), 455, 1991.
110. Herlihy, J. T. and Yu, B. P., Dietary manipulation of age-related decline in vascular smooth muscle function, *Am. J. Physiol.*, 238, H652, 1980.
111. Yu, B. P., Masoro, E. J., Murata, I., Bertrand, H. A., and Lynd, F. T., Life span study of SPF Fischer 344 male rats fed ad libitum or restricted diets: longevity, growth, lean body mass and disease, *J. Gerontol.*, 37, 130, 1982.
112. McCarter, R., Yu, B. P., and Radicke, D., Effects of caloric restriction on contraction of skeletal muscle, *Nutr. Rep. Int.*, 17, 339, 1978.
113. McCay, C. M., Maynard, L. A., Sperling, G., and Barnes, L. L., Retarded growth, lifespan, ultimate body size and age changes in the albino rat after feeding diets restricted in calories, *J. Nutr.*, 18, 1, 1939.
114. Merry, B. J. and Holehan, A. M., Onset of puberty and duration of fertility in rats fed a restricted diet, *J. Reprod. Fert.*, 57, 253, 1979.
115. Merry, B. J. and Holehan, A. M., Serum profiles of LH, FSH, testosterone and 5α-DHT from 21 to 1000 days in *ad libitum* fed and dietary restricted long-lived rats, *Exp. Gerontol.*, 16, 431, 1981.
116. Holehan, A. M. and Merry, B. J., The control of puberty in the dietary restricted rat, *Mech. Ageing Dev.*, 32, 179, 1985.
117. Koizumi, A., Wada, Y., Tsukada, M., Kamiyama, S., and Weindruch, R., Effects of energy restriction on mouse mammary tumor virus mRNA levels in mammary glands and uterus and on uterine endometrial hyperplasia and pituitary histology in C3H/SHN F1 mice, *J. Nutr.*, 120(11), 1401, 1990.
118. Nelson, J. F. and Felicio, L. S., Reproductive aging in the female: an etiological perspective, *Rev. Biol. Res. Aging,* 2, 251, 1985.
119. Lu, J. K. H., Changes in ovarian function and gonadotropin and prolactin secretion in aging female rats, in *Neuroendocrinology of Aging,* Meites, J., Ed., Plenum Press, New York, 1983, 103.
120. Holehan, A. M. and Merry, B. J., Modification of the oestrous cycle hormonal profile by dietary restriction, *Mech. Ageing Dev.*, 32, 63, 1985.
121. Holehan, A. M. and Merry, B. J., Follicular steroidogenesis in diet restricted rats with delayed reproductive ageing, in XIIIth Int. Congr. Gerontol., New York, July 12–17, 1985, 135.
122. Merry, B. J. and Holehan, A. M., The effect of dietary restriction on the endocrine control of reproduction, in *Biological Effects of Dietary Restriction,* 1st ed., Fishbein, L., Ed., Springer-Verlag, Berlin, 1991, 140.
123. Osborne, T. B. and Mendel, L. B., The resumption of growth after long continued failure to grow, *J. Biol. Chem.*, 23, 439, 1915.
124. Ball, Z. B., Barnes, R. H., and Visscher, M. B., The effects of dietary caloric restriction on maturity and senescence with particular reference to fertility and longevity, *Am. J. Physiol.*, 150, 511, 1947.
125. Visscher, M. B., King, J. T., and Lee, Y. C. P., Further studies on influence of age and diet upon reproductive senescence in strain A female mice, *Am. J. Physiol.*, 170, 72, 1952.
126. Segall, P. E. and Timiras, P. S., Age-related changes in thermoregulatory capacity of tryptophan deficient rats, *Fed. Proc.*, 34, 1975.
127. Segall, P. E., Timiras, P. S., and Walton, J. R., Low tryptophan diets delay reproductive aging, *Mech. Ageing Dev.*, 23, 245, 1983.
128. Everitt, A. V., The nature and measurement of aging, in *Hypothalamus, Pituitary and Aging,* Everitt, A. V. and Burgess, J. A., Eds., Charles C Thomas, Springfield, IL, 1976, 5.
129. Everitt, A. V., Seedsman, N. J., and Jones, F., The effects of hypophysectomy and continuous food restriction, begun at ages 70 and 400 days, on collagen aging, proteinuria, incidence of pathology and longevity in the male rat, *Mech. Ageing Dev.*, 12, 161, 1980.
130. Merry, B. J. and Holehan, A. M., Effects of caloric restriction on endocrine function during aging in rats, in *The Potential for Nutritional Modulation of Aging Processes,* Vol. 1, Ingram, D., Baker, G. T., and Shock, N. W., Eds., Food & Nutrition Press, Inc., Trumbull, CT, 1991, 157.

131. Sorrentino, S., Reitner, R. J., and Schalch, D. S., Interactions of the pineal gland, blinding and underfeeding on reproductive organ size and radioimmunoassayable growth hormone, *Neuroendocrinology,* 7, 105, 1971.
132. Quigley, K., Goya, R., Nachreiner, R., and Meites, J., Effects of underfeeding and refeeding on GH and thyroid hormone secretion in young, middle-aged, and old rats, *Exp. Gerontol.,* 25(5), 447, 1990.
133. Koizumi, A., Masuda, H., Wada, Y., Tsukada, M., Kawamura, K., Kamiyama, S., and Walford, R. L., Caloric restriction perturbs the pituitary-ovarian axis and inhibits mouse mammary tumor virus production in a high-spontaneous-mammary-tumor-incidence mouse strain (C3H/SHN), *Mech. Ageing Dev.,* 49(2), 93, 1989.
134. Tannenbaum, G. S., Epelbaum, J., Colle, E., Brazeau, P., and Martin, J. B., Antiserum to somatostatin reverses starvation induced inhibition of growth hormone but not insulin secretion, *Endocrinology,* 102, 1909, 1978.
135. Kalu, D. N., Hardin, R. R., Cockerham, R., Yu, B. P., Norling, B. K., and Egan, J. W., Lifelong food restriction prevents senile osteopenia and hyperparathyroidism in F344 rats, *Mech. Ageing Dev.,* 26, 103, 1984.
136. Kalu, D. N., Cockerham, R., Yu, B. P., and Roos, B., Lifelong dietary modulation of calcitonin levels in rats, *Endocrinology,* 113, 2010, 1983.
137. Herlihy, J. T., Stacy, C., and Bertrand, H. A., Long-term food restriction depresses serum thyroid hormone concentrations in the rat, *Mech. Ageing Dev.,* 53(1), 9, 1990.
138. Chowers, I., Einat, R., and Feldman, S., Effects of starvation on levels of corticotrophin releasing factor, corticotrophin and plasma corticosterone in rats, *Acta Endocrinol. (Copenhagen),* 61, 687, 1969.
139. Gallo, P. V. and Weinberg, J., Corticosterone rhythmicity in the rat: interactive effects of dietary restriction and schedule of feeding, *J. Nutr.,* 111, 208, 1981.
140. Sabatino, F., Masoro, E. J., McMahan, C. A., and Kuhn, R. W., Assessment of the role of the glucocorticoid system in aging processes and in the action of food restriction, *J. Gerontol.,* 46(5), B171, 1991.
141. Kingsley, T. R., Nekvasil, N. P., and Snyder, D. L., The influence of dietary restriction, germ-free status, and aging on adrenal catecholamines in Lobund-Wistar rats, *J. Gerontol.,* 46(4), B135, 1991.
142. Rosenfeld, M. G., Amora, S. G., Birnberg, N. G., Mermod, J.-J., Murdoch, G. H., and Evans, R. M., Calcitonin, prolactin and growth hormone gene expression as model systems for the characterization of neuroendocrine regulation, *Recent Prog. Horm. Res.,* 39, 305, 1983.
143. Reaven, E. and Reaven, G. M., Structure and function changes in the endocrine pancreas of aging rats with reference to the modulating effects of exercise and caloric restriction, *J. Clin. Invest.,* 68, 75, 1981.
144. Masoro, E. J., McCarter, R. J., Katz, M. S., and McMahan, C. A., Dietary restriction alters characteristics of glucose fuel use, *J. Gerontol.,* 47(6), B202, 1992.
145. Chung, M. H., Kasai, H., Nishimura, S., and Yu, B. P., Protection of DNA damage by dietary restriction, *Free Rad. Biol. Med.,* 12(6), 523, 1992.
146. Masoro, E. J., Katz, M. S., and McMahan, C. A., Evidence for the glycation hypothesis of aging from the food-restricted rodent model, *J. Gerontol.,* 44(1), B20, 1989.
147. Spindler, S. R., Grizzle, J. M., Walford, R. L., and Mote, P. L., Aging and restriction of dietary calories increases insulin receptor mRNA, and aging increases glucocorticoid receptor mRNA in the liver of female C3B10RF1 mice, *J. Gerontol.,* 46(6), B233, 1991.
148. Spindler, S. R., Crew, M. D., Mote, P. L., Grizzle, J. M., and Walford, R. L., Dietary energy restriction in mice reduces hepatic expression of glucose-regulated protein 78 (BiP) and 94 mRNA, *J. Nutr.,* 120(11), 1412, 1990.
149. Bertrand, H. A., Lynd, F. T., Masoro, E. J., and Yu, B. P., Changes in adipose mass and cellularity through the adult life of rats fed *ad libitum* or a life-prolonging restricted diet, *J. Gerontol.,* 35, 827, 1980.
150. Volicer, L., West, C. D., Chase, A. R., and Greene, L., Beta-adrenergic receptor sensitivity in cultured vascular smooth muscle cells: effect of age and dietary restriction, *Mech. Ageing Dev.,* 21, 283, 1983.
151. Stokkan, K. A., Reiter, R. J., Nonaka, K. O., Lerchl, A., Yu, B. P., and Vaughan, M. K., Food restriction retards aging of the pineal gland, *Brain Res.,* 545(1–2), 66, 1991.
152. Levin, P., Haji, M., Joseph, J. A., and Roth, G. S., Effect of aging on prolactin regulation of rat striatal dopamine receptor concentrations, *Life Sci.,* 32, 1743, 1983.
153. Roth, G. S., Ingram, D. K., and Joseph, J. A., Delayed loss of striatal dopamine receptors during aging of dietary restricted rats, *Brain Res.,* 300, 27, 1984.
154. Yeung, J. M. and Friedman, E., Effect of aging and diet restriction on monoamines and amino acids in cerebral cortex of Fischer-344 rats, *Growth Dev. Aging,* 55(4), 275, 1991.
155. Joseph, J. A., Whitaker, J., Ingram, D. K., and Roth, G. S., Dietary restriction retards age-related decrements in stereotypy following intrastriatal injection of DA-active agents, *Soc. Neurosci. Abstr.,* 9, 840, 1982.
156. London, E. D., Ingram, D. K., and Waller, S. B., Effects of dietary restriction on neurotransmitter synthetic enzymes and binding in aging brain, *Soc. Neurosci. Abstr.,* 9, 98, 1983.
157. Cutler, R. G., Superoxide dismutase, longevity and specific metabolic rate, *Gerontology,* 29, 113, 1983.
158. Makrides, S. C., Protein synthesis and degradation during aging and senescence, *Biol. Rev.,* 58, 343, 1983.
159. Mays, P. K., Macnulty, R. J., and Laurent, G. J., Age-related changes in rates of protein synthesis and degradation in rat tissues, *Mech. Ageing Dev.,* 59(3), 229, 1991.
160. Jacobus, S. and Gershon, D., Age-related changes in inducible mouse liver enzymes: ornithine decarboxylase and tyrosine aminotransferase, *Mech. Ageing Dev.,* 12, 311, 1980.
161. Reznick, A. Z., Lavie, L., Gershon, H. E., and Gershon, D., Age-associated accumulation of altered FDP aldolase B in mice. Conditions of detection and determination of aldolase half life in young and old animals, *FEBS Lett.,* 128, 221, 1981.
162. Ward, W. and Richardson, A., Effect of age on liver protein synthesis and degradation, *Hepatology,* 14(5), 935, 1991.
163. Wulf, J. H. and Cutler, R. G., Altered protein hypothesis of mammalian aging processes. I. Thermal stability of glucose-6-phosphate dehydrogenase in C57BL/6J mouse tissue, *Exp. Gerontol.,* 10, 101, 1975.
164. Ishigami, A. and Goto, S., Effect of dietary restriction on the degradation of proteins in senescent mouse liver parenchymal cells in culture, *Arch. Biochem. Biophys.,* 283(2), 362, 1990.
165. Harman, D., Aging: a theory based on free radical and radiation chemistry, *J. Gerontol.,* 11, 298, 1956.
166. Harman, D., Free radical theory of aging: consequences of mitochondrial aging, *Age,* 6, 86, 1983.
167. Laganiere, S. and Yu, B. P., Effect of chronic food restriction in aging rats. II. Liver cytosolic antioxidants and related enzymes, *Mech. Ageing Dev.,* 48, 221, 1989.
168. Laganiere, S. and Yu, B. P., Effect of chronic food restriction in aging rats. I. Liver subcellular membranes, *Mech. Ageing Dev.,* 48, 207, 1989.

169. Koizumi, A., Weindruch, R., and Walford, R. L., Influences of dietary restriction and age on liver enzyme activities and lipid peroxidation in mice, *J. Nutr.*, 117, 361, 1987.
170. Albrecht, R., Pelissier, M. A., Atteba, S., and Smaili, M., Dietary restriction decreases thiobarbituric acid-reactive substances generation in the small intestine and in the liver of young rats, *Toxicol. Lett.*, 63(1), 91, 1992.
171. Pieri, C., Falasca, M., Moroni, F., Recchioni, R., Marcheselli, F., Ioppolo, C., and Marmocchi, F., Antioxidant enzymes in erythrocytes from old and diet restricted old rats, *Boll. Soc. Ital. Biol. Sper.*, 66(10), 909, 1990.
172. Rao, G., Xia, E., Nadakavukaren, M. J., and Richardson, A., Effect of dietary restriction on the age-dependent changes in the expression of antioxidant enzymes in rat liver, *J. Nutr,* 120, 602, 1990.
173. Bhuyan, K. C., Bhuyan, D. K., and Podos, S. M., Lipid peroxidation in cataract of the human, *Life Sci.*, 38, 1463, 1986.
174. Taylor, A., Zuliani, A. M., and Hopkins, R. E., Moderate caloric restriction delays cataract formation in the Emory mouse, *FASEB J.*, 3, 1741, 1989.
175. Lee, D. W. and Yu, B. P., Modulation of free radicals and superoxide dismutases by age and dietary restriction, *Aging Milano,* 2(4), 357, 1990.
176. Ames, B. N. and Shigenaga, M. K., Oxidants are a major contributor to aging, *Ann. N.Y. Acad. Sci.*, 663, 85, 1992.
177. Youngman, L. D., Park, J. Y. K., and Ames, B. N., Protein oxidation associated with aging is reduced by dietary restriction of protein or calories, *Proc. Natl. Acad. Sci. U.S.A.*, 89(19), 9112, 1992.
178. Djuric, Z., Lu, M. H., Lewis, S. M., Luongo, D. A., Chen, X. W., Heilbrun, L. K., Reading, B. A., Duffy, P. H., and Hart, R. W., Oxidative DNA damage levels in rats fed low-fat, high-fat, or calorie-restricted diets, *Toxicol. Appl. Pharmacol.*, 115(2), 156, 1992.
179. Harris, S. B., Weindruch, R., Smith, G. S., Mickey, M. R., and Walford, R. L., Dietary restriction alone and in combination with oral ethoxyquin/2-mercaptoethylamine in mice, *J. Gerontol.*, 45(5), B141, 1990.
180. Richardson, A., Semsei, I., Rutherford, M. S., and Butler, J. A., Effect of dietary restriction on the expression of specific genes, *Fed. Proc.*, 46, 568, 1987.
181. Semsei, I., Rao, G., and Richardson, A., Changes in the expression of superoxide dismutase and catalase as a function of age and dietary restriction, *Biochem. Biophys. Res. Commun.*, 164(2), 620, 1989.
182. Richardson, A., Butler, J. A., Rutherford, M. S., Semsei, I., Gu, M.-Z., Fernandes, G., and Chiang, W.-H., Effect of age and dietary restriction on the expression of α_{2u}-globulin, *J. Biol. Chem.*, 262, 12821, 1987.
183. Chatterjee, B., Fernandes, G., Yu, B. P., Song, C., Kim, J. M., Demyan, W., and Roy, A. K., Calorie restriction delays age-dependent loss in androgen responsiveness of the rat liver, *FASEB J.*, 3(2), 169, 1989.
184. Song, C. S., Rao, T. R., Demyan, W. F., Mancini, M. A., Chatterjee, B., and Roy, A. K., Androgen receptor messenger ribonucleic acid (mRNA) in the rat liver: changes in mRNA levels during maturation, aging and calorie restriction, *Endocrinology,* 128, 349, 1991.
185. Fu, C. S., Harris, S. B., Wilhelmi, P., and Walford, R. L., Lack of effect of age and dietary restriction on DNA single-stranded breaks in brain, liver, and kidney of ($C_3H \times C_{57}BL/10)F_1$ mice, *J. Gerontol.*, 46(2), B78, 1991.
186. Weraarchakul, N., Strong, R., Wood, W. G., and Richardson, A., The effect of aging and dietary restriction on DNA repair, *Exp. Cell Res.*, 181(1), 197, 1989.
187. Srivastava, V. K. and Busbee, D. L., Decreased fidelity of DNA polymerases and decreased DNA excision repair in aging mice: effects of caloric restriction, *Biochem. Biophys. Res. Commun.*, 182(2), 712, 1992.
188. Mulinos, M. G. and Pomerantz, L., Pseudo-hypophysectomy, a condition resembling hypophysectomy produced by malnutrition, *J. Nutr.*, 19, 493, 1940.
189. Harrison, D. E., Archer, J. R., and Astle, C. M., The effect of hypophysectomy on thymic aging in mice, *J. Immunol.*, 129, 1982.
190. Segall, P. E. and Timiras, P. S., Pathophysiologic findings after chronic tryptophan deficiency in rats: a model for delayed growth and aging, *Mech. Ageing Dev.*, 5, 109, 1976.
191. Fernstrom, J. D. and Wurtman, R. J., Effect of chronic corn consumption on serotonin content of rat brain, *Nature,* 234, 62, 1971.
192. Segall, P. E., Interrelations of dietary and hormonal effects in aging, *Mech. Ageing Dev.*, 9, 515, 1979.
193. Walford, R. L., Introduction to special issue. Immunology and aging, *Gerontologia,* 18, 243, 1972.
194. Pierpaoli, W. and Sorkin, E., Relationship between thymus and hypophysis, *Nature,* 215, 834, 1967.
195. Burnett, F. M., *Intrinsic Mutagenesis: A Genetic Approach to Aging,* MTP, John Wiley & Sons, New York, 1974.
196. Roth, G. S., Ingram, D. K., and Cutler, R. G., Primate models for dietary restriction research, in *Biological Effects of Dietary Restriction,* Fishbein, L., Eds., Springer-Verlag, Berlin, 1991, 305.
197. Lane, M. A., Ingram, D. K., Cutler, R. G., Knapka, J. J., Barnard, D. E., and Roth, G. S., Dietary restriction in nonhuman primates: progress report on the NIA study, *Ann. N.Y. Acad. Sci.*, 673, 36, 1992.
198. Kemnitz, J. W., Weindruch, R., Roecker, E. B., Crawford, K., Kaufman, P. L., and Ershler, W. B., Dietary restriction of adult male rhesus monkeys: design, methodology, and preliminary findings from the first year of study, *J. Gerontol.*, 48(1), B17, 1993.
199. Weg, R. B., *Nutrition and the Later Years,* University of Southern California Press, Los Angeles, 1979,
200. Takata, H., Ishi, T., Suzuki, M., Seiguchi, S., and Iri, H., Influence of major histocompatibility complex region genes on human longevity among Okinawan Japanese centenarians and nonagenarians, *Lancet,* 2, 824, 1987.
201. Walford, R. L., Harris, S. B., and Gunion, M. W., The calorically restricted low-fat nutrient-dense diet in Biosphere 2 significantly lowers blood glucose, total leukocyte count, cholesterol, and blood pressure in humans, *Proc. Natl. Acad. Sci. U.S.A.*, 89(23), 11533, 1992.

IV
PREVENTION AND
REHABILIATION

25 AN AGENDA FOR HEALTHFUL AGING

Paola S. Timiras

1 BIOMEDICAL AND SOCIOECONOMIC INTERRELATIONS IN AGING

The major purpose of this book has been to present an overall, systematic view of physiologic aging, not only in humans but also, for comparison and elucidation, in experimental animals and cultured cell models. References to medical illness are intended primarily as a recognition of the increased incidence of disease with aging, and a few diseases have been selected for brief discussion as examples of how pathology influences physiologic adjustments with advancing age.

Limitations of time and space and lack of expertise by this writer have prevented incursions into areas not specifically biomedically oriented but nevertheless extremely important for the promotion and maintenance of the quality of life in the elderly. Thus, in drawing the physiologic profile of an aged individual, the *contribution of nonbiologic factors to good health status has been deliberately omitted,* with little attention for the contribution to physiologic decline of social losses, preexistent personality, and sociocultural variables. *Effective socioeconomic support networks and public health community programs do appear to play as important a role in determining the quality of life at all ages as medical care does.*

Biomedical progress in the U.S. and in all Western world countries has allowed us to live many years longer than our parents and ancestors. Preventive and rehabilitative strategies to maintain independence and active life-style contribute considerably to preserving some of the key functions (e.g., intellect, motility, sensory competence) in old age but fall short of assuring a good quality of life. We are now at the cutting edge of rapid biotechnologic advances and we should exploit, in full, our knowledge of these new techniques to treat the diseases of old age and to prolong and improve the lifespan. Some of the new directions of research, grouped under so-called life extension sciences, are briefly presented below.

Maintenance of the quality of life with old age is not merely a matter of improving physical fitness, but social, cultural, economic, and emotional factors must also be considered. We know that individuals, including old individuals, do not always see health as the most important issue in their lives. So far the increase in life expectancy witnessed in this century has been primarily attributed as much to improvements in hygiene, better nutrition, and the rising standard of living as to the successful eradication of a number of infectious diseases. Further breakthroughs in gerontology/geriatrics should be based on the effective articulation of biological measures with socioeconomic policy.

2 WELLNESS AND HEALTH PROMOTION

At all ages, the condition of well-being or wellness (a term considerably abused by faddists and alternative-health worshippers, but retained here for its original and simplest meaning) depends on good physical and mental health in a favorable environment. Senescence, in contrast, is characterized by declining physiological competence and increasing incidence and severity of disease, under the everyday stresses of life. Therefore, maintenance or restoration of wellness, with advancing age, depends both on strengthening physiological responses and eliminating disease.[1]

With considerable effort, some success has been achieved in the prevention and elimination of some diseases, particularly acute infections. However, some of these diseases, like tuberculosis, believed to be eradicated have now reappeared and have spread in vulnerable groups such as the elderly (Chapter 18). The increasing incidence of chronic degenerative diseases (e.g., Alzheimer's disease, Chapters 8 and 9, atherosclerosis, Chapters 16 and 17) in the aged population underlines the need for continuing research toward elimination of disease.

Much less, indeed very little, has been done to strengthen physiological competence and prevent, delay, or repair its decline with aging. This is an area that requires concerted effort as physiological integrity is the nexus not only for well-being but also for disease prevention. For an effective strengthening of physiologic competence, however, the molecular and cellular mechanisms of the aging process and their role in the regulation of systemic function must be understood. Therefore, maintenance of good health in the elderly requires a multifaceted approach.

2.1 Toward a New Image of Aging

As already discussed in Chapter 2, there are a number of erroneous myths and age-related stereotypes that pervade our culture. These myths obscure the facts about the later years of life and undermine the truth of a possible vigorous and healthy senescence: "At eighty I believe I am a far more cheerful person than I was at twenty or thirty. I most definitely would not want to be a teenager again. Youth may be glorious, but it is also painful to endure. Moreover, what is called youth is not youth; it is rather something like premature old age."[2] It is essential that these myths be disavowed for the real picture of physiologic aging to emerge.[1] A brief list will identify the most common and insidious of these myths:

- Old age starts at 65 years (although Picasso says, "One starts to get young at the age of sixty, and then it's too late").[3]
- Old age is a disease.

0-8493-8979-8/94/$0.00+$.50

TABLE 25-1

Factors Affecting the Elderly and Amenable to Multifactorial Interventions

Conditions	Underlying cause	Control
	Nutrition	
Reduced calcium absorption; loss of lean body mass and bone mass; low levels of vitamins K and D; muscle weakness and tetany	Inadequate caloric intake; little variety; excess processed food; lack of nutrition knowledge; alcohol, drugs; low income; social status; illness; physical difficulty in daily activities	Improved economics and education leading to adequate diet; improved physical condition leading to increased daily activities and greater social interactions
Hypertension; obesity; increased insulin resistance; high cholesterol; low HDL	Excess salt; excessive food intake; high caloric food intake; inactivity	Hormonal therapy; antihypertensive drugs; anticholesterol drugs
	Exercise	
Reduced muscle strength and endurance; loss of lean body mass; increased fat	Inactivity due to lack of stimulation in the environment; retirement	Education for need to exercise; improved social contacts; improved safety
Poor neurologic control; reduced coordination, balance; mental impairment	Confinement due to fear of crime; lack of knowledge for need to exercise	Treatment of underlying medical conditions
Reduced joint mobility, bone mass	Joint and bone discomfort; cardiovascular disease; respiratory disease	
	Drugs, Smoking, and Alcohol	
Polypharmacy; improper therapeutic drug dosage; drug addiction	Lack of knowledge of geriatric pharmacology; inaccurate diagnosis; noncompliance; unawareness of side effects; cost of drugs; lack of health insurance	Adequate diagnosis and treatment; surveillance of appropriate intake and side effects of drugs; economic support; rehabilitation from addiction
Respiratory insufficiency; loss of task capacity	Mass media; sex image; peer pressure	Stop smoking; support group
Impaired hepatic function; neurologic symptoms	Addiction; sex image; social condition; heredity	Stop drinking; peer and family support

Adapted from Minkler, M. and Pasick, R. J., *Wellness and Health Promotion for the Elderly*, Dychtwald, K., Ed., Aspen Systems Corp., Rockville, MD, 1986, 31. With permission.

- Old age always brings mental impairment.
- The elderly represent an homogenous group characterized by equal physiologic decline.
- Old people are poor.
- Old people are powerless.
- Nothing can be done to modify aging; future elderly will have the same problems as current elderly.
- The U.S. has always been a youth-focused culture.

These statements are incorrect and must be dispelled by intensive educational programs initiated at an early age, with emphasis on the following:

- Good health is important during early ages as a forerunner of good health in old age.
- Considerable advances have been achieved in prolongation of life expectancy and in prevention and treatment of many diseases, and similar advances may be expected in the future.
- Quality of life in old age depends on several complex factors, and its improvement must be reached by a concerted effort of biomedical as well as socioeconomic and educational measures.

Concepts of wellness are illustrated in Table 25-1, presenting selected common conditions, which affect the elderly and which depend on physical, emotional, and social circumstances, that can be approached by a multipronged attack.[4] Nutrition and exercise have been selected, as they are physiological parameters easily modified (Chapters 21 and 24); therapeutic drugs, as their use increases with aging (Chapter 23); and smoking and alcohol, as they are two commonly met abuses. It emerges clearly that there is no single approach to the promotion of wellness, but, rather, that a combination of medical, social, economic, and psychologic interventions is necessary.

Among the factors that can be modified to positively improve the quality of life are medical care, income, nutrition and housing, social integration, and education. The latter is worthy of greater emphasis than it has heretofore received. Inasmuch as aging must be viewed as a continuation of previous experiences, knowledge of physical make-up and environment is essential for programming life-styles capable of generating an ongoing satisfactory quality of life. Several programs are currently offered and more are planned to educate the elderly and increase their awareness of aging-related diseases and means of prevention and treatment. Useful as these educational programs may be, they belie the need for a broader view, i.e., beginning education about aging during the school years; it is at an early age that initiation of regimens for good health will yield the greatest benefit for better quality of future life.

2.2 Benefits of Continuing Learning

Another important role for education is to provide the skills to enable the individual to choose more effectively his/her position in society and to take advantage of opportunities and options. Indeed, "control of one's own destiny" rather than economic status may determine the quality of health and length of lifespan. It is possible that "formal" education is a key to improving control of one's destiny. While this is not a startling or original concept, formal education is rarely a top priority of any programs organized for the benefit of the elderly. A number of observations in support of this role of education in disease prevention and life improvement have been reviewed by Syme.[5] Only one example will be presented here to underline that socioeconomic status as a risk factor must be interpreted in a broader context than the mere restrictive difference of poverty versus nonpoverty. In this study, British civil servants were studied for the incidence of coronary heart disease and factors that might influence disease incidence. In all groups, after adjusting for the usual risk factors (serum cholesterol, smoking, obesity, etc.), the rate of coronary heart disease was three times higher in employees at the bottom than in those at the top of the civil service hierarchy.[6] Most interesting was the presence of a "gradient" among the different groups, with the high-level administrators having the lowest rate, followed by the professionals and executives with an intermediate rate, and the clerical workers having the highest rate. Clearly all subjects enjoyed a standard of living above poverty level, and therefore the differential incidence of the disease was to be ascribed to some still-unidentified aspect, other than income, of the social and work status. The concept that the lower the position of an individual in the social structure and the fewer the resources and training for options of type of work and style of life, the greater the incidence of disease must be supported by future research. If it proves to be of value, then we would have available, through strengthening education, an avenue for intervention that is more precise and understandable than simply suggesting that we intervene in "socioeconomic status".

According to a Japanese proverb *"Aging begins when we stop learning"*; indeed, the importance of continuing learning not only on the mental but also the overall physical well-being of the older person — as well as of persons of all ages — can never be overevaluated. Two recent retrospective and prospective surveys of elderly black and whites reported that "despite an overall decline in death rates in the United States between 1960 and 1989, poor and poorly educated people still die at higher rates than those with higher incomes or better education"[7] and that "among older black and white...the level of education has a substantially stronger relation to total life expectancy and active life expectancy than does race."[8]

The value of learning does not apply only to the young years when education results in the acquisition of humanistic, technical, or professional skills that will benefit the individual and the work force of the nation. The benefits extend to the entire lifespan. Hence, there are efforts in many countries to open the universities to all ages and provide learning facilities not only for training young students towards employment but also to old students for continuing learning and maintenance of good physical and mental health. Although, at present, very few universities in the U.S. have opened their doors to adult and old non–degree-seeking students, it is hoped that continuing studies demonstrating the importance of learning activity in reducing disease and prolonging active life expectancy will provide the necessary impetus (and finances) to make education a lifelong endeavor.

2.3 Health Programs for the Elderly

For the elderly, many programs have been and are being designed in the U.S. and most countries of the world, including the developing nations where the lifespan expectancy is rapidly rising and the percentage of elderly is increasing. Among the major programs capable of improving health, some strive to eliminate premature disease and to manage late degenerative disease — primarily the task of the geriatricians. Others focus on the improvement of funding, social policy, social security programs, establishment of federal and state health care insurance, and tax exemptions — primarily the task of policy makers. Still others emphasize physical and mental fitness by promoting programs in nutrition, exercise, personal safety, establishing senior citizen centers, centers for long-term care, and programs for stress management — primarily the task of the individual and the community.

A number of programs for the elderly concerned with health care, social policy, funding, senior centers, community-based health care, and education are continually being proposed, established, and tested. How effective they are and which health strategies are most likely to benefit the elderly remain to be clarified by vigorous and sound research. Compassion and desire to help the elderly are not sufficient qualifications to ensure success. Improvements in the rational care of the elderly and, incidentally, better cost-effectiveness will depend ultimately on improvement in governmental health policy and the concerted effort of educators, health care providers, community and family support, and the elderly themselves.

3 PHYSIOLOGIC APPROACH TO REHABILITATION

3.1 The "Rehabilitation Team"

Prevention of old-age disabilities improves the quality of life in old age, delays pathology, and prolongs the lifespan. Consequently, a number of preventive measures have been emphasized throughout this text. Key strategies include

- Improving life habits (diet exercise, continuing learning)
- Avoiding stress situations
- Providing appropriate medications

There are, however, situations when prevention is not sufficient or fails and disability and disease ensue. In this case, rehabilitation may play a crucial role in allowing the elderly to maintain independence and health. A brief listing of some rehabilitative measures is summarized here, representing a synopsis of a more comprehensive chapter included in the first edition of this text.[9]

Rehabilitation for the elderly, i.e., restoration to a former capacity, must take into consideration the heterogeneity of this age group:

- Some elderly continue to enjoy good health into old age. For these "successful agers", prevention (encompassing the strategies listed above) may be sufficient to ensure independence and health.
- Some elderly may be severely disabled and often present multisystem disease with medical, psychologic, and social problems. For these, rehabilitation is an important component of care and medical assistance to treat the disease and minimize the disabilities.

In view of the multisystem nature of the disease and disability of some of the elderly, efficient rehabilitation must be carried out by a "rehabilitation team" capable of assessing and restoring failing functions. This team should consist of individuals trained in various fields, not only medical but also social.

3.2 Functional Assessment

While medicine is traditionally concerned with disease, rehabilitation is concerned with the person as a whole. Inasmuch as disability represents a limitation of function, one of the roles of rehabilitation is to restore and/or maintain independence (Chapters 2, 3). To achieve this goal, the first step must assess the current physiologic competence/decrements of the individual, and this assessment must be carried out periodically, to pace the ongoing "dynamic process" of aging changes.

3.3 Levels of Rehabilitation

Rehabilitation programs are tailored to the physiologic or pathologic states taking into account the nature of the disability/disease:

- Its duration, whether acute or chronic
- Degree and severity of functional limitation(s)
- Medical and social facilities available at home, in the community, or hospital/nursing home

These factors will dictate the program to be followed with the necessary flexibility to match the progression of the aging process.

3.4 Therapeutic Rehabilitative Modalities

These modalities are complex and varied; they will just be listed here, and the reader is referred to several reviews for a complete presentation of the subject.[9-14]

1. *Physiologic effects of exercise training*. How beneficial physical exercise may be has already been underlined in previous chapters, particularly in relation to cardiovascular, skeletal neurologic, and mental rehabilitation. Possible benefits include
   - prolonged independence in the activities of day living due to increased muscular and cardiopulmonary fitness
   - improved general life-style
   - opportunity for social interaction with elevation of mood
   - lessened risk of certain medical disorders, including obesity, non–insulin-dependent diabetes, osteoporosis, coronary heart disease, hypertension, chronic obstructive heart disease (emphysema), and others
2. *Thermal therapy* is based on the use of various devices of heat conduction, convection, and radiation, as well as deep heat (short-wave diathermy). It offers a number of beneficial effects such as increased extensibility of collagen, increased blood flow, reduction of inflammatory processes, and reduction of pain.
3. *Electromagnetic therapy* is based on electrical stimulation to muscle, nerve, and bone, ultraviolet treatment of dermatologic lesions, and static magnetic field for pain management.
4. *Massage and manipulation* to improve muscle and joint motility. These procedures are generally delivered by a second party. They induce mechanical stimulation and improve blood flow.
5. *Use of devices, equipment, orthosis, and prosthesis*. These are employed to help in performing the activities of day living (e.g., wheelchairs, crutches, cane, orthopedic shoes, etc.)
6. *Pharmacologic tools,* as discussed in Chapter 23.

The effectiveness of restoring function by rehabilitative measures depends on several factors related to the individual who is rehabilitated and the environment in which he/she lives. While some of the strategies outlined above, either singly or in combination, are effective in temporarily improving physical and mental health, other interventions, as discussed in the next section, promise to achieve in the future more definitive and effective results.

4 FUTURE PERSPECTIVE IN BIOMEDICAL GERONTOLOGY: THE LIFE-EXTENSION SCIENCES*

The life-extension sciences include both classic and innovative biomedical areas such as interventive gerontology, cryonics, cloning, transplantation, resuscitation, artificial organs, and regeneration. Through the studied development and employment of these technologies, the course of aging and senescence, its ultimate culmination in morbidity and mortality, will be dramatically altered.

4.1 "Interventive" Gerontology

Experiments in *nutritional gerontology* indicate that the rate of aging can be modified in laboratory rodents (Chapter 24). Maximum lifespans of rats and mice, usually 3.5 years, have been extended to up to 5 years. Rats, usually infertile after 15 months, placed on a restricted dietary regimen early in life and maintained thereon for years have borne litters at ages approaching 3 years.[15] In addition, the age-related increase in tumor incidence, decline in homeostatic capability, and even the senescent deterioration of the coat have been delayed.[16] These findings show that mammalian aging, once believed to be somehow inexorable, can be manipulated in the laboratory.

Substantial efforts are currently underway to determine the mechanisms underlying delayed aging in underfed rodents. A comprehensive understanding of why food restriction produces its effects will provide the ability to even more profoundly influence the rate of aging, delay, and even reverse senescent changes, and prolong lifespan. It is expected then that efforts may extend beyond the laboratory and into the area of clinical medicine.

---

* This section (pages 314–316) contributed by P. E. Segall, Ph.D., BioTime Inc., Berkeley, CA.

Human growth hormone treatment has resulted in reversing certain signs of aging in frail elderly men (Chapter 21). These included significant increases in bone density and muscle mass, while the aging-related accumulation of adipose tissue was diminished.[17] The fact that hormone treatments can modify the rate of senescent changes, and even change their direction (such as seen in the augmentation of bone and reduction of skin adipose tissue), suggests that some forms of rejuvenation of the older body may be within the scope of physiological and pharmacological manipulation.

Parallel to these developments, advances in deep *hypothermia* and *cryonics* may facilitate prophylactic intervention in nonterminal situations. Surgical procedures are becoming available that permit hours of ice-cold bloodless perfusion and thereby may make possible surgical removal and specific organ-directed chemotherapeutic destruction of conventionally inoperable tumors, more effective transplantation of vital organs and tissues, and the implantation of a wide variety of prosthetic devices. The above strategies will allow us to more successfully counteract the effects of senescence.

4.2 Cryopreservation

Several species of frogs overwinter in the partially frozen state. These animals have been reported to survive for days, or even weeks, at –2°C or slightly below, with one third to two thirds of their body water frozen as ice. In order to restrict freezing to the extracellular compartment, and to avoid intracellular ice formations, these animals produce and circulate to their tissues large quantities of glucose or glycerol, which act as cryoprotectants. These amphibians also undergo partial dehydration, which further diminishes the formation of ice in the tissues.[18,19] Although ice is known to form in the abdomen, subcutaneously, and even in large blood vessels, ice formation within vital organs, such as the brain and heart, is minimal, if it even occurs at all.

Hamsters have been revived following partial freezing at temperatures between –1 and –3°C for 1 to 3 h. In these animals, as in the frogs, up to 50% of their body water (presumably the extracellular component) was frozen into ice. As these hamsters lacked cryoprotectant, they could not endure longer periods of freezing or lower temperatures.[20]

Hamsters,[21,22] dogs,[23,24] and even monkeys[25] can survive periods of total blood replacement with synthetic blood substitutes at temperatures very near the freezing point. A few hamsters have been revived following circulation of a blood substitute containing an amount of glycerol approaching the concentration of this cold-protecting chemical present in the blood of overwintering frogs. Although attempts are currently in progress to revive cryoprotected hamsters from partial freezing, this has not yet been possible. However, heartbeat has been restored in animals partially frozen overnight.[26,27]

When such revival from partial freezing becomes available in brain-dead humans, multiorgan transplant donors could be stored for periods long enough to allow maximum utilization of all their organs. This technology will provide for an increased availability of organs for transplantation. Advances in immunosuppression, already impressive since the discovery of cyclosporin A, and more recently FK-506, coupled with increased organ availability, will qualify older patients as transplant recipients (Chapters 2 and 3).

New forms of cryopreservation are now emerging. The most notable of these is vitrification, a technique in which tissues are saturated with high concentrations of chemical agents, subjected to pressures approaching and exceeding 1000 atmospheres, and frozen to cryogenic temperatures for indefinite storage. Mouse embryos have survived vitrification and liquid-nitrogen storage and have been brought back to term following implantation in surrogate mothers.[28] Workers in the field are currently attempting to extend this technique to the cryogenic storage of vital organs for transplantation.

One mammalian organ already capable of surviving cryopreservation in liquid nitrogen is the skin.[29] Full-thickness rat skin samples have been soaked in glycerol, cryopreserved for days or weeks at liquid-nitrogen temperature, and then transplanted to host rats with long-term viability.

The fact that a complex organ such as skin, with a multiplicity of cell types, can be revived after liquid-nitrogen storage suggests that the recovery of tissues, organs, organ systems, and whole organisms may be possible after extended cryogenic storage. The eventual goal is the reversible cryonic suspension of terminal patients or individuals with incurable degenerative disease, such as Alzheimer's disease, until a time when a cure is available.

Differentiated amphibian somatic cell nuclei, such as those from adult eye lens and adult red blood cells, have been transplanted into enucleated amphibian eggs, and the transplanted embryos have been raised to tadpole stage. When tadpole intestinal cells were used as the nuclear donor, a small number of transplant embryos reached adulthood and were capable of reproduction. The application of such cloning techniques as nuclear transplants in mammals has not progressed as far, but successful transplantation of nuclei from one-, two-, and eight-cell–stage embryos of mice, sheep, and cows have been reported.[30,31] More recently, cows have been produced by nuclear transplantation using nuclei harvested from 64-cell–stage blastocysts, as have been sheep from 120-cell blastocysts.[32]

4.3 Cloning

When mammalian nuclear-transplant technology produces the development of mature adults using nuclei transplanted from adult somatic cells, it may be possible to develop genetically identical bodies using single cells from an older patient. These "cloned" bodies could then be used as a source of young organs, which could be back-transplanted to replace those failing in the older individual. During the development of such a "parts clone", the prosencephalon could be removed and stored in liquid nitrogen. This would prevent the parts clone from developing into a sentient human being. After removal of the cells responsible for higher-brain formation, the embryo would develop into the ethical equivalent of a brain-dead individual. Once matured by rapid hormone-stimulated growth, it could serve as a tissue and organ donor to supply healthy youthful parts.

The embryonic nerve cells from the forebrain removed during early development can eventually be used to replace analogous cells damaged by aging. With such cloning technology, it may be possible to nearly completely "renovate" an aging individual using genetically identical and, therefore, immunologically compatible parts. Such a technique, using cells, tissues, and organs from a parts clone, could ultimately be used to rescue older individuals placed in cryonic suspension due to Alzheimer's disease, terminal cancer, or extreme debilitation secondary to advanced aging.

4.4 Critical-Care Interventions

Another means to increase lifespan in the aging population involves rapid progress in emergency and critical-care medicine. Cardiovascular and pulmonary disorders leading to terminal conditions are more prevalent in the older patient, responsible for a disproportionate number of emergency calls. Improvements in resuscitation technology will add years to their life expectancy as well as improve their quality of life by reducing morbidity from cerebral ischemia. Some recent innovations in this area include the use of calcium channel blockers to prevent arterial constriction in the microcirculation; removal of leukocytes in blood used for reperfusion, employing manitol to augment blood flow and decrease post-ischemic tissue swelling; and the exploration of hypothermia in revival following cardiac arrest.[33]

Coincidental with these advances is progress in the development of *artificial internal organs* such as the artificial kidney (renal dialysis), the artificial heart, and the extracorporeal membrane oxygenator.[34] Advances of techniques for direct sustenance of tissues through the use of total parenteral nutrition allow those whose digestive functions have been impaired to survive until these functions return.[35]

A revolution in *medical visualization technology,* including computer assisted tomography (CAT scanning), positron emission tomography (PET scanning), magnetic resonance imaging (MRI), and ultrasonics, coupled with the use of semi-micro multianalyses of physiological chemistry, allows early and accurate diagnosis of life-threatening situations and permits planning appropriate surgical, pharmacological, and other forms of intervention.

Newly found *growth factors,* along with drugs or even electric and electromagnetic fields, which some have indicated may spur regeneration, could lead to techniques for revitalizing aged, damaged, or diseased tissues and organs. A preliminary example of these is the use of pulsed electromagnetism to heal cracks in bones (common in older individuals often compromised by osteoporosis).[13,36] Let us not minimize the importance of improving the self-image as through continuing progress in plastic surgery, cosmetic dermatology, and pharmacology (such as the use of minoxidil to stimulate hair growth, or retene A for revitalizing aging skin).

5 IMMORTALITY AND RISK ASSESSMENT

Emphasis in achieving longevity and wellness in old age has, so far, focused primarily on prevention and treatment of disease and less often on enhancement of physiologic competence. Reducing risks and avoiding hazards represent yet another means, perhaps as effective as biomedical research, of achieving these goals, *Risk assessment* is an interesting and constructive first step in attempting to minimize mistakes (e.g., car accidents, cigarette smoking, radiation exposure) that may endanger the length and quality of life.[37] However, ranking of such risks seems to remain quite difficult[38] as is the perception of risk[39] and the measures taken for correction.[40,41] Thus, "a zero-risk society is not yet in the scientific cards".[42]

While some of the predictions for life extension may sound like science fiction, this century has produced such monumental changes that these predictions should not be considered outrageous. Seeking "immortality in the arduous old-fashioned way, doing good deeds and tak[ing] care of [one's] children"[42] remains a goal worthy of pursuit. It should not detract from efforts of continuing to improve and lengthen life. Aging and death remain, indeed, the last sacred enemies, particularly frustrating to 20th century man who has circled the moon, harnessed nuclear energy, artificially reproduced DNA, and significantly extended life expectancy. Such a person can be expected to continue to strive to improve the quality of life at all stages, including old age, as well as extend the duration of life, as the Russian writer Tolstoy[43] suggested when he wrote in a letter to a friend, "Don't complain about old age. How much good it has brought me that was unexpected and beautiful" and concluded, "that the end of old age and of life will be just as unexpectedly beautiful..." Here, again, we are emphasizing the biomedical; however, all of this is of little avail unless we also consider ethical, moral, economic, and cultural implications for the elderly population.[44] To be successful, aging depends on the cooperation of the individual, who is responsible for the well-being of his/her own body; on the community, including the family, who are responsible for a supportive network; and on the government, which must provide legal and monetary support for this segment of the population.

REFERENCES

1. Dychtwald, K., Ed., *Wellness and Health Promotion for the Elderly,* Aspen Systems Corp., Rockville, MD, 1986.
2. Miller, H., in *The Oxford Book of Ages,* Sampson, A. and Sampson S., Eds., Oxford University Press, Oxford, 1985, 151.
3. Picasso, P., in *The Oxford Book of Ages,* Sampson, A. and Sampson S., Eds., Oxford University Press, Oxford, 1985, 118.
4. Minkler, M. and Pasick, R. J., Health promotion in the elderly: a critical perspective on the past and future, in *Wellness and Health Promotion for the Elderly,* Dychtwald, K., Ed., Aspen Systems Corp., Rockville, MD, 1986, 31.
5. Syme, S. L., Social determinants of disease, *Ann. Clin. Res.,* 19, 44, 1986.
6. Rose, G. and Marmot, M. G., Social class and coronary heart disease, *Br. Heart J.,* 45, 13, 1981.
7. Pappas, G., Queen, S., Hadden, W., and Fisher, G., The increasing disparity in mortality between socioeconomic groups in the United States, 1960 and 1986, *N. Engl. J. Med.,* 329, 103, 1993.
8. Guralnik, J. M., Land, K. C., Blazer, D., Fillenbaum, G. G., and Branch, L. G., Educational status and active life expectancy among older blacks and whites, *N. Engl. J. Med.,* 329, 110, 1993.
9. Hong, C.-Z. and Tobis, J. S., Physiologic approach to rehabilitation medicine, in *Physiological Basis of Aging and Geriatrics,* Timiras, P. S., Ed., Macmillan, New York, 1988, 427.
10. Granger, C. V. and Greshan, G. E., Eds., *Functional Assessment in Rehabilitation Medicine,* Williams & Wilkins, Baltimore, 1984.
11. Tobis, J. S., Rehabilitation of the geriatric patients, in *Clinical Geriatrics,* 2nd ed., Rossman, I., Ed., Lippincott, Philadelphia, 1982.
12. Lehmann, J. F., Ed., *Therapeutic Heat and Cold,* 3rd ed., Williams & Wilkins, Baltimore, 1982.
13. Stillwell, G. K., Ed., *Therapeutic Electricity and Ultraviolet Radiation,* 3rd ed., Williams & Wilkins, Baltimore, 1983.

14. Tobis, J. S. and Hoehler, F, *Musculoskeletal Manipulation: Evolution of the Scientific Evidence,* Charles C Thomas, Springfield, IL, 1986.
15. Segall, P. E., Timiras, P. S., and Walton, J. R., Low tryptophan diets delay reproductive aging, *Mech. Ageing Dev.,* 23, 245, 1983.
16. Timiras, P. S., Hudson, D. B., and Segall, P. E., Lifetime brain serotonin: regional effects of age and precursor availability, *Neurobiol. Aging.,* 5, 235, 1984.
17. Rudman, D., Feller, A. G., Nagraj, H. S., Gergans, G. A., Lalitha, P. Y., Goldberg, A. F., Schlenker, R. A., Cohn, L., Rudman, I. W., and Matson, D. E., Effects of human growth hormone in men over 60 years old, *N. Engl. J. Med.,* 323, 1, 1990.
18. Storey, K. B. and Storey, J. M., Freeze tolerance and intolerance as strategies of winter survival in terrestrially-hibernating amphibians, *Comp. Biochem. Physiol.,* 83, 613, 1986.
19. Costanzo, J. P., Lee, R. E., Jr., and Wright, M. F., Cooling rate influences cryoprotectant distribution and organ dehydration in freezing wood frogs, *J. Exp. Zool.,* 261, 373, 1992.
20. Smith, A. U., Studies on golden hamsters during cooling to and rewarming from body temperature below 0°C, *Proc. R. Soc.,* 145, 391, 1956.
21. Waitz, H. D., Yee, H., Gan, S. C., et al., Reviving hamsters after asangineous hypothermic perfusion, *Cryobiology,* 21, 699, 1984.
22. Gan, S. C., Segall, P. E., Waitz, H. D., et al., Ice-cold blood substituted hamsters revived, *Fed. Proc.,* 44, 623, 1985.
23. Segall, P. E., Waitz, H. D., Sternberg, H., et al., Ice-cold bloodless dogs revived using protocol developed in hamsters, *Fed. Proc.,* 46, 1338, 1987.
24. Bailes, J. E., Leavitt, M. L., Teeple, E., Jr., Maroon, J. C., Shih, S.-R., Marquardt, M., El Rifai, A., and Manack, L., Ultraprofound hypothermia with complete blood substitution in a canine model, *J. Neurosurg.,* 74, 781, 1991.
25. Segall, P. E., Sternberg, H., Waitz, H. D., Bellport, V., Segall, J. M., Shermer, S., Kinoshita, G., Day, J., Parker-Boysen, C., and Breznock, E. M., Ice-cold blood-substituted baboons survive, *FASEB J.,* 7, A441, 1993.
26. Sternberg, H., Segall, P. E., Waitz, H. D., Bellport, V., and Segall, J. M., Partly-frozen overnight, thawed hamster's hearts beat, *FASEB J.,* 5, A1638, 1991.
27. Sternberg, H., Segall, P. E., and Waitz, H. D., Preserving organs using cold blood-substitutes, *FASEB J.,* 6, A1350, 1992.
28. Rall, W. F. and Fahy, G. M., Ice-free cryopreservation of mouse embryos at –196 degrees by vitrification, *Nature,* 313, 573, 1985.
29. Hirase, Y., Kojima, T., Takeishi, M., Hwang, K. H., and Tanaka, M., Transplantation of long-term cryopreserved allocutaneous tissue by skin graft or microsurgical anastomosis: experimental studies in the rat, *Plastic Reconstructive Surg.,* 91, 492, 1993.
30. McKinnell, R. G., *Cloning of Frogs, Mice, and Other Animals,* University of Minnesota Press, Minneapolis, 1985.
31. Willadsen, S. M., Nuclear transplantation in sheep embryos, *Nature,* 320, 63, 1986.
32. First, N. L. and Prather, R. S., Genomic potential in mammals, *Differentiation,* 48, 1, 1991.
33. Safar, P., Cerebral resuscitation after cardiac arrest: a review, *Circulation,* 74, IV138, 1986.
34. Cauwels, J. M., *The Body Shop: Bionic Revolution in Medicine,* C. V. Mosby, St. Louis, MO, 1986.
35. Irving, M., ABC of nutrition. Enteral and parenteral nutrition, *Br. Med. J.,* 291, 1404, 1986.
36. Barker, A. T., Dixon, R. A., Sharrard, W. J., and Sutchliffe, M. L., Pulsed magnetic field therapy for tibial non-union. Interim results of a double-blind trial, *Lancet,* 1, 994, 1984.
37. Wilson, R. and Crouch, E. A., Risk assessment and comparisons: an introduction, *Science,* 236, 267, 1987.
38. Ames, B. N., Magaw, R., and Gold, L. S., Ranking possible carcinogenic hazards, *Science,* 236, 271, 1987.
39. Slovic, P., Perception of risk, *Science,* 236, 280, 1987.
40. Russell, M. and Gruber, M., Risk assessment in environmental policy-making, *Science,* 236, 286, 1987.
41. Lave, L. B., Health and safety risk analyses: information for better decisions, *Science,* 236, 291, 1987.
42. Koshland, D. E., Immortality and risk assessment, *Science,* 236, 241, 1987.
43. Tolstoy, L., in *The Oxford Book of Ages,* Sampson, A. and Sampson S., Eds., Oxford University Press, Oxford, 1985, 148.
44. Wells, T., *Aging and Health Promotion,* Aspen Systems Corp., Rockville, MD, 1982.

INDEX

A

Absorption
  of drugs, 279
  of nutrients, 251–252
Accommodation, visual, 116–118
ACE inhibitors, 282
Acid–base balance, 244–245
Acquired immune deficiency syndrome, see AIDS
Activities of daily living (ADL), 16-17
Activity level, 28–29, see also Exercise
Acute renal failure, 239–240
Adaptation response, 140–144
Adrenal medulla, 138
Adrenergic synapses, 96
Adrenocortical effects, 134–138
Adrenocorticotropic hormone (ACTH), 135, 137–140
Advanced glycosylation endproducts (AGEs), 43
Afro-Americans
  bone aging in, 263
  demographics and, 9
Age, chronologic, 13
Age pigments (lipofuscins), 42–43, 53, 92
Age-related macular degeneration (ARMD), 121
Aging, see also specific conditions and disorders
  comparative physiology of, 11–13
  demography of, 7–11
  developmental aspects, 1–4
  differential in humans, 13–17
  and disease, 23–33, see also Diseases of aging
  versus disease, 23
  free radical theory, 61–72
  models of
    in vitro, 31, 67–68
    restricted feeding model, 296–305
  myths about, 311–312
  neuroendocrine control theory, 43–44
  normal vs. Werner's syndrome, 30
  and physiologic competence, 15
  pre-and postnatal, 1–2
  pre- vs. postovulatory, 161-162
  research goals, 3
  sociological aspects, 17–19, 311–314
  stages of maturity and, 2–3
  study of, 29–33
  successful vs. usual, 13
  terminology of, 3–4
  theories of
    antagonistic pleiotropy, 41
    cellular, 42–43
    dysdifferentiation, 41–42
    genetic/environmental, 37
    molecular, 37–41
    neuroendocrine, 133–134
    system level theories, 43–44
AIDS, 31, 82–83, 111
Aldosterone, 136
α-tocopherol (vitamin E), 63
Alveolar function, 226
Alzheimer's disease, 92–93, 94
  and apoliprotein E4, 223
  differential diagnosis, 109
  etiology, 91, 110–111
  management, 111
  neuronal loss vs. normal aging, 91
  and olfaction, 128
  pathogenesis, 109–110
  study models for, 31–33
Amadori products, 43
Amino acids, free radical effects, 64–65
Amyloid, 53
Amyloidoses, 53–54
  and Alzheimer's disease, 111
Analgesics, 282
Anemia, 266
  Fanconi's, 65
Angina pectoris, 209–211
Angiography, 208
Animal experiments, 30–31
Anosmia, 125–126
Antacid drugs, 250
Antagonistic pleiotropy, 41
Anthropologic aspects, 9–11
Anticholinergic drugs, 281
Anticoagulant drugs, 281
Antidiuretic hormone (ADH; vasopressin), 140, 237–238
Antigen-antibody system, 75–85 see also Immune system
Antioxidants, 62–63, 65–67, 71
  bilirubin as, 255–256
Apathetic hyperthyroidism, 186
Apolipoproteins, 217–220
  and Alzheimer's disease, 223
Arterial structure, 200–201
Arteriosclerosis, 199, see also Atherosclerosis
Arteritis, giant-cell, 266
Arthritis
  osteo-, 264–265
  rheumatoid, 265–266
Ascorbic acid (vitamin C), 63
Assessment
  of carbohydrate metabolism, 192–193
  of endocrine function, 134
  functional, 314
  geriatric, 16–17
  of hepatic/biliary function, 254–255
  of menopausal age, 149–150
  of physiologic age in humans, 15–16
  of plasma lipoprotein levels, 215–216
  of renal function, 235–238
  of thyroid function, 181–182
Atherosclerosis
  epidemiology, 199–200
  etiology and pathogenesis, 200–204
  and lipoprotein metabolism
    assay methods, 215–216
    cellular interactions, 221–222
    enzyme activity in, 220–221
    hyperlipoproteinemia, 222–223
    nomenclature, 215
    types and pathogenetic roles, 216–220
  progression, 204–208
  study methods, 208
  theories of, 208–209
Auditory changes, 121–124
Autoimmune disease, 78
  diabetes (type 1), 195, 196

and dietary restriction, 291
and free radical effects, 71
of joints, 265–266
thyroid, 183

B

Balance, 104
Basal metabolic rate (BMR), 12, 65–66, 179–187, see also Thyroid gland
and dietary restriction, 291–292
Benign prostatic hypertrophy, 243–244
β blockers, 281
Beta carotene, 63
Biliary system, 253–255
Bilirubin as antioxidant, 255–256
Biologic aspects, see Physiologic aging
Biologic clocks, 104, 292
Biomarkers, 13
Biomedical and socioeconomic issues, 311–316
Biomedical gerontology, 314–315
Birth defects, see Developmental abnormalities
Blacks, see Afro-Americans
Bladder function, 153, 240–243
Bleeding, gastrointestinal, 250
Blood cell disorders, 230–231
Blood flow, 201, see also Atherosclerosis
Blood pressure, 135, 211–213, 269–271
B lymphocytes, 84
BMR, see Basal metabolic rate (BMR)
Body/brain weight, 11–12, 14–15
Bone
calcium and, 259–262
fractures, 263
hormonal regulation of metabolism, 262
parathyroid disorders, 262-263
remodeling of, 259–260
strength of, 260
structure and function, 259
Bone spurs (osteophytes), 264
Brain weight, 90

C

Calcitonin, 187–188
Calcium channel blockers, 281
Calcium metabolism and bone, 261–262
Caloric restriction, see Diet; Restricted feeding model
Calorigenesis and thyroid function, 185
Cancer
and dietary restriction, 286–289
and free radical effects, 69–70
gastric, 250–251
and immune function, 79
Carbohydrate metabolism, 191–196, see also Diabetes mellitus
Cardiac block disorders, 270
Cardiac dysrhythmias, 270
Cardiac muscle, 268–271, 282
Cardiovascular system, see also Atherosclerosis
and atherosclerosis, 200–209
and coronary heart disease (CHD), 209
in diabetes mellitus, 196
diseases of
of cardiac muscle, 268–271
coronary heart disease (CHD), 209–211
hypertension, 211–213
menopausal changes, 153–156
terminology, 199–200
Carotenes, 63
Cartilage, see Joint disease
Catalases, 63
Cataracts, 120, 196
Cell death, 52–54
Cell doubling capacity assays, 17
Cell injury, 52–54
Cellular aging
cell injury/cell death, 52–54
in cytoplasm, 49–50
in enzyme activity, 51–52
erythrocyte as model, 231
functional and organizational, 47–49
human fibroblast studies, 55–56
in membranes, 49
transplant studies, 54–55
variability of, 47
Cellular defenses, 23–24, see also Immune system
Cellular theories of aging
age pigments (lipofuscins), 42–43
crosslinking theory, 43
wear and tear, 42
Cervical changes, 153
Cholesterol, 185–186, 215–222, see also Atherosclerosis; Lipid metabolism
Cholinergic synapses, 96
Chromaffin cells, 138
Chromosomal abnormalities, 160–161
Chronic renal failure, 239
Chylomicrons, 215–216
Cigarette smoke, see Smoking
Circadian rhythms, 105, 292
Cloning, 315–316
CNS plasticity, 100
Codon restriction theory, 37–38
Collagen (connective tissue), 43, 276–277, 296
Color perception, 120
Compression of morbidity, 26–27
Connective tissue, 43, 276–277, 296
Constipation, 253
Continence, see Incontinence
Contraceptive drugs, 150
Coronary heart disease (CHD), 153–155, 209–211, see also Atherosclerosis
and atherosclerosis, 199–209
in diabetes mellitus, 196
Corticotropin-releasing hormone (CRH), 135, 136–140
Creatinine clearance testing, 236–237
Critical-care issues, 316
Crosslinking theory of aging, 43
Cryopreservation, 315
Cu/Zn SOD-, see Superoxide dismutase (SOD)
Cytoplasmic changes in aging, 49–50

D

Death, 11
versus immortality, 4
Death rate
and cardiovascular disease, 25–26
compression of morbidity concept, 26–27
coronary heart disease (CHD), 209
and diseases of aging, 24–28
lifespan comparisons, 24
Defecation disorders, 252–253
Degenerative aspects of aging, 47–56, see also Cell death; Cellular aging
Degenerative joint disease, 264–265
Dehydration, cellular, 52

Dementia
AIDS-related, 111
senile, see Alzheimer's disease; Senile dementia
Demographic aspects
ethnicity and understudied groups, 9
international comparisons, 10–11
life tables, 7–8
maximum human lifespan, 11
primitive and ancient populations, 10
sex differences, 9
survival curves, 8–9
Dendritic loss, 91
Dental changes, 248
Dermis, 274
Developmental abnormalities
and age at parity, 160–161
Developmental aspects
pre- and postnatal, 1–2
stages of maturity and aging, 2–3
Diabetes mellitus
and accelerated aging, 194–195
anatomic and physiological considerations, 191–192
assessment, 192–193
and dietary restriction, 299
and insulin resistence, 193
in native Americans, 9
type 1 (insulin deficiency), 195
type 2 (non–insulin dependent), 195
Dialysis treatment, 239-240
Diarrhea, 253
Diet
and age at menopause, 150
and free radical formation, 66–67
and gastrointestinal function, 247
human studies, 305
and malabsorption syndromes, 251–252
in peptic ulcer disease, 250
primate studies, 304–305
restricted feeding (dietary restriction) model, 285-305, see also Restricted feeding model
Digestion, see Gastrointestinal system
Disability defined, 16
Diseases of aging
as cause of death, 24–25
characterization of, 23
as age-related, 27
as multiple pathology, 26–27
experimental induction of, 29–33
eye diseases, 120–121
and free radical effects, 69–71
hematopoietic disorders, 230–231
and immune (cellular) defenses, 23–24
oral cavity and teeth, 248
pathology vs. normal aging, 23
reproductive system
female, 149–164, see also Menopause
male, 243–244
respiratory disorders, 225–230
thyroid disorders, 186–187
treatment goals in, 28
vs. disuse, 28-29
Disposable soma hypothesis, 42
Disuse, 28–29, 228, 270
Diverticulitis/diverticulosis, 252
DNA
and aging, 37–42, see also Molecular theories of aging
and dietary restriction, 302
free radical effects, 65
Dopaminergic pathways, 96–99
Down's syndrome, 65, 160–161
Drug therapy
adverse reactions/side effects, 282–283
and cardiac activity, 271
and kidney damage, 241
and peptic ulcer formation, 250
and pharmacodynamics, 281
and pharmacokinetics, 279–281
and senile dementia, 108
and urinary continence, 242–243
Dry mouth (xerostomia), 248
Duodenal ulcers, 249–251
Duodenum, 249–251, see also Gastrointestinal system
Dysdifferentiation, 41–42
Dysphagia, 248–249
Dysrhythmias, 270

E

EEG changes, sleep/wakefulness, 105
Elderly persons
self-awareness of disease in, 27–28
societal value, 17–19
Electroencephalography (EEG), see EEG changes
Emphysema, 229
Endocrine–immunologic theory of aging, 44
Endocrine system
adaptation mechanisms, 140–144
adrenal cortex, 134–138
adrenal medulla, 138
assessment, 134
and dietary restriction, 302–304
endocrine pancreas and carbohydrate metabolism, 191-196
and neuroendocrine function, 133–134
parathyroid gland, 187, 261–263
pituitary gland, 138–140
reproductive system
female, 151–164, see also Menopause
male, 171–175
thyroid gland and basal metabolism, 179–187
Endoplasmic reticulum, 49
Environmental/genetic interactions, 37
Enzyme activity
in aging process, 50–52
in lipoprotein metabolism, 220–221
Epidermis, 273–274
Error theory of aging, 39–40
Erythrocyte function, 230–231
Essential hypertension, 211–213
Estrogen replacement therapy, 154-158
Ethnicity, 9
Exercise
and cardiac output, 269
and free radical formation, 68
functional benefits of, 29
in rehabilitation, 314
and respiratory changes, 227–228
and skeletal muscle aging, 267, 268
Eye diseases, 120–121
Eyes, 115–121, see also Vision

F

Falls, 104, 263
Fanconi's anemia, 65
Fatty change, 52–53
Fecal incontinence, 252–253
Fecundity, 12

Fertility
  female, 158–164
  male, 171–172
Fibroblast growth factor, 208
Fluid/electrolyte balance, 244–245
Follicle-stimulating hormone (FSH), 174
Fractional sodium excretion test, 239
Fractures, 263
Free radicals, 43, see also Antioxidants; Oxidants
  bilirubin as antioxidant, 255–256
  cellular and tissue effects, 63–65
  chemistry and biochemistry, 61–63
  and dietary restriction, 292–293, 301–302
  and genetic disorders, 65
  interventive strategies, 71
  lifespan effects, 65–68
  and lung damage, 228
  susceptibility and related diseases, 68–71
Freezing (cryopreservation), 315

G

Gabanergic synapses, 96
Gait and balance, 103–104
Gastric ulcers, 249–250
Gastrointestinal system
  biliary system and, 255–256
  exocrine pancreas, 253–254
  fecal incontinence, 252–253
  small and large intestines, 251–253
  stomach and duodenum, 249–251
  swallowing and pharyngoesophageal function, 248–249
  teeth, gums, and oral mucosa, 248
Gene exression, and dietary restriction, 302
General adapation syndrome, 141–145
Gene regulation theory of aging, 40–41
Genetic disease
  and dietary restriction, 289–290
  and free radical activity, 65
Genetic/environmental interactions, 2, 37
Genitourinary system, see also Reproductive system
  lower urinary tract/incontinence, 240–243
  renal function/renal failure, 235–240
  in women, 152
Geriatric assessment, 16–18
  in vitro measurements, 17
  and screening programs, 17
Gerontology as science, 2–3
Giant-cell arteritis, 266
Gingivae (gums), 248
Glaucoma, 120–121
Glomerular funtion testing, 235–237
Glucagon, and cardiac activity, 271
Glucagon disorders, 194, see also Diabetes mellitus
Glucocorticoids, 134–135, see also Steroids
  and dietary restriction, 299
  and neuronal damage, 143–144
Glucose intolerance, 191–196, see also Diabetes mellitus
  and dietary restriction, 299
Glucose tolerance test, 192–193
Glutathione peroxidase, 63
Glycogen-storage diseases, 53
Glycosylation, 43
Golgi complex, 49-50
Gonadotropin, testicular stimulation by, 173–174
Gout, 264–265
Growth factors in atherosclerosis, 208
Growth hormone, 140
  and dietary restriction, 298
  and skeletal muscle, 267–268
Growth hormone–inhibiting hormone (somatostatin), 140, 191
Growth hormone–releasing hormone (GRH), 140
Growth period duration, 12

H

Handedness, 149
Health care system, 28, 311–314
Health promotion, 311–314
Hearing changes, 121–124
Heart muscle, 268–271
Hematopoietic disorders, 230–231
Hemodialysis, 241
Hemoglobin, 64, 231
Hepatic function, 254
Heredity, 2, see also Genetic factors
High-density lipoproteins (HDLs), 154, 186, 206, 208-209, 215–223
Homeless people, 9
Hormonal system, see Endocrine system and individual organs
Hospitalization, 27
Hot flushes/flashes, 150–152
Human skin fibroblast culture assays, 17
Hydroxl radicals, see Free radicals
Hyperlipoproteinemia, 222–223
Hypertension, 211–213
  and dietary restriction, 289–290
  essential, 211
Hyperthyroidism, 186–187
  apathetic, 186
Hypophysis (pituitary gland), 138–140
Hyposmia, 125–126
Hypotension, postural, 212, 269, 281
Hypothalamic-pituitary-adrenocortical axis, 136
Hypothalamic regulation, 183
  of male sexual/reproductive function, 174–175
  of menopause, 152
Hypothalamo-pituitary-thyroid axis, 179–187, see also Thyroid gland
Hypothyroidism, 187
  as protective mechanism, 186

I

Immortality, 3, 316
  selective, 12–13
Immune system, 43
  in AIDS
  autoimmune disease, 71, 265–266, 291
  B lymphocytes and antibody procution, 84
  cellular constituents, 80–81
  decline with aging, 23–24
  in diabetes mellitus, 196
  and dietary restriction, 290–291, 304
  endocrine–immunologic theory of aging, 44
  free radical effects, 64, 69
  and free radicals, 65
  reversibility of age-related dysfunction, 84–85
  structure and function, 75–79
  thymus and thymosins, 79–80
  T lymphocytes, 81–84
Immunity
  cell-mediated, 78
  humoral, 84
Incontinence
  fecal, 252–253
  urinary, 240–243
Infarction
  cerebral, 93–94

gastrointestinal, 250
and multi-infarct dementia, 109
Instrumental activities of daily living (IADL), 16–17
Insulin, 191–193, 299, see also Diabetes mellitus
Integumentary system, 273–277
Intestinal absorption, 251–252
Intestinal system, 251–253
Ionizing radiation, 66
Islets of Langerhans, 191–196, see also Diabetes mellitus; Insulin
Isoenzymes, 51

J

Joints, see also Musculoskeletal system
age-related changes, 263–264
diseases
autoimmune arthritides, 265
giant-cell arteritis, 266
osteoarthritis and gout, 264–265
polymyalgia rheumatica, 266
structure, 263

K

Kidneys, 235–240, see also Genitourinary system

L

Lactate dehydrogenase (LDH), 51
LAG-1 gene, 41
Learning and rehabilitation, 313–314
Lecithin: cholesterol transferase (LCAT), 220
Leukoplakia, 248
Lewy bodies, 97
Life cycle and aging, 1–4
Life expectancy, see Lifespan
Life extension sciences, 314–316
Lifespan, see also Longevity
biological phases of, 40
and erythrocyte function, 230–231
and free radical effects, 66–67
and immune function, 79
international comparative, 10–11
maximum human, 11
and metabolic rate, 65–66
reproductive, 158–164
stages, 1–2
and stem cell turnover, 80–81
and thyroid function, 186
Lifespan extension, 71
Life tables, 7–8
Limbic system, 105–106
Lipid metabolism
and atherosclerosis, 205–209, 215–223
and menopause, 154–155
and thyroid function, 185–186
Lipid peroxidation, see Free radicals
Lipofuscins (age pigments), 42–43, 53, 92
Lipoprotein lipase (LPL), 220–221
Liver, 254–256
and drug absorption, 280
Longevity, 7–15, see also Lifespan
physiologic correlates of, 11–12
selective, 11
selective immortality concept, 12–13
Low-density lipoproteins (LDLs), 154, 186, 206, 208-209, 215–223
Lower urinary tract, 242–243
Lungs and lung disorders, 225–230, see also Pulmonary disorders
Luteinizing hormone, see Gonadotropin

M

Macrophages, 83
Macular degeneration, 121
Major histocompatibility complex, 75–76
Malabsorption syndromes, 251–252
Marital status, and age at menopause, 150
Maturity, 2–3
Meissner's corpuscles, 125, 274
Membrane effects, 49, 64
Memory, 106–107
Menopause
endocrinology menopause/postmenopause, 149–150
estrogen deprivation
and cardiovascular system, 153–156
and genitourinary organs, 152–153
and skeletal system, 156–157
and skin, 153
estrogen replacement therapy, 157–158
evolutionary aspects, 147
fertility patterns and, 158–164
functional characteristics, 147–149
symptoms of, 150–152
Metabolic rate, see Basal metabolic rate (BMR)
MHC antigens, 75–76, see also Immune system
Micturition, see Genitourinary tract; Incontinence, urinary
Mineralocorticoids, 135–136
Miosis, senile, 118
Mitochondrial effects, 50, 63–64, 68–69
Molecular aging, 47–56, see also Cell death; Cellular aging
Molecular theories of aging
codon restriction, 37–38
conceptual basis, 37
error theory, 39–40
gene regulation theory, 40–41
somatic mutation, 39
Mortality, 24–26, see also Death rate
compression of, 26–27
Motor changes, 103–104
Multi-infarct dementias, 109
Multiple pathology concept, 26
Muscles, 266–268, see also Musculoskeletal system
Musculoskeletal system
bones, 259–263
cardiac muscle, 268–271
joints, 263–266
skeletal muscle, 266–268
Mutagenesis, and ionizing radiation, 66
Myoglobin, 64
Myths about aging, 311–312

N

Native Americans, 9
Natural killer (NK) cells, 83–84, 291
Necrosis, 54, see also Cell death
Nerve growth factor, 100
Nervous system, see also Sensory system and individual senses
biochemical changes, 94–99
and CNS plasticity, 100
and dietary restriction, 299–300, 302–304
functional changes
memory, 106–107
motor: gait and balance, 103–104
senile dementias, 107–111
sleep/wakefulness, 104–106
metabolic changes, 99–100
sensory aging and, 115–129
structural changes, 89–94

Neural growth factor, 111
Neuritic plaques, 93
Neuroendocrine control theory of aging, 43–44, 302–303
Neuroendocrine function, see Endocrine system; Nervous system
Neurofibrillary tangles, 92–93
Neuronal damage, glucocorticoids and stress in, 143–144
Neurons
  free radical damage, 68
  loss of, 90–91, 92
  types and functions, 90
Neuropeptide Y, 174–175
Neurotransmitters
  as hormones, 134
  and memory, 107
Neurotropic agents, 100
NGF, see Neural growth factor
Nitric oxide (NO), 94–95
Norepinephrine, see Neurotransmitters
NSAIDs, 282
Nucleic acids, see DNA

O

Obesity, 193–194
Ocular changes, 115–121, see also Vision
Olfaction, 126–128
Oocyte depletion, 159–16
Oral cavity and teeth, 248
Oral contraceptives, 150
Organ transplants, 78–79
Orthostatic (postural) hypertension, 212, 269, 281
Osteitis deformans (Paget's disease), 263
Osteoarthritis, 264–265
Osteophytes (bone spurs), 263–264
Osteoporosis, 155–156
Ovaries, 153
Oviducts, 153
Oxidation, 61–62, see also Free radicals

P

Pacinian corpuscles, 125, 274
Paget's disease (osteitis deformans), 263
Pain-relieving drugs, 282
Pancreas
  endocrine, 191–196, 299, see also Diabetes mellitus; Insulin
  exocrine, 253–254
Parathyroid gland, 187–188
  and bone disorders, 262–263
  and bone metabolism, 262
Parathyroid hormone (PTH), 187–188
Parity, and age at menopause, 150
Parkinsonism, 281
Parkinson's disease, 96–99, 281
  drug-induced, 98
  and free radical effects, 70–71
Peptic ulcers, 250
Pharmacodynamics, 281–283
Pharmacokinetics, 279–281
Pharmacologic treatment, see Drug therapy
Physical fitness, see Exercise
Physiologic aging
  assessment of physiologic age, 15–16
  comparative, 11–13
  competency issues, 15
  differential, 13–17
  geriatric assessment, 16–19
  heterogeneity of, 13–14
  successful vs. usual, 13
Physiologic competence, 15
Pituitary gland, 137–140, 182, see also Thyroid gland
Plasma lipoproteins, 215–223, see also Atherosclerosis; Lipid metabolism
Platelets and atherosclerosis, 207–208
Pleiotropy, antagonistic, 41
Pneumonia, 229–230
Polymyalgia rheumatica, 266
Postural hypotension, 269, 281
Prebycusis, 121–123
Prenatal development, 1–2
Presbyopia, 117–118
Pressure sores, 275
Prevention, 311–313
Progeria/progeroid syndromes, 30, 40, 65
Prostate gland, 243–244
Protein synthesis, and dietary restriction, 294–296, 300–301
Psychological aspects, of menopause, 152
Pulmonary disorders
  aging-associated changes, 225–229
  emphysema, 229
  oxygen toxicity (free radicals), 228
  and physiological "battering," 225
  pneumonia, 229–230
  and smoking, 229
  tuberculosis, 230

R

Race
  and age at menopause, 150
  and bone aging, 263
Radiation, ionizing and free radical formation, 66
Rehabilitation, 313–314
Renal disease
  acute renal failure, 239–240
  chronic renal failure, 240
Renal function
  age-related changes, 235–238
  and drug elimination, 280–281
  renal failure, 238–240
Reproductive lifespan, 158–164
Reproductive system
  and dietary restriction, 296–298
  female
    anatomic and physiologic considerations, 149–150
    loss of fertility, 158–162
    in menopause, 147–149, 150–158
    rodent studies, 162–164
  male, 171–176
    anatomic and physiological aspects, 171
    prostate gland, 243–244
    sexual function, 175–176
    testicular changes, 171–175
Respiratory system, 225–230
Restricted feeding model
  age of initiation, 285–286
  dietary composition, 285
  effects
    on genetic disease models, 289–290
    on immunologic response, 290–291
    on tumor incidence, 286–289
  metabolic response to
    collagen, 296
    core body temperature, 292
    enzyme activity, 292–293
    lipid peroxidation, 293–294
    mitochondrial recovery, 292
    protein synthesis, 294–296

physiologic response
neurology, 299–300
nonreproductive endocrinology, 298–299
proposed mechanism of action, 300–304
reproduction, 296–298
Reticular activating system, 105–106
Retinal changes, 118–119
Rheumatoid arthritis, 265–266
Risk assessment, 316
Risk management, 311–314

S

Safety issues, falls and lack of balance, 104
SDAT (senile dementia of Alzheimer type), see Alzheimer's disease
Sedative drugs, 281
Self-awareness of disease, 27–28
Selye's stress theory, 140–145
Senile dementia
of Alzheimer type, 109–111, see Alzheimer's disease
definitions and prevalence, 107–108
multi-infarct, 109
Senile macular degeneration, 121
Senile miosis, 118
Sensory system, 274
hearing, 121–124
olfaction and taste, 126–130
somatic sensation, 124–126
vision, 115–121
Serotonergic synapses, 96
Serotonin, see Neurotransmitters
Sex differences, in physiologic aging, 9
Sex hormones, 135, 136
male, 171–175
and menopause, 156–158, 161–150
Sexual function, male, 175–176
Skeletal system, 155–156, 259–263, see also Musculoskeletal system
Skin and skin appendages, 273–275
Skin changes, 153
Sleep/wakefulness, 104–106
Smell, 126–128
Smoking
and age at menopause, 150–151
and emphysema, 229
and free radical formation, 62, 66
and oral disease, 248
Smooth muscle cell proliferation theory, 209
SMP-2 gene, 41
Socioeconomic issues, 311–316
Somatic mutation theory, 39
Somatic sensory changes, 124–126
Somatostatin (GIH), 140, 191
Speech perception, 123–124
Statistical methods, 17
Stem cells, 80–81
Steroids, see also Glucocorticoids
ovarian, see Estrogen replacement therapy
testicular, 172–175
Stomach, 249–251, see also Gastrointestinal system
Stress, see Environmental stress
adaptation syndrome, 3, 140–143
adrenocortical response, 136–137
and Alzheimer's disease, 111
cellular defenses, 23–24
cigarette smoke and free radical formation, 62
glucocorticoids and neuronal damage, 143–144
neuroendocrine response to, 145
Superoxide dismutase (SOD), 41, 50, 62, 65
Survival curves, 8–9
Swallowing difficulty, 248–249
Synaptic changes, 91–92
System level theories
endocrineimmunologic theory, 44
immunology theory, 43
neuroendocrine control theory, 43–44

T

Tardive dyskinesia, 281
Taste, 129–130
Teeth and gums, 248
Terminology of aging, 3
Testosterone, 171–175
T-helper cells, 82
Thermoregulation, 185
Thrombogenic theory, 209
Thrombophlebitis, 269
Thrombosis, 207, 281, see also Atherosclerosis
Thymic polypeptide factors, 80
Thymosins, 79–80
Thymus gland, 79–80
Thyroid gland
abnormal states in aging, 186–187
and cardiac activity, 271
and dietary restriction, 298–299
disorders and Alzheimer's disease, 111
effects of thyroid hormones, 184–186
hypothalamo-pituitary-thyroid axis, 179–181
regulatory changes, 182–184
secretory changes, 181–182
structural changes, 181
Thyroid hormone receptors, 183–184
Thyroid nodules, 181
T-killer cells, 82
T lymphocytes, 81–84, 290, see also Immune system
Touch, 124–126
Transplantation of organs, 78–79, 241, 316
Tryptophan, 304–305
TSH (thyroid-stimulating hormone; hypothalmic releasing hormone), 182–183
T-suppressor cells, 82
Tuberculosis, 230
Tumorigenicity, see Cancer
Tumor incidence, and dietary restriction, 286–287

U

Ulcers
gastric and peptic, 249–251
pressure, 275
Underfeeding, 295–305, see also Restricted feeding model
Urethra, 153
Uric acid, 62, 63
Urinary incontinence, 240–243
Uterine changes, 153, 161–162

V

Vaginal changes, 153
Vascular changes, see Circulatory changes
Vasopressin (ADH), 140, 237–238
Ventilatory changes, see Pulmonary disorders
Very-low-density lipoproteins (VLDLs), 215–223, see also Atherosclerosis
Vibrotactile sensations, 124–125
Vision
anatomic aspects, 115–119
functional aspects, 119–120

Vitamin C, 63
Vitamin E, 63, 66, 67
Vulval changes, 152–153

W

Wear and tear theory, 42
Wellness, 311–313
Werner's syndrome, 30

X

Xanthines and cardiac activity, 271
Xerostomia (dry mouth), 248